The Making, Shaping and Treating of Steel

11th Edition

Steelmaking and Refining Volume

Copyright © 1998, 2012
Association for Iron & Steel Technology (AIST)
(Formerly published by the AISE Steel Foundation)
186 Thorn Hill Road
Warrendale, PA 15086 USA

All rights reserved.

No part of this publication may be reproduced,
stored in a retrieval system, or transmitted,
in any form or by any means,
electronic, mechanical, photocopying, recording, or otherwise,
without the prior permission of the Association for Iron & Steel Technology.

Library of Congress Catalog Card Number: 98–73477

The Association for Iron & Steel Technology makes no warranty, expressed or implied,
and no warranty as to the merchantability, fitness for any particular purpose
or accuracy of any information contained in this publication.
The user of any information contained herein assumes full responsibility for
such use and AIST, the editor and
the authors of this volume shall have no liability therefor.
The use of this information for any specific application should be based upon the
advice of professionally qualified personnel after independent verification
by those personnel of the suitability of the information for such use.
No license under any third party patents or other proprietary interest is expressly
or impliedly granted by publication of the information contained herein.
Furthermore, in the event of liability arising out of this publication,
consequential damages are excluded.

ISBN: 0-930767-02-0

Printed in the United States of America.

Preface

With the publication of the 10th edition of *The Making, Shaping and Treating of Steel* in 1985, the Association of Iron and Steel Engineers assumed total responsibility for the future of this prestigious document from the U.S. Steel Corporation. In 1998, the Association of Iron and Steel Engineers transferred all rights to *The Making, Shaping and Treating of Steel* to The AISE Steel Foundation. Readers of the 11th edition will obviously note the most dramatic change in technology and style of presentation since the book's inception in 1919.

In 1995, The AISE Steel Foundation formed an MSTS Steering Committee to oversee the creation of the 11th edition, and this committee looked out at a vastly different steel industry than that of the 10th edition. Hence, a new publication concept was deemed necessary, and this concept had to be consistent with the massive changes in steel industry economics that had occurred during the 1980s and early 1990s. These changes were occasioned by restructuring, downsizing, and wholesale implementation of new and improved technology. In turn, these changes produced major increases in labor productivity, huge reductions in energy consumption, and vastly improved yields. Concomitant with these improvements, the steel marketplace saw the introduction of a host of new and improved products.

Given the backdrop of the industry's transformation, the Steering Committee deemed a revision to the 10th edition in its current format to be impractical, and therefore decided the 11th edition would be a series of separate volumes dealing with specific subjects. These initial volumes, along with their scheduled publication dates, are:

> *Ironmaking Volume* (1999)
> *Steelmaking and Refining Volume* (1998)
> *Casting Volume* (2000)
> *Flat Products Volume* (2001)
> *Long Products Volume* (2002)

The separate volume concept was implemented by selecting Volume Chairpersons who were recognized as world leaders in their respective fields of technology. These leaders, in turn, recruited a team of top-notch authors to create the individual chapters. The leaders and expert auhors, many with backgrounds in the Association of Iron and Steel Engineers and the Iron and Steel Society, came from individual steel companies, the steel industry supplier base, and several universities with close associations with the steel industries. Thus, for the first time, the MSTS represents a broad and diverse view of steel technology as seen from various vantage points within industry and academe.

Despite all the changes to be found in the 11th edition, the MSTS Steering Committee has held on to certain traditions. One such tradition has been to provide to a wide audience (or readership) within the steel industry a basic reference containing the current practices and latest technology used in the making, shaping, and treating of steel. The primary readership targets are university students (technical knowledge), steel producers (training and technology implementation), and customers and suppliers (technical orientation and reference). As noted by the author of the 1st edition in 1919, "the book was written for . . . (those) . . . who are seeking self-instruction." The 11th edition attempts to maintain that tradition by incorporating technical information at several different levels of complexity and detail, thereby offering information of value to a wide-ranging readership.

The *Ironmaking Volume* and the *Steelmaking and Refining Volume,* both being published in the same year, contain common information on physical chemistry and kinetics, refractories, industrial gases, and fuels and water to make each book self-sufficient. The *Ironmaking Volume* includes descriptions of the newly emerging field of alternative iron production, and the *Steelmaking and Refining Volume* includes updated information on EAF technology and secondary refining, and new information on alternatives to conventional steelmaking. The *Casting Volume*, to be published in 1999, will include new information on near-net-shape and strip casting, as well as updated information on ingot teeming and conventional continuous casting.

The AISE Steel Foundation, which is dedicated to the advancement of the iron and steel industry of North America through training, publications, research, electronic resources and other related programs of benefit to the industry, receives the benefits of all sales of this publication.

In closing, the MSTS Steering Committee wants to personally thank all of the authors who have contributed their time and expertise to make the 11th edition a reality.

Allan M. Rathbone
Chairman, MSTS Steering Committee
Honorary Chairman, TheAISESteelFoundation

About the Editor

Richard J. Fruehan received his B.S. and Ph.D. degrees from the University of Pennsylvania and was an NSF post-doctoral scholar at Imperial College, University of London. He then was on the staff of the U.S. Steel Laboratory until he joined the faculty of Carnegie Mellon University as a Professor in 1980. Dr. Fruehan organized the Center for Iron and Steelmaking Research, an NSF Industry/University Cooperative Research Center, and is a Co-Director. The Center currently has twenty-seven industrial company members from the U.S., Europe, Asia, South Africa and South America. In 1992 he became the Director of the Sloan Steel Industry Study which examines the critical issues impacting a company's competitiveness and involves numerous faculty at several universities. Dr. Fruehan has authored over 200 papers, two books on steelmaking technologies and co-authored a book on managing for competitiveness, and is the holder of five patents. He has received several awards for his publications, including the 1970 and 1982 Hunt Medal (AIME), the 1982 and 1991 John Chipman Medal (AIME), 1989 Mathewson Gold Medal (TMS-AIME), the 1993 Albert Sauveur Award (ASM International), and the 1976 Gilcrist Medal (Metals Society UK), the 1996 Howe Memorial Lecture (ISS of AIME); he also received an IR100 Award for the invention of the oxygen sensor. In 1985 he was elected a Distinguished Member of the Iron and Steel Society. He served as President of the Iron and Steel Society of AIME from 1990–91. He was the Posco Professor from 1987 to 1997 and in 1997 he was appointed the U.S. Steel Professor of the Materials Science and Engineering Department of Carnegie Mellon University.

About the Authors

Keith J. Barker is Manager of Technology—Steelmaking and Continuous Casting for USX Engineers and Consultants, Inc., a subsidiary of U.S. Steel Corp., located in Pittsburgh, Pa. He received his B.S. and M.S. degrees in metallurgical engineering from Lehigh University. He has held various positions, during his 24 year career with U.S. Steel, in both the Research and Development Engineering departments. Prior to his current position he was involved in the project development and implementation of most of the capital improvements for U.S. Steel, since 1983, in the areas of steelmaking, ladle metallurgy and continuous casting.

Charles D. Blumenschein, P.E., D.E.E., is Senior Vice President of Chester Engineers, where he manages the Science and Technology Division. He received both his B.S. degree in civil engineering and his M.S. degree in sanitary engineering from the University of Pittsburgh. He has extensive experience in industrial water and wastewater treatment. At Chester Engineers, he is responsible for wastewater treatment projects, groundwater treatment investigations, waste minimization studies, toxic reduction evaluations, process and equipment design evaluations, assessment of water quality based effluent limitations, and negotiation of NPDES permit limitations with regulatory agencies. His experience includes conceptual process design of contaminated groundwater recovery and treatment systems; physical/chemical wastewater treatment for the chemical, metal finishing, steel, and non-ferrous industries; as well as advanced treatment technologies for water and wastewater recycle systems. He has actively negotiated effluent limitations for numerous industrial clients and has served as an expert witness in litigation matters. In addition, he has authored several publications addressing various wastewater treatment technologies and the implications of environmental regulations governing industry.

Ben Bowman has been Senior Corporate Fellow at the UCAR Carbon Co. Technical Center in Parma, Ohio, since 1993. Before that he had spent 22 years in the European headquarters of UCAR, located in Geneva, Switzerland, as customer technical service manager for arc furnace technology. After obtaining a Ph.D. in arc physics from the University of Liverpool in 1965, he commenced his involvement with arc furnaces at the Arc Furnace Research Laboratory of British Steel. He continues to study arc furnaces.

Allen H. Chan is Manager of AOD Process Technology for Praxair, Inc. He received his B.S., M.S., and Ph.D. degrees in metallurgical engineering and materials science from Carnegie Mellon University. Since joining Praxair, he has also worked in applications research and development and market development for the steel and foundry industries. His interests include high temperature physical chemistry, process development, and process modeling.

Richard J. Choulet is currently working as a steelmaking consultant to Praxair. He graduated in 1958 with a B.S. degree in metallurgical engineering from Purdue University. He previously worked in Research and Development for Inland Steel and Union Carbide, in the steel refining area. Since 1970 he has worked as a steelmaking consultant for Union Carbide (now Praxair), primarily on development and commercialization of the AOD process. He has extensive experience and has co-authored several papers and patents in the field of stainless steel refining.

Dennis J. Doran is Market Development Manager for Primary Metals in the Basic Industry Group of Nalco Chemical Co. He received his B.S. in metallurgy and materials science from Carnegie Mellon University in 1972 and an MBA from the University of Pittsburgh in 1973. Prior to joining Nalco in sales in 1979, he was employed by Vulcan Materials in market research and business development for their Metals Div. and by Comshare, Inc. in sales and technical support of computer timeshare applications. His area of expertise involves the interaction of water with process, design, cooling and environmental considerations in iron and steelmaking facilities. His responsibilities include technical, marketing, and training support for the steel industry and non-ferrous market segments. Technical support activities have included travel in North America, Asia, and Australia. He is a member of the Iron and Steel Society and AISE, and is a member of AISE Subcommittee No. 39 on Environmental Control Technologies.

Raymond F. Drnevich is the Manager of Process Integration for Praxair, Inc. Process integration focuses on developing industrial gases supply system synergies with iron and steelmaking technologies as well as technologies used in the chemical, petrochemical, and refining industries. He received a B.S. in chemical engineering from the University of Notre Dame and an M.S. in water resources engineering from the University of Michigan. In his 27 years at Praxair he has authored or co-authored more than 20 technical papers and 20 patents dealing with the production and use of industrial gases.

Peter C. Glaws is currently a Senior Research Specialist at The Timken Co. Research Center in Canton, Ohio, He received his B.S. in metallurgical engineering at Lafayette College and both his M.S. and Ph.D. degrees in metallurgical engineering and materials science from Carnegie Mellon University. He was a Postdoctoral Fellow at the University of Newcastle in New South Wales, Australia before joining The Timken Co. in 1987. His research interests include the physical chemistry of steelmaking and process modeling.

Daniel A. Goldstein received a B.S. degree in mechanical engineering from the Universidad Simon Bolivar in Caracas, Venezuela in 1987. He then joined a Venezuelan mini-mill steel producer, where he worked in production planning. In 1992 he enrolled at Carnegie Mellon University, sponsored by the Center for Iron and Steelmaking Research. His research work at CMU, done under the supervision of Prof. R. J. Fruehan and Prof. Bahri Ozturk, focused on nitrogen reactions in electric and oxygen steelmaking. He received his M.S. and Ph.D. degrees in materials science and engineering from Carnegie Mellon University in 1994 and 1996, respectively. He then joined Homer Research Laboratories at Bethlehem Steel Corporation as a Research Engineer working for the Steelmaking Group. He recently received the 1997 Jerry Silver Award from the Iron and Steel Society.

David H. Hubble was involved in refractory research, development and application for 34 years with U.S. Steel Corp. and continued as a consultant another five years following his retirement. Following graduation from Virginia Polytechnical Institute as a ceramic and metallurgical engineer, he was involved in all phases of steel plant refractory usage and facility startups in both domestic and foreign environments. He is the author of numerous papers and patents and has been involved in various volunteer activities since his retirement.

Ronald M. Jancosko is Executive Vice President and Partner of Vulcan Engineering Co. He received B.S. degrees in chemistry and biology from John Carroll University and has been a member of AISE since 1987. Vulcan Engineering designs and supplies special application steel mill equipment and processes for primary steelmaking throughout the world. In addition to working with Vulcan Engineering, he also is a steel industry consultant with Iron Technologies, Inc.

Jesus Jimenez is an Associate Research Consultant at the U.S. Steel Technical Center. He received a B.S. in chemical engineering from the Universidad Autonoma de Coahuila in Mexico and an

M.S. in metallurgical engineering from the University of Pittsburgh. His principal research interests are hot metal desulfurization, oxygen steelmaking (BOF, Q-BOP and combined blowing processes) and degassing. He was named as a Candidate for National Researcher by the National System of Researchers in Mexico in 1984.

Jeremy A.T. Jones is currently Vice President of Business Development for the Steelmaking Technology Division of AG Industries. He received his B.S. and M.S. degrees in chemical engineering from Queen's University at Kingston, Ontario, Canada, in 1983 and 1985, respectively. Following several years at Hatch Associates Ltd., he held key positions at Nupro Corp. and at Ameristeel. In September 1995, he joined Bechtel Corp. as principal engineer for iron and steel projects worldwide. In March 1998 he joined AG Industries in his current position. His previous consulting roles have involved many international assignments focused on both ferrous and non-ferrous process technologies, and included process plant improvements, review and development of environmental systems, development of process control systems and plant start-ups. Recently, he has focused on EAF technologies under development and alternative iron feedstocks, including new ironmaking technologies. He is a regular presenter at both AISE and ISS training seminars and has authored over 50 papers in the field of EAF steelmaking. He is currently chairman of the ISS Continuing Education Committee, and also sits on the ISS Advanced Technology Committee and the ISS Energy and Environment Committee.

G. J. W. (Jan) Kor received a Ph.D. in metallurgical engineering from the University of London, Imperial College of Science and Technology in 1967. He started his career in the steel industry with Hoogovens in the Netherlands. In 1968 he joined U.S. Steel Corp.'s Edgar C. Bain Laboratory for Fundamental Research in Monroeville, PA. His work there resulted in a number of papers in the areas of physical chemistry of iron and steelmaking, casting and solidification, as well as processing of ferroalloys. In 1986 he became a Scientist at the Technology Center of The Timken Co., where he was primarily involved in the application and implementation of basic technologies in steelmaking, ladle refining and casting. He retired from The Timken Co. in 1997.

Peter J. Koros currently is Senior Research Consultant for the LTV Steel Co. at the Technology Center in Independence, Ohio. He obtained a B.S. in metallurgical engineering at Drexel University and both S.M. and Sc.D. degrees in metallurgy at M.I.T. He joined Jones and Laughlin Steel Corp., a predecessor of LTV Steel, and held positions in Research and Quality Control. He was responsible for the development work in injection technology for desulfurization of hot metal and steel at Jones and Laughlin Steel Corp. and served on the AISI-DOE Direct Steelmaking Program. Dr. Koros has over 70 publications, seven U.S. patents, and has organized numerous conferences and symposia. He has been elected Distinguished Member by the Iron and Steel Society and Fellow by ASM International.

Peter A. Lefrank received his B.S., M.S. and Ph.D. degrees in chemical engineering from the University of Erlangen-Nuremberg in Germany. He has held technical management positions with graphite manufacturers in Europe and in the U.S. As an entrepreneur, he has founded the Intercarbon Engineering firm engaging in design, modernization and improvement of graphite production processes and plants. He has studied electrode consumption processes in EAFs extensively, and has developed proposals to improve electrode performance, specifically for DC operations. Worldwide, he is considered a leading specialist in the area of development, manufacturing, and application of graphite electrodes for EAF steelmaking. He is currently working as an international consultant to the SGL Carbon Corp.

Antone Lehrman is a Senior Development Engineer for LTV Steel Co. at the Technology Center in Independence, Ohio. He received a B.E. degree in mechanical engineering at Youngstown State University in 1970 and worked for Youngstown Sheet & Tube Co. and Republic Steel Corp. prior to their merger with Jones and Laughlin Steel Corp. His entire career has been focused in the energy and utility field of steel plant operations. He held the positions of Fuel Engineer, Boiler Plant Supervisor, and others prior to joining the corporate Energy Group in 1985.

Ronald J. Marr has over 30 years experience in the application, installation, wear mechanisms, and slag reactions of basic refractories. He has worked, taught, and published extensively in the areas of electric arc furnace, ladle, ladle furnace, and AOD/VOD process slag control and refractory design. After receiving a B.S. in ceramic engineering from Alfred University, he was employed in the laboratories of both General Refractories and Martin-Marietta Refractories prior to joining Baker Refractories in 1975, where he is currently Projects Manager—Research and Development.

Charles J. Messina is Director of Bulk Gas Sales in Cleveland, Ohio for Praxair, Inc. He received his M.S. in process metallurgy from Lehigh University in 1976. He also holds degrees in mechanical engineering and business administration. In 1976 he began his career in the steel industry at the U.S. Steel Research Laboratory, where he worked on steelmaking applications and process control; in 1981 he was transferred to the Gary Works. In 1983 he joined the Linde Division of Union Carbide as technology manager of the AOD process. He joined PennMet, located in Ridgway, PA, in 1985 as vice president of operations and returned to Praxair, Inc. in 1986 as process manager of steelmaking and combustion. He was named sales manager, bulk gases in 1990 and was named technology manager, primary metals in 1992. His work included development BOF slag splashing, EAF post-combustion and Praxair's coherent jet technology.

Timothy W. Miller is currently Supervisor of Steelmaking and Casting at Bethlehem Steel Corp.'s Homer Research Laboratories. He graduated from Rensselaer Polytechnic Institute with a B.Met.Eng. After working for General Electric in the development of alloys for nuclear reactors, he obtained an M.Met.Eng at RPI. He then worked at Bethlehem's Homer Labs for several years before moving into steelmaking production at the Lackawanna Plant, where he was General Supervisor of the BOF and Supervisor of Steelmaking Technology. He transferred to the Bethlehem Plant as Supervisor of Steelmaking Technology when the Lackawanna Plant was shut down. Later, at the Bethlehem Plant he headed all areas of technology for the plant. He returned to Homer Research Laboratories as a Steelmaking Consultant when the Bethlehem Plant was shut down. Over the years he has acquired much experience in steelmaking and long bar rolling and finishing.

Claudia L. Nassaralla is currently Assistant Professor at Michigan Technological University and is an ISS Ferrous Metallurgy Professor. She began her education in Brazil and moved to the U.S. in 1986, where she received her Ph.D. in metallurgical and materials engineering from Carnegie Mellon University. Before joining Michigan Technological University in 1993, Dr. Nassaralla was a Senior Research Engineer at the U.S. Steel Technical Center and was the U.S. Steel representative on the Technical Board of the AISI Direct Steelmaking Program. Her principal research interests are on applications of physical chemistry and kinetics to the development of novel processes for recycling of waste materials in the metal industry.

Balaji (Bal) V. Patil is Manager, Process Research and Development at the Technical Center of Allegheny Ludlum Steel, a Division of Allegheny Teledyne Company. He received his Bachelor of Technology degree in metallurgical engineering from Indian Institute of Technology, Mumbai, India. He pursued his graduate studies at Columbia University in New York City. He has an M.S. in mineral engineering and a Doctor of Engineering Science in chemical metallurgy. After a brief employment with Cities Service Company (later acquired by Occidental Petroleum), he joined Allegheny Ludlum Corp. in 1976. His areas of expertise include raw material selection, EAF melting, BOF and AOD steelmaking, ladle treatment as well as continuous casting. He is a member of the Iron & Steel Society, The Metallurgical Society and ASM International.

John R. Paules is currently General Manager at Ellwood Materials Technologies, a division of the Ellwood Group, Inc. He received B.S. and M.Eng. degrees in metallurgical engineering from Lehigh University, and he is a registered Professional Engineer. He previously worked at Bethlehem Steel Corp., Stratcor Technical Sales, and Berry Metal Co. Involved with the technology of steel production for over 20 years, he has authored numerous publications and patents in the fields of steelmaking and new product development.

About the Authors

Robert O. Russell is the Manager of Refractories at LTV Steel Co. and is a member of ISS, AISI (past chairman), American Ceramic Society (fellow), and ASTM. He has won the prestigious American Ceramic Society Al Allen award for best refractory paper (two times), and the Charles Herty Award from the Iron & Steel Society. He was conferred with the T. J. Planje - St. Louis Refractories Award for distinguished achievement in the field of refractories. In his 36 years at LTV Steel, he has authored over twenty papers on refractories for BOFs, steel ladles, degassers and on steelmaking raw materials. Mr. Russell has seven patents related to phosphate bonding, slagmaking, and refractory compositions and design. He received his formal education from Miami University (Ohio) with A.B. and M.S. degrees in geology.

Nicholas Rymarchyk is Vice President for Berry Metal Co. and is responsible for all sales and marketing activities for oxygen lances in steelmaking processes. He received a B.S. in mechanical engineering from Geneva College. In 1966, he began his career in steelmaking at the U.S. Steel Applied Research Laboratory, where he worked in the Structural Mechanics Group. In his 32 years at Berry Metal Co. he has authored several technical papers and has 25 patents dealing with oxygen lance design for primary steelmaking.

Ronald J. Selines is a Corporate Fellow at Praxair, Inc. and is responsible for efforts to develop and commercialize new industrial gas based iron and steelmaking process technology. He received an Sc.D. in metallurgy and materials science from MIT in 1974, has been actively involved in iron and steelmaking technology for the past 24 years, and has authored 11 publications and 12 patents in this field.

Alok Sharan received his Bachelor Technology degree in metallurgical engineering from the Indian Institute of Technology at Kanpur in 1989. In 1993 he completed his Ph.D. in materials science and engineering from Carnegie Mellon University. He joined Bethlehem Steel Corp. in 1994 to work in the Steelmaking Group at Homer Research Laboratories. He has published several papers in the area of steel processing and also has a patent. He was the recipient of the Iron and Steel Society's Frank McKune award for the year 1998.

Steven E. Stewart is District Account Manager in Northwest Indiana for Nalco Chemical Co. He received a B.S. in biology from Indiana University in 1969 and received an M.S. in chemistry from Roosevelt University in Chicago. He has specialized in industrial water treatment during his 22 year career with Nalco. He has had service responsibility in all of the major steel manufacturing plants in Northwest Indiana. He has experience in power generation plants, cooling water systems, and wastewater treatment plants. He has been responsible for the startup and implementation of numerous automated chemical control and monitoring systems during his career.

E. T. Turkdogan, a Ph.D. graduate of the University of Sheffield, was appointed in 1950 as Head of the Physical Chemistry Section of the British Iron and Steel Research Association, London. In 1959, he was invited to join U.S. Steel Corp. as an Assistant Director of research at the Edgar C. Bain Laboratory for Fundamental Research, Monroeville, PA, as it was known prior to 1972. Subsequently he became a Senior Research Consultant at the Research Center of U.S. Steel. Upon retirement from USX Corp. in 1986, he undertook a private consultancy business entailing a wide range of industrial and research and development technologies, including technical services to law firms. He published approximately 200 papers in the fields of chemical metallurgy, process thermodynamics and related subjects, authored 13 patents and contributed to chapters of numerous reference books on pyrometallurgy. He authored three books: *Physical Chemistry of High Temperature Technology* (1980), *Physicochemical Properties of Molten Slags and Glasses* (1983) and *Fundamentals of Steelmaking* (1996). He received numerous awards from the British and American metallurgical institutes and in 1985 was awarded the Degree of Doctor of Metallurgy by the University of Sheffield in recognition of his contributions to the science and technology of metallurgy. He was further honored by a symposium held in Pittsburgh in 1994, which was sponsored by USX Corp. and the Iron and Steel Society of AIME. He is a Fellow of the Institute of Materials (U.K.), a Fellow of The Minerals, Metals and Materials Society and a Distinguished Member of the Iron and Steel Society.

H. L. Vernon is Market Manager—Iron and Steel for Harbison-Walker Refractories Co. and has 29 years experience in the refractories industry. He has held previous management positions with H-W in research, field sales, marketing, customer service, and technical services. Prior to working for H-W, he was employed as a quality control metallurgist with Armco Steel in Houston, Texas. He has a B.S. in metallurgy from Case Institute of Technology, Cleveland, Ohio, and an MBA degree from Pepperdine University. He has authored several technical papers, most recently "Electric Furnace Refractory Lining Management" at the ISS 1997 Electric Furnace Conference.

Acknowledgments

This volume of the 11th edition of *The Making, Shaping and Treating of Steel* would not have been possible if not for the oversight and guidance of the members of the MSTS Steering Committee. Their efforts, support and influence in shaping this project to its completion are greatly appreciated, and they are recognized here:

>Allan Rathbone, Chairman, MSTS Steering Committee, U.S. Steel Corp., General Manager, Research (Retired)
>Brian Attwood, LTV Steel Co., Vice President, Quality Control and Research
>Michael Byrne, Bethlehem Steel Corp., Research Manager
>Alan Cramb, Carnegie Mellon University, Professor
>Bernard Fedak, U.S. Steel Corp., General Manager, Engineering
>Frank Fonner, Association of Iron and Steel Engineers, Manager, Publications
>Richard Fruehan, Carnegie Mellon University, Professor
>David Hubble, U.S. Steel Corp., Chief Refractory Engineer (Retired)
>Dennis Huffman, The Timken Co., Manager—Steel Product Development
>Lawrence Maloney, Association of Iron and Steel Engineers, Managing Director
>David Matlock, Colorado School of Mines, Professor
>Malcolm Roberts, Bethlehem Steel Corp., Vice President–Technology and Chief Technology Officer
>David Wakelin, LTV Steel Co., Manager, Development Engineering, Primary

Oversight of the project was also provided by The AISE Steel Foundation Board of Trustees, and they are recognized here:

>Timothy Lewis, 1998 Chairman, Bethlehem Steel Corp., Senior Advisor
>James Anderson, Electralloy, President
>Bernard Fedak, U.S. Steel Corp., General Manager, Engineering
>Steven Filips, North Star Steel Co., Executive Vice President–Steelmaking Operations
>William Gano, Charter Manufacturing Co., President and Chief Operating Officer
>J. Norman Lockington, Dofasco, Inc., Vice President–Technology
>Lawrence Maloney, Association of Iron and Steel Engineers, Managing Director
>Rodney Mott, Nucor Steel–Berkeley, Vice President and General Manager
>R. Lee Sholley, The Timken Co., General Manager–Harrison Steel Plant
>Thomas Usher, USX Corp., Chairman, The AISE Steel Foundation Board of Trustees, and Chief Executive Officer
>James Walsh, AK Steel Corp., Vice President–Corporate Development

Many hours of work were required in manuscript creation, editing, illustrating, and typesetting. The diligent efforts of the editor and the authors, many of whom are affiliated with other technical societies, are to be commended. In addition, the strong support and contributions of the AISE staff deserve special recognition.

The AISE Steel Foundation is proud to be the publisher of this industry classic, and will work to keep this title at the forefront of technology in the years to come.

Lawrence G. Maloney
Managing Director, AISE
Publisher and Secretary/Treasurer, The AISE Steel Foundation

Pittsburgh, Pennsylvania
July 1998

Table of Contents

Preface — v
About the Editor — vii
About the Authors — ix
Acknowledgments — xv

Chapter 1 Overview of Steelmaking Processes and Their Development — 1

  1.1 Introduction — 1

  1.2 Historical Development of Modern Steelmaking — 1
    1.2.1 Bottom-Blown Acid or Bessemer Process — 2
    1.2.2 Basic Bessemer or Thomas Process — 4
    1.2.3 Open Hearth Process — 4
    1.2.4 Oxygen Steelmaking — 7
    1.2.5 Electric Furnace Steelmaking — 8

  1.3 Evolution in Steelmaking by Process — 10

  1.4 Structure of This Volume — 12

Chapter 2 Fundamentals of Iron and Steelmaking — 13

  2.1 Thermodynamics — 13
    2.1.1 Ideal Gas — 13
    2.1.2 Thermodynamic Laws — 14
    2.1.3 Thermodynamic Activity — 18
    2.1.4 Reaction Equilibrium Constant — 23

  2.2 Rate Phenomena — 24
    2.2.1 Diffusion — 24
    2.2.2 Mass Transfer — 26
    2.2.3 Chemical Kinetics — 39
    2.2.4 Mixed Control — 47

  2.3 Properties of Gases — 49
    2.3.1 Thermochemical Properties — 49

    2.3.2 Transport Properties 55
    2.3.3 Pore Diffusion 57
2.4 Properties of Molten Steel 60
    2.4.1 Selected Thermodynamic Data 60
    2.4.2 Solubility of Gases in Liquid Iron 61
    2.4.3 Iron-Carbon Alloys 64
    2.4.4 Liquidus Temperatures of Low Alloy Steels 69
    2.4.5 Solubility of Iron Oxide in Liquid Iron 69
    2.4.6 Elements of Low Solubility in Liquid Iron 70
    2.4.7 Surface Tension 72
    2.4.8 Density 75
    2.4.9 Viscosity 75
    2.4.10 Diffusivity, Electrical and Thermal Conductivity, and Thermal Diffusivity 76
2.5 Properties of Molten Slags 79
    2.5.1 Structural Aspects 79
    2.5.2 Slag Basicity 80
    2.5.3 Iron Oxide in Slags 81
    2.5.4 Selected Ternary and Quaternary Oxide Systems 81
    2.5.5 Oxide Activities in Slags 84
    2.5.6 Gas Solubility in Slags 89
    2.5.7 Surface Tension 95
    2.5.8 Density 98
    2.5.9 Viscosity 100
    2.5.10 Mass Diffusivity, Electrical Conductivity and Thermal Conductivity 101
    2.5.11 Slag Foaming 102
    2.5.12 Slag Models and Empirical Correlations for Thermodynamic Properties 104
2.6 Fundamentals of Ironmaking Reactions 104
    2.6.1 Oxygen Potential Diagram 104
    2.6.2 Role of Vapor Species in Blast Furnace Reactions 105
    2.6.3 Slag-Metal Reactions in the Blast Furnace 109
2.7 Fundamentals of Steelmaking Reactions 118
    2.7.1 Slag-Metal Equilibrium in Steelmaking 119
    2.7.2 State of Reactions in Steelmaking 123
2.8 Fundamentals of Reactions in Electric Furnace Steelmaking 132
    2.8.1 Slag Chemistry and the Carbon, Manganese, Sulfur and
          Phosphorus Reactions in the EAF 132
    2.8.2 Control of Residuals in EAF Steelmaking 134
    2.8.3 Nitrogen Control in EAF Steelmaking 135
2.9 Fundamentals of Stainless Steel Production 136
    2.9.1 Decarburization of Stainless Steel 136
    2.9.2 Nitrogen Control in the AOD 138
    2.9.3 Reduction of Cr from Slag 139
2.10 Fundamentals of Ladle Metallurgical Reactions 140
    2.10.1 Deoxidation Equilibrium and Kinetics 140
    2.10.2 Ladle Desulfurization 147
    2.10.3 Calcium Treatment of Steel 150
2.11 Fundamentals of Degassing 151
    2.11.1 Fundamental Thermodynamics 151
    2.11.2 Vacuum Degassing Kinetics 152

Chapter 3 Steel Plant Refractories — 159

- 3.1 Classification of Refractories — 159
  - 3.1.1 Magnesia or Magnesia–Lime Group — 160
  - 3.1.2 Magnesia–Chrome Group — 163
  - 3.1.3 Siliceous Group — 164
  - 3.1.4 Clay and High-Alumina Group — 166
  - 3.1.5 Processed Alumina Group — 169
  - 3.1.6 Carbon Group — 170
- 3.2 Preparation of Refractories — 172
  - 3.2.1 Refractory Forms — 172
  - 3.2.2 Binder Types — 173
  - 3.2.3 Processing — 176
  - 3.2.4 Products — 177
- 3.3 Chemical and Physical Characteristics of Refractories and their Relation to Service Conditions — 178
  - 3.3.1 Chemical Composition — 178
  - 3.3.2 Density and Porosity — 179
  - 3.3.3 Refractoriness — 181
  - 3.3.4 Strength — 182
  - 3.3.5 Stress-Strain Behavior — 185
  - 3.3.6 Specific Heat — 186
  - 3.3.7 Emissivity — 187
  - 3.3.8 Thermal Expansion — 188
  - 3.3.9 Thermal Conductivity and Heat Transfer — 190
  - 3.3.10 Thermal Shock — 194
- 3.4 Reactions at Elevated Temperatures — 194
- 3.5 Testing and Selection of Refractories — 206
  - 3.5.1 Simulated Service Tests — 206
  - 3.5.2 Post-Mortem Studies — 212
  - 3.5.3 Thermomechanical Behavior — 213
- 3.6 General Uses of Refractories — 215
  - 3.6.1 Linings — 215
  - 3.6.2 Metal Containment, Control and Protection — 217
  - 3.6.3 Refractory Use for Energy Savings — 222
- 3.7 Refractory Consumption, Trends and Costs — 224

Chapter 4 Steelmaking Refractories — 227

- 4.1 Refractories for Oxygen Steelmaking Furnaces — 227
  - 4.1.1 Introduction — 227
  - 4.1.2 Balancing Lining Wear — 228
  - 4.1.3 Zoned Linings by Brick Type and Thickness — 230
  - 4.1.4 Refractory Construction — 231
  - 4.1.5 Furnace Burn-In — 235
  - 4.1.6 Wear of the Lining — 235
  - 4.1.7 Lining Life and Costs — 238
- 4.2 BOF Slag Coating and Slag Splashing — 239
  - 4.2.1 Introduction — 239

4.2.2 Slag Coating Philosophy	239
4.2.3 Magnesia Levels and Influences	239
4.2.4 Material Additions	240
4.2.5 Equilibrium Operating Lining Thickness	240
4.2.6 Other Refractory Maintenance Practices	241
4.2.7 Laser Measuring	241
4.2.8 Slag Splashing	241
4.3 Refractories for Electric Furnace Steelmaking	243
4.3.1 Electric Furnace Design Features	243
4.3.2 Electric Furnace Zone Patterns	244
4.3.3 Electric Furnace Refractory Wear Mechanisms	247
4.3.4 Conclusion	248
4.4 Refractories for AOD and VOD Applications	248
4.4.1 Background	248
4.4.2 AOD Refractories	249
4.4.3 VOD Refractories	258
4.4.4 Acknowledgments	261
4.5 Refractories for Ladles	262
4.5.1 Function of Modern Steel Ladle	262
4.5.2 Ladle Design	265
4.5.3 Ladle Refractory Design and Use	268
4.5.4 Ladle Refractory Construction	276
4.5.5 Refractory Stirring Plugs	277
4.5.6 Refractory Life and Costs	281
4.6 Refractories for Degassers	285

Chapter 5 Production and Use of Industrial Gases for Iron and Steelmaking — 291

5.1 Industrial Gas Uses	291
5.1.1 Introduction	291
5.1.2 Oxygen Uses	292
5.1.3 Nitrogen Uses	294
5.1.4 Argon Uses	295
5.1.5 Hydrogen Uses	296
5.1.6 Carbon Dioxide Uses	296
5.2 Industrial Gas Production	297
5.2.1 Introduction	297
5.2.2 Atmospheric Gases Produced by Cryogenic Processes	298
5.2.3 Atmospheric Gases Produced by PSA/VSA/VPSA Membranes	302
5.2.4 Hydrogen Production	305
5.2.5 Carbon Dioxide Production	305
5.3 Industrial Gas Supply System Options and Considerations	306
5.3.1 Introduction	306
5.3.2 Number of Gases	306
5.3.3 Purity of Gases	307
5.3.4 Volume of Gases	307
5.3.5 Use Pressure	307
5.3.6 Use Pattern	307
5.3.7 Cost of Power	307

  5.3.8 Backup Requirements   307
  5.3.9 Integration   307

 5.4 Industrial Gas Safety   307
  5.4.1 Oxygen   308
  5.4.2 Nitrogen   308
  5.4.3 Argon   308
  5.4.4 Hydrogen   309
  5.4.5 Carbon Dioxide   309

Chapter 6 Steel Plant Fuels and Water Requirements   311

 6.1 Fuels, Combustion and Heat Flow   311
  6.1.1 Classification of Fuels   311
  6.1.2 Principles of Combustion   312
  6.1.3 Heat Flow   326

 6.2 Solid Fuels and Their Utilization   329
  6.2.1 Coal Resources   330
  6.2.2 Mining of Coal   336
  6.2.3 Coal Preparation   339
  6.2.4 Carbonization of Coal   341
  6.2.5 Combustion of Solid Fuels   341

 6.3 Liquid Fuels and Their Utilization   344
  6.3.1 Origin, Composition and Distribution of Petroleum   345
  6.3.2 Grades of Petroleum Used as Fuels   347
  6.3.3 Properties and Specifications of Liquid Fuels   348
  6.3.4 Combustion of Liquid Fuels   351
  6.3.5 Liquid-Fuel Burners   351

 6.4 Gaseous Fuels and Their Utilization   352
  6.4.1 Natural Gas   353
  6.4.2 Manufactured Gases   353
  6.4.3 Byproduct Gaseous Fuels   356
  6.4.4 Uses for Various Gaseous Fuels in the Steel Industry   358
  6.4.5 Combustion of Various Gaseous Fuels   360

 6.5 Fuel Economy   363
  6.5.1 Recovery of Waste Heat   364
  6.5.2 Minimizing Radiation Losses   366
  6.5.3 Combustion Control   366
  6.5.4 Air Infiltration   367
  6.5.5 Heating Practice   368

 6.6 Water Requirements for Steelmaking   368
  6.6.1 General Uses for Water in Steelmaking   368
  6.6.2 Water-Related Problems   371
  6.6.3 Water Use by Steelmaking Processes   372
  6.6.4 Treatment of Effluent Water   379
  6.6.5 Effluent Limitations   385
  6.6.6 Boiler Water Treatment   395

Chapter 7 Pre-Treatment of Hot Metal   413

 7.1 Introduction   413

 7.2 Desiliconization and Dephosphorization Technologies   413

7.3 Desulfurization Technology ... 416
   7.3.1 Introduction ... 416
   7.3.2 Process Chemistry ... 417
   7.3.3 Transport Systems ... 421
   7.3.4 Process Venue ... 422
   7.3.5 Slag Management ... 423
   7.3.6 Lance Systems ... 424
   7.3.7 Cycle Time ... 426
   7.3.8 Hot Metal Sampling and Analysis ... 426
   7.3.9 Reagent Consumption ... 426
   7.3.10 Economics ... 427
   7.3.11 Process Control ... 427

7.4 Hot Metal Thermal Adjustment ... 427

7.5 Acknowledgments ... 428

7.6 Other Reading ... 428

Chapter 8 Oxygen Steelmaking Furnace Mechanical Description and Maintenance Considerations   431

8.1 Introduction ... 431

8.2 Furnace Description ... 431
   8.2.1 Introduction ... 431
   8.2.2 Vessel Shape ... 433
   8.2.3 Top Cone-to-Barrel Attachment ... 434
   8.2.4 Methods of Top Cone Cooling ... 435
   8.2.5 Vessel Bottom ... 438
   8.2.6 Types of Trunnion Ring Designs ... 438
   8.2.7 Methods of Vessel Suspension ... 439
   8.2.8 Vessel Imbalance ... 445
   8.2.9 Refractory Lining Design ... 446
   8.2.10 Design Temperatures ... 448
   8.2.11 Design Pressures and Loading ... 451
   8.2.12 Method of Predicting Vessel Life ... 457
   8.2.13 Special Design and Operating Considerations ... 458

8.3 Materials ... 460

8.4 Service Inspection, Repair, Alteration and Maintenance ... 460
   8.4.1 BOF Inspection ... 460
   8.4.2 BOF Repair and Alteration Procedures ... 462
   8.4.3 Repair Requirements of Structural Components ... 463
   8.4.4 Deskulling ... 464

8.5 Oxygen Lance Technology ... 465
   8.5.1 Introduction ... 465
   8.5.2 Oxidation Reactions ... 465
   8.5.3 Supersonic Jet Theory ... 466
   8.5.4 Factors Affecting BOF Lance Performance ... 468
   8.5.5 Factors Affecting BOF Lance Life ... 469
   8.5.6 New Developments in BOF Lances ... 470

8.6 Sub-Lance Equipment ... 471

Chapter 9 Oxygen Steelmaking Processes — 475

- 9.1 Introduction — 475
  - 9.1.1 Process Description and Events — 475
  - 9.1.2 Types of Oxygen Steelmaking Processes — 476
  - 9.1.3 Environmental Issues — 477
  - 9.1.4 How to Use This Chapter — 477
- 9.2 Sequences of Operations—Top Blown — 478
  - 9.2.1 Plant Layout — 478
  - 9.2.2 Sequence of Operations — 478
  - 9.2.3 Shop Manning — 486
- 9.3 Raw Materials — 489
  - 9.3.1 Introduction — 489
  - 9.3.2 Hot Metal — 489
  - 9.3.3 Scrap — 491
  - 9.3.4 High Metallic Alternative Feeds — 491
  - 9.3.5 Oxide Additions — 493
  - 9.3.6 Fluxes — 494
  - 9.3.7 Oxygen — 495
- 9.4 Process Reactions and Energy Balance — 496
  - 9.4.1 Refining Reactions in BOF Steelmaking — 496
  - 9.4.2 Slag Formation in BOF Steelmaking — 498
  - 9.4.3 Mass and Energy Balances — 499
  - 9.4.4 Tapping Practices and Ladle Additions — 503
- 9.5 Process Variations — 504
  - 9.5.1 The Bottom-Blown Oxygen Steelmaking or OBM (Q-BOP) Process — 504
  - 9.5.2 Mixed-Blowing Processes — 507
  - 9.5.3 Oxygen Steelmaking Practice Variations — 512
- 9.6 Process Control Strategies — 515
  - 9.6.1 Introduction — 515
  - 9.6.2 Static Models — 515
  - 9.6.3 Statistical and Neural Network Models — 516
  - 9.6.4 Dynamic Control Schemes — 517
  - 9.6.5 Lance Height Control — 519
- 9.7 Environmental Issues — 519
  - 9.7.1 Basic Concerns — 519
  - 9.7.2 Sources of Air Pollution — 519
  - 9.7.3 Relative Amounts of Fumes Generated — 521
  - 9.7.4 Other Pollution Sources — 522
  - 9.7.5 Summary — 522

Chapter 10 Electric Furnace Steelmaking — 525

- 10.1 Furnace Design — 525
  - 10.1.1 EAF Mechanical Design — 525
  - 10.1.2 EAF Refractories — 545
- 10.2 Furnace Electric System and Power Generation — 551
  - 10.2.1 Electrical Power Supply — 551
  - 10.2.2 Furnace Secondary System — 554
  - 10.2.3 Regulation — 555

10.2.4 Electrical Considerations for AC Furnaces	557
10.2.5 Electrical Considerations for DC Furnaces	560
10.3 Graphite Electrodes	562
10.3.1 Electrode Manufacture	562
10.3.2 Electrode Properties	564
10.3.3 Electrode Wear Mechanisms	564
10.3.4 Current Carrying Capacity	569
10.3.5 Discontinuous Consumption Processes	569
10.3.6 Comparison of AC and DC Electrode Consumption	572
10.3.7 Development of Special DC Electrode Grades	575
10.4 Gas Collection and Cleaning	577
10.4.1 Early Fume Control Methods	577
10.4.2 Modern EAF Fume Control	579
10.4.3 Secondary Emissions Control	583
10.4.4 Gas Cleaning	586
10.4.5 Mechanisms of EAF Dust Formation	590
10.4.6 Future Environmental Concerns	590
10.4.7 Conclusions	594
10.5 Raw Materials	594
10.6 Fluxes and Additives	595
10.7 Electric Furnace Technology	597
10.7.1 Oxygen Use in the EAF	597
10.7.2 Oxy-Fuel Burner Application in the EAF	598
10.7.3 Application of Oxygen Lancing in the EAF	601
10.7.4 Foamy Slag Practice	604
10.7.5 CO Post-Combustion	605
10.7.6 EAF Bottom Stirring	615
10.7.7 Furnace Electrics	617
10.7.8 High Voltage AC Operations	617
10.7.9 DC EAF Operations	618
10.7.10 Use of Alternative Iron Sources in the EAF	621
10.7.11 Conclusions	622
10.8 Furnace Operations	622
10.8.1 EAF Operating Cycle	622
10.8.2 Furnace Charging	623
10.8.3 Melting	624
10.8.4 Refining	624
10.8.5 Deslagging	626
10.8.6 Tapping	627
10.8.7 Furnace Turnaround	627
10.8.8 Furnace Heat Balance	628
10.9 New Scrap Melting Processes	629
10.9.1 Scrap Preheating	629
10.9.2 Preheating With Offgas	630
10.9.3 Natural Gas Scrap Preheating	630
10.9.4 K-ES	631
10.9.5 Danarc Process	634
10.9.6 Fuchs Shaft Furnace	635
10.9.7 Consteel Process	642
10.9.8 Twin Shell Electric Arc Furnace	645
10.9.9 Processes Under Development	648

Chapter 11 Ladle Refining and Vacuum Degassing — 661

- 11.1 Tapping the Steel — 662
  - 11.1.1 Reactions Occurring During Tapping — 662
  - 11.1.2 Furnace Slag Carryover — 663
  - 11.1.3 Chilling Effect of Ladle Additions — 664

- 11.2 The Tap Ladle — 665
  - 11.2.1 Ladle Preheating — 665
  - 11.2.2 Ladle Free Open Performance — 667
  - 11.2.3 Stirring in Ladles — 669
  - 11.2.4 Effect of Stirring on Inclusion Removal — 672

- 11.3 Reheating of the Bath — 673
  - 11.3.1 Arc Reheating — 673
  - 11.3.2 Reheating by Oxygen Injection — 675

- 11.4 Refining in the Ladle — 677
  - 11.4.1 Deoxidation — 677
  - 11.4.2 Desulfurization — 680
  - 11.4.3 Dephosphorization — 683
  - 11.4.4 Alloy Additions — 685
  - 11.4.5 Calcium Treatment and Inclusion Modification — 687

- 11.5 Vacuum Degassing — 693
  - 11.5.1 General Process Descriptions — 694
  - 11.5.2 Vacuum Carbon Deoxidation — 694
  - 11.5.3 Hydrogen Removal — 698
  - 11.5.4 Nitrogen Removal — 701

- 11.6 Description of Selected Processes — 705
  - 11.6.1 Ladle Furnace — 705
  - 11.6.2 Tank Degasser — 705
  - 11.6.3 Vacuum Arc Degasser — 705
  - 11.6.4 RH Degasser — 708
  - 11.6.5 CAS-OB Process — 709
  - 11.6.6 Process Selection and Comparison — 710

Chapter 12 Refining of Stainless Steels — 715

- 12.1 Introduction — 715

- 12.2 Special Considerations in Refining Stainless Steels — 720

- 12.3 Selection of a Process Route — 721

- 12.4 Raw Materials — 723

- 12.5 Melting — 724
  - 12.5.1 Electric Arc Furnace Melting — 724
  - 12.5.2 Converter Melting — 725

- 12.6 Dilution Refining Processes — 725
  - 12.6.1 Argon-Oxygen Decarburization (AOD) Converter Process — 725
  - 12.6.2 K-BOP and K-OBM-S — 726
  - 12.6.3 Metal Refining Process (MRP) Converter — 727
  - 12.6.4 Creusot-Loire-Uddeholm (CLU) Converter — 727
  - 12.6.5 Krupp Combined Blowing-Stainless (KCB-S) Process — 728
  - 12.6.6 Argon Secondary Melting (ASM) Converter — 728

12.6.7 Sumitomo Top and Bottom Blowing Process (STB) Converter … 729
12.6.8 Top Mixed Bottom Inert (TMBI) Converter … 729
12.6.9 Combined Converter and Vacuum Units … 729

12.7 Vacuum Refining Processes … 729

12.8 Direct Stainless Steelmaking … 730

12.9 Equipment for EAF-AOD Process … 732
12.9.1 Vessel Size and Shape … 732
12.9.2 Refractories … 733
12.9.3 Tuyeres and Plugs … 733
12.9.4 Top Lances … 733
12.9.5 Gases … 734
12.9.6 Vessel Drive System … 734
12.9.7 Emissions Collection … 735

12.10 Vessel Operation … 735
12.10.1 Decarburization … 735
12.10.2 Refining … 737
12.10.3 Process Control … 737
12.10.4 Post-Vessel Treatments … 738

12.11 Summary … 738

Chapter 13 Alternative Oxygen Steelmaking Processes … 743

13.1 Introduction … 743

13.2 General Principles and Process Types … 743

13.3 Specific Alternative Steelmaking Processes … 745
13.3.1 Energy Optimizing Furnace (EOF) … 746
13.3.2 AISI Continuous Refining … 748
13.3.3 IRSID Continuous Steelmaking … 749
13.3.4 Trough Process … 752
13.3.5 Other Steelmaking Alternatives … 753

13.4 Economic Evaluation … 755

13.5 Summary and Conclusions … 757

Index … **761**

Chapter 1

Overview of Steelmaking Processes and Their Development

R. J. Fruehan, Professor, Carnegie Mellon University

1.1 Introduction

This volume examines the basic principles, equipment and operating practices involved in steelmaking and refining. In this introductory chapter the structure of this volume is briefly described. Also the evolution of steelmaking processes from about 1850 to the present is given along with statistics on current production by process and speculation on future trends.

For the purpose of this volume steelmaking can be roughly defined as the refining or removal of unwanted elements or other impurities from hot metal produced in a blast furnace or similar process or the melting and refining of scrap and other forms of iron in a melting furnace, usually an electric arc furnace (EAF). Currently most all of the hot metal produced in the world is refined in an oxygen steelmaking process (OSM). A small amount of hot metal is refined in open hearths, cast into pigs for use in an EAF or refined in other processes. The major element removed in OSM is carbon which is removed by oxidation to carbon monoxide (CO). Other elements such as silicon, phosphorous, sulfur and manganese are transferred to a slag phase. In the EAF steelmaking process the chemical reactions are similar but generally less extensive.

After treating the metal in an OSM converter or an EAF it is further refined in the ladle. This is commonly called secondary refining or ladle metallurgy and the processes include deoxidation, desulfurization and vacuum degassing. For stainless steelmaking the liquid iron-chromium-nickel metal is refined in an argon-oxygen decarburization vessel (AOD), a vacuum oxygen decarburization vessel (VOD) or a similar type process.

In this volume the fundamental physical chemistry and kinetics relevant to the production of iron and steel is reviewed. Included are the critical thermodynamic data and other data on the properties of iron alloys and slags relevant to iron and steelmaking. This is followed by chapters on the support technologies for steelmaking including fuels and water, the production of industrial gases and the fundamentals and application of refractories. This volume then describes and analyzes the individual refining processes in detail including hot metal treatments, oxygen steelmaking, EAF steelmaking, AOD and VOD stainless steelmaking and secondary refining. Finally future alternatives to oxygen and EAF steelmaking are examined.

1.2 Historical Development of Modern Steelmaking

In the 10th edition of *The Making Shaping and Treating of Steel*[1] there is an excellent detailed review of early steelmaking processes such as the cementation and the crucible processes. A new discussion of these is not necessary. The developments of modern steelmaking processes such as the Bessemer, open hearth, oxygen steelmaking and EAF have also been chronicled in detail in the 10th edition. In this volume only a summary of these processes is given. For more details the reader is referred to the 10th edition or the works of W.T. Hogan[2,3].

1.2.1 Bottom-Blown Acid or Bessemer Process

This process, developed independently by William Kelly of Eddyville, Kentucky and Henry Bessemer of England, involved blowing air through a bath of molten pig iron contained in a bottom-blown vessel lined with acid (siliceous) refractories. The process was the first to provide a large scale method whereby pig iron could rapidly and cheaply be refined and converted into liquid steel. Bessemer's American patent was issued in 1856; although Kelly did not apply for a patent until 1857, he was able to prove that he had worked on the idea as early as 1847. Thus, both men held rights to the process in this country; this led to considerable litigation and delay, as discussed later. Lacking financial means, Kelly was unable to perfect his invention and Bessemer, in the face of great difficulties and many failures, developed the process to a high degree of perfection and it came to be known as the acid Bessemer process.

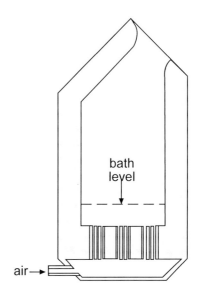

Fig. 1.1 Principle of the bottom blown converter. The blast enters the wind box beneath the vessel through the pipe indicated by the arrow and passes into the vessel through tuyeres set in the bottom of the converter.

The fundamental principle proposed by Bessemer and Kelly was that the oxidation of the major impurities in liquid blast furnace iron (silicon, manganese and carbon) was preferential and occurred before the major oxidation of iron; the actual mechanism differs from this simple explanation, as outlined in the discussion of the physical chemistry of steelmaking in Chapter 2. Further, they discovered that sufficient heat was generated in the vessel by the chemical oxidation of the above elements in most types of pig iron to permit the simple blowing of cold air through molten pig iron to produce liquid steel without the need for an external source of heat. Because the process converted pig iron to steel, the vessel in which the operation was carried out came to be known as a converter. The principle of the bottom blown converter is shown schematically in Fig. 1.1.

At first, Bessemer produced satisfactory steel in a converter lined with siliceous (acid) refractories by refining pig iron that, smelted from Swedish ores, was low in phosphorus, high in manganese, and contained enough silicon to meet the thermal needs of the process. But, when applied to irons which were higher in phosphorus and low in silicon and manganese, the process did not produce satisfactory steel. In order to save his process in the face of opposition among steelmakers, Bessemer built a steel works at Sheffield, England, and began to operate in 1860. Even when low phosphorus Swedish pig iron was employed, the steels first produced there contained much more than the admissible amounts of oxygen, which made the steel "wild" in the molds. Difficulty also was experienced with sulfur, introduced from the coke used as the fuel for melting the iron in cupolas, which contributed to "hot shortness" of the steel. These objections finally were overcome by the addition of manganese in the form of spiegeleisen to the steel after blowing as completed.

The beneficial effects of manganese were disclosed in a patent by R. Mushet in 1856. The carbon and manganese in the spiegeleisen served the purpose of partially deoxidizing the steel, which part of the manganese combined chemically with some of the sulfur to form compounds that either floated out of the metal into the slag, or were comparatively harmless if they remained in the steel.

As stated earlier, Bessemer had obtained patents in England and in this country previous to Kelly's application; therefore, both men held rights to the process in the United States.

The Kelly Pneumatic Process Company had been formed in 1863 in an arrangement with William Kelly for the commercial production of steel by the new process. This association included the Cambria Iron Company; E.B.Ward; Park Brothers and Company; Lyon, Shord and Company; Z.S. Durfee and , later, Chouteau, Harrison and Vale. This company, in 1864, built the first commercial Bessemer plant in this country, consisting of a 2.25 metric ton (2.50 net ton) acid lined vessel erected at the Wyandotte Iron Works, Wyandotte, Michigan, owned by Captain E.B. Ward. It may be mentioned that a Kelly converter was used experimentally at the Cambria Works, Johnstown, Pennsylvania as early as 1861.

As a result of the dual rights to the process a second group consisting of Messrs. John A. Griswold and John F. Winslow of Troy, New York and A. L. Holley formed another company under an arrangement with Bessemer in 1864. This group erected an experimental 2.25 metric ton (2.50 net ton) vessel in Troy, New York which commenced operations on February 16, 1865. After much litigation had failed to gain for either sole control of the patents for the pneumatic process in America, the rival organizations decided to combine their respective interests early in 1866. This larger organization was then able to combine the best features covered by the Kelly and Bessemer patents, and the application of the process advanced rapidly.

By 1871, annual Bessemer steel production in the United States had increased to approximately 40,800 metric tons (45,000 net tons), about 55% of the total steel production, which was produced by seven Bessemer plants.

Bessemer steel production in the United States over an extended period of years remained significant; however, raw steel is no longer being produced by the acid Bessemer process in the United States. the last completely new plant for the production of acid Bessemer steel ingots in the United States was built in 1949.

As already stated, the bottom blown acid process known generally as the Bessemer Process was the original pneumatic steelmaking process. Many millions of tons of steel were produced by this method. From 1870 to 1910, the acid Bessemer process produced the majority of the world's supply of steel.

The success of acid Bessemer steelmaking was dependent upon the quality of pig iron available which, in turn, demanded reliable supplies of iron ore and metallurgical coke of relatively high purity. At the time of the invention of the process, large quantities of suitable ores were available, both abroad and in the United States. With the gradual depletion of high quality ores abroad (particularly low phosphorus ores) and the rapid expansion of the use of the bottom blown basic pneumatic, basic open hearth and basic oxygen steelmaking processes over the years, acid Bessemer steel production has essentially ceased in the United Kingdom and Europe.

In the United States, the Mesabi Range provided a source of relatively high grade ore for making iron for the acid Bessemer process for many years. In spite of this, the acid Bessemer process declined from a major to a minor steelmaking method in the United States and eventually was abandoned.

The early use of acid Bessemer steel in this country involved production of a considerable quantity of rail steel, and for many years (from its introduction in 1864 until 1908) this process was the principal steelmaking process. Until relatively recently, the acid Bessemer process was used principally in the production of steel for buttwelded pipe, seamless pipe, free machining bars, flat rolled products, wire, steel castings, and blown metal for the duplex process.

Fully killed acid Bessemer steel was used for the first time commercially by United States Steel Corporation in the production of seamless pipe. In addition, dephosphorized acid Bessemer steel was used extensively in the production of welded pipe and galvanized sheets.

1.2.2 Basic Bessemer or Thomas Process

The bottom blown basic pneumatic process, known by several names including Thomas, Thomas-Gilchrist or basic Bessemer process, was patented in 1879 by Sidney G. Thomas in England. The process, involving the use of the basic lining and a basic flux in the converter, made it possible to use the pneumatic method for refining pig irons smelted from the high phosphorus ores common to many sections of Europe. The process (never adopted in the United States) developed much more rapidly in Europe than in Great Britain and, in 1890, European production was over 1.8 million metric tons (2 million net tons) as compared with 0.36 million metric tons (400,000 net tons) made in Great Britain.

The simultaneous development of the basic open hearth process resulted in a decline of production of steel by the bottom blown basic pneumatic process in Europe and, by 1904, production of basic open hearth steel there exceeded that of basic pneumatic steel. From 1910 on, the bottom blown basic pneumatic process declined more or less continuously percentage-wise except for the period covering World War II, after which the decline resumed.

1.2.3 Open Hearth Process

Karl Wilhelm Siemens, by 1868, proved that it was possible to oxidize the carbon in liquid pig iron using iron ore, the process was initially known as the "pig and ore process."

Briefly, the method of Siemens was as follows. A rectangular covered hearth was used to contain the charge of pig iron or pig iron and scrap. (See Fig.1.2) Most of the heat required to promote the chemical reactions necessary for purification of the charge was provided by passing

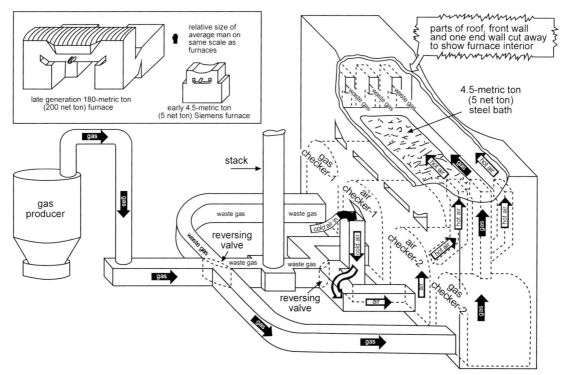

Fig. 1.2 Schematic arrangement of an early type of Siemens furnace with about a 4.5 metric ton (5 net ton) capacity. The roof of this design (which was soon abandoned) dipped from the ends toward the center of the furnace to force the flame downward on the bath. Various different arrangements of gas and air ports were used in later furnaces. Note that in this design, the furnace proper was supported on the regenerator arches. Flow of gas, air and waste gases were reversed by changing the position of the two reversing valves. The inset at the upper left compares the size of one of these early furnaces with that of a late generation 180 metric ton (200 net ton) open hearth.

burning fuel gas over the top of the materials. The fuel gas, with a quantity of air more than sufficient to burn it, was introduced through ports at each end of the furnace, alternately at one end and then the other. The products of combustion passed out of the port temporarily not used for entrance of gas and air, and entered chambers partly filled with brick checkerwork. This checkerwork, commonly called checkers, provided a multitude of passageways for the exit of the gases to the stack. During their passage through the checkers, the gases gave up a large part of their heat to the brickwork. After a short time, the gas and air were shut off at the one end and introduced into the furnace through the preheated checkers, absorbing some of the heat stored in these checkers The gas and air were thus preheated to a somewhat elevated temperature, and consequently developed to a higher temperature in combustion than could be obtained without preheating. In about twenty minutes, the flow of the gas and air was again reversed so that they entered the furnace through the checkers and port used first; and a series of such reversals, occurring every fifteen or twenty minutes was continued until the heat was finished. The elements in the bath which were oxidized both by the oxygen of the air in the furnace atmosphere and that contained in the iron ore fed to the bath, were carbon, silicon and manganese, all three of which could be reduced to as low a limit as was possible in the Bessemer process. Of course, a small amount of iron remains or is oxidized and enters the slag.

Thus, as in all other processes for purifying pig iron, the basic principle of the Siemens process was that of oxidation. However, in other respects, it was unlike any other process. True, it resembled the puddling process in both the method and the agencies employed, but the high temperatures attainable in the Siemens furnace made it possible to keep the final product molten and free of entrapped slag. The same primary result was obtained as in the Bessemer process, but by a different method and through different agencies, both of which imparted to steel made by the new process properties somewhat different from Bessemer steel, and gave the process itself certain metallurgical advantages over the older pneumatic process, as discussed later in this section.

As would be expected, many variations of the process, both mechanical and metallurgical, have been worked out since its original conception. Along mechanical lines, various improvements in the design, the size and the arrangement of the parts of the furnace have been made. Early furnaces had capacities of only about 3.5–4.5 metric tons (4–5 net tons), which modern furnaces range from about 35–544 metric tons (40–600 net tons) in capacity, with the majority having capacities between about 180–270 metric tons (200–300 net tons).

The Siemens process became known more generally, as least in the United States, as the open hearth process. The name "open hearth" was derived, probably, from the fact that the steel, while melted on a hearth under a roof, was accessible through the furnace doors for inspection, sampling, and testing.

The hearth of Siemens' furnace was of acid brick construction, on top of which the bottom was made up of sand, essentially as in the acid process of today. Later, to permit the charging of limestone and use of a basic slag for removal of phosphorus, the hearth was constructed with a lining of magnesite brick, covered with a layer of burned dolomite or magnesite, replacing the siliceous bottom of the acid furnace. These furnaces, therefore, were designated as basic furnaces, and the process carried out in them was called the basic process. The pig and scrap process was originated by the Martin brothers, in France, who, by substituting scrap for the ore in Siemens' pig and ore process, found it possible to dilute the change with steel scrap to such an extent that less oxidation was necessary.

The advantages offered by the Siemens process may be summarized briefly as follows:

1. By the use of iron ore as an oxidizing agent and by the external application of heat, the temperature of the bath was made independent of the purifying reactions, and the elimination of impurities could be made to take place gradually, so that both the temperature and composition of the bath were under much better control than in the Bessemer process.

2. For the same reasons, a greater variety of raw materials could be used (particularly scrap, not greatly consumable in the Bessemer converter) and a greater variety of products could be made by the open hearth process than by the Bessemer process.

3. A very important advantage was the increased yield of finished steel from a given quantity of pig iron as compared to the Bessemer process, because of lower inherent sources of iron loss in the former, as well as because of recovery of the iron content of the ore used for oxidation in the open hearth.

4. Finally, with the development of the basic open hearth process, the greatest advantage of Siemens' over the acid Bessemer process was made apparent, as the basic open hearth process is capable of eliminating phosphorus from the bath. While this element can be removed also in the basic Bessemer (Thomas-Gilchrist) process, it is to be noted that, due to the different temperature conditions, phosphorus is eliminated before carbon in the basic open hearth process, whereas the major proportion of phosphorus is not oxidized in the basic Bessemer process until after carbon in the period termed the afterblow. Hence, while the basic Bessemer process requires a pig iron with a phosphorus content of 2.0% or more in order to maintain the temperature high enough for the afterblow, the basic open hearth process permits the economical use of iron of any phosphorus content up to 1.0%. In the United States, this fact was of importance since it made available immense iron ore deposits which could not be utilized otherwise because of their phosphorus content, which was too high to permit their use in the acid Bessemer or acid open hearth process and too low for use in the basic Bessemer process.

The open hearth process became the dominant process in the United States. As early as 1868, a small open hearth furnace was built at Trenton, New Jersey, but satisfactory steel at a reasonable cost did not result and the furnace was abandoned. Later, at Boston, Massachusetts, a successful furnace was designed and operated, beginning in 1870. Following this success, similar furnaces were built at Nashua, New Hampshire and in Pittsburgh, Pennsylvania, the latter by Singer, Nimick and Company, in 1871. The Otis Iron and Steel Company constructed two 6.3 metric ton (7 net ton) furnaces at their Lakeside plant at Cleveland, Ohio in 1874. Two 13.5 metric ton (15 net ton) furnaces were added to this plant in 1878, two more of the same size in 1881, and two more in 1887. All of these furnaces had acid linings, using a sand bottom for the hearths.

The commercial production of steel by the basic process was achieved first at Homestead, Pennsylvania. The initial heat was tapped March 28, 1888. By the close of 1890, there were 16 basic open hearth furnaces operating. From 1890 to 1900, magnesite for the bottom began to be imported regularly and the manufacture of silica refractories for the roof was begun in American plants. For these last two reasons, the construction of basic furnaces advanced rapidly and, by 1900, furnaces larger than 45 metric tons (50 net tons) were being planned.

While the Bessemer process could produce steel at a possibly lower cost above the cost of materials, it was restricted to ores of a limited phosphorus content and its use of scrap was also limited. The open hearth was not subject to these restrictions, so that the annual production of steel by the open hearth process increased rapidly, and in 1908, passed the total tonnage produced yearly by the Bessemer process. Total annual production of Bessemer steel ingots decreased rather steadily after 1908, and has ceased entirely in the United States. In addition to the ability of the basic open hearth furnace to utilize irons made from American ores, as discussed earlier, the main reasons for proliferation of the open hearth process were its ability to produce steels of many compositions and its ability to use a large proportion of iron and steel scrap, if necessary. Also steels made by any of the pneumatic processes that utilize air for blowing contain more nitrogen than open hearth steels; this higher nitrogen content made Bessemer steel less desirable than open hearth steel in some important applications.

With the advent of oxygen steelmaking which could produce steel in a fraction of the time required by the open hearth process, open hearth steelmaking has been completely phased out in the United States. The last open hearth meltshop closed at Geneva Steel Corporation at Provo, Utah in 1991. Worldwide there are only a relative few open hearths still producing steel.

1.2.4 Oxygen Steelmaking

Oxygen steelmaking has become the dominant method of producing steel from blast furnace hot metal. Although the use of gaseous oxygen (rather than air) as the agent for refining molten pig iron and scrap mixtures to produce steel by pneumatic processes received the attention of numerous investigators from Bessemer onward, it was not until after World War II that commercial success was attained.

Blowing with oxygen was investigated by R. Durrer and C. V. Schwarz in Germany and by Durrer and Hellbrugge in Switzerland. Bottom-blown basic lined vessels of the designs they used proved unsuitable because the high temperature attained caused rapid deterioration of the refractory tuyere bottom; blowing pressurized oxygen downwardly against the top surface of the molten metal bath, however, was found to convert the charge to steel with a high degree of thermal and chemical efficiency.

Plants utilizing top blowing with oxygen have been in operation since 1952–53 at Linz and Donawitz in Austria. These operations, sometimes referred to as the Linz-Donawitz or L-D process were designed to employ pig iron produced from local ores that are high in manganese and low in phosphorus; such iron is not suitable for either the acid or basic bottom blown pneumatic process utilizing air for blowing. The top blown process, however, is adapted readily to the processing of blast furnace metal of medium and high phosphorus contents and is particularly attractive where it is desirable to employ a steelmaking process requiring large amounts of hot metal as the principal source of metallics. This adaptability has led to the development of numerous variations in application of the top-blown principle. In its most widely used form, which also is the form used in the United States, the top blown oxygen process is called the basic oxygen steelmaking process (BOF for short) or in some companies the basic oxygen process (BOP for short).

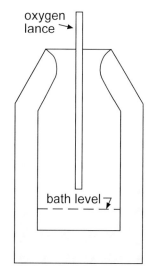

Fig. 1.3 Principle of the top blown converter. Oxygen of commercial purity, at high pressure and velocity is blown downward vertically into surface of bath through a single water cooled pipe or lance.

The basic oxygen process consists essentially of blowing oxygen of high purity onto the surface of the bath in a basic lined vessel by a water cooled vertical pipe or lance inserted through the mouth of the vessel (Fig. 1.3).

A successful bottom blown oxygen steelmaking process was developed in the 1970s. Based on development in Germany and Canada and known as the OBM process, or Q-BOP in the United States, the new method has eliminated the problem of rapid bottom deterioration encountered in earlier attempts to bottom blow with oxygen. The tuyeres (Fig. 1.4), mounted in a removable bottom,

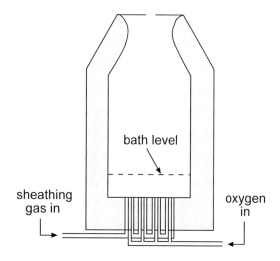

Fig. 1.4 Schematic cross-section of an OBM (Q-BOP) vessel, showing how a suitable gas is introduced into the tuyeres to completely surround the stream of gaseous oxygen passing through the tuyeres into the molten metal bath.

are designed in such a way that the stream of gaseous oxygen passing through a tuyere into the vessel is surrounded by a sheath of another gas. The sheathing gas is normally a hydrocarbon gas such as propane or natural gas. Vessel capacities of 200 tons and over, comparable to the capacities of typical top blown BOF vessels, are commonly used.

The desire to improve control of the oxygen pneumatic steelmaking process has led to the development of various combination blowing processes. In these processes, 60–100% of the oxygen required to refine the steel is blown through a top mounted lance (as in the conventional BOF) while additional gas (such as oxygen, argon, nitrogen, carbon dioxide or air) is blown through bottom mounted tuyeres or permeable brick elements. The bottom blown gas results in improved mixing of the metal bath, the degree of bath mixing increasing with increasing bottom gas flow rate. By varying the type and flow rate of the bottom gas, both during and after the oxygen blow, specific metallurgical reactions can be controlled to attain desired steel compositions and temperatures. There are, at present many different combination blowing processes, which differ in the type of bottom gas used, the flow rates of bottom gas that can be attained, and the equipment used to introduce the bottom gas into the furnace. All of the processes, to some degree, have similar advantages. The existing combination blowing furnaces are converted conventional BOF furnaces and range in capacity from about 60 tons to more than 300 tons. The conversion to combination blowing began in the late 1970s and has continued at an accelerated rate. Further details of these processes are given in Chapter 9.

Two other oxygen blown steelmaking, the Stora-Kaldo process and the Rotor process, did not gain wide acceptance.

1.2.5 Electric Furnace Steelmaking

In the past twenty years there has been a significant growth in electric arc furnace (EAF) steelmaking. When oxygen steelmaking began replacing open hearth steelmaking excess scrap became available at low cost because the BOF melts less scrap than an open hearth. Also for fully developed countries like the United States, Europe and Japan the amount of obsolete scrap in relationship to the amount of steel required increased, again reducing the price of scrap relative to that of hot metal produced from ore and coal. This economic opportunity arising from low cost scrap and the lower capital cost of an EAF compared to integrated steel production lead to the growth of the mini-mill or scrap based EAF producer. At first the mini-mills produced lower quality long products such as reinforcing bars and simple construction materials. However with the advent of thin slab casting a second generation of EAF plants has developed which produce flat products. In the decade of the 1990's approximately 15–20 million tons of new EAF capacity has been built or planned in North America alone. As discussed later and in Chapter 10 in detail, the EAF has evolved and improved its efficiency tremendously. Large quantities of scrap substitutes such as direct reduced iron and pig iron are now introduced in the EAF as well as large quantities of oxygen.

It has been said that arc-type furnaces had their beginning in the discovery of the carbon arc by Sir Humphrey Davy in 1800, but it is more proper to say that their practical application began with the work of Sir William Siemens, who in 1878 constructed, operated and patented furnaces operating on both the direct arc and indirect arc principles.

At this early date, the availability of electric power was limited and its cost high; also, carbon electrodes of the quality required to carry sufficient current for steel melting had not been developed. Thus the development of the electric melting furnace awaited the expansion of the electric power industry and improvement in carbon electrodes.

The first successful commercial EAF was a direct arc steelmaking furnace which was placed in operation by Heroult in 1899. The Heroult patent stated in simple terms, covered single-phase or multi-phase furnaces with the arcs in series through the metal bath. This type of furnace, utilizing three phase power, has been the most successful of the electric furnaces in the production of steel. The design and operation of modern electric arc furnaces are discussed in Chapter 10.

In the United States there were no developments along arc furnace lines until the first Heroult furnace was installed in the plant of the Halcomb Steel Company, Syracuse New York, which made its first heat on April 5, 1906. This was a single phase, two electrode, rectangular furnace of 3.6 metric tons (4 net tons) capacity. Two years later a similar but smaller furnace was installed at the Firth-Sterling Steel Company, McKeesport, Pennsylvania, and in 1909, a 13.5 metric ton (15 net ton) three phase furnace was installed in the South Works of the Illinois Steel Company. The latter was, at that time, the largest electric steelmaking furnace in the world, and was the first round (instead of rectangular) furnace. It operated on 25-cycle power at 2200 volts and tapped its first heat on May 10, 1909.

From 1910 to 1980 nearly all the steelmaking EAFs built had three phase alternating current (AC) systems. In the 1980s single electrode direct current (DC) systems demonstrated some advantages over the conventional AC furnaces. In the past 15 years a large percentage of the new EAFs built were DC. Commercial furnaces vary in size from 10 tons to over 300 tons. A typical state-of-the-art furnace is 150–180 tons, has several natural gas burners, uses considerable oxygen (30m^3/ton), has eccentric bottom tapping and often is equipped with scrap preheating. A schematic of a typical AC furnace is shown in Fig. 1.5. The details concerning these furnaces and their advantages are discussed in detail in Chapter 10.

Another type of electric melting furnace, used to a certain extent for melting high-grade alloys, is the high frequency coreless induction furnace which gradually replaced the crucible process in the production of complex, high quality alloys used as tool steels. It is used also for remelting scrap from fine steels produced in arc furnaces, melting chrome-nickel alloys, and high manganese scrap, and, more recently, has been applied to vacuum steelmaking processes.

The induction furnace had its inception abroad and first was patented by Ferranti in Italy in 1877. This was a low frequency furnace. It had no commercial application until Kjellin installed and

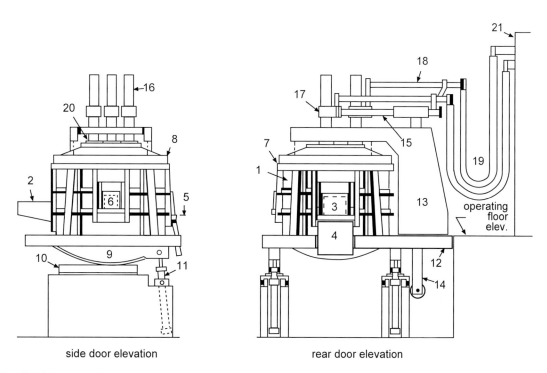

Fig. 1.5 Schematic of a typical AC electric arc furnace. Elements are identified as follows:
1. shell
2. pouring spout
3. rear door
4. slag apron
5. sill line
6. side door
7. bezel ring
8. roof ring
9. rocker
10. rocker rail
11. tilt cylinder
12. main (tilting) platform
13. roof removal jib structure
14. electrode mast stem
15. electrode mast arm
16. electrode
17. electrode holder
18. bus tube
19. secondary power cables
20. electrode gland
21. electrical equipment vault

operated one in Sweden. The first large installation of this type was made in 1914 at the plant of the American Iron and Steel Company in Lebanon, Pennsylvania, but was not successful. Low frequency furnaces have operated successfully, especially in making stainless steel.

A successful development using higher frequency current is the coreless high frequency induction furnace. The first coreless induction furnaces were built and installed by the Ajax Electrothermic Corporation, who also initiated the original researches by E.F. Northrup leading to the development of the furnace. For this reason, the furnace is often referred to as the Ajax-Northrup furnace.

The first coreless induction furnaces for the production of steel on a commercial scale were installed at Sheffield, England, and began the regular production of steel in October, 1927. The first commercial steel furnaces of this type in the United States were installed by the Heppenstall Forge and Knife Company, Pittsburgh, Pennsylvania, and were producing steel regularly in November, 1928. Each furnace had a capacity of 272 kilograms (600 pounds) and was served by a 150 kVA motor-generator set transforming 60 hertz current to 860 hertz.

Electric furnace steelmaking has improved significantly in the past twenty years. The tap to tap time, or time required to produce steel, has decreased from about 200 minutes to as little as 55 minutes, electrical consumption has decreased from over 600 kWh per ton to less than 400 and electrical consumption has been reduced by 70%. These have been the result of a large number of technical developments including ultra high power furnaces, long arc practices using foamy slags, the increased use of oxygen and secondary refining. With new EAF plants using scrap alternatives to supplement the scrap charge and the production of higher quality steels, EAF production may exceed 50% in the United States and 40% in Europe and Japan by the year 2010.

Electric furnaces of other various types have been used in the production of steel. These include vacuum arc remelting furnaces (VAR), iron smelting furnaces and on an experimental basis plasma type melting and reheating furnaces. Where appropriate these are discussed in detail in this volume.

1.3 Evolution in Steelmaking by Process

The proportion of steel produced by the major processes for the United States and the World are given in Fig. 1.6 and Fig. 1.7, respectively. The relative proportions differ widely from country to country depending on local conditions and when the industry was built.

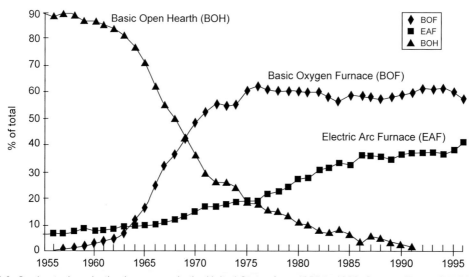

Fig. 1.6 Crude steel production by process in the United States from 1955 to 1996. Source: International Iron and Steel Institute.

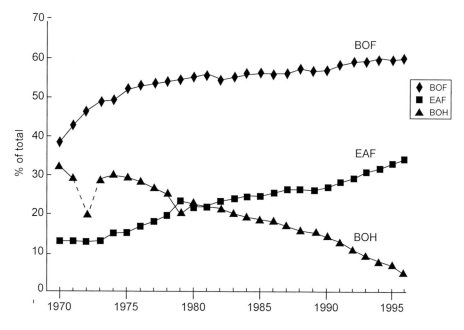

Fig. 1.7 Crude steel production by process for the World from 1970 to 1996. Source: International Iron and Steel Institute.

For the United States in 1955 nearly 90% of the steel produced was in by the basic open hearth processes. The last new open hearth shop was opened in 1958, approximately three years after the first BOF. Starting in about 1960 the BOF began to replace the BOH. By 1975 BOF production reached about 62% of the total. The remaining open hearth plants were either completely abandoned or converted to EAF plants. The last open hearth in the United States at Geneva Steel was closed in 1991. Also as the BOF melts less scrap than the BOH, electric furnace production grew because of the surplus and relatively low cost of scrap. For most of the period from 1970 to 1990 scrap costs were significantly lower than the cost of producing hot metal. Due to the increased price of scrap costs and more efficient blast furnace operation, in the past few years the costs of hot metal and scrap have become similar. Nevertheless electric furnace production continues to grow in the United States increasing by over 15 million tons of production in the the 1990s.

World production has followed a generally similar trend, however, the patterns in different regions vary greatly. For Japan and more recently Korea and Brazil, where the steel industry was built or completely rebuilt after 1955, the BOF was the dominant process earlier. In other regions such as the CIS (former Soviet Union), Eastern Europe and India open hearths were extensively used through the 1980s. The choice of steelmaking process will continue to depend on local conditions but a gradual growth in EAF production will continue. Much of the new production will be in smaller developing countries and the scrap will be supplemented with other forms of iron including direct reduced iron and pig iron.

In general the EAF share of production worldwide will continue to grow. However scrap alone can never supply all of the iron requirements. For at least 30 years the blast furnace-BOF steelmaking route will be required. In 20 to 30 years, the scrap/DRI-EAF share of production may grow to 50% worldwide.

In the future the BOF and EAF processes will continue to evolve. In particular more gaseous oxygen will be used in the EAF reducing electric energy consumption and be operated more as a hybrid process between the BOF and EAF. Processes other than the BOF and EAF, which are discussed in Chapter 13 may be commercialized. However even by 2010 or 2020 it is doubtful any will be significant and attain 10% of total production.

1.4 Structure of This Volume

Following this introductory chapter, a major chapter on the fundamentals of iron and steelmaking are given in Chapter 2, including the basic thermodynamics and kinetics along with reference data on metal and slag systems. The next four chapters deal with the production and use of major materials for steelmaking such as refractories, industrial gases and fuels and utilities.

The first refining operations in steelmaking are hot metal treatments, which are described in Chapter 7. This is followed by oxygen steelmaking, discussed in Chapters 8 and 9. In this edition, the electric furnace chapter (Chapter 10) has been greatly expanded to reflect the advances in this process in the last decade. Ladle and other secondary refining processes, including the AOD for stainless steel production, are dealt with in Chapters 11 and 12. Finally, other and developing future steelmaking processes are examined in Chapter 13.

References

1. *The Making, Shaping and Treating of Steel*, 10th ed., AISE, Pittsburgh, PA, 1985.

2. W.T.Hogan, *The Development of American Heavy Industry in the Twentieth Century*, Fordham University, New York, NY, 1954.

3. W.T.Hogan, *Economic History of the Iron and Steel Industry in the United States,* Heath, Lexington, MA, 1971.

Chapter 2

Fundamentals of Iron and Steelmaking

E.T. Turkdogan, Consultant: Pyrometallurgy & Thermochemistry
R.J. Fruehan, Professor, Carnegie Mellon University

There have been tremendous improvements in iron and steelmaking processes in the past twenty years. Productivity and coke rates in the blast furnace and the ability to refine steel to demanding specifications have been improved significantly. Much of this improvement is based on the application of fundamental principles and thermodynamic and kinetic parameters which have been determined. Whereas, many future improvements will be forthcoming in steelmaking equipment, process improvements resulting from the application of fundamental principles and data will likewise continue.

In this chapter the basic principles of thermodynamics and kinetics are reviewed and the relevant thermodynamic data and properties of gases, metals and slags relevant to iron and steelmaking are presented. These principles and data are then applied to ironmaking, steelmaking and secondary refining processes. These principles and data are also used in subsequent chapters in this volume.

In writing this chapter, an attempt has been made to limit the discussion to an average level suitable for the students of metallurgy pursuing graduate or post-graduate education as well as for those with some scientific background engaged in the iron and steel industry. It is assumed that the reader has some basic knowledge of chemistry, physics and mathematics, so that the chapter can be devoted solely to the discussion of the chemistry of the processes.

2.1 Thermodynamics

2.1.1 Ideal Gas

A gas which obeys the simple gas laws is called an ideal gas satisfying the following relation:

$$PV = nRT \qquad (2.1.1)$$

where n is the number of mols and R the universal molar gas constant.

For one mol of an ideal gas at 273.16K and 1 atm pressure, the value of the molar gas constant is:

$$R = \frac{1 \times 22.414}{273.16} = 0.08205 \text{ l atm mol}^{-1} \text{ K}^{-1}$$

For pressure in Pa ($\equiv Nm^{-2} \equiv Jm^{-3}$) and volume in m^3,

$$R = \frac{1.01325 \times 10^5 \times 22.414 \times 10^{-3}}{273.16} = 8.314 \text{ J mol}^{-1} \text{ K}^{-1}$$

In a gas mixture containing $n_1, n_2, n_3...$ number of mols of gases occupying a volume V at a total pressure P, the partial pressures of the constituent gaseous species are as given below.

$$p_1 = \frac{n_1}{n_1 + n_2 + n_3 + ...} \times P \qquad (2.1.2)$$

$$P = p_1 + p_2 + p_3 + ... \qquad (2.1.3)$$

The following equations are for a given mass of gas at constant pressure, volume and temperature:

Constant pressure (isobaric) $\qquad \dfrac{V_1}{T_1} = \dfrac{V_2}{T_2} = \dfrac{V_3}{T_3}... \qquad (2.1.4)$

Constant volume (isochoric) $\qquad \dfrac{P_1}{T_1} = \dfrac{P_2}{T_2} = \dfrac{P_3}{T_3}... \qquad (2.1.5)$

Constant temperature (isothermal) $\qquad P_1V_1 = P_2V_2 = P_3V_3... \qquad (2.1.6)$

Generally speaking, deviation from the ideal gas equation becomes noticeable with easily liquefiable gases and at low temperatures and high pressures. The behavior of gases becomes more ideal with decreasing pressure and increasing temperature. The nonideality of gases, the extent of which depends on the nature of the gas, temperature and pressure, is attributed to two major causes: (1) van der Waals' forces and (2) chemical interaction between the different species of gas molecules or atoms.

2.1.2 Thermodynamic Laws

2.1.2.1 The First Law

The first law of thermodynamics is based on the concept of conservation of energy. When there is interaction between systems, the gain of energy of one of the systems is equal to the loss of the other system. For example, the quantity of heat required to decompose a compound into its elements is equal to the heat generated when that compound is formed from its elements.

2.1.2.1.1 Enthalpy (heat content) The internal energy of a system includes all forms of energy other than the kinetic energy. Any exchange of energy between a system and its surroundings, resulting from a change of state, is manifested as heat and work.

When a system expands against a constant external pressure P, resulting in an increase of volume ΔV, the work done by the system is

$$w = P\Delta V = P(V_B - V_A)$$

Since this work is done by the system against the surroundings, the system absorbs a quantity of heat q and the energy E of the system increases in passing from state A to state B.

$$\Delta E = E_B - E_A = q - P\Delta V = q - P(V_B - V_A)$$

Upon re-arranging this equation, we have

$$(E_B + PV_B) - (E_A + PV_A) = q$$

The quantity E + PV is represented by a single symbol H, thus

$$\Delta H = q = (E_B + PV_B) - (E_A + PV_A) \qquad (2.1.7)$$

The function H is known as enthalpy or heat content.

There are two fundamental thermochemical laws which express the first law specifically in terms of enthalpy. The first principle derived by Lavoisier and Laplace (1780) states that "the quantity of heat required to decompose a compound into its elements is equal to the heat evolved when that compound is formed from its elements"; i.e. the heat of decomposition of a compound is numerically equal to its heat of formation, but of opposite sign. The second principle is that discovered by Hess (1840); it states that "the heat of reaction depends only on the initial and final states, and not on the intermediate states through which the system may pass."

2.1.2.1.2 Heat Capacity The heat capacity of a substance is defined as the quantity of heat required to raise the temperature by one degree. The heat capacity of 1 g of a substance is called the specific heat. The heat capacity of 1 g-molecule (abbreviated as mol) is called the molar heat capacity.

The variation of energy, at constant volume, and of enthalpy, at constant pressure, with temperature gives the heat capacity of the system, thus

$$C_V = \left(\frac{\partial E}{\partial T}\right)_V \tag{2.1.8}$$

$$C_p = \left(\frac{\partial H}{\partial T}\right)_p \tag{2.1.9}$$

For an ideal gas the difference between the molar heat capacities at constant pressure, C_p, and constant volume, C_V, is equal to the molar gas constant.

$$C_p - C_V = R \tag{2.1.10}$$

Because of experimental convenience, the heat capacity is determined under conditions of constant pressure (usually atmospheric).

From the temperature dependence of heat capacity at constant pressure, the enthalpy change is obtained by integrating equation 2.1.9.

$$H_{T_2}^\circ - H_{T_1}^\circ = \int_{T_1}^{T_2} C_p dT \tag{2.1.11}$$

Above 298 K, the temperature dependence of C_p is represented by:

$$C_p = a + bT - cT^{-2} \tag{2.1.12}$$

$$\Delta H = \int_{298}^{T} \left(a + bT - cT^{-2}\right) dT \tag{2.1.13}$$

where the coefficients, a, b and c are derived from C_p calorimetric measurements at different temperatures.

In recent compilations of thermochemical data, the H values are tabulated at 100 K intervals for the convenience of users.

2.1.2.1.3 Standard State The enthalpy is an extensive property of the system, and only the change in heat content with change of state can be measured. A standard reference state is chosen for each element so that any change in the heat content of the element is referred to its standard state, and this change is denoted by ΔH°.

The natural state of elements at 25°C and 1 atm pressure is by convention taken to be the reference state. On this definition, the elements in their standard states have zero heat contents.

The heat of formation of a compound is the heat absorbed or evolved in the formation of 1 g-mol of the compound from its constituent elements in their standard states, denoted by ΔH_{298}°.

2.1.2.1.4 Enthalpy of Reaction The change of enthalpy accompanying a reaction is given by the difference between the enthalpies of the products and those of the reactants.

For an isobaric and isothermal reaction,

$$A + B = C + D \tag{2.1.14}$$

the enthalpy change is given by:

$$\Delta H = (\Delta H°C + \Delta H°D) - (\Delta H°A + \Delta H°B) \tag{2.1.15}$$

By convention, H is positive (+) for endothermic reactions, i.e. heat absorption, and H is negative (–) for exothermic reactions, i.e. heat evolution.

Temperature effect:

$$\Delta H°_T = \Sigma \Delta H°_{298}(\text{products}) - \Sigma \Delta H°_{298}(\text{reactants}) + \int_{298}^{T} \left[\Sigma C_p(\text{products}) - \Sigma C_p(\text{reactants})\right] dT \tag{2.1.16}$$

$$\Delta H°_T = \Delta H°_{298} + \int_{298}^{T} (\Delta C_p) dT \tag{2.1.17}$$

The following are some examples of the special terms of the heat of reaction.

Enthalpy or heat of formation	Fe + ½O$_2$	→	FeO
Heat of combustion	C + O$_2$	→	CO$_2$
Heat of decomposition	2CO	→	C + CO$_2$
Heat of calcination	CaCO$_3$	→	CaO + CO$_2$
Heat of fusion (melting)	Solid	→	Liquid
Heat of sublimation	Solid	→	Vapor
Heat of vaporization	Liquid	→	Vapor
Heat of solution	Si(l)	→	[Si] (dissolved in Fe)

2.1.2.1.5 Adiabatic Reactions When a reaction occurs in a thermally insulated system, i.e. no heat exchange between the system and its surroundings, the temperature of the system will change in accordance with the heat of reaction.

As an example, let us consider the internal oxidation of unpassivated direct reduced iron (DRI) in a stockpile, initially at 25°C. The enthalpy of reaction at 298K is

$$Fe + \tfrac{1}{2}O_2 \to FeO, \quad \Delta H°_{298} = -267 \text{ kJ mol}^{-1} \tag{2.1.18}$$

The heat balance calculation is made for 1000 kg Fe in the stockpile with 150 kg FeO formed in oxidation. The heat absorbed by the stockpile is $150 \times 10^3/72 \times 267$ kJ and the temperature rise is calculated as follows:

$$Q = [n_{Fe}(C_p)_{Fe} + n_{FeO}(C_p)_{FeO}](T - 298)$$
$$n_{Fe} = 17{,}905 \text{ g-mol for 1000 kg Fe}$$
$$n_{FeO} = 2087.7 \text{ g-mol for 150 kg FeO}$$
$$C_p(Fe) = 0.042 \text{ kJ mol}^{-1}\text{K}^{-1}$$
$$C_p(FeO) = 0.059 \text{ kJ mol}^{-1}\text{K}^{-1}$$
$$\therefore Q = 557{,}416 = (752 + 123)(T - 298)$$

With this adiabatic reaction, the stockpile temperature increases to T = 935K (662°C).

The moisture in the stockpile will react with iron and generate H$_2$ which will ignite at the elevated stockpile temperature. This has been known to happen when DRI briquettes were not adequately passivated against oxidation.

2.1.2.2 The Second Law

The law of dissipation of energy states that all natural processes occurring without external interference are spontaneous (irreversible processes). For example, heat conduction from a hot to a cold part of the system. The spontaneous processes cannot be reversed without some change in the system brought about by external interference.

2.1.2.2.1 Entropy The degree of degradation of energy accompanying spontaneous, hence irreversible, processes depends on the magnitude of heat generation at temperature T and temperatures between which there is heat flow.

The quantity q/T is a measure of degree of irreversibility of the process, the higher the quantity q/T, the greater the irreversibility of the process. The quantity q/T is called the increase in entropy. In a complete cycle of all reversible processes the sum of the quantities $\Sigma q/T$ is zero.

The thermodynamic quantity, entropy S, is defined such that for any reversible process taking place isothermally at constant pressure, the change in entropy is given by

$$dS = \frac{dH}{T} = \frac{C_p}{T} dT = C_p \, d(\ln T) \qquad (2.1.19)$$

2.1.2.3 The Third Law

The heat theorem put forward by Nernst (1906) constitutes the third law of thermodynamics: 'the entropy of any homogeneous and ordered crystalline substance, which is in complete internal equilibrium, is zero at the absolute zero temperature.' Therefore, the integral of equation given above has a finite value at temperature T as shown below.

$$S_T = \int_0^T C_p \, d(\ln T) \qquad (2.1.20)$$

The entropy of reaction is

$$\Delta S = \Sigma S(\text{products}) - \Sigma S(\text{reactants}) \qquad (2.1.21)$$

and the entropy of fusion at the melting point T_m is

$$\Delta S_m = \frac{\Delta H_m}{T_m} \qquad (2.1.22)$$

2.1.2.4 Gibbs Free Energy

From a combined form of the first and second laws of thermodynamics, Gibbs derived the free energy equation for a reversible process at constant pressure and temperature.

$$G = H - TS \qquad (2.1.23)$$

The Gibbs free energy is also known as the chemical potential.

When a system changes isobarically and isothermally from state A to state B, the change in the free energy is

$$G_B - G_A = \Delta G = \Delta H - T\Delta S \qquad (2.1.24)$$

During any process which proceeds spontaneously at constant pressure and temperature, the free energy of the system decreases. That is, the reaction is thermodynamically possible when $\Delta G < 0$. However, the reaction may not proceed at a perceptible rate at lower tempertures, if the activation energy required to overcome the resistance to reaction is too high. If $\Delta G > 0$, the reaction will not take place spontaneously.

As in the case of enthalpy, the free energy is a relative thermodynamic property with respect to the standard state, denoted by $\Delta G°$.

The variation of the standard free energy change with temperature is given by:

$$\Delta G°_T = \Delta H°_{298} + \int_{298}^T \Delta C_p \, dT - T\Delta S°_{298} - T \int_{298}^T \frac{\Delta C_p}{T} dT \qquad (2.1.25)$$

2.1.2.4.1 Generalization of Entropy Change of Reaction

1. When there is volume expansion accompanying a reaction, i.e. gas evolution, at constant pressure and temperature the entropy change is positive, hence, ΔG decreases with an increasing temperature.

$$C + CO_2 = 2CO$$
$$\Delta G° = 166{,}560 - 171.0T \text{ J} \tag{2.1.26}$$

2. When there is volume contraction, i.e. gas consumed in the reaction, at constant pressure and temperature the entropy change is negative, hence ΔG increases with an increasing temperature.

$$H_2 + \tfrac{1}{2}S_2 = H_2S$$
$$\Delta G° = -91{,}600 + 50.6T \text{ J} \tag{2.1.27}$$

3. When there is little or no volume change the entropy change is close to zero, hence temperature has little effect on ΔG.

$$C + O_2 = CO_2$$
$$\Delta G° = 395{,}300 - 0.5T \text{ J} \tag{2.1.28}$$

2.1.2.4.2 Selected Free Energy Data For many reactions, the temperature dependence of $\Delta H°$ and $\Delta S°$ are similar and tend to cancel each other, thus the nonlinearity of the variation of $\Delta G°$ with the temperature is minimized. Using the average values of $\Delta H°$ and $\Delta S°$, the free energy equation is simplified to

$$\Delta G° = \Delta H° - \Delta S°T \tag{2.1.29}$$

The standard free energies of reactions encountered in ferrous metallurgical processes can be computed using the free energy data listed in Table 2.1.

2.1.3 Thermodynamic Activity

The combined statement of the first and second laws for a system doing work only against pressure gives the following thermodynamic relation.

$$dG = VdP - SdT \tag{2.1.30}$$

At constant temperature $G = VdP$ and for 1 mol of an ideal gas $V = RT/P$; with these substituted in equation (2.1.30) we obtain

$$dG = RT\,\frac{dP}{P} = RT\,d(\ln P) \tag{2.1.31}$$

Similarly, for a gas mixture

$$dG_i = RT\,d(\ln p_i) \tag{2.1.32}$$

where p_i is the partial pressure of the ith species in the gas mixture, and G_i partial molar free energy.

In a homogeneous liquid or solid solution, the thermodynamic activity of the dissolved element is defined by the ratio

$$a_i = \left(\frac{\text{vapor pressure of component (i) in solution}}{\text{vapor pressure of pure component}}\right)_T \tag{2.1.33}$$

In terms of solute activity, the partial molar free energy equation is

$$dG_i = RT\,d(\ln a_i) \tag{2.1.34}$$

Integration at constant temperature gives the relative partial molar free energy in solution

$$G_i = RT\,\ln a_i \tag{2.1.35}$$

In terms of the relative partial molar enthalpy and entropy of solution

$$\overline{G}_i = \overline{H}_i - \overline{S}_i T \tag{2.1.36}$$

which gives

$$\ln a_i = \frac{\overline{H}_i}{RT} - \frac{\overline{S}_i}{R} \tag{2.1.37}$$

or

$$\log a_i = \frac{\overline{H}_i}{2.303\,RT} - \frac{\overline{S}_i}{2.303\,R} \tag{2.1.38}$$

2.1.3.1 Solutions

A solution is a homogeneous gas, liquid or solid mixture, any portion of which has the same state properties. The composition of gas solution is usually given in terms of partial pressures of species in equilibrium with one another under given conditions. For liquid solutions, as liquid metal and slag, the composition is given in terms of the molar concentrations of components of the solution.

The atom or mol fraction of the component i in solution is given by the ratio

$$N_i = \frac{n_i}{\Sigma n}$$

where n_i is the number of g-atoms or mols of component i per unit mass of solution, and n the total number of g-atoms or mols. Since the metal and slag compositions are reported in mass percent, n_i per 100g of the substance is given by the ratio

$$n_i = \frac{\%i}{M_i}$$

where M_i is the atomic or molecular mass of the component i.

Noting that the atomic mass of iron is 55.85g, the atom fraction of solute i in low alloy steels is given by a simplified equation

$$N_i = \frac{\%i}{M_i} \times 0.5585 \tag{2.1.39}$$

In low alloy steelmaking, the composition of slag varies within a relatively narrow range, and the total number of g-mol of oxides per 100g of slag is within the range $\Sigma n = 1.6 \pm 0.1$. With this simplification, the mol fraction of the oxide in the slag is given by

$$N_i = \frac{\%i}{1.6\,M_i} \tag{2.1.40}$$

2.1.3.1.1 Ideal Solutions – Raoult's Law: The solutions are said to be ideal, if the activity is equal to the mol or atom fraction of the component i in solution,

$$a_i = N_i \tag{2.1.41}$$

A thermodynamic consequence of Raoult's law is that the enthalpy of mixing for an ideal solution, $H^{M,id}$, is zero. Substituting $a_i = N_i$ and $H^{M,id} = 0$ in the free energy equation gives for the entropy of formation of an ideal solution.

$$S^{M,id} = R(N_1 \ln N_1 + N_2 \ln N_2 + N_3 \ln N_3 + ...) \tag{2.1.42}$$

Table 2.1 The Standard Free Energies of Formation of Selected Compounds from Compiled Thermochemical Data

Notations: < > solid, { } liquid, ( ) gas, d decomposition, m melting, b boiling.

	$\Delta G° = \Delta H° - \Delta S°T$			Temp.Range °C
	$-\Delta H°$ kJ mol^{-1}	$-\Delta S°$ J mol^{-1}K^{-1}	$\Delta G°$ ±kJ	
<Al> = {Al}	−10.8	11.5	0.2	660m
2{Al} + 3/2(O$_2$) = <Al$_2$O$_3$>	1683.2	325.6	8	660–1700
{Al} + 1/2(N$_2$) = <AlN>	328.3	115.5	4	660–1700
<C> + 2(H$_2$) = (CH$_4$)	91.0	110.7	2	25–2000
<C> + 1/2(O$_2$) = (CO)	114.4	−85.8	2	25–2000
<C> + (O$_2$) = (CO$_2$)	395.3	−0.5	2	25–2000
<Ca> = {Ca}	−8.5	7.7	0.5	842m
{Ca} = (Ca)	153.6	87.4	0.5	842–1500b
{Ca} + 1/2(O$_2$) = <CaO>	900.3	275.1	6	842–1500b
{Ca} + 1/2(S$_2$) = <CaS>	548.1	103.8	4	842–1500b
<CaO> + <Al$_2$O$_3$> = <CaAl$_2$O$_4$>	19.1	−17.2	8	25–1605m
<CaO> + (CO$_2$) = <CaCO$_3$>	161.3	137.2	4	25–880d
2<CaO> + <SiO$_2$> = <Ca$_2$SiO$_4$>	118.8	−11.3	10	25–1700
<CaO> + <SiO$_2$> = <CaSiO$_3$>	92.5	2.5	12	25–1540m
<Cr> = {Cr}	−16.9	7.9	–	1857m
2<Cr> + 3/2(O$_2$) = <Cr$_2$O$_3$>	1110.3	247.3	2	900–1650
<Fe> = {Fe}	−13.8	7.6	1	1537m
0.947<Fe> + 1/2(O$_2$) = <Fe$_{0.947}$O>	263.7	64.3	4	25–1371m
{Fe} + 1/2(O$_2$) = {FeO}	225.5	41.3	4	1537–1700
3<Fe> + 2(O$_2$) = <Fe$_3$O$_4$>	1102.2	307.4	4	25–1597m
2<Fe> + 3/2(O$_2$) = <Fe$_2$O$_3$>	814.1	250.7	4	25–1500
<Fe> + 1/2(S$_2$) = <FeS>	154.9	56.9	4	25–988m
{Fe} + 1/2(O$_2$) + <Cr$_2$O$_3$> = <FeCr$_2$O$_4$>	330.5	80.3	2	1537–1700
2<FeO> + <SiO$_2$> = <Fe$_2$SiO$_4$>	36.2	21.1	4	25–1220m
(H$_2$) + 1/2(O$_2$) = (H$_2$O)	247.3	55.9	1	25–2000
(H$_2$) + 1/2(S$_2$) = (H$_2$S)	91.6	50.6	1	25–2000
3/2(H$_2$) + 1/2(N$_2$) = (NH$_3$)	53.7	32.8	0.5	25–2000
{K} = (K)	−84.5	82.0	0.5	63–759b
(K) + <C> + 1/2(N$_2$) = {KCN}	171.5	93.5	16	622–1132b
{KCN} = 1/2(KCN)$_2$	109.2	76.7	4	622–1132b
<Mg> = {Mg}	−9.0	9.7	0.5	649m
{Mg} = (Mg)	129.6	95.1	2	649–1090b
(Mg) + 1/2(O$_2$) = <MgO>	759.4	202.6	10	1090–2000
(Mg) + 1/2(S$_2$) = <MgS>	539.7	193.0	8	1090–2000
2<MgO> + <SiO$_2$> = <Mg$_2$SiO$_4$>	67.2	4.3	8	25–1898m
<MgO> + <SiO$_2$> = <MgSiO$_3$>	41.1	6.1	8	25–1577m
<MgO> + (CO$_2$) = MgCO$_3$)	116.3	173.4	8	25–402d
<Mn> = {Mn}	−14.6	9.6	1	1244m
<Mn> + 1/2(O$_2$) = <MnO>	391.9	78.3	4	25–1244m
{Mn} + 1/2(O$_2$) = <MnO>	406.5	87.9	4	1244–1700
{Mn} + 1/2(O2) = {MnO}**	352.2	61.5	4	1500–1700

** supercooled liquid below the melting point 1785°C

<Mn> + 1/2(S$_2$) = <MnS>	277.9	64.0	4	25–1244m
{Mn} + 1/2(S$_2$) = <MnS>	292.5	73.6	4	1244–1530m
{Mn} + 1/2(S$_2$) = {MnS}	265.0	66.1	4	1530–1700
<MnO> + <SiO$_2$> = <MnSiO$_3$>	28.0	2.8	12	25–1291m
<Mo> = {Mo}	−27.8	9.6	6	2620m
<Mo> + (O$_2$) = <MoO$_2$>	578.2	166.5	12	25–2000
<Mo> + 3/2(O$_2$) = (MoO$_3$)	359.8	59.4	20	25–2000

Table 2.1 (continued)

	$-\Delta H°$ kJ mol^{-1}	$-\Delta S°$ J mol^{-1}K^{-1}	$\Delta G°$ ±kJ	Temp.Range °C
$1/2(N_2) + 3/2(H_2) = (NH_3)$	53.7	116.5	0.5	25–2000
$1/2(N_2) + 1/2(O_2) + (NO)$	–90.4	–12.7	0.5	25–2000
$1/2(N_2) + (O_2) + (NO_2)$	–32.3	63.3	1	25–2000
{Na} = (Na)	–101.3	87.9	1	98–883b
(Na) + <C> + $1/2(N_2)$ = {NaCN}	152.3	83.7	16	833–1530b
2(Na) + $1/2(O_2)$ = {Na$_2$O}	518.8	234.7	12	1132–1950d
<Nb> = {Nb}	–26.9	9.8	–	2477m
2<Nb> + $1/2(N_2)$ = <Nb$_2$N>	251.0	83.3	16	25–2400m
<Nb> + $1/2(N_2)$ = <NbN>	230.1	77.8	16	25–2050m
2<Nb> + $5/2(N_2)$ = <Nb$_2$O$_5$>	1888.2	419.7	12	25–1512m
<Ni> = {Ni}	–17.5	10.1	2	1453m
<Ni> + $1/2(O_2)$ = <NiO>	235.6	86.1	2	25–1984m
<Ni> + $1/2(S_2)$ = <NiS>	146.4	72.0	6	25–600
3<Ni> + (S$_2$) = <Ni$_3$S$_2$>	331.5	163.2	8	25–790m
$1/2(S_2) + (O_2) = (SO_2)$	361.7	72.7	0.5	25–1700
<Si> = {Si}	–49.3	30.0	2	1412m
{Si} + $1/2(O_2)$ = (SiO)	154.7	–52.5	12	1412–1700
<Si> + (O$_2$) = <SiO$_2$>	902.3	172.9	12	400–1412m
{Si} + (O$_2$) = <SiO$_2$>	952.5	202.8	12	1412–1723m
<Ti> = {Ti}	–18.6	9.6	–	1660m
<Ti> + $1/2(N_2)$ = <TiN>	336.3	93.3	6	25–1660m
<Ti> + (O$_2$) = <TiO$_2$>	941.0	177.6	2	25–1660m
<V> + {V}	–22.8	10.4	–	1920m
<V> + $1/2(N_2)$ = <VN>	214.6	82.4	16	25–2346d
2<V> + $3/2(O_2)$ = <V$_2$O$_3$>	1202.9	237.5	8	25–2070m
{Zn} = (Zn)	–118.1	100.2	1	420–907b
(Zn) + $1/2(O_2)$ = <ZnO>	460.2	198.3	10	907–1700
{Zn} + $1/2(S_2)$ = <ZnS>	277.8	107.9	10	420–907b
(Zn) + $1/2(S_2)$ = (ZnS)	–5.0	30.5	10	1182–1700
<Zr> = {Zr}	–20.9	9.8	–	1850m
<Zr> + $1/2(N_2)$ = <ZnN>	363.6	92.0	16	25–1850m
<Zr> + (O$_2$) = <ZrO$_2$>	1092.0	183.7	16	25–1850m
<Zr> + (S$_2$) = <ZrS$_2$>	698.7	178.2	20	25–1550m
<ZrO$_2$> + <SiO$_2$> = <ZrSiO$_4$>	26.8	12.6	20	25–1707m

* References to the compiled thermochemical data used in deriving $\Delta H°$ and $\Delta S°$ values are given in Ref. 27 cited in Section 2.2.2.4

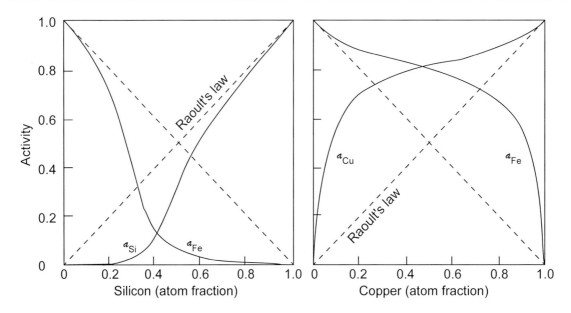

Fig. 2.1 Activities in liquid Fe–Si and Fe–Cu alloys at 1600°C showing strong negative and positive departures from Raoult's law.

2.1.3.1.2 Nonideal Solutions Almost all metallic solutions and slags exhibit nonideal behavior. Depending on the chemical nature of the elements constituting a solution, the activity vs composition relation deviates from Raoult's law to varying degrees, as demonstrated in Fig. 2.1 for liquid Fe–Si and Fe–Cu systems at 1600°C.

2.1.3.2 Activity Coefficient

The activity coefficient of solute i is defined by the ratio

$$\gamma_i = \frac{a_i}{N_i} \quad (2.1.43)$$

If the activity is relative to the pure component i, it follows from Raoult's law that as $N_i \rightarrow 1, \gamma \rightarrow 1$.

2.1.3.2.1 Henry's Law for Dilute Solutions In infinitely dilute solutions, the activity is proportional to the concentration

$$a_i = \gamma_i^\circ N_i \quad (2.1.44)$$

The approach to Henry's law at dilute solutions is demonstrated in Fig. 2.2 for the activity of carbon (relative to graphite) in austenite at 1000°C. The austenite containing 1.65%C ($N_C = 0.072$) is saturated with graphite at 1000°C for which the carbon activity is one relative to graphite.

Since Henry's Law is valid at infinite dilution only, the ration γ_i/γ_i° is used as a measure of departure from Henry's Law for finite solute contents in dilute solutions. For solute concentration in terms of mass percent, Henry's activity coefficient is defined by the ratio

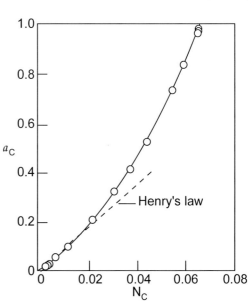

Fig. 2.2 Activity of carbon in austenite (relative to graphite) at 1000°C, demonstrating deviation from Henry's law.

Fundamentals of Iron and Steelmaking

$$f_i = \frac{\gamma_i}{\gamma_i^\circ} \qquad (2.1.45)$$

such that $f_i \to 1$ when $\%i \to 0$.

2.1.3.2.2 Interaction Coefficients Over several mass percentages of the solute, the composition dependence of activity coefficient f_i of solute i in binary systems is represented by the following relation in terms of mass %i.

$$\log f_i = e_i^i [\%i] \qquad (2.1.46)$$

where e is called the solute interaction coefficient. For multi-component solutions, the following summation is used

$$\log f_i = e_i^i [\%i] + \Sigma e_i^j [\%j] \qquad (2.1.47)$$

where e_i^j is the effect of the alloying element j on the activity coefficient of solute i.

2.1.3.3 Conversion from One Standard State to Another

In steelmaking processes we are concerned with reactions involving dissolved elements at low concentrations in liquid steel. Therefore, it is convenient to express the solute activity relative to Henry's law. The free energy change accompanying the isothermal transfer of solute from state A to state B is

$$\Delta G = RT \ln\left(\frac{a_B}{a_A}\right) \qquad (2.1.48)$$

Taking a pure component for state A, i.e. $a_A = 1$ and Henry's law for state B, i.e. $a_B = \gamma_i^\circ N_i$

$$\Delta G = RT \ln(\gamma_i^\circ N_i) \qquad (2.1.49)$$

For one mass percent solute in iron,

$$N_i \approx \frac{0.5585}{M_i} \qquad (2.1.50)$$

where M_i is the atomic mass (g) of the solute. Assuming that Henry's law holds at 1 mass %, the standard free energy of solution of pure component i in iron at 1 mass % is

$$\Delta G_s = RT \ln\left(\frac{0.5585}{M_i} \gamma_i^\circ\right) \qquad (2.1.51)$$

Since Henry's law is valid at infinite dilution only, appropriate correction must be made for non-ideal behavior when using the above equation.

2.1.4 Reaction Equilibrium Constant

Let us consider the following reaction equilibrium occurring at constant temperature and pressure.

$$mM + nN = uU + vV \qquad (2.1.52)$$

The state of equilibrium is defined by the following thermodynamic relation involving the activities of reactants M and N and the activities of the products U and V.

$$K = \frac{(a_U)^u (a_V)^v}{(a_M)^m (a_N)^n} \qquad (2.1.53)$$

where the equilibrium constant K can be derived from the standard free energy change accompanying the reaction thus,

$$\Delta G^\circ = -RT \ln K \qquad (2.1.54)$$

Since $\Delta G°$ is a function of temperature only, the equilibrium constant is also a function of temperature only.

$$\ln K = -\frac{\Delta G°}{RT} \qquad (2.1.55)$$

In terms of enthalpy and entropy changes

$$\ln K = -\frac{\Delta H°}{RT} + \frac{\Delta S°}{R} \qquad (2.1.56)$$

In exothermic reactions ΔH is negative, therefore the equilibrium constant K decreases with an increasing temperature. The temperature effect on K is the opposite in endothermic reactions.

For $\Delta G°$ in J, R = 8.314 J mol^{-1}K^{-1} and substituting log for ln,

$$\log K = -\frac{\Delta G°}{19.144\,T} \qquad (2.1.57)$$

As an example, let us consider the following reaction equilibrium between liquid slag and low alloy steel.

$$(MnO) + [C] = [Mn] + CO(g) \qquad (2.1.58)$$

where the parenthesis ( ) and [ ] represent respectively the oxides dissolved in slag and elements dissolved in iron. From the free energy data in Table 2.1 and those in Table 2.8 in section 2.4 for the dissolution of graphite and liquid manganese in liquid iron, the standard free energy change accompanying reaction 2.1.58 is

$$\Delta G° = 230{,}560 - 147.32T\,J \qquad (2.1.59)$$

which gives for the equilibrium constant K

$$\log K = -\frac{12{,}043}{T} + 7.695 \qquad (2.1.60)$$

$$K = \frac{[\%Mn]\,p_{CO}}{(a_{MnO})[\%C]} \qquad (2.1.61)$$

where p_{CO} is in atm, the oxide activity is with respect to pure molten MnO and at low concentrations, the mass percentages of Mn and C in the steel are essentially equivalent to their activities, i.e. f_{Mn} and f_C are close to unity.

For a steelmaking slag of basicity 3.0 containing 5% MnO, $a_{MnO} = 0.088$; at 1600°C the equilibrium constant is K = 18.42 for which equation 2.1.61 gives for the slag-metal equilibrium [%Mn]/[%C] = 1.62 at 1 atm pressure of CO.

2.2 Rate Phenomena

There are many different facets of rate phenomena involving homogeneous or heterogeneous chemical reactions, mass transfer via atomic or molecular diffusional processes, viscous flow, thermal and electrical conduction and so on. The concepts of diffusion, mass transfer and chemical kinetics presented briefly in this section of Chapter 2, are confined to cases which are relevant to the study of the rates of reactions in the iron and steelmaking processes.

2.2.1 Diffusion

2.2.1.1 Fick's Diffusion Laws

The first law is for the steady state diffusion.

The quantity of diffusing substance which passes per unit time through unit area of a plane perpendicular to the direction of diffusion, known as the flux J, is proportional to the concentration gradient of the diffusing substance

$$J = -D\frac{dC}{dx} \quad (2.2.1)$$

The coefficient D is the diffusivity of the substance in the medium; C is the concentration of the substance per unit volume and x the distance in the direction of diffusion.

The second law is for the nonsteady state diffusion.

The rate of accumulation of diffusing substance in a given volume element is the difference between the inward and outward flux. In other words, what goes in and does not come out, stays there.

The rate of concentration change $-dC/dt$ resulting from flux over a distance dx is

$$J = -\frac{dC}{dt}dx \quad (2.2.2)$$

Hence, the change in flux with distance is

$$\frac{dJ}{dx} = -\frac{dC}{dt} \quad (2.2.3)$$

Invoking the first law and re-arranging gives the second law.

$$\frac{dJ}{dx} = -\frac{d}{dx}\left(D\frac{dC}{dx}\right) \quad (2.2.4)$$

The solution of this equation depends on the geometry and on the boundary conditions of the medium in which the dissolved substance is diffusing.

2.2.1.2 Chemical and Self-Diffusivities

There are two kinds of diffusivities:

(i) The chemical diffusivity is the coefficient of diffusion defined by Fick's law $D_i = -J_i(\partial C/\partial x)$ for the single diffusing species i.

(ii) The self-diffusivity is the coefficient of diffusion measured in a homogeneous medium with the radioactive or stable isotope (tracer) of the diffusing species. The self-diffusion is a consequence of random movements of atoms or molecules.

For an ideal solution, or an infinitely dilute solution, we have the equality

$$D_i = D_i^* \quad (2.2.5)$$

For nonideal solutions in general (ionic or metallic), Onsager and Fuoss[1] derived the following relation for the chemical diffusivity corrected for departure from ideal behavior.

$$D_i = D_i^*\left(\frac{\partial \ln a_i}{\partial \ln C_i}\right) = D_i^*\left(1 + \frac{\partial \ln \gamma_i}{\partial \ln C_i}\right) \quad (2.2.6)$$

where a_i is the activity and γ_i the activity coefficient of the single diffusing species i.

2.2.1.3 Types of Diffusional Processes

There are various types of diffusional processes, depending on the medium in which the diffusion occurs.

2.2.1.3.1 Diffusion in Molten Slags and Glasses
Because of the ionic nature of the medium, the diffusion in molten slags and glasses is by the ions. The ionic conduction is a purely random jump process. For the limiting case of purely free and random movements of ions, the self-diffusivity of the ionic species i is related to the electrical conductivity λ_i by the Nernst-Einstein[2] equation

$$D_i^* = \frac{RT}{F^2 Z_i^2 C_i} \lambda_i \qquad (2.2.7)$$

where Z_i is the valence and C_i the concentration (mol cm^{-3}) of the species i and $F = 96,489$ C mol^{-1} the Faraday constant.

Reference may be made to a previous publication by Turkdogan[3] on the review of interrelations between ionic diffusion and conduction in molten slags.

2.2.1.3.2 Diffusion in Porous Media The rate of a heterogeneous reaction between a gas and a porous medium is much affected by the counter-diffusive flow of gaseous reactants and products through the pores of the medium. There are two types of effective gas diffusivities, depending on the pore structure of the medium:

(i) Molecular diffusivity in large pores and at ordinary to high pressures.

(ii) Knudsen diffusivity in small pores and at low pressures.

The pore diffusion is discussed in more detail in Section 2.3.

2.2.1.3.3 Eddy Diffusivity The mass transfer occurring within a turbulent fluid flow is characterized by the diffusion coefficient called eddy diffusivity which is a function of the time-averaged and fluctuating velocity components in turbulent flow. This subject is covered in detail in a text book on fluid phenomena by Szekely[4].

2.2.2 Mass Transfer

In high temperature processes as in pyrometallurgy, the rates of interfacial chemical reactions are in general much faster than the rates of transfer of the reactants and reaction products to and from the reaction site. The formulations of rate equations for transport controlled reactions vary considerably with the physical properties and type of fluid flow over the surface of the reacting condensed phase. Then, there are formulations of rate equations for different regimes of gas bubbles in liquid metal and slag. For a comprehensive discussion of the transport controlled rate phenomena in metallurgical processes, reference may be made to textbooks by Szekely and Themelis[5] and by Geiger.[6]

There are numerous computer software packages, as for example PHOENICS, FLUENT, CFDS-FLOWS3D, GENMIX, TEACH(2D, 3D), 2E/FIX, SOLA/VOF, METADEX(R) and FID, which are used in the steel industry in the process and design engineering of plant facilities and in the control of steelmaking processes. References may also be made to the publications by Szekely et al.[7-9] on the mathematical and physical modelling of metallurgical processes involving fluid flow, heat transfer, mass transfer and gas-slag-metal reactions.

There are of course many other publications on this subject which cannot all be cited here as the subject matter is well outside the scope of this book.

The rate phenomena discussed in this chapter are on chosen subjects which are pertinent to various aspects of ferrous-pyrometallurgical processes, in addition to oxygen steelmaking, ladle refining and degassing.

2.2.2.1 Parabolic Rate of Oxidation of Iron

The scale forming in the oxidation of iron, or low alloy steels, consists of three layers of oxides: wustite on the iron surface followed by magnetite then hematite on the outer surface of the scale. From several experimental studies it was found that the average wustite:magnetite:hematite thickness ratios were about 95:4:1. In iron oxides, the diffusivity of iron is greater than the oxygen diffusivity. The rate of oxidation is controlled primarily by diffusion of iron through the wustite layer from the iron-wustite interface to the wustite-gas or wustite-magnetite interface, then through the thin layers of magnetite and hematite.

$$X^2 = \lambda t \quad (2.2.8)$$

where λ is the parabolic rate constant, cm^2 (scale) s^{-1}

If the measurement of the rate of oxidation is made by the thermogravimetric method, then the parabolic rate constant k_p would be in units of $(gO)^2 cm^{-4} s^{-1}$. From the compositions and densitites of the oxides, with the relative thickness ratios of wustite:magnetite:hematite = 95:4:1, the values of k_p and λ are related as follows.

$$k_p \, (gO)^2 \, cm^{-4} \, s^{-1} = 1.877\lambda, \, cm^2 \, (scale) \, s^{-1} \quad (2.2.9)$$

Many experimental studies have been made of the rate of oxidation of iron in air and oxygen at temperatures 600 to 1300°C. The temperature dependence of the parabolic rate constant is shown in Fig. 2.3; the references to previous studies denoted by different symbols are given in a paper by Sheasby et al.[10]

From theoretical considerations, Wagner[11] derived the following equation for the parabolic rate constant in terms of the activity of oxygen and self-diffusivities of the mobile species in the scale. For the case of wustite with $D^*_{Fe} \gg D^*_O$ Wagner's equation is simplified to

$$\lambda = 2 \int_{a'_O}^{a''_O} \frac{N_O}{N_{Fe}} D^*_{Fe} \, d(\ln a_O) \quad (2.2.10)$$

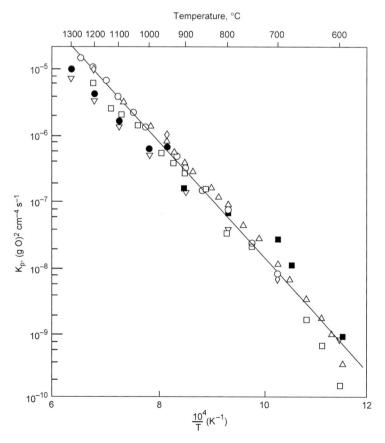

Fig. 2.3 Temperature dependence of the parabolic rate constant for oxidation of iron in air or oxygen with or without H_2O. *From Ref. 10.*

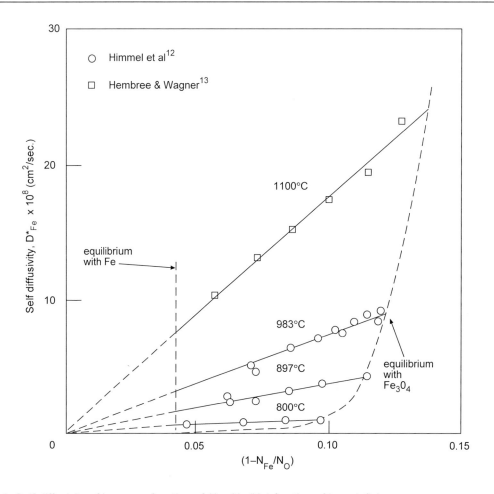

Fig. 2.4 Self-diffusivity of iron as a function of $(1 - N_{Fe}/N_O)$ fraction of iron deficiency.

where a'_O and a''_O are oxygen activities at the iron-wustite and wustite-magnetite interfaces, respectively. The diffusion of iron ions in the nonstoichiometric phase wustite Fe_xO, is via the cation vacant sites in the wustite lattice. It is for this reason that the self-diffusivity of iron increases with an increasing O/Fe atom ratio in wustite, i.e. with increasing iron deficiency, as shown in Fig. 2.4

The values of $k_p = 1.877\lambda$ calculated from the measured tracer diffusivity of iron using Wagner's equation, are consistent with the average experimental values in Fig. 2.3 which is represented by the following equation.

$$\text{Log } k_p = -\frac{8868}{T} + 0.9777 \tag{2.2.11}$$

2.2.2.2 Oxidation of Carbon in CO_2–CO Mixtures

Most forms of carbon are porous, therefore the rate of oxidation, i.e. gasification of carbon, is decisively affected by the pore structure, internal pore surface area, effective gas diffusivity in the pores and particle size.

In a critical review of the oxidation of carbon, Walker et al.[14] gave a detailed and a comprehensive account of (up to 1959) experimental and theoretical studies on this subject. In the late 1960s, Turkdogan and co-workers[15–17] pursued further experimental work on the oxidation of several types of carbon in CO_2–CO mixtures at temperatures up to 1300°C and at pressures of 0.01 to 10 atm. A few salient features of their findings are briefly given here.

Depending on the particle size, pore structure, temperature and gas pressure, there is either complete internal burning or external burning of carbon. The rate equations are of the following forms for these two limiting cases:

(i) For small porous carbon particles and at low temperatures and pressures, there is rapid counter current pore diffusion of CO_2 and CO, and the rate of oxidation is controlled by the chemical kinetics on the pore walls of the carbon. For this limiting case of complete internal burning, the rate equation is

$$\ln(1-F) = -\phi S C_i t \tag{2.2.12}$$

where

F = the mass fraction of carbon oxidized,
ϕ = specific isothermal rate constant of chemical reaction per unit area of the pore wall,
S = connected internal pore surface area per unit mass $cm^2 mol^{-1}$,
C_i = molar concentration of CO_2 per cm^3 in the gas stream,
t = oxidation time

(ii) For large carbon particles and at high temperatures and pressures, the gas diffusion is confined to the pore mouths on the surface of the carbon particle. For this limiting case of external burning the rate equation is

$$1-(1-F)^{1/3} = \frac{(\phi \rho S D_e)^{1/2}}{\rho r_o} C_i t \tag{2.2.13}$$

where

D_e = effective CO_2–CO pore diffusivity,
ρ = molar bulk density of the carbon,
r_o = initial radius of carbon particle.

The effects of the particle size and gas pressure on the rate of oxidation of electrode graphite is shown in Fig. 2.5

A mathematical analysis was made by Tien and Turkdogan[18] to formulate the pore diffusion effect on partial internal burning of relatively large carbon particles. The experimentally determined initial rates of oxidation of coke and graphite spheres ($\approx$ 2 cm dia.) in 1:1 CO_2:CO mixture at 1 atm pressure are compared in Fig. 2.6 with the

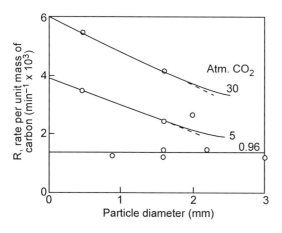

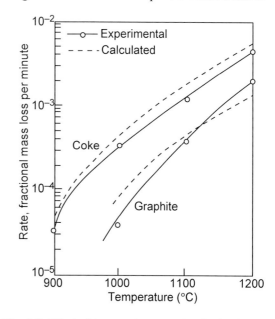

Fig. 2.5 Effect of particle size on the rate of oxidation of electrode graphite in pure carbon dioxide at 1000°C. *From Ref. 16.*

Fig. 2.6 Effect of temperature on rate of oxidation of coke and graphite (1.9 and 2.2 cm dia. spheres, respectively) in 50:50 CO_2:CO at 1 atm. *From Ref. 18.*

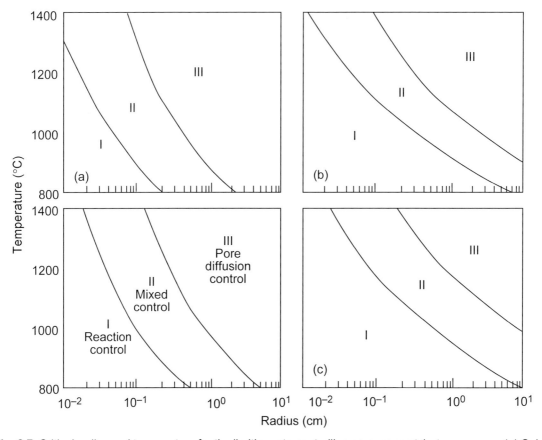

Fig. 2.7 Critical radius and temperature for the limiting rate controlling processes at 1 atm pressure. (a) Coke, 100% CO_2; (b) Coke, 1:1 $CO:CO_2$; (c) Graphite, 100% CO_2; (d) Graphite, 1:1 $CO:CO_2$. *From Ref. 18.*

values calculated from the mathematical analysis, which is summarized in Fig. 2.7. In each diagram, the lower curve is for 80% internal burning; therefore, in the region below this curve there is almost complete pore diffusion. The upper curve is for 20% internal burning; therefore, in the region above this curve, the pore diffusion control predominates.

Reference should also be made to a paper by Aderibigbe and Szekely[19] for a more detailed mathematical formulation of the rate of pore-diffusion and reaction controlled gasification of metallurgical coke in CO_2–CO mixtures.

2.2.2.3 Reduction of Iron Oxides

In view of its practical importance to the understanding and control of ironmaking processes, a great deal of research has been done on the gaseous reduction of iron oxides and iron ores. Because of the porous nature of iron oxides and the reduction products, the interpretation of the reduction rate data is inherently complex.

The formation of product layers during the gaseous reduction of dense sintered hematite and magnetite pellets or natural dense iron ore particles is a well-known phenomenon, as shown in Fig. 2.8.

In several studies made in the early 1960s[20–22] it was found that the thickness of the reduced iron layer, encasing the iron oxide core of the pellet, increased linearly with the reduction time. The measured rates were interpreted in terms of the rate-controlling chemical reaction at the iron wustite interface; the diffusive fluxes of gases through the porous layers were assumed to be relatively fast. On the other hand, Warner[23] and Spitzer *et al.*[24] have expressed the view that the rate of gaseous reduction is much affected by the gaseous diffusional processes, e.g. the gas-film resistance at the

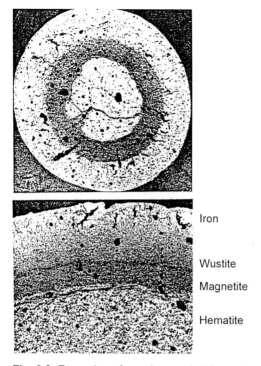

Fig. 2.8 Formation of reaction product layers in the gaseous reduction of hematite pellets.

pellet surface and particularly the resistance to diffusion in the porous product layers. The rate measurements made by Turkdogan and Vinters[25] on the reduction of hematite pellets in H_2–H_2O and CO–CO_2 mixtures have clearly demonstrated that the rate-controlling effects of gas diffusion into the pores of the oxide granules or pellets and through the porous iron layer dominate the reaction kinetics.

The sketch in Fig. 2.9 demonstrates three limiting rate controlling processes as outlined below.

(a) With fine granules there is internal reduction producing rosettes or platelets of metallic iron within the oxide particle; the rate in terms of mass fraction reduced, F, is independent of particle size.

(b) With large and dense oxide particles and at high temperatures, the gas diffusion is slow and the reaction is confined essentially to pore mouths on the outer surface of the particle, bringing about the development of a porous iron layer around the pellet. Because of the layer formation, this mode of reduction is often called topochemical reduction. In the early stages of reduction, the porous iron layer is sufficiently thin for rapid gas diffusion, therefore the initial rate of reduction is controlled jointly by (i) gas diffusion into the pore mouths of the oxide and (ii) reaction on the pore walls of the wustite.

$$\frac{dF}{dt} \propto \frac{\sqrt{\phi S D_e}}{r} \qquad (2.2.14)$$

(c) When the porous iron layer becomes sufficiently thick, the rate of reduction will be controlled essentially by the counter current gas diffusion (H_2–H_2O and CO–CO_2) for which the limiting rate equation, for a given temperature is as follows.

$$\left[3 - 2F - 3(1-F)^{2/3}\right] = Y = \frac{D_e}{\rho r^2}\left(\frac{p_i - (p_i)_{eq}}{RT}\right)t + C \qquad (2.2.15)$$

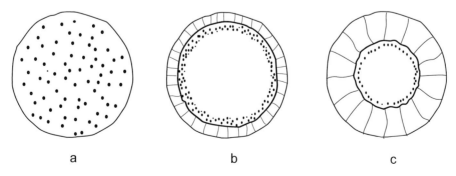

Fig. 2.9 Schematic representation of (a) uniform internal reduction of wustite to iron, (b) limiting mixed control (partial internal reduction) and (c) diffusion in porous iron as the rate-controlling step.]

where p_i is the H_2 or CO partial pressure in the gas stream and $(p_i)_{eq}$ that for the iron-wustite equilibrium and C a constant (a negative number) that takes account of all early time departures from the assumed boundary conditions for this limiting case.

(d) At high reduction temperatures, with large oxide pellets and low velocity gas flows, the rate of reduction is controlled primarily by mass transfer in the gas-film layer at the pellet surface. In this limiting case the rate is inversely proportional to the square of the particle diameter.

The experimental data are given Fig. 2.10 showing the particle size effect on the initial rate of reduction of hematite granules or pellets in hydrogen.

When the thickness of the reduced porous iron layer exceeds 1 mm, the subsequent rate of reduction is controlled primarily by gas diffusion through the porous iron layer. The reduction data plotted in accord with equation 2.2.15 usually give elongated S-shaped curves as in Fig. 2.11 for 15 mm diameter spheroidal hematite ore reduced in H_2 at 1 atm. From about 50% to 95% or 99% reduction, data are well represented by straight lines. The effective H_2–H_2O diffusivities in the pores of the iron layer are derived from the slopes of the lines; details are given in Section 2.3.3.

The effect of reaction temperature on the pore surface area of iron and wustite formed by the reduction of hematite is shown in Fig. 2.12. The higher the reduction temperature, the smaller the internal pore surface area, i.e. the coarser the pore structure. The iron oxides reduced at temperatures below 800°C are known to be pyrophoric, which is a consequence of a fine pore structure with a large pore surface area as noted from the data in Fig. 2.12.

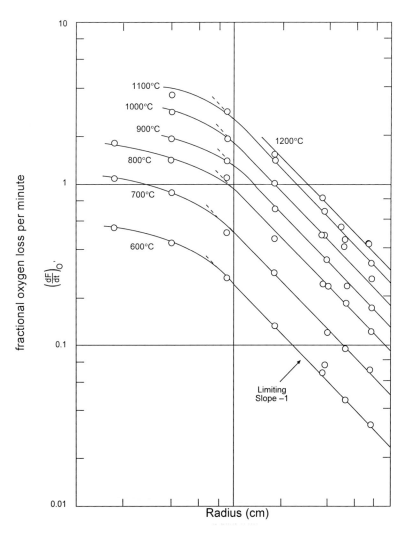

Fig. 2.10 Initial rate in hydrogen at 0.96 atm. as a function of particle radius at indicated temperatures. *From Ref. 25.*

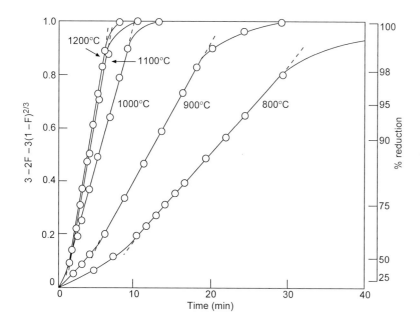

Fig. 2.11 Diffusion plot of reduction data for 15 mm diameter spheroidal hematite ore. *From Ref. 25.*

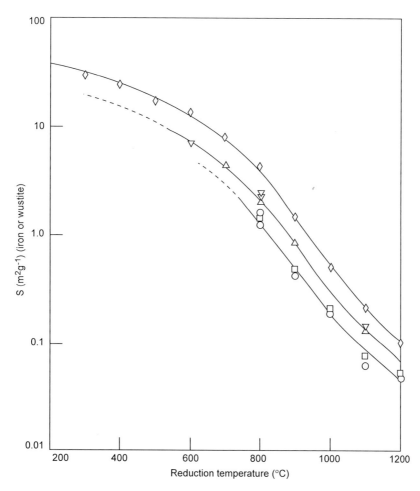

Fig. 2.12 Connected internal pore surface area of iron and wustite formed by reduction of hematite ores' A and B; (a) ore A reduced to iron in H_2 (◊); (b) ore B reduced to iron in H_2(∇) and reduced to wustite in an H_2–H_2O mixture (△); (c) ore B reduced to iron in a CO–CO_2 mixture (□) and reduced to wustite in another CO–CO_2 mixture (○). *From Ref. 25.*

2.2.2.4 Mass Transfer with Gas Bubbles in Oxygen Steelmaking

In most practical applications of gas injection into liquids, we are concerned mainly with rates of reactions in the swarm of bubbles. Mass-transfer phenomena in the swarm of bubbles have been studied extensively with aqueous solutions; accumulated knowledge is well documented in a review paper by Calderbank.[26] Also, several studies have been made of the bubble size and velocity of rise of single bubbles in liquid metals. Szekely presented a comprehensive review of these studies in his book on *Fluid Flow Phenomena in Metals Processing*.[4]

In his recent book *Fundamentals of Steelmaking*, Turkdogan[27] made an assessment of the size range of dispersed gas bubbles and the corresponding mass transfer in liquid steel for the conditions prevailing in the BOF and Q-BOP steelmaking processes. In this assessment, the bubbles are assumed to be generated from a multitude of points across the bottom of the vessel, as though the entire vessel bottom were in fact one big porous plug.

2.2.2.4.1 Gas Holdup and Superficial Gas Velocity The volume fraction of gas holdup (ε) is given by the following ratio:

$$\varepsilon = \frac{V_b}{V_b + V_s} = 1 - \frac{V_s}{HA} \qquad (2.2.16)$$

where

V_b = transitory bubble volume in the bath at any time,
V_s = steel volume in the vessel,
H = average bath depth,
A = average bath cross sectional area.

The velocity of bubble rise, U_b, in a swarm of bubbles is given by the ratio

$$U_b = \frac{U_s}{\varepsilon} = \frac{\dot{V}}{\varepsilon A} \qquad (2.2.17)$$

where

$\dot{V}$ = gas flow rate,
U_s = superficial gas velocity.

In the studies of Calderbank and co-workers[26] with the air-water system, the gas was injected into the column of water through a sieve plate on the bottom of the column. The fractional holdup of gas was reported as shown in Fig. 2.13. The units in F factor are U_s, m s^{-1} and ρ_g (the gas density), kg m^{-3}.

The dot-dash curve is taken to represent the average values. It appears that at $F \geq 1$, the gas holdup reaches an essentially constant value of $\varepsilon = 0.6$.

In the estimation of fractional gas holdup in the steel bath, during decarburization or argon stirring, the gas flow rates taken are for the melt temperature and average gas bubble pressure. The values used are for 1600°C and 1.5 atm so that $\dot{V}_T$(m^{3}s^{-1}) = 4.57 × $\dot{V}_o$(Nm^{3}s^{-1}); for these conditions the gas densities are 0.27 kg m^{-3} for N_2 and CO and 0.39 kg m^{-3} for Ar; an average value is used, ρ_g = 0.33 kg m^{-3}.

The average bath cross sectional area A in BOF or Q-BOP vessel increases with an increasing gas flow rate. Keeping in mind the inner dimensions of the BOF or Q-BOP vessel for 200-240 ton heats, the area values used are: A = 22 m^2 at $\dot{V}_T$ = 60 m^{3}s^{-1} decreasing to A = 16 m^2 at $\dot{V}_T$ = 1 m^{3}s^{-1}. On this basis, the following equation is obtained for the superficial gas velocity U_s as a function of $\dot{V}_T$.

$$U_s\left(\text{m s}^{-1}\right) = \frac{\dot{V}_T}{14.55 + 0.089 \dot{V}_T} \qquad (2.2.18)$$

The volume fraction gas holdup and bubble velocity of rise calculated using equation 2.2.18 and Fig. 2.13 are plotted in Fig. 2.14 as a function of the gas flow rate. The gas holdup in the slag layer is a more complex phenomenon, because of slag foaming and injection of large amounts of steel droplets into the slag layer during decarburisation with oxygen blowing.

2.2.2.4.2 Estimation of Bubble Size for Uniformly Dispersed Bubbles in the Steel Bath Many estimates were made in the past of gas bubble diameters in steelmaking processes, the estimates varying over a wide range from 1 to 8 cm or more. Bubbles with diameters >1cm acquire a spherical cap shape as they rise in liquids. The apparent diameter is that of a sphere which has a volume equivalent to that of the spherical cap shaped bubble.

Calderbank and co-workers[26] and Leibson et al.[28] have found that in aqueous solutions the bubble size becomes smaller with an increasing gas flow rate, ultimately reaching a minimum size of about 0.45 cm diameter at large gas flow rates. Relatively large gas bubbles in motion are subject to deformation and ultimately to fragmentation into smaller bubbles. The drag force exerted by the liquid on a moving bubble induces rotational and probably a turbulent motion of the gas within the bubble.

This motion creates a dynamic pressure on the bubble surface; when this force exceeds the surface tension, bubble breakup occurs. Because of the large difference between the densitites of the gas and liquid, the energy associated with the drag force is much greater than the kinetic energy of the gas bubble. Therefore, the gas velocity in the bubble will be similar to the bubble velocity. On the basis of this theoretical reasoning, Levich[29] derived the following equation for the critical bubble size as a function of bubble velocity.

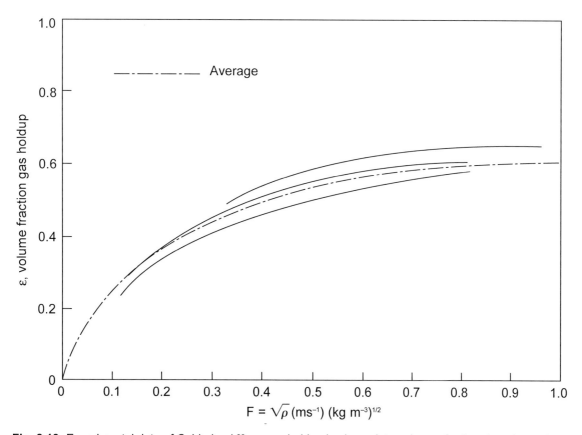

Fig. 2.13 Experimental data of Calderbank[26] on gas holdup in sieve plate columns for the air-water system.

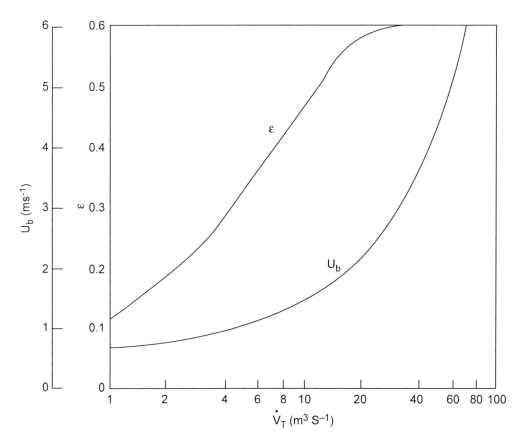

Fig. 2.14 Gas holdup and bubble velocity as a function of gas flow rate. *From Ref. 27.*

$$d_c = \left(\frac{3}{C_d \rho_g \rho_\ell^2}\right)^{1/3} \frac{2\sigma}{U_b^2} \quad (2.2.19)$$

where σ is the surface tension and C_d the drag coefficient, which is close to unity. As the bubble size decreases with increasing bubble velocity, the gas circulation within the bubble diminishes, thus becoming less effective in bubble fragmentation. Another view to be considered is that, as the gas holdup increases with increasing gas flow rate, the liquid layer separating the bubbles becomes thinner. Consequently, the dynamic gas pressure exerted on the bubble surface becomes nullified by similar forces exerted in the neighboring bubbles, hence the cessation of bubble fragmentation at high values of ε.

The bubble fragmentation with increasing ε, hence increasing U_b, is calculated for bubbles in water with $\sigma = 0.072$ Nm^{-1} and in liquid steel with (a) $\sigma = 1.7$ Nm^{-1} in the presence of O < 4 ppm and 40 ppm S, and (b) $\sigma = 1.3$ Nm^{-1} in the presence of 600 ppm O and 120 ppm S. The results are shown in Fig. 2.15. As noted earlier, the limiting bubble size in air-water system is d = 0.45 cm which intersects the fragmentation curve at $\varepsilon = 0.41$, depicting the cessation of bubble fragmentation.

In an attempt to estimate the minimum bubble size in the swarm of bubbles in liquid steel, Turkdogan proposed the following hypothesis: 'the surface energy of bubbles per unit mass of liquid which the bubbles displace, are equal for all liquids.' Thus, surface energy E_b of dispersed bubbles of uniform size per unit mass of liquid which they displace is given by:

$$E_b = \frac{6\varepsilon}{1-\varepsilon}\left(\frac{\sigma}{\rho d}\right) \text{ J kg}^{-1} \text{ liquid} \quad (2.2.20)$$

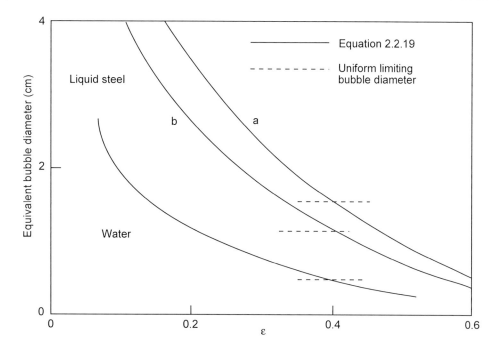

Fig. 2.15 Calculated bubble diameter for bubble fragmentation mechanism equation 2.2.19, compared with uniform limited diameter at high values for water and liquid steel: (a) for O < 4 ppm and 40 ppm S (b) for 600 ppm O and 120 ppm S. *From Ref. 27.*

This conjectured statement leads to the following similarity relation for non-foaming inviscid liquids.

$$\frac{\sigma_1}{\rho_1 d_1} = \frac{\sigma_2}{\rho_2 d_2} = \cdots \text{J kg}^{-1} \text{ liquid} \qquad (2.2.21)$$

With $\sigma_1 = 0.072$ Jm^{-2} and $d_1 = 0.45$ cm for gas-water system and $\sigma_2 = 1.3$ Jm^{-2} and $\rho_2 = 6940$ kg m^{-3}, we obtain $d_2 = 1.16$ cm for the limiting bubble diameter in liquid steel containing ≈ 600 ppm O and 120 ppm S. For the melt containing O < 4 ppm and 40 ppm S, $d_2 = 1.52$ cm. These limiting values intersect the fragmentation curves for liquid steel at about $\varepsilon = 0.40$, similar to that for water.

A uniform constant bubble size and essentially constant $\varepsilon = 0.6$ at high gas flow-rates, means a constant number of bubbles per unit volume of the emulsion. Noting that for the close-packed (fcc) arrangement of bubbles, the packing fraction (gas holdup) is $\varepsilon = 0.74$, the thickness of the liquid film at the nearest approach of the bubbles is derived from the following equality of the bubble number.

$$\frac{0.6}{d^3} = \frac{0.74}{(d + \delta/2)^3} \qquad (2.2.22)$$

Noting that for water the experimental data give $d = 0.45$ cm and that estimated for liquid steel $d = 1.16$ cm, equation 2.2.22 gives $\delta = 0.65$ mm for bubbles in aqueous solutions and $\delta = 1.68$ for bubbles in liquid steel at the limiting value of $\varepsilon = 0.6$ for non-foaming inviscid liquids.

2.2.2.4.3 Rate Equation for Transport Controlled Gas Bubble Reactions in Liquid Steel
For mass-transfer controlled reactions of gas bubbles in liquid steel, the rate equation has the following form:

$$\ln \frac{\%X - \%X_e}{\%X_o - \%X_e} = -S_o k_m t \qquad (2.2.23)$$

where X_o and X_e are the initial and gas-metal equilibrium concentrations of the diffusing reactant in the melt, k_m the mass-transfer coefficient and S_o the bubble surface area which is in terms of ε.

$$S_o = \frac{6\varepsilon}{d}, m^2/m^3 \text{ gas-melt emulsion} \tag{2.2.24}$$

Bubble surface area with respect to unit mass of liquid is:

$$S_m = \frac{6\varepsilon}{1-\varepsilon}\left(\frac{1}{\rho d}\right), m^2/kg \text{ liquid} \tag{2.2.24a}$$

From many studies of gas bubble reactions in nonfoaming aqueous solutions, the following formulation have been derived for the liquid-phase mass-transfer coefficient for the regime of surface-renewal at the gas-liquid interface.[26]

$$k_m = 1.28\left(\frac{DU_b}{d}\right)^{1/2} \tag{2.2.25}$$

where D is the diffusivity of the reactant in the liquid-film boundary layer.

In the presence of surface active solutes such as oxygen and sulphur in liquid steel, the bubble surface will be less mobile hence the rate constant k_m will be somewhat less than that given by equation 2.2.25 for the mobile surface, which is a necessary condition for surface-renewal. On the other hand, the surface active solutes decrease the bubble diameter, hence increase the bubble surface area for a given gas holdup. It appears that the product $S_o k_m$ may not be too sensitive to the presence of surface active solutes in the liquid.

The rate constant $S_o k_m$ for the transport controlled reaction is obtained from the combination of equations 2.2.17, 2.2.24 and 2.2.25 as given below for an average solute diffusivity $D = 5 \times 10^{-9}$ m²s⁻¹.

$$S_o k_m = \frac{5.43 \times 10^{-4} (\varepsilon U_s)^{1/2}}{d^{3/2}} \tag{2.2.26}$$

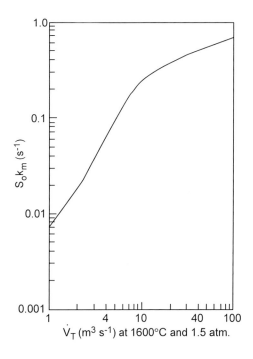

Fig. 2.16 Calculated rate constant for mass-transfer controlled reaction of gas bubbles in 220 ± 20 ton steel bath in the BOF or OBM (Q-BOP) vessel. From Ref. 27.

The rate constant thus calculated is plotted in Fig. 2.16 against the gas flow rate $\dot{V}_T$ at 1600°C and for an average gas pressure of 1.5 atm in the steel bath for 220 ± 20 ton heats.

2.2.2.4.4 Maximum Rate of Degassing of Liquid Steel with Argon Purging If the rates of diffusional processes and chemical reactions are sufficiently fast, the argon bubbles traversing the melt will be saturated with N_2 or H_2 to the corresponding equilibrium values for the concentrations present in the melt. It is for this limiting case that the following rate equation is applicable.

Denoting N and H by X, and using molar contents in the gas bubbles,

$$\dot{n}_{X_2} = \frac{d(\text{ppm} X)}{dt} \frac{10^{-6}}{M} \times 10^6 \text{ g-mol } X_2/\text{min ton} \tag{2.2.27}$$

$$\dot{n}_{Ar} = \frac{\dot{V}}{22.414 \times 10^{-3}} \text{ g-mol Ar/min ton} \tag{2.2.28}$$

where M is the molecular mass of X_2 and $\dot{V}$ the argon blowing rate in Nm³ min⁻¹t⁻¹. In the X_2-saturated argon bubbles for $\dot{n}_{Ar} \gg \dot{n}_{X_2}$

$$p_{X_2} = \frac{\dot{n}_{X_2}}{\dot{n}_{Ar}} \bar{P} = \frac{d(\text{ppm X})}{dt} \frac{22.414 \times 10^{-3}}{M\dot{V}} \bar{P} \qquad (2.2.29)$$

where $\bar{P}$ is the average bubble pressure in the melt.

Since the gas bubbles are in equilibrium with the melt

$$p_{X_2} = \frac{[\text{ppm X}]^2}{K^2} = \frac{d(\text{ppm X})}{dt} \frac{22.414 \times 10^{-3}}{M\dot{V}} \bar{P} \qquad (2.2.30)$$

where K is the equilibrium constant. The integration of the above rate equation gives with X_o being the initial concentration,

$$\frac{1}{\text{ppm X}} - \frac{1}{\text{ppm X}_o} = \frac{M\dot{V}}{22.414 \times 10^{-3} K^2 \bar{P}} t \qquad (2.2.31)$$

For the average liquid steel temperature of 1650°C at turn down, the values of the equilibrium constants are as given below.

$$\left. \begin{array}{l} \text{For } N_2: K = 459 \\ \text{For } H_2: K = 27.2 \end{array} \right\} \text{with } p_{X_2} \text{ in atm}$$

For an average gas bubble pressure of $\bar{P} = 1.5$ atm in the melt, the following rate equations are obtained.

$$\frac{1}{\text{ppm N}} - \frac{1}{\text{ppm N}_o} = 0.00395 \dot{V} t \qquad (2.2.32)$$

$$\frac{1}{\text{ppm H}} - \frac{1}{\text{ppm H}_o} = 0.0804 \dot{V} t \qquad (2.2.33)$$

As discussed later in detail, the maximum rates are not observed in steelmaking processes; the rates are limited by mass transfer and or chemical kenetics. Nevertheless equations 2.2.32 and 2.2.33 indicate, even for the limiting case, large quantities of purging gas are required and the processes are normally not feasible.

2.2.3 Chemical Kinetics

The theory of the absolute reaction rates is based on the concept of the formation of an activated complex as an intermediate transition state, which has an infinitesimally short life time of the order of 10^{-15} second. For an indepth study of the theory of reaction kinetics, to which outstanding contributions were made by Eyring and co-workers, reference may be made to the classical text books by Glasstone et al.[30] and by Hinshelwood.[31]

As illustrated in Fig. 2.17, there is a change in energy profile accompanying the reaction that involves the formation and decomposition of an activated complex. While the change in free energy accompanying reaction is $\Delta G° < 0$, the activation energy $\Delta G^* > 0$.

In almost all pyrometallurgical processes, we are concerned with heterogeneous reactions involving an interface between two reacting

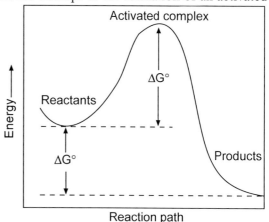

Fig. 2.17 Change in energy profile during the formation and decomposition of the activated complex involved in the reaction.

phases, e.g. solid-gas, solid-liquid, liquid-gas and two immiscible liquids (slag-liquid metal). For the case of a fast rate of transport of reactants and products to and from the reaction site, the rate is controlled by a chemical reaction occurring in the adsorbed layer at the interface. The reaction between adsorbed species L and M on the surface producing product Q occurs via the formation of an activated complex (LM)*.

$$L + M = (LM)^* \rightarrow Q$$
$$\{\text{stage I reactants}\} \quad \{\text{activated complex}\} \quad \{\text{stage II products}\} \quad (2.2.34)$$

The theory of the absolute rates states that the activated complex is in equilibrium with the reactants for which the equilibrium constant K* for constant temperature is

$$K^* = \frac{a^*}{a_L \, a_M} \quad (2.2.35)$$

where a's are the thermodynamic activities.

Next to be considered is the specific rate of decomposition of the activated complex to the overall reaction product Q, represented by

$$\frac{dn}{dt} = \left(\frac{kT}{h}\right) \Gamma_o \, \theta^* \quad (2.2.36)$$

where

- dn/dt = the reaction rate, mol cm^{-2}s^{-1},
- k = the Boltzmann constant, 1.380×10^{-23} J K^{-1},
- h = the Planck constant, 6.626×10^{-34} J s,
- T = temperature, K,
- Γ_o = total number of adsorption sites on the surface, 10^{15} mole cm^{-2},
- θ^* = fractional coverage by the activated complex.

For single site occupancy by the activated complex in the adsorbed layer, the activity of the complex is represented by

$$a^* = \varphi^* \frac{\theta^*}{1 - \theta} \quad (2.2.37)$$

where θ is the total fractional occupancy of the sites by the adsorbed species and φ^* is the activity coefficient of the complex in the chemisorbed layer. Combining equations 2.2.35 - 2.2.37 gives the rate of forward reaction in terms of the absolute reaction rate theory.

$$R_f = \frac{dn}{dt} = \left(\frac{kT}{h}\right) \Gamma_o \left(\frac{K^*}{\varphi^*}\right)(1 - \theta)\{a_L \, a_M\} \quad (2.2.38)$$

The thermodynamics of the chemisorbed layer at the interface, i.e. value of Γ_o, K* and φ are not known, therefore the isothermal rate equation is given in a simplified general form thus

$$R_f = \Phi_f (1 - \theta)\{a_L \, a_M\} \quad (2.2.39)$$

where Φ_f is the isothermal rate constant of the forward reaction.

As the reaction progresses, concentrations of the reactants L and M decrease while the concentration of the product Q increases. Because of these composition changes and the influence of the reverse reaction Q→L + M, the rate decreases with an increasing reaction time. The rate of the reverse reaction is represented by

$$R_r = -\Phi_r (1 - \theta)\{a_Q\} \quad (2.2.40)$$

where Φ_r is the rate constant of the reverse reaction. Therefore, the net overall rate of reaction is

$$\frac{dn}{dt} = \Phi_f (1-\theta)\{a_L a_M\} - \Phi_r (1-\theta)\{a_Q\} \qquad (2.2.41)$$

When the rates of forward and reverse reactions are the same, i.e. $dn/dt = 0$, the reaction is said to be at an equilibrium state. It follows that the ratio of the rate constants Φ_f/Φ_r is the equilibrium constant of the reaction.

$$K = \frac{\Phi_f}{\Phi_r} = \left(\frac{a_Q}{a_L a_M}\right)_{eq} \qquad (2.2.42)$$

In terms of a single rate constant, the net reaction rate is formulated as

$$\frac{dn}{dt} = \Phi_f (1-\theta)\{a_L a_M - (a_L a_M)_{eq}\} \qquad (2.2.43)$$

where $(a_L a_M)_{eq}$ is the equilibrium value for the activity a_Q in stage II at any given reaction time.

For a given surface coverage, the temperature dependence of the rate constant is represented by

$$\Phi = \left[\Phi_o \exp\left(\frac{-\Delta H^*}{RT}\right)\right]_\theta \qquad (2.2.44)$$

where Φ_o is a pre-exponential constant and ΔH^* is the apparent heat of activation for the reaction at a given site fillage θ.

Since the activation energy ΔG^* is always positive, the enthalpy of activation for the reaction is also positive. As shown in Fig. 2.18 in the plot $\ln \Phi$ vs $1/T$ the slope of the line gives ΔH^*. It should be noted that since the thermodynamic quantities for the activated complex (K^*/φ^*) in the chemisorbed layer are not known, the ΔH^* derived from the rate measurements over a sufficiently large temperature range, is the apparent heat of activation.

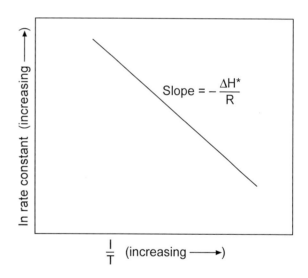

Fig. 2.18 Effect of temperature on rate constant for reaction involving an activated complex.

Another aspect of the kinetics of interfacial reactions is the rate of chemisorption or desorption at the reaction surface as given by the following theoretical equation for an uncontaminated surface.

$$\Phi_f = (2\pi M_i RT)^{-1/2} \exp\left(\frac{-\Delta H^*}{RT}\right) \qquad (2.2.45)$$

where M_i is the molecular mass of the adsorbed species. This equation is transformed to the following form for the maximum rate of vaporization, i.e. free vaporization, from an uncontaminated surface at low pressures

$$\text{Rate}_{max} = \frac{p_i}{\sqrt{2\pi M_i RT}} \qquad (2.2.46)$$

where p_i is the vapor pressure in atm, for which the equation is reduced to the following form.

$$\text{Rate}_{max} \text{ (g-mol cm}^{-2}\text{ s}^{-1}) = 44.3\, p_i\, (MT)^{-1/2} \qquad (2.2.47)$$

2.2.3.1 Examples of Experiments on Rates of Interfacial Reactions

2.2.3.1.1 Nitrogen Transfer Across Iron Surface The rate of reaction of gaseous nitrogen with liquid and solid iron, in the presence of surface active alloying elements, has been studied by many investigators since the late 1950s. It was in the late 1960s[32–34] that the rate controlling reaction mechanism was resolved, which was further confirmed in the subsequent experimental work done by Fruehan and Martonik[35], Byrne and Belton[36], and Glaws and Fruehan[37].

In particular the latter investigators used an isotope exchange technique which measured the rate of dissociation of the N_2 molecule. They demonstrated the rate of dissociation was the same as the rate of absorption of nitrogen, providing strong evidence that the absorption rate is controlled by dissociation.

When the rate of reaction is not hindered by slow nitrogen transport to and from the gas-metal interface, the rate of nitrogenation

$$N_2(g) = N_2^* \rightarrow 2[N] \tag{2.2.48}$$

is controlled either by the rate of chemisorption or dissociation of N_2 molecules on the metal surface. The rate of reverse reaction, i.e denitrogenation, is of course a second order type with respect to nitrogen dissolved in the metal. The equation below represents the rate of nitrogen transfer from gas to liquid iron

$$\frac{d[\%N]}{dt} = \frac{100 A}{\rho V} \Phi_f (1-\theta) \left\{ p_{N_2} - \left(p_{N_2}\right)_{eq} \right\} \tag{2.2.49}$$

where ρ is the density of liquid iron, A the surface area of the melt on which the nitrogen stream is impinging and V the volume of the melt. The rate constant Φ_f, in units of g N cm^{-2}min^{-1}atm^{-1}N$_2$, is for the forward reaction 2.2.48. The equilibrium partial pressure $(p_{N_2})_{eq}$ corresponding to the nitrogen content of the melt at the reaction time t, is that given by the equilibrium constant K for nitrogen solubility.

$$\left(p_{N_2}\right)_{eq} = \frac{[\%N]^2}{K} \tag{2.2.50}$$

With this substitution the isothermal rate equation is

$$\frac{d[\%N]}{dt} = \frac{100 A}{\rho V} \Phi_f (1-\theta) \left\{ p_{N_2} - \frac{[\%N]^2}{K} \right\} \tag{2.2.51}$$

For constant N_2 pressure and temperature, the integration of equation 2.2.51 gives for %N = 0 at t = 0,

$$\ln \frac{K p_{N_2} + [\%N]}{K p_{N_2} - [\%N]} = 2 p_{N_2} \frac{100 A}{\rho V} \Phi_f (1-\theta) t \tag{2.2.52}$$

Byrne and Belton made an accurate determination of the rate constant Φ_f for reaction of N_2 with high purity iron and Fe-C alloys at 1550–1700°C, by measuring the rate of $^{15}N \rightarrow ^{14}N$ isotope exchange that occurs on the iron surface, as represented by

$$\log \Phi_f = \frac{-6340 \pm (710)}{T} + 1.85 (\pm 0.38) \tag{2.2.53}$$

where the rate constant Φ_f is in units of g N cm^{-2}min^{-1}atm^{-1}. The apparent heat of activation = 121.4 kJ mol^{-1} is much lower than the value expected for the rate of dissociation of N_2. As pointed out by Byrne and Belton, the rate of chemisorption of N_2 is presumably controlling the reaction mechanism.

The surface active elements dissolved in iron, e.g. O, S, Se, Te, are known to lower the rate of nitrogen transfer across the iron surface. On the basis of the experimental rate data with liquid iron containing O and S, as given in various publications (Refs. 33–38) and the surface tension data, the effects of O and S on the fraction of vacant sites, $1-\theta$, in the chemisorbed layer may be represented by

$$1-\theta = \frac{1}{1+260\left(\%O + \frac{\%S}{2}\right)} \quad (2.2.54)$$

which is a slightly simplified form of the equation that was derived by Byrne and Belton.

For the chemical reaction-controlled nitrogen removal from liquid iron (or steel) in reduced pressures or in an inert gas stream with very low N_2 pressure

$$2[N] \to N_2 (g) \quad (2.2.55)$$

the integrated form of the rate equation is

$$\frac{1}{\%N} - \frac{1}{\%N_\circ} = \frac{100 A}{\rho V} \Phi_r (1-\theta) t \quad (2.2.56)$$

where $\%N_o$ is the initial nitrogen content and Φ_r the rate constant $= \Phi_f/K$.

The solubility of N_2 in liquid iron or low alloy steel is given by

$$\log K = \frac{[\%N]^2}{p_{N_2}(\text{atm})} = -\frac{376}{T} - 2.48 \quad (2.2.57)$$

Combining this with equation 2.2.53 gives for the rate constant Φ_r in gN cm^{-2}min^{-1}%N^{-1}

$$\log \Phi_r = \frac{-5964}{T} + 4.33 \quad (2.2.58)$$

Now let us consider the chemical-reaction controlled rate of denitrogenation. It will be assumed that the rate is controlled by only chemical kinetics. In actuality it is controlled by mass transfer and chemical kinetics in series which is discussed later in detail. From equation 2.2.58 the rate for 1600°C $\Phi_r = 0.233$ g N cm^{-2}s^{-1}%N^{-1}. For liquid steel containing 600 ppm O and 120 ppm S, $(1-\theta) = 0.055$ from equation 2.2.54. For the gas flow rate $\dot{V}_T = 34$ m^{3}s^{-1}, $\varepsilon = 0.6$ and the bubble surface area is $S_m = 1.11$ cm^{2}g^{-1} liquid steel. Inserting these numbers in equation 2.2.56 for the limiting case of $\%N_e \approx 0$, gives

$$\frac{1}{\%N} - \frac{1}{\%N_o} = 100 \times 1.11 \times 0.233 \times 0.055 t = 1.42 t \quad (2.2.59)$$

In 60 seconds time of purging, 0.0030% N_o will be reduced only to 0.0024% N. In actual fact, the argon bubbles traversing the melt will contain some N_2; for example, if the nitrogen partial pressure in gas bubbles were 0.001 atm N_2, the equilibrium content in the melt would be 0.0015% N_e. Therefore, with $\%N_e > 0$, the nitrogen removal will be much less than that calculated above for the chemical-reaction controlled rate for the hypothetical limiting case of $\%N_e \approx 0$.

The foregoing predication from the rate equation is in complete accord with the practical experience that even at high rates of argon purging of the melt, as in OBM(Q-BOP), there is no perceptible nitrogen removal primarily because of the presence of surface active solutes, oxygen and sulfur, in the steel bath.

2.2.3.1.2 CO_2 and H_2O Reaction with Fe–C Melts Another example of chemical kinetics controlling reaction rates is the decarburization of Fe–C melts.

$$\underline{C} + CO_2 = 2CO \quad (2.2.60)$$

Sain and Belton[39] and later Mannion and Fruehan[40] showed that the rate was affected by surface active elements and was consistent with CO_2 dissociation on the surface controlling the rate.

Nagasaka and Fruehan[41] measured the rate of decarburization by H_2O.

$$H_2O + \underline{C} = CO + H_2 \qquad (2.2.61)$$

They also found that the rate was controlled by chemical kinetics and most likely dissociation of H_2O on the surface. For both the CO_2 and H_2O reaction, the rate constant could be expressed by the relationship

$$k_C = \frac{k_p}{1+K_S[\%S]} + k_r \qquad (2.2.62)$$

where k_C, k_p and k_r are the rate constants, k_p is the rate constant for pure iron and k_r is the residual rate at high sulfur contents respectively and K_s is the absorption coefficient for sulfur on liquid iron. The first term is based on the analysis by Belton[42] and the residual rate which is relatively small is discussed by Nagasaka et al in detail.[41] The overall rate constants are given as a function of sulfur content in Fig. 2.19.

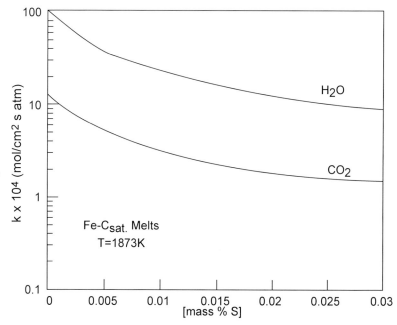

Fig. 2.19 Rates of dissociation of CO_2 and H_2O on iron as a function of sulfur content. *From Ref. 41.*

2.2.3.1.3 Chemical Kinetics of Carbon Oxidation in CO_2–CO Mixtures Examples of the experimental data on the initial rate of internal oxidation of granular (0.5 mm dia.) electrode graphite and metallurgical coke are given in Figs. 2.20 and 2.21, reproduced from a previous publication.[16]

In a subsequent publication, Turkdogan[27] re-assessed the interpretation of the rate data on the assumption that both CO and CO_2 are chemisorbed on the pore walls of the carbon and the rate of oxidation is due to the rate of dissociation of chemisorbed CO_2. For the limiting case of complete internal burning, the rate equation will be in the following form

$$\text{Rate} = \frac{\Phi' p_{CO_2}}{1 + k_1 p_{CO} + k_2 p_{CO_2}} t \qquad (2.2.63)$$

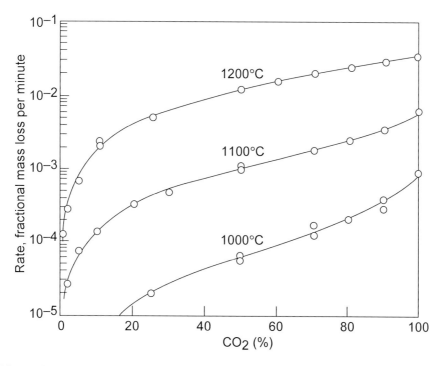

Fig. 2.20 Effects of temperature and gas composition on the rate of oxidation of electrode graphite granules (~0.5 mm dia.) in CO_2–CO mixtures at 0.96 atm total pressure. *From Ref. 16.*

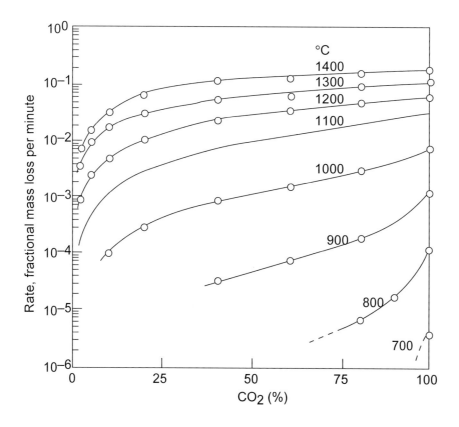

Fig. 2.21 Effect of temperature and gas composition on the rate of oxidation of granular metallurgical coke (~0.5 mm dia.) in CO_2–CO mixtures at 0.96 atm total pressure. *From Ref. 16.*

where Φ' is the rate constant (min^{-1}atm^{-1}CO$_2$) for a given temperature and a particular type of carbon; the constants k_1 and k_2 are associated with the chemisorption of CO and CO$_2$. It should be noted that in compliance with the notation in equation 2.2.64 for the ideal monolayer, the constants k_1 and k_2 are reciprocals of the respective activity coefficients of the absorbed species, i.e. $k_1 = 1/\varphi_{CO}$ and $k_2 = 1/\varphi_{CO_2}$; the fraction of vacant sites being $(1 - \theta) = (1 - \theta_{CO} - \theta_{CO_2})$.

The fraction of surface sites θ_i covered by the adsorbed species is related to the solute activity as given in the following Langmuir adsorption isotherm.

$$a_i = \varphi_i \frac{\theta_i}{1 - \theta_i} \tag{2.2.64}$$

From the cited experimental data for the oxidation of graphite and coke the following equations are obtained for the temperature dependence of Φ', $\varphi_{CO} = 1/k_1$ and $\varphi_{CO_2} = 1/k_2$.

For electrode graphite:

$$\log \Phi'\left(\min^{-1} \text{atm}^{-1} CO_2\right) = -\frac{16{,}540}{T} + 10.75 \tag{2.2.65}$$

$$\log \varphi_{CO} (\text{atm}) = -\frac{8719}{T} + 4.84 \tag{2.2.66}$$

$$\log \varphi_{CO_2} (\text{atm}) = -\frac{590}{T} - 0.072 \tag{2.2.67}$$

For metallurgical coke:

$$\log \Phi'\left(\min^{-1} \text{atm}^{-1} CO_2\right) = -\frac{16{,}540}{T} + 11.37 \tag{2.2.68}$$

$$\log \varphi_{CO} (\text{atm}) = -\frac{2117}{T} + 0.27 \tag{2.2.69}$$

$$\log \varphi_{CO_2} (\text{atm}) = \frac{3840}{T} - 3.45 \tag{2.2.70}$$

It should be noted that these equations fit the measured initial rates of oxidation within a factor of about 1.5 for temperatures 800 to 1200°C and for CO$_2$ pressures 0.03 to 3 atm. For easy comparison of the reactivities of electrode graphite and metallurgical coke, the numerical values are given below from the above equations for 900 and 1200°C.

Temperature °C	Electrode graphite			Metallurgical coke		
	Φ'	φ_{CO}	φ_{CO_2}	Φ'	φ_{CO}	φ_{CO_2}
900	4.5 x 10^{-4}	0.0025	0.27	1.9 x 10^{-3}	0.029	0.67
1200	3.3 x 10^{-1}	0.083	0.34	1.4 x 10^0	0.068	0.14

After about 3 to 5 percent of initial oxidation, the pore surface area of the coke samples were found to be four to five times greater than the graphite samples; this is consistent with the Φ' values for coke being greater than for graphite by a similar factor. The extent of CO adsorption on the pore walls of electrode graphite or coke is greater than the CO$_2$ adsorption.

There are variations in the reported values of the apparent heat of activation for oxidation of the electrode (or reactor grade) graphite in CO$_2$. For example, Gulbransen et al.[43] found $\Delta H^* = 368$ kJ mol^{-1} while according to Blackwood's work [44] $\Delta H^* = 260$ kJ mol^{-1}; in the present case $\Delta H^* = 317$ kJ mol^{-1}. Reference should be made also to the papers of Ergun,[45] Hedden and Lowe,[46] and by Grabke[47] for various other interpretations of the kinetics of oxidation of carbons.

Aderibigbe and Szekely[48] also investigated the oxidation of metallurgical coke in CO$_2$–CO mixtures at 850 to 1000°C. Their rate constants K_1 ($\equiv \Phi'$) are similar to those given by equation 2.2.68 their estimate of the apparent heat of activation being $\Delta H^* = 249\pm47$ kJ mol^{-1}.

Recently Story and Fruehan[49] examined the oxidation of carbons in CO–CO_2 gas mixtures at high temperatures (>1300°C) under conditions of high external mass transfer. The gas mixture was jetted at high velocities, reducing the resistance to mass transfer at the surface. They were in limiting case of external oxidation and determined the parameter ($\phi \rho S\, D_e$) from the data, and then using published values of S and computed values of D_e, computed the rate constant. The results are compared to the rates measured at lower temperatures in Fig. 2.22.

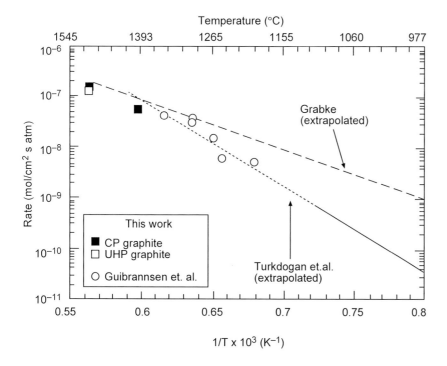

Fig. 2.22 Arrhenius plot of graphite–CO_2 data.

2.2.4 Mixed Control

In many metallurgical operations the rate is not controlled by a single reaction step such as mass transfer or chemical kinetics but rather two steps in series such as mass transfer followed by chemical kinetics. By equating the fluxes for each of the reaction steps it is possible to obtain expressions for the overall rate, or a series of expressions, which can be solved numerically.

2.2.4.1 Mixed Control for Gas Phase Mass Transfer and Chemical Kinetics

In general the temperature dependence of chemical kinetics is greater than for gas phase mass transfer. Consequently at low temperatures chemical kinetics is predominantly controlling the rate while at high temperatures mass transfer dominates. However for many reactions over a large range of conditions including temperature, fluid flow conditions and the concentration of surface active elements which effect the chemical rate, the rate is controlled by both processes in series.

For example, the rate of dissociation of CO_2 and H_2O on liquid iron is controlled by both gas phase mass transfer and chemical kinetics in series over a large range of temperature and fluid flow conditions. In this case the pressure of the reacting gas at the surface is not the bulk composition. The flux of the reacting gas and the rate of the reaction are given by 2.2.71 and 2.2.72 respectively,

$$J = \frac{m_i}{RT}\left(p_i^B - p_i^s\right) \qquad (2.2.71)$$

$$R = k\left(p_i^s - p_i^e\right) \qquad (2.2.72)$$

where

m_i = the gas phase mass transfer coefficient for i,
p_i^B = the pressure of i in the bulk phase,
p_i^s = the pressure of i at the surface,
p_i^e = the equilibrium pressure of i,
k = the chemical rate constant.

Equations 2.2.71 and 2.2.72 assume that the chemical rate is first order and the reaction is equilmolar in terms of the moles of reactant and product gases or the pressure of the reaction gas is low. As it is not possible to accumulate component i at the surface, the processes are assumed to be at steady state and the flux and chemical rate are equated. It is then possible to calculate the pressure at the surface and the rate in terms of the bulk composition. Further simplification gives

$$R = \frac{1}{\frac{1}{k} + \frac{RT}{m_i}} \left(p_i^B - p_i^e \right) \quad (2.2.73)$$

Equation 2.2.73 is analogous to an electrical current with two resistances in series where k and m_i/RT are analogous to conductivities. The first term in equation 2.2.73 is often referred to as the observed or overall rate constant.

For example Nagasaka and Fruehan[41] measured the rate of dissociation of H_2O on liquid Fe-C-S alloys from 1673 to 1873°K for varying sulfur contents. At 1873K and 0.1% S the measured overall rate constant was 2.5×10^{-4} mole/cm² s atm for conditions where the mass transfer coefficient was 100 cm/s. For this case the chemical rate was computed to be 3.9×10^{-4} mole/cm² s atm.

Similarly the nitrogen and carbon dioxide reactions can be controlled by gas phase mass transfer and chemical kinetics and equations 2.2.71 – 2.2.73 can be applied.

2.2.4.2 Mixed Control for Chemical Kinetics and Liquid Phase Mass Transfer

For many conditions in steelmaking and refining the rate of pick up or removal of nitrogen from liquid iron is controlled by chemical kinetics and liquid phase mass transfer in series. Fruehan, Lally and Glaws[50] developed a mixed control model and applied it to several metallurgical processes. For nitrogen gas reacting with iron the flux of nitrogen for mass transfer and the chemical reaction are given by 2.2.74 and 2.2.75.

$$J_N = \frac{m \rho}{100}\left[\%N^s - \%N \right] \quad (2.2.74)$$

$$J_N = k\left(p_{N_2} - \frac{[\%N^s]^2}{K} \right) \quad (2.2.75)$$

where

J_N = flux of nitrogen (g/cm² s),
m = mass transfer coefficient for nitrogen (cm/s),
k = chemical rate constant (g/cm² s atm),
$\%N^s$ = nitrogen content at the surface,
K = equilibrium constant for nitrogen reaction with liquid iron,
ρ = density of iron (g/cm³).

Since no nitrogen can be accumulated at the surface it is possible to equate the two flux equations, 2.2.74 and 2.2.75, and solve for the nitrogen content at the surface. The equation is a quadratic and is solved numerically for small time increments for specific conditions.. The surface nitrogen content is then used to compute the rate of nitrogen absorption or removal. This technique was applied to nitrogen absorption into liquid iron when bubbling nitrogen in the ladle and typical results are shown in Fig. 2.23. The rate of nitrogen pick up for low sulfur steel is considerably faster than for high sulfur steel, as sulfur decreases the chemical rate. Below 0.005% S the rate does not increase further as liquid phase mass transfer then dominates and it is not affected by sulfur content.

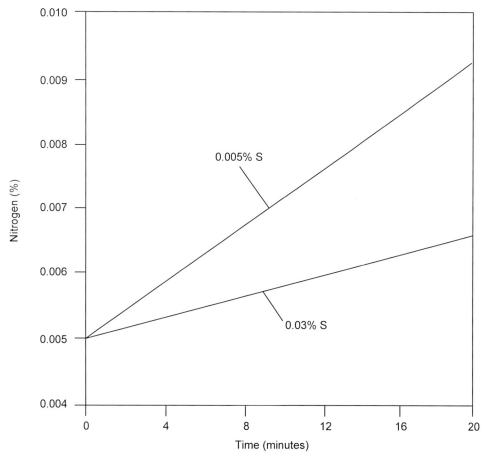

Fig 2.23 Rate of nitrogen pick-up in 200 metric tons of steel containing 0.005 and 0.030% S, using 10 scfm of N_2. *From Ref. 50.*

2.3 Properties of Gases

Of importance to the iron and steelmaking and related pyrometallurgical processes, the selected thermochemical and transport properties of gases are presented in this section.

2.3.1 Thermochemical Properties

2.3.1.1 Heat Content

The molar heat contents of simple selected gases of interest to pyrometallurgical processes are plotted in Fig. 2.24 over the temperature range 300–1900K.

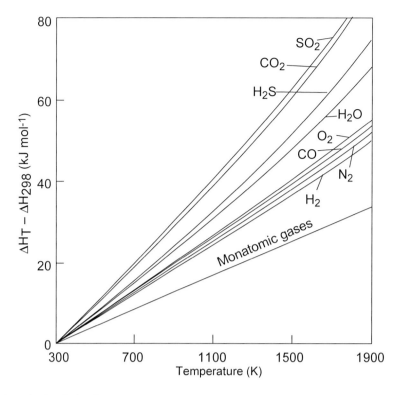

Fig. 2.24 Heat contents of simple gases.

2.3.1.2 Equilibrium States in Gas Mixtures

When assessing the state of reactions involving gas mixtures, the partial pressures of the gaseous species should be evaluated from the reaction equilibrium data for all the pertinent gas reactions. The principle of the computation is demonstrated by the following example for the gas mixture CO–CO_2–H_2O–CH_4.

This example is in relation to reheating a reformed gas in the direct iron ore reduction processes. If there is complete gas equilibrium at the reheating temperature and pressure, the activity of carbon in the equilibrated gas mixture must satisfy the following three basic reaction equilibria:

$$2CO = C + CO_2 \tag{2.3.1}$$

$$K_1 = \frac{a_C \, p_{CO_2}}{p_{CO}^2} \tag{2.3.2}$$

$$H_2 + CO = C + H_2O \tag{2.3.3}$$

$$K_2 = \frac{a_C \, p_{H_2O}}{p_{H_2} \, p_{CO}} \tag{2.3.4}$$

$$CH_4 = C + 2H_2 \tag{2.3.5}$$

$$K_3 = \frac{a_C \, p_{H_2}^2}{p_{CH_4}} \tag{2.3.6}$$

For the predominant gaseous species the following equalities are derived from the mass balance for (ΣC), (ΣO) and (ΣH).

$$\frac{\Sigma C}{\Sigma O} = \left(\frac{p_{CO_2} + p_{CO} + p_{CH_4}}{2p_{CO_2} + p_{CO} + p_{H_2O}}\right)_i = \left(\frac{p_{CO_2} + p_{CO} + p_{CH_4}}{2p_{CO_2} + p_{CO} + p_{H_2O}}\right)_e \quad (2.3.7)$$

$$\frac{\Sigma C}{\Sigma H} = \left(\frac{p_{CO_2} + p_{CO} + p_{CH_4}}{2p_{H_2O} + 2p_{H_2} + 4p_{CH_4}}\right)_i = \left(\frac{p_{CO_2} + p_{CO} + p_{CH_4}}{2p_{H_2O} + 2p_{H_2} + 4p_{CH_4}}\right)_e \quad (2.3.8)$$

where i and e indicate partial pressures in the ingoing and equilibrated gas mixtures, respectively. The equation below gives the total pressure P of the equilibrated gas mixture.

$$(p_{CO} + p_{CO_2} + p_{H_2} + p_{H_2O} + p_{CH_4})_e = P \quad (2.3.9)$$

The partial pressures of gaseous species and the activity of carbon in the equilibrated gas mixture are computed by simultaneous solution of equations 2.3.1 to 2.3.9 with known values of the equilibrium constants K_1, K_2, and K_3.

Examples are given in Fig. 2.25 showing the calculated carbon activities for 4 atm total pressure and for an initial gas mixture containing 73% H_2, 18% CO, 8% CO_2, 1% CH_4 to which 0.2% to 6.0% H_2O has been added. The carbon deposition is imminent when its activity in the equilibrated gas exceeds unity, with respect to graphite. With the addition of 0.2% H_2O to this gas mixture, there should be no carbon deposition in the equilibrated gas mixture at temperatures below 512°C and above 720°C. In an earlier study of the reduction of iron oxides in a similar gas mixture, carbon deposition was observed at all temperatures below 1000°C (see Ref. 51), indicating lack of complete gas equilibrium even at 1000°C. These experimental findings suggest that carbon deposition may occur during reheating of reformed natural gas in industrial operations, even when the calculated carbon activity is less than unity in the gas mixture under equilibrium conditions.

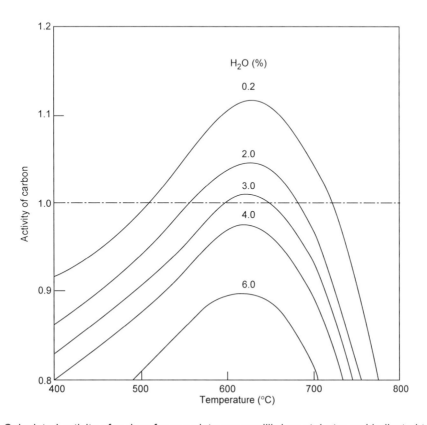

Fig. 2.25 Calculated activity of carbon for complete gas equilibrium at 4 atm and indicated temperatures for reformed natural gas containing 73% H_2, 18% CO, 8% CO_2 and 1% CH_4 to which an indicated amount of H_2O is added. *From Ref. 52.*

The activity of carbon cannot be calculated for a nonequilibrated gas mixture. However, for the purpose of identifying the reactions that may be responsible for carbon deposition, we may calculate carbon activities for individual reactions on the assumption that there is no change in the composition of the inlet gas upon heating. As is seen from the example given in Fig. 2.26 the carbon activities for individual reactions differ considerably from those calculated for complete gas equilibrium. For complete gas equilibrium, the carbon activity is below unity at temperatures below 540 and above 700°C, while for the individual reactions the carbon activity is below unity only within the temperature range 880–1060°C. Therefore, in the nonequilibrated gas, there can be no carbon deposition only within the temperature range 880–1060°C. The temperature at which carbon deposition may start depends on the relative rates of reactions 2.3.1 and 2.3.3: below 730°C if reaction 2.3.1 is fast, and below 880°C if reaction 2.3.3 is fast.

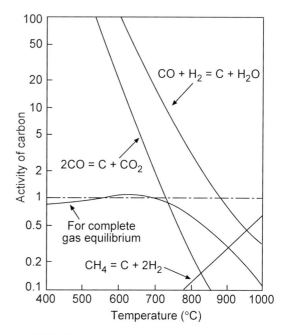

Fig 2.26 Calculated activity of carbon for complete gas equilibrium at 4 atm. compared with the activity of carbon for individual reactions in a reformed inlet gas containing 2% H$_2$O. *From Ref. 52.*

On the basis of the foregoing reasoning, we may compute a diagram of % H$_2$O in the inlet reformed gas versus temperature, as in Fig. 2.27, to delineate the regions where (I) carbon will not deposit, (II) carbon may deposit, and (III) carbon will deposit. With increasing pressure, regions II and III move to higher temperatures and I to higher concentrations of H$_2$O, and chances of carbon deposition become greater. From the equilibrium constant for reaction 2.3.3, we can also compute the critical p_{H_2O}/p_{CO} ratio for possible carbon deposition at any temperature and partial pressure of H$_2$ in the inlet gas. The possibility of carbon deposition in region II depends much on the catalytic behavior of the inner surface of the reheater tubes.

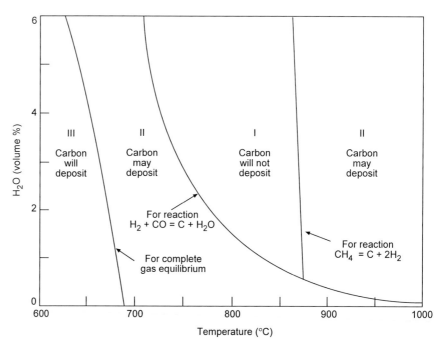

Fig. 2.27 Regions of carbon deposition at reheating temperatures of reformed natural gas at 1 atm containing indicated amounts of H$_2$O. *From Ref. 52.*

2.3.1.3 Catalytic Decomposition of Carbon Monoxide

The catalytic decomposition of carbon monoxide has been the subject of much study, because of the occurrence of carbon deposition in many processes, e.g., blast furnaces, the Stelling process, the Fischer-Tropsch process, nuclear reactors, and so on. References to early work on this subject are listed in an annotate bibliography by Donald[53]. The consensus is that iron, cobalt, and nickel are the most effective catalysts for the decomposition of carbon monoxide. Kehrer and Leidheiser[54] reported virtually no catalytic effect of copper, silver, chromium, molybdenum, palladium, and rhenium.

In a comprehensive study of carbon deposition in CO–H_2 mixtures at temperatures of 450–700°C, Walker et al.[55] observed that the C/H atom ratio in the deposit increased from 10 to 50 with increasing reaction temperature and increasing CO/H_2 ratio in the gas. They also found that the properties of carbon deposit were affected by the amount accumulated on the catalyst, e.g., with increasing thickness of the carbon layer, the crystallinity became poorer, the surface area increased, the electrical conductivity decreased, and the C/H_2 ratio increased. Another important finding was that carbon deposition ultimately ceased when most of the iron was converted to cementite. Upon hydrogen treatment, cementite decomposed to iron and graphite, but the regenerated iron lost most of its reactivity as a catalyst. Similar observations were made by Turkdogan and Vinters[51] in subsequent studies of the catalytic decomposition of carbon monoxide in the presence of porous iron.

It is generally agreed that iron, not cementite, catalyzes the decomposition of carbon monoxide. The chemisorption of H_2 and CO on the surface of iron is believed to approach equilibrium rapidly. The carbon formed by reactions 2.3.1 and 2.3.3 migrates across the surface to a nucleating center where cementite and free carbon are deposited. Based on x-ray diffraction and electron microscopic studies, Ruston et al.[56] suggested that although the decomposition of carbon monoxide was catalyzed by iron, an iron carbide Fe_7C_3 formed as an intermediate step in the decomposition of the activated reaction product to graphite.

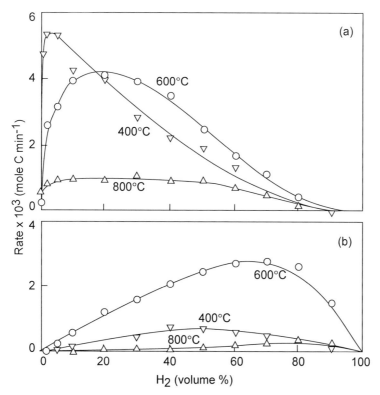

Fig. 2.28 Effect of gas composition (H_2–CO) on rate of carbon deposition on 660-mg porous iron granules at 1 atm (a) by reaction $2CO \rightarrow C + CO_2$ and (b) by reaction $CO + H_2 \rightarrow C + H_2O$. *From Ref. 57.*

As is seen from the experimental results of Olsson and Turkdogan[57] in Fig. 2.28, the rate of carbon deposition on an iron catalyst in CO–H_2 mixtures is a complex function of temperature and gas composition. From the measured rate of carbon deposition and the measured CO_2/H_2O in the exhaust gas, the rates of reactions 2.3.1 and 2.3.3 were determined. These measured rates are shown in Fig. 2.29 as a function of gas composition at 400, 600 and 800°C. The hydrogen appears to have a dual role: at low concentrations the hydrogen catalyzes reaction 2.3.1, and at high concentrations of hydrogen reaction 2.3.3 contributes directly to carbon deposition.

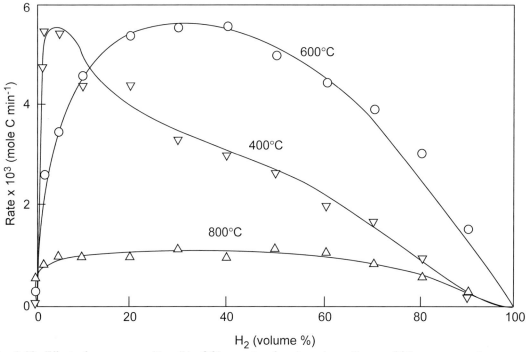

Fig. 2.29 Effect of gas composition (H_2–CO) on rate of carbon deposition on 660–mg porous iron granules at atmospheric pressure and indicated temperatures. *From Ref. 57.*

Although H_2O is expected to retard or inhibit carbon deposition by virtue of reverse reaction 2.3.4, it has been found[57] that an addition of 0.5%–2% H_2O to CO–CO_2 mixtures enhances the rate of carbon deposition at 400–600°C. A similar catalytic effect of H_2O has been observed for carbon deposition from methane in the presence of nickel[58]. We might mention in passing that the rate of carbon formation by the disproportionation of carbon monoxide is much faster than the pyrolysis of methane under similar conditions.[59] Ever since the first observations made by Byrom[60] and soon after by Carpenter and Smith[61], it has become generally recognized that sulfur in the gas retards or inhibits the catalytic action of iron in the decomposition of carbon monoxide. This is due to the coating of iron surface with cementite, which does not decompose readily in the presence of sulfur. In fact, in the Stelling process[62] the iron ore is reduced to cementite in a CO–CO_2 atmosphere without the formation of free carbon, probably because of the presence of sulfur in the system.

The nitrogen-bearing gaseous species such as NH_3 and $(CN)_2$ are also known to retard the decomposition of carbon monoxide.[62–63] Again, this may be attributed to extended metastability of cementite in the presence of nitrogen. The retarding effect of sulfur and nitrogen on carbon deposition is similar to sluggish graphitization of steel (by decomposition of cementite) in the presence of sulfur and nitrogen in solution in steel.

The carburization of iron in gas mixtures containing H_2 and CO following reactions similar to 2.3.1 and 2.3.3 occurs on iron surfaces. These reactions are relevant to the early stages of metal dusting and were studied by Fruehan and Martonik.[64] They also found that the rate of reaction 2.3.3 is faster than 2.3.1 and contributes significantly to the carburization of iron[64] as shown in Fig.2.30. When CO is replaced by H_2 the rate increases significantly. The H_2–CO reaction rate constant is approximately tenfold greater than that for CO for carburization.

2.3.2 Transport Properties

The molecular transfer of mass, momentum and energy are interrelated transport processes of diffusion under a concentration gradient, viscous flow in a velocity gradient and heat conduction in a thermal gradient.

The derivation of the transport properties from the rigorous kinetic theory of gases, described in depth by Chapman and Cowling[65] and Hirschfelder *et al.*[66] is based on the evaluation of the intermolecular energy of attraction, the collision diameter and the collision integral involving the dynamics of a molecular encounter, hence the intermolecular force law.

For a brief description of the theoretical equations for the calculation of the transport properties of gases, reference may be made to a previous publication.[52] For the present purpose, it is sufficient to give selected numerical data on the transport properties of a few gases, relevant to ironmaking and steelmaking processes.

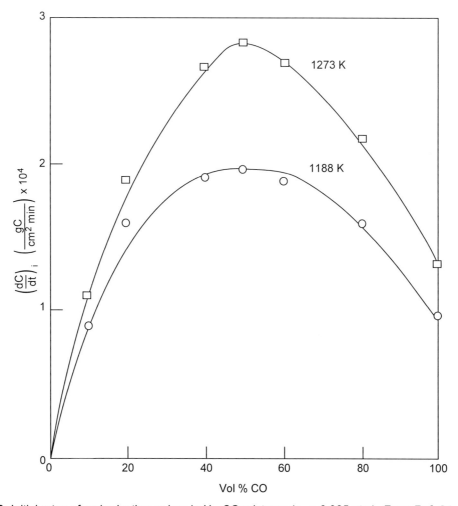

Fig. 2.30 Initial rates of carburization or iron in H_2–CO mixtures (p_T = 0.965 atm). *From Ref. 64.*

Table 2.2 Diffusivity D at 1 atm Pressure: $cm^2s^{-1} \equiv 10^{-4} m^2s^{-1}$

Gas	Temperature, K			
	300	1000	1500	2000
H_2	1.46	10.82	21.07	33.90
Ar	0.18	1.46	2.88	4.56
N_2	0.20	1.59	3.05	4.98
O_2	0.22	1.67	3.27	5.30
CO	0.20	1.63	3.20	5.19
CO_2	0.11	0.93	1.87	3.00
H_2O	0.28	2.83	5.79	9.57

Table 2.3 Viscosity η (poise) $\equiv$ g cm^{-1} s^{-1} $\equiv$ 0.1N s m^{-2} $\equiv$ 0.1J m^{-3}s

Gas	Temperature, K			
	300	1000	1500	2000
H_2	0.89	1.95	2.53	3.06
Ar	2.50	5.34	6.98	8.27
N_2	1.77	4.07	5.11	6.30
O_2	2.07	4.82	6.20	7.64
CO	1.79	4.15	5.34	6.53
CO_2	1.49	3.85	5.01	6.05
H_2O	1.57	4.77	6.44	7.84

Table 2.4 Thermal Conductivity $\kappa \times 10^4$, J cm^{-1}s^{-1}K^{-1} $\equiv \times 10^2$, kg m s^{-3}K^{-1} $\equiv \times 10^2$, Wm^{-1}K^{-1}

Gas	Temperature, K			
	300	1000	1500	2000
H_2	17.45	39.96	53.87	67.64
Ar	1.97	4.18	5.44	6.44
N_2	2.50	6.18	8.15	10.53
O_2	2.57	6.68	9.02	11.62
CO	2.53	6.35	8.57	10.96
CO_2	1.62	5.49	7.71	9.95
H_2O	3.84	13.55	20.20	26.93

2.3.2.1 Interrelations Between Transport Properties

2.3.2.1.1 Viscosity/Thermal Conductivity

For monatomic gases,

$$\kappa = \frac{15\,R}{4\,M}\eta \qquad (2.3.10)$$

where
- R = 8.314 J mol^{-1}K^{-1} (molar gas constant),
- M = molecular mass, kg mol^{-1},
- η = viscosity, Nsm^{-2} $\equiv$ Jm^{-3}s.

For polyatomic gases,

$$\kappa = \frac{15R}{4M}\eta\left\{\frac{4}{15}\frac{C_V}{R} + \frac{3}{5}\right\} \quad (2.3.11)$$

where C_V is the molar heat capacity at constant volume and

$$\left\{\frac{4}{15}\frac{C_V}{R} + \frac{3}{5}\right\}$$

is the Eucken[67] correction factor for thermal conductivity of polyatomic gases.

For monatomic gases $C_V = 3R/2$; with this substitution, equation 2.3.11 is reduced to equation 2.3.10. In terms of the molar heat capacity at constant pressure, $C_V = C_p - R$, equation 2.3.11 is transformed to the following form.

$$\kappa = \left(C_p + \frac{5R}{4}\right)\frac{\eta}{M} \quad (2.3.12)$$

2.3.2.1.2 Thermal Diffusivity/Thermal Conductivity The thermal diffusivity, D^T, is analogous to mass diffusivity and is given by the ratio

$$D^T = \frac{\kappa}{\rho C_P} \quad (2.3.13)$$

where ρ is the molar density of the gas.

2.3.2.1.3 Temperature and Pressure Effects: The diffusivity, viscosity and thermal conductivity increase with an increasing temperature, thus

$$D \propto T^{3/2} \quad (2.3.14)$$
$$\eta \propto T^{1/2} \quad (2.3.15)$$
$$\kappa \propto T^{1/2} \quad (2.3.16)$$

The viscosity and thermal conductivity are independent of pressure; however, the diffusivity is inversely proportional to pressure.

2.3.2.1.4 Molecular Mass Effect

$$D \propto M^{-1/2} \quad (2.3.17)$$
$$\eta \propto M^{1/2} \quad (2.3.18)$$
$$\kappa \propto M^{-1} \quad (2.3.19)$$

2.3.3 Pore Diffusion

When the pores are small enough such that the mean free path of the molecules is comparable to the dimensions of the pore, diffusion occurs via the collision of molecules with the pore walls. Knudsen[68] showed that the flux involving collision of molecules with reflection from the surface of the capillary wall is represented by the equation

$$J = -\frac{2}{3}\bar{v}r\frac{dC}{dx} \quad (2.3.20)$$

where $\bar{v}$ is the mean thermal molecular velocity = $(8kT/\pi m)^{1/2}$, r the radius of capillary tube and dC/dx the concentration gradient along the capillary. For r in cm and the Knudsen diffusivity D_K in cm^{2}s^{-1}

$$D_K = 9.7 \times 10^3 \, r \left(\frac{T}{M}\right)^{1/2} \quad (2.3.21)$$

The Knudsen diffusivity being independent of pressure but molecular diffusivity inversely proportional to pressure, it follows that in any porous medium, the Knudsen diffusion predominates at low pressures where the mean free path is larger than the dimensions of the pore.

For the mixed region of molecular and Knudsen diffusion in capillaries, Bosanquet[69] derived the following expression based on the random walk model in which the successive movements of molecules are terminated by collision with the capillary wall or with other molecules:

$$\frac{1}{D} = \frac{1}{D_{Ki}} - \frac{1}{D_{ii}} \qquad (2.3.22)$$

where D_{ii} is the molecular self-diffusivity. For a porous medium of uniform pore structure with pores of equal size, the Bosanquet interpretation formula gives for the effective diffusivity of component i

$$D_{ei} = \frac{\varepsilon}{\tau} \frac{D_{12} D_{Ki}}{D_{12} + D_{Ki}} \qquad (2.3.23)$$

where ε is the volume fraction of connected pores, τ the tortuosity factor, D_{12} the molecular diffusivity for a binary mixture 1–2 and D_{ki} the Knudsen diffusivity of component i for a given uniform pore radius r.

2.3.3.1 Examples of Experimental Data

The experimental data are shown in Fig. 2.31 for effective diffusivity He–CO_2 at 20°C. As discussed in Section 2.2.2.3 the pore structure of iron becomes finer at a lower reduction temperature. It is for this reason that the Knudsen diffusion effect at lower pressures becomes more pronounced for porous iron reduced at 800°C as compared to that reduced at 1000°C. More detailed information is given in the previous publications on the subject of gas diffusion in porous media.[51, 70]

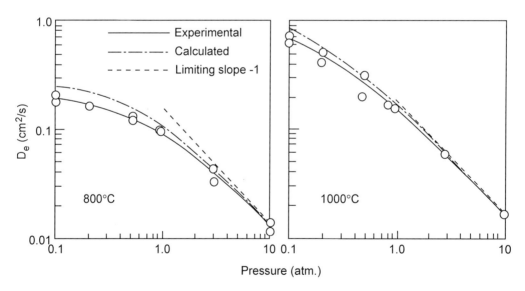

Fig. 2.31 Pressure dependence of effective diffusivity He–CO_2 at 20°C in iron reduced from hematite ore in hydrogen at indicated temperature. *From Ref. 70.*

It is seen from the data in Fig. 2.32 that the effective diffusivities for H_2–H_2O measured directly or calculated from the pore structure considerations, agree well with those derived from the rate measurements within the regime of pore-diffusion control, i.e. from the slopes of the lines in Fig. 2.11. The temperature has a small effect on gas diffusivities. A marked effect of temperature on the effective diffusivity seen in Fig. 2.32 is due mainly to the coarseness of the pore structure of iron when the oxide is reduced at higher temperatures.

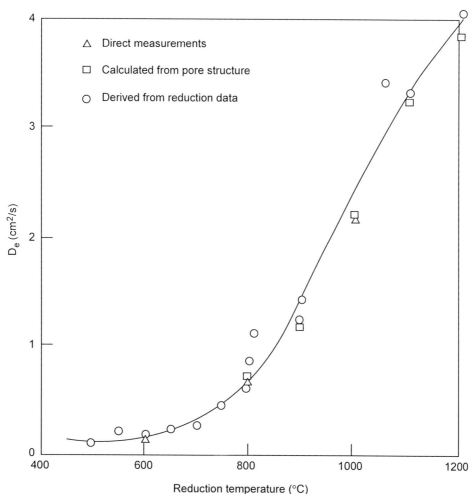

Fig. 2.32 H_2–H_2O effective diffusivity derived from data for rate of reduction of hematite to iron is compared with that obtained by direct measurements and calculated from pore structure. *From Ref. 70.*

In the case of internal burning of carbon, the pore structure becomes coarser as the gasification progresses, resulting in an increase in the effective gas diffusivity. A few examples of the measured CO–CO_2 effective diffusivities for electrode graphite are given in Table 2.5, reproduced from Ref. 17.

Table 2.5 Measured CO–CO_2 Effective Diffusivities for Electrode Graphite

Temperature	% oxidized	D_e(CO–CO_2), cm² s⁻¹
500	0	0.006
500	3.8	0.016
500	7.8	0.020
500	14.6	0.045
700	0	0.009
700	3.8	0.023
700	7.8	0.029
700	14.6	0.067
800	0	0.012
900	0	0.014

Recently Story and Fruehan[49] demonstrated the effect of Knudsen diffusion in the oxidation of carbon in CO_2–Ar and CO_2–He gas mixtures. Normally one would expect the rate to be significantly faster in CO_2–He and CO_2–Ar due to their higher molecular diffusivities. In general the diffusivity is given by 2.3.22. Knudsen diffusion is important when the pore diameter is smaller than the mean free path of the molecules, where the molecules strike the pore walls and diffusivity does not depend on the molecule size. For ultra pure graphite the overall rate constant for external oxidation, (SDe) as discussed in Section 2.2.2.2, was the same for CO_2–He and CO_2–Ar gas mixtures. For this particular material the pore size was small and Knudsen diffusion predominates. On the other hand for coke with large pores the rate is significantly faster with He as the dilution gas as the pores are large and molecular diffusion dominates.

2.4 Properties of Molten Steel

2.4.1 Selected Thermodynamic Data

The thermodynamic properties of liquid iron alloys have been studied extensively, both experimentally and theoretically. Most of these studies were made over a period from 1950 through 1970. For the present purpose, only a selected set of activity data need to be considered, in relation to the study and understanding of the chemistry of iron and steelmaking reactions.

2.4.1.1 Solute Activity Coefficients

For low solute contents, as in low-alloy steels, the activity is defined with respect to Henry's law and mass percent of the solute

$$a_i = f_i \times [\%i] \tag{2.4.1}$$

such that $f_i \to 1.0$ as $[\%i] \to 0$.

Up to several percent of the solute content, $\log f_i$ increases or decreases linearly with an increasing solute concentration.

$$\log f_i = e_i[\%i] \tag{2.4.2}$$

The proportionality factor e_i is known as the interaction coefficient.

The binary interaction coefficients e_i^i for dilute solutions in liquid iron are listed in Table 2.6, taken from the data compiled by Sigworth and Elliott.[71]

Table 2.6 Values of e_i^i for Dilute Solutions in Liquid Iron at 1600°C. *From Ref. 71*

Element i	e_i^i	Element i	e_i^i
Al	0.045	O	–0.10
C	0.18	P	0.062
Cu	0.023	S	–0.028
Cr	0.0	Si	0.11
Mn	0.0	Ti	0.013
Ni	0.0	V	0.015

In multicomponent melts, as in alloy steels, the activity coefficient of solute i is affected by the alloying elements for which the formulation is

$$\log f_i = e_i^i[\%i] + \sum e_i^j[\%j] \tag{2.4.3}$$

where e_i^j is the interaction coefficient of i as affected by the alloying element j.

$$e_i^j = \frac{d \log f_i}{d[\%j]} \tag{2.4.4}$$

Selected interaction coefficients in dilute solutions of ternary iron base alloys for C, H, N, O and S at 1600°C are given in Table 2.7.

Table 2.7 Selected Interaction Coefficients in Dilute Solutions of Ternary Iron Base Alloys for C, H, N, O and S at 1600°C. From Ref. 71.

Element j	e_C^j	e_H^j	e_N^j	e_O^j	e_S^j
Al	0.043	0.013	−0.028	−3.9	0.035
C	0.14	0.06	0.13	−0.13	0.11
Cr	−0.024	−0.002	−0.047	−0.04	−0.011
Mn	−0.012	−0.001	−0.02	−0.021	−0.026
N	0.13	0	0	0.057	0.007
O	−0.34	−0.19	0.05	−0.20	−0.27
P	0.051	0.011	0.045	0.07	0.029
S	0.046	0.008	0.007	−0.133	−0.028
Si	0.08	0.027	0.047	−0.131	0.063

For the mass concentrations of carbon and silicon above 1%, as in the blast furnace and foundry irons, the following values of f_S^j should be used.

Mass % C or Si:	2.0	2.5	3.0	3.5	4.0	4.5	5.0
f_S^C:	1.79	2.14	2.53	3.05	3.74	4.56	5.75
f_S^{Si}:	1.37	1.50	1.64	1.78	1.95	2.10	2.32

2.4.1.2 Free Energies of Solution in Liquid Iron

For the solution of element X_i in liquid iron at mass % X_i,

$$X_i (\text{pure}) = [X_i] \,(1 \text{ mass \%})$$

the free energy of solution is

$$\Delta G_S = RT \ln\left(\frac{0.5585}{M_i}\gamma_i^\circ\right) \quad (2.4.5)$$

where M_i is the atomic mass (g-atom) and γ_i° the activity coefficient (with respect to pure element) at infinite dilution (%$X_i \rightarrow 0$).

For ΔG_S in J mol^{-1} and substituting log for ln,

$$\Delta G_S = 19.144\, T \log\left(\frac{0.5585}{M_i}\gamma_i^\circ\right) \quad (2.4.6)$$

The free energies of solution of various elements in liquid iron are listed in Table 2.8.

2.4.2 Solubility of Gases in Liquid Iron

Diatomic gases such as O_2, S_2, N_2 and H_2 dissolve in liquid and solid metals in the atomic form

$$\tfrac{1}{2} X_2(g) = [X] \quad (2.4.7)$$

for which the isothermal equilibrium constant is

$$K = \frac{[\%X]}{(p_{X_2})^{1/2}} \quad (2.4.8)$$

For ideal solutions, the concentration of X is directly proportional to the square root of the equilibrium gas partial pressure; this is known as the Sievert's law.

Table 2.8 Free Energies of Solution in Liquid Iron for 1 mass %: (g) gas, (l) liquid, (s) solid. *From Ref. 71.*

Element i	γ_i^o	ΔG_s, Jmol⁻¹
Al(l)	0.029	−63,178 − 27.91T
C(gr)	0.57	22,594 − 42.26T
Co(l)	1.07	1,004 − 38.74T
Cr(s)	1.14	19,246 − 46.86T
Cu(l)	8.60	33,472 − 39.37T
½H₂(g)	−	36,377 + 30.19T
Mg(g)	−	−78,690 + 70.80T
Mn(l)	1.30	4,084 − 38.16T
½N₂(g)	−	3,599 + 23.74T
Ni(l)	0.66	−20,920 − 31.05T
½O₂(g)	−	−115,750 − 4.63T
½P₂(g)	−	−122,173 − 19.25T
½S₂(g)	−	−135,060 + 23.43T
Si(l)	0.0013	−131,500 − 17.24T
Ti(s)	0.038	−31,129 − 44.98T
V(s)	0.10	−20,710 − 45.61T
W(s)	1.20	31,380 − 63.60T
Zr(s)	0.043	−34,727 − 50.00T

2.4.2.1 Solubilities of H₂, N₂ and O₂

For the solute content in ppm (by mass) and the gas pressure in atm., the temperature dependence of the equilibrium constants for H₂, N₂ and O₂ solubilities are given in Table 2.9 from the compiled data cited in Ref. 27.

Table 2.9 Equilibrium Constants of Solubilities of H₂, N₂ and O₂ in Liquid Iron.

$$\log \frac{[\text{ppm H}]}{(p_{H_2})^{1/2}} = -\frac{1900}{T} + 2.423 \quad (2.4.9)$$

$$\log \frac{[\text{ppm N}]}{(p_{N_2})^{1/2}} = -\frac{188}{T} + 2.760 \quad (2.4.10)$$

$$\log \frac{[\text{ppm O}]}{(p_{O_2})^{1/2}} = \frac{6046}{T} + 4.242 \quad (2.4.11)$$

2.4.2.2 Solubility of Gaseous Oxides

2.4.2.2.1 Solubility of CO Carbon monoxide dissolves in liquid iron (steel) by dissociating into atomic carbon and oxygen

$$CO(g) = [C] + [O] \quad (2.4.12)$$

for which the equilibrium constant (which is not too sensitive to temperature) is given by the following equation for low alloy steels containing less than 1% C.

$$K = \frac{[\%C][\text{ppm O}]}{p_{CO}(\text{atm})} = 20 \tag{2.4.13}$$

At higher carbon contents, a correction should be made to K regarding the activity coefficients f_O^C (log $f_O^C = -0.13[\%C]$) and f_C^C(log $f_C^C = 0.18[\%C]$).

For the reaction

$$CO_2(g) = CO(g) + [O] \tag{2.4.14}$$

$$K = \frac{p_{CO}}{p_{CO_2}}[\text{ppm O}] = 1.1 \times 10^4 \text{ at } 1600°C \tag{2.4.15}$$

For 800 ppm O in low carbon steel at tap, the equilibrium ratio p_{CO}/p_{CO_2} is 13.75. For this state of equilibrium, the gas mixture contains 6.8% CO_2 and 93.2% CO.

2.4.2.2.2 Solubility of H_2O From the free energy of formation of water vapor and the solubilities of hydrogen and oxygen in liquid iron, the following equilibrium constant is obtained for the reaction of water vapor with liquid iron for 1600°C.

$$H_2O(g) = 2[H] + [O] \tag{2.4.16}$$

$$K = \frac{[\text{ppm H}]^2[\text{ppm O}]}{p_{H_2O}(\text{atm})} = 1.77 \times 10^6 \text{ at } 1600°C \tag{2.4.17}$$

The hydrogen and oxygen contents of low alloy liquid steel in equilibrium with H_2–H_2O mixtures at 1 atm pressure and 1600°C are shown in Fig. 2.33.

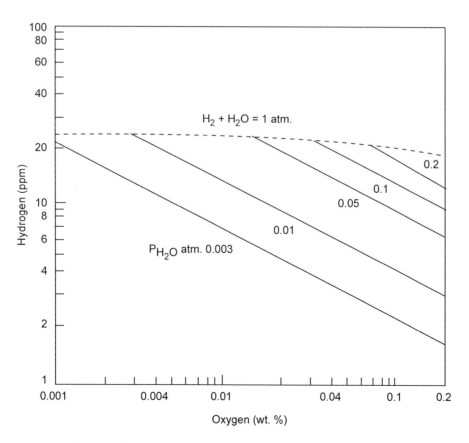

Fig. 2.33 Concentrations of hydrogen and oxygen in liquid iron at 1600°C in equilibrium with indicated compositions of H_2–H_2O mixtures.

2.4.2.2.3 Solubility of SO_2 From the free energy of formation of SO_2 and the solubilities of gaseous sulfur and oxygen in liquid iron, the following equilibrium constant is obtained for the reaction of SO_2 with liquid iron at 1600°C.

$$SO_2(g) = [S] + 2[O] \qquad (2.4.18)$$

$$K = \frac{[\%S][\%O]^2}{p_{SO_2}(\text{atm})} = 1558 \text{ at } 1600°C \qquad (2.4.19)$$

For the concentrations of sulfur and oxygen present in liquid steel, it is seen that the corresponding equilibrium pressure of SO_2 is infinitesimally small. It is for this reason that no sulfur can be oxidized to SO_2 during steelmaking with oxygen blowing.

2.4.3 Iron-Carbon Alloys

2.4.3.1 Fe-C Phase Equilibrium Diagram

Since carbon is one of the most important ingredients of steel, the study of the iron-carbon phase equilibrium diagram has received much attention during the past several decades. The phase diagram is given in Fig. 2.34.

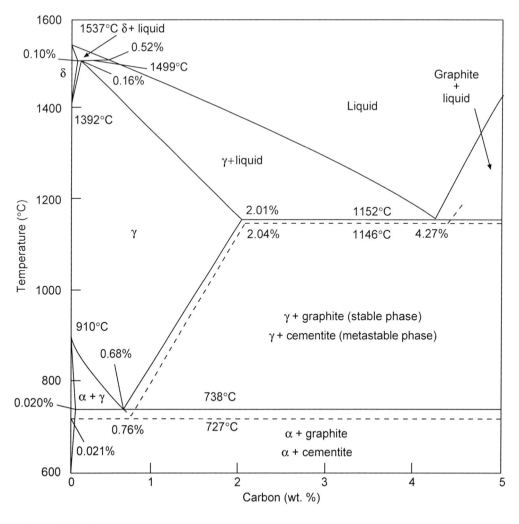

Fig. 2.34 Iron-carbon phase equilibrium diagram. Dashed lines represent phase boundary for metastable equilibrium with cementite.

There are three invariants in this system; peritectic at 1499°C, eutectic at 1152°C and eutectoid at 738°C. The phase boundaries shown by broken lines are for the metastable equilibrium of cementite, Fe_3C, with austenite. During heat treatment of steel if sufficient time is allowed, iron-carbon alloys containing austenite and cementite decompose to austenite and graphite. The solubilitites of graphite and of cementite in α-iron below the eutectoid temperature are given by the following equations

$$\log \%C \text{ (graphite)} = -\frac{5250}{T} + 3.53 \qquad (2.4.20)$$

$$\log \%C \text{ (cementite)} = -\frac{3200}{T} + 1.50 \qquad (2.4.21)$$

The solubility of graphite in pure liquid iron is well established through many independent studies; the experimental data are summarized by the equation

$$[\%C] = 1.30 + 2.57 \times 10^{-3} T(°C) \qquad (2.4.22)$$

If atom fraction N_C is used, the same set of data can be represented by the following equation in terms of $\log N_C$ and the reciprocal of the absolute temperature.

$$\log N_C = -\frac{560}{T} - 0.375 \qquad (2.4.23)$$

The effect of silicon, phosphorus, sulfur, manganese, cobalt and nickel on the solubility of graphite in molten iron was determined by Turkdogan et al.[72, 73] and graphite solubility in iron-silicon and iron-manganese melts by Chipman et at.[74, 75] Similar measurements with iron-chromium melts were made by Griffing and co-workers.[76] The experimental data are given graphically in Fig. 2.35 for 1500°C; the solubility at other temperatures can be estimated from this plot by using the temperature coefficient given in equation 2.4.22 for binary iron-carbon melts. In the iron-sulfur-carbon system there is a large miscibility gap. For example, at 1500°C the melt separates into two liquids containing phase (I) 1.8% S and 4.24% C and phase (II) 26.5% S and 0.90% C.

2.4.3.2 Activity Coefficient of Carbon

The activity of carbon in liquid iron was measured in many independent studies. The data compiled and re-assessed by Elliott et al.[77] are given in Figs. 2.36 and 2.37 as activity coefficients γ_C and f_C for two different standard states.

2.4.3.3 Peritectic Reaction

The peritectic reaction occurring in the early stages of solidification of low carbon steels is of particular importance in the continuous casting of steel. The peritectic region of the Fe–C system is shown on a larger scale in Fig. 2.38. As the temperature decreases within the two phase region, δ+liquid, the carbon contents of δ-iron and residual liquid iron increase. At the peritectic temperature 1499°C, δ-iron containing 0.10% C reacts with liquid iron containing 0.52% C to form γ-iron with 0.16% C.

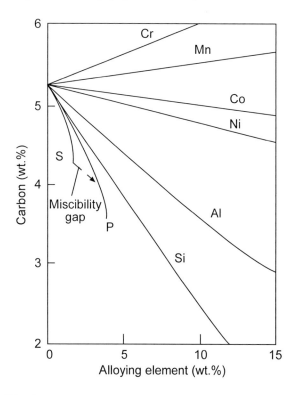

Fig. 2.35 Solubility of graphite in alloyed iron melts at 1500°C.

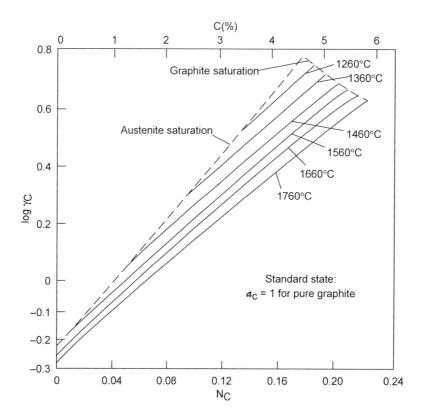

Fig. 2.36 Activity coefficient (γ_C) of carbon in liquid iron.

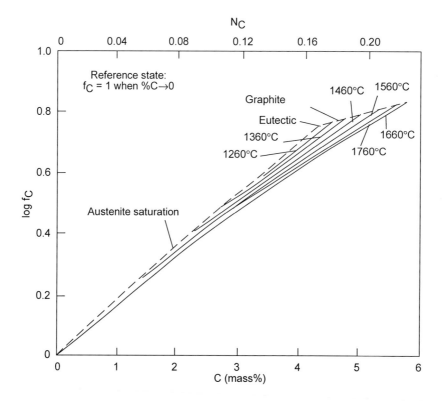

Fig. 2.37 Activity coefficient (f_C) of carbon in liquid iron.

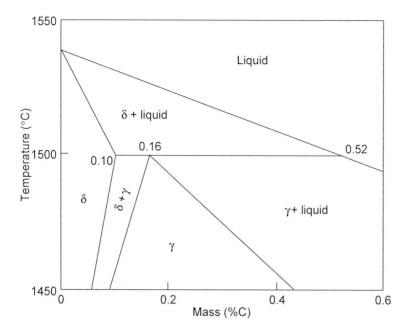

Fig. 2.38 Peritectic region of binary iron-carbon system.

The X-ray diffraction data give for the densities of iron-carbon alloys: 7.89 g cm^{-3} for δ–Fe with 0.10% C and 8.26 g cm^{-3} for γ–Fe with 0.16% C. Hence, the δ to γ phase transformation is accompanied by 4.7% volume shrinkage. Because of this shrinkage, the thin solidified shell in the mould of the caster will contract, producing a gap between the shell surface and mould wall. This situation leads to an uneven surface which is in partial contact with the mould wall, hence resulting in reduced heat flux at contracted areas. A reduced solidification growth rate and a nonuniform shell with thin spots, lowers the resistance of the steel to cracking which my cause a breakout in the mould. This phenomenon was well demonstrated experimentally by Singh and Blazek[78] using a bench scale caster; this subject is discussed further in a paper by Wolf and Kurz.[79]

In low alloy steels containing 0.10 to 0.16% C, the solid/liquid ratio at the peritectic invariant is higher than for steels containing more than 0.16% C. Therefore, due to the peritectic reaction, low alloy steels with 0.10 to 0.16% C are more susceptible to the development of surface cracks in continuous casting than steels with higher carbon contents.

2.4.3.3.1 Effect of Alloying Elements on Peritectic Invariant
As shown schematically in Fig. 2.39 for the ternary Fe–C–X, or multicomponent alloy steels, the peritectic reaction occurs over a temperature and composition range. From the experimental data on the liquidus and solidus temperatures of alloy steels, the carbon and temperature equivalents of alloying elements have been evaluated in three independent investigations.[80–82]

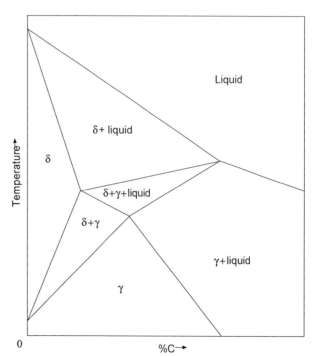

Fig. 2.39 Schematic representation of phase boundaries in low alloy steels as pseudo binary Fe–C system.

The carbon equivalents of the alloying elements for the peritectic reaction

$$C(\delta - Fe) + C\,(l - Fe) = C\,(\gamma - Fe) \qquad (2.4.24)$$

are usually formulated as follows

$$\text{Liquid phase:}\quad \Delta\%C = 0.52 + \Sigma\Delta C_{P\ell}^{X}[\%X] \qquad (2.4.25)$$

$$\text{Delta phase:}\quad \Delta\%C = 0.10 + \Sigma\Delta C_{P\delta}^{X}[\%X] \qquad (2.4.26)$$

The changes in peritectic temperatures are formulated as

$$\text{Liquid phase:}\quad \Delta T = 1499°C + \Sigma\Delta T_{P\ell}^{X}[\%X] \qquad (2.4.27)$$

$$\text{Delta phase:}\quad \Delta T = 1499°C + \Sigma\Delta T_{P\delta}^{X}[\%X] \qquad (2.4.28)$$

The coefficients ΔC_P^X and ΔT_P^X can be positive or negative, depending on the alloying element X. The peritectic temperature and carbon coefficients determined by Yamada et al.[81] are listed in Table 2.10 with minor numerical adjustments to some of the parameters in accord with the empirical correlations given in Fig. 2.40, reproduced from Ref. 27.

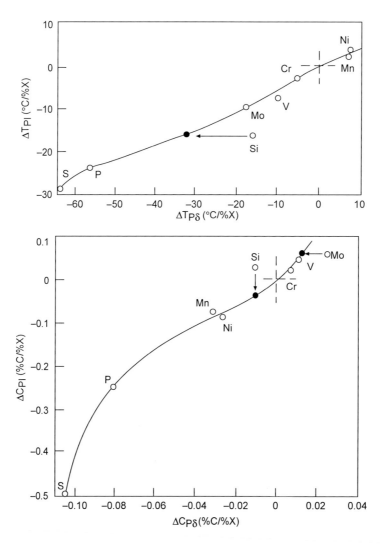

Fig. 2.40 Interrelations between peritectic temperature and composition changes. *From Ref. 27.*

Table 2.10 Peritectic Temperature and Composition Parameters in Accord with the Curves in Fig. 2.40.

Alloying Element	Composition range, wt.%	Δ°C/%X		Δ%C/%X	
		$\Delta T_{P\ell}$	$\Delta T_{P\delta}$	$\Delta C_{P\ell}$	$\Delta C_{P\delta}$
Cr	<1.5	−3	−5	+0.022	+0.006
Mn	<1.5	+3	+7	−0.085	−0.029
Mo	<1.5	−10	−18	+0.055	+0.012
Ni	<3.5	+4	+9	−0.082	−0.027
P	<0.05	−24	−56	−0.250	−0.080
S	<0.03	−28	−64	−0.500	−0.105
Si	<0.6	−16	−32	−0.035	−0.010
V	<1.0	−6	−11	+0.045	+0.010

The phase boundaries in the peritectic region of the Fe–C–X alloys, projected on the Fe–C–X composition diagram, are shown schematically in Fig. 2.41 for alloys where X is: (a) ferrite stabiliser and (b) austenite stabiliser. In the Fe–C–Mn (or Ni) system, the peritectic reaction occurs at all compositons between the peritectic regions of the binaries Fe–C and Fe–Mn (Ni) systems. In the Fe–C alloys with ferrite stabilisers, the peritectic reaction will not occur beyond a certain concentration of the alloying element X as shown in Fig. 2.41. This limiting case applies only to high alloy steels.

2.4.4 Liquidus Temperatures of Low Alloy Steels

The liquidus temperatures of low alloy steels are derived from the binary Fe–X systems on the assumption that the coefficients $\alpha = \Delta T/\%X$ are additive in their effects on the melting point of iron.

Below 0.5% C, where the solidification begins with the formation of delta (δ) iron, the following equation would apply

$$\text{Liquidus T (°C)} = 1537 - 73.1[\%C] + \Sigma\alpha[\%X] \quad (2.4.29)$$

For the carbon contents within the range 0.5 to 1.0% C, where the solidification begins with the formation of gamma (γ) iron, the following equation is recommended.

$$\text{Liquidus T (°C)} = 1531 - 61.5[\%C] + \Sigma\alpha[\%X] \quad (2.4.30)$$

The same coefficients α are used in both equations.

Alloying element X	Coefficient α, °C/%X
Al	−2.5
Cr	−1.5
Mn	−4.0
Mo	−5.0
Ni	−3.5
P	−30.0
Si	−14.0
S	−45.0
V	−4.0

2.4.5 Solubility of Iron Oxide in Liquid Iron

Subsequent to earlier studies by various investigators, Taylor and Chipman[83] made the most reliable measurement of the oxygen solubility in liquid iron in equilibrium with essentially pure liquid iron oxide at temperatures of 1530–1700°C. In a later study, Distin *et al.*[84] extended the solubility measurements up to 1960°C. The solubility data are represented by the equation

$$\log[\%O]_{sat.} = -\frac{6380}{T} + 2.765 \quad (2.4.31)$$

The oxygen content of liquid iron oxide in equilibrium with liquid iron decreases with an increasing temperature as given below and reaches the stoichiometric composition (22.27%) at about 2000°C.

Temperature °C	%O in iron	%O in liquid iron oxide
1527 (eutectic)	0.16	22.60
1785	0.46	22.40
1880	0.63	22.37
1960	0.81	22.32

2.4.6 Elements of Low Solubility in Liquid Iron

A few elements of low solubility in liquid iron play some role in the steelmaking technology; a brief comment on the chemistry of such elements in liquid iron is considered desirable.

2.4.6.1 Lead

The break out of the furnace lining is often blamed on the presence of lead in the melt. Small amounts of metal trapped in the crevices of the lining are likely to get oxidized subsequent to tapping. The lead oxide together with iron oxide will readily flux the furnace lining, hence widening the cracks which ultimately leads to failure of the lining. The solubility of lead in liquid iron is sufficiently high that in normal steelmaking practice there should be no accumulation of lead at the bottom of the melt, except perhaps in the early stages of the melting of lead-containing scrap.

At steelmaking temperatures the vapor pressure of lead is about 0.5 atm and the solubility[85] in liquid iron is about 0.24% Pb at 1500°C increasing to about 0.4% Pb at 1700°C. The free-machining leaded steels contain 0.15 to 0.35% Pb. Evidently lead added to such steels is in solution in the metal prior to casting and precipitates as small lead spheroids during the early stages of freezing.

2.4.6.2 Calcium

The boiling point of calcium is 1500°C and its solubility in liquid iron is very low. Sponseller and Flinn[86] measured the solubility of calcium in iron at 1607°C for which the calcium vapor pressure is 1.69 atm. Under these conditions at 1607°C, 0.032% Ca is in solution. That is, the solubility at 1 atm pressure is 189 ppm Ca. They also investigated the effect of some alloying elements on the solubility; as seen from the data in Fig. 2.42, C, Si, Ni and Al increase markedly the calcium solubility in liquid iron. In melts saturated with CaC_2 the solubility of calcium of course decreases with an increasing carbon content.

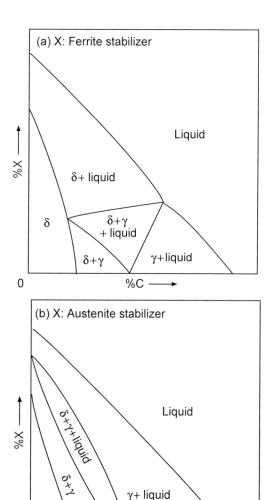

Fig. 2.41 Phase boundaries in the peritectic region of the Fe-C-X alloys projected on the composition diagram.

2.4.6.3 Magnesium

The solubility of magnesium in iron-carbon alloys determined by Trojan and Flinn[87], is shown is Fig. 2.43 as functions of temperature and carbon content. Subsequently, Guichelaar et al.[88], made similar measurements with liquid Fe-Si-Mg alloys. Their data have been used in numerous studies to derive the equilibrium relations for the solubility of magnesium in liquid iron. However, there are some variations in the interpretation of the above mentioned experimental data. A reassessment of these experimental data is considered desirable.

The equilibrium relation for the solubility of Mg (in units of mass % atm^{-1}) is represented by

$$Mg(g) = [Mg]$$

$$K_{Mg} = \frac{[\%Mg] f_{Mg}}{P_{Mg}} \qquad (2.4.32)$$

where f_{Mg} is the activity coefficient affected by the alloying elements. In the experiments with the Fe–C–Mg melts coexistent with liquid Mg, the latter contained less than 2 percent Fe, therefore, the Mg vapor pressure prevailing in the reactor would be essentially the same as that for pure Mg for which the following is obtained from the data of Guichelaar, et al.

$$\log P^o_{Mg}(atm) = -\frac{6730}{T} + 4.94 \qquad (2.4.33)$$

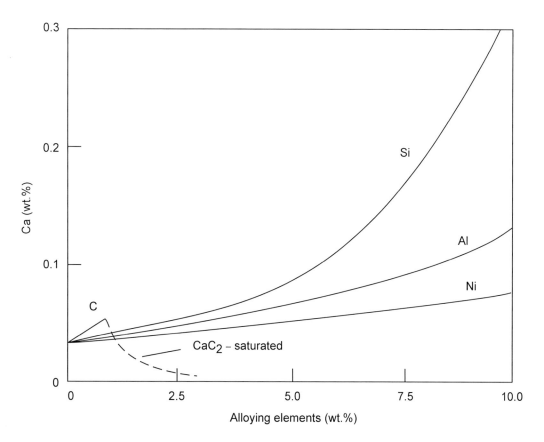

Fig. 2.42 Effect of alloying elements on the solubility of liquid calcium in liquid iron at 1607°C, corresponding to 1.69 atm pressure of calcium vapor. *From Ref. 87.*

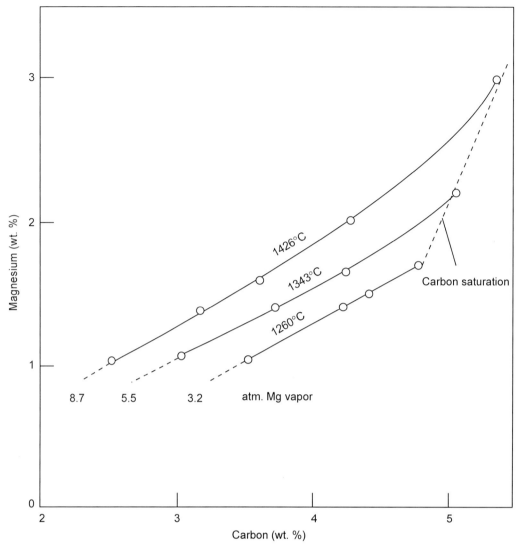

Fig. 2.43 Solubility of magnesium in liquid iron–carbon alloys at indicated temperatures and pressures of magnesium vapor. *From Ref. 88.*

2.4.7 Surface Tension

The experimental data for the surface tension of liquid iron and its binary alloys have been compiled recently by Keene.[89] For purified liquid iron, the average value of the surface tension at temperature T(°C) is represented by

$$\sigma_{Fe} = (2367 \pm 500) - 0.34(°C), mNm^{-1} \qquad (2.4.34)$$

Keene derived the following weighted average limiting values of (mNm^{-1}) for dilute solutions of X in Fe-X binary alloys.

Fe–C:	Virtually no effect of C on σ_{Fe}
Fe–Ce:	$\sigma = \sigma_{Fe} - 700[\%Ce]$
Fe–Mn:	$\sigma = \sigma_{Fe} - 51[\%Mn]$; $\partial\sigma/\partial T = -0.22$
Fe–N:	$\sigma = \sigma_{Fe} - 5585[\%N]$
Fe–P:	$\sigma = \sigma_{Fe} - 25[\%P]$
Fe–S:	(Discussed later)
Fe–Si:	$\sigma = \sigma_{Fe} - 30[\%Si]$; $\partial\sigma/\partial T = -0.25$

Gibbs' exact treatment of surface thermodynamics gives, for fixed unit surface area and constant temperature and pressure,

$$d\sigma = -RT \sum_i^k \Gamma_i d(\ln a_i) \qquad (2.4.35)$$

where Γ_i is the surface excess concentration of the ith component and a_i its activity. For a ternary system, equation 2.4.35 is reduced to

$$d\sigma = -RT(\Gamma_2 d(\ln a_2) + \Gamma_3 d(\ln a_3)) \qquad (2.4.36)$$

Since carbon dissolved in iron has virtually no effect on the surface tension of liquid iron, for the ternary system Fe–C–S, equation 2.4.36 is simplified to

$$d\sigma = -RT\Gamma_s d(\ln a_s) \qquad (2.4.37)$$

As is seen from the compiled data in Fig. 2.44, the experimental results of various investigators for the Fe–S melts are in close agreement. References to these data are given in an earlier publication.[90]

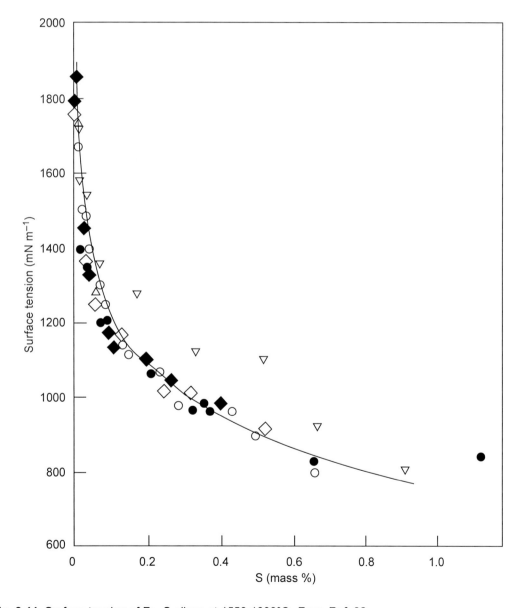

Fig. 2.44 Surface tension of Fe–S alloys at 1550-1600°C. *From Ref. 90.*

The classical work of Kozakevitch[91] on the surface tension of Fe–C–S melts at 1450°C are reproduced in Fig. 2.45

The points read off from the curves in Figs. 2.44 and 2.45 are plotted in Fig. 2.46 as σ versus log a_S. It is seen that for a_S (%S x f_S) > 0.01, σ is a linear function of log a_S, a limiting case for almost complete surface coverage with chemisorbed S. The shaded area represents the data of Selcuk and Kirkwood[92] for 1200°C.

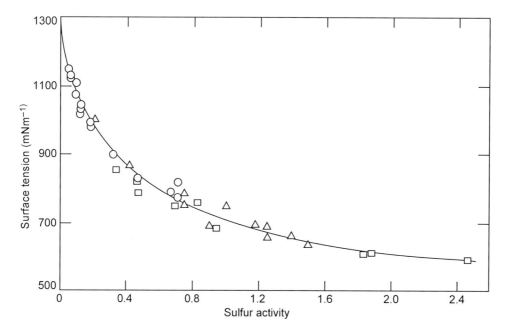

Fig. 2.45 Surface tension of liquid Fe–C–S alloys at 1450°C. *From Ref. 91.*

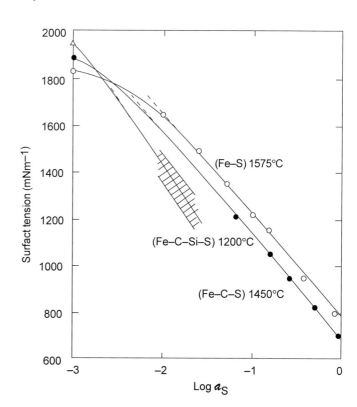

Fig. 2.46 Surface tension of Fe–S, Fe–C–S and Fe–C–Si–S alloys related to sulfur activity; a_S→%S when %C and %Si→0. *From Ref. 90.*

Numerous studies have been made of the surface tension of the Fe–O melts. The more recent experimental data[93,94] are reproduced in Fig. 2.47. At oxygen activities a_O (%O x f_O) > 0.1, there is essentially complete surface coverage with the chemisorbed oxygen.

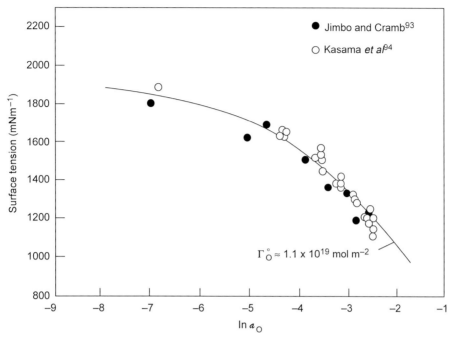

Fig. 2.47 Effect of oxygen on surface tension of liquid iron at 1550°C.

2.4.8 Density

The temperature dependence of the densities of liquid iron, nickel, cobalt, copper, chromium, manganese, vanadium and titanium are given by the following equations as a linear function of temperature in °C in g cm^{-3}:

Iron:	$8.30 - 8.36 \times 10^{-4}T$
Nickel:	$9.60 - 12.00 \times 10^{-4}T$
Cobalt:	$9.57 - 10.17 \times 10^{-4}T$
Copper:	$9.11 - 9.44 \times 10^{-4}T$
Chromium:	$7.83 - 7.23 \times 10^{-4}T$
Manganese:	$7.17 - 9.30 \times 10^{-4}T$
Vanadium:	$6.06 - 3.20 \times 10^{-4}T$
Titanium:	$4.58 - 2.26 \times 10^{-4}T$

The specific volume and density of liquid iron-carbon alloys are given in Fig. 2.48 for various temperatures. It should be noted that the density of the liquid in equilibrium with austenite does not change much over the entire liquidus range.

2.4.9 Viscosity

The viscosity is a measure of resistance of the fluid to flow when subjected to an external force. As conceived by Newton, the shear stress ε, i.e. force per unit area, causing a relative motion of two adjacent layers in a fluid is proportional to the velocity gradient du/dz, normal to the direction of the applied force

$$\varepsilon = \eta \frac{du}{dz} \tag{2.4.38}$$

where the proportionality factor η is viscosity of the fluid (liquid or gas).

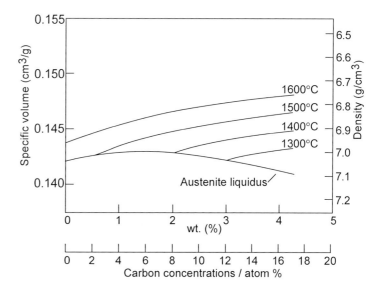

Fig. 2.48 Density of iron-carbon alloys. *From Ref. 95.*

Viscosity unit: poise (g cm⁻¹s⁻¹) ≡ 0.1N s m⁻².

The viscosities of Fe–C alloys determined by Barfield and Kitchener[96] are given in Fig. 2.49. In the iron-carbon melts the coefficient of viscosity is essentially independent of composition within the range 0.8 to 2.5% C; above 2.5% C, the viscosity decreases continously with an increasing carbon content.

2.4.10 Diffusivity, Electrical and Thermal Conductivity, and Thermal Diffusivity

The diffusivity is an exponential function of temperature,

$$D = D_o \exp\left(-\frac{E}{RT}\right) \quad (2.4.39)$$

where D_o is a constant for a given solute and E the activation energy (enthalpy) for the diffusion process.

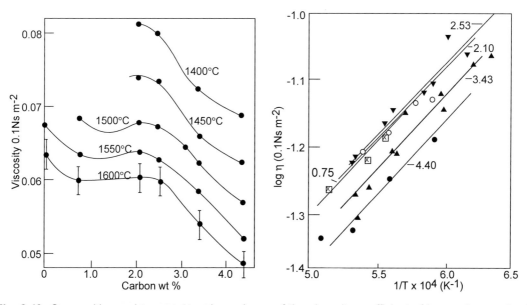

Fig. 2.49 Composition and temperature dependence of the viscosity coefficient of iron-carbon melts. *From Ref. 96.*

The general trend in the temperature dependence of solute diffusivity in solid iron is shown in Fig. 2.50. The interstitial elements, e.g. O, N, C, B, H have diffusivities much greater than the substitutional elements. Because of larger interatomic spacing, the diffusivities of interstitials in bcc-iron are greater and the heats of activations (~ 85 kJ) are smaller than those in the fcc-iron with E ~ 170 kJ. With the substitutional elements, the heat of activation is within 210 to 250 kJ for bcc-iron and within 250 to 290 kJ for fcc-iron.

In liquid iron alloys, diffusivities of elements are within 10^{-5} to 10^{-4} cm^{2}s^{-1} with E within 15 to 50 kJ.

Some solute diffusivity data for liquid iron and iron-carbon alloys are given in Table 2.11.

Table 2.11 Selected Solute Diffusivities in Liquid Fe-C Alloys

Diffusing element	Concentration mass %	Medium	Temp. range °C	D cm²sec⁻¹	D$_o$ cm²sec⁻¹	E kJ
C	0.03	Fe	1550	7.9×10⁻⁵	–	–
C	2.1	Fe	1550	7.8×10⁻⁵	–	–
C	3.5	Fe	1550	6.7×10⁻⁵	–	–
Co	Dilute sol.	Fe	1568	4.7×10⁻⁵	–	–
Co	Dilute sol.	Fe	1638	5.3×10⁻⁵	–	–
Fe	–	Fe–4.6%C	1240–1360	–	4.3×10⁻³	51
Fe	–	Fe–2.5%C	1340–1400	–	1.0×10⁻²	66
H	Dilute sol.	Fe	1565–1679	–	3.2×10⁻³	14
Mn	2.5	gr.satu.Fe	1300–1600	–	1.93×10⁻⁴	24
N	Dilute sol.	Fe	1600	1.1×10⁻⁴	–	–
N	Dilute sol.	Fe–0.15%C	1600	5.6×10⁻⁵	–	–
O	Dilute sol.	Fe	1600	5.0×10⁻⁵	–	–
P	Dilute sol.	Fe	1550	4.7×10⁻⁵	–	–
S	<0.64	gr.satu.Fe	1390–1560	–	2.8×10⁻⁴	31
S	~1	Fe	1560–1670	–	4.9×10⁻⁴	36
Si	<2.5	Fe	1480	2.4×10⁻⁵	–	–
Si	<1.3	Fe	1540	3.8×10⁻⁵	–	–
Si	1.5	gr.satu.Fe	1400–1600	–	2.4×10⁻⁴	34

The electrical conductivity λ in the units of Ω^{-1}cm^{-1} is the reciprocal of the electrical resistivity. The electrical conductivity of liquid low alloy steel is about $\lambda = 7140$ Ω^{-1}cm^{-1} at steelmaking temperatures.

From Fick's law, the thermal conductivity is defined by

$$\kappa = -\frac{Q}{\partial T/\partial x} \qquad (2.4.40)$$

where Q is heat flux, energy per unit per unit time, $\partial T/\partial x$ the temperature gradient normal to the direction of heat flow and κ the thermal conductivity.

Units of κ: J cm^{-1}s^{-1}K^{-1} ≡ 10^{-2} kg m s^{-3}K^{-1} ≡ 10^{-2} Wm^{-1}K^{-1}

The effect of temperature on the thermal conductivities of iron, carbon steels and high alloy steels are shown in Fig. 2.51.

Analogous to mass diffusivity, the thermal diffusivity is defined as

$$\alpha = \frac{\kappa}{\rho C_p} \text{ cm}^2\text{s}^{-1} \qquad (2.4.41)$$

where ρ is density and C_p molar heat capacity. In metals, the electrons migrate at much faster rates than the atoms; therefore the thermal diffusivity is much greater than the mass diffusivity.

For low–alloy steels at 1000°C:

$\alpha \approx 0.04$ cm^{2}s^{-1}
$D \approx 3 \times 10^{-12}$ cm^{2}s^{-1} (substitutional)
$D \approx 3 \times 10^{-7}$ cm^{2}s^{-1} (interstitial)

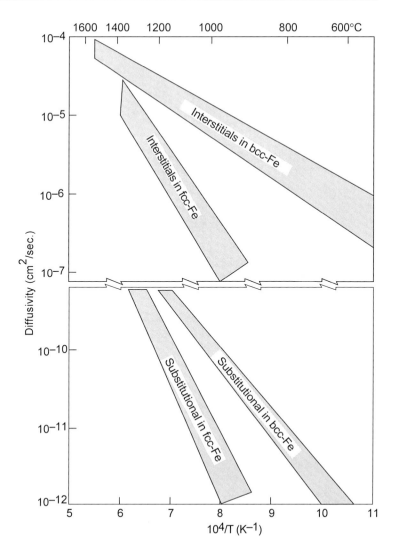

Fig. 2.50 Range of diffusivities of interstitial and substitutional elements in bcc and fcc iron.

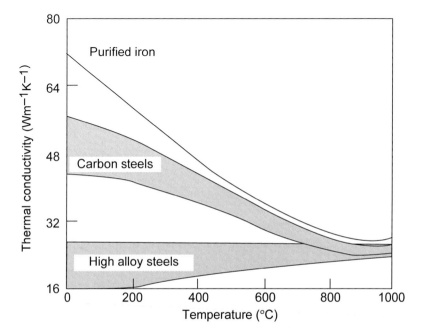

Fig. 2.51 Temperature dependence of thermal conductivity of purified iron and iron alloys.

2.5 Properties of Molten Slags

2.5.1 Structural Aspects

Molten slags are ionic in nature consisting of positively charged ions known as cations, and negatively charged complex silicate, aluminate and phosphate ions known as anions.

The fundamental building unit in solid silica and molten silicates is the silicate tetrahedron SiO_4^{4-}. Each silicon atom is tetrahedrally surrounded by four oxygen atoms and each oxygen atom is bonded to two silicon atoms. The valency of silicon is +4 and that of oxygen is –2, therefore the silicate tetrahedron has 4 negative charges.

The addition of metal oxides such as FeO, CaO, MgO, . . . to molten silica brings about a breakdown of the silicate network, represented in a general form by the reaction

$$\left(-Si-O-Si-\right) + MO \rightarrow 2\left(-Si-O\right)^- + M^{2+} \qquad (2.5.1)$$

The cations are dispersed within the broken silicate network. In $MO-SiO_2$ melts the atom ratio $O/Si > 2$, therefore part of the oxygen atoms are bonded between two silicon atoms and part to only one silicon atom. Partial depolymerization of the silicate network with the addition of a metal oxide MO is illustrated in Fig. 2.52.

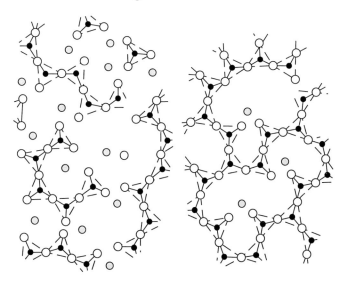

Fig. 2.52 Schematic representation of depolymerization of the silicate network with the dissolution of metal oxides in silicate melts.

In highly basic slags with molar ratio $MO/SiO_2 > 2$, the silicate network completely breaks down to individual SiO_4 tetrahedra intermixed with cations M^{2+} and some oxygen ions O^{2-}.

At low concentrations, Al_2O_3 behaves like a network-modifying oxide and forms aluminum cations Al^{3+}. At high concentrations, the aluminum enters the tetrahedral structure isomorphous with silicon. This process may be schematically represented by the reaction

$$\left(-Si-O-Si-\right) + MAlO_2 \rightarrow \left(-Si-O-Al-O-Si-\atop O\right)^- + M^+ \qquad (2.5.2)$$

The cation M^+ is located in the vicinity of Al–O bonding to preserve the local charge balance. At low concentrations of phosphorus in steelmaking slags, the phosphate ions PO_4^{3-} are incorporated in the silicate network. In steelmaking slags, the sulfur exists as a sulfide ion S^{2-}. The sulfate ions SO_4^{2-} exist in slags only under highly oxidizing conditions and in the absence of iron or any other oxidizable metal.

Although molten slags are ionized, the slag composition can be represented in terms of the constituent oxides, e.g. CaO, FeO, SiO$_2$, P$_2$O$_5$. The thermodynamic activity of an ion in the slag cannot be determined. However, the activity of an oxide dissolved in molten slag, forming M^{2+} and O^{2-} ions, can be determined experimentally and the following equality can be written

$$MO \rightarrow M^{2+} + O^{2-} \tag{2.5.3}$$

$$\frac{a_{MO}}{(a_{MO})^\circ} = \frac{(a_{M^{2+}} a_{O^{2-}})}{(a_{M^{2+}} a_{O^{2-}})^\circ} \tag{2.5.4}$$

where the superscript ° refers to the standard state which is usually pure solid or liquid oxide.

2.5.2 Slag Basicity

For steelmaking slags of low phosphorus content, the slag basicity has traditionally been represented by the mass concentration ratio

$$V = \frac{\%CaO}{\%SiO_2} \tag{2.5.5}$$

For slags containing high concentrations of MgO and P$_2$O$_5$, as in some laboratory experiments, the basicity may be defined by the following mass concentration ratio, with the assumption that on a molar basis the concentrations of CaO and MgO are equivalent. Similarly, on a molar basis 1/2 P$_2$O$_5$, i.e. PO$_{2.5}$, is equivalent to SiO$_2$.

$$B = \frac{\%CaO + 1.4 \times \%MgO}{\%SiO_2 + 0.84 \times \%P_2O_5} \tag{2.5.6}$$

For slags containing MgO < 8% and P$_2$O$_5$ < 5%, the basicity B is essentially directly proportional to V.

$$B = 1.17V \tag{2.5.7}$$

Another measure of slag basicity is the difference between the sum of the concentrations of basic oxides and acidic oxides.

$$(\%CaO + \%MgO + \%MnO) - (\%SiO_2 + P_2O_5 + TiO_2) \tag{2.5.8}$$

This formulation of slag basicity is not used very often.

For the calcium aluminate type of slags used in steel refining in the ladle furnace, the slag basicity used in some German and Japanese publications, is defined by the ratio

$$\frac{\%CaO}{\%SiO_2 \times \%Al_2O_3} \tag{2.5.9}$$

However, such a ratio becomes meaningless at low concentrations of either SiO$_2$ or Al$_2$O$_3$. For the ladle furnace slag the basicity may be defined by the following mass concentration ratio, on the assumption that on a molar basis Al$_2$O$_3$ is equivalent to SiO$_2$.

$$B_{LF} = \frac{\%CaO + 1.4 \times \%MgO}{\%SiO_2 + 0.6 \times \%Al_2O_3} \tag{2.5.10}$$

The slag basicities as defined above are for the compositions of molten slags. In practice, the steelmaking slags often contain undissolved CaO and MgO. The chemical analyses of such slag samples without correction for undissolved CaO and MgO, will give unrealistic basicities which are much higher than those in the molten part of the slag.

There has been a trend in recent years to relate some physiochemical properties of slags, such as sulfide capacity, phosphate capacity, carbide capacity, etc., to optical slag basicity. The optical basicity is discussed briefly in Section 2.5.12.

2.5.3 Iron Oxide in Slags

Iron oxide dissolves in slags in two valency states: divalent iron cations Fe^{2+} and trivalent iron cations Fe^{3+}. The ratio Fe^{3+}/Fe^{2+} depends on temperature, oxygen potential and slag composition; this is discussed later in this section. In the formulation of the equilibrium constants of slag-metal reactions and the thermodynamic activities of oxides in slags, the total iron dissolved in the slag as oxides is usually converted to the stoichiometric formula FeO and denoted by Fe_tO, thus

$$\%Fe_tO = \%FeO \text{ (analyzed)} + 0.9 \times \%Fe_2O_3 \text{ (analyzed)} \qquad (2.5.11)$$

or

$$\%Fe_tO = 1.286 \times \%Fe \text{ (total as oxides)} \qquad (2.5.12)$$

For the ease of understanding, the subscript t will be omitted in all the subsequent equations and diagrams.

2.5.4 Selected Ternary and Quaternary Oxide Systems

Most steelmaking slags consist primarily of CaO, MgO, SiO_2 and FeO. In low-phosphorus steelmaking practices, the total concentration of these oxides in liquid slags is in the range 88 to 92%. Therefore, the simplest type of steelmaking slag to be considered is the quaternary system CaO–MgO–SiO_2–FeO.

First let us consider the ternary system CaO·SiO_2–FeO; the liquidus isotherms of this system is shown in Fig. 2.53. The isothermal section of the composition diagram in Fig. 2.54 shows the phase equilibria at 1600°C.

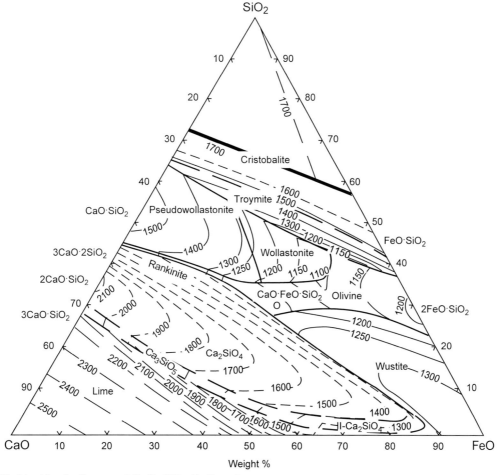

Fig. 2.53 Liquidus isotherms of CaO–SiO_2–FeO system.

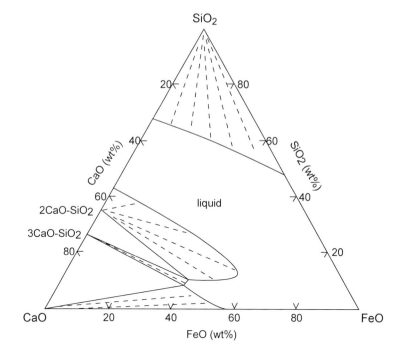

Fig. 2.54 Phase equilibrium in the system CaO–SiO$_2$–FeO in equilibrium with liquid iron at 1600°C.

There are four two-phase regions where, as depicted by dotted lines, the melt is saturated with SiO$_2$, 2CaO·SiO$_2$, 3CaO·SiO$_2$ or CaO; two three-phase regions (2CaO·SiO$_2$ + 3CaO·SiO$_2$ + liquid) and (3CaO·SiO$_2$ + CaO + liquid); and one liquid phase region.

Magnesia is another important ingredient of steelmaking slags, which are invariably saturated with MgO to minimize slag attack on the magnesia refractory lining of the furnace.

The effect of MgO on the solubility of calcium silicates and calcium oxide is shown in Fig 2.55 for the system (CaO + MgO)–SiO$_2$–FeO, in equilibrium with liquid iron at 1600°C.

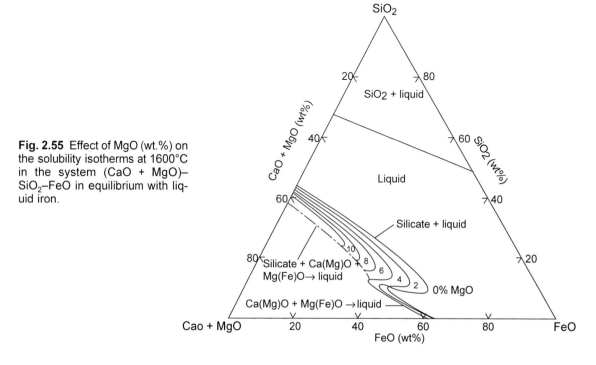

Fig. 2.55 Effect of MgO (wt.%) on the solubility isotherms at 1600°C in the system (CaO + MgO)–SiO$_2$–FeO in equilibrium with liquid iron.

The broken-line curve delineates the region of saturation of molten slag with solid calcium (magnesium) silicates and solid magnesio-wustite (MgO–FeO solid solution). Effects of the concentrations of MgO and FeO on the solubility of CaO in $2CaO \cdot SiO_2$-saturated slags are shown in Fig. 2.56.

Tromel et al.[97] have made a detailed study of the solubility of MgO in iron–calcium silicate melts in equilibrium with liquid iron at 1600°C. Below the dotted curve BACD in Fig. 2.57 for double saturations, the curves for 10 to 60% FeO are the MgO solubilities in the slag.

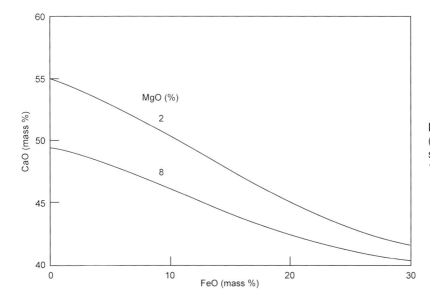

Fig. 2.56 Solubility of CaO in (CaO–MgO–SiO$_2$–FeO) slags saturated with $2CaO \cdot SiO_2$ at 1600°C.

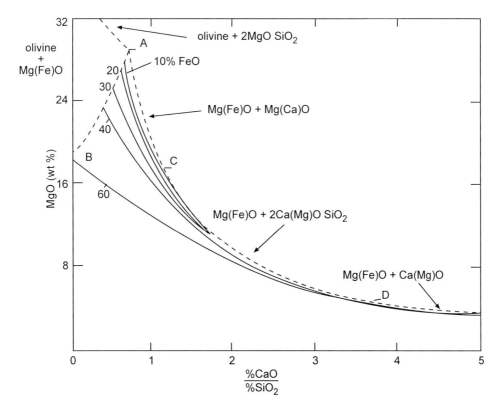

Fig. 2.57 Solubility of MgO, as magnesio–wustite, in the system CaO–MgO–SiO$_2$–FeO at 1600°C as a function of slag basicity and FeO concentration. *From Ref. 97.*

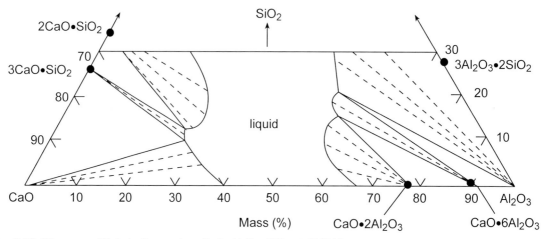

Fig. 2.58 Phase equilibria in the system CaO–Al$_2$O$_3$–SiO$_2$ at 1600°C.

Pertinent to the compositions of neutral ladle slags, the phase equilibria in part of the system CaO–Al$_2$O$_3$–SiO$_2$ at 1600°C is shown in Fig. 2.58.

2.5.5 Oxide Activities in Slags

In this section the activity data are given for a few ternary and multicomponent systems which are closely related to the iron and steelmaking slags.

2.5.5.1 Activities in CaO – FeO–SiO$_2$ System

The oxide activities in the CaO–FeO–SiO$_2$ melts relevant to oxygen steelmaking in equilibrium with liquid iron at about 1550°C are given in Fig. 2.59. The iso-activity curves in the left diagram represent the experimentally determined activities of iron oxide with respect to liquid FeO.[98–99] The activities of CaO and SiO$_2$ with respect to solid oxides were calculated from the FeO activities by Gibbs-Duhem integration.[99]

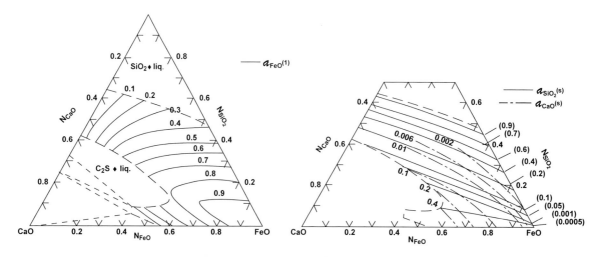

Fig. 2.59 Activities of FeO(l) (experimental), CaO(s) and SiO$_2$(s) (by Gibbs-Duhem integration) in CaO–FeO–SiO$_2$ melts in equilibrium with liquid iron at 1550°C. From Refs. 98, 99.

Recently new iron smelting processes have been developed and are briefly discussed here. The slags are CaO–SiO$_2$–FeO–Al$_2$O$_3$–MgO (saturated) with low FeO contents (<5%). Knowledge of the activity of FeO in these slags is critical and has been recently measured by Liu et al.[100] The activity coefficient of FeO in these slags is given in Fig. 2.60 and is approximately 3.5. It increases slightly with basicity as shown in Fig. 2.61 but is nearly constant with a value of 3.5.

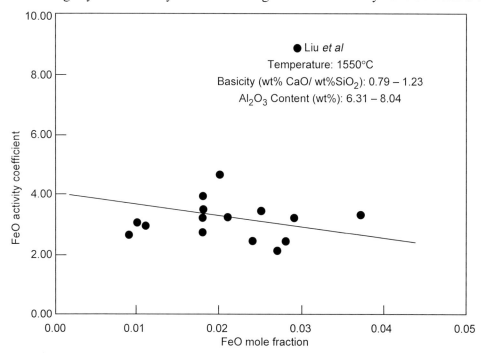

Fig. 2.60 The activity of FeO in CaO–SiO$_2$–Al$_2$O$_3$–FeO–MgO (saturated) slags relevant to iron smelting. *From Ref. 100.*

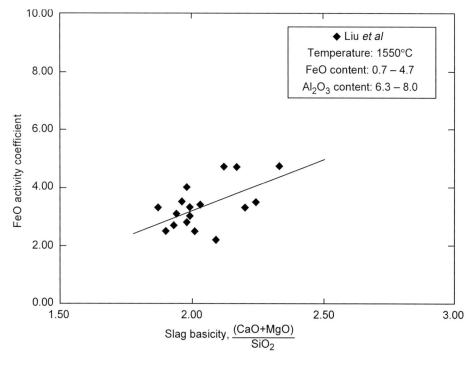

Fig. 2.61 The effect of slag basicity on the activity coefficient of FeO in CaO–SiO$_2$–Al$_2$O$_3$–FeO–MgO (saturated) slags relevant to iron smelting. *From Ref. 100.*

2.5.5.2 Activities in CaO–Al$_2$O$_3$–SiO$_2$ System

Rein and Chipman[101] measured the activity of silica in the CaO–Al$_2$O$_3$–SiO$_2$ system; from these experimental data they calculated the activities of CaO and Al$_2$O$_3$ by Gibbs-Duhem integration. The salient features of these oxide activities at 1600°C, with respect to solid oxides, are shown in Fig. 2.62 for the mass ratios of CaO/Al$_2$O$_3$ = 2/3 and 3/2; the compositions of ladle slags are well within the range given in Fig. 2.62.

Rein and Chipman also measured the activity of silica in the quaternary melts CaO–MgO–Al$_2$O$_3$–SiO$_2$. The activity coefficients of SiO$_2$ derived from these data are given in Fig. 2.63 for melts containing 10% MgO and 0, 10 or 20% Al$_2$O$_3$.

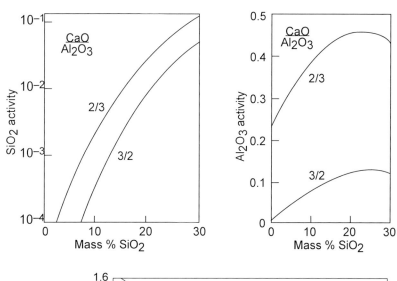

Fig. 2.62 Silica and alumina activities, with respect to solid oxides, in CaO–Al$_2$O$_3$–SiO$_2$ melts at 1600°C, derived from experimental data of Rein and Chipman.[101]

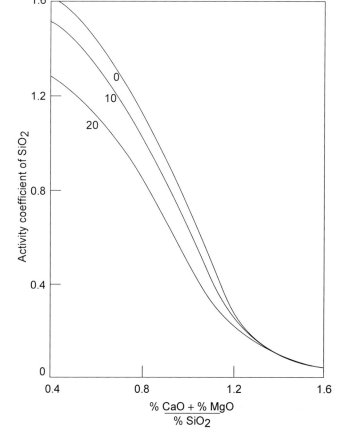

Fig. 2.63 Activity coefficient of SiO$_2$ with respect to pure solid oxide, in CaO–MgO–Al$_2$O$_3$–SiO$_2$ melts at 1600°C, containing 10% MgO and 0, 10 or 20% Al$_2$O$_3$, derived from experimental data of Rein and Chipman[101].

The ratio of the activities $(a_{Al_2O_3})^{1/3}/a_{CaO}$ within the entire liquid composition range up to 30% SiO_2 is shown in Fig. 2.64 reproduced from Ref. 27.

2.5.5.3 Activities in MnO–Al$_2$O$_3$–SiO$_2$ System

In the deoxidation of steel with the ladle addition of silicomanganese and aluminum together, the deoxidation product is molten manganese aluminosilicate with the mass ratio MnO/SiO$_2$ at about 1:1 and containing 10 to 45% Al$_2$O$_3$. The activities of oxides in the MnO–Al$_2$O$_3$–SiO$_2$ melts at 1550 and 1650°C were computed by Fujisawa and Sakao[102] from the available thermochemical data on the system. They also determined experimentally the activities of MnO and SiO$_2$ by the selected slag-metal equilibrium measurements and found a close agreement with the computed data.

The activities of Al$_2$O$_3$ and SiO$_2$, with respect to solid oxides, are plotted in Fig. 2.65 for melts with mass ratio of MnO/SiO$_2$ = 1. For melts containing up to 30% Al$_2$O$_3$ the activity of MnO remains essentially unchanged at about 0.1 then decreases to about 0.05 at 40% Al$_2$O$_3$.

2.5.5.4 Activity Coefficient of FeO in Slags

With hypothetical pure liquid FeO as the standard state, the activity of iron oxide is derived from the concentration of dissolved oxygen in liquid iron that is in equilibrium with the slag. For the reaction equilibrium

$$FeO(l) = Fe + [O] \tag{2.5.13}$$

$$K_O = \frac{[a_O]}{a_{FeO}} \tag{2.5.14}$$

where $a_O = [\%O]f_O$, using $\log f_O = -0.1 \times [\%O]$. The temperature dependence of the equilibrium constant K_O is given below.

$$\log K_O = -\frac{5370}{T} + 2.397 \tag{2.5.15}$$

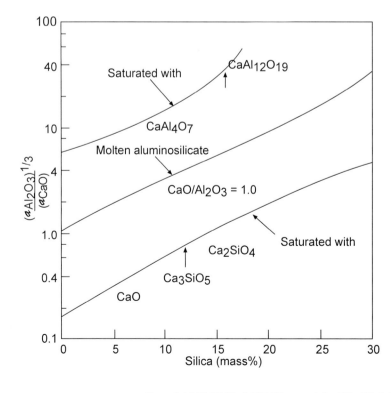

Fig. 2.64 Effect of slag composition on the activity ratio $(a_{Al_2O_3})^{1/3}/a_{CaO}$ for the system CaO–Al$_2$O$_3$–SiO$_2$ at 1600°C. *From Ref. 27.*

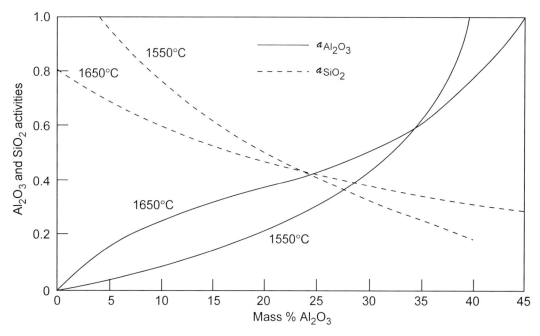

Fig. 2.65 Al_2O_3 and SiO_2 activities in $MnO–Al_2O_3–SiO_2$ system for mass ratio $MnO/SiO_2 = 1$, derived from data compiled by Fujisawa & Sakao[102].

In the past a wide variety of formulations were generated to represent the composition dependence of the iron oxide activity or activity coefficient in complex slag. On a recent reassessment of this property of the slag, Turkdogan[103] came to the conclusion that, within the limits of uncertainty of the experimental data on slag-metal reaction equilibrium, there is a decisive correlation between the activity coefficient of FeO and the slag basicity as shown in Fig. 2.66. The γ_{FeO} reaches a peak at a basicity of about B = 1.8. It should be pointed out once again that the concentration of iron oxide is for total iron as oxides in the slag represented by the stoichiometric formula FeO.

As discussed in Ref. 103, the experimental data used in deriving the relation in Fig. 2.66 are for simple and complex slags which differ considerably in their compositions thus: in mass percent, CaO 0–60, MgO 0–20, SiO_2 0–35, P_2O_5 0–20, FeO 5–40, MnO 0–15.

2.5.5.5 Activity of MnO in Slags

The activity coefficient ratio $\gamma_{FeO}/\gamma_{MnO}$, derived from the slag-metal equilibrium data cited in Ref. 103, varies with slag basicity as shown in Fig. 2.67. There is a sharp decrease in the ratio of the activity coefficients as the basicity B increases from 1.5 to 2.0. At basicities above 2.5, the ratio $\gamma_{FeO}/\gamma_{MnO}$ is essentially constant at about 0.63. The curve for γ_{MnO} in Fig. 2.66 is that derived from the combination of the curve for γ_{FeO} with the ratio in Fig. 2.67.

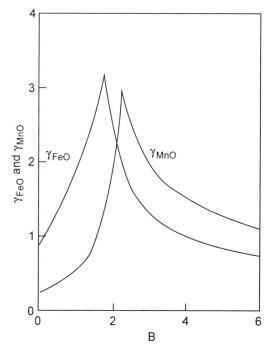

Fig. 2.66 Effect of slag basicity B on the activity coefficients of FeO and MnO, with respect to pure liquid oxides, in simple and complex slags at temperatures of 1550 to 1700°C. *From compiled data in Ref. 103.*

2.5.6 Gas Solubility in Slags

2.5.6.1 Solubility of H₂O

In acidic melts, H₂O vapor reacts with double bonded oxygen and depolymerizes the melt, thus

$$\left(-\overset{|}{\underset{|}{Si}}-O-\overset{|}{\underset{|}{Si}}-\right) + H_2O = 2\left(-\overset{|}{\underset{|}{Si}}-OH\right)$$

In basic melts, H₂O reacts with free oxygen ions

$$(O^{2-}) + H_2O = 2(OH)^- \quad (2.5.16)$$

Both for acidic and basic melts the overall reaction is represented by

$$(O^*) + H_2O = 2(OH^*) \quad (2.5.17)$$

where O^* represents double or single bonded oxygen, or O^{2-}, and OH^* is single bonded to silicon or as a free ion. The equilibrium constant for a given melt composition is

$$C_{OH} = \frac{(\text{ppm } H_2O)}{(p_{H_2O})^{1/2}} \quad (2.5.18)$$

where P_{H_2O} is the vapor partial pressure.

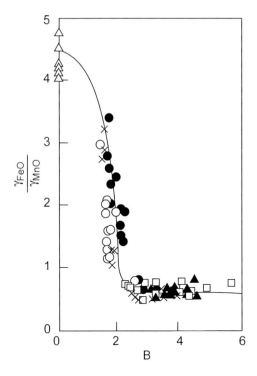

Fig. 2.67 Variation of activity coefficient ratio with basicity. *From Ref. 103.*

The solubilities of H₂O (in units of mass ppm H₂O) at 1 atm pressure of H₂O vapor in CaO–FeO–SiO₂ at 1550°C, measured by Iguchi *et al.*[104] are given in Fig. 2.68.

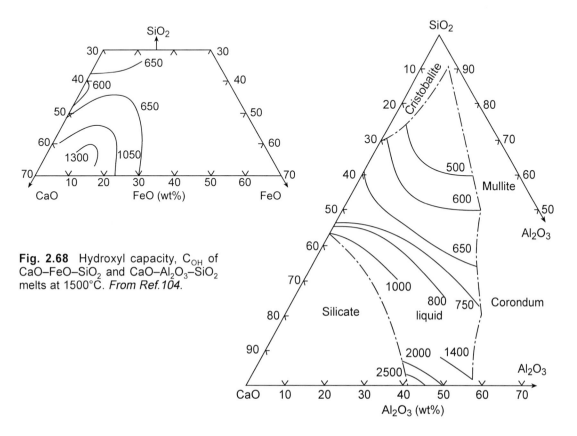

Fig. 2.68 Hydroxyl capacity, C_{OH} of CaO–FeO–SiO₂ and CaO–Al₂O₃–SiO₂ melts at 1500°C. *From Ref. 104.*

2.5.6.2 Solubility of N_2

Nitrogen dissolves in molten slags as a nitride ion N^{3-} only under reducing conditions

$$\tfrac{1}{2}N_2(g) + \tfrac{3}{2}(O^{2-}) = (N^{3-}) + \tfrac{3}{4}O_2(g) \qquad (2.5.19)$$

for which the equilibrium constant (known as nitride capacity) is

$$C_N = (\%N)\frac{p_{O_2}^{3/4}}{p_{N_2}^{1/2}} \qquad (2.5.20)$$

Many studies have been made of the solubility of nitrogen in $CaO-Al_2O_3$ and $CaO-Al_2O_3-SiO_2$ melts in the 1970s. These were reviewed in a previous publication[3]. Reference should be made also to a subsequent work done by Ito and Fruehan[105] on the nitrogen solubility in the $CaO-Al_2O_3-SiO_2$ melts. They showed that the nitride capacity of the aluminosilicate melts increases with a decreasing activity of CaO.

2.5.6.2.1 Nitrogen Dissolution into Slags The nitride capacity in certain slag systems increases and in others decreases with basicity. According to equation 2.5.19 one would expect it to increase with increasing slag basicity, i.e. increasing concentration of free oxygen ion O^{2-}. The reason for this apparent contradiction is that nitrogen can enter the slag by replacing single bonded oxygen ions in the SiO_2 network. This phenomena is explained in detail by Ito and Fruehan.[105]

Briefly, spectroscopic research indicates that the nitrogen is combined with the network former in silicate melts which can be represented by

$$(2O^-) + \tfrac{1}{2}N_2 = (N^-) + \tfrac{1}{2}(O^{2-}) + \tfrac{3}{4}O_2 \qquad (2.5.21)$$

where O^- is a nonbridging system in the network. If SiO_2 is the network former equation 2.5.21 corresponds to

$$\left(O^- - Si - O^-\right) + \tfrac{1}{2}N_2 = \left(Si = N^-\right) + \tfrac{1}{2}(O^{2-}) + \tfrac{3}{4}O_2 \qquad (2.5.22)$$

In this case the nitride capacity is given by

$$C_{N^{3-}} = \frac{(\%N^{3-})(p_{O_2})^{3/4}}{(p_{N_2})^{1/2}} = K\frac{a_{O^-}}{f_{N^-}(a_{O^{2-}})^{1/2}} \qquad (2.5.23)$$

where K is an equilibrium constant for reaction 2.5.21, f_i is an activity coefficient for species i. In this case N^{3-} does not mean a free nitride ion but the nitrogen analyzed as a nitride which is associated with the network former. Similar reactions can be written for bridging oxygen, but then reactions are less likely. In either case the nitride capacity decreases with oxygen ion activity, i.e. increasing slag basicity.

For the $CaO-SiO_2$, $CaO-Al_2O_3$, and $CaO-SiO_2-Al_2O_3$ systems CaO is the basic component. Therefore the equations predict that a plot of the logarithm of nitride capacity versus the logarithm of the activity of CaO should yield a straight line with a slope of $-\tfrac{1}{2}$ as shown in Fig. 2.69.

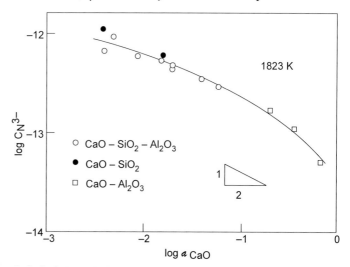

Fig. 2.69 The relation between the nitride capacity and the activity of lime for $CaO-SiO_2-Al_2O_3$ slag at 1823 K. Ref. 105.

For highly basic oxide systems the nitrogen enters as a free nitride ion and the simple relationship given by 2.5.19 is valid. In this case the free nitride concentration and the nitride capacity is proportional to the oxygen ion activity to the 3/2 power $(a_{O^{2-}})^{3/2}$. Min and Fruehan[106] have shown that the nitride capacity does increase with basicity for highly basic slags, particularly when there is no strong network former such as SiO_2 present. In theory the nitride capacity should decrease with basicity when the nitrogen is primarily in the network and increase when it is a free nitride. This was observed for the $BaO-B_2O_3$ and $CaO-B_2O_3$ system as shown in Fig. 2.70.

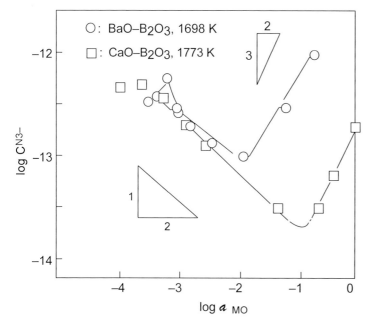

Fig. 2.70 Nitride capacity of the slags as a function of the activity of CaO or BaO. *From Ref. 106.*

2.5.6.2.2 Nitrogen Removal by Slags

For Al-killed steels the corresponding equilibrium partial pressure of oxygen is

$$p_{O_2} \text{ (atm)} = \frac{1.22 \times 10^{-16}}{[\%Al]^{4/3}} \text{ at } 1600°C \tag{2.5.24}$$

From the solubility data for nitrogen we have

$$\left(p_{N_2}\right)^{1/2} \text{ (atm)}^{1/2} = 21.9[\%N] \text{ at } 1600°C \tag{2.5.25}$$

Substituting these in the equation for C_N gives

$$C_N = \frac{(\%N)}{[\%N][\%Al]} \times 5.3 \times 10^{-14}$$

In lime-rich aluminate ladle slags of low SiO_2 content, the lime activity is $a_{CaO} = 0.5$ for which $C_N = 10^{-13}$. These give the following equilibrium relation

$$\frac{(\%N)}{[\%N]} = 1.9 \times [\%Al]$$

Dissolved [%Al]	(%N) / [%N]
0.005	0.0095
0.02	0.038
0.06	0.114
0.10	0.190

The equilibrium nitrogen distribution ratio between slag and steel is very low even at high aluminum contents, well above the practical range. The slag/steel mass ratio in the ladle is also low, about 1/100. For these reasons, liquid steel in the ladle cannot be de-nitrogenized by the slag that is usable in industry in steel refining.

As noted from the recent experimental studies of Fruehan and co-workers,[106–108] that even with fluxes of high nitride capacities (as with alkaline earth borates) it is not practically possible to remove nitrogen from the aluminum-killed steel in the ladle. The highest nitride capacity was for a slag containing about 22% CaO – 27% BaO – 48% Al_2O_3 – 3% TiO_2 and was about 5×10^{-12}. For this case for a steel containing 0.06% Al the nitrogen distribution ratio is about 5.5 which is still not high enough for effective nitrogen removal.

2.5.6.3 Solubility of S_2

Sulfur-bearing gases dissolve in molten slags as sulfide ions (S^{2-}) under reducing conditions, and as sulfate ions (SO_4^{2-}) under highly oxidizing conditions. In steelmaking the oxygen potential is not high enough for the solution of sulfur as sulfate ions, therefore we need to consider only the sulfide reaction.

Whether the sulfur-bearing species is primarily H_2S or SO_2, there is a corresponding equilibrium value of p_{S_2} depending on the temperature and gas composition. It is convenient to consider the reaction in a general form as

$$\tfrac{1}{2}S_2(g) + (O^{2-}) = (S^{2-}) + \tfrac{1}{2}O_2(g) \tag{2.5.26}$$

For a given slag composition the equilibrium relation is represented by

$$C_S = (\%S)\left(\frac{p_{O_2}}{p_{S_2}}\right)^{1/2} \tag{2.5.27}$$

where p's are equilibrium gas partial pressures. The equilibrium constant C_S is known as the sulfide capacity of the slag. The value of C_S depends on slag composition and temperature.

Experimentally determined sulfide capacities of binary oxide melts are shown in Fig. 2.71. References to experimental data are given in Ref. 3.

For the slag-metal system, the sulfur reaction is formulated in terms of the activities ($\cong$ concentrations) of sulfur and oxygen dissolved in the steel.

$$[S] + (O^{2-}) = (S^{2-}) + [O] \tag{2.5.28}$$

For low-alloy steels

$$k_S = \frac{(\%S)}{[\%S]}[\%O] \tag{2.5.29}$$

where the equilibrium constant k_S depends on slag composition and temperature.

2.5.6.3.1 Conversion of p_{O_2}/p_{S_2} to [%O]/[%S] The free energies of solution of O_2 and S_2 in liquid low alloy steel are given below; see Table 2.1 in section 2.1.2.4.2.

$$\tfrac{1}{2}S_2 = [S]; \quad \Delta G_s = -135{,}060 + 23.43T J \tag{2.5.30}$$

$$\tfrac{1}{2}O_2 = [O]; \quad \Delta G_o = -115{,}750 - 4.63T J \tag{2.5.31}$$

For the reaction equilibrium

$$\tfrac{1}{2}S_2 = [O] = \tfrac{1}{2}O_2 + [S]$$

the standard free energy change is

$$\Delta G° = \Delta G_s - \Delta G_o = -19{,}310 + 28.06T J \tag{2.5.32}$$

$$\log\left(\frac{p_{O_2}}{p_{S_2}}\right)^{1/2}\frac{[\%S]}{[\%O]} = \frac{1009}{T} - 1.466 \qquad (2.5.33)$$

For steelmaking temperatures an average value of K is 0.133; with this conversion factor the following is obtained.

$$\left(\frac{p_{O_2}}{p_{S_2}}\right)^{1/2} = 0.133 \frac{[\%O]}{[\%S]} \qquad (2.5.34)$$

With this substitution the values of C_S are converted to k_S.

$$7.5 \times C_S = k_S = \frac{(\%S)}{[\%S]}[\%O] \qquad (2.5.35)$$

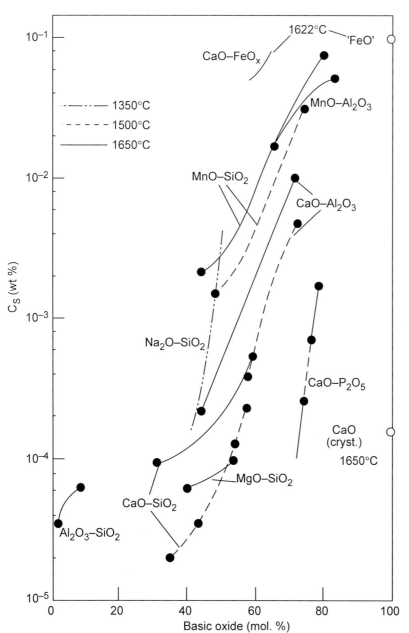

Fig. 2.71 Sulfide capacities of binary oxide melts. *From Ref. 3.*

A slag of high k_S value and steel deoxidation to low levels of [O] are necessary conditions for steel desulfurization. The subject is discussed in more detail later.

2.5.6.4 Solubility of O_2

Oxygen dissolves in molten slags by oxidizing the divalent iron ions to the trivalent state.

$$\text{Gas–slag reaction:} \quad \tfrac{1}{2}O_2(g) + 2(Fe^{2+}) = 2(Fe^{3+}) + (O^{2-}) \quad (2.5.36)$$

$$\text{Slag–metal reaction:} \quad Fe° + 4(Fe^{3+}) + (O^{2-}) = 5(Fe^{2+}) + [O] \quad (2.5.37)$$

These reactions provide the mechanism for oxygen transfer from gas to metal through the overlaying slag layer. Partly for this reason, the steel reoxidation will be minimized by maintaining a low concentration of iron oxide in the ladle slag, tundish and mold fluxes.

Examples are given in Fig. 2.72 of variations of the ratio Fe^{3+}/Fe^{2+} with slag composition of melts co-existing with liquid iron, reproduced from a paper by Ban-ya and Shim[109].

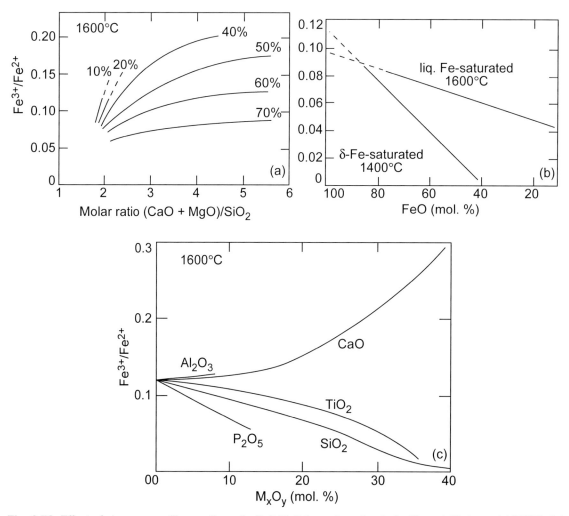

Fig. 2.72 Effect of slag composition on the ratio Fe^{3+}/Fe^{2+} in melts saturated with metallic iron at 1600°C: (a) CaO–MgO–SiO$_2$–FeO melts at indicated molar concentrations of total iron oxide FeO; (b) MgO–SiO$_2$–FeO melts; (c) pseudobinary FeO–M$_x$O$_y$ melts. *From Ref. 109.*

2.5.7 Surface Tension

Surface tensions measured by Kozakevitch[110] are given in Fig. 2.73 for binary melts with iron oxide, and in Fig. 2.74 for FeO–MnO–SiO$_2$ and FeO–CaO–SiO$_2$ melts at 1400°C. For additional data on surface tension of a wide variety of slags and mold fluxes, reference may be made to a review paper by Mills and Keene[111].

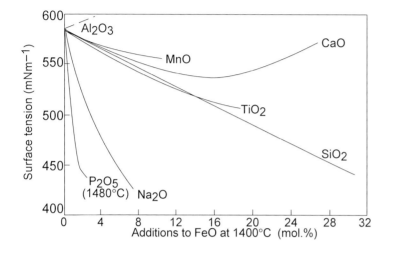

Fig. 2.73 Surface tensions of binary iron oxide melts at 1400°C. *From Ref. 110.*

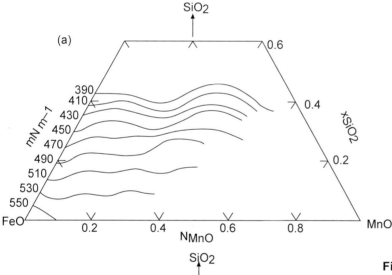

Fig. 2.74 Isosurface tension curves in (a) FeO-MnO-SiO$_2$ and (b) FeO-CaO-SiO$_2$ melts saturated with iron at 1400°C. *From Ref. 110.*

2.5.7.1 Interfacial Tension

The slag-metal interfacial tensions have values between those for the gas-slag and gas-metal surface tensions. Consequently, the addition of surface active elements to liquid iron lowers the slag-metal interfacial tension.

Effects of sulfur and oxygen dissolved in iron on the interfacial tension between liquid iron and $CaO-Al_2O_3-SiO_2$ melts at 1600°C, determined by Gaye et al.[112], are shown in Figs. 2.75 and 2.76; compared to the surface tensions of Fe-S and Fe-O melts. The effect of oxygen on the interfacial tension is greater than sulfur, e.g. $\sigma_i = 600$ mN m^{-1} with $a_O = 0.05$ (0.05%) while at $a_S = 0.05$ (0.05%), $\sigma_i = 1000$ mN m^{-1}.

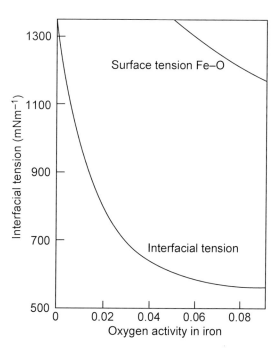

Fig. 2.75 Effect of oxygen in iron on the interfacial tension between liquid iron and $CaO-Al_2O_3-SiO_2$ melts, determined by Gaye et al.[112] is compared to the surface tension of Fe-O melt at 1600°C.

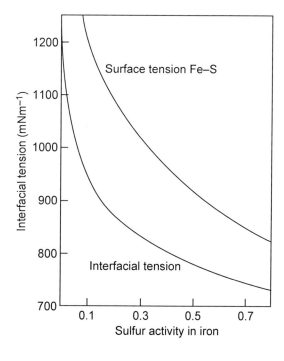

Fig. 2.76 Effect of sulfur in iron on the interfacial tension between liquid iron and $CaO-Al_2O_3-SiO_2$ melts, determined by Gaye et al.[112] is compared to the surface tension of Fe-S melts at 1600°C.

As shown by Ogino et al.[113] Fig. 2.77, a single curve describes adequately the effect of oxygen in iron on the interfacial tension between liquid iron and a wide variety of simple and complex slags, including those containing Na_2O and CaF_2. In the case of slags containing iron oxide, a decrease in interfacial tension with an increasing iron oxide content is due entirely to the corresponding increase in the oxygen content of the iron.

Recently Jimbo and Cramb[114] accurately measured the interfacial tension between liquid iron and CaO–Al_2O_3–SiO_2 slags. As shown in Fig. 2.78 their results are slightly higher than those of Gaye et al.

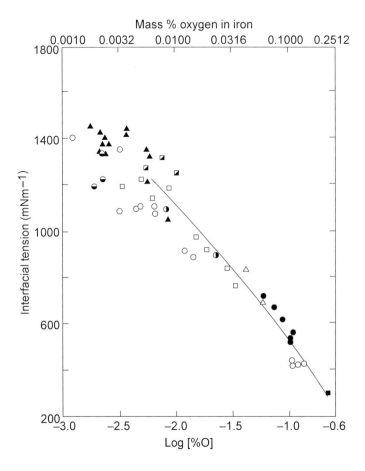

Fig. 2.77 General relation between the oxygen content of iron and the interfacial tension between the metal and various slag systems at 1580°C. *From Ref. 113.*

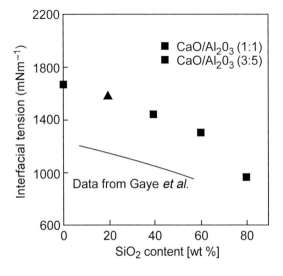

Fig. 2.78 Interfacial tension between CaO–Al_2O_3–SiO_2 and Fe at 1600°C. *From Ref. 114.*

2.5.8 Density

Many repetitive measurements of slag densities have been made. Only selected references are given on the density data cited here for steelmaking type of slags. Since the density of silica (2.15 g cm^{-3} at 1700°C) is much lower than the densities of other metal oxide components of slags, densities of slags will decrease with an increasing silica content.

The density data for binary silicates are given in Fig. 2.79. The data in Fig. 2.80 are for $CaO–MgO–Al_2O_3–SiO_2$ melts, relevant to neutral slags for steel refining in the ladle.

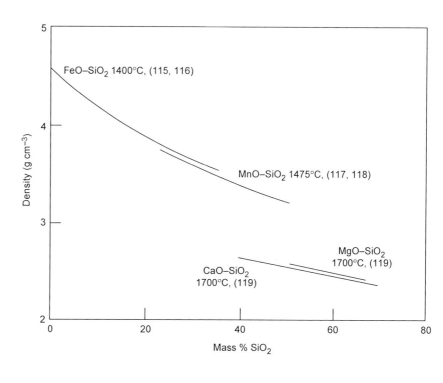

Fig. 2.79 Densities of binary silicate melt. *From Refs. 115–119.*

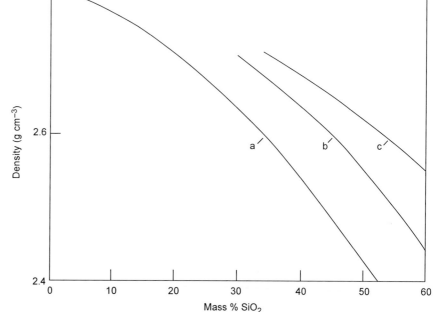

Fig. 2.80 Densities of $CaO–MgO–Al_2O_3–SiO_2$ melt at 1550°C, using data in Refs; 120–122. (a) 0% MgO, CaO/Al_2O_3 = 1; (b) 0% MgO, 5% Al_2O_3; (c) 5% MgO, 5% Al_2O_3.

The density data in Fig. 2.81 compiled by Mills and Keene[111] are for simple and complex slags containing FeO, CaO, MgO, SiO$_2$ and P$_2$O$_5$. Since the densities of FeO–SiO$_2$ and MnO–SiO$_2$ are essentially the same, the average of the data in Fig. 2.81 is represented by the following equation in terms of (%FeO + %MnO).

$$\rho, \text{gcm}^{-3} = 2.46 + 0.018 \times (\%\text{FeO} + \%\text{MnO}) \tag{2.5.38}$$

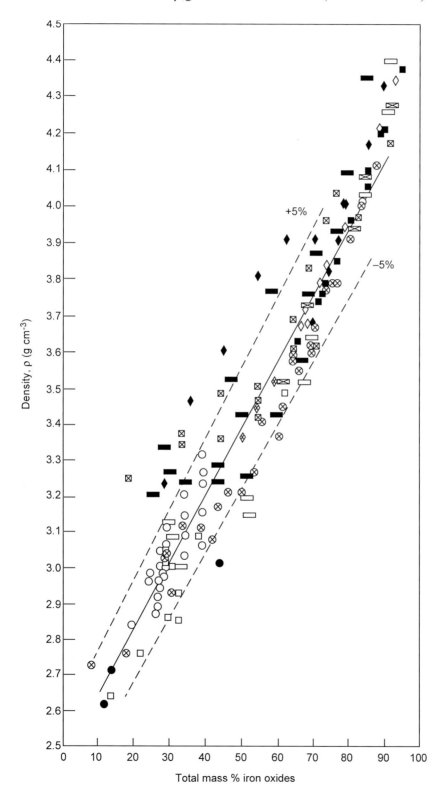

Fig. 2.81 Densities of simple and complex slags containing iron oxide at about 1400°C. *From Ref. 111.*

2.5.9 Viscosity

The size of the silicate and aluminosilicate network in molten slags becomes larger with increasing SiO_2 and Al_2O_3 contents, hence their mobility decreases resulting in a higher viscosity. The addition of metal oxides or an increase in temperature leads to the breakdown of the $Si(Al)O_4$ network, resulting in lower melt viscosity.

Machin and Yee[123] made an extensive study of the viscosity of CaO–MgO–Al_2O_3–SiO_2 melts at temperatures of 1350 to 1500°C. The data in Fig. 2.82 are for the ternary system at 1500°C. The isokoms are approximately parallel to the binary side Al_2O_3–SiO_2, indicating that Al_2O_3–SiO_2 are isomorphous in their effect on the slag viscosity. The isokoms in Fig. 2.83 are for the quaternary system with 35% and 50% SiO_2. In this case, the isokoms are approximately parallel to the binary side CaO–MgO, indicating that Ca^{2+} and Mg^{2+} cations have similar effects on the breakdown of the aluminosilicate network.

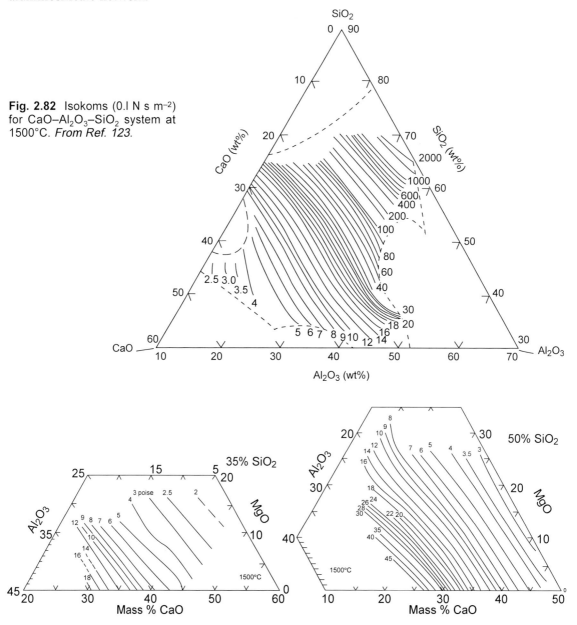

Fig. 2.82 Isokoms (0.1 N s m⁻²) for CaO–Al_2O_3–SiO_2 system at 1500°C. *From Ref. 123.*

Fig. 2.83 Isokoms (0.1 N s m⁻²) for CaO–MgO–Al_2O_3–SiO_2 system at 1500°C for melts containing 35% and 50% SiO_2. *From Ref. 123.*

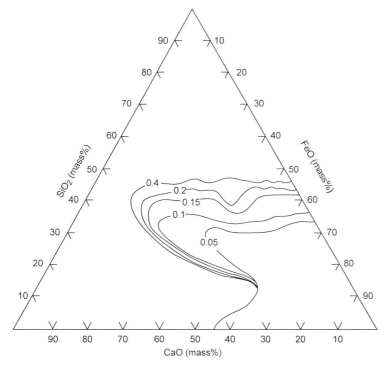

Fig. 2.84 Viscosity (N s m^{-2}) of CaO–FeO–SiO$_2$ melts in 1400°C. From Ref. 110.

Viscosities of steelmaking slags are well represented by the experimental data of Kozakevitch[110] given in Fig. 2.84

In the study of viscosities of mold fluxes for continuous casting, the experimental results have been represented as a function of temperature using the relation.

$$\eta = AT \exp(B/T) \qquad (2.5.39)$$

where A and B are functions of slag composition. For the composition range (wt %) 33–56% SiO$_2$, 12–45% CaO, 0–11% Al$_2$O$_3$, 0–20% Na$_2$O and 0–20% CaF$_2$, an interpolation formula has been derived for the parameters A and B as a function of the mole fractions of the constituents as given below.

$$\ln A = -17.51 - 35.76(Al_2O_3) + 1.73(CaO) + 5.82(CaF_2) + 7.02(Na_2O) \qquad (2.5.40)$$

$$B = 31{,}1140 - 68{,}833(Al_2O_3) - 23{,}896(CaO) - 46{,}351(CaF_2) - 39{,}519(Na_2O) \qquad (2.5.41)$$

where A is in units of 0.1 N s m^{-2} K^{-1} (poise/deg.) and B in degrees Kelvin.

2.5.10 Mass Diffusivity, Electrical Conductivity and Thermal Conductivity

2.5.10.1 Mass Diffusivity

Because of the ionic nature of molten slags, the diffusive mass transfer is by ions. The ionic diffusivities are measured using the radioactive tracer elements dissolved in an oxidized form in the melt. Typical examples of ionic diffusivities in slags at 1600°C are given below.

Ion	D_i^*, cm^2/s
Si^{4+}, O^{2-}	$4 \times 10^{-7} - 1 \times 10^{-6}$
Al^{3+}	$\approx 1 \times 10^{-6}$
Ca^{2+}, Mg^{2+}, Fe^{2+}	$6 \times 10^{-6} - 1 \times 10^{-5}$
S^{2-}	$\approx 4 \times 10^{-6}$

Since the electroneutrality has to be maintained, diffusion of a cation is accompanied by diffusion of the oxygen ion. The diffusion that occurs in the dissolution of a solid oxide in the slag is controlled by the mobility of the O^{2-} ion which is smaller than the divalent cations.

2.5.10.2 Electrical Conductivity

The electrical current in molten slags is carried by the cations. However, in slags containing high concentrations of FeO or MnO (> 70%) the electronic conduction becomes the dominant mechanism.

The ionic conductivity λ_i is theoretically related to the self diffusivity of the ionic species i by the Nernst-Einstein equation

$$D_i^* = \frac{RT}{F^2 Z_i^2 C_i} \lambda_i \qquad (2.5.42)$$

where
- F = Faraday constant; 96,489 C mol^{-1},
- Z_i = valency of ion i,
- C_i = concentration of ion i, mol cm^{-3},
- λ_i = specific conductivity, Ω^{-1}cm^{-1}.
  - For steelmaking slags: λ = 0.5–1.5 Ω^{-1}cm^{-1}.
  - For ladle slags: λ = 0.4–0.7 Ω^{-1}cm^{-1}.

The electrical conductivity increases with an increasing slag basicity and increasing temperature.

2.5.10.3 Thermal Conductivity

Because of the presence of iron oxide, the metallurgical slags are opaque to infrared radiation, therefore the heat conduction is primarily thermal.

Thermal conductivity of slags and mold fluxes are in the range 0.5 to 1.2 Wm^{-1}K^{-1}. From experimental data the following approximate empirical relation has been found

$$\kappa (Wm^{-1} K^{-1}) = 1.8 \times 10^{-5} V^{-1} \qquad (2.5.43)$$

where V is the molar volume = M/ρ, m^{3}mol^{-1}.

2.5.11 Slag Foaming

Slag foaming plays an important role in many steelmaking processes. In oxygen steelmaking excessive foaming can lead to slopping. On the other hand controlled foaming in the electric arc furnace is desirable to protect the refractories from the electrical arc radiation. In the direct ironmaking processes such as DIOS and AISI Direct Steelmaking, excessive foaming can cause operational problems. However some foam is desirable to help capture the energy from post combustion. Slag foaming should not be confused with simple gas bubbling or hold up. Whenever gas passes through a liquid, the liquid expands due to the presence of the gas. However in some liquids a stable foam develops which consists of foam bubbles on the top of an unfoamed liquid.

Early work on foaming was limited to a qualitative understanding of this complex phenomenon. Due to the added importance of foaming in the EAF and iron bath systems Ito and Fruehan[124,125] and other researchers developed a quantitative measure of foaming and measured the foaming characteristics of many important slag systems. The foam index (Σ) was defined by

$$\Sigma = \frac{\Delta h}{V_g^s} \qquad (2.5.44)$$

$$V_g^s = \frac{Q}{A} \qquad (2.5.45)$$

where
- Δ = the increase in the height of the slag,
- V_g^s = the superficial gas velocity,
- Q = the gas flow rate,
- A = the area of the vessel.

It follows that the volume of foam is approximately given by

$$Vf = Q\Sigma \qquad (2.5.46)$$

The foam index has the units of time and represents the average traveling time of the gas through the foam.

The foam can also be characterized by the time for the foam to decay. The average foam life (τ) is defined by the first order decay

$$\ln \frac{h}{h_o} = -\frac{t}{\tau} \qquad (2.5.47)$$

where h and h_o are the foam height and the foam height when the gas flow stops, respectively. It can be shown that for an ideal foam, in which the gas fraction of the foam is constant, that the foam index (Σ) and foam life (τ) are equal.

The foam index for $CaO-SiO_2-FeO-Al_2O_3$ system is shown in Fig. 2.85. The foam index decreases initially with increasing basicity because the viscosity is decreasing. However when the solubility limit is reached the foam index increases with basicity because the second phase particles increase the bulk viscosity of the slag.

Zhang and Fruehan[126] later demonstrated that the foam index on stability increased with decreasing bubble size and developed a general correlation demonstrated in Fig. 2.86 from which the foam index for many complex systems would be desired.

$$\Sigma = \frac{\eta^{1.2}}{\sigma^{0.2} \rho D_B^{0.9}} \qquad (2.5.48)$$

where

η = slag viscosity,
σ = surface tension,
ρ = slag density,
D_B = foam bubble diameter.

Takumetsu et al[127] showed that in bath smelting processes char could reduce foaming significantly. Zhang and Fruehan[128] examined the fundamentals of slag control by carbonaceous materials. Due to surface tension phenomenon the slag foam bubbles collapse and form longer bubbles resulting in a less stable foam. Carbonaceous materials will cause the foam to collapse while others such as alumina particles or iron oxide pellets do not due to the different wetting characteristics with the slag.

The fundamentals of slag foaming have been applied to a variety of processes such as oxygen and EAF steelmaking, bath smelting and ladle processing. The fundamental principles allow for reasonable predictions of foaming in these processes.

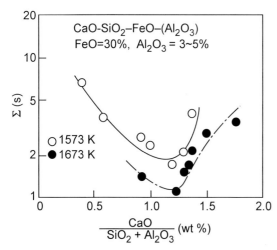

Fig. 2.85 The foam index of $CaO-SiO_2-FeO-Al_2O_3$ slags illustrating the effect of second phase particles. *From Ref. 126.*

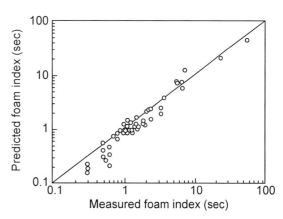

Fig. 2.86 Comparison between the measured foam index and that predicted by the correlation equation 2.5.48. From Ref 126.

2.5.12 Slag Models and Empirical Correlations for Thermodynamic Properties

There have been several attempts to develop models and correlations for slags, from which thermodynamic properties such as activities and capacities can be predicted. Simple models such as a regular solution have been used.[129] However because of the ionic nature and structure of slags they are generally inadequate. Recently Gaye and Lehmann reviewed the current status of structural models.[130]

The most successful models have been developed by Gaye and coworkers[131] at IRSID and Blander et al.[132] In the IRSID model the slag is represented using an oxygen sublattice and a cationic sublattice. The structure is defined in terms of cells composed of a central oxygen surrounded by two cations. Two parameters are used to calculate the energy of the system; the energies of formation of asymmetric cells and the interaction energies limited to one parameter per couple of cations. These parameters are computed from the known thermodynamics of the binary systems. It is beyond the scope of this chapter to describe the model in greater detail. The model has been used successfully to estimate the phase diagrams and thermodynamic activities in multi-component systems as well as sulfide capacities. Whereas this model is complex and requires binary data, to date it is the best model available for estimating thermodynamic properties of slags.

In recent years various capacities such as the sulfide and phosphate capacities have been correlated to a term called the optical basicity. This concept has been recently reviewed by Summerville et al.[133] Briefly each oxide is assigned a value for its optical basicity which is related to the Pauling electronegativity. By definition the optical basicity of CaO is taken as unity. Components such as Na_2O have optical basicities greater than one while acidic oxides such as SiO_2 have values less than one. The optical basicity of a slag is then computed from a weighted average of the components in the slag. The optical basicity has been used to correlate a large number of studies on sulfide and other capacities.[133, 134] Whereas the optical basicity is an interesting concept, the correlations are not sufficiently accurate to predict capacities for industrial purposes. For example the deviations between the correlation and actual data is 25–50% in some cases. This is not accurate enough for computing sulfur equilibrium in industrial applications. However the optical basicity correlation maybe useful for making crude estimates of capacities for which no data exists or to predict trends when changing slag compositions slightly.

2.6 Fundamentals of Ironmaking Reactions

Reference may be made to a book entitled "The Iron Blast Furnace", authored by Peacey and Davenport.[135] In this book, the theoretical, experimental and operational data have been put together about the blast furnace with particular emphasis on the interrelations between countercurrent heat and mass transfer between gases and solids in the blast furnace stack. Reference may also be made to the earlier studies of Rist and co-workers[136–140] on the heat and mass transfer and reduction reactions in the stack.

In this section, the following aspects of the blast furnace reactions are considered: oxygen potential diagram, role of vapor species in blast furnace reactions, slag-metal equilibrium, reaction mechanisms and operational slag-metal data.

2.6.1 Oxygen Potential Diagram

The reducibility of a metal oxide relative to other oxides, or the oxidisability of a metal relative to other metals, can readily be assessed from the free energy data.

For the oxidation reaction involving pure metals and metal oxides in their standard states, i.e. $a_M = 1$, $a_{MO} = 1$

$$2M + O_2 = 2MO \qquad (2.6.1)$$

the isothermal equilibrium constant is

$$K = \frac{1}{p_{O_2}} \qquad (2.6.2)$$

where p_{O_2} is the equilibrium oxygen partial pressure for which the standard state is 1 atm at the temperature under consideration. The standard free energy change is

$$\Delta G° = -RT \ln K = RT \ln p_{O_2} \qquad (2.6.3)$$

which is also called the oxygen potential.

In an earlier study, Richardson and Jeffes[141] compiled the free energy data on metal oxides which were then available, and presented the data, Fig. 2.87, as an oxygen potential diagram.

The oxides for which the oxygen potential lines are above that of CO may be reduced by carbon. As the affinity of the metals for oxygen increases, i.e. $\Delta G°$ decreases, the temperature of reduction of the oxides by carbon increases.

For easy conversion of oxygen potentials to the corresponding values of p_{O_2} or to the equilibrium ratios of H_2/H_2O and CO/CO_2, appropriate scales are included in the enclosed oxygen potential diagram for various oxides of metallurgical interest.

Scale for p_{O_2}:

Lines drawn from the point O on the ordinate for the absolute zero temperature through the points marked on the right hand side of the diagram give the isobars. For example, for the Fe–FeO equilibrium the oxygen potential is –82 kcal (–343 kJ) at 1200°C. By drawing a line passing through this point and the point O, the oxygen partial pressure of about 7×10^{-13} atm is read off the log p_{O_2} scale.

Scale for CO/CO_2:

Draw the line from point C to a point on the oxygen potential; the extension of this line gives the corresponding equilibrium log CO/CO_2 ratio.

Scale for H_2/H_2O:

Same as above by using point H.

These diagrams can be used to predict CO/CO_2 or H_2/H_2O ratios as well as the pressure of oxygen in equilibrium with numerous metals and their oxides. They are particularly useful for predicting the CO/CO_2 ratio in a blast furnace or the H_2/H_2O ratio in direct reduction with natural gas. It also gives an indication of the relative reduceability of the oxides. It should be remembered that the diagrams are for unit activity of the metal. If the metal is in solution, it is easier to reduce. For example, if the activity of Si is 10^{-3} relative to pure Si in liquid iron, the equilibrium oxygen pressure is 10^3 times higher. Similarly, if the activity of SiO_2 is lowered in a slag it is more difficult to reduce.

2.6.2 Role of Vapor Species in Blast Furnace Reactions

Because of the high temperatures involved, vaporization plays a significant role in pyrometallurgical reactions. In a reactor such as a blast furnace, the formation and condensation of various vapor species is responsible for the recycle of some of the elements between the high and low temperature regions of the furnace.

The countercurrent flow in the blast furnace of solids from low- to high-temperature zones and of gases from high- to low-temperature zones brings about a cyclic process of vaporization and condensation which has a decisive influence on the overall operation of the furnace.

2.6.2.1 Vapor Species SiO, SiS and CS

It is now well established that the silicon and sulfur in the coke are transferred to the slag and iron in the blast furnace via the vapor species SiO, SiS, CS, and other minor sulfur-bearing species. The earlier and recent studies of quenched blast furnaces have revealed that while the sulfur content of the slag increases with the descent of the burden in the bosh, the sulfur and silicon contents of the metal droplets reach maxima and then decrease as they pass through the slag layer. This phenomenon of vapor-phase mass transfer has been substantiated by the experimental work of Tsuchiya *et* al,[142] Turkdogan et al [143] and Ozturk and Fruehan[144] under conditions partly simulating the bosh region of the blast furnace.

In the presence of carbon and depending on temperature and activity of silica in the slag or coke, the SiO vapor is generated by one of the following reactions

$$SiO_2 \text{ (in coke ash or slag)} + C \rightarrow SiO + CO \tag{2.6.4}$$

$$SiC + CO \rightarrow SiO + 2C \tag{2.6.5}$$

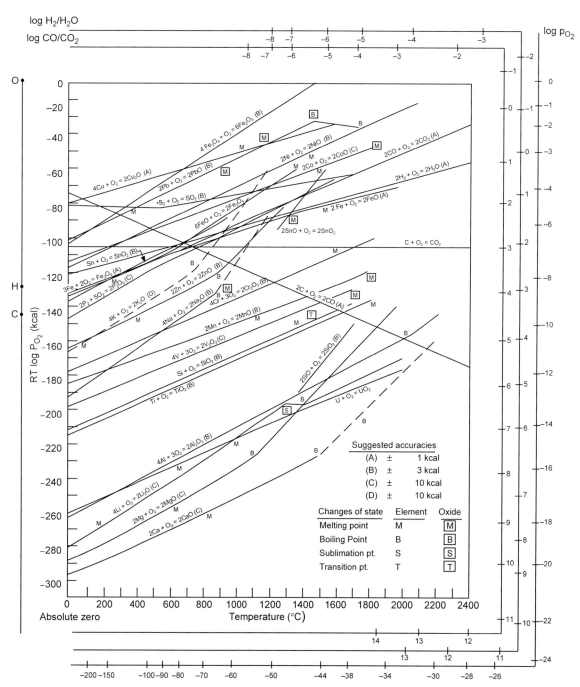

Fig. 2.87 Oxygen potential diagram. *From Ref. 141.*

Fundamentals of Iron and Steelmaking

At 1 bar pressure of CO, $a_{SiO_2} = 1$ and above about 1500°C, SiO_2 is converted to SiC and SiO is generated by reaction 2.6.5; for $a_{SiO_2} = 0.1$, reaction 2.6.5 applies only at temperatures above about 1610°C. In the experiments cited,[143] the SiO generated by passage of CO through a coke bed at temperatures of 1600 to 1900°C, were found to be close to the equilibrium values for reaction 2.6.5. Volatile SiS is generated by reactions of the type with the coke ash:

$$CaS \text{ (in coke ash)} + SiO \rightarrow SiS + CaO \quad (2.6.6)$$

$$FeS \text{ (in coke ash)} + SiO + C \rightarrow SiS + CO + Fe \quad (2.6.7)$$

The reaction of these vapor species with metal and slag may be represented by the equations:

$$\text{Metal} \begin{cases} SiO + [C] = [Si] + CO \\ SiS = [Si] + [S] \end{cases} \quad (2.6.8)$$

$$\text{Slag} \begin{cases} SiO + \tfrac{1}{2}O^{2-} + \tfrac{1}{2}C = \left(-\overset{|}{\underset{|}{Si}}-O^-\right) + \tfrac{1}{2}CO \\ SiS + \tfrac{3}{2}O^{2-} + \tfrac{1}{2}C = \left(-\overset{|}{\underset{|}{Si}}-O^-\right) + (S^{2-}) + \tfrac{1}{2}CO \end{cases} \quad (2.6.9)$$

Ozturk and Fruehan,[144] in carefully controlled experiments, demonstrated that the rate of the SiO reactions with metal and slag are controlled by gas phase mass transfer and are rapid.

Coke samples taken from the tuyere zone of the blast furnace usually contain 12 to 16% ash which has a basicity $(CaO + MgO)/SiO_2$ of about 0.6; for a typical blast furnace slag, the basicity is about 1.5. Therefore, the partial pressures of SiO and sulfur-bearing species in the gas near the coke surface are expected to differ from those interacting with the slag. For the purpose of comparison, calculations are made for assumed local equilibrium at 1500°C for the systems gas-slag and gas-coke ash; the results are given in Table 2.12. The amount of ash in the tuyere coke and its composition suggests that, although most of the silica and sulfur in the coke ash are removed from the furnace by the slag and liquid iron, there is significant recycle of silicon and sulfur in the bosh region by the vaporization and condensation of these species on the coke particles.

Table 2.12 Calculated Equilibrium Vapor Pressures in bar for the Systems Gas–Slag and Gas–Coke Ash at 1500°C and 1 bar Pressure of CO.

Vapor	$B = 1.5$, $a_{SiO_2} = 0.05$ 2%(S), $C_S = 3 \times 10^{-4}$, (wt%) *	$B = 0.6$, $a_{SiO_2} = 0.3$ 5%(S), $C_S = 10^{-5}$, (wt%)*
SiO	2.6×10^{-4}	1.5×10^{-3}
SiS	3.9×10^{-5}	1.7×10^{-2}
S_2	8.6×10^{-9}	4.9×10^{-5}
CS	5.5×10^{-5}	4.3×10^{-3}
COS	5.5×10^{-6}	1.2×10^{-4}

* C_S(wt%) is the sulfide capacity of the slag.

2.6.2.2 Alkali Recycle and Removal by Slag

The alkalies constitute another dominant vapor species in the bosh and stack regions of the blast furnace. Many studies have been made of the alkali recycle in the blast furnace via the process of vaporization and condensation; the subject is well covered in the references cited.[145–147] General consensus evolved from elementary thermodynamic calculations is that the alkali silicates in the ore and coke ash decompose at elevated temperatures in the lower part of the bosh and the combustion zone. In part the alkali vapors carried away with the ascending gas react with the slag and can therefore be removed from the furnace; part are converted to alkali cyanides and carbonates and deposited on the burden and the refractory lining of the furnace stack. Some new thoughts are presented in the following discussion on the sequence of reactions and vapor species that may govern the alkali recycle in the blast furnace.

As an example, let us calculate the equilibrium vapor pressure of potassium for the reduction reaction:

$$(K_2O) + C = 2K(g) + CO(g) \qquad (2.6.10)$$

The computed equilibrium values in Table 2.13 are for a blast furnace slag containing 0.5% K_2O and having a basicity of B = 1.5: K is the equilibrium constant for reaction 2.6.10 from the thermochemical data and a_{K_2O} is the activity of K_2O in the slag, relative to the hypothetical solid K_2O, from the experimental data of Steiler.[148]

If the potassium input with the blast furnace burden is 4 kg K/t–HM and 80% of it is removed by the slag, about 0.8 kg K/t–HM will be in the vapor phase carried away by the furnace gas. For a typical total gas volume of 2300 nm³/t–HM, the partial pressure of the total potassium vapor species would be 7.2 x 10^{-4} bar (for a total gas pressure of 3.6 bar in the bosh region). Comparison of this total potassium pressure in the gas with those in Table 2.13 for the gas-slag equilibrium indicates that the slag particles will pick up potassium from the gas only at temperatures below 1200°C. The ash in the coke samples taken from the tuyere zone contains 2 to 4% ($Na_2O + K_2O$) which is several fold of that in the original coke ash. Because of the low basicity of coke ash, the amount of alkali therein is expected to be 5 to 10 times greater than that found in the slag. In the upper part of the bosh, the slag and coke will pick up alkalies from the gas. As the temperature of the slag and coke increases during descent in the bosh, they will emit alkali vapors to the gas phase. However, the partial pressure of potassium vapor in the gas would probably be below that given in Table 2.13 for the gas-slag equilibrium. Nevertheless, the numerical examples cited suggest that much of the alkalies accumulated in the bosh and hearth zone of the blast furnace, via recycling, are in the gas phase.

Table 2.13 Equilibrium Vapor Pressure of Potassium for Reaction 2.6.10 with Slag having 0.5% K_2O and B = 1.5.

Temperature (°C)	K, bar	a_{K_2O}	p_K, bar
1200	3.10 x 10^5	2.1 x 10^{-12}	7.1 x 10^{-4}
1300	2.69 x 10^6	5.5 x 10^{-12}	3.4 x 10^{-3}
1400	1.81 x 10^7	1.4 x 10^{-11}	1.4 x 10^{-2}
1500	9.78 x 10^7	3.5 x 10^{-11}	2.6 x 10^{-1}

According to a mass spectrometric study of vaporization of potassium cyanide by Simmons et. al,[149] the equilibrium pressures of vapor species are in the order $p_{(KCN)_2} > p_{(KCN)} > p_{(KCN)_3}$. For atmospheric pressure of nitrogen and at graphite saturation, the following are the equilibrium ratios of the vapor species: at the melting point of KCN (635°C), $(KCN)_2$: KCN : K = 2 : 1 : 0.76 and at the boiling point (1132°C), $(KCN)_2$: KCN : K = 4.6: 1 : 0.19. It is all too clear that $(KCN)_2$ is the dominant alkali-bearing species in the blast furnace. The sodium in the burden undergoes cyclic reactions similarly to potassium; however, the concentration of sodium-bearing species are about one-tenth of the potassium-bearing species.

One method of minimizing the ascent of alkali vapors to the stack, hence reducing the alkali recycle, is by operating the furnace with a slag of low basicity, particularly when the alkali input is high. The slag of low basicity, however, will have an adverse effect on the composition of the hot metal produced; it will result in low manganese, high silicon and high sulfur in the metal. For a given preferred basicity, the slag mass per ton of hot metal may have to be increased when there is an increase in the alkali input to the furnace. An increase in amount of slag will increase the coke rate, hence will increase the alkali input. Obviously, with a burden of high alkali input, a compromise has to be made in adjusting slag basicity and slag mass so that the alkali recycle would be low.

Even with a low alkali input some alkali buildup will occur. To prevent excessive accumulation of alkalies, which lead to scaffolding, gas channeling, furnace upsets and so on, periodic additions of gravel (silica) or olivine (magnesium silicate) are made to the blast furnace charge. The silica reacts with alkali carbonates and cyanides, thus facilitates the discharge of the accumulated alkalies from the furnace by the slag. Practical experience also shows that periodic addition of calcium chloride to the burden also facilitates the removal of alkalies by the slag.

2.6.2.3 Ammonia and Hydrogen Cyanide in Blast Furnace Stack

The contaminants in the wash water for the iron blast-furnace off-gas are in the ranges 10–300 mg NH_3/l, 1–30 mg CN_t (total)/l, and 0.3–15 mg phenols/l. In terms of ppm by volume in the stack gas, the concentrations are usually in the ranges 3000–4000 ppm NH_3 and 150–250 ppm CN_t. As observed by Pociecha and Biczysko,[150] for example, most of the cyanide in the de-dusted furnace off gas is in the form of hydrogen cyanide. They also found that the blast furnace dust contains only a small amount of solid cyanides (presumably alkali cyanides) in comparison with the amount of hydrogen cyanide present in the de-dusted top gas. The same contaminants in the ferromanganese blast furnace off gas are about an order of magnitude greater than those in the iron blast furnace.

The experimental work of Turkdogan and Josephic,[151] under conditions partly simulating the blast furnace stack, substantiated the validity of this mechanism of generation of ammonia and hydrogen cyanide in the stack, by the following reactions.

$$2K(Na)CN + 3H_2O = (K, Na)_2CO_3 + 2NH_3 + C \quad (2.6.11)$$

$$NH_3 + CO = HCN + H_2O \quad (2.6.12)$$

The amounts of ammonia and hydrogen cyanide generated will increase with an increase in the concentration of water vapor in the stack gas. Other reactions will occur with these volatile species generating some amino-nitro phenols, e.g. $NH_2(NO_2)C_6H_3OH$. The equilibrium concentrations of these volatile organic species become greater at lower temperatures.

2.6.3 Slag-Metal Reactions in the Blast Furnace

Slag-metal reactions in the hearth of the blast furnace control the chemistry of the hot metal. In particular, the silicon, sulfur and manganese contents depend upon these reactions. In this section the equilibrium and reaction mechanisms are presented.

2.6.3.1 Slag-Metal Equilibrium for Blast Furnace Reactions

From the thermodynamics of the slag systems and of the various solutes in iron such as sulfur, silicon and manganese, it is possible to predict the equilibrium relationship for blast furnace reactions.

The three-phase reactions to be considered are for graphite-saturated melts at 1 atm pressure of carbon monoxide, which is close to the partial pressure of CO in the blast furnace bosh and hearth zones.

2.6.3.1.1 Silicon Reaction The reaction equilibrium to be considered is

$$(SiO_2) + 2[C] = [Si] + 2CO(g) \quad (2.6.13)$$

$$K_{Si} = \frac{[\%Si] f_{Si}}{(a_{SiO_2})} \left(\frac{p_{CO}}{a_C}\right)^2 \quad (2.6.14)$$

where the SiO_2 and C activities are with respect to pure solid silica and graphite, respectively.

The temperature dependence of the equilibrium constant is from the free energy data

$$\log K_{Si} = -\frac{30,935}{T} + 20.455 \quad (2.6.15)$$

Substituting $a_{SiO_2} = (\%SiO_2)\gamma_{SiO_2}/60 \times 1.65$ and for the activity coefficient $f_{Si} = 15$ in graphite-saturated iron containing Si < 2%, the equilibrium metal/slag silicon distribution ratio is represented by

$$\frac{[\%Si]}{(\%SiO_2)} p_{CO}^2 = 6.73 \times 10^{-4} \, \gamma_{SiO_2} \, K_{Si} \quad (2.6.16)$$

The equilibrium silicon distribution ratios computed previously, using the γ_{SiO_2} values for the blast furnace type slags (CaO, 10% MgO, 10% Al$_2$O$_3$, SiO$_2$), are given in Fig. 2.88. The dotted curves are derived from other data discussed later. For the blast furnace type slags, the basicity is usually defined by the mass ratio (%CaO + %Mg)/%SiO$_2$.

2.6.3.1.2 Manganese Reaction
For the reaction

$$(MnO) + [C] = [Mn] + CO(g) \quad (2.6.17)$$

$$K_{Mn} = \frac{[\%Mn] f_{Mn}}{(a_{MnO})} \frac{p_{CO}}{a_C} \quad (2.6.18)$$

$$\log K_{Mn} = -\frac{15,090}{T} + 10.970 \quad (2.6.19)$$

For the graphite-saturated iron, the activity coefficient $f_{Mn} = 0.8$, and inserting $a_{MnO} = (\%MnO) \gamma_{MnO}/71 \times 1.65$, the equilibrium manganese distribution ratio is represented by

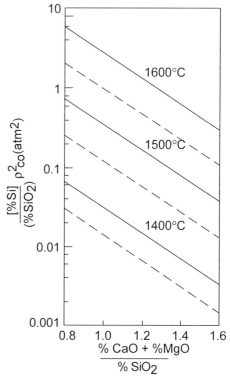

Fig. 2.88 Silicon distribution ratio for graphite–saturated melts for blast furnace type slags containing about 10% MgO and 10% Al$_2$O$_3$. Lines – – – are derived from the data in Fig. 2.90 and equation (2.6.23).

$$\frac{[\%Mn]}{(\%MnO)} p_{CO} = 1.07 \times 10^{-2} \gamma_{MnO} \, K_{Mn} \quad (2.6.20)$$

Abraham et al.[152] measured the MnO activity in CaO–Al$_2$O$_3$–SiO$_2$– MnO melts containing Al$_2$O$_3$ < 20% and MnO < 8%. The values of γ_{MnO} derived from their data are seen in Fig. 2.89 to increase with an increasing slag basicity. The equilibrium manganese distribution ratio for graphite saturated melts thus evaluated are given in Fig. 2.90.

Fig. 2.89 Activity coefficient of MnO, relative to pure solid oxide, in CaO–Al$_2$O$_3$–SiO$_2$–MnO slags, with MnO < 8%, at 1500 to 1650°C, derived from the data of Abraham, Davies and Richardson[152].

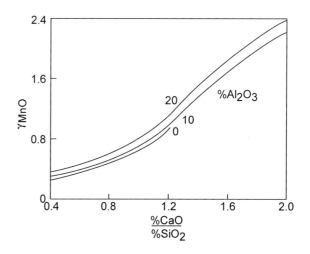

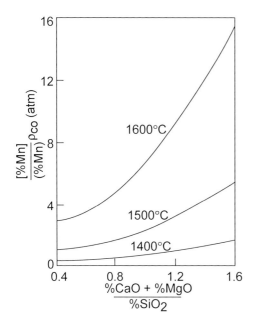

Fig. 2.90 Manganese distribution ratio for graphite-saturated melts derived for blast furnace type slags containing about 10% MgO and 10% Al$_2$O$_3$.

2.6.3.1.3 Si–Mn Coupled Reaction

The silicon-manganese coupled reaction

$$2(MnO) + [Si] = (SiO_2) + 2[Mn] \tag{2.6.21}$$

between graphite-saturated iron and blast furnace type slags has been investigated by several experimental studies.[143,153,154]

$$K = \left(\frac{[\%Mn]}{(\%MnO)}\right)^2 \frac{(\%SiO_2)}{[\%Si]} = 0.17 \frac{K_{Mn}^2}{K_{Si}} \frac{\gamma_{MnO}^2}{\gamma_{SiO_2}} \tag{2.6.22}$$

The combination of equations 2.6.16 and 2.6.20 gives the following equilibrium constant for reaction (2.6.21)

The variation of the equilibrium relation K_{MnSi} with slag basicity is shown in Fig. 2.91. The dotted line for 1400 to 1600°C is calculated from the thermochemical data, i.e. from the computed equilibrium data in Figs. 2.88 and 2.90. The difference of the calculated equilibrium dotted line from the average experimental values by a factor of 2.7 ($\equiv$ ~15 kJ) is due to the accumulated uncertainties in the free energy data on MnO and SiO$_2$ as well as ΔG_s for solutions of Mn and Si in graphite-saturated iron.

The experimental data in Fig. 2.91 for graphite saturated melts at 1400 – 1600°C may be represented by the following equation.

$$\log K_{MnSi} = 2.8\left(\frac{\%CaO + \%MgO}{\%SiO_2}\right) - 1.16 \tag{2.6.23}$$

To be consistent with the experimental data in Fig. 2.91, the dotted curves are derived from the reliable equilibrium data in Fig. 2.90 and the experimental data represented by equation (2.6.23) for the Si–Mn coupled reaction in graphite saturated melts.

In a study of slag-metal equilibria between CaO–MnO–SiO$_2$ slags and graphite-saturated manganese-base alloys, Turkdogan and Hancock[155] found that the ratios Si/SiO$_2$ and Mn/MnO varied with slag basicity in a manner similar to that for iron base alloys. Their results for 1400°C are shown in Fig. 2.92; the dot-dash line is for the iron-base alloys, reproduced from Fig. 2.91.

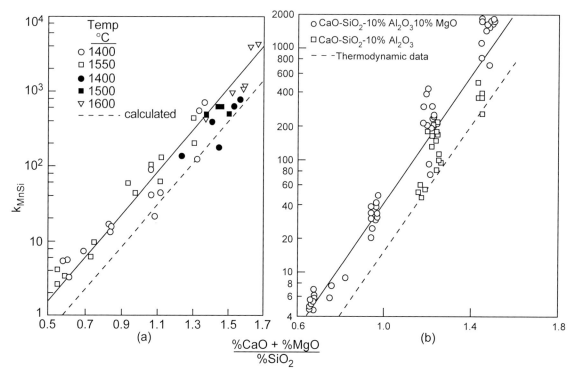

Fig. 2.91 Experimental data showing the equilibrium relation K_{MnSi} with basicity of the blast furnace type slags at graphite saturation: (a) reproduced from Ref. 143 and (b) from Ref. 154; lines---derived from equation 2.6.23 and Fig. 2.90.

2.6.3.1.4 Si–S and Mn–S Coupled Reactions

The silicon-sulfur coupled reaction is represented by

$$[S] + \tfrac{1}{2}[Si] + (CaO) = (CaS) + \tfrac{1}{2}(SiO_2) \tag{2.6.24}$$

for which the isothermal equilibrium constant is

$$K_{SiS} = \frac{(\%S)}{[\%S]} \left\{ \frac{(\%SiO_2)}{[\%Si]} \right\}^{1/2} \frac{1}{(\%CaO)} \tag{2.6.25}$$

From an analysis of the most available experimental data, Turkdogan et al.[143] developed a relation for K_{SiS} shown in Fig. 2.93. For the range 1400 to 1600°C, the temperature dependence of K_{SiS} is represented by the equation

$$\log K_{SiS} = \frac{6327}{T} - 4.43 + 1.4 \left(\frac{\%CaO + \%MgO}{\%SiO_2} \right) \tag{2.6.26}$$

From the experimental data represented by equations 2.6.23 and 2.6.26, the following equation is obtained for the temperature dependence of the equilibrium relation for the manganese-sulfur coupled reaction in graphite-saturated melts.

$$\log K_{MnS} = \log \frac{(\%S)}{[\%S]} \frac{(\%MnO)}{[\%Mn]} \frac{1}{(\%CaO)} = \frac{6327}{T} - 3.85 \tag{2.6.27}$$

The equilibrium sulfur distribution ratio as a function of slag basicity in graphite-saturated melts shown in Fig. 2.94.

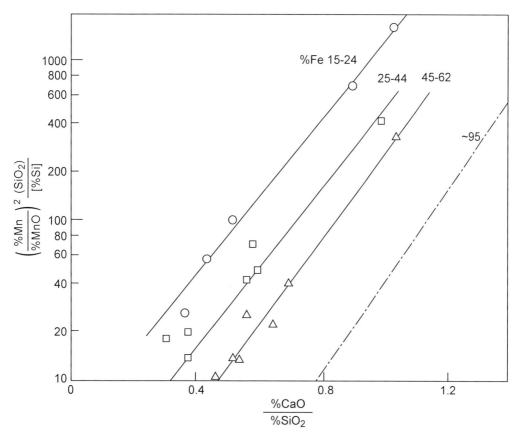

Fig. 2.92 Product of manganese and silicon ratios for graphite–saturated ferromanganese melts, of indicated iron contents, equilibrated with CaO–MnO–SiO$_2$ slags at 1400°C and 1 atm CO, derived from the data of Turkdogan and Hancock.[155]

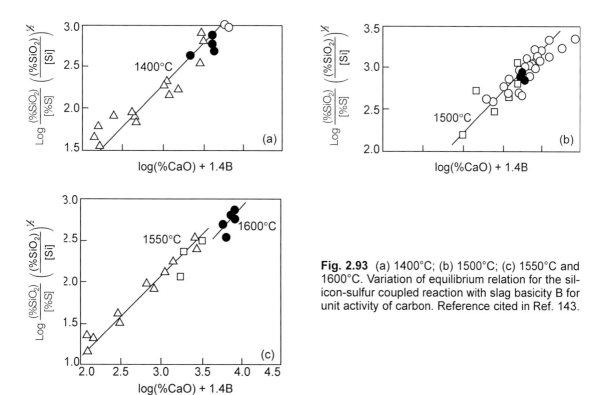

Fig. 2.93 (a) 1400°C; (b) 1500°C; (c) 1550°C and 1600°C. Variation of equilibrium relation for the silicon-sulfur coupled reaction with slag basicity B for unit activity of carbon. Reference cited in Ref. 143.

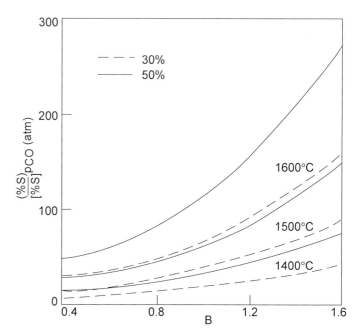

Fig. 2.94 Sulfur distribution ratio for graphite-saturated melts as a function of slag basicity. *From Ref. 143.*

2.6.3.1.5 Si–Ti Coupled Reaction Delve et al.[156] have studied the silicon-titanium coupled reaction in graphite-saturated melts using blast furnace type slags:

$$(TiO_2) + [Si] = (SiO_2) + [Ti] \quad (2.6.28)$$

Their experimental results, for temperatures of 1500 and 1600°C and basicities of 1 to 2, may be summarized by the following equation.

$$K_{SiTi} = \frac{[\%Ti]}{(\%TiO_2)} \frac{(\%SiO_2)}{[\%Si]} \quad (2.6.29)$$

$$\log K_{SiTi} = 0.46 \frac{\%CaO}{\%SiO_2} + 0.39 \quad (2.6.30)$$

2.6.3.2 Mechanisms of Slag-Metal Reactions

2.6.3.2.1 Experimental Work Experiments were made by the authors[143] to simulate the events in the upper part of the blast furnace hearth where metal droplets pass through the slag layer. Solid pieces (4 to 5 mm dia.) of graphite-saturated iron containing Si and S were dropped on the surface of a pool of remelted blast furnace slag (65 g) contained in a graphite tube (20 mm dia.) at 1480°C; an argon atmosphere was maintained over the melt. Twenty pieces of metal (7 g total) were dropped one at a time at 10–20 s intervals, within 7–10 min total time of the experiment. The metal pieces melted rapidly and descended a 10 cm deep column of molten slag as individual metal droplets. After the last piece was dropped the melt was rapidly cooled. The residence time of the droplets in the slag column is estimated to be about 1–2 s, if there is no appreciable buoyancy effect due to gas evolution. The metal droplets collected in a well at the bottom of the slag column; metal and slag were subsequently analyzed for manganese, silicon and sulfur.

With the addition of MnO to the slag, there is silicon transfer from metal to slag. As is seen from the experimental results in Fig. 2.95, the extent of silicon oxidation increases with increasing %MnO in the slag via the Si–Mn coupled reaction

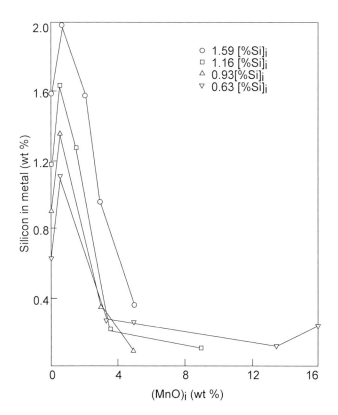

Fig. 2.95 Change in silicon content of metal droplets passing through a 10 cm deep slag column. *From Ref. 143.*

$$2(MnO) + [Si] \rightarrow 2[Mn] + (SiO_2) \qquad (2.6.31)$$

In addition, there is also MnO reduction by carbon in the metal droplets.

$$(MnO) + [C] \rightarrow [Mn] + CO \qquad (2.6.32)$$

With the slags containing MnO > 3%, the metal droplets desiliconized down to 0.1 – 0.3% Si. Within this region, the silicon and manganese distribution ratios are within the ranges $(\%SiO_2)/[\%Si]$ = 120 to 350 and $[\%Mn]/(\%MnO)$ = 0.7 to 1.8. These ratios give K_{MnSi} values below the equilibrium value of 1096 derived from equation 2.6.23 for the slag basicity of 1.5.

The effect of MnO content of the slag on the extent of desulfurization of metal droplets during descent in the slag column is shown in Fig. 2.96. The shaded area within the range 0.010 to 0.035% S corresponds to the state of slag-metal equilibrium represented by equation 2.6.26, for metal droplets containing 0.1 to 0.3% Si and the slag of basicity 1.5 containing 41% CaO, 35% SiO$_2$ and 1.5% S.

The extremely fast rates of reactions observed in the foregoing experiments may be attributed to the presence of a thin gas film (consisting of the vapor species CO, SiO, SiS and Mn) around the metal droplets, which provides the reaction path for the transfer of reactants to and from slag and metal via the gas phase.

2.6.3.3 Blast Furnace Tap Chemistry Data

In a previous study[157] an assessment was made of the state of slag-metal reactions in the blast furnace hearth from an analysis of the daily average compositions of metal and slag samples as normally recorded in the plants. The plant data used were those reported by Okabe et al.[158] from Chiba Works of Kawasaki Steel and the daily average data for a period of a month from U.S. Steel plants.

In the plant data used, the melt temperatures at tap are mostly in the range 1500 ± 25°C. The average slag compositions in mass percent are in the range 38–42% CaO, 8–10% MgO, 34–38% SiO$_2$, 10–12% Al$_2$O$_3$, 0.5–1.0% MnO, 2% S and minor amounts of other oxides. The slag basicity is about

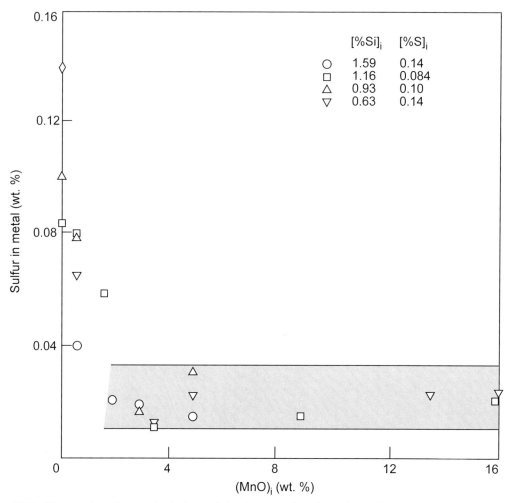

Fig. 2.96 Change in sulfur content of metal droplets passing through a 10 cm deep slag column; shaded region is for slag-metal equilibrium at 1480°C. *From Ref. 143.*

1.4 ± 0.15. The hot metal compositions in the tap stream are in the range: 0.4–0.8% Mn, 0.5–1.5% Si, 0.02–0.05% S, 5% C and other usual impurities.

The silicon, manganese and sulfur distribution ratios from plant data are scattered within the shaded areas in Fig. 2.97. The dotted lines are for the slag-metal equilibria at 1 atm pressure of CO, which is close to the CO partial pressure in the hearth zone of the blast furnace. It should be noted that the equilibrium line for the Si/SiO$_2$ ratio is based on equation 2.6.16 representing the experimental data and the equilibrium relation in Fig. 2.88, i.e. dotted curves. It is only the silicon distribution ratio that is scattered about the gas-slag-metal equilibrium line for the three-phase reaction

$$(SiO_2) + 2[C] = [Si] + 2CO(g) \tag{2.6.33}$$

Values of K_{MnSi} scattered within the shaded area in Fig. 2.98 are below the equilibrium line (dotted). It should be pointed out that in the upper part of the slag layer, the iron oxide content will be higher than manganese oxide. Therefore, the silicon in metal droplets will be oxidized more readily by iron oxide in the upper part of the slag layer, resulting in the observed non-equilibrium state for the Si–Mn coupled reaction in equation 2.6.31.

With iron ore of low basicity in the blast furnace burden, the slag basicity will be low and furthermore, the iron oxide content in the upper part of the slag layer is expected to be higher. It is

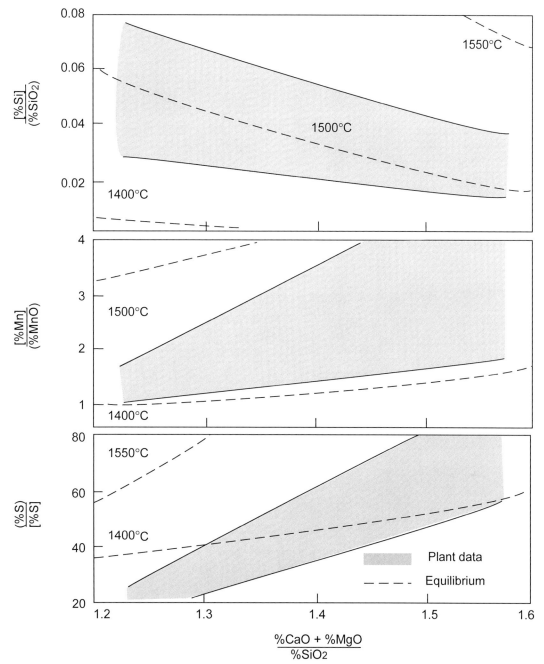

Fig. 2.97 Si, Mn and S distribution ratios are compared with the equilibrium data for graphite-saturated melts at 1 atm CO. *From Ref. 159.*

presumably for this reason that a low slag basicity, the K_{MnSi} values are much lower than the equilibrium values for the Si–Mn couple reaction.

In an earlier study of plant data from a ferromanganese blast furnace in the UK,[160] it was found that there were departures from equilibrium for the Si oxidation by MnO in the slag in a manner similar to that observed for the iron blast furnace.

The plant data in Fig. 2.99 for the sulfur distribution ratio are scattered about the equilibrium line for the Mn–S coupled reaction at 1500°C, represented by equation 2.6.27. However, the ratios

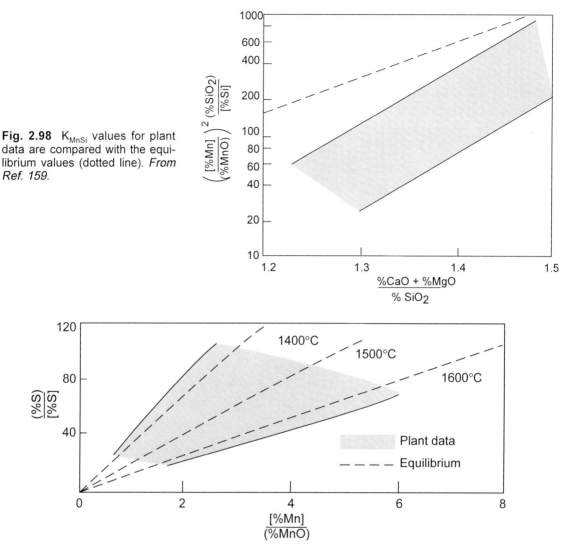

Fig. 2.98 K_{MnSi} values for plant data are compared with the equilibrium values (dotted line). *From Ref. 159.*

Fig. 2.99 The blast furnace data are compared with the slag–metal equilibrium values for the Mn–S coupled reaction. *From Ref. 159.*

(%S)/[%S] in the plant data are lower than the equilibrium values for the Si–S coupled reaction represented by equation 2.6.26. Nevertheless, despite departures from equilibrium, the silicon and sulfur contents of the blast furnace iron change in a systematic manner, i.e. high silicon/low sulfur and low silicon/high sulfur.

It should be borne in mind that because of fluctuations in the operating conditions in the blast furnace, there will be variations in temperature and the extent of reaction of metal droplets during descent through the slag layer. Consequently, the metal droplets of varying composition collecting in the stagnant hearth will bring about the composition and temperature stratification that is common to all blast furnaces.

2.7 Fundamentals of Steelmaking Reactions

The oxygen steelmaking and electric-arc furnace steelmaking processes are described in detail in Chapters 9 and 10 respectively. The practical and technical aspects of the steel refining in the ladle furnace and vacuum degassing are described in Chapter 11. In this section of Chapter 2, the discussion of steelmaking reactions will be confined to an assessment of the reaction mechanisms and the state of slag-metal reactions at the time of furnace tapping.

2.7.1 Slag-Metal Equilibrium in Steelmaking

The reaction equilibria in the liquid steel-slag systems have been extensively studied, both experimentally and theoretically by applying the principles of thermodynamics and physical chemistry. In a recent reassessment of the available experimental data on steel-slag reactions[103], it became evident that the equilibrium constants of slag-metal reactions vary with the slag composition in different ways, depending on the type of reaction. For some reactions the slag basicity is the key parameter to be considered; for another reaction the key parameter could be the mass concentration of either the acidic or basic oxide components of the slag.

2.7.1.1 Oxidation of Iron

In steelmaking slags, the total number of g-mols of oxides per 100 g of slag is within the range 1.65 ± 0.05. Therefore, the analysis of the slag-metal equilibrium data, in terms of the activity and mol fraction of iron oxide, can be transposed to a simple relation between the mass ratio [ppm O]/(%FeO) and the sum of the acidic oxides $\%SiO_2 + 0.84 \times \%P_2O_5$ as depicted in Fig. 2.100(a). The experimental data used in the diagram are those cited in Ref. 103. There is of course a corollary relation between the ratio [ppm O]/(%FeO) and the slag basicity as shown in Fig. 2.100(b).

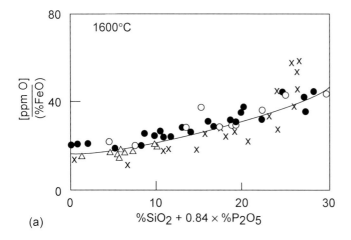

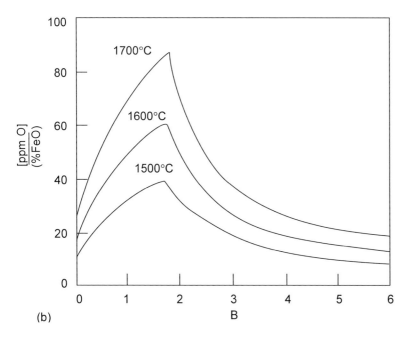

Fig. 2.100 Equilibrium ratio [ppm O]/(%FeO) related to (a) SiO_2 and P_2O_5 contents and (b) slag basicity; experimental data are those cited in Ref. 103.

2.7.1.2 Oxidation of Manganese

For the FeO and MnO exchange reaction involving the oxidation of manganese in steel, formulated below,

$$(FeO) + [Mn] = (MnO) + [Fe] \qquad (2.7.1)$$

the equilibrium relation may be described in terms of the mass concentrations of oxides

$$K'_{FeMn} = \frac{(\%MnO)}{(\%FeO)[\%Mn]} \qquad (2.7.2)$$

where the equilibrium relation K'_{FeMn} depends on temperature and slag composition.

The values of K'_{FeMn} derived from the equilibrium constant for reaction 2.7.1, given in Ref. 27 and the activity coefficient ratios $\gamma_{FeO}/\gamma_{MnO}$ in Fig. 2.67, are plotted in Fig. 2.101 against the slag basicity. In BOF, OBM(Q-BOP) and EAF steelmaking, the slag basicities are usually in the range 2.5 to 4.0 and the melt temperature in the vessel at the time of furnace tapping in most practices is between 1590 and 1630°C for which the equilibrium K'_{FeMn} is 1.9 ± 0.3. The plant analytical data for tap samples give K'_{FeMn} values that are scattered about the indicated slag-metal equilibrium values.

Morales and Fruehan[161] have recently determined experimentally the equilibrium constant K'_{FeMn} for reaction 2.7.1 using MgO-saturated calcium silicate melts. Their values of K'_{FeMn} are plotted in Fig. 2.102 together with some data from the studies of Chipman et al.[162] and Suito et al.[163]. The broken line curve is reproduced from Fig. 2.101. Resolution of observed differences in the values of K'_{FeMn} awaits future studies.

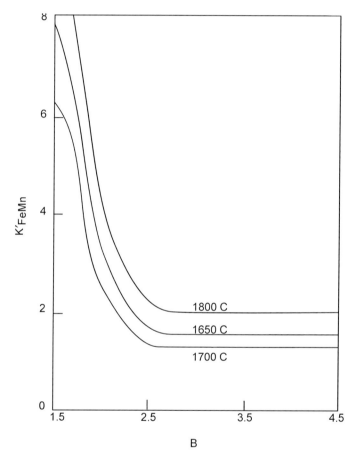

Fig. 2.101 Equilibrium relation in equation 2.7.2 related to slag basicity. *From Ref. 103.*

2.7.1.3 Oxidation of Carbon

With respect to the slag-metal reaction, the equilibrium relation for carbon oxidation would be

$$(FeO) + [C] = CO + [Fe] \qquad (2.7.3)$$

$$K_{FC} = \frac{p_{CO}\,(atm)}{[\%C]\,a_{FeO}} \qquad (2.7.4)$$

$$\log K_{FC} = -\frac{5730}{T} + 5.096 \qquad (2.7.5)$$

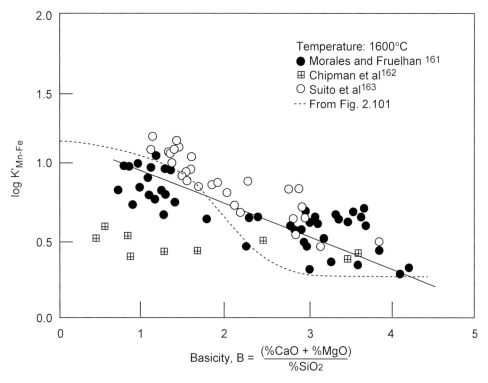

Fig. 2.102 Experimental values of K'_{FeMn} measured recently by Morales and Fruehan[161].

For 1600°C, $\gamma_{FeO} = 1.3$ at slag basicity of B = 3.2 and $p_{CO} = 1.5$ atm (average CO pressure in the vessel), we obtain the following equilibrium relation between the carbon content of steel and the iron oxide content of slag.

$$K_{FC} = 108.8$$

$$a_{FeO} = 1.3\, N_{FeO} \approx \frac{1.3}{72 \times 1.65}(\%FeO) = 0.011 \times (\%FeO) \quad (2.7.6)$$

$$(\%FeO)[\%CO] = 1.25$$

2.7.1.4 Oxidation of Chromium

There are two valencies of chromium (Cr^{2+} and Cr^{3+}) dissolved in the slag. The ratio Cr^{2+}/Cr^{3+} increases with an increasing temperature, decreasing oxygen potential and decreasing slag basicity. Under steelmaking conditions, i.e. in the basic slags and at high oxygen potentials, the trivalent chromium predominates in the slag. The equilibrium distribution of chromium between slag and metal for basic steelmaking slags, determined by various investigators, is shown in Fig. 2.103; slope of the line represents an average of these data.

$$\frac{(\%Cr)}{[\%Cr]} = (0.3 \pm 0.1) \times (\%FeO) \quad (2.7.7)$$

2.7.1.5 Oxidation of Phosphorus

It was in the late 1960s that the correct formulation of the phosphorus reaction was at last realized, thus

$$[P] + \tfrac{5}{2}[O] + \tfrac{3}{2}(O^{2-}) = (PO_4^{3-}) \quad (2.7.8)$$

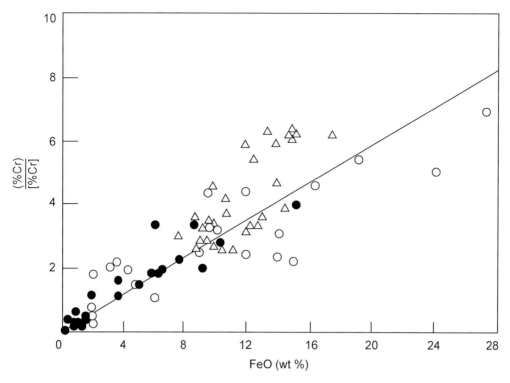

Fig. 2.103 Variation of chromium distribution ratio with the iron oxide content of slag, in the (Δ) open hearth[164] and (o) electric arc furnace[165] at tap is compared with the results of laboratory experiments[166](•).

At low concentrations of [P] and [O], as in most of the experimental melts, their activity coefficients are close to unity, therefore mass concentrations can be used in formulating the equilibrium relation K_{PO} for the above reaction.

$$K_{PO} = \frac{(\%P)}{[\%P]}[\%O]^{-5/2} \qquad (2.7.9)$$

The equilibrium relation K_{PO}, known as the phosphate capacity of the slag, depends on temperature and slag composition.

From a reassessment of all the available experimental data, discussed in detail in Ref. 103, it was concluded that CaO and MgO components of the slag, had the strongest effect on the phosphate capacity of the slag. Over a wide range of slag composition and for temperatures of 1550 to 1700°C, the steel-slag equilibrium with respect to the phosphorus reaction may be represented by the equation

$$\log K_{PO} = \frac{21,740}{T} - 9.87 + 0.071 \times BO \qquad (2.7.10)$$

where BO = %CaO + 0.3 (%MgO).

2.7.1.6 Reduction of Sulfur

The sulfur transfer from metal to slag is a reduction process as represented by this equation

$$[S] + (O^{2-}) = (S^{2-}) + [O] \qquad (2.7.11)$$

for which the state of slag-metal equilibrium is represented by

$$K_{SO} = \frac{(\%S)}{[\%S]}[\%O] \qquad (2.7.12)$$

As is seen from the plots in Fig. 2.104, the sulfide capacities of slags, K_{SO}, measured in three independent studies are in general accord. The effect of temperature on K_{SO} is masked by the scatter in the data. The concentration of acidic oxides, e.g. $\%SiO_2 + 0.84 \times \%P_2O_5$, rather than the slag basicity seems to be better representation of the dependence of K_{SO} on the slag composition.

2.7.2 State of Reactions in Steelmaking

There are numerous versions of oxygen steelmaking such as top blowing (BOF, BOP, LD, etc.), bottom blowing (OBM, Q-BOP, etc.) and combined blowing (K-OBM, LBE, etc.); these are described in detail in Chapter 10. The state of the refining reactions depend to some degree on the process. Bottom blowing in general provides better slag-metal mixing and reactions are closer to equilibrium. In the following section the state of reactions for top blowing (BOF) and bottom blowing (OBM) is given. The state of reactions for combined or mixed blowing processes would be between these two limiting cases. The analytical plant data used in this study were on samples taken from the vessel at first turndown from the BOP and Q-BOP shops of U.S. Steel. These plant data were acquired through the kind collaboration of the USS research personnel at the Technical Center.

2.7.2.1 Decarburization and FeO Formation

The most important reaction in steelmaking is decarburization. It not only determines the process time but also the FeO content of the slag, affecting yield and refining. When oxygen is injected into an oxygen steelmaking furnace a tremendous quantity of gas is evolved, forming a gas-metal-slag emulsion which is three to four times greater in volume than the non-emulsified slag and metal. The chemical reactions take place between the metal droplets, the slag and gas in the emulsion. These reactions have been observed in the laboratory using x-ray techniques indicating in many cases, the gas phase (primarily CO) separates the slag and metal, and that gaseous intermediates play a role in decarburization.

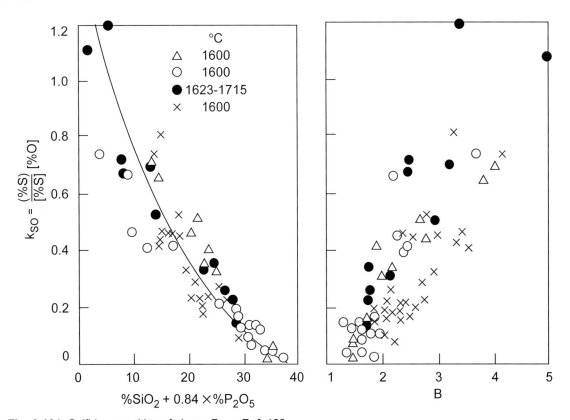

Fig. 2.104 Sulfide capacities of slags. *From Ref. 103.*

When oxygen first contacts a liquid iron-carbon alloy it initially reacts with iron according to reaction 2.7.13, even though thermodynamically it favors its reaction with carbon. This is due to the relative abundance of iron in comparison to carbon. Carbon in the liquid metal then diffuses to the interface reducing the FeO by reaction 2.7.14. The net reaction is the oxidation of carbon, reaction 2.7.15.

$$Fe + \tfrac{1}{2}O_2 = FeO \tag{2.7.13}$$

$$FeO + \underline{C} = CO + Fe \tag{2.7.14}$$

$$\underline{C} + \tfrac{1}{2}O_2 = CO \tag{2.7.15}$$

However, carbon is only oxidized as fast as it can be transferred to the surface.

At high carbon contents the rate of mass transfer is high such that most of the FeO formed is reduced and the rate of decarburization is controlled by the rate of oxygen supply:

$$\frac{d\%C}{dt} = -\frac{\dot{N}_{O_2} M_C 100}{W}(f+1) \tag{2.7.16}$$

where

$\dot{N}_{O_2}$ = the flow rate of oxygen in moles,
M_C = the molecular weight of carbon (12),
W = the weight of steel,
f = the fraction of the product gas which is CO; the remainder is CO_2 and f is close to unity (0.8 to 1).

Below a critical carbon content the rate of mass transfer is insufficient to react with all the injected oxygen. In this case the rate of decarburization is given by[168]:

$$\frac{d\%C}{dt} = -\frac{\rho}{W}(\%C - \%C_e)\sum_i m_i A_i \tag{2.7.17}$$

where

ρ = density or steel,
$\%C_e$ = the equilibrium carbon with the slag for reaction 2.7.14 and is close to zero,
m_i = the mass transfer coefficient for the specific reaction site,
A_i = the metal-FeO surface area for the specific reaction site.

For top blown processes the reaction takes place between the metal droplets ejected into the emulsion and the FeO in the slag. For bottom blowing there is less of an emulsion and the reaction takes place at the interface of the metal bath and the rising bubbles which have FeO associated with them. The critical carbon is when the rates given by 2.7.16 and 2.7.17 are equal and is typically about 0.3% C.

The actual values of m_i and A_i are not known, consequently an overall decarburization constant (k_C) can be defined and the rate below the critical carbon content is given by 2.7.19 and 2.7.20:

$$k_C = \frac{\rho}{W}\sum_i m_i A_i \tag{2.7.18}$$

$$\frac{d\%C}{dt} = -k_C(\%C - \%C_e) \tag{2.7.19}$$

or,

$$\ln\frac{(\%C - \%C_e)}{(\%C_C - \%C_e)} = -k_C(t - t_C) \tag{2.7.20}$$

where t_c is the time at which the critical carbon content is obtained. The equilibrium carbon content ($\%C_e$) is close to zero. However in actual steelmaking there is a practical limit of about 0.01 to 0.03 for $\%C_e$.

The value of k_C increases with the blowing rate since the amount of ejected droplets and bubbles increases. Also, k_c decreases with the amount of steel, (W), as indicated by 2.7.19. In actual processes, the blowing rate is proportional to the weight of steel and, therefore, k_C has similar values in most oxygen steelmaking operations. For top blowing, k_C is about 0.015 s^{-1} and for bottom blowing 0.017 s^{-1}. The values of the critical carbon content are also similar ranging from 0.2 to 0.4% C. The rate of decarburization for a typical steelmaking process is shown in Fig. 2.105. At the initial stage when silicon is being oxidized, which will be discussed later, the rate of decarburization is low. The value of k_C is higher for bottowm blowing because mixing is more intensive and the reaction occurs at the interface of the rising bubbles as well as the slag-metal emulsion.

Below the critical carbon content, when the rate of mass transfer of carbon is insufficient to reduce all of the FeO formed and, therefore, the FeO content of the slag increases rapidly. In actual processes, initially, some FeO forms because it has a low activity in the slag resulting in about 5-10 % FeO in the slag. The FeO remains constant until the critical carbon content and then the FeO increases rapidly. The amount of FeO can be computed from a mass balance for oxygen. Specifically the oxgyen not used for carbon, silicon or manganese oxidizes iron to FeO. The moles of FeO in the slag at anytime (N_{FeO}) is given by:

$$N_{FeO} = \int_0^t 2\left[\dot{N}_{O_2} - \frac{d\%C}{dt}\frac{W(1+f)}{M_C\,100}\right]dt - \left(N_{O_2}^{Si} + N_{O_2}^{Mn}\right) \qquad (2.7.21)$$

where $N_{O_2}^{Si}$ and $N_{O_2}^{Mn}$ are the moles of oxygen consumed in oxidizing Si and Mn. Since the rate of decarburizaiton is slightly higher for bottom and mixed blowing these processes have lower FeO contents.

At all levels of turndown carbon, the iron oxide content of BOF slag is about twice that of OBM (Q-BOP) slag (Fig. 2.106). The dotted line depicts the slag-metal equilibrium value of $(\%FeO)[\%C] \approx 1.25$.

As depicted in Fig. 2.107, the square root correlation also applies to the product $(\%FeO)[\%C]$ for low carbon contents, with marked departure form this empirical correlation at higher carbon contents. The slopes of the lines for low carbon contents are given below.

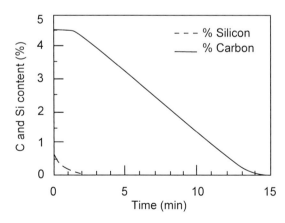

Fig. 2.105 Computed carbon and silicon contents of steel in oxygen steelmaking processes.

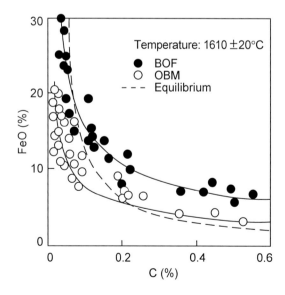

Fig. 2.106 Iron oxide-carbon relations in BOF and OBM (Q-BOP) are compared with the average equilibrium relation (-------). *From Ref. 103.*

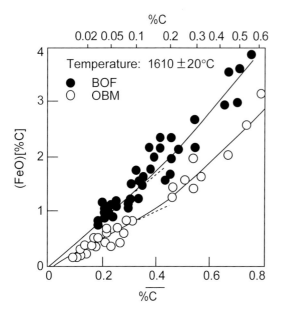

Fig. 2.107 Variation of product (FeO)[%C] with carbon content of steel at first turndown. *From Ref. 103.*

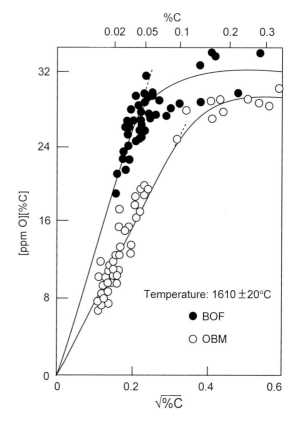

Fig. 2.108 Variation of product [ppm O][%C] with carbon content of steel at first turndown. *From Ref. 103.*

BOF with C < 0.10%,
$$(\%FeO)\sqrt{\%C} = 4.2 \pm 0.3 \quad (2.7.22)$$

OBM (Q - BOP) with C < 0.10%,
$$(\%FeO)\sqrt{\%C} = 2.6 \pm 0.3 \quad (2.7.23)$$

2.7.2.2 Oxygen-Carbon Relation

Noting from the plot in Fig. 2.108, there is an interesting correlation between the product [ppm O][%C] and the carbon content of the steel. The square root correlation holds up to about 0.05% C in BOF and up to 0.08% C in OBM(Q-BOP) steelmaking. At low carbon contents, the oxygen content of steel in the BOF practice is higher than that in OBM steelmaking. For carbon contents above 0.15% the product [ppm O] [%C] is essentially constant at about 30 ± 2, which is the equilibrium value for an average gas (CO) bubble pressure of about 1.5 atmosphere in the steel bath.

At low carbon levels in the melt near the end of the blow, much of the oxygen is consumed by the oxidation of iron, manganese and phosphorus, resulting in a lower volume of CO generation. With the bottom injection of argon in the BOF combined-blowing practice, and the presence of hydrogen in the gas bubbles in OBM, the partial pressure of CO in the gas bubbles will be lowered in both processes when the rate of CO generation decreases. A decrease in the CO partial pressure at low carbon contents will be greater in OBM than in BOF practices, because the hydrogen content of gas bubbles in OBM is greater than the argon content of gas bubbles in BOF. It is presumably for this reason that the concentration product [O][C] in OBM steelmaking is lower than in the BOF combined-blowing practice, particularly at low carbon levels.

The non-equilibrium states of the carbon-oxygen reaction at low carbon contents in BOF and OBM(Q-BOP) are represented by the folowing empirical relations.

BOF with C < 0.05%,

$$[\text{ppm O}]\sqrt{\%C} = 135 \pm 5 \qquad (2.7.24)$$

OBM (Q - BOP) with C < 0.08%,

$$[\text{ppm O}]\sqrt{\%C} = 80 \pm 5 \qquad (2.7.25)$$

2.7.2.3 Desiliconization

Silicon is oxidized out of hot metal early in the process. The oxidation reaction supplies heat and the SiO_2 reacts with the CaO to form the slag.

$$\underline{Si} + \tfrac{1}{2}O_2 = (SiO_2) \qquad (2.7.26)$$

The thermodynamics of the reaction indicate virtually all of the Si is oxidized. The rate is controlled by liquid phase mass transfer, represented by the following equation:

$$\frac{d\%Si}{dt} = -\frac{A\rho m_{Si}}{W}[\%Si - \%Si_e] \qquad (2.7.27)$$

where

m_{Si} = mass transfer coefficient for silicon,
$\%Si_e$ = the silicon content in the metal in equilibrium with the slag and is close to zero.

As with decarburization, an overall rate parameter for Si can be defined, k_{Si}, similar to k_C. Within our ability to estimate k_{Si} it is about equal to k_C since A, W, and are the same and mass transfer coefficients vary only as the diffusivity to the one half power. Therefore 2.7.27 simplifies to:

$$\ln\frac{\%Si}{\%Si°} = -k_{Si}\,t \qquad (2.7.28)$$

where $\%Si°$ is the initial silicon and k_{Si} is the overall constant for Si and is approximately equal to k_C. The rate for Si oxidadtion is shown in Fig. 2.105.

2.7.2.4 Manganese Oxide–Carbon Relation

At low carbon contents, the ratio [%Mn]/(%MnO) is also found to be proportional to $\sqrt{\%C}$, represented as given below.

BOF with C < 0.10%,

$$\frac{[\%Mn]}{(\%MnO)}\frac{1}{\sqrt{\%C}} = 0.1 \pm 0.02 \qquad (2.7.29)$$

OBM (Q - BOP) with C < 0.10%,

$$\frac{[\%Mn]}{(\%MnO)}\frac{1}{\sqrt{\%C}} = 0.2 \pm 0.02 \qquad (2.7.30)$$

2.7.2.5 FeO–MnO–Mn–O Relations

From the foregoing empirical correlations for the non-equilibrium states of reactions involving the carbon content of steel, the relations obtained for the reaction of oxygen with iron and manganese are compared in the table below with the equilibrium values for temperatures of $1610 \pm 20°C$ and slag basicities of $B = 3.2 \pm 0.6$.

	BOF C < 0.05%	OBM(Q-BOP) C < 0.08%	Values for slag-metal equilibrium
$\dfrac{[\text{ppm O}]}{(\%\text{FeO})}$	32 ± 4	32 ± 5	26 ± 9
$\dfrac{[\%\text{Mn}][\text{ppm O}]}{(\%\text{MnO})}$	13.6 ± 3.2	16.1 ± 2.6	18 ± 6
$\dfrac{(\text{MnO})}{(\%\text{FeO})[\%\text{Mn}]}$	2.6 ± 0.5	2.2 ± 0.2	1.9 ± 0.2

It is seen that the concentration ratios of the reactants in low carbon steel describing the states of oxidation of iron and manganese, are scattered about the values for the slag-metal equilibrium. However, as indicated by the plant data in Fig. 2.108, the oxidation of iron and manganese are in the non-equilibrium states for high carbon contents in the steel at turndown. Although the concentration product [O][C] is close to the equilibrium value for an average CO pressure of about 1.5 atm in the steel bath, the concentrations of iron oxide and manganese oxide in the slag are above the equilibrium values for high carbon contents in the melt.

It is concluded from these observations that:

(i) at low carbon contents the equilibrium state of iron and manganese oxidation controls the concentration of dissolved oxygen

(ii) at high carbon contents it is the CO–C–O equilibrium which controls the oxygen content of the steel.

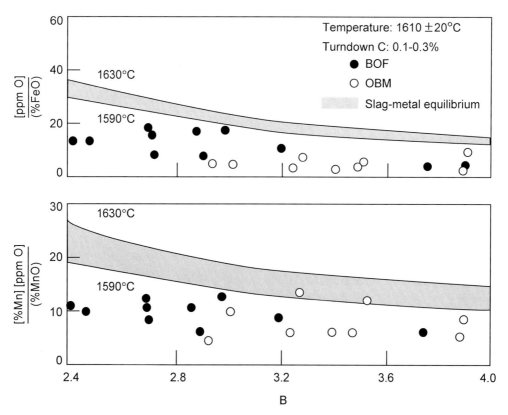

Fig. 2.108 Non-equilibrium states of oxidation of iron and manganese at high carbon contents in steel. *From Ref. 103.*

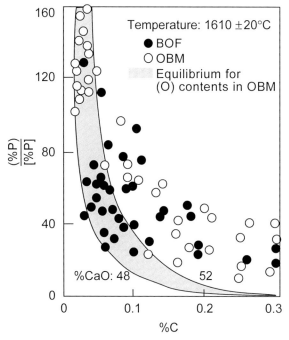

Fig. 2.109 Slag/metal phosphorus distribution ratios at first turndown in BOF and OBM(Q–BOP) practices; slag–metal equilibrium values are within the hatched area for Q–BOP. *From Ref. 103.*

2.7.2.6 State of Phosphorus Reaction

The slag/metal phosophorus distribution ratios at first turndown in BOF and OBM (Q-BOP) steelmaking are plotted in Fig. 2.109 against the carbon content of the steel. The plant data are for the turndown temperatures of $1610 \pm 20°C$ and slags containing $50 \pm 2\%$ CaO and $6 \pm 2\%$ MgO. The shaded area is for the slag-metal equilibrium for the above stated conditions in OBM(Q-BOP), based on the empirical [O][C] relations, i.e. for C < 0.08%, [ppm O] $\sqrt{\%C}$ = 80 and for C > 0.15%, [ppm O][%C] = 30.

Noting that the oxygen contents of the steel at low carbon levels are greater in BOF steelmaking, the equilibrium phosphorus distribution ratios will likewise be greater in BOF than in (OBM) Q-BOP steelmaking. For example, at 0.05% C and about 600 ppm O in BOF at turndown the average equilibrium value of (%P)/[%P] is about 200 at 1610°C, as compared to the average value of 60 in OBM(Q-BOP) at 0.05% C with about 360 ppm O.

Below 0.04% C the phosphorus distribution ratios in OBM(Q-BOP) are in general accord with the values for slag-metal equilibrium. However, at higher carbon contents the ratios (%P)/[%P] are well above the equilibrium values. In the case of BOF steelmaking below 0.1% C at turndown, the ratios (%P)/[%P] are much lower than the equilibrium values. On the other hand, at higher carbon contents the phosphorus distribution ratios are higher than the equilibrium values as in the case of OBM(Q-BOP).

The effect of temperature on the phosphorus distribution ratio in OBM(Q-BOP) is shown in Fig. 2.110 for melts containing 0.014% to 0.022% C with BO = $52 \pm 2\%$ in the slag.

2.7.2.7 State of Sulfur Reaction

A highly reducing condition that is required for extensive desulfurization of steel is opposite to the oxidizing condition necessary for steel making. However, some desulfurization is achieved during oxygen blowing for decarburization and dephosphorization. As seen from typical examples of the BOF and OBM(Q-BOP) plant data in Fig. 2.111, the state of steel desulfurization at turndown, described by the expression [%O](%S)/[%S], is related to the SiO_2 and P_2O_5 contents of the slag. Most of the points for OBM(Q-BOP) are within the hatched area for the slag-metal equilibrium reproduced from Fig. 2.104. However, in the case of BOF steelmaking the slag/metal sulfur distribution ratios at turndown are about one-third or

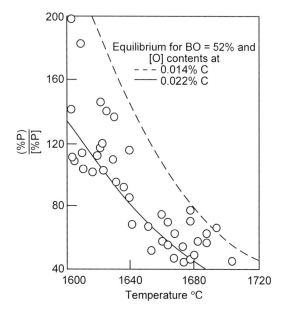

Fig. 2.110 Effect of turndown temperature on the slag/metal phosphorus distribution ratios in OBM (Q-BOP) for turndown carbon contents of 0.014 to 0.022% C; curves are for slag–metal equilibrium. *From Ref. 103.*

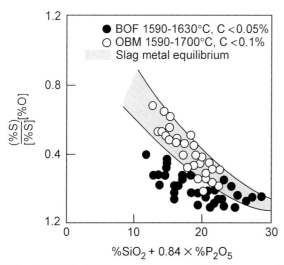

Fig. 2.111 Equilibrium and non–equilibrium states of sulphur reaction in OBM(Q-BOP) and BOF steelmaking. *From Ref. 103.*

one-half of the slag-metal equilibrium values. At higher carbon contents, e.g. C > 0.1%, the sulfur distribution ratios in both processes are below the slag-metal equilibrium values.

2.7.2.8 Hydrogen in BOF and OBM(Q-BOP) Steelmaking

Hydrogen and nitrogen are removed from the steel bath during oxygen blowing by CO carrying off H_2 and N_2.

$$2\underline{H}(\text{in steel}) \rightarrow H_2(g) \quad (2.7.31)$$

$$2\underline{N}(\text{in steel}) \rightarrow N_2(g) \quad (2.7.32)$$

There is however continuous entry of both hydrogen and nitrogen into the bath by various means. In both BOF and OBM(Q-BOP) there is invariably some leakage of water from the water cooling system of the hood into the vessel. In the case of OBM(Q-BOP), the natural gas (CH_4) used as a tuyere coolant is a major source of hydrogen.

The hydrogen content of steel in the tap ladle, measured by a probe called HYDRIS, is less than 5 ppm in BOF steelmaking and 6 to 10 ppm H in OBM(Q-BOP) steelmaking. Because natural gas (CH_4) is used as a tuyere coolant in the OBM, the hydrogen content of steel made in this vessel is always higher than that in the BOF vessel. In both practices the re-blow will always increase the hydrogen content of the steel. A relatively small volume of CO evolved during the re-blow cannot overcome the hydrogen pickup from various sources.

2.7.2.9 The Nitrogen Reaction

As demonstrated, the decarburization reaction is reasonably well understood as well as the slag-metal re-fining reactions which approach equilibrium. The removal of nitrogen is complex and depends on numerous operating variables. With the advent of the production of high purity interstitial free (IF) steels the control of nitrogen has become of great importance. For years there have been "rules of thumb" for achieving low nitrogen. For example, using more hot metal and less scrap in the charge or with combined blowing processes, switching the stirring gas from nitrogen to argon have resulted in lower nitrogen contents.

The sources of nitrogen in oxygen steelmaking include the hot metal, scrap, impurity nitrogen in the oxygen, and nitrogen in the stirring gas. Nitrogen from the atmosphere is not a major factor unless at first turndown a correction or reblow is required, in which case the furnace fills up with air which is entrained into the metal when the oxygen blow restarts, resulting in a significant nitrogen pickup of 5 to 10 ppm.

Goldstein and Fruehan[169] developed a comprehensive model to predict the nitrogen reaction in steelmaking. Nitrogen is removed by diffusing to the CO bubbles in the emulsion, or to the bubbles in the bath in bottom blowing. The nitrogen atoms combine to form N_2, the rate of which is controlled by chemical kinetics and the nitrogen gas is removed with the CO bubbles. Both mass transfer and chemical kinetics contribute to the overall rate of removal.

$$\underline{N}(\text{metal}) = \underline{N}(\text{surface}) \quad (2.7.33)$$

$$2\underline{N}(\text{surface}) = N_2 \quad (2.7.34)$$

The mixed control model presented previously is the basis for the model. There are, however, several complications. For example, all of the rate parameters are functions of temperature. The sulfur content of the metal changes with time and the oxygen content at the surface is considerably different from the bulk concentration. Consequently the temperature and surface contents of oxygen and sulfur must be known as functions of time. Also, scrap is melting during the process changing the composition of the metal. These details are dealt with and discussed in detail elsewhere.[169]

An example of the results of the model calculations are shown in Fig. 2.112 and Fig. 2.113. In Fig. 2.112 the nitrogen contents are shown for two hypothetical cases for a 200 ton oxygen steelmaking converter blowing oxygen at 800 Nm³/min. In one case the charge contains 80% heavy scrap, 16 cm thick containing 50 ppm nitrogen; in the second case there is no scrap but cooling is achieved by the use of a 15% addition of DRI containing 20 ppm nitrogen. For the case of the heavy scrap, it melts late in the process releasing its nitrogen after most of the CO is generated, causing the nitrogen content to increase to 30 ppm at the end of the blow. With no scrap the final nitrogen is significantly lower at 20 ppm. The model also indicates if ore, which has no nitrogen, is used for cooling, levels of 10 ppm nitrogen can be obtained. In comparing the bulk and surface nitrogen contents, it is found that the surface concentration is only slightly less than the bulk, indicating that the nitrogen reaction is primarily controlled by chemical kinetics. In particular, towards the end of the blow the oxygen content at the surface is high, retarding the rate. This model is useful for optimizing the process for nitrogen control. For example, the effect of oxygen purity, the time of switching the stirring gas from N_2 to argon, the metal sulfur content and the amount and size of scrap can be determined. The effect of switching the stirring gas from N_2 to Ar is shown in Fig. 2.113.

2.7.2.10 General Considerations

The analyses of plant data on slag and metal samples taken at first turndown have revealed that there are indeed equilibrium and non-equilibrium states of reactions in oxygen steelmaking. In all the reactions considered, the carbon content of steel is found to have a decisive effect on the state of slag-metal reactions.

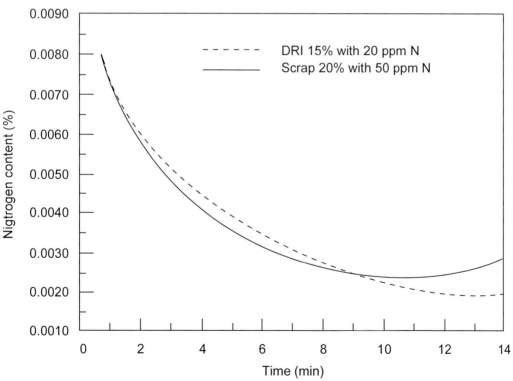

Fig. 2.112 Computed nitrogen content in a 200 ton oxygen steelmaking converter for 20% heavy scrap – 80% light scrap and DRI at a blowing rate of 800 Nm³/min O_2. *From Ref. 167.*

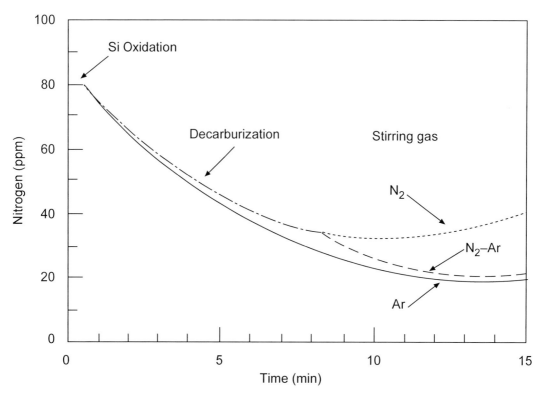

Fig. 2.113 The effect of bottom stirring practice on nitrogen removal in the BOF with N_2 during the entire blow, Ar during the entire blow, and with a switch from N_2 to Ar at 50% through the blow. *From Ref. 167.*

Considering the highly dynamic nature of steelmaking with oxygen blowing and the completion of the process in less than 20 minutes, it is not surprising that the slag-metal reactions are in the non-equilibrium states in heats with high carbon contents at turndown. With regard to low carbon heats all the slag-metal reactions are close to the equilibrium states in bottom blown processes. In the case of BOF steelmaking however, the states of steel dephosphorization and desulfurization are below the expected levels for slag-metal equilibrium. As we all recognize it is of course the bottom injection of lime, together with oxygen, that brings about a closer approach to the slag-metal equilibrium in OBM(Q-BOP) as compared to the BOF practice, particularly in low carbon heats.

2.8 Fundamentals of Reactions in Electric Furnace Steelmaking

The EAF process is described in detail in Chapter 10. The slag-metal equilibrium for EAF refining reactions are similar to those for oxygen steelmaking discussed in detail in Section 2.7. The state of reactions in EAF steelmaking are similar but, in general, are slightly further from equilibrium due to less stirring and slag-metal mixing as compared to oxygen steelmaking. In this section, the slags and the state of the carbon, sulfur and phosphorous reactions are briefly discussed. The nitrogen reaction is considerably different and , along with control of residuals, is more critical and consequently is discussed in more detail.

2.8.1 Slag Chemistry and the Carbon, Manganese, Sulfur and Phosphorus Reactions in the EAF

The state of slag-metal reactions at the time of furnace tapping discussed here is based on about fifty heats acquired from EAF steelmaking plants. The data considered are for the grades of low alloy steels containing 0.05 to 0.20% C and 0.1 to 0.3% Mn plus small amounts of alloying elements.

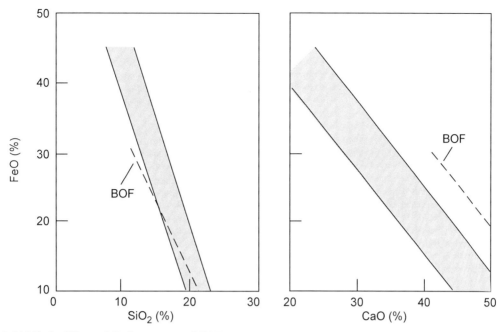

Fig. 2.114 FeO, SiO$_2$ and CaO contents of EAF slags at tap are compared with the BOF slags. *From Ref. 27.*

The iron oxide, silica and calcium oxide contents of slags are within the hatched areas shown in Fig. 2.114. The silica contents are slightly higher than those in oxygen steelmaking slags. On the other hand the CaO contents of EAF slags are about 10% lower than those in the BOF slags. This difference is due to the higher concentrations of Al$_2$O$_3$, Cr$_2$O$_3$ and TiO$_2$ in the EAF slags; 4 to 12% Al$_2$O$_3$, 1 to 4% Cr$_2$O$_3$ and 0.2 to 1.0% TiO$_2$. In these slags the basicity ratio B is about 2.5 at 10% FeO, increasing to about 4 at 40% FeO.

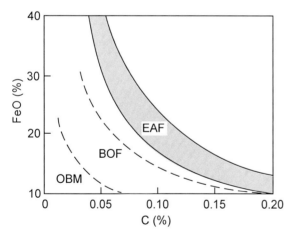

Fig. 2.115 Relation between %FeO in slag and %C in steel at tap in EAF steelmaking is compared with the relations in BOF and OBM(Q–BOP) steelmaking. *From Ref. 27.*

As is seen from the plant data within the hatched area in Fig. 2.115, the iron oxide contents of EAF slags are much higher than those of the BOF slags for the same carbon content at tap. However, the product [%C][ppm O] at tap is about 26 ± 2 which is slightly lower than that for oxygen steelmaking.

For the slag-metal reaction involving [Mn], (FeO) and (MnO), the equilibrium relation K'_{FeMn} = (%MnO)/(FeO)[%Mn] is 1.8 ± 0.2 at 1625 ± 15°C and the basicity B > 2.5. The EAF data are scattered about K'_{FeMn} = 1.8 ± 0.4, which is near enough to the slag-metal equilibrium similar to those in oxygen steelmaking.

For steels containing 0.07 to 0.09% C in the furnace at tap, the estimated dissolved oxygen contents will be about 370 to 290 ppm O.

For slags containing 38–42% CaO and 5–7% MgO, the equilibrium constant K_{PO} for the phosphorus reaction, given in equations 2.7.10, is in the range 2.57×10^4 to 4.94×10^4 at 1625°C. For the tap carbon contents and slag composition ratios will be in the range 4 to 13. The EAF plant data show ratios (%P)/[%P] from 15 to 30. Similar to the behavior in BOF steelmaking, the state of phosphorus oxidation to the slag in EAF steelmaking is greater than would be anticipated from the equilibrium considerations for carbon, hence oxygen contents of the steel at tap.

The state of steel desulfurization, represented by the product {(%S)/[%S]} × (%FeO), decreases with increasing contents of SiO$_2$ and P$_2$O$_5$ in the slag, as shown in Fig. 2.116. The EAF plant data within the shaded area are below the values for the slag-metal equilibrium. This non-equilibrium state of the sulfur reaction in EAF steelmaking is similar to that observed in BOF steelmaking.

General indications are that the states of slag-metal reactions at tap in EAF steelmaking are similar to those noted in oxygen steelmaking at carbon contents above 0.05% C. That is, the slag/metal distribution ratio of manganese and chromium are scattered about the equilibrium values; the phosphorus and sulfur reactions being in non-equilibrium states. Also, the iron oxide content of the slag is well above the slag-metal equilibrium values for carbon contents of steel at tap. On the other hand, the product [%C][ppm O] = 26 ± 2 approximately corresponds to the C–O equilibrium value for gas bubble pressures of 1.3 ± 0.2 atm in the EAF steel bath.

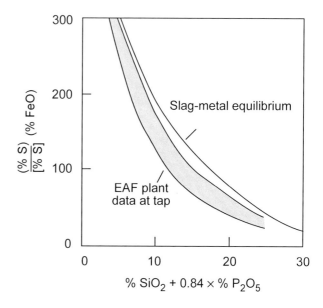

Fig. 2.116 Slag/metal sulfur distribution ratios at tap in EAF steelmaking are compared with the slag-metal equilibrium values. *From Ref. 27.*

2.8.2 Control of Residuals in EAF Steelmaking

The growth of EAF steelmaking in the 1990's was in the production of higher grades of steel, including flat rolled and special bar quality steels. These steels require lower residuals (Cu, Ni, Sn, Mo, etc.) than merchant long products normally produced from scrap in an EAF. It is possible to use high quality scrap such as prompt industrial scrap, but this is expensive and not always available. The solution to the problem is the use of direct reduced iron such as DRI, HBI, iron carbide or other scrap substitutes such as pig iron.

The control of residuals in EAF steelmaking is based on a simple mass balance. Residual elements such as copper will report almost entirely to the steel. The control of residuals can be demonstrated in a very simple example. Consider the copper content of steel produced from two grades of scrap, Number 1 bundles and shredded scrap, and DRI/HBI. It is critical to know the yield of each charge material and its copper content, which is not always easy to assess accurately. There are also secondary effects such as increased slag levels. Nevertheless, this simple example demonstrates the general concepts.

Material	Yield (%)	Charge (kg)	Fe Yield (kg)	Cu Content (wt. %)	Cu (kg)
#1 Bundles	95	363	345	0.05	0.17
Shredded	95	500	475	0.25	1.19
DRI/HBI	90	200	180	0.0	0
Totals		1063	1000		1.36

Cu content in the steel will be 0.136%

2.8.3 Nitrogen Control in EAF Steelmaking

The nitrogen content of steels produced in the EAF is generally higher than in oxygen steelmaking primarily because there is considerably less CO evolution, which removes nitrogen from steel. The usual methods of controlling nitrogen are by carbon oxidation and the use of direct reduced iron or pig iron. The control of nitrogen in the EAF is discussed in recent publications by Goldstein and Fruehan.[170,171] They developed a model to predict the rate of nitrogen removal similar to the one presented earlier for oxygen steelmaking. The removal of nitrogen is shown for a 100 ton EAF as a function of oxygen usage in Fig. 2.117. Nitrogen removal decreases once the carbon content falls below approximately 0.3%C, as most of the oxygen is then reacting with Fe and is therefore not producing CO. Starting at a higher initial carbon allows for more CO evolution and reduces the activity of oxygen, which retards the rate of the nitrogen reaction.

The other method of reducing nitrogen is the use of DRI or HBI. These materials reduce nitrogen primarily through dilution. There was a belief that the CO evolved from these products also removed nitrogen. However it has been demonstrated that the CO evolved from the reaction of C and FeO in the HBI/DRI is evolved at lower temperatures (1000°C) and is released while heating or in the slag phase. Since it does not pass through the metal it does not remove nitrogen. The CO does dilute any N_2 in the furnace atmosphere.

The nitrogen removal using 25% and 50% DRI/HBI is shown in Fig. 2.118. The primary effect is simple dilution.

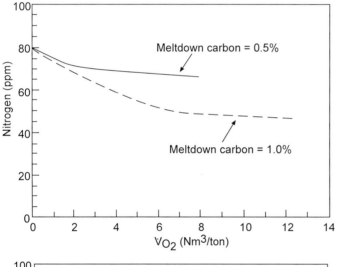

Fig. 2.117 Change in bulk nitrogen concentrations during EAF steelmaking. *From Ref. 171.*

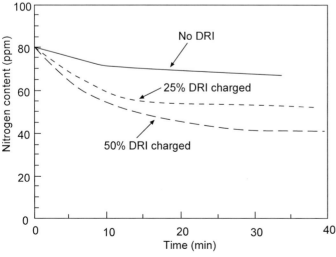

Fig. 2.118 Effect of DRI on nitrogen content during EAF steelmaking. From *From Ref. 171.*

2.9 Fundamentals of Stainless Steel Production

There are numerous processes to produce stainless steel including the AOD (argon oxygen decarburization), VOD (vacuum oxygen decarburization), and other variations of these processes. All of the processes are based on the reduction of the CO pressure to promote the oxidation of carbon in preference to chromium. The processes and reactions are described in detail in Chapter 12. This section will be limited to the reaction mechanisms and fundamentals of decarburization, nitrogen control and slag reduction.

2.9.1 Decarburization of Stainless Steel

Stainless steels can not be easily produced in an EAF or oxygen steelmaking converter as under normal conditions Cr will be readily oxidized in preference to C at low carbon contents. This leads to excessive Cr yield loss and necessitates the use of high cost low carbon ferrochrome rather than lower cost high carbon ferrochrome.

The critical carbon content at which Cr is oxidized rather than carbon can be computed based on the following reaction:

$$Cr_2O_3 + 3\underline{C} = 3CO + 2\underline{Cr} \quad (2.9.1)$$

The equilibrium constant is given by:

$$K = \frac{p_{CO}^3 \, a_{Cr}^2}{a_{Cr_2O_3} \, f_C^3 \, [\%C]^3} \quad (2.9.2)$$

where f_C is the activity coefficient of carbon and a_i is the activity of species i. The reaction could be written in terms of Cr_3O_4 and the results would be similar. In Fig. 2.119 the equilibrium for reaction 2.9.1 is given as a function of p_{CO} for an 18% Cr steel.

The critical carbon content is defined as the carbon content below which Cr is oxidized. The critical carbon increases with chromium content or activity ($a_{Cr}^{2/3}$) and decreasing temperature.

However decarburization in stainless steelmaking is not controlled by equilibrium but rather reaction kinetics. It was found that for the AOD process to be effective the gas had to be injected deep in the steel bath. Fruehan[172] demonstrated in laboratory experiments that when oxygen is injected into an Fe–Cr–C bath the oxygen initially oxidizes Cr. This lead to the following reaction mechanism and model for decarburization.

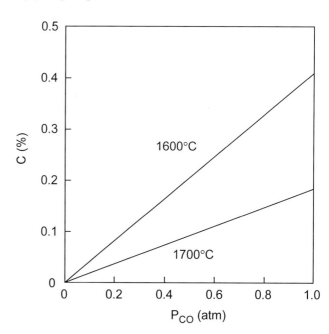

Fig. 2.119 Critical carbon content for an 18% Cr steel as a function of CO pressure.

When oxygen initially contacts an Fe–Cr–C bath it primarily oxidizes Cr to Cr_2O_3. As the chrome oxide particles rise through the bath with the gas bubbles, carbon diffuses to the surface and reduces the oxide according to the reaction 2.9.1. The rate of the reaction is controlled by mass transfer of carbon. The reaction also takes place with the top slag. The rate equation can be expressed as:

$$\frac{d\%C}{dt} = -\frac{\rho}{W} \sum_i m_i A_i \left[\%C - \%C_i^e\right] \quad (2.9.3)$$

where ρ is the density of steel, W is the weight of the metal, m_i are the mass transfer coefficients, A_i are the surface areas, $\%C_i^e$ is the equilibrium carbon content, and the subscript i refers to the individual reaction sites such as the rising bubble and the top slag.

The equilibrium carbon content is determined from the equilibrium for reaction 2.9.1 which is a function of temperature, chromium content and local CO pressure. The CO pressure in turn depends on the rate of decarburization and the Ar or N_2 in the gas and the total pressure. The details of the calculation are given elsewhere.

Equation 2.9.3 is valid when mass transfer of carbon is limiting. At high carbon contents mass transfer of carbon is sufficient to consume all the oxygen and the rate of decarburization is simply given by the mass balance based on oxygen flow rate.

The Cr loss to the slag can be computed from the mass balance for oxygen and is given by

$$\Delta[\%Cr] = \frac{4 M_{Cr}}{3 W 10^{-2}} \dot{N}_{O_2} t - \frac{10^{-2} W}{2 M_C} \Delta[\%C] \qquad (2.9.4)$$

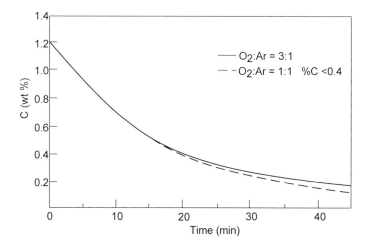

Fig. 2.120 Calculated rate of decarburization for an 18-8 stainless steel. *From Ref. 172.*

where

M_{Cr}, M_C = molecular weights of Cr and C,
W = weight of steel,
$\dot{N}_{O_2}$ = molar flow rate of oxygen,
$\Delta[\%C]$ = change in carbon content.

Simplified calculations for the rates of decarburization and Cr oxidation of an 18–8 stainless steel are presented in Fig. 2.120 and Fig. 2.121. These calculations indicate that switching the O_2/Ar ratio at 0.4% C slightly increases the rate of decarburization but significantly reduces Cr oxidation.

This model is the basis of process control models for the AOD, VOD and other similar processes. These are discussed in Chapter 12 in more detail.

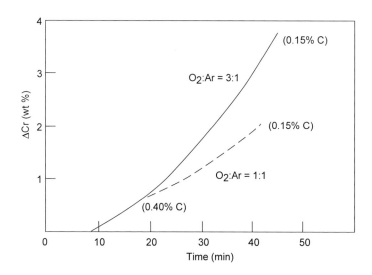

Fig. 2.121 Calculated rate of Cr oxidation for an 18–8 type stainless steel. *From Ref. 172.*

2.9.2 Nitrogen Control in the AOD

It is desirable to use nitrogen (N_2) in the AOD in place of Ar because of its lower cost. Also, for some steels it is desirable to alloy nitrogen by blowing N_2 gas. A model for nitrogen control was developed by one of the authors based on fundamental principles and is briefly described below.

The nitrogen control model presented here is based on the mixed control model for chemical kinetics and mass transfer in series given in section 2.2.4. Equation 2.9.5 requires knowledge of the pressure of N_2 which is given by

$$p_{N_2} = \frac{W R T P_T}{100 M_{N_2}} \frac{d\%N}{dt} + \frac{\dot{N}_{N_2}}{\dot{N}_{CO} + \dot{N}_{N_2} + \dot{N}_{Ar}} P_T \tag{2.9.5}$$

where

$\dot{N}_i$ = molar flow rate of species i,
W = weight of steel,
P_T = total pressure,
M_{N_2} = molecular weight of N_2 (28).

The rate of CO evolution is determined by the rate of decarburization, equations 2.9.3 and 2.9.4, and consequently it is necessary to include the decarburization model given in the previous section.

The rate of dissociation of N_2 depends on the Cr and S contents. Glaws and Fruehan[173] measured the rate for Fe–Cr–Ni–S alloys and found that Cr increased the rate while sulfur decreased the rate. The mass transfer parameter can be estimated from basic principles but requires knowledge of the bubble size; alternatively, it can be determined with a limited quantity of plant data by determining the best value of mA, the mass transfer coefficient times the surface area. Basic principles can be used to estimate m reasonably accurately using equations 2.2.74 and 2.2.75, thereby allowing for an estimation of A. Results of the model are presented in Fig. 2.122 and 2.123.

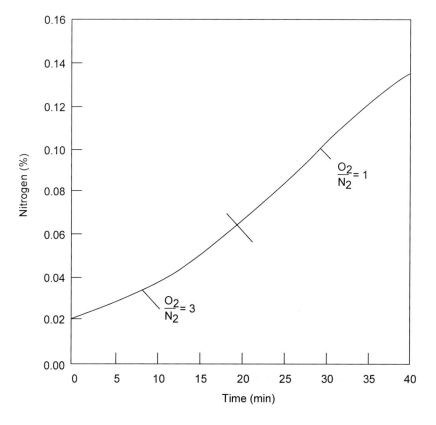

Fig. 2.122 Nitrogen pickup in 75 metric tons of 18–8 stainless steel containing 0.03% S using $O_2/N_2 = 3$ and $O_2/N_2 = 1$ gas mixtures, and with total flow of 0.85 m³/s at 1600°C. *From Ref. 50.*

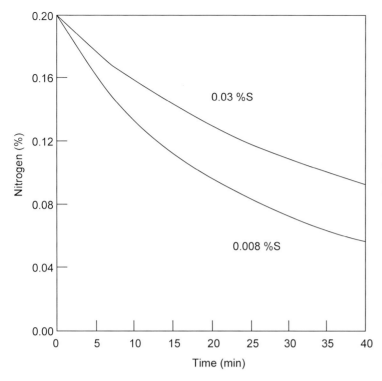

Fig. 2.123 Rate of nitrogen removal from 75 metric tons of 18–8 stainless steel containing 0.005 and 0.03% S using 0.85 m³/s of O_2 and Ar at 1600°C. *From Ref. 50.*

2.9.3 Reduction of Cr from Slag

In stainless steel production some chromium is oxidized to slag, approximately 3% for an 18% Cr steel. It is necessary to recover this Cr by adding a reductant, usually Si, (as ferrosilicon) or aluminum, which reacts with Cr_2O_3. For example, for Si the reaction is given by

$$3\underline{Si} + 2(Cr_2O_3) = 3(SiO_2) + 4\underline{Cr} \tag{2.9.6}$$

The equilibrium distribution of chromium between slag and metal is given in Fig. 2.124 as a function of Si content and slag chemistry. From Fig. 2.124 and a mass balance from reaction 2.9.6, it is possible to compute the Si required for the desired Cr reduction.

The typical compositions of AOD slags after decarburization and after silicon reduction are given in Table 2.14. As is seen from the experimental data in Fig. 2.124, the higher the slag basicity and higher the silicon content of steel, the lower is the equilibrium slag/metal distribution of chromium, i.e. the greater the chromium recovery from the slag.

Table 2.14 Ranges of AOD Slag Composition after Decarburization and after Silicon Reduction.

	Composition, wt.%	
	After decarburization	After silicon reduction
FeO	4–6	1–2
MnO	4–8	1–3
SiO_2	12–18	30–40
Al_2O_3	18–22	3–8
CaO	8–15	33–43
MgO	7–15	10–20
Cr_2O_3	20–30	1–3

Aluminum is a stronger reductant and nearly all will react to reduce the Cr_2O_3.

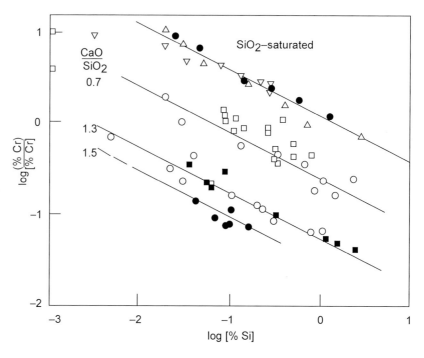

Fig. 2.124 Equilibrium slag/metal chromium distribution varying with the concentration of silicon in iron coexisting with chromium oxide containing CaO–Al$_2$O$_3$–SiO$_2$ slags at temperatures of 1600 to 1690°C. See Ref. 3 for references to experimental data.

2.10 Fundamentals of Ladle Metallurgical Reactions

Several books and conferences have been devoted to ladle or secondary metallurgy. The processes and fundamentals are described in detail in Chapter 11. In this section only the fundamentals of deoxidation, desulfurization and inclusion modification are discussed.

2.10.1 Deoxidation Equilibrium and Kinetics

There are primarily three elements used in steel deoxidation:

(1) Mn as low or high C ferro alloy,
(2) Si as low or high C ferro alloy or as silico manganese alloy,
(3) Al of approximately 98% purity.

2.10.1.1 Deoxidation with Fe/Mn

When the steel is partially deoxidized with Mn, the iron also participates in the reaction, forming liquid or solid Mn(Fe)O as the deoxidation product.

$$\left.\begin{array}{l}[Mn]+[O] \to MnO \\ [Fe]+[O] \to FeO\end{array}\right\} \text{liquid or solid } Mn(Fe)O \quad (2.10.1)$$

The state of equilibrium of steel with the deoxidation product Mn(Fe)O is shown in Fig. 2.215.

2.10.1.2 Deoxidation with Si/Mn

Depending on the concentrations of Si and Mn added to steel in the tap ladle, the deoxidation product will be either molten manganese silicate or solid silica.

$$\left.\begin{array}{l}[Si]+2[O] \to SiO_2 \\ [Mn]+[O] \to MnO\end{array}\right\} \text{molten } xMnO \cdot SiO_2 \text{ or solid } SiO_2 \quad (2.10.2)$$

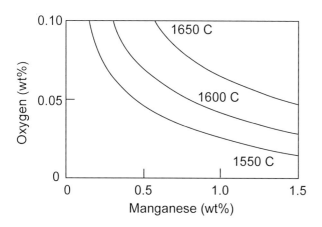

Fig. 2.125 Manganese and oxygen contents of iron in equilibrium with solid FeO–MnO deoxidation product.

From the experimental work of various investigators[174–176] the following equilibrium relation is obtained for the Si/Mn deoxidation reaction.

$$[Si] + 2(MnO) = 2[Mn] + (SiO_2) \tag{2.10.3}$$

$$K_{MnSi} = \left(\frac{[\%Mn]}{a_{MnO}}\right)^2 \frac{a_{SiO_2}}{[\%Si]} \tag{2.10.4}$$

$$\log K = \frac{1510}{T} + 1.27 \tag{2.10.5}$$

where the oxide activities are relative to pure solid oxides. For high concentrations of silicon (> 0.4%) the activity coefficient f_{Si} should be used in the above equation, thus $\log f_{Si} = 0.11 \times \%Si$.

The activities of MnO in manganese silicate melts have been measured by Rao and Gaskell.[177] Their results are in substantial agreement with the results of the earlier work by Abraham et al.[152] The activity of the oxides (relative to solid oxides) are plotted in Fig. 2.126. For liquid steel containing Mn < 0.4% the deoxidation product is a MnO-rich silicate with FeO < 8%; therefore the activity data in Fig. 2.126 can be used together with equation 2.10.4 in computing the equilibrium state of the Si/Mn deoxidation as given in Fig. 2.127. The deoxidation product being either solid silica or molten manganese silicate depends on temperature, Si and Mn contents, as shown in Fig. 2.127.

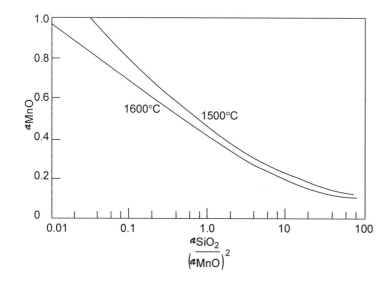

Fig. 2.126 Activities in MnO–SiO$_2$ melts with respect to solid oxides. *From Ref. 27.*

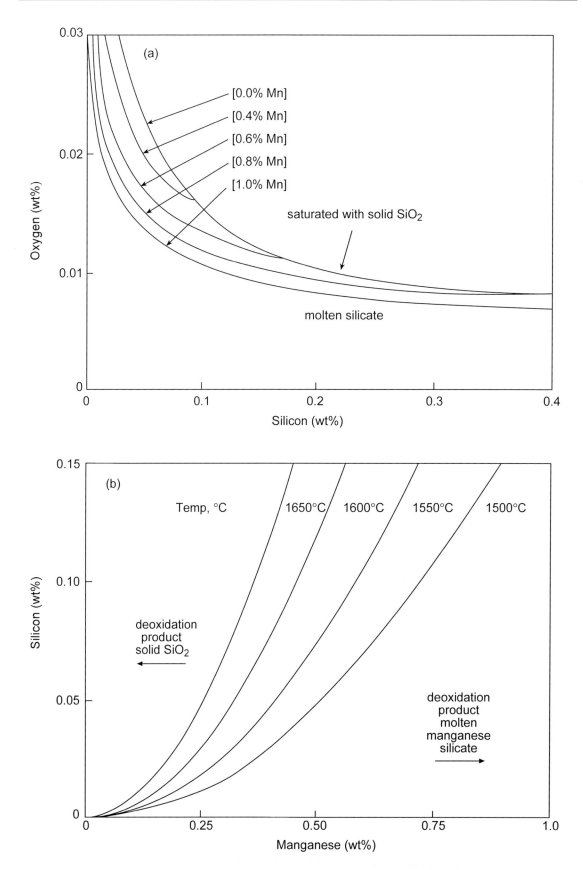

Fig. 2.127 Equilibrium relations for deoxidation of steel with silicon and manganese at 1600°C. *From Ref. 27.*

At silica saturation, the deoxidation is by silicon alone, for which the equilibrium relation for unit SiO_2 activity is reduced to

$$\log\left[\%Si\right]\left[ppm\ O\right]^2 = -\frac{30,410}{T} + 19.59 \qquad (2.10.6)$$

2.10.1.3 Deoxidation with Si/Mn/Al

Semi-killed steels with residual dissolved oxygen in the range 40 to 25 ppm are made by deoxidizing steel in the tap ladle with the addition of a small amount of aluminum together with silicomanganese, or a combination of ferrosilicon and ferromanganese. In this case, the deoxidation product is molten manganese aluminosilicate having a composition similar to $3MnO \cdot Al_2O_3 \cdot 3SiO_2$. With a small addition of aluminum, e.g. about 35 kg for a 220 to 240 ton heat together with Si/Mn, almost all the aluminum is consumed in this combined deoxidation with Si and Mn. The residual dissolved aluminum in the steel will be less than 10 ppm. For the deoxidation product $3MnO \cdot Al_2O_3 \cdot 3SiO_2$ saturated with Al_2O_3, the silica activities are 0.27 at 1650°C, 0.17 at 1550°C and decreasing probably to about 0.12 at 1500°C. Using these activity data the deoxidation equilibria are calculated for Al/Si/Mn; these are compared in Fig. 2.128 with the residual ppm O for the Si/Mn deoxidation at the same concentrations of Mn and Si.

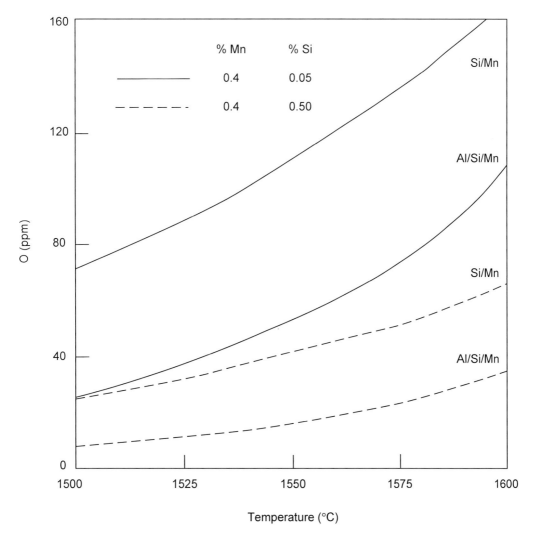

Fig. 2.128 Deoxidation equilibria with Si/Mn compared with Al/Si/Mn for the deoxidation product saturated with Al_2O_3. *From Ref. 27.*

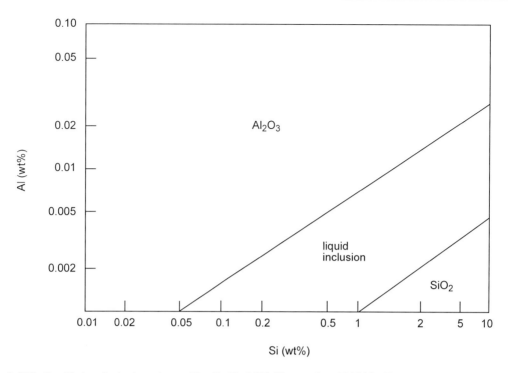

Fig. 2.129 Equilibrium inclusions for an Fe–Al–Si–1.0% Mn steel at 1600°C. *From Ref. 178.*

When intentionally using some Al for deoxidation or unintentionally with Al being in the ferrosilicon the equilibrium deoxidation product can be solid Al_2O_3, molten manganese silicate or solid SiO_2. To avoid clogging of continuous casting nozzles it is desirable to have liquid inclusions. The equilibrium deoxidation product is given as a function of Al and Si contents in Figure 2.129. As seen in this figure even small amount of Al in solution will lead to Al_2O_3 inclusions which can clog nozzles.

2.10.1.4 Deoxidation with Al

Numerous laboratory experiments have been made on the aluminum deoxidation of liquid iron using the EMF technique for measuring the oxygen activity in the melt. The equilibrium constants obtained from independent experimental studies, cited in Ref. 179 agree within about a factor of two. An average value for the equilibrium constant is given below.

$$Al_2O_3(s) = 2[Al] + 3[O] \tag{2.10.7}$$

$$k = \frac{[\%Al]^2 [ppm\ O \times f_O]^3}{a_{Al_2O_3}} \tag{2.10.8}$$

$$\log K = -\frac{62,680}{T} + 31.85 \tag{2.10.9}$$

The alumina activity is with respect to pure solid Al_2O_3. The effect of aluminum on the activity coefficient of oxygen dissolved in liquid steel is given by $\log f_O = 3.9 \times [\%Al]$. At low concentration of aluminum, $f_{Al} \sim 1.0$.

It should be noted that in the commercial oxygen sensors the electrolyte tip is MgO-stabilized zirconia. At low oxygen potentials as with aluminum deoxidation, there is some electronic conduction in the MgO-stabilized zirconia which gives an EMF reading that is somewhat higher than Y_2O_3 or ThO_2 stabilized zirconia where the electronic conduction is negligibly small. In other words, for a given concentration of Al in the steel the commercial oxygen sensor, without correction for partial electronic conduction, registers an oxygen activity that is higher than the true equilibrium value. To be consistent with the commercial oxygen sensor readings, the following apparent equilibrium constant may be used for reaction 2.10.7 for pure Al_2O_3 as the reaction product.

$$\log K_a = -\frac{62,680}{T} + 32.54 \qquad (2.10.10)$$

Another point to be clarified is that in the commercial oxygen sensor system, the EMF reading of the oxygen activity is displayed on the instrument panel in terms of ppm O, as though the activity coefficient $f_O = 1.0$ in the Al-killed steel. If the deoxidized steel contains 0.05% Al, the apparent oxygen activity using equation 2.10.10 will be (ppm O × f_O) = 3.62; noting that at 0.05% Al, f_O = 0.64, the apparent concentration of dissolved oxygen will be 3.62/0.64 = 5.65 ppm O.

When the Al-killed steel is treated with Ca–Si the alumina inclusions are converted to molten calcium aluminate. For the ratio %CaO/Al_2O_3 = 1:1, the activity of Al_2O_3 is 0.064 with respect to pure Al_2O_3 at temperatures in the range 1500–1700°C. The apparent equilibrium relations, consistent with the readings of commercial oxygen sensors, are shown in Fig. 2.130 for the deoxidation products: pure Al_2O_3 and molten calcium aluminate with %CaO/%Al_2O_3 = 1:1.

2.10.1.5 Silicon and Titanium Equilibrium in Aluminum Deoxidized Steel

When steels are deoxidized with aluminum and also contain silicon and titanium slag metal equilibrium is established for these elements according to the reactions.

$$3SiO_2 + 4\underline{Al} = 2(Al_2O_3) + 3\underline{Si} \qquad (2.10.11)$$

$$3TiO_2 + 4\underline{Al} = 2(Al_2O_3) + 4\underline{Ti} \qquad (2.10.12)$$

In principle the equilibrium distribution ratio %Si/(%SiO_2) and %Ti/(%TiO_2) can be computed from basic thermodynamics. The equilibrium ratios observed in practice are shown in Figure 2.131 and 2.132.

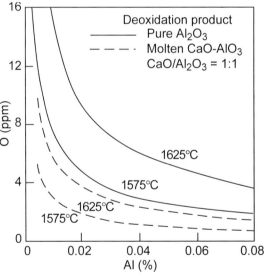

Fig. 2.130 Deoxidation with aluminum in equilibrium with Al_2O_3 or molten calcium aluminate with CaO/Al_2O_3 = 1:1. *From Ref. 27.*

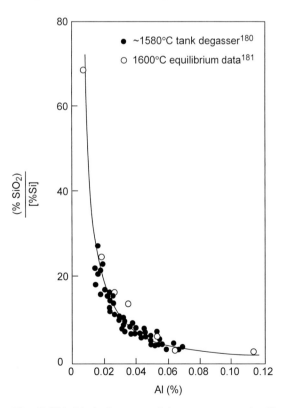

Fig. 2.131 Tank degasser data are compared with the experimental equilibrium data for aluminum reduction of silica from lime-saturated calcium aluminate melts containing SiO_2 < 5%. *From Ref. 27.*

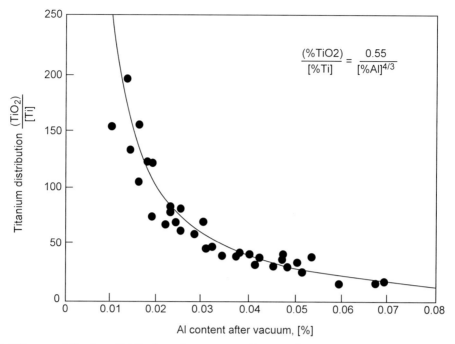

Fig. 2.132 Silicon and titanium distribution after vacuum. *From Ref. 178.*

2.10.1.6 Rate Phenomena in Deoxidation

The equilibrium states for the common deoxidation reactions pertaining to steelmaking conditions have been established reasonably well. However, the rate phenomena concerning the deoxidation reactions is complex and is discussed in detail in a review paper by Turkdogan.[182] A brief mention of the state of our knowledge is adequate for the present purpose.

There are three basic consecutive steps involved in the deoxidation reaction; namely, formation of critical nuclei of the deoxidation product in a homogeneous medium, progress of deoxidation resulting in growth of the reaction products and their flotation from the melt.

Turkdogan has shown that because of the high interfacial tension between liquid iron and oxide and silicate inclusions, a high supersaturation of the reactants in the metal is needed for spontaneous nucleation of the deoxidation products as predicted from the theory of homogeneous nucleation. In estimating the supersaturation ratio likely to be achieved under practical conditions, homogeneous solution of deoxidizers in the steel was assumed. However, the dissolution of added deoxidizers in liquid steel takes a finite time during which certain regions of the melt are expected to be very rich in solute concentration:; in these regions the solution is sufficiently supersaturated locally for homogeneous nucleation of the deoxidation product. Owing to the agitation in the ladle, the nuclei thus formed are considered to be distributed in the melt, soon after the addition of deoxidizers. Another source of nuclei is, of course, the thin oxide layer on the surface of particles of solid deoxidizers added to steel.

A generally accepted view is that the deoxidation reactions at steelmaking temperatures are fast relative to other rate-controlling processes responsible for the growth and ultimate flotation of inclusions.

The rate phenomena is deoxidation is complex because of the side effects caused by the interplay of several variables which cannot readily be accounted for in mathematical simulations of the deoxidation process. However, certain important deductions can be made from the results of several conceptual analyses based on simplified models and those of experimental observations.

1. The number of nuclei (z) formed at the time of addition of deoxidizers is of the order of $z = 10^7/cm^3$ or higher.

2. The diffusion-controlled deoxidation reaction is essentially complete within a few seconds when $z > 10^6/cm^3$.

3. The deoxidation reaction may cease prematurely in parts of the melt depleted of nuclei or oxide inclusions.

4. The inclusion size during deoxidation is in the range 1 to 40 μm.

5. In laboratory experiments with inductively stirred melts (~5 cm deep) most of the oxide inclusions float out of the melt in 5 to 10 minutes.

6. The growth by collision and coalescence of ascending inclusions does not seem possible under the conditions of laboratory experiments with unstirred or moderately stirred melts.

These observations are not mutually consistent. One possible explanation perhaps is that the nuclei formed at the time of dissolution of deoxidizers are unevenly distributed in molten steel. In parts of the melt where the number of nuclei is small, e.g. $10^4/cm^3$ to $10^5/cm^3$, the inclusions 20 to 40 μm in size rapidly float out of the melt, presumably prior to the deoxidation reaction. In parts of the melt containing about 10^8 nuclei/cm^3, the inclusions grow only to a micron size and ascend in the melt with a creeping velocity. Convection currents or other means of stirring eventually bring about more uniform distribution of these small inclusions. The particles thus brought to the parts of the melt where the deoxidation reaction was incomplete, due to early depletion of inclusions, bring about further deoxidation, growth and flotation.

Under practical conditions of deoxidation during filling of the ladle, there is sufficient stirring that some inclusion growth may take place by collision and coalescence; also stirring brings about a motion in the melt such that the inclusions could get attached to the surface of the ladle lining and caught by the slag layer. Controlled gas stirring at low flow rates for 4–8 minutes is common practice to enhance inclusion removal.

2.10.2 Ladle Desulfurization

It is possible to desulfurize aluminum killed steels in the ladle or ladle furnace, using CaO based slags, to less than 20 ppm S. The chemical reaction can be written as

$$3(CaO) + 2\underline{Al} + 3\underline{S} = 3(CaS) + Al_2O_3 \quad (2.10.13)$$

The equilibrium sulfur distribution ratio can be calculated from the thermodynamics of reaction 2.10.13. In terms of the ionic reaction the reaction is

$$\tfrac{2}{3}\underline{Al} + \underline{S} + (O^{2-}) = (S^{2-}) + \tfrac{1}{3}(Al_2O_3) \quad (2.10.14)$$

for which the equilibrium ratio is

$$K_{SA} = \frac{(\%S)}{[\%S]}[\%Al]^{-\tfrac{2}{3}} \quad (2.10.15)$$

The value of K_{SA} for CaO–Al_2O_3 and CaO–Al_2O_3–SiO_2–MgO slags is shown in Figure 2.133. The sulfur distribution ratios are shown in Fig. 2.134 and Fig. 2.135 for the stated Al contents. From the data presented and a mass balance for sulfur it is possible to compute the final equilibrium sulfur contents.

The slags will absorb sulfur until CaS forms decreasing the amount of dissolved CaO and decreasing K_{SA}. The solubility of CaS is given in Fig. 2.136.

Steels deoxidized with Si are difficult to desulfurize because the oxygen potential is significantly higher than for Al killed steels. For these steels desulfurization is limited to 10 to 20%. Calcium carbide is an effective desulfurizer because it decreases the oxygen potential and the resulting CaO then is able to desulfurize.

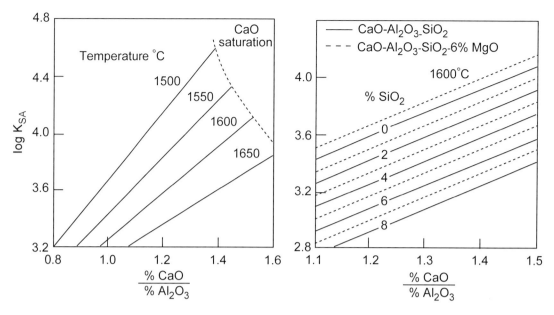

Fig. 2.133 Effects of temperature and slag composition on the equilibrium relation K_{SA} for the calcium-magnesium aluminosilicate melts. *From Ref. 27.*

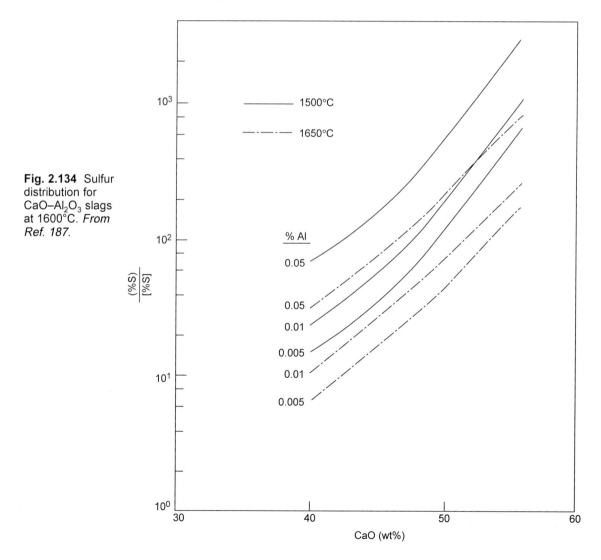

Fig. 2.134 Sulfur distribution for $CaO-Al_2O_3$ slags at 1600°C. *From Ref. 187.*

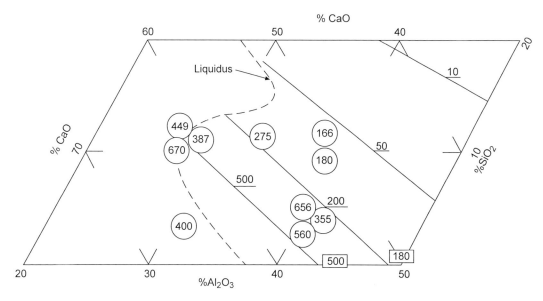

Fig. 2.135 Sulfur distribution for CaO–Al$_2$O$_3$–SiO$_2$ slags at 1600°C steel containing 0.03% Al. *From Ref. 178.*

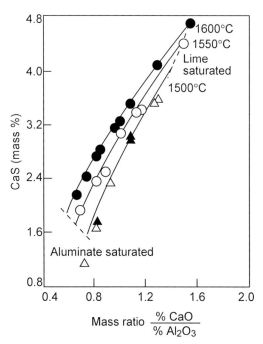

Fig. 2.136 Solubility of CaS in calcium aluminate melts related to the mass ratio %CaO/%Al$_2$O$_3$. *From Ref. 27.*

Desulfurization is controlled by liquid phase mass transfer. Since the sulfur distribution ratio is high sulfur mass transfer in the metal is the primary rate controlling process. For this case the rate of desulfurization is given by the following[178].

$$K_{SA} = \frac{(\%S)}{[\%S]}[\%Al]^{-2/3} \qquad (2.10.16)$$

$$[\%S^e] = \frac{(\%S_t)}{L_S} \qquad (2.10.17)$$

$$(\%S_t) = \frac{W_m}{W_s}\left([\%S_\circ] - [\%S_t]\right) \qquad (2.10.18)$$

where

m = mass transfer coefficient,
A = surface area,
W_m = weight of metal,
W_s = weight of steel,
ρ = density of steel,
L_S = sulfur distribution ratio,
$(\%S_t)$ = sulfur content of the slag,
$[\%S_t]$ = sulfur content of the metal,
$[\%S^e]$ = equilibrium sulfur content of the metal,
$[\%S_\circ]$ = initial sulfur content of the metal.

Equations 2.10.16 through 2.10.18 can be solved and the rate is given by

$$\frac{\ln\left\{1 + \frac{1}{L_S}\left(\frac{W_m}{W_s}\right)\left[\frac{\%S_t}{\%S_\circ}\right] - \frac{1}{L_S}\left(\frac{W_m}{W_s}\right)\right\}}{1 + \frac{1}{L_S}\left(\frac{W_m}{W_s}\right)} = \frac{mA\rho}{W_s}t \qquad (2.10.19)$$

Desulfurization increases with stirring rate, which increases m and more importantly A by providing slag-metal mixing, increased values of L_s, and higher slag volumes. It has been shown that 90% desulfurization can be achieved in 10–15 minutes of intense stirring.

2.10.3 Calcium Treatment of Steel

Calcium is usually employed in ladle metallurgy as Ca or CaSi cored wire or by injecting CaSi powder. Calcium is highly reactive and it could deoxidize, desulfurize, modify oxide inclusions or modify sulfide inclusions. It should only be used in deoxidized steels because it is too expensive to be used as a deoxidizer.

Since often the primary purpose of a calcium injection into the steel bath is to convert solid Al_2O_3 inclusions to liquid calcium aluminates to prevent Al_2O_3 from clogging casting nozzles, it is necessary to know under what conditions it will react with the inclusions or simply react with sulfur. There have been numerous thermodynamic calculations to predict the conditions for Al_2O_3 inclusion modification. Several of these required knowledge of the thermodynamics of Ca in steel, which is not accurately known. The conditions can be computed based on the thermodynamic properties of the inclusions themselves.

Consider the following reaction equilibrium in the calcium treated steel,

$$3(CaS) + (Al_2O_3) = 3(CaO) + 2\underline{Al} + 3\underline{S} \qquad (2.10.20)$$

where CaS is that formed by the reaction of calcium and sulfur, (Al_2O_3) is an alumina rich inclusion such as Al_2O_3, $CaO \cdot Al_2O_3$, etc and (CaO) represents an inclusion richer in CaO. The results of these calculations are given in Fig. 2.137 in which the inclusion stability is given as a function of Al and S contents[183]. Below the CA curve $CaO \cdot Al_2O_3$ is the stabler oxide and below the $C_{12}A_7$ liquid curve, liquid calcium aluminates form. For example if Ca is added to a steel containing 0.04% Al and 0.015% S the alumina inclusion will be converted to solid $CaO \cdot Al_2O_3$ and calcium there will react to form CaS. These inclusions will clog casting nozzles. For effective inclusion modification the sulfur content should be below 0.01% for a 0.04% Al steel.

Calcium also helps eliminate MnS inclusions which form during solidification. For a steel low in sulfur which contains liquid calcium aluminate inclusions, much of the remaining sulfur will be absorbed during cooling and solidification by these inclusions as they are excellent desulfurizers. Also, if any dissolved Ca is remaining, although the authors believe this will be very small, it could react with sulfur on

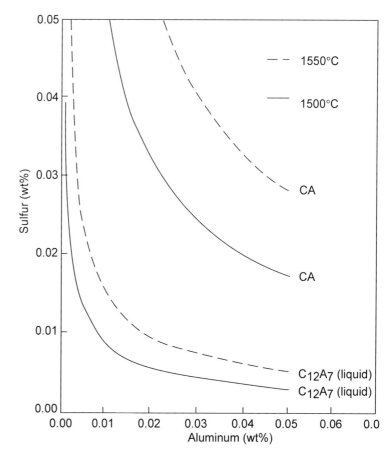

Fig. 2.137 Univariant equilibrium for CaS (a_{CaS} = 0.74) and $C_{12}A_7$ or CA as a function of %Al and %S at 1550 and 1500° C.

the inclusions. The net result is a duplex inclusion consisting of a liquid calcium aluminate with a CaS (MnS) rim. This is preferable to stringer MnS which can cause brittle fracture.

Effective calcium treatment is a critical issue and is discussed in Chapter 12 in more detail.

2.11 Fundamentals of Degassing

During the past two decades there has been an increase in vacuum degassing to reduce hydrogen and nitrogen contents. However the major increase in vacuum degassing is for decarburization to low carbon contents (40 ppm) which are required for good formability such as for interstitial free (IF) steels. Vacuum degassing is discussed in detail in Chapter 11. Briefly there are two major types of degassers: circulating, such as RH and RH–OB, and non-recirculating such as ladle and tank degassers. In some cases oxygen is used to enhance the reactions, examples of these are RH–OB (oxygen blowing) and VOD (vacuum oxygen decarburization).

Details on the types of degassers and the processes are given in Chapter 11. In this section only the fundamental aspects of the reactions will be discussed.

2.11.1 Fundamental Thermodynamics

In vacuum degassing hydrogen, nitrogen, carbon, and oxygen can be removed by the following reactions.

$$\underline{H} = \tfrac{1}{2}H_2 \quad (2.11.1)$$

$$\underline{N} = \tfrac{1}{2}N_2 \quad (2.11.2)$$

$$\underline{C} + \underline{O} \,(FeO, \tfrac{1}{2}O_2) = CO + (Fe) \quad (2.11.3)$$

Hydrogen and nitrogen are dissolved in steel and removed by forming diatomic molecules. The thermodynamics are given in Section 2.4.

Carbon is removed by reaction with oxygen as dissolved oxygen, FeO in slag or gaseous oxygen (O_2) to form CO. Due to the reduced pressure the formation of CO is favored. This is shown schematically in Figure 2.138 as a plot of the equilibrium carbon and oxygen contents at 1 and 0.2 atmosphere pressure. In RH and other processes with oxygen the carbon reacts with oxygen and follows the indicated reaction path. Theoretically one atom of oxygen requires one atom of dissolved oxygen. However in actual processes there are other sources of oxygen such as leaks and unstable oxides in the slag (FeO and MnO). In oxygen assisted processes the reaction path is indicated as RH–OB (or VOD). Initially gaseous oxygen removes carbon, the oxygen blow is terminated, and dissolved oxygen reacts with carbon.

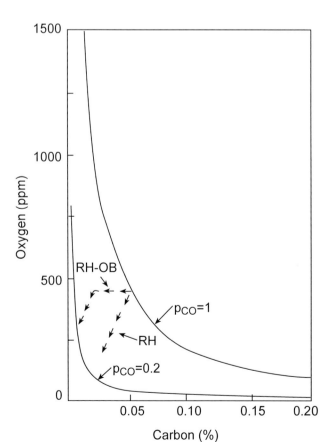

Fig. 2.138 Carbon–oxygen equilibrium as a function of CO pressure. *From Ref. 184.*

2.11.2 Vacuum Degassing Kinetics

In recirculating systems H_2, N_2 and CO are formed at two reaction sites: the rising argon bubbles and the metal-vacuum interface above the melt. The model and plant data given by Bannenberg[185] et al indicates that the Ar bubbles are the major reaction surface area for reactions at high Ar rates. Recent work by Uljohn and Fruehan[186] indicate at lower Ar flow rates both sites should be considered. As the argon bubbles burst at the surface they generate a large number of metal droplets and surface area. In recirculating systems the reactions include the rising argon bubbles, the metal droplets and homogenous or heterogeneous nucleated bubbles in the bath.

The hydrogen reaction is simply controlled by liquid phase mass transfer of hydrogen. Hydrogen diffusivity is high and consequently the reaction is fast. For non recirculating systems the reactions increase with argon stirring in a complex manner. Due to the expansion of the gas bubbles and the dependence of bubble size on operating parameters, the rate equations are complex and the reader is referred to Ref. 184 for details.

It should be noted that the reaction is limited not by H_2 gas in the vessel but rather H_2O. Even at very low H_2O pressures the equilibrium hydrogen content for reaction 2.11.4 is relatively high.

$$3H_2O + 2\underline{Al} = Al_2O_3 + 6\underline{H} \qquad (2.11.4)$$

For example, for $p_{H_2O} = 2 \times 10^{-6}$ atm and an aluminum content of 0.03%, the equilibrium hydrogen content is approximately 1 ppm.

The nitrogen reaction is controlled by mass transfer of nitrogen and chemical kinetics in series and the mixed control model given in Section 2.2.4 can be applied. Again the reactions are complex and the details are given elsewhere. However, the rate depends on the argon flow rate and sulfur content; sulfur is surface active and retards the chemical reaction as discussed in Section 2.2.3. Examples of calculated rates of nitrogen are given in Figs. 2.139 and 2.140. Nitrogen can only be removed effectively at low sulfur contents and high Ar bubbling rates.

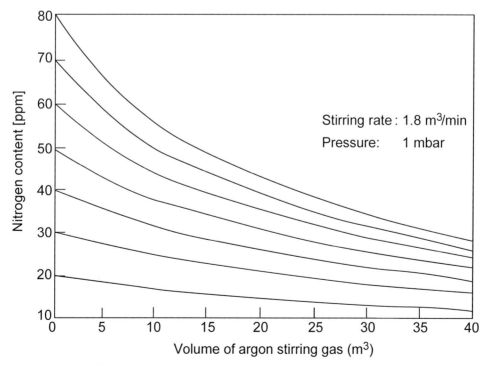

Fig. 2.139 Decrease of nitrogen content during vacuum for a sulfur content of 10 ppm. *From Ref. 185.*

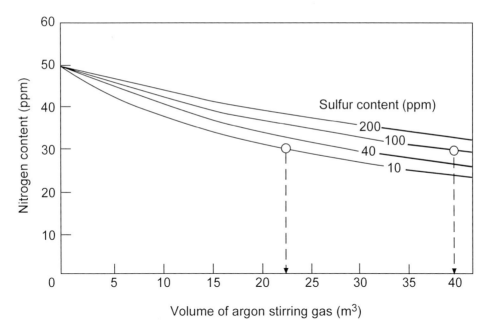

Fig. 2.140 Influence of sulfur content on the denitrogenisation. *From Ref. 185.*

The rates in recirculating systems (RH) are complicated by the fact they also depend on the circulation rate into the RH unit. For example the rate of decarburization in the ladle is given by

$$\frac{d\,\%C}{dt} = -\frac{Q}{W}\left[\%C - \%C_{RH}\right] \tag{2.11.5}$$

where

Q = metal recirculation rate,
W = the weight of steel,
$\%C_{RH}$ = the carbon content in the RH.

The change of carbon in the RH is given by

$$\frac{d\,\%C_{RH}}{dt} = -k\left[\%C_{RH} - \%C_e\right] + \frac{Q}{W_{RH}}\left[\%C - \%C_{RH}\right] \tag{2.11.6}$$

where k is the decarburization rate constant, $\%C_e$ is the equilibrium carbon content and W_{RH} is the weight of steel in the RH. An approximate solution to equation 2.11.6 is given by

$$\ln\left(\frac{\%C - \%C_e}{\%C_\circ - \%C_e}\right) = -Kt \tag{2.11.7}$$

where

$\%C_\circ$ = the initial carbon content,
K = the overall reaction rate constant given by

$$K = \frac{Q}{W}\left(\frac{k\,W_{RH}}{k\,W_{RH} + Q}\right) \tag{2.11.8}$$

Therefore the rate can be increased by increasing Q, W_{RH} and k. For low circulation rates the rate is controlled by the circulation rates. Consequently newer RH units have large snorkels and high circulation rates.

The hydrogen reaction in RH units are given by similar expressions. The major difference is that k_{RH} for hydrogen is larger because of faster mass transfer of hydrogen. Nitrogen removal in RH units is generally slow and generally less than 5 ppm is removed.

References

1. L. Onsager and R. M. Fuoss, *J. Phys. Chem.* 36 (1932): 2687.
2. A. Einstein, *Ann. Phys (Leipzig)* 17 (1905): 54; *Z. Elektrochem* (1908): 14,235.
3. E. T. Turkdogan, "Physicochemical Properties of Molten Slags and Glasses," The Metals Society (now The Institute of Materials), London, 1983.
4. J. Szekely, *Fluid Flow Phenomena in Metals Processing* (New York: Academic Press, 1979).
5. J. Szekely and N. J. Themelis, *Rate Phenomena in Process Metallurgy* (New York: Wiley, 1971).
6. G. H. Geiger and D. R. Poirier, *Transport Phenomena in Metallurgy* (Reading, MA: Addison-Wesley, 1973).
7. J. Szekely, J. W. Evans, and J. K. Brimacombe, *The Mathematical and Physical Modelling of Primary Metals Processing Operations* New York: Wiley, 1988).
8. J. Szekely, G. Carlsson, and L. Helle, *Ladle Metallurgy* (Berlin: Springer, 1989).
9. J. Szekely and O. J. Ilegbusi, *The Physical and Mathematical Modelling of Tundish Operations* (Berlin: Springer, 1989).
10. J. S. Sheasby, W. E. Boggs, and E. T. Turkdogan, *Metal Science* 18 (1984): 127.
11. C. Wagner, *Z. Phys. Chem., Abt B* 21 (1933), 25 (in "Atom Movements," ASM (1950): 153.
12. L. Himmel, R. F. Mehl, and C. E. Birchenall, *Trans. AIME* 197 (1953): 827.
13. P. Hembree and J. B. Wagner, *Trans. AIME* 245, (1969): 1547.
14. P. L. Walker, F. Rusinko, and L. G. Austin, *Advan. Catalysis* 21 (1959): 33.
15. E. T. Turkdogan et al., *Carbon* 6 (1968): 467.
16. E. T. Turkdogan and J. V. Vinters, *Carbon* 7 (1969): 101; 8 (1970): 39.
17. E. T. Turkdogan, R. G. Olsson, and J. V. Vinters, *Carbon* 8 (1970): 545.
18. R. H. Tien and E. T. Turkdogan, *Carbon* 8 (1970): 607.
19. D. A. Aderibigbe and J. Szekely, *Ironmaking and Steelmaking* 9 (1982): 32.
20. W .M. McKewan, *Trans. TMS-AIME* 218 (1960): 2.
21. J. M. Quets, M. E. Wadsworth, and J. R. Lewis, *Trans. TMS-AIME* 218 (1960): 545: 221 (1961): 1186.
22. N. J. Themelis and W. H. Gauvin, *A.E. Ch. E.J.* 8, (1962): 437; *Trans. TMS-AIME* 227 (1963): 290.
23. N. A. Warner, *Trans. TMS-AIME* 230 (1964): 163.
24. R .H. Spitzer, F. S. Manning, and W. O. Philbrook, *Trans. TMS-AIME* 236 (1966): 726.
25. E. T. Turkdogan and J. V. Vinters, *Metall. Trans.* 2 (1971): 3175; 3 (1972): 1561.
26. P. H. Calderbank, *Chem. Eng. (London)* 212 (1967): CE 209.
27. E. T. Turkdogan, "Fundamentals of Steelmaking," The Institute of Materials, London, 1996.
28. I. Leibson et al., *A.I.Ch.E.J.* 2, (1956): 296.
29. V. G. Levich, *Physicochemical Hydrodynamics* (Englewood Cliffs: Prentice-Hall, 1962).
30. S. Glasstone, K. J. Laidler, and H. Eyring, The Theory of Rate Processes (New York: McGraw-Hill, 1941).
31. C. N. Hinshelwood, Kinetics of Chemical Change, (London: Oxford University Press, 1942).
32. T. Fuwa, S. Ban-ya, and T. Shinohara, *Tetsu-to-Hagane* 53 (1967): S328.
33. M. Inouye and T. Choh, *Trans. Iron Steel Inst. Japan* 8 (1968): 134.
34. K. Mori and K. Suzuki, *Trans. Iron Steel Inst. Japan* 10 (1970): 232.
35. R. J. Fruehan and L. J. Martonik, *Metall. Trans. B* 11B (1980): 615.
36. M. Byrne and G. R. Belton, *Metall. Trans. B* 14B (1983): 441.
37. P. Glaws and R. J. Fruehan, *Metall Trans. B* 16B (1985): 551.
38. S. Ban-ya et al., *Tetus-to-Hagane* 60 (1974): 1443.
39. G. Sain and G. R. Belton, *Metall Trans. B* 5B (1974): 1027–32; 7B (1976): 403–7.
40. F. Manmion and R. J Fruehan, *Metall Trans. B,* 20B (1989): 853.
41. T. Nagasaka and R. J. Fruehan, *Metall Trans. B,* 25B (1994): 245.
42. G. R. Belton, *Metall Trans. B,* 7B (1976): 35–42.
43. E. A. Gulbransen, K. F. Andrew, and F. A. Brassort, *Carbon* 2 (1965): 421.
44. J. D. Blackwood, *Aust. J. Appl. Sci.* 13 (1962): 199.

45. S. Ergun, *J. Phys. Chem.* 60 (1956): 480.
46. K. Hedden and A. Lowe, *Carbon* 5 (1967): 339.
47. H. J. Grabke, *Ber. Bunsenges. Physik Chem.* 70 (1966): 664.
48. D. A. Aderibigbe and J. Szekely, *Ironmaking & Steelmaking* 8 (1981): 11.
49. S. Story and R. J. Fruehan, To be submitted to Metall Trans., 1997
50. R. J. Fruehan, B. Lally, and P. C. Glaws, *Trans. ISS, I&SM,* (April 1987), 31.
51. E. T. Turkdogan and J. V. Vinters, *Metall. Trans.* 5 (1974): 11.
52. E. T. Turkdogan, *Physical Chemistry of High Temperature Technology* (New York: Academic Press, 1980).
53. H. J. Donald, "An Annotated Bibliography," Mellon Inst. Ind. Res. Pittsburgh, PA. 1956.
54. J. V. Kehrer and H. Leidheiser, *J. Phys. Chem.* 58, (1954): 550.
55. W .R. Walker, J. F. Rakszawski, and G. R. Imperial, *J. Phys. Chem.* 63 (1959): 133, 140.
56. W. R. Ruston et al., *Carbon* 7 (1969): 47.
57. R. G. Olsson and E. T. Turkdogan, *Metall. Trans.* 5 (1974): 21.
58. J. Macak, P. Knizek, and J. Malecha, *Carbon* 16 (1978): 111.
59. E. R. Gilliland and P. Harriott, *Ind. Eng. Chem.* 46 (1954): 2195.
60. T. H. Byrom, *J. Iron Steel Inst.* 92 (1915): 106.
61. H. C. H. Carpenter and C. C. Smith, *J. Iron Steel Inst.* 96 (1918): 139.
62. O. Stelling, *J. Met.* 10 (1958): 290.
63. H. Schenck and W. Maachlanka, *Arch. Eisenhuttenwes.*31 (1960): 271.
64. R. J. Fruehan and L. J. Martonik, *Metall Trans.* 4 (1973): 2129.
65. S. Chapman and T. G. Cowling, *The Mathematical Theory of Non-uniform Gases,* 3rd ed. (London: Cambridge University Press, 1970).
66. J. O. Hirschfelder, C. F. Curtiss, and R. B. Bird, *Molecular Theory of Gases and Liquids,* (New York: Wiley, 1954).
67. A. Eucken, *Physik.Z.* 14 (1913): 324.
68. M. Knudsen, *Ann. Phys. (Leipzig)* 28 (1909): 75.
69. C. H. Bosanquet, Br. TA Rep. BR–507, September 1944.
70. E. T. Turkdogan, R. G. Olsson, and J. V. Vinters, *Metall. Trans.* 2 (1971): 3189.
71. G. K. Sigworth and J. F. Elliott, *Met. Sci.* 8 (1974): 298.
72. E. T. Turkdogan and L. E. Leake, *J. Iron and Steel Inst.* 179 (1955): 39.
73. E. T. Turkdogan and R. A. Hancock, 179 (1955): 155; 183 (1956): 69.
74. J. Chipman et al., *Trans. ASM.* 44 (1952): 1215.
75. J. Chipman and T. P. Floridis, *Acta Met* 3 (1955): 456.
76. N. R. Griffing, W. D. Forgeng, and G. W. Healy, *Trans. AIME* 224 (1962): 148.
77. J. F. Elliott, M. Gleiser, and V. Ramakrishna, *Thermochemistry for Steelmaking,* vol. II (Reading, MA: Addison-Wesley, 1963).
78. S. N. Singh and K. E. Blazek, *J. Metals (AIME)* 26(10) (1974): 17.
79. M. M. Wolf and W. Kurz, *Metall.Transaction B* 12B (1981) 85.
80. A. A. Howe, *Applied Scientific Research* 44 (1987): 51.
81. H. Yamada, T. Sakurai, and T. Takenouchi, *Tetsu-to-Hagane* 76 (1990): 438.
82. M. M. Wolf, in "1st European Conference on Continuous Casting", p. 2.489. Florence, Italy, Sept. 23–25, 1991.
83. C. R. Taylor and J. Chipman, *Trans. AIME* 154 (1943): 228.
84. P .A. Distin, S. G. Whiteway, and C. R. Masson, *Can. Metall. Q.* 10 (1971): 13.
85. A. E. Lord and N. A. Parlee, *Trans. Met. Soc. AIME* 218 (1960): 644.
86. D. L. Sponseller and R. A. Flinn, *Trans. Met. Soc. AIME* 230 (1964): 876.
87. P. K. Trojan and R. A. Flinn, *Trans. ASM 54* (1961): 549.
88. P. J. Guichelaar et al., *Metall. Trans.* 2 (1971): 3305.
89. B. J. Keene, A survey of extant data for the surface tension of iron and its binary alloys, NPL Rep. DMA(A) 67, June 1983.
90. E. T. Turkdogan, in *Foundry Processes—Their Chemistry and Physics*, ed. S. Katz and C. F. Landefeld (New York: Plenum Press, 1988), 53.
91. P. Kozakevitch, in *Surface Phenomena of Metal* (London: Chem. Ind., 1968), 223.
92. E. Selcuk and D. H. Kirkwood, *J. Iron and Steel Inst.* 211 (1973): 134.

93. I. Jimbo and A. W. Cramb, *Iron Steel Inst. Japan International* 32 (1992): 26.
94. A. Kasama et al., *Can. Met. Quart.* 22 (1983): 9.
95. L. D. Lucas, *Compt. Rend.* 248 (1959): 2336.
96. R. N. Barfield and J. A. Kitchener, *J. Iron and Steel Inst.* 180 (1955): 324.
97. G. Tromel et al., *Arch. Eisenhuttenwes.* 40 (1969): 969.
98. C. R. Taylor and J. Chipman, *Trans. AIME* 159 (1943): 228.
99. M. Timucin and A. E. Morris, *Metall. Trans* 1 (1970): 3193.
100. S. H. Liu, A. Morales, and R. J. Fruehan, To be submitted to Metall Trans. B, 1997.
101. R. H. Rein and J. Chipman, *Trans. Met. Soc. AIME* 233 (1965): 415.
102. T. Fujisawa and H. Sakao, *Tetsu-to-Hagane* 63 (1977): 1494, 1504.
103. "Ethem T. Turkdogan Symposium", pp. 253–269. The Iron and Steel Society of AIME, May 15–17, 1994.
104. Y. Iguchi, S. Ban-ya and T. Fuwa, *Trans. Iron and Steel Inst. Japan* 9 (1969): 189.
105. K. Ito and R. J. Fruehan, *Metall. Trans. B* 19B (1988): 419.
106. D. J. Min and R. J. Fruehan, *Metall Trans. B* 21B (1990): 1025.
107. H. Suito and R. Inoue, *Tetsu-to-Hagane* 73 (1987): S246.
108. K. Nomura, B. Ozturk, and R. J. Fruehan, *Metall Trans. B* 22B (1991): 783.
109. S. Ban-ya and J. D. Shim, *Can. Metall. Q.* 21 (1982): 319.
110. P. Kozakevitch, *Rev. Metall.* 46 (1949): 505, 572.
111. K. C. Mills and B. J. Keene, *Inter. Met. Rev.* 26 (1981): 21.
112. H. Gaye et al., *Can. Metall. Q.* 23 (1984): 179.
113. K. Ogino et al., *Trans. Iron and Steel Inst. Japan* 24 (1984): 522.
114. I. Jimbo and A. W. Cranb, The Sixth International Iron and Steel Congress, Naguya, ISIJ, 1990, 499.
115. D. R. Gaskell, A. McLean, and R. G. Ward, *Trans. Faraday Soc.* 65 (1969): 1498.
116. Y. Skiraishi et al., *Trans. Japan Inst. Met.* 19 (1978): 264.
117. Yu M. et al., *Soob A Gruz. SSR* 32(1) (1963): 117.
118. L. Segers, A. Fontana, and R. Winand, *Electrochem. Acta.* 23 (1978): 1275.
119. J. W. Tomlinson, M. S. R. Heynes, and J. O'M. Bockris, *Trans. Faraday Soc.* 54 (1958): 1822.
120. L. R. Barrett and A. G. Thomas, *J. Glas Technol.* (1959): 179s.
121. H. Winterhager, L. Greiner, and R. Kammel, *Forschungsberichte des Landes* (Nordrhein-Westfalen, Nr. 1630: Westdeutscher Verlag, 1966).
122. E. V. Krinochkin, K. T. Kurochin, and P. V. Umrikhin, *Fiz. Khim. Poverkh.* (Kiev: Naukova Dumba, Yavlenii Rasp., 1971) ,179–83.
123. J. S. Machin and T. B. Yee, *J. Am. Cerm. Soc.* 32 (1948): 200; 37 (1954) 177.
124. K. Ito and R. J. Fruehan, *Metall Trans. B* 20B (1989): 509.
125. K. Ito and R.J. Fruehan, *Metall Trans. B* 20B (1989): 515.
126. Y. Zhang and R. J. Fruehan, *Metall Trans. B* 26B (1995): 811.
127. T. Tokumetsu et al., Process Technology Conference Proceedings, Toronto, ISS, 1989.
128. Y. Zhang and R. J. Fruehan, *Metall Trans. B.* 26B (1995): 803.
129. S. Ban-ya, *ISIJ International* 33 (1993): 2–11.
130. H. Gaye and J. Lehmann, *Molten Slags, Fluxes, and Salts Conference,* Sydney, 1997, IS Warrendale, PA, 27–34.
131. H. Gaye and J. Welfinger, *Proc. Second International Symposium on Slags and Fluxes*, 1984, TMS Warrendale, PA 357–375.
132. M. Blander, A Pelton and G. Erikssen, *4th Internation Conference on Slags and Fluxes*, 1992, ISIJ, 56–60.
133. I. D. Sommerville, A. McLean and Y. D. Yang, Molten Slags, Fluxes, and Salts Conference, Sydney, 1997, ISS Warrendale PA, 375–383.
134. D. J. Sosinsky and I. D. Sommerville, *Metall Trans. B.* 17B (1986): 331–7.
135. J. G. Peacey and W. G. Davenport, *The Iron Blast Furnace* (Oxford: Pergamon Press, 1979)>
136. A. Rist and G. Bonnivard, *Rev. Met.* 60 (1963): 23.
137. A. Rist and N. Meysson, *Rev. Met.* 61 (1964): 121.
138. N. Meysson, J. Weber, and A. Rist, *Rev. Met.* 61 (1964): 623
139. A. Rist and N. Meysson, *Rev. Met.* 62 (1965): 995.
140. N. Meysson, A. Maaref, and A. Rist, *Rev. Met.* 62 (1965): 1161.

141. F. D. Richardson and J. H. E. Jeffes, *J. Iron and Steel Inst.* 160 (1948): 261.
142. N. Tsuchiya, M. Tokuda, and M. Ohtani, *Metall. Trans* B7 (1976): 315.
143. E. T. Turkdogan, G. J. W. Kor, and R. J. Fruehan, *Ironmaking and Steelmaking* 7 (1980): 268.
144. B. Ozturk and R. J. Fruehan, *Metall Trans. B* 16B (1985): 121.
145. N. Standish and W.-K. Lu, eds., *Alkalies in Blast Furnaces* (Ontario: McMaster University Press, 1973).
146. K. P. Abraham and L. J. Staffansson, *Scand. J. Metall.* 4 (1975): 193.
147. J. Davies, J. T. Moon and F. B. Traice, *Ironmaking and Steelmaking* 5 (1978): 151.
148. J. M. Steiler, "Etude Thermodynamique des Laitiers Liquides des Systems K_2O–SiO_2 et K_2O–CaO–SiO_2–Al_2O_3," IRSID, PCM–RE 646, July 1979.
149. L. L. Simmons, L. R. Lowden and T. C. Ehlert, *J. Phys. Chem.* 81 (1977): 709.
150. Z. Pociecha and J. Biczysko, in Seminar on Problems of Air and Water Pollution Arising in the Iron and Steel Industry, Leningrad, Aug. 1971, CI.
151. E. T. Turkdogan and P .H. Josephic, *Ironmaking and Steelmaking* 11 (1984): 192.
152. K. L. Abraham, M. W. Davies, and F. D. Richardson, *J. Iron and Steel Inst.* 19 (1960): 82.
153. E. W. Filer and L. S. Darken, *Trans. AIME* 194 (1952): 253.
154. W. Oelsen, *Arch. Eisenhuttenwes* 35 (1964): 699, 713, 1039, 1115.
155. E. T. Turkdogan and R. A. Hancock, *Trans. Inst. Min. Metall.* 67 (1957–58): 573.
156. F. D. Delve, H. W. Meyer, and H. N. Lander, in *Phys. Chem. Proc. Metallurgy,* ed. G. R. St. Pierre (New York: Interscience, 1961).
157. E. T. Turkdogan, *Metall. Trans. B* 9B (1978): 163.
158. K. Okabe et al., Kawasaki Steel Tech. Res. Lab. Rept, pp. 1–43, May 4, 1974 (English Translation BISI 13657); pp. 1–11 November 27, 1974 (English Translation 13658).
159. E. T. Turkdogan, *Trans. Iron and Steel Inst. Japan* 24 (1984): 591.
160. E. T. Turkdogan, *J. Iron and Steel Inst.* 182 (1956): 74.
161. A. Morales and R. J. Fruehan, *Metall Trans. B,* To be published 1997.
162. J. Chipman et al., *Trans. AIME* 188 (1950): 341.
163. H. Suito and R. Inoue, *Trans. ISIJ* 24 (1984): 301.
164. P. Bremer, *Stahl u. Eisen* 71 (1951): 575.
165. E. Aukrust, P. J. Koros, and H. W. Meyer, *J. Met.* 18 (1966): 433.
166. R. V. Pathy and R. G. Ward, *J. Iron and Steel Inst.* 202 (1964): 995.
167. R. J. Fruehan, in *Advanced Physical Chemistry for Process Metallurgy,* ed. N. Sano, wKLW and P. Ribaud (London: Academic Press, 1977).
168. R. J. Fruehan, *Ironmaking and Steelmaking* 1 (1976): 33.
169. D. Goldstien and R. J. Fruehan, Submitted to *Metall Trans. B*, 1997.
170. R. J. Fruehan and D. Goldstien, 5th European Electric Steel Congress, 1995, Paris, Revue de Metallurgic Paris.
171. D. Goldstien and R. F. Fruehan, Submitted to *Metall Trans. B*, 1997.
172. R. J Fruehan, *Ironmaking and Steelmaking* 3 (1976): 158.
173. P. C. Glaws and R. J. Fruehan, *Metall. Trans. B* 17B (1986): 317.
174. F. Korber and W. Oelsen, *Mitt. Kaiser-Wilhelm Inst. Eisenforsch* 17 (1935): 231.
175. S. Matoba, K. Gunji and T. Kuwana, *Tetsu-to-Hagane* 45 (1959): 229.
176. J. Chipman and T. C. Pillay, *Trans. Met. Soc. AIME* 221 (1961): 1277
177. B. K. D. P. Rao and D. R. Gaskell, *Metall. Trans. B.* 12B (1981): 311.
178. R. J. Fruehan, "*Ladle Metallurgy Principles and Practices*, ISS Warrendale PA, 1985.
179. D. Janke and W. A. Fischer, *Arch. Eisenhuttenwes*, 47 (1976): 195.
180. N. Bannenberg and B. Bergmann, *Stahl u. Eisen*, 112(2) (1992): 57.
181. B. Ozturk and E. T. Turkdogan, *Metal Science*, 18 (1984): 306.
182. E. T. Turkdogan, *J. Iron and Steel Inst.* 210 (1972): 21.
183. K. Larsen and R. J. Fruehan, *Trans. ISS, I&SM*, July 1990, 45.
184. R. J. Fruehan, *"Vacuum Degassing of Steel",* ISS Warrendale PA, 1990.
185. N. Bannenberg, B. Bergmann, H. Wagner and H. Gaye, Proceeding of 6th International Iron and Steel Congress Nagoya, 1990, ISIJ, 3, 603.
186. H. Uljohn and R. J. Fruehan, To be submitted into Trans ISS, 1997.
187. E. T. Turkdogan, Arc. Eisenhuttenwes., 54 (1983) :1.

Chapter 3

Steel Plant Refractories

D.H. Hubble, Chief Refractory Engineer, U.S. Steel Corp. (Retired)

3.1 Classification of Refractories

Refractories are the primary materials used by the steel industry in the internal linings of furnaces for making iron and steel, in vessels for holding and transporting metal and slag, in furnaces for heating steel before further processing, and in the flues or stacks through which hot gases are conducted. At the risk of oversimplification, they may, therefore, be said to be materials of construction that are able to withstand temperatures from 260–1850°C (500–3400°F).

As seen in Fig. 3.1, the melting point of refractory materials in the pure state varies from 1815–3315°C (3300°–6000°F). Refractories in service can tolerate only small amounts of melting

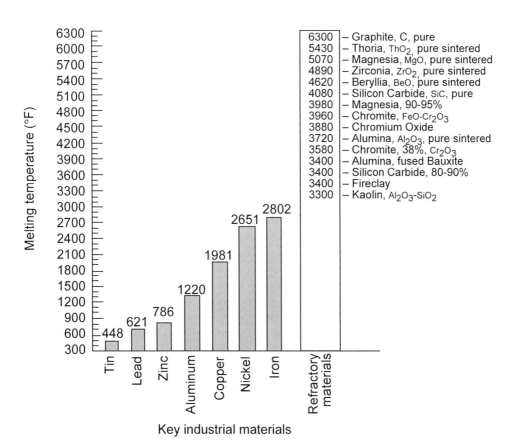

Fig. 3.1 Refractory and industrial materials melting point chart.

(1–5%) without loss of their important structural characteristics. Subsequent discussion will show, however, that the use of many such materials is limited by factors such as cost or instability in certain atmospheres. Also, fluxes present in the initial impure refractory and/or encountered in service can seriously reduce these melting points.

Refractories are expensive, and any failure in the refractories results in a great loss of production time, equipment, and sometimes the product itself. The type of refractories also will influence energy consumption and product quality. Therefore, the problem of obtaining refractories best suited to each application is of supreme importance. Economics greatly influence these problems, and the refractory best suited for an application is not necessarily the one that lasts the longest, but rather the one which provides the best balance between initial installed cost and service performance. This balance is never fixed, but is constantly shifting as a result of the introduction of new processes or new types of refractories. History reveals that refractory developments have occurred largely as the result of the pressure for improvement caused by the persistent search for superior metallurgical processes. The rapidity with which these ever recurring refractory problems have been solved has been a large factor in the rate of advancement of the iron and steel industry. To discuss the many factors involved in these problems and to provide information helpful to their solution are the objectives of this chapter.

Refractories are also vital in the safe operation of the processes and must not expose personnel to hazardous conditions during their manufacture, installation, use or during disposal following their use.

Refractories may be classified in a number of ways. From the chemical standpoint, refractory substances, in common with matter in general, are of three classes; namely, acid, basic, and neutral. Theoretically, acid refractories should not be used in contact with basic slags, gases or fumes whereas basic refractories can be best used in contact with a basic chemical environment. Actually, for various reasons, these rules are often violated. Hence, the time honored chemical classification is largely academic, and of little value as a guide to actual application. Also, the existence of a truly neutral refractory may be doubted. Classifications by use, such as blast furnace refractories or refractories for oxygen steelmaking, are generally too broad and are constantly subject to revision.

For our purposes, refractories will be classified with reference to the raw materials used in their preparation and to the minerals predominating after processing for use. This classification is believed to offer the best possibility for a clear understanding of the origin and nature of steel plant refractories.

3.1.1 Magnesia or Magnesia–Lime Group

This group includes all refractories made from synthetic magnesites and dolomite. These constitute the most important group of refractories for the basic steelmaking processes. All these materials are used primarily as a source of magnesia (MgO).

3.1.1.1 Magnesia

Modern high-purity magnesias are produced in well controlled processes. The principal sources of magnesias are brines (often deep well type) and seawater. Magnesium hydroxide, $Mg(OH)_2$, is precipitated from these sources by reaction with calcined dolomite or limestone; one source uses a novel reactor process. The resultant magnesium hydroxide slurry is filtered to increase its solids content. The filter cake can then be fed directly to a rotary kiln to produce refractory grade magnesia, but more commonly now the filter cake is calcined at about 900–1000°C (1650–1830°F), usually in multiple-hearth furnaces, to convert the magnesium hydroxide to active magnesia. This calcined magnesia is then briquetted or pelletized for firing into dense refractory-grade magnesia, usually in shaft kilns which reach temperatures around 2000°C (3630°F). The end product is sintered magnesia.

Fused magnesia is produced by melting a refractory grade magnesia or other magnesia precursor in an electric arc furnace. The molten mass is then removed from the furnace, cooled, and broken up to begin its path for use in refractories.

The impurities in magnesia are controlled by the composition of the original source of the magnesia (brine or seawater), the composition of the calcined dolomite or limestone, and the processing techniques. In particular the amounts and ratio of CaO and SiO_2 are rigorously controlled, and the B_2O_3 is held to very low levels. The end results are high-grade refractory magnesias which are ready for processing into refractory products. Tables 3.1 through 3.4 show the compositions of different grades of magnesia.

Table 3.1 Selected Sintered Magnesias Produced in North America (high CaO/SiO_2 ratio)

Reference Code	SM-1	SM-2	SM-3
Chemical Analysis (wt%)			
CaO	2.2	2.2	0.8
SiO_2	0.7	0.35	0.1
Al_2O_3	0.1	0.20	0.1
Fe_2O_3	0.2	0.20	0.1
B_2O_3	0.015	0.02	0.005
MgO	96.7	96.3	98.8
CaO/SiO_2 ratio	3.1	6.3	8.0
Bulk Density (kg/m³)	3400	3420	3400
Average Crystallite Size (μm)	~ 80	~ 90	~ 100

Table 3.2 Selected Sintered Magnesias Produced in Europe and the Middle East (high CaO/SiO_2 ratio)

Reference Code	SM-4	SM-5	SM-6
Chemical Analysis (wt%)			
CaO	1.90	0.65	0.7
SiO_2	0.20	0.15	0.03
Al_2O_3	0.05	0.06	0.03
Fe_2O_3	0.20	0.50	0.04
B_2O_3	0.015	0.008	0.005
MgO	97.5	98.5	99.2
CaO/SiO_2 ratio	9.5	4.3	23.3
Bulk Density (kg/m³)	3440	3450	3430–3450
Average Crystallite Size (μm)	150–160	~ 100	70–90

From Ref. 2.

Table 3.3 Selected Fused Magnesia Produced in North America

Reference Code	FM-1	FM-2
Chemical Analysis (wt%)		
CaO	0.95	1.8
SiO_2	0.30	0.5
Al_2O_3	0.11	0.2
Fe_2O_3	0.16	0.5
B_2O_3	0.0025	0.01
MgO	98.50	97.0
CaO/SiO_2 ratio	3.2	3.6
Bulk Density (kg/m³)	3530	3500
Average Crystallite Size (μm)	~ 780	~ 800

From Ref. 2.

Table 3.4 Selected Sintered Magnesias Produced in North America (low CaO/SiO_2 ratio)

Reference Code	SM-7	SM-8	SM-9
Chemical Analysis (wt%)			
CaO	0.7	0.80	0.9
SiO_2	0.7	0.60	2.1
Al_2O_3	0.1	0.20	0.15
Fe_2O_3	0.2	0.20	0.35
B_2O_3	0.1	0.10	0.23
MgO	98.2	98.05	96.2
CaO/SiO_2 ratio	1.0	1.3	0.4
Bulk Density (kg/m³)	3330	3340	3260
Average Crystallite Size (μm)	50-60	~ 60	~ 40

From Ref. 2.

High purity is quite important because MgO has high refractoriness and good resistance to basic slags. Minimizing the total impurities content in magnesias is quite important because impurities affect refractoriness and performance. A high CaO/SiO_2 ratio, preferably 2:1 or slightly higher, is optimum for maintaining high refractoriness in magnesias. As the total impurities are reduced to about 1% or less, the CaO/SiO_2 ratio has less and less significance. High density reduces infiltration and dissociation of magnesia grain by slag. Large crystallite size (best achieved in fused magnesias) provides less surface area for slag attack. Low B_2O_3 content ensures high strength at elevated temperatures for burned brick. Low lime to SiO_2 ratios are required in certain products where MgO is used with other raw materials (such as Cr_2O_3) or for maximum resistance to hydration.

3.1.1.2 Dolomite

The natural double carbonate dolomite ($CaCO_3 \cdot MgCO_3$) can be converted to refractory dolomite ($CaO \cdot MgO$) by high temperature firing. A limited number of dolomite deposits exists in the world with satisfactory uniformity, purity, and calcining behavior to be processed into high purity, refractory dolomite at a reasonable cost. High purity dolomite is greater than 97% CaO + MgO and

0.5–3% impurities. Most high purity dolomite deposits are difficult to calcine and sinter to high density and usually require special methods to yield acceptable refractory grade dolomite. Silica, iron oxide and alumina are the most common impurities in high purity dolomite. See Table 3.5.

Table 3.5 Chemical and Physical Properties of High Purity, Dead-Burned Dolomite

Chemical Analysis (wt%)		Physical Properties	
Al_2O_3	0.45	Bulk Density (kg/m^3)	3250
Fe_2O_3	0.90	Closed pore volume (%)	1.3
SiO_2	0.70		
MgO	41.20		
CaO	56.70		

From Ref. 2.

Dolomite has excellent refractoriness and is thermodynamically very stable in contact with steel or steelmaking slags. Note in Table 3.6 that CaO is the most stable of the common refractory oxides at steelmaking temperatures.

Table 3.6 The Free Energies of Formation for Various Refractory Oxides

	ΔG @ 1600 °C $\left(\dfrac{Kcal}{mole\ O_2}\right)$
$2\ Ca + O_2 = CaO$	–205
$^4/_3\ Al + O_2 = {}^2/_3\ Al_2O_3$	–175
$2\ Mg + O_2 = 2\ MgO$	–170
$Si + O_2 = SiO_2$	–140
$^4/_3\ Cr + O_2 = {}^2/_3\ Cr_2O_3$	–110

From Ref. 2.

The free lime portion of the dead burned dolomite can react with atmospheric moisture which causes the material to powder and crumble. The degree of hydration under set conditions of time, temperature, and relative humidity is dependent upon the proportion of lime and impurities contained in the material and upon the density of the grain achieved during the dead burning process.

In practice, with modern packaging materials and techniques together with other means of protecting the products, the storage of dolomite products can be extended.

3.1.2 Magnesia–Chrome Group

Chrome ores, or chromites, often called chrome enriched spinels, are naturally occurring members of the spinel mineral group. These materials are all characterized by relatively high melting points, good temperature stability (particularly in thermal union with magnesite) and moderate thermal expansion characteristics.

Chrome ores are covered by the formula $(Mg \cdot Fe^{+2})(Cr,Al,Fe^{+3})_2O_4$, where magnesium can substitute for iron, and aluminum for chromium. Accessory minerals, called gangue, are often associated with these ores. Gangue minerals contribute residual silica, lime and additional magnesia to the chrome ore. Control, placement and quantity of gangue minerals is as important to finished refractory properties as the basic chemical composition of the individual ore. Ores are often upgraded by mechanical separation techniques to minimize impurities. Thus, chrome concentrates with low gangue impurities, as shown in Table 3.7, are often the article of commerce used in refractories.

Pure Cr_2O_3 chemically separated from chromite can be used in combination with lower cost oxides to produce specific refractory properties.

In addition to basic refractory raw materials composed of dead burned magnesia and chrome ore as starting materials, other magnesite-chrome combinations are also a part of this series. A magnesite-chrome group of raw materials exists, including co-sintered magnesite-chrome, fused magnesite-chrome and synthetic picrochromite—a combination of magnesite and chromic oxide. Typical magnesite-chrome combinations appear in Table 3.8.

Table 3.7 Examples of Refractory-Grade Chrome Ores (typical data)

Source	Philippines −10 concentrates	South Africa	Turkey
Typical Chemistry			
Silica (SiO_2)	3.20%	0.70%	3-4%
Alumina (Al_2O_3)	28.10	14.80	20-21
Iron Oxide (Fe_2O_3)	15.20	28.60	15-16
Lime (CaO)	0.28	0.05	0.30
Magnesia (MgO)	17.80	10.80	18.00
Chromic Oxide (Cr_2O_3)	35.00	47.10	41.42
Size	12% min. +14 mesh 20% max. −65 mesh	−20+150	−10+100 or Lump
Structure	Hard, Massive	Friable	

Table 3.8 Examples of Magnesite-Chrome Raw Materials (typical data)

Source	U. Kingdom Co-Sintered	South Africa Fused Mag. Chrome	Canada/U.S. Fused Mag. Chrome
Typical Chemistry			
Silica (SiO_2)	1.15 %	1.60 %	1.2 %
Alumina (Al_2O_3)	5.70	7.20	6.1
Titania (TiO_2)	0.20	0.20	0.3
Iron Oxide (Fe_2O_3)	12.50	10.70	13.0
Lime (CaO)	0.80	0.60	0.4
Magnesia (MgO)	61.50	59.90	60.1
Chromic Oxide (Cr_2O_3)	18.00	18.80	18.6
Bulk Specific Gravity (g/cm³)	3.53	3.75	3.8

From Ref. 3.

Table 3.9 Examples of Refractory-Grade Silicas (typical data)

Type	Natural Silica (Quartzite)	Fused Silica
Typical Chemistry		
Silica (SiO_2)	99.70 %	99.60 %
Alumina (Al_2O_3)	0.09	0.20
Titania (TiO_2)	0.01	—
Iron Oxide (Fe_2O_3)	0.09	0.03
Lime (CaO)	0.03	0.04
Magnesia (MgO)	0.01	
Total Alkalies	0.02+	0.01
Bulk Specific Gravity (g/cm³)	2.33	2.20
Major Minerals		
Quartz	X	
Vitreous or Glassy		X

From Ref. 3.

3.1.3 Siliceous Group

3.1.3.1 Natural

Natural silica occurs primarily as the mineral quartz. Heat treatment of silica in the manufacture of refractory products can result in the formation of a number of different crystalline forms. Depending on the thermal history applied, silica refractories may contain various mineral assemblages of quartz, cristobalite, and tridymite. Silica raw materials used in the manufacture of conventional silica refractories, as shown in Table 3.9, must contain high silica (99% or above SiO_2) and low impurity levels, particularly alumina and alkalies, which can act as fluxes during firing of the refractory and reduce overall refractoriness of the end product.

3.1.3.2 Fused Silica

Fused silica is produced by actual fusion of specially selected, very high grade silica sands in electric arc, electrical resistance, or other furnace procedures. Crystalline raw material is converted into an amorphous glass, or fused silica. Properties of this fused raw material vary considerably from those of the original quartz sand, in particular fused silica has very low thermal expansion. Fused silica products exhibit low thermal conductivity, high purity and excellent resistance to thermal shock.

3.1.3.3 Silicon Carbide

Commercial silicon carbide (SiC) used as a refractory raw material is manufactured by abrasive grain producers in electric furnaces from a mixture of coke and silica sand. The finished material is extremely hard (9.1 MOH's scale) with high thermal conductivity and good strength at elevated temperatures, as well as very good resistance to thermal shock. Silicon carbide dissociates at 2185°C (3965°F) and oxidizes slowly in air, but is relatively stable under reducing conditions. The material is serviceable at 1535–1650°C (2800–3000°F) for many applications. Table 3.10 shows some analysis of refractory grade silicon carbides.

Table 3.10   Examples of Refractory-Grade Silicon Carbides

Type	Coarse, 6/F Low Iron	200/F Low Iron	10/F
Typical Chemistry			
Silica (SiO_2)	3.50%	1.90%	2.40%
Alumina (Al_2O_3)	0.50	0.40	0.76
Titania (TiO_2)	0.05	0.06	0.07
Iron Oxide (Fe_2O_3)	0.24	0.45	0.93
Lime (CaO)	0.12	0.20	0.49
Magnesia (MgO)	0.02	0.03	0.29
Alkalies	<0.04	<0.04	—
Silicon Carbide (SiC)	94.90	96.80	91.70
Total Carbon	29.10	29.20	30.70
Free Carbon	0.60	0.30	3.20

From Ref. 3.

3.1.3.4 Zircon/Zirconia

Zircon, or zirconium silicate ($ZrO_2 \cdot SiO_2$), is a naturally occurring raw material having excellent refractoriness. Specific gravity (4.5–4.6 g/cm³) is unusually high compared to most refractory materials.

Major zircon sources include the natural sands of western Australia, eastern Florida, South Africa's northern Natal east coast, the European Economic Community countries and the Peoples Republic of China. Zircon usually is found with other heavy mineral sands, most notably titania minerals. Table 3.11 shows properties of zircon.

Zirconium oxide (ZrO_2) is produced commercially as the naturally occurring mineral baddeleyite. The refractory industry has been a major growth area for zirconia. The relatively high melting point of baddeleyite, along with superior resistance to corrosion and erosion, make zirconia an ideal component for several refractory systems. Zirconia in the natural state occurs in the monoclinic crystal phase. When heated, zirconia undergoes a phase change (to tetragonal) and a volume shrinkage of about 5%; with additional increases in temperature, the cubic form is stable at 2350°C (4260°F). To counteract these deleterious phase changes, zirconia may be stabilized to the cubic phase with small amounts of calcia, magnesia, or yttria, with the result that stability, thermal shock

Table 3.11 Refractory-Grade Zircons (typical data)

Type	Zircon Sand	Zircon −200 Mesh	Zircon −400 Mesh
Typical Chemistry (Calcined Basis)			
Silica (SiO_2)	32.60%	33.60%	33.70%
Alumina (Al_2O_3)	0.80	0.50	0.50
Titania (TiO_2)	0.20	0.10	0.20
Iron Oxide (Fe_2O_3)	0.03	0.02	0.01
Lime (CaO)	0.03	0.03	0.03
Magnesia (MgO)	0.03	0.02	0.02
Zirconia (Zr_2O)	66.20	65.60	65.60
Screen Analysis			
+200 Mesh	—	2	0
+325 Mesh	—	11	1
+400 Mesh	—	7	5
−400 Mesh	—	79	95

From Ref. 3.

Table 3.12 Refractory-Grade Zirconia (typical data)

Type	Baddelevite (natural mineral concentrate)	Zirconia Fused (lime stabilized)
Typical Chemistry (Calcined Basis)		
Silica (SiO_2)	0.36%	0.37%
Alumina (Al_2O_3)	0.01	0.50
Titania (TiO_2)	0.30	0.31
Iron Oxide (Fe_2O_3)	0.20	0.09
Lime (CaO)	0.01	3.98
Magnesia (MgO)	0.05	0.27
Zirconia (Zr_2O)	99.20	94.46
Soda (Na_2O)	0.02	0.02
Potash (K_2O)	<0.01	<0.01
Lithia (Li_2O)	<0.01	<0.01
Loss on Ignition	0.11	—

From Ref. 3.

resistance, and hot load properties are enhanced in the final product. Zirconia is available in the natural, stabilized, or fused state, the latter often as a mixture with alumina, silica, or other compatible oxides. Table 3.12 shows typical properties of zirconia.

3.1.4 Clay and High-Alumina Group

3.1.4.1 Clays

Although clays were among the first raw materials used to make refractories, their usage has diminished as demands placed on modern refractories have necessitated better performing materials to replace them. Nevertheless, clays are still an important material in the industry.

Clays may be used as binders, plasticizers, or as aggregates for producing refractories. The properties of some domestic clays are shown in Table 3.13.

3.1.4.2 Bauxitic Kaolins

Several other types of 50–70% alumina raw materials are also used in refractories. Bauxitic kaolins, or bauxite clay combinations, represent another class of natural aluminum silicates used in refractory manufacture. Major U.S. commercial deposits of aluminous refractory raw materials are found in southeastern Alabama, and in east-central Georgia. These materials, shown in Table 3.14, are mined, blended and fired in rotary kilns to yield a versatile line of calcined aluminum silicates, ranging from about 50% to over 70% alumina content.

Table 3.13 Some Examples of Refractory Clays (typical data)

Locale Type	Missouri Flint	Missouri Plastic	Kentucky Mine Run	Georgia Kaolin	Ball Clay	Bentonite
Typical Chemistry **(Calcined Basis)**						
Silica (SiO_2)	50.0%	59.5%	51.4%	52.1%	62.5%	67.7%
Alumina (Al_2O_3)	45.3	32.3	42.5	44.5	31.4	20.4
Titania (TiO_2)	2.4	1.6	2.6	1.8	1.5	0.2
Iron Oxide (Fe_2O_3)	0.8	2.2	1.8	0.6	2.5	4.5
Lime (CaO)	0.2	0.4	0.2	0.3	0.5	1.1
Magnesia (MgO)	0.2	0.8	0.5	0.1	0.7	2.3
Alkalies (total)	0.6	3.1	1.4	0.3	0.7	3.5
L.O.I. (dry basis)	—	8.3	12.7	13.7	10.3	5.4
B.S.G. (g/cm³)	2.57–2.62					
Bulk Density (Pcf)	148	130	143	—	—	—
Linear Change, after Cone 15 (in./ft.)	0.5–1.1	0.75–1.25	0.75–1.0	—	—	—
PCE (Cone)	34+	29–31	32½–33	34+	31½	13+

From Ref. 3.

Table 3.14 Domestic U.S. Bauxitic Kaolins

Typical Chemistry	Ucal 50	Ucal 60	Ucal 70
Silica (SiO_2)	47.30%	36.90%	25.70%
Alumina (Al_2O_3)	49.20	59.20	70.00
Titania (TiO_2)	2.40	2.60	3.20
Iron Oxide (Fe_2O_3)	1.00	1.10	1.00
Lime (CaO)	0.02	0.02	0.02
Magnesia (MgO)	0.04	0.05	0.04
Alkalies (total)	0.08	0.11	0.07
Specific Gravity (g/cm³)	2.60	2.70	2.78
Pyrometric Cone Equivalent	35–36	37–38	38–39

From Ref. 3.

3.1.4.3 Sillimanite

Andalusite, sillimanite and kyanite comprise the water-free, natural aluminum silicate varieties of minerals known as the sillimanite group. Andalusite and kyanite are the more common commercial materials. These minerals are normally about 60% alumina, with the balance composed primarily of silica with minor iron and titania impurities, Table 3.15. Andalusite and sillimanite have several important characteristics; when heated at high temperatures, the refractory mineral mullite ($3Al_2O_3 \cdot 2SiO_2$) is formed. Complete mullite occurs at 1300–1400°C (2372–2552°F). This mineral is a key component of many high-alumina materials.

3.1.4.4 Bauxite

Bauxite in the crude state is a naturally occurring group of minerals composed primarily of either gibbsite ($Al_2O_3 \cdot 3H_2O$), diaspore, or boehmite [AlO(OH)], and various types of accessory clays. Refractory grade calcined bauxites are a specific form as found in Table 3.16. These are produced from low iron, low silica materials in rotary kiln calcining operations or down-draft kilns. Calcining temperatures are in the 1400–1800°C (2550–3275°F) range. Crude bauxite is converted to the

Table 3.15 Examples of Andalusite and Kyanites

Source	S. Africa A	Andalusite S. Africa B	French A	Kyanite United States
Typical Chemistry				
Silica (SiO_2)	38.00%	38.10%	43.80%	39.90%
Alumina (Al_2O_3)	60.30	61.90	53.00	56.00
Titania (TiO_2)	0.17	0.16	0.25	1.80
Iron Oxide (Fe_2O_3)	0.72	0.59	1.20	1.60
Lime (CaO)	0.11	0.06	0.20	0.04
Magnesia (MgO)	0.09	0.11	0.15	0.04
Soda (Na_2O)	0.06	0.06	0.20	
Potash (K_2O)	0.22	0.13	0.20	0.16 total
Lithia (Li_2O)	0.02	0.01	—	
L.O.I. (dry Basis)	0.44	0.28	0.08	0.90
Screen Analysis				
+6 Mesh	17%	77%		35 mesh or
–6 +10 Mesh	34	14		100 mesh ×
–10 +16 Mesh	44	9		down
–16 Mesh	5	—		

From Ref. 3.

Table 3.16 Refractory-Grade Calcined Bauxites (typical data)

Source	Guyana, S.A.	China	Brazil	United States (Diaspore)
Typical Chemistry (Calcined Basis)				
Silica (SiO_2)	6.50%	5.60%	9.0–10.0%	16.8%
Alumina (Al_2O_3)	88.00	87.50	85.0–87.0	75.7
Titania (TiO_2)	3.25	3.64	1.9–2.3	3.8
Iron Oxide (Fe_2O_3)	2.00	1.56	1.6–2.1	0.9
Lime (CaO)	0.02	0.06	<0.2	0.3
Magnesia (MgO)	0.02	0.17	<0.2	0.5
Alkalies (total)	0.01	0.62	<0.1	2.3
Loss in Ignition	0.25	trace	<0.1	trace
Bulk Specific Gravity (g/cm³)	3.10 min.	3.20	3.15–3.20	2.74

From Ref. 3.

minerals corundum (Al_2O_3) and mullite ($3Al_2O_3 \cdot 2S_iO_2$) — both very refractory components. Important features of bauxites are maximum alumina values (85% or more desired), maximum bulk specific gravity, and minimum impurities such as iron oxide, titania, alkalies (Na_2O, K_2O and Li_2O) and alkaline earths (CaO and MgO). Major sources for calcined, refractory grade bauxite are Guyana, South America, and the Peoples Republic of China.

3.1.5 Processed Alumina Group

Several types of chemically and thermally processed aluminas are used in refractories. These include calcined, tabular, and fused alumina all from the Bayer process as illustrated in Fig. 3.2 and Fig. 3.3.

In general, calcined aluminas are used to promote refractory binding during manufacture or use whereas tabular or fused products form very stable aggregates. Tabular alumina is formed by calcination at 1925°C (3500°F), whereas fused alumina is more dense after total melting and rapid solidification. Tables 3.17 through 3.19 show properties of these alumina grades.

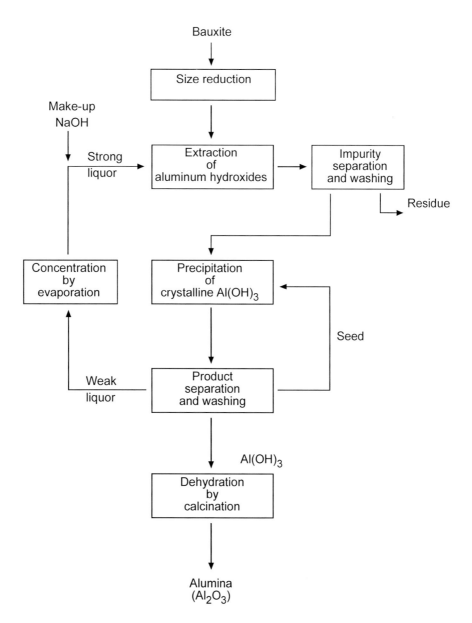

Fig. 3.2 Schematic of Bayer process.

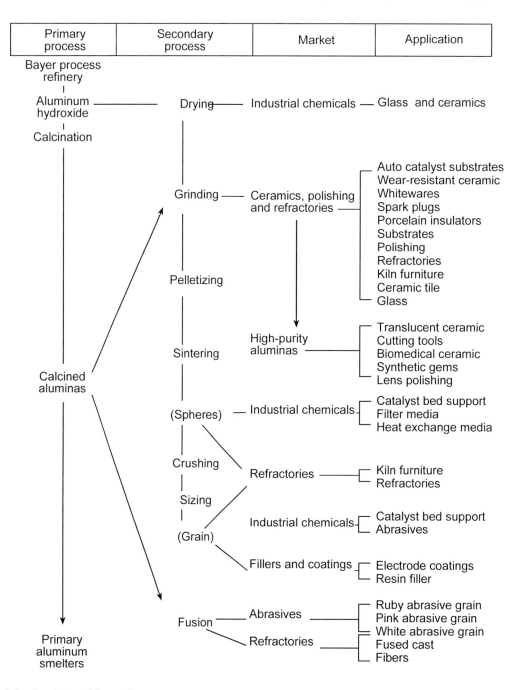

Fig. 3.3 Products of Bayer Process.

3.1.6 Carbon Group

Various carbon forms are used to an ever increasing extent in refractories. For example, modern refractories use various graphite forms in combination with oxides to impart special properties. The graphite may be synthetic in nature as produced by heating calcined petroleum coke to 3000°C (5400°F) or may be natural graphite(s) from China, Mexico, Canada etc. Some all-carbon or all-graphite refractories may be produced for applications in highly reducing atmospheres.

Table 3.17 Calcined Aluminas (typical data)

Source	Hydrated	Intermediate Soda Calcined A	Intermediate Soda Fine Ground Calcined B	Low Soda Super Ground Calcined C
Typical Chemistry				
Silica (SiO_2)	0.01%	0.03%	0.05%	0.04%
Alumina (Al_2O_3)	65.00	99.40	99.80	99.70
Titania (TiO_2)	—	0.03	0.01	<0.01
Iron Oxide (Fe_2O_3)	0.004	0.04	0.02	0.04
Lime (CaO)	—	0.06	0.20	0.04
Magnesia (MgO)	—	0.02	0.03	0.01
Total Alkalies (Na_2O)	0.15	0.19	0.23	0.08
Loss on Ignition (LOI)	34.50	0.21	0.10	0.03
Ultimate Crystal Size (μm)	—	3 to 5	—	1.50
Surface Area (m^2/g)	0.15	0.50	—	1.00

From Ref. 3.

Table 3.18 Tabular Alumina (typical data)

Source	United States	United States
Typical Chemistry		
Silica (SiO_2)	0.04%	0.04%
Alumina ($Al2O_3$)	99.70+	99.80+
Titania (TiO_2)	0.01	0.01
Iron Oxide (Fe_2O_3)	0.06	0.03
Lime (CaO)	0.04	0.01
Magnesia (MgO)	<0.001	0.02
Boron Oxide (B_2O_3)	<0.001	—
Soda (Na_2O)	0.16	0.05
Bulk Specific Gravity (g/cm^3)	3.55	3.54

From Ref. 3.

Table 3.19 Fused Aluminas (typical data)

Typical Chemistry	Brown[1]	White[2]	White[3]
Silica (SiO_2)	0.40%	0.04%	0.14%
Alumina (Al_2O_3)	96.70	99.50	99.41
Titania (TiO_2)	2.52	—	<0.01
Iron Oxide (Fe_2O_3)	0.10	0.10	0.02
Lime (CaO)	0.05	0.05	0.02
Magnesia (MgO)	0.18	—	0.02
Soda (Na_2O)	0.10	0.30	0.39
Potash (K_2O)	—	—	0.01
Bulk Specific Gravity (g/cm^3)	3.75	3.77	3.50*

[1]General Abrasives – Treibacher, Inc. [2]Washington Mills [3]C.E. Minerals *bulk density

From Ref. 3.

Generally, graphites are used in refractories in order to reduce the wetting characteristics of the refractory material with respect to slag corrosion and to increase the thermal conductivity which will result in better thermal shock resistance. In oxide-carbon refractories, the carbon content may range anywhere from as low as 4–5% up to as high as 30–35%. Note that as the graphitic content increases, the thermal conductivity of the refractory increases, but the density of the refractory decreases. This result is primarily due to the fact that the density of graphite is much less than the density of the other refractory materials being used. There are other contrasting differences in the morphology of the graphite as compared to the other refractory materials. The graphite materials, which are used in refractories, are commonly of a flaky structure; therefore, these flakes do not lend themselves to the same particle packing phenomena as do granular particles.

Table 3.20 shows the properties of several types of graphite. Flake graphite is commonly purified to extremely high carbon contents of 99 wt% carbon or higher. This purification utilizes both a chemical process and a thermal process; these steps have a significant effect on the price of the materials.

Table 3.20 Typical Properties of Graphite

	Amorphous	Flake	High crystalline	Primary artificial	Secondary artificial
Carbon (wt%)	81.00	90.00	96.70	99.90	99.00
Sulfur (wt%)	0.10	0.10	0.70	0.001	0.01
True Density (kg/m^3)	2310	2290	2260	2250	2240
Graphite Content (wt%)	28.0	99.9	100.0	99.9	92.3
d-Spacing (002) (nm)	33.61	33.55	33.54	33.55	33.59
Ash True Density (kg/m^3)	2680	2910	2890	2650	2680
Resistivity (ohm-m)	0.00091	0.00031	0.00029	0.00035	0.00042
Morphology	Granular	Flaky	Plates Needles Granular	Granular	Granular

From Ref. 2.

3.2 Preparation of Refractories

3.2.1 Refractory Forms

Refractories are produced in two basic forms: preshaped objects and unformed compositions in granulated, plastic forms or spray mixes. The preformed products are called bricks and shapes. The unformed products, depending on composition and application, are categorized as specialties or monolithics. Refractory shapes, found in Fig. 3.4, may range from simple nine inch brick to tapered 30 inch brick, or complex tubes or rods.

Specialties or monolithics are refractories that cure to form a monolithic, integral structure after application. These include products known as plastics, ramming mixes, castables, pumpables, pourables, spray mixes, gunning mixes and shotcreting. The mortars used to install brick and shapes constitute a category of the specialty refractory classification.

Mortars are available in compositions that either approximate the brick they are bonding together, or are chosen so that their thermal expansion will be similar to the brick with which they are used. The goal is to achieve a lining that comes as close as possible to being a monolithic and continuous refractory structure.

As the name implies, plastic refractories are ready-to-use materials that are installed by tamping or ramming. After drying, either the heat from firing the equipment or a chemical binder converts the plastic material to a solid, monolithic structure. Plastic refractories are available in both clay and nonclay compositions.

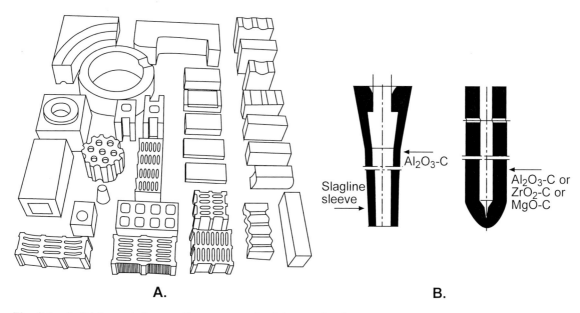

Fig. 3.4 A. Bricks and shapes; B. more complex tubes and rods.

Castables, or refractory concretes, are predominately dry, granular refractory mixes designed to be mixed on site with water and capable of curing to a stable dimensional form through hydraulic or chemical setting. Castables are particularly suited to the molding of special shapes and parts at the installation site. They can be used for forming complete furnace linings, and other unique shapes. They can be applied by pouring, pumping, troweling, gunning, and shotcreting. They have the advantage of being readily usable at the operating temperature of the equipment after hydraulic or chemical setting and removal of all moisture has taken place.

Spray mixes are made from a variety of refractory compositions. The common feature of these materials is that they contain sufficient water for transport via pumps and for spray application onto a furnace wall or ceiling. A set accelerator is added in sufficient quantity such that the mix sets rapidly.

Gunning mixes comprise a variety of specialty refractory compositions that develop a solid shape by air drying, hydraulic setting, or heat curing. The principal requirements are that they can be blown into position by air pressure through a lance or nozzle, but must adhere on impact and build up to the desired lining thickness. They are used for patch-type repairs, especially inside empty, hot furnaces.

Fig. 3.5 is a greatly simplified flowsheet illustrating the various methods of refractory manufacture and the resultant products, and classifies raw materials as either calcined, uncalcined (raw), or binders. Calcined materials have been fired to remove moisture and volatiles and to densify the material to minimize subsequent in-service shrinkage and reaction. The calcining temperature will range from 1093–3315°C (2000–6000°F). Raw or uncalcined materials are cheaper to use than calcined materials and are used to impart desirable characteristics such as plasticity or volume expansion to certain refractories. Binders are used to impart strength to the refractory during manufacture or in service.

3.2.2 Binder Types

3.2.2.1 Temporary Binder

Temporary binders include paper byproducts, sugar, or certain clays. Their function is to improve handling strength during manufacture.

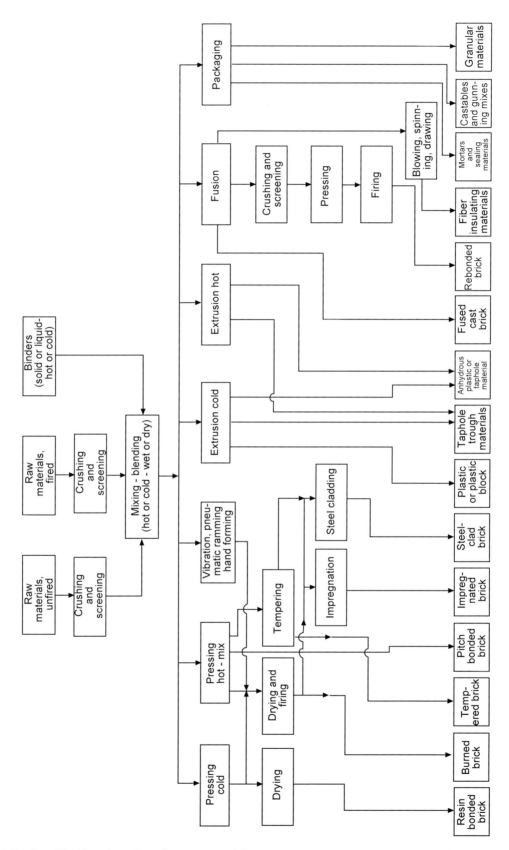

Fig 3.5 Simplified flowsheet for refractory materials.

Table 3.21 Dry Binder Ingredients for Ramming Mixes, Plastics and Patching Plasters

Boric Acid
Colloidal Aluminas
Colloidal Silicas
Goulac
Hydrated Aluminas
Methyl Cellulose
Mono-Aluminum Phosphates
Petroleum Pitches
Resins
Silica Fume
Sodium Aluminum Phosphates
Sodium Poly Phosphates
Sodium Silicate
Starches
Synthetic Clays
Waxes

3.2.2.2 Chemical Binders

Chemical binders impart strength during manufacture, after manufacture, or on installation as a monolithic material. Tables 3.21 and 3.22 show some of the binders which might be used for monolithic materials depending on the application and properties desired.

3.2.2.3 Cement Binders

Cement binders set hydraulically when mixed with water. The primary binders of this type used in refractories are the calcium-aluminate cements which set rapidly and are able to retain some of their bonding strength to intermediate temperatures. The cement(s) used in refractories are $CaO \cdot Al_2O_3$ type cements which develop strength more rapidly than portland cements (near maximum strength in one day as opposed to 7–10 days) and are capable of retaining strength to higher temperatures than Portland cements. Table 3.23 shows several grades of refractory cements.

Table 3.22 Liquid Binder Ingredients for Ramming Plastics and Patching Mixes

Aqueous	Non-Aqueous
Alginates	Coal Tars
Chromic Acid	Oils
Colloidal Silica Solution	Petroleum Tars
Glutrin Resins	
Mono Aluminum Phosphates	
Molasses	
Phosphoric Acids	
Potassium Silicates	
Resins	
Sodium Aluminum Phosphates	
Sodium Silicates	
Water	

3.2.2.4 Organic Binders

Organic binders include tars, pitches, or resins for use in reducing atmospheres where the carbon residuals impart bonding strength or act to inhibit alteration.

The coal tar pitches, derived as by-products of coke oven operation, were the preferred binders for refractories up to the mid 1970s. Around 1978, transition was made from coal tar to petroleum pitches. Today most organic binders are resin based so they can be processed cold to minimize environmental problems.

Table 3.23 Chemical Composition of Calcium-Aluminate Cements

	A	B	C	D	E
Al_2O_3	39.0	47.0	58.5	72.5	79.0
$TiO_2 + Fe_2O_3$	17.8	7.4	1.7	0.25	0.3
CaO	38.5	34.3	33.5	26.0	18.0
SiO_2	4.5	7.9	5.2	0.35	0.2
MgO	1.0	0.9	0.1	—	0.4
Alkalies + SO_3	0.25	2.1	0.4	0.4	0.5

Phenolic resins, the most important synthetic binders for refractory materials, are condensation products of phenol and formaldehyde. These resins are differentiated according to whether they represent novolacs or self curing resoles. The versatility of phenolic resins is derived from the various bonding functions that they can provide. Among other things phenolic resins can provide an intermediate bonding function associated with a thermosetting range of properties during production of refractory bricks and, at a later stage, are capable of forming (polymeric) carbon by pyrolytic decomposition. In the form of bonding carbon, these results contribute to the quality of the products.

3.2.3 Processing

A knowledge of the various steps in refractory processing is very important in understanding the behavior of steelplant refractories. As shown in Figure 3.5, all refractories use crushing, sizing and mixing or blending as the first steps in their manufacture. These steps produce the proper particle sizings necessary for the desired product density and strength. In sizing a mix to produce dense brick, for example, the raw materials are crushed and screened to produce some desired particle size range as illustrated below:

Screen Size		
mm	Tyler Mesh Size	wt% of Mix
4.6 through 1.65	4 through 10	20
1.65 through 1.17	10 through 14	30
1.17 through 0.83	14 through 20	5
0.83 through 0.30	20 through 48	5
0.30 through 0.15	48 through 100	5
0.15 through 0.07	100 through 200	5
0.07 through 0.04	200 through 325	10
Below 0.04	Below 325	20

The crushing and screening techniques used are increasingly complex, including vibratory screening equipment and air classification techniques.

Mixing and blending steps range from the simple addition of water to clay, to hot mixing of preheated aggregates with selected resin or other anhydrous binders. Special sequences of combining raw materials, the time of mixing, and the use of high energy mixing equipment are used to obtain uniform mixing and equal distribution of additives.

Fig. 3.5 illustrates some (but not all) of the forming methods that are used after mixing and blending. The most widely used manufacturing method involves cold pressing of the grain-sized and blended mix to produce a dense refractory shape. Power or dry pressing of the mix into a shape is done on hydraulic or mechanical presses capable of forming the moist material (2–5% water) at pressures of 34.5 to 103.4 MPa (5 to 15 ksi). The degree of compaction obtained in this pressure range depends on plasticity and particle sizing, but most high quality brick are pressed to the point where further pressure would produce laminations or internal cracking. The pressing chamber or mold may be evacuated or de-aired to increase density and prevent laminations resulting from entrapped air. Dry pressing lends itself to a wide variety of materials and can produce a wide range of properties. Certain products may be pressed hot and these materials are usually plasticized with liquid pitch.

Shapes may also be formed by applying pressure by other means such as vibration, pneumatic ramming, hand molding, or isostatic pressing. Many brick and special shapes in fire clay compositions are also formed by extrusion followed by low pressure pressing (the stiff mud repress process). In this process, more plastic mixes (10–15% water) are forced through a die by a power driven auger, cut into slugs, and then pressed to shape. This process usually involves de-airing during extrusion.

Hot extrusion of pitch or other anhydrous bonded materials may also be used; however, this is mainly for monolithic materials.

In limited cases, raw materials are fused in very high temperature electric furnaces and cast into larger ingots in graphite molds. These ingots can subsequently be cut to the desired shape, or may be broken and crushed into a refractory raw material for use in conventional powder pressed brick or for use in monoliths. In still another process, molten refractory may be blown, drawn, or spun into fibers for subsequent use in forming mats, blankets, or boards.

Many refractory materials are used in bulk form. Sized, granular refractories may be used in dry form or mixed with water at the plant site before installation by casting or gunning. Wet extruded material may be packaged to avoid drying and shipped to the plant site ready for application by ramming into place as a large monolithic structure. Wet bonding mortars may be shipped in sealed containers ready for use.

As shown in Fig. 3.5 many products are prefired before shipment. The purpose of firing is to produce dimensionally stable products having specific properties. Firing in modern refractory plants is accomplished in continuous or tunnel kilns.

In tunnel kiln firing, which is usually preceded by tunnel drying, the unfired brick loaded on small cars are passed slowly through a long tunnel shaped refractory lined structure, divided successively into preheating, firing and cooling zones, generally taking three to five days for the trip. This time will vary widely, however, with the product being fired. Products of combustion from the fuel burned in the firing zone pass into the preheating zone (countercurrent to the direction of travel of the cars onto which the brick are stacked) and give up their heat to the oncoming loads of brick. Some refractories are also fired in batch or periodic kilns where two to four week cycles are used for heating, cooling, and loading and unloading kilns.

Temperatures of firing are important regardless of the type of kiln used, because both the quality and properties of the brick may be affected. The final properties and behavior of most brick can be modified by firing them in an oxidizing or a reducing atmosphere. By controlling the rate of heating and the maximum soaking temperature and soaking time, change in the crystalline structure can be effected, which in turn can also affect the service performance of the brick. In general, the objectives in firing are to (a) drive off hygroscopic, combined water, and CO_2; (b) bring about desired chemical changes such as oxidizing iron and sulfur compounds, and organic matter, etc; (c) effect transformations of the mineral constituents and convert them to the most stable forms; and (d) effect necessary combinations and vitrification of bonding agents. Firing temperatures vary from as low as 1093°C (2000°F) for certain fireclay materials to over 1770°C (3200°F) for some basic products.

Certain refractories with carbon binders or containing oxidizable constituents may be indirectly fired inside muffles to prevent oxidation or may be packed in coke or graphite during firing for the same purpose. One grade of carbon refractory is hot pressed by electrically heating it during pressing. This accomplishes the forming and thermal treatment of the refractory in a single step. Nitrogen or other special atmospheres may be used to impart special binding phases such as silicon nitride.

Manufacturing processes for making lightweight or insulating brick aim for high porosity, preferably with a fine pore structure. This is accomplished by mixing a bulky combustible substance, like sawdust or ground cork, or volatile solid, such as napthalene, with the wet batch, by forcing air into the wet plastic mass, or by mixing into the batch reagents which will react chemically to form a gas and a product not injurious to the brick. In firing such brick, the combustible or volatile material is eliminated and the remaining refractory structure is rigidized. Low density, pre-expanded aggregate may also be used to make products by conventional brickmaking methods.

Some processing after the fired brick are produced may also be performed. For example the brick may be steelcased for use in applications where oxidization of the steel case between brick serves to weld or hold the brick together. Many brick types are also impregnated by placing the fired brick into vacuum tanks and introducing liquid pitch or resins into the brick pore structure. This treatment results in formation of a carbon phase in service which has highly beneficial effects in some applications.

3.2.4 Products

With the large number of raw material types, refractory forms, and manufacturing techniques a multitude of refractory products are produced. A significant number of them are currently used in the iron and steel industry. This chapter and the one that follows will present considerable more detail regarding the specific uses of refractories.

3.3 Chemical and Physical Characteristics of Refractories and their Relation to Service Conditions

The foregoing discussions have indicated that there is a wide variety of refractories from the standpoint of raw materials, overall composition, and method of manufacture. The requirements for refractories are equally diverse. Analysis of service conditions in iron and steelmaking in general shows that refractories are required to withstand:

1. A wide range of temperature, up to 2200°C (4000°F).
2. Sudden changes in temperature; high tensile stresses accompanying these rapid temperature changes cause thermal shock which result in cracking or fracturing.
3. Low levels of compressive stresses at both high and low temperatures.
4. Abrasive forces at both high and low temperatures.
5. The corrosive action of slags, ranging from acidic to basic in character.
6. The action of molten metals, always at high temperatures and capable of exerting great pressures and buoyant forces.
7. The action of gasses, including CO, SO_2, Cl, CH_4, H_2O, and volatile oxides and salts of metals. All are capable of penetrating and reacting with the refractory.
8. As a refractory is being subjected to one or more of the previously stated conditions, it usually functions as a highly effective insulator, or may also be required to be a conductor or absorber of heat depending on its application. The refractory also must perform without exposing workers or environments to unsafe or unhealthy conditions at all times.

As any particular service environment usually involves more than one of the above factors, predetermining the life of a refractory is a complex process involving information on physical and thermal properties as determined by laboratory testing, analysis of the effect of service or process conditions and media on the refractory, and a knowledge of the fundamental reactions between refractories and the various contaminants encountered in service. In this section the physical and chemical characteristics of selected refractories as measured in a variety of laboratory tests will be described as a general guide to understanding the complex nature of these materials in relation to their service environments.

3.3.1 Chemical Composition

As described in the section on refractory classification, the raw materials used in making refractories differ appreciably and result in materials with a wide range of compositions. It must be emphasized that these are unaltered refractories before use and not refractories that have been chemically changed in service. Refractories have the unique ability to withstand alteration by penetration, contamination, and/or reaction in service and still function as reliable engineering materials. Section 3.4 will describe the reactions between refractories and their environments. As a general rule, however, recent trends in refractory development require refractories with the minimum content of impurities, and these impurities are deliberately decreased during raw material or product processing. The following describes the undesirable constituents (originally present or from contamination) in several types of refractories. It should be noted that many other impurities (such as PbO, ZnO, B_2O_3, etc.) which are undesirable in all refractories because of their low melting points have not been shown.

Refractory Type	Undesirable Impurities
Silica	Al_2O_3 alkali TiO_2
Fireclay—all types	alkali, iron oxide, CaO, MgO
High alumina—all types	SiO_2, iron oxide, CaO, MgO
Magnesia-chrome—all types	SiO_2, iron oxide
Magnesia—all types	SiO_2, Al_2O_3, iron oxide
Carbon	alkali, iron oxide

The importance of composition will be described in Section 3.4 where phase diagrams will be used to indicate the reactions in refractories and their environment at elevated temperatures.

3.3.2 Density and Porosity

Refractory density and porosity are among the most misunderstood and yet useful characteristics of steelplant refractories. It must be appreciated that most refractories are not fully dense but deliberately contain both open and closed pores. Fully dense refractory materials with their inherently low tensile strengths can not resist the temperature gradients in normal service at constant temperatures (for example across a wall of a furnace or a tube with steel inside and ambient conditions outside). Refractory materials are also commonly made as heterogeneous structures to promote resistance to thermal cracking. The volume of pores are measured by immersion of a refractory specimen in a liquid or in vacuum. (The measurement of total volume is based on Archimedes' principle that states that a body submerged in a fluid will weigh less than its actual weight by an amount equal to the weight of the displaced fluid. Knowing that difference in weight and the density of the fluid, the volume of the submerged body is easily calculated. The volume of open pores is measured by the amount of liquid absorbed by the sample.)

Fig. 3.6 illustrates the difference between apparent and closed porosity and Fig. 3.7 shows the steps in determining these properties using a vacuum pressure technique. These simple measurements are somewhat useful in comparing like refractory products but most useful as quality control measurements for the consistency of manufacture for a given refractory brand.

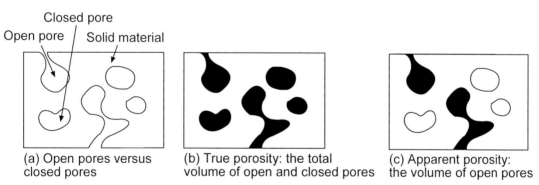

Fig. 3.6 Schematic representation of: (a) open pore, and closed pore, and sample volume for (b) true porosity, and (c) apparent porosity. From Ref. 2.

The size and quantity of pores in solid materials are found by measuring the quantity of mercury that can be forced into the pores of the material under study at various pressures. Since mercury does not usually wet the materials, the mercury will not penetrate the openings among the cluster of particles unless force is applied. Thus, the mercury will penetrate into pores in strict relationship with pressure.

Pore size not only defines the size distribution of the pores of a refractory material, but also defines the size of the bonds formed in the porous matrix and between the porous matrix and dense coarse particles. The size of each bonding area has an important effect on the critical tradeoff in refractories properties. A large number of smaller bond areas can blunt crack propagation and increase crack propagation resistance, thereby improving thermal-shock resistance. However, a smaller number of bonds with a larger cross-sectional area are slower to dissolve and can enhance resistance to slag corrosion.

Fig. 3.8 shows typical pore size distribution results for several refractories. Note the small pore sizes and differences between these products.

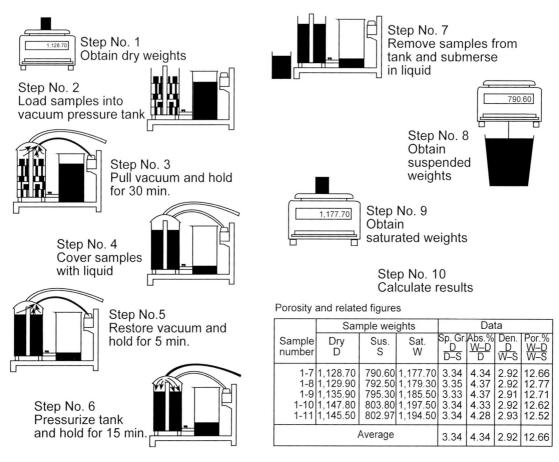

Fig. 3.7 Vacuum pressure technique for measuring porosity in ten steps. From Ref. 2.

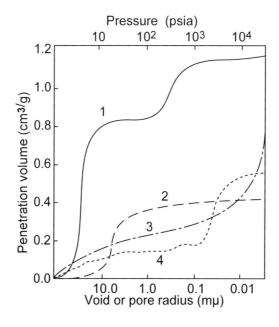

Fig. 3.8 Typical results of pore size distribution in shaped bricks. From Ref. 2.

3.3.3 Refractoriness

The use of the classic PCE, pyrometric cone equivalent test, has little usefulness for today's refractories. (This test measures a relative softening point.) The relative refractoriness of modern refractories is more meaningfully measured using tests for deformation under load and/or creep. Table 3.24 shows comparable results in a short time load test.

Table 3.24 Typical Results of Load Test as Indicated by 24 lb/in² Load Testing

Type of brick	Results of load testing
Fireclay	
Superduty	1.0–3.0% subsidence after heating at 2640°F
High-duty	Withstands load to <2640°F
Low-duty	Withstands load to <2640°F
High-alumina	
60%class	0.1–0.5% subsidence after heating at 2640°F
70%class	0.4–1.0% subsidence after heating at 2640°F
85%class	0.2–0.8% subsidence after heating at 2640°F
90%class	0.0–0.4% subsidence after heating at 3200°F
Corundum class	0.1–1.0% subsidence after heating at 2900°F
Silica (superduty)	Withstands load to 3060°F
Basic	
Magnesite, fired	Withstands load to >3200°F
Magnesite-chrome, fired	Withstands load to 2700°F
Magnesite-chrome, unburned	Withstands load to 2950°F
Chrome, fired	Withstands load to 2800°F
Chrome-magnesite, fired	Withstands load to 3020°F
Chrome-magnesite, unburned	Withstands load to 3020°F
Silicon carbide	Withstands load to 2800°F
Zircon	0.01–0.8% subsidence after heating at 2900°F

From Ref. 2.

Like most structural materials, refractories experience creep behavior, Fig. 3.9, when exposed to high temperatures (0.5 × melting temperature). Most refractories show two characteristic stages of creep. In the first stage, called primary creep, the rate of subsidence declines gradually with time. In the secondary stage, called steady state, the rate of subsidence is constant. At very high temperatures, steady state creep is sometimes followed by tertiary creep region where the rate of subsidence accelerates and leads to catastrophic failure or creep rupture. Primary creep is generally short in duration, while secondary creep can occur over a long term.

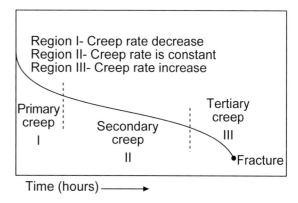

Fig. 3.9 Conventional creep curve. From Ref. 2.

Therefore, secondary creep usually provides a more meaningful comparison of refractories. Secondary creep is the parameter determined by the method described above.

Fig. 3.10 shows creep test results on several high alumina brick. Note that the low-alkali (low impurity) 60% Al_2O_3 brick has superior creep resistance to high alumina brick with more alkali.

Such load and creep results are useful in many refractory applications. For example in a blast furnace stove checker setting where loads and temperatures are known, Fig. 3.11, such information can be readily used for design purposes.

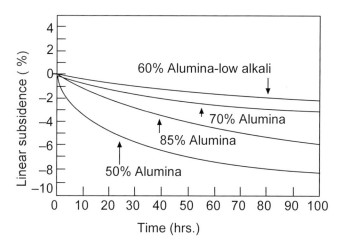

Fig. 3.10 Creep measurement of various high-alumina refractories under 25 psi load at 1425°C (2600°F) for 0–100hrs. Note the excellent creep resistance of 60% alumina low alkali brick. From Ref. 3.

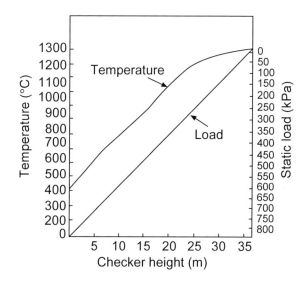

Fig. 3.11 Temperature and load distribution in checker setting for 1315°C (2400°F) top temperature.

3.3.4 Strength

The cold strength(s) of refractories are commonly measured in either compression, Fig. 3.12 or transversely in a modulus of rupture test, Fig. 3.13.

These cold strength values give some measure of the consistency of the refractory product and its ability to function during shipment and installation.

Hot strength values are also determined by similar procedures with the refractory specimen contained in a furnace which may also provide for a protective atmosphere surrounding the specimen. Fig. 3.14 shows the simplest hot modulus of rupture apparatus.

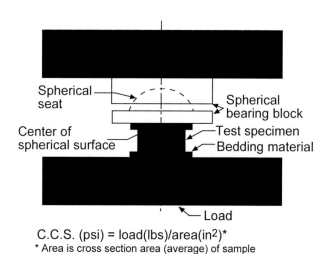

C.C.S. (psi) = load(lbs)/area(in^2)*
* Area is cross section area (average) of sample

Fig. 3.12 Cold crushing strength test apparatus. From Ref. 2.

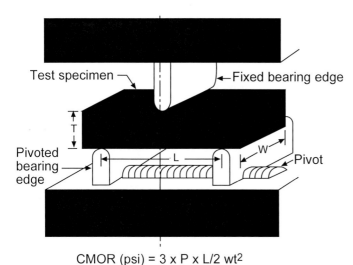

Fig. 3.13 Cold modulus of rupture test apparatus. From Ref. 2.

CMOR (psi) = 3 × P × L/2 wt²

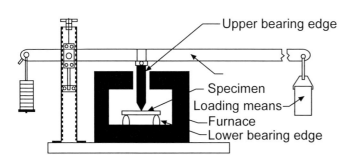

Fig. 3.14 Cross section of a typical hot modulus of rupture apparatus. From Ref. 2.

Table 3.25 Physical Properties of Refractory Brick

Type of brick	Density (lb/ft³)	Apparent porosity (%)	Cold crushing strength (lb/in²)	Modulus of rupture (lb/in²)
Fireclay				
Superduty	144–148	11.0–14.0	1800–3000	700–1000
High-duty	132–136	15.0–19.0	4000–6000	1500–2200
Low-duty	130–136	10–25	2000–6000	1800–2500
High-alumina				
60% class	156–160	12–16	7000–10,000	2300–3300
70% class	157–161	15–19	6000–9000	1700–2400
85% class	176–181	18–22	8000–13,000	1600–2400
90% class	181–185	14–18	9000–14,000	2500–3000
Corundum class	185–190	18–22	7000–10,000	2500–3500
Silica (superduty)	111–115	20–24	4000–6000	600–1000
Basic				
Magnesite, fired	177–181	15.5–19.0	5000–8000	2600–3400
Magnesite-chrome, fired	175–179	17.0–22.0	4000–7000	600–800
Magnesite-chrome, unburned	185–191	—	3000–5000	800–1500
Chrome, fired	195–200	15.0–19.00	5000–8000	2500–3400
Chrome-magnesite, fired	189–194	19.0–22.0	3500–4500	1900–2300
Chrome-magnesite, unburned	200–205	—	4000–6000	800–1500
Magnesite-carbon	170–192	9.0–13.0	—	1000–2500
Dolomite	165–192	5.0–20.0	1500–3500	500–2500
Fused cast magnesite-chrome	205–245	1.0–15.0	900–1400	6000–8000
Silicon carbide	160–166	13.0–17.0	9000–12,000	3000–5000
Zircon	225–232	19.5–23.5	7000–11,000	2300–3300

From Ref. 3.

Table 3.25 shows some typical cold properties for several types of refractories and the range of properties between and within these groups are obvious. As previously stated, these cold properties are most useful in rating the consistency of quality of particular refractory products.

The hot strengths of refractories will vary significantly with temperature and with other parameters such as furnace atmosphere. Fig. 3.15 and 3.16 illustrate the complex nature of refractory hot strength for two types of refractories. Fig. 3.15 shows the strength of periclase-chrome refractories as it varies with temperature and is affected by both composition and initial firing temperature during manufacture. Fig. 3.16 shows the strength at a single test temperature as it is influenced by small changes in the impurity (SiO_2) level and the ratio of CaO to SiO_2 in the refractory.

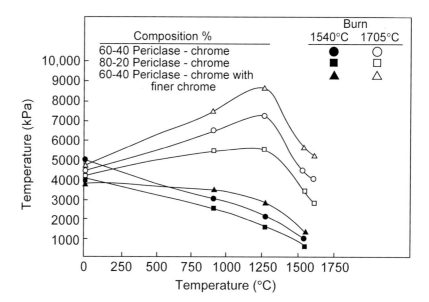

Fig. 3.15 Strength-temperature relationship for refractories of the indicated compositions.

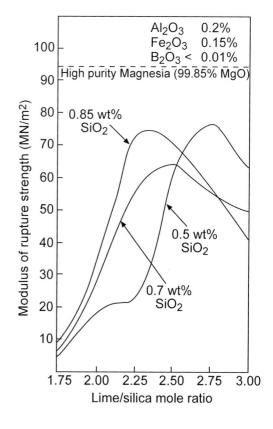

Fig. 3.16 Effects of lime/silica ratio on the hot modulus of rupture at 1500°C (2732°F) of burned magnesia bricks containing a different silica level. From Ref. 2.

3.3.5 Stress-Strain Behavior

When a refractory is subjected to a mechanical load, it will compress. This behavior may be quantified by the following equation:

$$\varepsilon = \frac{\sigma}{E} \tag{3.3.1}$$

where:

ε = strain (dimensionless),
σ = stress psi (MPa),
E = Modulus of elasticity, psi (MPa).

Strain is equal to the amount of compression divided by the original length.

$$\varepsilon = \frac{\Delta L}{L} \tag{3.3.2}$$

Stress is the force applied per unit area.

$$\sigma = \frac{F}{A} \tag{3.3.3}$$

The modulus of elasticity, or Young's Modulus, of a refractory is constant for a given material and temperature.

The stress-strain behavior of a refractory material is determined by a method which is similar to that used to measure hot crushing strength. A cylindrical sample is heated uniformly to the test temperature and compressed using a mechanical testing machine. While loading the sample, the change in its height is monitored by an electrical transducer which is connected by sapphire sensing rods with the top and bottom of the sample. The data are used to create a stress versus strain curve.

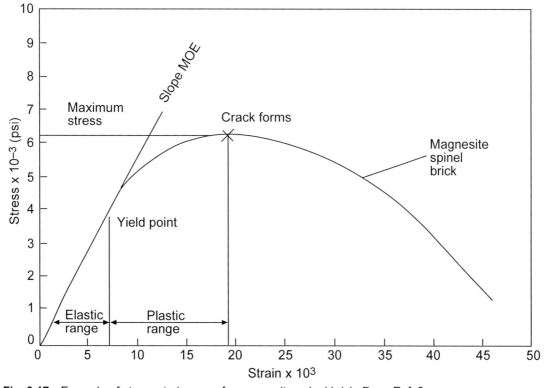

Fig. 3.17 Example of stress-strain curve for magnesite-spinel brick. From Ref. 3.

Table 3.26 Static Modulus of Elasticity (approximate values)

Type of brick	Modulus of Elasticity at 2000°F (psi)
Silica	9.0×10^5
50% alumina	6.9×10^5
60% alumina	17.0×10^5
Magnesite-alumina-spinel	14.0×10^5
Magnesite-chrome	12.0×10^5
Magnesite	9.3×10^5
Magnesite-tar impregnated	21.0×10^5
Magnesite-tar bonded	14.0×10^5
Magnesite-carbon (20%)	4.4×10^5
Magnesite-carbonn (15%)	4.0×10^5

From Ref. 3.

Fig. 3.17 illustrates a typical stress-strain curve. The initial portion of the curve is usually linear and is called the elastic range. The slope of the curve over this range gives the modulus of elasticity. The point at which the stress-strain curve becomes non-linear is called the yield point. Straining the material beyond this point results in permanent deformation; this portion of the curve is known as the plastic range. Further straining of the material brings failure. Some materials do not show plastic behavior at certain temperatures, but instead fail after elastic straining. Modulus of elasticity values of a number of refractories are shown in Table 3.26.

An understanding of the stress-strain behavior of refractories at elevated temperatures is important in nearly all applications. The shell or superstructure of a vessel usually acts to restrain the thermal expansion of the lining. Proper lining design may require gaps in the lining (thermal expansion allowance) during installation to prevent catastrophic stress buildup in the lining during heating. However, the expansion allowance must also be designed to prevent instability of the lining. The engineer, too, must be concerned with the stresses which the expanding refractories induce in the vessel shell or superstructure. Section 3.5 on the selection of refractories will further describe analysis techniques for refractory behavior under stress.

3.3.6 Specific Heat

The specific heat of a refractory material indicates the amount of heat necessary to change its temperature by a given amount. The specific heat is the amount of heat in British Thermal Units (BTU's) which must be absorbed by one pound of material to raise its temperature by one degree Fahrenheit. Table 3.27 gives the specific heats of several types of refractory brick and refractory minerals.

Table 3.27 Specific Heats of Refractory Brick and Minerals

Mean Specific Heats between 32°F, and T°F, in BTU per pound per °F

T°F	Fireclay Brick	Silica Brick	Magnesite Brick	Chrome Brick	Forsterite Brick	T°F	Mullite $3Al_2O_3 \cdot 2SiO_2$	Cristobalite SiO_2	Periclase MgO	Corundum Al_2O_3
32	0.193	0.169	0.208	0.170	0.180	32	0.184	0.165	0.208	0.171
200	0.199	0.188	0.219	0.176	0.200	200	0.192	0.183	0.227	0.196
400	0.206	0.211	0.232	0.182	0.216	400	0.214	0.204	0.240	0.214
600	0.212	0.229	0.242	0.188	0.230	600	0.223	0.239	0.251	0.226
800	0.220	0.238	0.251	0.194	0.240	800	0.229	0.251	0.257	0.235
1000	0.227	0.246	0.258	0.199	0.246	1000	0.233	0.258	0.262	0.242
1200	0.234	0.252	0.263	0.204	0.250	1200	0.237	0.264	0.267	0.248
1400	0.241	0.256	0.268	0.208	0.254	1400	0.240	0.268	0.270	0.252
1600	0.248	0.260	0.273	0.212	0.258	1600	0.242	0.271	0.274	0.257
1800	0.253	0.264	0.278	0.216	0.262	1800	0.245	0.273	0.277	0.260
2000	0.258	0.268	0.283	0.220	0.266	2000	0.247	0.275	0.280	0.264
2200	0.262	0.272	0.288	0.222	0.270	2200	0.249	0.277	0.282	0.267
2400	0.266	0.276	0.293	0.224	0.274	2400	0.251	0.278	0.285	0.270
2600	0.269	0.279	0.297	0.226	0.278	2600	0.253	0.279	0.288	0.273
						2800	0.255	0.280	0.290	0.276
						3000	0.256	0.281	0.292	0.279
						3200	0.258	—	0.294	0.282

From Ref. 3.

The specific heat values for refractory materials are important in many applications because the amount of heat stored in the lining during heating is often significant. In furnaces which are cycled, large amounts of heat are alternately stored during heating stages and lost to the surroundings during cooling stages. Applications for which high specific heat is desirable include blast furnace stoves. Stoves are specifically designed to absorb and store heat from hot waste gases. The stored heat is subsequently used to preheat combustion air.

3.3.7 Emissivity

Emissivity is the relative power of a surface to emit heat by radiation. It is expressed as a fraction of the emissivity of an ideal black body. Such black body radiation is the maximum possible, but it is never achieved by actual materials. A good radiator is an equally good absorber of heat. A good reflector, obviously, is a poor absorber, and consequently a poor radiator. A perfect reflector, which also does not exist, would have an emissivity of zero. Materials do not radiate equally well at all wavelengths. The ability to radiate at a particular wavelength is referred to as the monochromatic emissivity. Total emissivity refers to heat radiation over the entire spectrum of wavelengths. The total emissivity of most refractories decreases somewhat with an increase in temperature.

Emissivities dictate the amount of heat which is radiated across a gap in a refractory structure. The equation used to calculate this quantity is

$$Q_r = \sigma \left(\frac{T_1^4 - T_2^4}{\frac{1}{\varepsilon_1} + \frac{1}{\varepsilon_2} - 1} \right) \qquad (3.3.4)$$

where:

Q_r = heat radiated across gap, Wm^{-2} (BTU hr^{-1} ft^{-2}),
σ = Stefan-Boltzmann constant, W/m^{-2} K^{-4} (BTU hr^{-1} ft^{-2} $°R^{-4}$),
T_1 = temperature of hotter surface of gap, K (°R),
T_2 = temperature of cooler surface of gap, K (°R),
ε_1 = emissivity of hotter surface of gap, dimensionless,
ε_2 = emissivity of cooler surface of gap, dimensionless.

In the above equation, the temperatures must be expressed using an absolute temperature scale. The temperature in Kelvins is K = °C + 273; the temperature in degrees Rankin is °R = °F + 460. The Stefan-Boltzmann constant is 5.670×10^{-9} Wm^{-2} K^{-4} (1.714×10^{-9} BTU hr^{-1} ft^{-2} R^{-4})

The emissivity of the outermost component of a vessel determines, to a large extent, the amount of heat radiated to the surroundings. The loss to the environment by unobstructed radiation is given by:

$$Q_r = \varepsilon \sigma (T_1^4 - T_2^4) \qquad (3.3.5)$$

where:

Q_r = heat radiated to surroundings, Wm^{-2} (BTU hr^{-1} ft^{-2}),
σ = Stefan-Boltzmann constant, W/m^{-2} K^{-4} (BTU hr^{-1} ft^{-2} $°R^{-4}$),
ε = emissivity of outer surface, dimensionless,
T_1 = temperature of outer surface, K (°R),
T_2 = temperature of surroundings, K (°R).

In most applications, the outer surface is a metallic shell such as carbon steel plate. the emissivity of the shell, in this case, is affected by the degree of oxidation. The emissivity tends to increase as the amount of oxidation increases.

Emissivity values for a number of refractories and materials commonly used for shells are listed in Table 3.28

Table 3.28 Representative Values of Total Emissivity

Material	Temperature °F						
	200	400	800	1600	2000	2400	2800
Fireclay Brick	.90	(.90)	(.90)	.81	.76	.72	.68
Silica Brick	(.90)	—	—	.82–.65	.78–.60	.74–.57	.67–.52
Chrome-Magnesite Brick	—	—	—	.87	.82	.75	.67
Chrome Brick	(.90)	—	—	.97	.975	—	—
High-Alumina Brick	.90	.85	.79	(.50)	(.44)	—	—
Mullite Brick	—	—	—	.53	.53	.62	.63
Silicon Carbide Brick	—	—	—	.92	.89	.87	.86
Carbon Steel	.80	—	—	—	—	—	—
Aluminum	.20	—	—	—	—	—	—

(Figures in parentheses were obtained by extrapolation of test data.)

From Ref. 3.

3.3.8 Thermal Expansion

All refractories expand on heating in a manner related to their compositions. Fig. 3.18 shows classic examples of the expansion of divergent refractory types It is obvious that the shape of the expansion curve with temperature and the maximum magnitude of the expansions differ appreciably. Compressible materials or voids are used for relieving such thermal expansion in any refractory construction because growth of several inches can obviously be experienced. Refractories with linear or near linear thermal expansion generally require considerably less care than those with non-linear expansion characteristics. A classic example of a refractory that requires care during heatup is silica brick, which after firing consists of a carefully controlled balance of various mineral forms of the compound SiO. Fig. 3.19 shows the thermal expansion of the various silica mineral forms which make slow heatup through various critical ranges necessary. As a result of the complex mineral makeup of silica brick, large constructions using these brick (such as coke ovens) are heated and cooled at very slow rates and may require several weeks to reach operating temperatures. Once the critical temperature ranges are passed, however, silica brick exhibit low and uniform thermal expansion and may be thermally cycled without damage between 1000–1538°C (1800–2800°F)

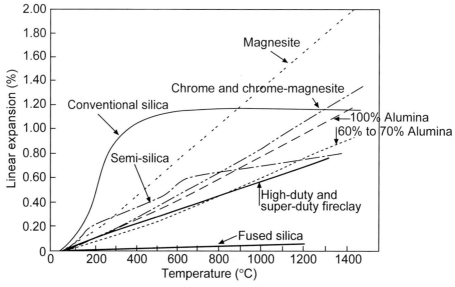

Fig. 3.18 Typical curves of linear expansion of various types of refractories.

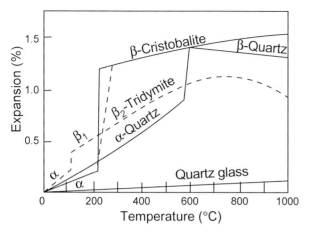

Fig. 3.19 Thermal expansion of silica minerals.

In general, the thermal expansion behaviors of unfired refractories are more complex than those of their fired counterparts. During initial heating, dramatic expansions or contractions may occur in an unfired material as a result of changes in bonding structure, changes in mineralogy, and sintering effects.

The thermal expansion characteristics of a number of cement-bonded refractories during initial heat-up are shown in Fig. 3.20 These materials show shrinkage over the temperature range 205–315°C (400–600°F) which is associated with thermal decomposition of the cement. The amount of shrinkage is determined by the quality and amount of cement. At temperatures above 980 to 1090°C (1800 to 2000°F), additional shrinkage occurs as a result of sintering. The underlying thermal expansion is determined by the characteristics of the aggregate. The shrinkages which take place during initial heat-up to 1430°C (2600°F) are permanent in nature and commonly are on the order of 0.2 to 1.5%.

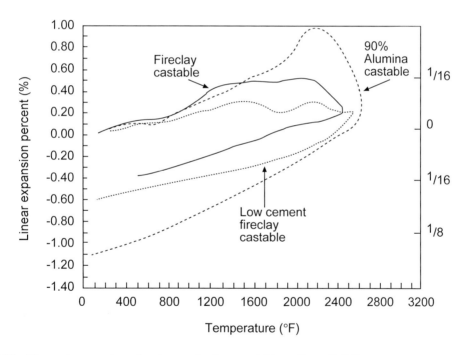

Fig. 3.20 Thermal expansion of various refractory castables. From Ref. 3.

Thermal expansion curves for a number of magnesite-carbon brick are shown in Fig. 3.21. Through the temperature range 370–540°C (700–1000°F), the formation of a glassy carbon bond tends to densify the structure and shrinkage results. In materials which contain metals, expansive reactions take place at higher temperatures. Brick which contain magnesium show a dramatic thermal expansion over the range 540–705°C (1000–1300°F). Materials containing aluminum show gradual acceleration of expansion above about 760°C (1400°F). Thermal expansion at high temperatures increases significantly with increases in metals content. After heating to 1430°C (2600°F), magnesite-carbon brick show permanent expansion of 0.1 to 1.0%.

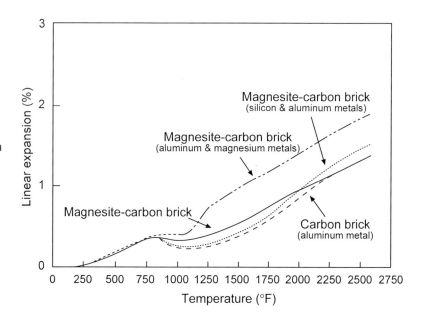

Fig. 3.21 Thermal expansion of magnesite-carbon brick. From Ref. 3.

3.3.9 Thermal Conductivity and Heat Transfer

As insulators, refractory materials have always been used to conserve heat, and their resistance to heat flow is a prime selection factor in many applications. Fig. 3.22 shows thermal conductivity curves for several refractory types ranging from dense refractories to insulating brick. Some refractories (for example, carbon or silicon carbide) have appreciably higher conductivities (up to 43.26 $Wm^{-1}K^{-1}$ or 300 BTU $hr^{-1}ft^{-2}°F$ in^{-1}) whereas others are available with conductivities lower than 0.14 $Wm^{-1}K^{-1}$ or 1 BTU $hr^{-1}ft^{-2}°F$ in^{-1} (for example, block insulation or refractory fiber forms).

Using measured conductivity values, heat transfer losses through single or multiple component refractory walls can be calculated using the general formula:

$$\frac{Q}{A} = \frac{T_1 - T_2}{\frac{L_1}{K_1} + \frac{L_2}{K_2} + \ldots \frac{L_n}{K_n}} \quad (3.3.6)$$

where:

Q/A = heat loss expressed as Wm^2 (BTU $hr^{-1}ft^{-2}$),
T_1 = temperature of the hotter surface, °C (°F),
T_2 = temperature of the cooler surface, °C (°F),
$L_1, L_2 \ldots L_n$ = thickness of each material, (in.),
$K_1, K_2 \ldots K_n$ = thermal conductivity of each material, $Wm^{-1}K^{-1}$ (BTU in. $ft^{-2}hr^{-1}°F^{-1}$).

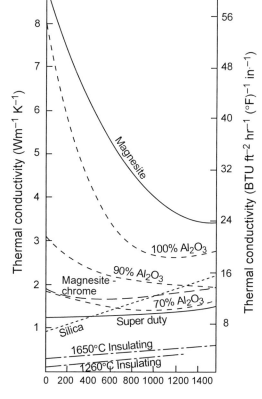

Fig. 3.22 Typical thermal conductivity curves for various refractory brick.

Such calculations are now rapidly made using computer simulations where calculated heat transfer rates can be balanced with loss from the outer refractory surface by radiation and natural or forced convection. Moreover, where once it was a slow mathematical process to determine even the steady-state heat times, the advent of computers has made the rapid determination of the heat transfer data for even transient conditions routine. Although it might seem that every construction should be designed for minimum heat losses, this is not always the case, and care must be taken in some situations including the following:

(1) The hot face refractories in a particular application must be able to withstand the higher temperatures that will result when layers of highly insulating backup materials are added.

(2) Other refractory properties must be suitable for the environment. For example, most insulating materials will not stand direct exposure to metal or slag, and backup materials may be subject to attack by vaporized process components (alkali, sulfur compounds, acids) or their condensates. Gas channeling through permeable materials must also be considered to prevent hot spots on shells.

(3) Insulation increases the depth of penetration and chemical attack on the hot-face layer.

A wide variety of very insulating fiber based refractories are available in bulk, blanket, board, or shaped forms. In such products, thermal conductivity is highly dependent on product density as shown in Fig. 3.23.

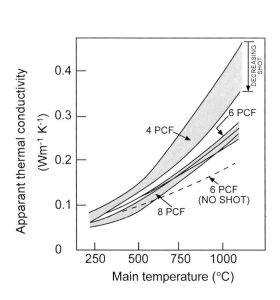

Fig. 3.23 Thermal conductivity as a function of temperature for various fiber materials.

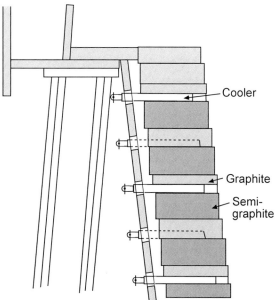

Fig. 3.24 Use of ultra-conductivity materials in a typical bosh construction.

In certain applications, very high conductivity refractories are used to cool the refractory lining and stabilize it against further refractory wear. Fig. 3.24 for example shows a blast furnace bosh using graphite and semi-graphite materials with conductivities of 70–80 and 30–35 $Wm^{-1}K^{-1}$, respectively, at 1000°C. (Note these values in comparison to the more conventional refractories shown in Fig. 3.22). In such designs, heat extracted through the copper coolers enables the refractories to last many times longer than uncooled linings. Similarly, a composite construction with silicon carbide in a blast furnace stack as shown in Fig. 3.25 controls shell temperatures and allows the formation of a stable lining. In this case the cooling plate density is desirable as varying from none to very close or dense.

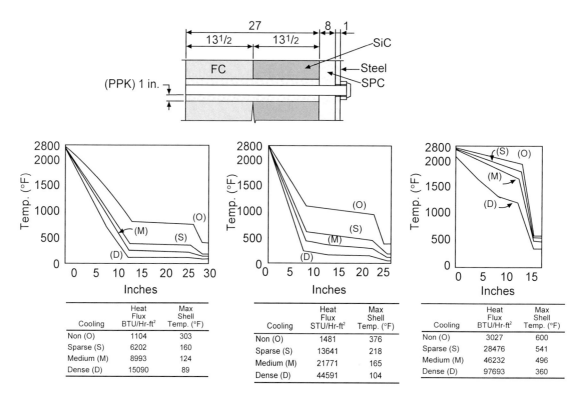

Fig. 3.25 Temperature profile for lining B.

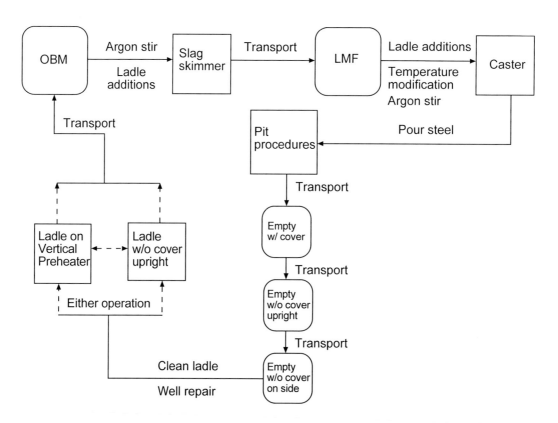

Fig. 3.26 Ladle cycling procedure.

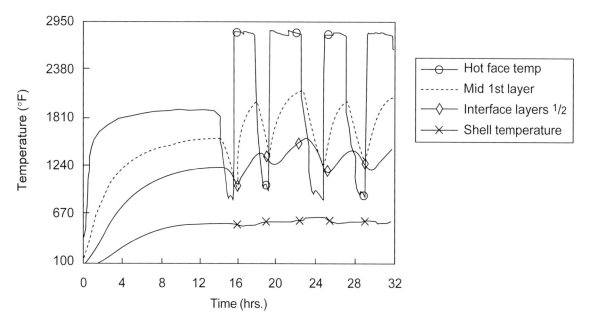

Fig. 3.27 Transient phenomena in ladle cycling.

Many steelplant refractory applications also never reach thermal equilibrium, and dynamic heat transfer calculations must be employed in the analysis of such applications. For example, a typical steel ladle will cycle several times daily as shown in Fig. 3.26. During such cycles, the temperature at the hot face and to a lesser extent the entire lining, will show cyclic behavior, as displayed in Fig. 3.27.

Importantly, remember that while heat flow at steady state is governed by thermal conductivity k (first Fourier's law), the material property required in the transient case is thermal diffusivity λ (second Fourier's law).

Thermal diffusivity λ is the equivalent of a diffusion coefficient in the Fourier equation for heat flow and is a measure of how fast a heat pulse is transmitted through a solid. Thermal diffusivity is related to both thermal conductivity and heat capacity through the relationship:

$$\lambda = \frac{k}{\rho C_p} \tag{3.3.7}$$

where:

k = thermal conductivity,
ρ = density
C_p = heat capacity,
λ = $m^2 s^{-1}$.

Thermal diffusivity is most commonly determined by the heat-pulse method where a heat pulse is applied to one side of a disc or short cylinder and the time period for the temperature of the opposite face to reach one-half its final value is recorded. The heat pulse is commonly provided by a power laser flash.

Table 3.29 shows some diffusivity values for selected products. The relationship among density, heat capacity, thermal conductivity and thermal diffusivity should be borne in mind, so if one property is unkown, the others can be reqadily calculated.

Table 3.29 Thermal Diffusivity of Different Products, in 10^{-6} m²s⁻¹

	25°C	500°C	1000°C	1500°C
Alumina > 99.5%	10.7	2.6	1.8	1.3
Alumina, 90%	6.2	1.7	1.2	0.9
Magnesia, 30% porous	~12	~4	~1.5	—
Spinel	~6	~2	~1.2	—
Zirconia (cubic)	0.7	0.6	0.5	0.6
Silicon carbide (dense)	40–50	15–18	11–13	7–9
Graphite ⊥ c	250–1000	50–150	30–75	—
Graphite ∥ c	2.8–5.4	0.4–0.8	0.2–0.5	—

From Ref. 2

3.3.10 Thermal Shock

Thermal shock resistance is a complex issue when investigating refractory system design. Thermal shock or spalling is caused by thermal stresses which develop from uneven rates of expansion and contraction within the refractory, caused by rapid temperature changes or high inherent temperature gradients in a refractory. A qualitative prediction of the resistance of materials to fracture by thermal shock can be expressed by the factor:

$$\frac{ks}{\alpha E} \qquad (3.3.8)$$

where:
 k = thermal conductivity,
 s = tensile strength,
 α = coefficient of thermal expansion,
 E = modulus of elasticity.

The higher the value of the factor in Eq. 3.3.8, the higher the predicted thermal shock resistance of the material. Some measure of shock resistance can be obtained using the work of fracture measurements at elevated temperatures to predict the energy to propagate a slow moving crack. Other thermal shock measurement techniques usually involve thermal cycle tests as described in Section 3.5.

3.4 Reactions at Elevated Temperatures

The foregoing discussion of the high temperature behavior of refractories emphasizes physical factors, but refractory behavior also depends greatly on high temperature reactions occurring not only within the the refractories themselves but between refractories and contaminants encountered in service.

Phase equilibrium diagrams have proved to be invaluable guides to understanding service reactions and the influence of composition on refractory properties. As excellent compilations of diagrams have been published (see references at end of chapter), only a few are reproduced here. However, it should be recognized that these diagrams are not without limitations in their use to predict or explain refractory behavior. For example, the various systems have been explored using simple combinations of pure oxides and represent equilibrium conditions, while refractories are rarely pure and seldom in equilibrium, either as manufactured or in service. Because of this complex chemical nature of refractories, information is often needed on reactions involving so many oxides that the usefulness of phase equilibrium information on systems involving three or four oxides is minimal. The diagrams also give no information on such significant matters as viscosity of the liquids formed or the rates at which reactions proceed.

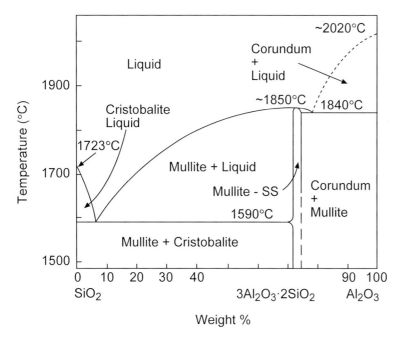

Fig. 3.28 Phase diagram of the Al_2O_3–SiO_2 system.

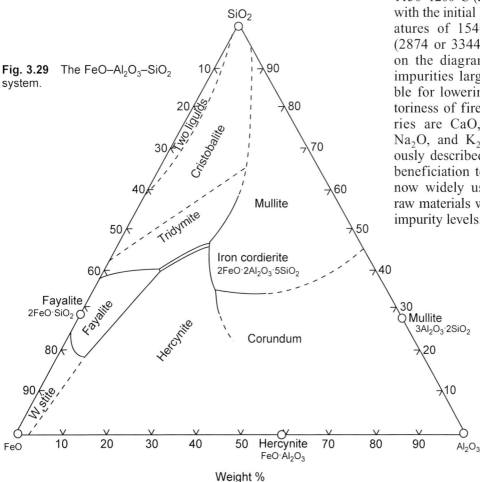

Fig. 3.29 The FeO–Al_2O_3–SiO_2 system.

Fig. 3.28 shows the Al_2O_3–SiO_2 system which applies to silica, fireclay, and high alumina refractories. It will be noted that the lowest temperature at which any liquid is developed in the system is 1590°C (2894°F), while those compositions more aluminous than mullite ($3Al_2O_3 \cdot 2SiO_2$) or above 71.8% Al_2O_3 develop liquids at quite high temperatures. It is obvious that useful refractories can be made of the pure oxides Al_2O_3 or SiO_2. The pronounced effect of the impurities present in most commercial refractories in this system can be appreciated by comparing the temperatures at which they deform initially under load, 1150–1200°C (2100–2200°F), with the initial liquid temperatures of 1540 or 1840°C (2874 or 3344°F) indicated on the diagram. The oxide impurities largely responsible for lowering the refractoriness of fireclay refractories are CaO, MgO, FeO, Na_2O, and K_2O. As previously described, mining and beneficiation techniques are now widely used to obtain raw materials with minimum impurity levels.

Fig. 3.29 shows the FeO–Al$_2$O$_3$–SiO$_2$ system. Here it is seen that the formation of some liquid can be expected even below about 1095°C (2000°F) with Al$_2$O$_3$–SiO$_2$ refractories and that very damaging amounts will be formed at the higher temperatures common to iron and steel processes. This is particularly true as iron oxide bearing liquids are characteristically very fluid.

Fig. 3.30 is the diagram of the CaO–Al$_2$O$_3$–SiO$_2$ system, which is most applicable to reactions of fireclay refractories with blast furnace slags and indicates superior resistance for higher Al$_2$O$_3$ products in such environments. This system has also been useful in predicting behavior of silica brick, which will be discussed later.

The ternary phase equilibrium diagrams for K$_2$O or Na$_2$O reactions with Al$_2$O$_3$ and SiO$_2$ are reproduced in Fig. 3.31 and Fig. 3.32. It is evident that the refractoriness of alumina-silica refractories will be seriously affected by very small amounts of Na$_2$O, less than 1% being sufficient to lower the temperature of initial liquid formation to less than about 1095°C (2000°F), while approximately 10% is sufficient to completely liquefy the more siliceous alumina-silica compositions at about 1205°C (2200°F). Potassium oxide (K$_2$O) has a similar effect in amounts up to 10%.

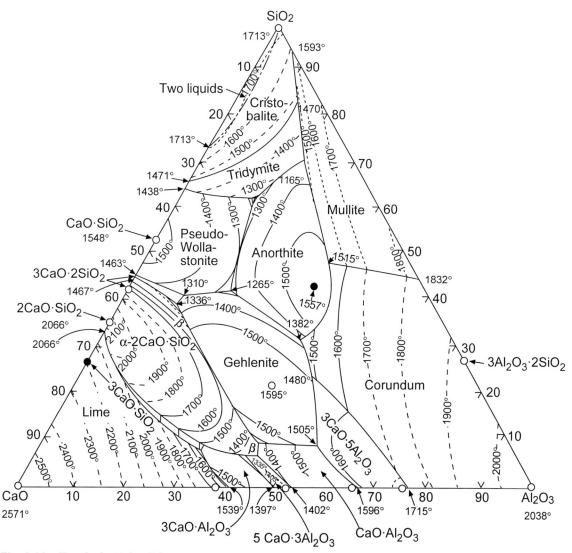

Fig. 3.30 The CaO–Al$_2$O$_3$–SiO$_2$ system.

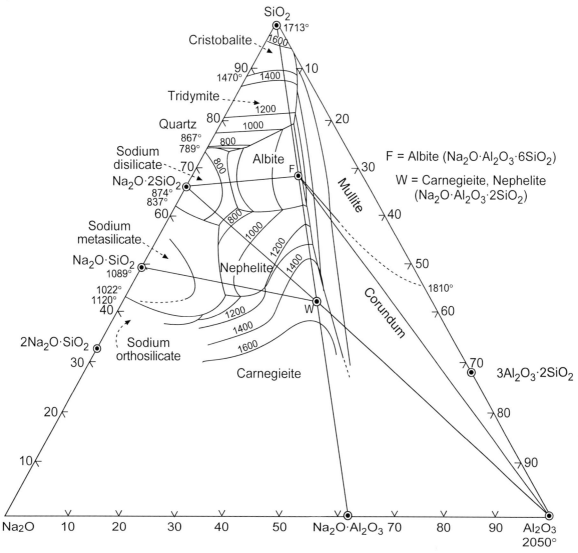

Fig. 3.31 The K$_2$O–Al$_2$O$_3$–SiO$_2$ system.

Low levels of impurities are far more critical in silica brick than fireclay brick, and alkalies are the worst offenders. The amounts of Na$_2$O, K$_2$O, and Al$_2$O$_3$ required to lower the melting point of pure silica from about 1725°C to about 1675°C (about 3140°F to about 3050°F) are, respectively 1.4, 1.9, and 3.1%.

Another very deleterious fluxing agent for fireclay refractories is MnO, as can ben seen in the diagram of the MnO–Al$_2$O$_3$–SiO$_2$ system in Fig. 3.33. Note that this system is quite similar to the FeO–Al$_2$O$_3$–SiO$_2$ system.

Because the raw materials for silica brick lack both a natural bond and a high melting point, care must be taken that the required bonding addition has the minimum effect on refractoriness. Fig. 3.34 shows the CaO–SiO$_2$ system and explains why lime is universally used for this purpose. With additions of CaO to SiO$_2$, the melting temperature remains unchanged between 1 and 27.5% CaO, due to the formation of two immiscible liquids. No such phenomenon occurs in the Al$_2$O$_3$–SiO$_2$ system, and by referring again to the CaO–Al$_2$O$_3$–SiO$_2$ system, it is found that only a small amount of Al$_2$O$_3$ is required to destroy the CaO–SiO$_2$ immiscibility. In fact the effect of minor increments

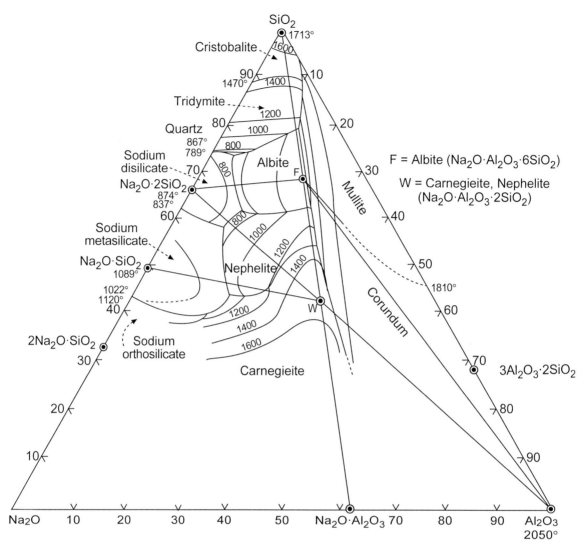

Fig. 3.32 The Na_2O–Al_2O_3–SiO_2 system.

of Al_2O_3 on the liquid development of silica brick is such that the temperature of failure under a load of 172 kPa (25 psi) will decrease approximately 5°C (10°F) for each 0.1% increase in Al_2O_3 in the 0.3 to 1.2% range of Al_2O_3 between super duty and conventional silica brick. Fig. 3.35 of the FeO–SiO_2 system shows that FeO, like CaO, also forms two immiscible liquids when added to SiO_2, thus greatly increasing the tolerance of silica brick for FeO. Furthermore, as with CaO, a small amount of Al_2O_3 can eliminate this immiscibility.

As atmospheric conditions in steelplant furnaces may range from highly reducing to highly oxidizing, the form of iron oxides present may vary from FeO to Fe_2O_3. Accordingly, Fig. 3.36 shows the system FeO–Fe_2O_3–SiO_2 and is of considerable importance in understanding the behavior of silica brick in service. Thus, it is seen that the lowest-melting liquids occur from reaction of FeO and SiO_2, and that at temperatures in the range of 1455–1665°C (2650–3030°F) less liquid, and a less siliceous liquid, will be produced with either $FeO \cdot Fe_2O_3$, or Fe_2O_3 than with FeO, due to the greater extent of the two liquid region under oxidizing conditions.

The two principal refractory oxides considered to be basic are magnesia (MgO) and calcia (CaO). Magnesia is noted for its tolerance to iron oxides. As shown in Fig. 3.37, MgO and FeO form a

Steel Plant Refractories

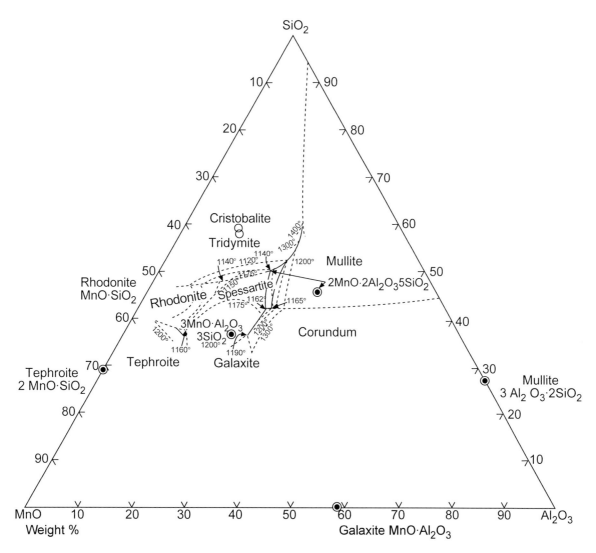

Fig. 3.33 The MnO–Al$_2$O$_3$–SiO$_2$ system.

continuous series of solid solutions which have high refractoriness even with very high FeO contents. Under oxidizing conditions, magnesia is even more tolerant to iron oxide. Magnesia and iron oxide form the refractory compound magnesioferrite (MgO·Fe$_2$O$_3$) which contains 80 wt% Fe$_2$O$_3$. Magnesioferrite forms solid solutions with magnetite (FeO·Fe$_2$O$_3$) at higher iron oxide contents and with magnesia at lower iron oxide contents. On the other hand, calcia is more reactive with iron oxide, forming low melting calcium ferrites such as dicalcium ferrite (2CaO·Fe$_2$O$_3$) that melts incongruently at ~1440°C (~2620°F). Also, calcia is subject to hydration and disruptive disintegration on exposure to atmospheric conditions and cannot be used in refractory shapes made by conventional procedures. It is evident, therefore, that magnesia is the more useful basic refractory oxide and forms the base for all types of basic refractories including those made from magnesite, olivine, dead burned dolomite, and magnesite and chrome ore.

Magnesia-bearing refractories, regardless of type, contain accessory refractory oxides and encounter other refractory oxides in service which exert an important influence on their performance. Fig. 3.38 shows the reactions and phase assemblages in the MgO–CaO–SiO$_2$ system. In the high MgO portion of this system, the principal mineral is of course always periclase. The accessory

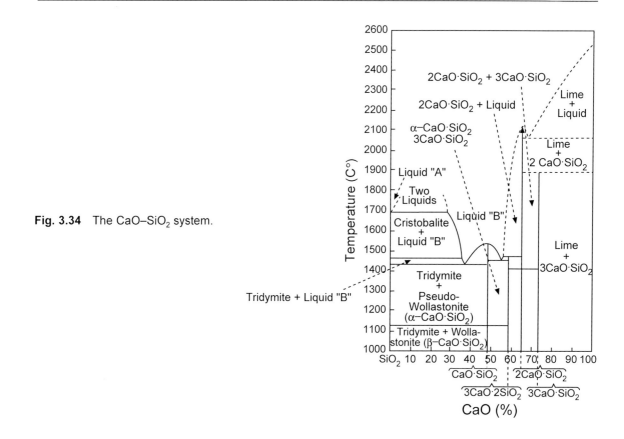

Fig. 3.34 The CaO–SiO$_2$ system.

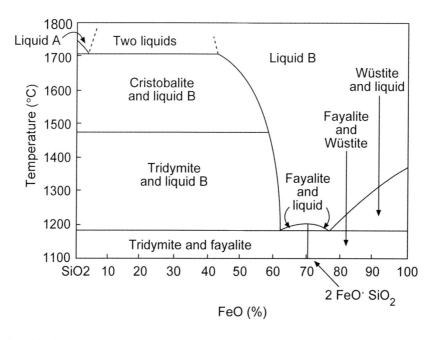

Fig. 3.35 The FeO–SiO$_2$ system.

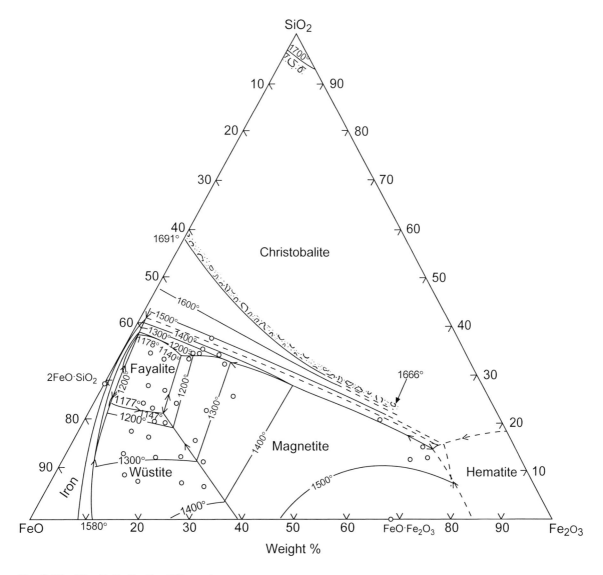

Fig. 3.36 The FeO–Fe$_2$O$_3$–SiO$_2$ system.

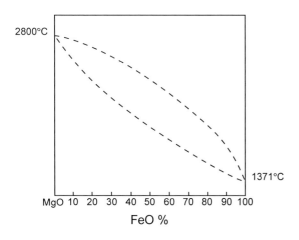

Fig. 3.37 The MgO–FeO system.

silicate bonding minerals, however, will vary considerably depending on the ratio of CaO to SiO$_2$. Table 3.30 presents a summary of the compounds present with periclase as affected by the CaO/SiO$_2$ weight ratio and their approximate melting points. The type of refractory bonding preferred will depend on the intended application. The fosterite bond, which occurs at CaO/SiO$_2$ ratios less than 0.93, is desired in some basic brick to prevent excessive formation of monticellite and merwinite. These minerals form low temperature liquids and therefore have poor high temperature load carrying ability. Refractory bonding phases with high melting points can also

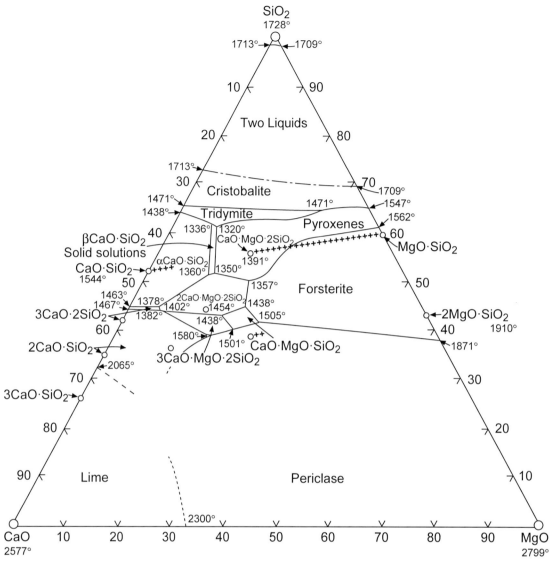

Fig. 3.38 The MgO–CaO–SiO$_2$ system.

be obtained at CaO/SiO$_2$ ratios above 1.86 where refractory dicalcium or tricalcium silicates are present. In recent years, in addition to the control of CaO/SiO$_2$ ratio, increased emphasis has been placed on decreasing the amounts of lime and silica present to obtain the maximum advantage of the properties of nearly pure magnesia. This has been accomplished largely through the use of the improved synthetic magnesites.

The high temperature reactions of refractories made from magnesia and chrome ore are under constant study. The properties of refractories made from these two raw materials are excellent because of the tendency of each to minimize the major weaknesses of the other constituent. Chrome ore consists of a solid solution of chrome spinels (Mg,Fe)O·(Cr,Al,Fe)$_2$O$_3$ with appreciable amounts of gangue silicates. At high temperatures, the gangue silicate in chrome ores is responsible for poor resistance to deformation under load and the iron oxides, when alternately oxidized and reduced, cause expansion and contraction, often causing disintegration. Also, chrome spinel shows considerable growth or bursting when reacted with iron oxide at high temperatures as a result of the formation of solid solutions of magnetite (FeO·Fe$_2$O$_3$) and other spinels. With additions of magnesia

Table 3.30 Mineral Phases in Equilibrium with Periclase (MgO) in the MgO–CaO–SiO$_2$ System

Weight Ratio CaO/SiO$_2$	Minerals Present	Composition	Approximate Melting Temperature (°C)
less than 0.93	Forsterite	2MgO•SiO$_2$	1900
	Monticellite	CaO•MgO•SiO$_2$	1490*
0.93	Monticellite	CaO•MgO•SiO$_2$	1490*
0.93 to 1.40	Monticellite	CaO•MgO•SiO$_2$	1490*
	Merwinite	3CaO•MgO•2SiO$_2$	1575*
1.40	Merwinite	3CaO•MgO•2SiO$_2$	1575*
1.40 to 1.86	Merwinite	3CaO•MgO•2SiO$_2$	1575*
	Dicalcium silicate	2CaO•SiO$_2$	2130
1.86	Dicalcium silicate	2CaO•SiO$_2$	2130
1.86 to 2.80	Dicalcium silicate	2CaO•SiO$_2$	2130
	Tricalcium silicate	3CaO•SiO$_2$	1900**
2.80	Tricalcium silicate	3CaO•SiO$_2$	1900**
More than 2.80	Tricalcium silicate	3CaO•SiO$_2$	1900**
	Lime	CaO	2565

*Incongruent melting
**Stable only between 1900° and 1250°C.
Dissociation below and above these temperature into 2Ca$_2$O•SiO and CaO.

to chrome ore, however, the gangue silicates are converted on firing during manufacture or in service to the more refractory phases such as forsterite or dicalcium silicate, and the iron oxide to the spinel MgO·Fe$_2$O$_3$ by co-diffusion of Fe$_2$O$_3$ and MgO between the magnesia and chrome spinel. MgO·Fe$_2$O$_3$ is more resistant to deterioration in cyclic oxidizing and reducing conditions than the iron oxides in the original chrome ore. The addition of still greater amounts of magnesia to chrome ore improves significantly the resistance of the refractory to iron oxide bursting because of the greater affinity of magnesia for iron oxide as compared with Cr$_2$O$_3$. The addition of chrome ore to magnesia on the other hand improves the resistance of magnesia to thermal spalling through an apparent stress relief in an otherwise rigid structure.

As in magnesia refractories, the CaO/SiO$_2$ ratio exerts an important influence on the phases present in composite refractories of magnesia and chrome ore. At CaO/SiO$_2$ ratios less than 1.86, the primary phases between MgO, CaO, and SiO$_2$ are the same as that previously discussed with the sesquioxides Cr$_2$O$_3$, Al$_2$O$_3$, and Fe$_2$O$_3$ combined with MgO and FeO to form spinel solid solutions. At higher CaO/SiO$_2$ ratios, the sesquioxides form low melting compounds with CaO.

As in magnesia refractories, efforts have been made to replace the silicate bond in refractories made from magnesia and chrome ore. The most noteworthy of these efforts to date is the use of firing temperatures above those normally employed to produce a low silica burned basic refractory with so-called direct bonding of the magnesia. Fig. 3.39 shows photomicrographs of the conventional silicate and these direct bonds, respectively. Direct-bonded brick have usually high hot strengths, several times that of conventional brick of similar composition.

Although the phase assemblages previously discussed indicate the general combinations of iron oxides in basic refractories, it is equally important to consider the influence of furnace temperature and atmosphere on the oxidation state of the iron present in various compositions including the oxides MgO, Cr$_2$O$_3$, Al$_2$O$_3$, FeO, and Fe$_2$O$_3$. It has been shown, for example, that in mixtures consisting originally of Fe$_2$O$_3$ and MgO·Fe$_2$O$_3$, dissociation of the Fe$_2$O$_3$ is accompanied by solution of the dissociation product magnetite (FeO·Fe$_2$O$_3$) in MgO·Fe$_2$O$_3$ until a single spinel phase is

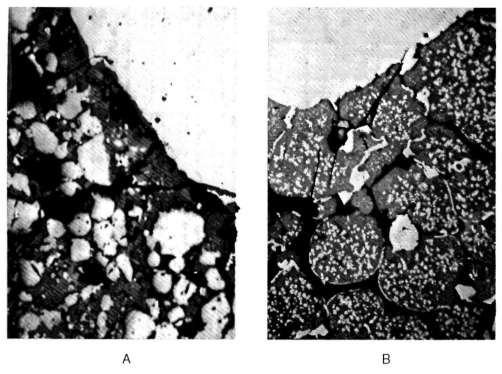

Fig. 3.39 Photomicrographs showing the structure of basic brick with different bonds. Photomicrograph A is a magnesite-chrome ore composition in which the bond between the large white chrome spinel grains and rounded magnesia crystals is principally the dark silicate. Photomicrograph B shows a magnesite-chrome composition having less silicate and a direct bond between white chrome spinel and periclase (MgO) containing multiple white magnesioferrite inclusions. Reflected light, 200X.

formed. In this same region, it has been shown that the spinel to sesquioxide transition temperature decreases as the magnesia content increases, but at the same time magnesia stabilizes Fe_2O_3 at higher temperatures. Chrome oxide (Cr_2O_3), on the other hand, has an opposite effect to magnesia in that it increases the sesquioxide to spinel transition temperature and lowers the degree of dissociation of Fe_2O_3 at higher temperatures. In mixtures consisting initially of $MgO \cdot Fe_2O_3$ and MgO, dissociation of Fe_2O_3 proceeds with a decrease in the amount of spinel and solution of iron oxide in periclase.

With the advent of carbon bonded refractories and oxide carbon refractories, the study of oxide reducing, gas-solid reactions has become an important concern for refractory technologists. Three main subjects are important: the straight or the direct oxidation of carbon (and graphite) by air; the reduction of magnesia (and other oxides) by carbon (graphite); and the role of antioxidants.

The rate of the magnesia-carbon reaction below is mainly affected, in addition to temperature, by the purity and the crystal size of the magnesia grain.

$$MgO(s) + C(s) \rightarrow Mg(g) + CO(g) \qquad (3.4.1)$$

This reaction reduces the magnesia to Mg vapor, which migrates to the surface, oxidizes, and forms a dense MgO zone or is incorporated in the furnace atmosphere. The dense zone may have a positive effect on the resistance to slag attack and oxidation resistance, but, on the other hand, the loss of refractory ingredients due to the MgO–C reaction results in a porous matrix behind the hot face which can be easily penetrated by slag at a later stage.

Antioxidants as metals (Mg, Al, Si) or carbides (SiC, B_4C) are often used to improve refractory properties such as oxidation resistance. Metals are added as antioxidants due to the fact that the

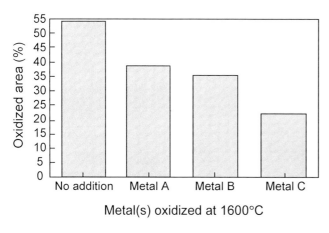

Fig. 3.40 Effect of metal(s) addition on resistance to oxidation. From Ref. 2.

affinity of some metals to oxygen, at the critical temperature range, is stronger than that of carbon. Each metal, or metal combination, has a different oxidation resistance characteristic as shown in Fig. 3.40. By protecting the graphite and the bond network from oxidation, and through the formation of carbides and nitrites, the metals improve the elevated temperature strength, in air, and under reducing atmospheres; see Fig. 3.41. Moreover, once the graphite is oxidized, there is no barrier preventing the magnesia grain from being washed into the slag; therefore, by protecting the graphite from oxidation, the metals also improve the resistance to slag attack.

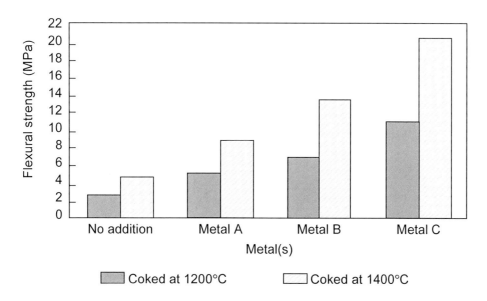

Fig. 3.41 Effect of metal(s) addition on coked strength. From Ref. 2.

Results of thermodynamic calculations of magnesia carbon refractories are summarized in Table 3.31. In this table Zone I is at the refractory hot face and Zone V is at the cold face. The oxides are to be found in a decarburized zone at the hot face. In the next zone, metal oxidation should be occurring together with the formation of secondary carbon from CO present inside of the brick matrix. This secondary carbon could produce extremely low porosity and a low permeable zone that precludes gas penetration and slag infiltration into the brick matrix. When CO is completely used up and only N_2 is available, the latter gas will form AlN and Mg_3N_2, respectively.

Carbides of Al, Si, and perhaps Mg can be formed by direct reaction between the respective metal and carbon further from the hot face. Thermodynamically, the effectiveness of the metals as antioxidants should be in the order Si–Al–Mg, least to most. Experimental data showed, however, that the order is reversed and does not follow oxygen affinity to the metals.

Table 3.31 Reaction Products for Three systems: MgO + Al + C, MgO + Mg + C, and MgO + Si + C at Different Temperatures Throughout a BOF Lining

Zone I (CO_2, CO, N_2)		
MgO + MA MA = MgO Al_2O_3 (Spinel)	MgO	MgO + M_2S M_2S = 2MgO SiO_2
Zone II (CO, N_2)		
> 2080K: MgO + AlN + C < 2080 K: MgO + MA + C	MgO	> 1800K: MgO + M_2S + C < 1800K: MgO + SiC + C
Zone III (N_2)		
MgO + AlN + C	> 1600K: MgO + Mg + C < 1600K: MgO + Mg_3N_2 + C	> 1700K: MgO + SiC + C < 1700K: MgO + Si_3N_4 + C
Zone IV **(Neutral Atmosphere)**		
> 1380K: MgO + Mg + C < 1380 MgO + Al_4C_3 + C	MgO + Mg + C	> 1810K: MgO + M_2S + Mg + C < 1810K: MgO + SiC + C
Zone V **(Unchanged Material)**		
MgO + Al + C	MgO + Mg + C	MgO + Si + C

From Ref. 2.

3.5 Testing and Selection of Refractories

3.5.1 Simulated Service Tests

Section 3.3 described various tests for single refractory properties. Because most refractories wear by a complex combination of mechanisms, many simulated tests have been developed to study refractory behavior for specific applications.

One simple example includes failure of refractories in a special gaseous environment such as carbon monoxide. Carbon monoxide disintegration occurs as carbon from the CO gas deposits around iron concentrations in a refractory:

$$2CO \rightarrow CO_2 + C \qquad (3.5.1)$$

The iron concentrations catalyze this reaction, producing carbon buildups and growth on iron locations which may fracture the refractory behind the hot face (the above reaction is highly temperature sensitive). Simulative tests have been designed to predict resistance to carbon monoxide disintegration in all types of refractories by heating the refractory for long periods in CO gas. Fig. 3.42 shows refractory samples after such tests. Although refractories without iron will not be affected by this reaction, the total removal of iron from most natural raw material is not economically practical. A more practical solution involves proper handling of raw materials and processes to remove ternary iron and minimize iron contamination from grinding. Control of the form of iron in the fired aggregate in monolithic materials or fired products has led to the development of special low iron refractories for blast furnace linings. Grinding equipment improvements and magnetic

Fig. 3.42 Specimen rating after the carbon monoxide disintegration test.

separation have further been used to minimize contamination. Examples of other simple simulated service tests include testing for exposure to alkalis, hydration, sulfidation, or oxidation.

Many more complex simulative tests exist where various failure mechanisms may be simulated at one time. For example, refractory samples may be exposed to slags in a rotating furnace heated with an oxygen-gas torch as shown in Fig. 3.43 (a), (b), and (c). In this test refractory wear from chemical reaction with the molten slag, erosion by slag, thermal cycling, and other effects can be measured for various times, temperatures, and atmospheres.

Similiar slag tests are conducted using a metal induction furnace with the specimen as part of the lining. In this case, wear by both metal and slag can be obtained on the same refractory specimens, as illustrated in Fig. 3.44.

The results from such tests can usually be correlated with the overall composition, with specific additives to the refractory and with the overall physical characteristics of the refractory. These effects are illustrated in Fig. 3.45 through Fig. 3.47. The slag resistance of various basic brick are

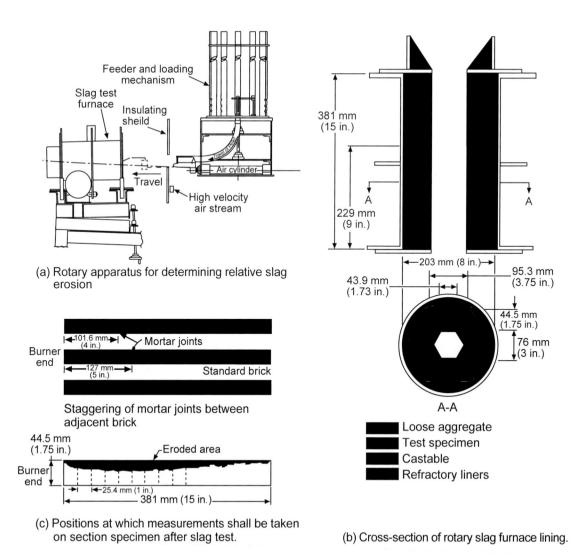

Fig. 3.43 (a) rotary apparatus for determining relative slag erosion, (b) cross section of rotary slag furnace lining, (c) positions at which measurements shall be taken on section specimen after slag test.

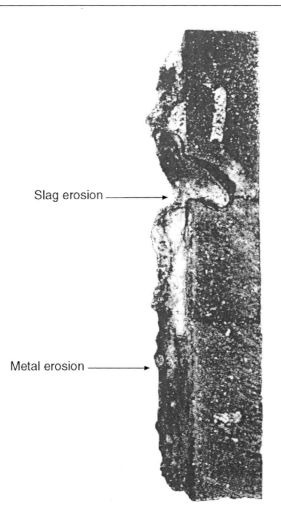

Fig. 3.44 Sectioned sample after slag test showing high erosion (1.27 in^2).

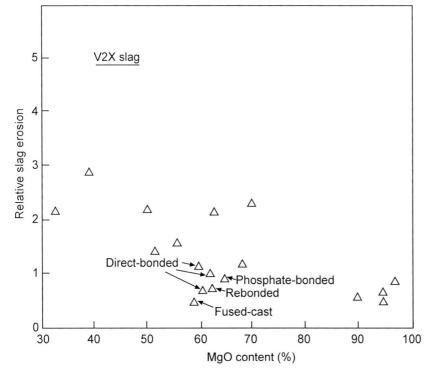

Fig. 3.45 Relation of relative slag erosion in rotary test to MgO content for various basic brick.

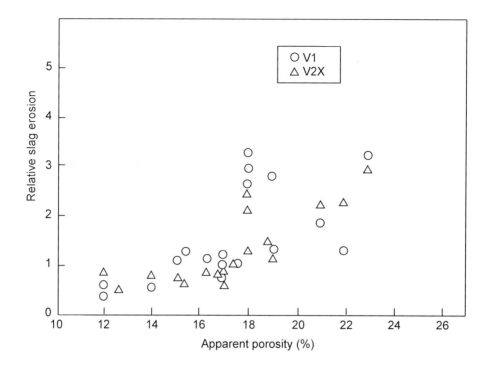

Fig. 3.46 Effect of porosity on relative slag erosion in rotary test of various basic brick with indicated slag compositions.

shown to be related to brick composition (MgO content) and physical structure (apparent porosity) and the slag resistance of a magnesia carbon brick is related to the type of metal antioxidant used.

Refractory specimens may also be exposed to erosion and thermal shock by dipping or spinning in steel baths or by pouring steel on or through them. For example, Fig. 3.48 shows steel pouring tubes after testing in steel while they were rotated at various speeds to simulate steel flowing through the exit ports.

It should always be remembered that any of the simulated refractory tests fail to some degree to duplicate service conditions. Fig. 3.49 is an effort to fully characterize the slag-refractory reaction involving wear by corrosion, penetration and dissolution.

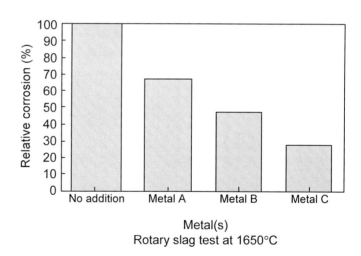

Fig. 3.47 Effect of metal(s) addition on corrosion resistance. From Ref. 2.

What makes corrosion resistance so difficult to measure is the fact that this characteristic is also a matter of the composition of the slags and their fluidity or viscosity, surface tension, wettability and work of adhesion, thermal conductivity, mass conductivity and density.

The extrinsic properties of both the refractory and the slag are also important parameters to define shape and size of the bricks, lining design, nature of the joints, expansion joints, and method of refractory installation, on one hand; fluid dynamics, heat and mass transport phenomena at the

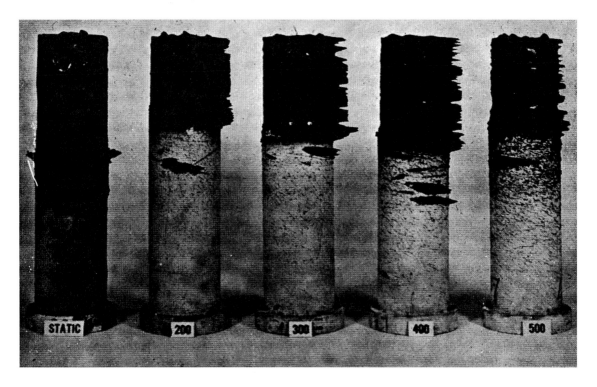

Fig. 3.48 One brand of pouring tubes after testing of various speeds for one hour in C 13450 (AISI 1345) steel.

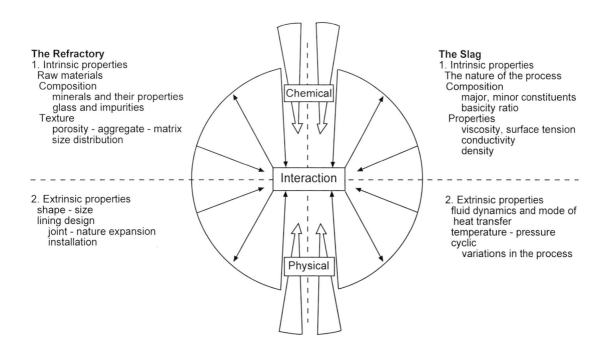

Fig. 3.49 Parameters influencing the slag-refratory reaction. Case of pure dissolution. From Ref. 2.

refractory-slag interface, temperature, pressure, temperature gradients, concentration gradients, variability, and the cyclic nature of the process, on the other hand.

In porous refractories, corrosion-dissolution is an accompaniment of penetration (chemical invasion) into pores, but also in all openings: at the lining level, in open joints, gaps, and cracks; at the brick level, in connected networks of voids so the texture of refractories and the method of installation are of paramount importance.

Both physical penetration and chemical invasion are favored by effective slag-solid wetting and by low viscosity of the slag. Therefore, the slag chemical composition is also of paramount importance.

Slag corrosion always begins with the dissolution of either the matrix material and/or the aggregates themselves, or, in other words, by the interaction of some minor or major constituents of the refractories with the liquid slag. Dissolution means either the solubility of a solid phase into a liquid phase, or a chemical reaction leading to the disappearance of the original solid phase. Therefore, knowing whether the product is liquid or solid at the prevailing temperature is very important.

3.5.2 Post-Mortem Studies

To improve refractory life, the mechanisms by which the refractories are consumed must be fully understood. A considerable and continuing effort has therefore been made to thoroughly analyze used refractories and to integrate this information into improved refractory products. In its simplest form, used refractories are examined by chemical analyses and by microscopic methods. Fig. 3.50 shows results using this approach on two types of brick taken from a BOF furnace after service. In the magnesite brick without pitch (burned product), CaO, FeO, and SiO_2 from the slag have penetrated the brick to a depth of 25 to 75 mm (1 to 3 in.) and altered the microstructure and properties of the refractory. Such a refractory may not wear uniformly due to structural spalling as well as hot face corrosion and erosion. In a similar magnesite brick made with pitch or having retained carbon, penetration has been restricted to an area only a few millimetres thick by the presence of carbon in the brick. The carbon minimizes wetting and causes the formation of a dense impervious layer due to limited MgO volatilization. A carbon bearing refractory of this type wears uniformly by corrosion and erosion. The brick without carbon has been severely altered by slag penetration, diffusion, and reaction with the brick components, whereas penetration has been held to 1 to 2 mm in the carbon bearing brick. Unfortunately, carbon penetration inhibitors cannot be used for all applications because the carbon is easily oxidized in some oxidizing processes.

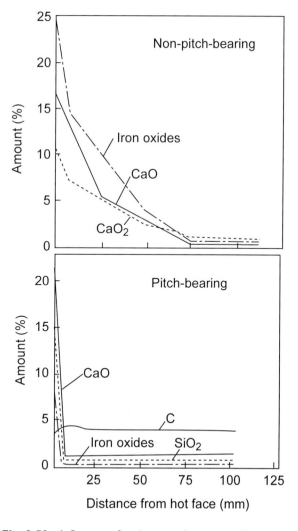

Fig. 3.50 Influence of carbon on slag penetration into magnesite brick in a BOF lining.

Steel Plant Refractories

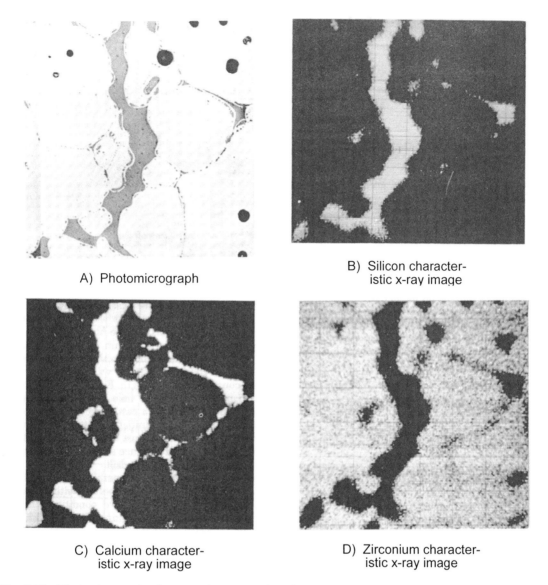

A) Photomicrograph

B) Silicon characteristic x-ray image

C) Calcium characteristic x-ray image

D) Zirconium characteristic x-ray image

Fig. 3.51 Photomicrograph of center of cross sectioned sensor button showing predominately large grains with glassy material at grain boundaries and characteristc x-ray images (600X) showing distribution of elements.

Many more sophisticated techniques can be used as part of post-mortem examinations including x-ray, scanning electron microscope, or electron probe techniques. Fig. 3.51 illustrated the use of the electron probe to indicate the distribution of impurities in a zirconia refractory. The concentration of impurities in the grain boundary area is typical of refractory material and helps to explain why small quantities of such impurities often produce significant detrimental effects including a significant loss in refractoriness.

3.5.3 Thermomechanical Behavior

Studies of the thermomechanical behavior of refractories and refractory lining systems are being used to an increased extent. The availability of larger personal and professional computers and sophisticated software has resulted in increased use of such techniques in refractory design and failure analysis. Compressive static stress/strain data along with compressive creep data, both at realistic stress levels, are required to characterize the mechanical behavior of refractories. Refractory lining systems are subjected primarily to strain controlled loadings (restrained thermal expansion) and to a much lesser degree stress controlled loadings. As a result, ultimate compressive strain is a better parameter for ranking and selecting the strongest refractory and not the ultimate compressive stress. Methods for measuring strain and creep were briefly described in Section 3.3.5.

The refractory criteria selected for use in these analysis techniques are becoming increasingly realistic. Fig. 3.52 shows calculated results using older elastic modulus data on refractories which gave unrealistically high shell stress values in a steel ladle. These values could be reduced to more realistic (measured) values by adjusting the modulus of elasticity values.

The latest analysis techniques require static compressive stress-strain data and compressive creep data at reasonably high stress levels. Fig. 3.53 shows stress-strain behavior of a high alumina refractory with and without mortar joints, which serve to reduce the stress level for any degree of strain.

Stress analysis can be used to understand and modify observed refractory behavior.

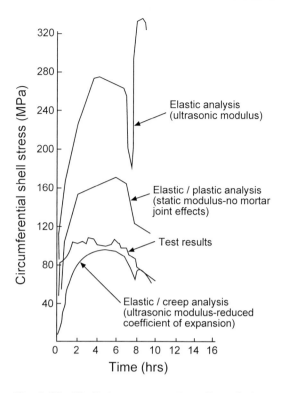

Fig. 3.52 Shell stress versus time. *From Ref. 4.*

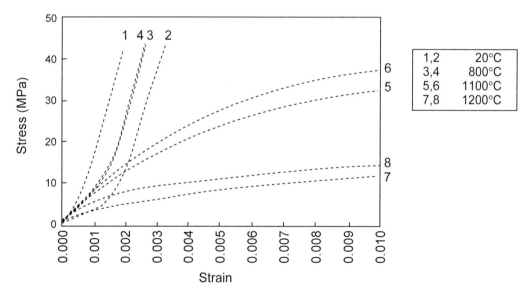

Fig. 3.53 Static compressive stress-strain behavior of a 70% alumina brick with a 1.6 mm brick mortar joint. From Ref. 2

Fig. 3.54 shows the radial compressive load of the working lining against the safety lining that results from the thermal expansion of the ladle refractory brick in an oval shaped ladle. Note that the force imposed by the working lining on the safety lining is greatest in the radial section of the ladle. This compressive force diminishes in the flatter section of the trunnion. Cooling of the refractory between heats causes an inverted thermal gradient through the ladle lining with the brick hot face at a lower temperature than the intermediate temperatures deeper in the lining, as illustrated in Fig. 3.55.

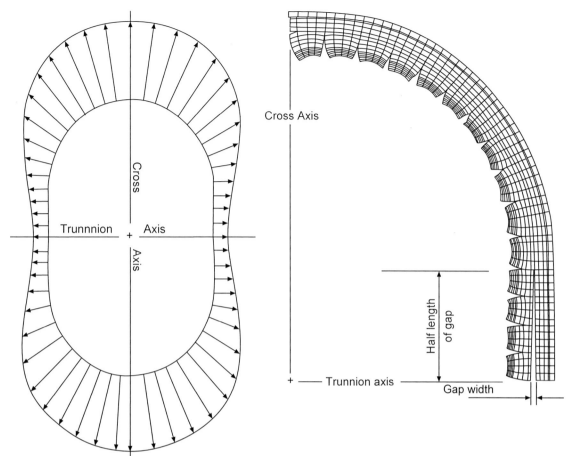

Fig. 3.54 Illustration of radial compressive load of working lining against the safety lining due to thermal expansion of the working lining. Note the compressive forces are lowest at the trunnion. *From Ref. 5.*

Fig. 3.55 Finite element analysis of ladle quarter section after cooling between heats. Note the densification at the hot face due to heating, opening of joints due to shrinkage during cooling, and gap formation at the trunion. *From Ref. 5.*

These mechanisms help explain the problems experienced by oval type ladles in the ladle straight sections. Varying material expansion under load characteristics can reduce such problems.

3.6 General Uses of Refractories

Other chapters of this book cover the uses of refractories in specific applications. This section is intended to offer only a flavor of refractory uses along with general comments.

3.6.1 Linings

The most common type of installation involves a refractory lining to contain molten metal and slag. Such linings are made of several types of refractories to resist specific conditions and usually consist of a working or hot face lining and a backup lining of material which may include an insulating material to reduce shell temperatures. Fig. 3.56 shows a typical lining in a steel ladle used in modern steelmaking to transport and treat molten steel and slag. Note the use of high alumina castable (85–95% Al_2O_3) in the metal portion of the ladle with basic (MgO–chrome or MgO–carbon) brick in the area which must resist slag. The conditions to which refractories in ladles are subjected are very severe as the ladle will undergo several process cycles daily and may be used for complex

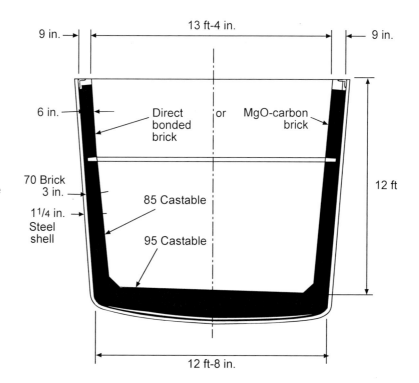

Fig. 3.56 Typical steel ladle lining.

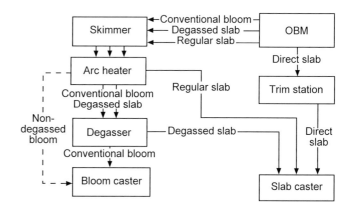

Fig. 3.57 Shop process routes.

ladle refining, arc reheating, degassing, or other treatments. Fig. 3.57 shows the different cycles to which ladles are exposed in one steelmaking shop which produces various slab and bloom heats.

Life of the working lining will range from as little as 20 heats for some slag lines to over 100 heats for the barrel or metal containing zone. Increased ladle wear will result from factors such as:

- higher tap temperatures,
- localized tap stream erosion,
- increased overall steel holding time,
- corrosion by superheated artificial slags during arc heating,
- thermomechanical cracking associated with ladle movements,
- more thermal cycling.

Table 3.32 Properties of Various MgO–C Brick for Ladle Slag Lines

Property	Better Resistance Obtained by
Arc Resistance	More graphite
	Higher purity graphite
Oxidation Resistance	Metal(s)
Hot Strength	Metal(s)
	Metals and higher purity graphite
Corrosion Resistance	High purity sintered grain
	High purity graphite
	Fused grain
	Combinations of above

Table 3.32 shows various changes which might be made in MgO–carbon brick to improve their life at the slag line. The exact brick selected will be based on the most economic materials in balance with the life of the remaining components of the ladle.

Slag line life will depend strongly on the extent of arc reheating and the manner in which it is conducted. Fig. 3.58 shows the extent to which slag is superheated depending on the thickness of the slag layer and the manner in which it is stirred with lances, porous plugs, or by electromagnetic force. Increased superheat can increase refractory erosion 2–4 times the normal rates.

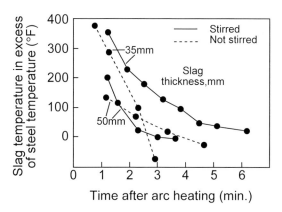

Fig. 3.58 Effect of slag thickness and stirring on slag superheat.

3.6.2 Metal Containment, Control and Protection

Fig. 3.59 shows a simple concept of the various refractories used in the modern casting process to contain, control and protect the steel between the steel ladle (top) and caster.

Various plates, rods, tubes, and linings are employed and their properties vary widely. All of these refractories must present the ultimate in

Table 3.33 Typical Properties of Sliding Gate Plates—Conventional Type

Type	85% Alumina	90% Alumina	High Alumina, > 92% Al_2O_3	Alumina/Chrome**	MgO
Apparent porosity (%)	16	17	15	16	18
Bulk density (g/cm³)	2.88	3.00	3.15	2.95	2.91
Cold crushing strength (kg/cm²)	1700	1500	1500	1500	ND*
Thermal expansion (% at 1000°C)	0.65	0.67	0.78	0.57	ND*
MOR 2500°F (Pa×10⁵)	68	82	82	82	109
Major composition (%)					
Al_2O_3	87	90	95	87	
SiO_2	11	8	3	9	
C***		2–3	2–3	2–3	2–3
Cr_2O_3				3	
MgO					95
Recommended applications	Mild, general	General	Throttling tundish	General, severe	Low C, billet

*ND – No data availlable.
**Replaces part of Al_2O_3.
***Carbon from tar-impregnated which may be followed by partial coking.

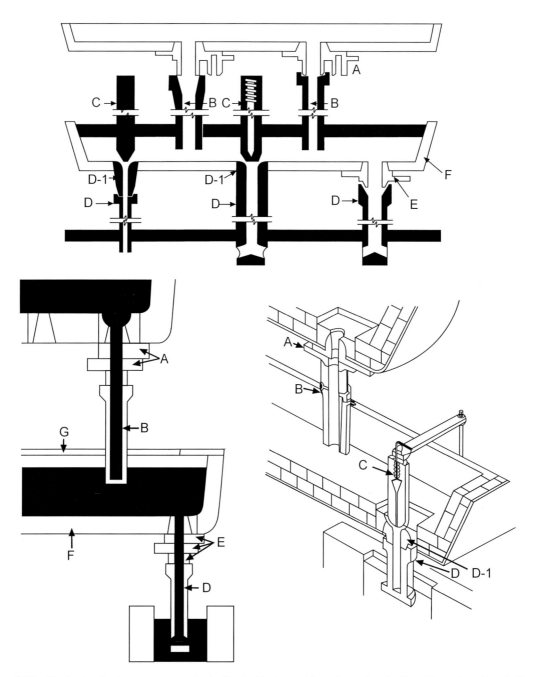

Fig. 3.59 Typical refractory components in the ladle to mold system. A = ladle slide gates, B = ladle to tundish shrouds, C = Tundish stopper rods, D = tundish to mold shrouds with associated nozzles (D–1), E = tundish slide gates, F = tundish linings, and G = tundish covers.

properties, such as thermal shock resistance, as they are exposed to instant temperature changes from ambient to steelmaking conditions. These refractories also must not erode or allow air infiltration to prevent deterioration to steel quality.

Tables 3.33 and 3.34 show properties of slide gate refractories.

The materials are designed to function in either ladle and/or tundish gates as generally illustrated in Fig. 3.60 and Fig. 3.61.

Table 3.34 Properties of Typical Higher Carbon Gate Plates

Property	Alumina–Carbon-A	Alumina–Carbon-B	Alumina–Zirconia Carbon
Composition (%)			
Al_2O_3	89–92	68–75	70–75
SiO_2	4–8	3–6	3–5
C	3–6	10–15	7–12
ZrO_2			6–9
Density (g/cm³)	3.15–3.20	2.70–2.80	3.00–3.10
Porosity (%)	10–12	11–13	9–11

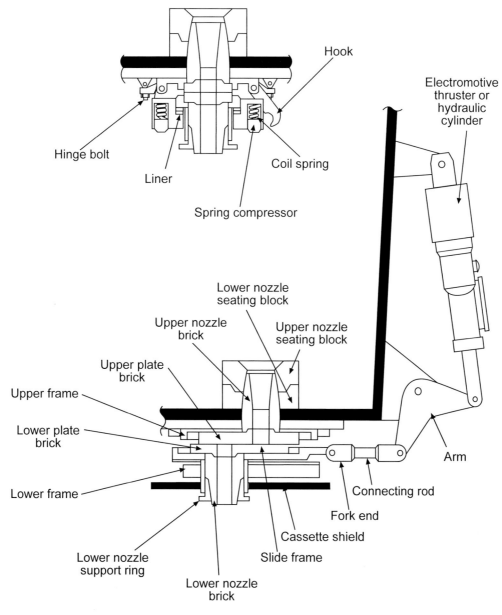

Fig. 3.60 Typical ladle with system.

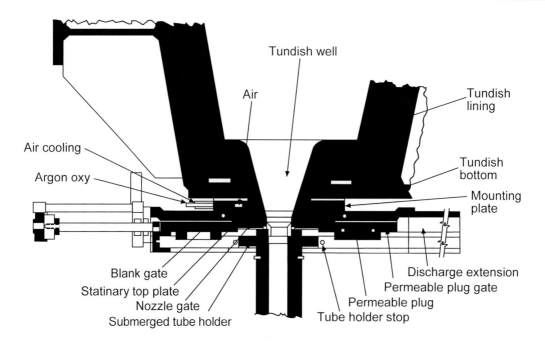

Fig. 3.61 Tundish valve showing submerged pouring tube in position.

The sliding ladle gate uses air-cooled stainless steel springs to maintain a constant sealing pressure on the sliding refractory interfaces. A valve mechanism permits rapid opening or closing to replace or inspect refractory components. Refractories for ladle gates must have excellent resistance to thermal shock and metal erosion and maintain a smooth surface during use. Depending upon steel grade, plate life may be single or multiple heats without replacement or inspection. Consistency is vital because of the severe danger to personnel and equipment.

The tundish nozzle is a critical link in the continuous-casting system because it must deliver a constant and controlled rate of flow of steel to the mold with minimum stream flare to minimize spatter, spray and atmospheric oxidation. Because the ferrostatic head remains substantially constant in the tundish throughout casting of much of each heat, the bore of the nozzle must remain at a constant diameter throughout the cast.

Tundish rods and ladle or tundish shrouds are made from isostatically pressed oxide-carbon combinations as illustrated in Table 3.35. These materials must have the ultimate resistance to thermal shock and erosion from steel and slag in the case of ladle to tundish shrouds or from steel and mold powders in the case of tundish to mold shrouds. Ladle to tundish shrouds may have lives of single or multiple heats whereas the tundish to mold shroud may be changed in mid-heat if clogging occurs from the buildup of deoxidation products such as Al_2O_3. Argon gas injection is usually supplied in one or more areas of the tundish to mold control system to minimize Al_2O_3 formation and/or buildup as shown in Fig. 3.62. This requires that a specific portion of the refractory be produced with a sufficient permeability to direct the inert gas flow.

Refractory usage in the tundish proper continues to evolve as the tundish is required to remove and store undesirable materials in the tundish slag or on baffles or filters in the tundish. The refractory inner linings are thin consumable layers applied between strings of heats of like composition, as in Fig. 3.63. After each string of heats, the tundish and skull are dumped and a fresh refractory layer applied. The fresh layer, such as a gunnable MgO layer ~1–2 in. thick, provides a completely clean material to avoid any possible steel contamination. A monolithic safety lining and insulation provide for safety and low shell temperatures. The length of the string of heats is largely independent of the refractory as it depends on allowable slag buildup in the tundish and the need for a particular steel order.

Table 3.35 Properties of Refractories for Shrouds and Tundish Rods

Area Used	Composition (%)						Apparent Porosity (%)	Bulk Density (g/cm³)
	Al_2O_3	SiO_2	C	ZrO_2	MgO	Other		
Base Al_2O_3								
Ladle-to-tundish shroud,	50–56	14–18	26–33			0–5	15–18	2.30–2.44
tundish-to-mold shroud, rod for tundish.	42–44	18–24	24–31	0–3		0–2	12–18	2.20–2.35
Fused silica								
ladle-to-tundish shroud, tundish-to-mold shroud.	≤ 0.3	≥ 99					12–17	1.82–1.90
ZrO_2–C								
Slag line or nozzle of	1–2	5–7	7–10	75–80		0–2	12–16	3.57–4.18
tundish-to-mold shroud, head of rod.			20–25	67–74		0–2	15–18	3.20–3.60
MgO–C								
Nozzle of tundish to mold		0–2	5–9		85–92		15–18	2.46–2.51
shroud, head of rod		12-16	10–20		58-78		14-17	2.25-2.50

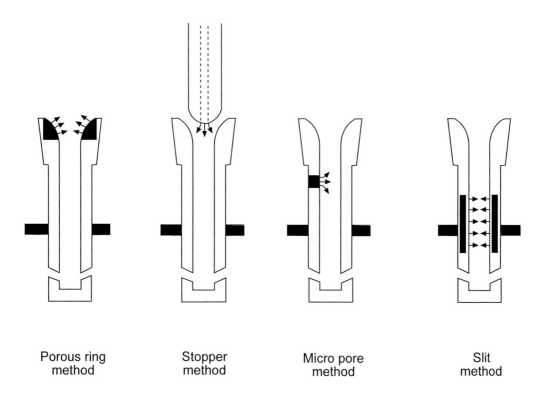

Porous ring method   Stopper method   Micro pore method   Slit method

Fig. 3.62 Inert gas bubbling methods used to decrease alumina clogging problems.

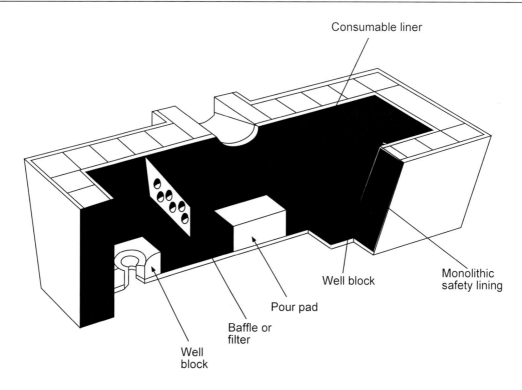

Fig. 3.63 Consumable liner in tundish for clean steel practice and to facilitate skull removal.

3.6.3 Refractory Use for Energy Savings

Although refractories always provide a degree of thermal insulation, there are some steel plant applications where these criteria are more important than containment or steel quality protection.

In slab reheating furnaces, refractories are used over the areas of the skid pipe system not in contact with the slab to decrease energy usage, as illustrated in Fig. 3.64. Special refractory metal anchoring systems or interlocking refractory shapes are used in this application to resist vibration and facilitate skid insulation replacement. Fig. 3.65 shows some examples of methods of holding refractory tile on to a skid pipe system.

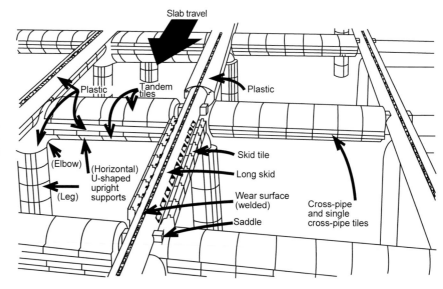

Fig. 3.64 Identification of the major components of a skid pipe system.

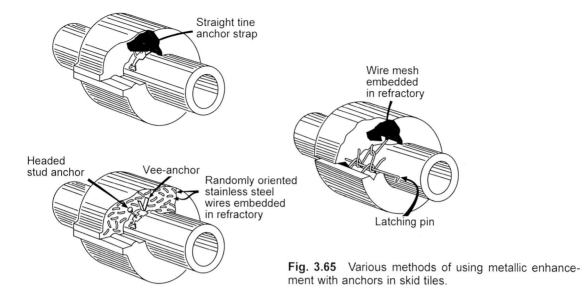

Fig. 3.65 Various methods of using metallic enhancement with anchors in skid tiles.

In recent years, increased use has been made of various forms of fiber refractory materials in many types of reheating furnaces because of their good insulating properties and light weight. Such materials are not suitable in contact with slag or in areas requiring abrasion resistance, but are suitable for many less severe areas. Fig. 3.66 shows a plate mill continuous-annealing furnace lined with fiber blanket held in place with metallic anchors. Refractory type anchors may be used in such a system to extend the useful temperature of fibers. Fibers may also be used in stackbonded or modular form where the fibers are folded or fastened to a steel backup plate attached directly to the steel super-structure. Veneers of fiber are also being evaluated over existing refractory linings to reduce energy consumption. In all cases, the economies of fiber materials must be carefully evaluated in comparison to other insulating materials including insulating brick, insulating castables and mineral wool or other insulating blocks.

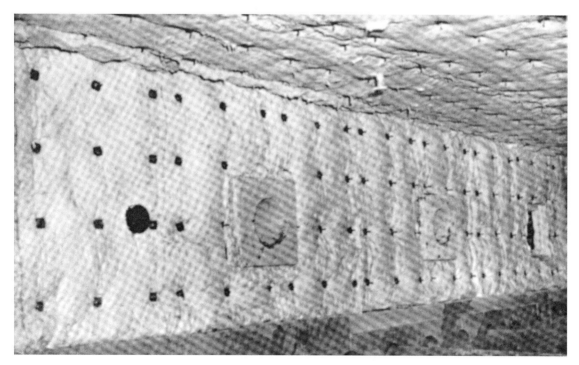

Fig. 3.66 Plate mill continuous annealing furnace with fiber blanket wallpaper construction.

Fibers offer added advantages when operating equipment under cyclic conditions because of their low heat storage characteristics.

3.7 Refractory Consumption, Trends, and Costs

Some general comments on refractory consumption, costs and trends are made here for illustrative purposes. Because this type of information changes rapidly with changes in the steelmaking process, changes in refractory technology, changes in economic conditions, and changes in environmental factors, care must be taken to gather updated information for any specific use. Also, comparing such data between countries (or even companies) must be done only after an understanding in the wide differences in conditions. Consideration must be given regarding integrated and nonintegrated shops and between the types of steel products being produced.

The types of refractories used in recent years trend toward more use of monolithic and special shaped products with less standard brick being used. This trend reflects the drive for both lower labor costs and the changes in the process to produce higher quality steel products. Tables 3.36 and 3.37 show the distribution of refractory product types for an integrated and nonintegrated plant in

Table 3.36 Breakdown of U.S. Refractory Usage—Integrated Plant

Refractory Type	Percentage In Indicated Form		
	Brick	Shapes	Monolithic
All	49	15	36
Silica	5	25 [†]	
Fireclay	15		25
High alumina	19	58	30
Al$_2$O$_3$–SiC–C	10		20
Burned basic	22		
Magnesia carbon	25*	2	
Other MgO		5 [‡]	23 [§]
Miscellaneous	3	10	2
Total	100	100	100

* All types including tempered, impregnated, graphite bearing.
† Mainly fused-silica shrouds and coke oven shapes.
‡ Mainly tundish boards.
§ Mainly basic gun materials.

Table 3.37 Breakdown of U.S. Refractory Usage—Nonintegrated Plant

Refractory Type	Percentage In Indicated Form		
	Brick	Shapes	Monolithic
All	41	15	44
Silica		25 [†]	
Fireclay	11		12
High alumina	30	55	21
Al$_2$O$_3$–SiC–C		5	
Burned basic	15		
Magnesia carbon	40*	2	
Other MgO		5 [‡]	65 [§]
Miscellaneous	4	8	2
Total	100	100	100

* Mainly graphite bearing.
† Mainly fused-silica shrouds.
‡ Mainly tundish boards.
§ Mainly basic gun materials.

Steel Plant Refractories

Table 3.38 Major Areas of High-Alumina Refractory Usage in the Steel Plant

Process Area	Application	Type of High Alumina Product Used	Desirable Refractory Features	Future U.S. Directions
Blast furnace	1. Furnace lining	90% Al_2O_3, Al_2O_3–Cr_2O_3 brick, 60–70% Al_2O_3 brick, often tar-impregnated	Shock and alkali resistance, hot strength	More nonoxide SiC, graphite-type brick
	2. Stove—hot/air system	Creep-resistant brick (60–80)	Creep resistance, volume stability	Longer life/high quality Al_2O_3 types
	3. Cast house (troughs, runners, spouts)	Al_2O_3–SiC–C or high-alumina castables, plastics, ram mixes	Volume stability, metal and slag erosion resistance	Increased Al_2O_3–SiC–C types as metal treatment increases
	4. Hot-metal transfer/treatment (torpedo ladles, mixers, lances, stirrers)	Burned and impregnated high-Al_2O_3 products (70–80) Al_2O_3–SiC–C brick for metal treatment(s)	Metal and slag erosion resistance Cycle resistance	
Steelmaking	1. Runners EAF	Al_2O_3–SiC–C monolithics	Metal and slag resistance	Increased use replacing older material types
	2. Delta sections, smoke rings–EAF	Low-moisture high-alumina castables	Shock resistance	
Ladles	1. Linings	Brick (70–80)	Thermal shock, metal and slag resistance	Increased castable use. Higher alumina with clean-steel practices.
	2. Pads-blocks	Low-moisture castables (70–95)		Increased use with clean-steel practices. Increased use over brick or other materials
	3. Lances	Low-moisture castables preformed	Thermal shock and erosion resistance	
	4. Permeable plugs	Low-moisture castables preformed Fired shapes	Permeability, resistance to erosion	
Tundish	1. Safety Linings	Low-moisture castable	Volume stability	Increased use over brick or other materials
	2. Covers	Low-moisture castable	Volume stability	
Shrouds	1. Ladle-to-tundish	Al_2O_3–C	Shock resistance, metal and flux erosion resistance	Increased use with decreased fused-silica use
	2. Tundish-to-mold	Al_2O_3–C		
Slide gates	Ladle and tundish	Fired impregnated Al_2O_3 Al_2O_3–Cr_2O_3 Al_2O_3–C	Shock resistance, metal erosion resistance	Increased use of alumina-carbon types. Repair of plates
Reheating furnaces	Linings	High-alumina plastics, castables, preformed shapes, fibers	Volume stability, insulating effects	Continued emphasis on energy savings and ease of installation

Table 3.39 Projected USA Refractory Usage in Indicated Plant

Area	Consumption (kg/ton)	
	Integrated Plant	Nonintegrated Plant
Coke oven	0.40	
Blast furnace	0.45	
Cast house	2.55	
Hot-metal transfer and treatment	2.25	
Steelmaking	3.05	4.00
Steel ladles & ladle metallurgy	3.60	4.05
Gates or rods/tubes	0.65	1.05
Tundish	1.25	1.65
Finishing	1.25	1.25
Total	15.45	12.00

the United States. Each of these product types show complex trends in refractory use and some of these will be obvious by reading other chapters which highlight refractory uses in specific cases. Table 3.38, for example, shows some of the trends in the use of high Al_2O_3 products in the steelplant. Iron or steel processing changes have driven such trends as increased use of high Al_2O_3 materials in blast furnace casthouses which now cast on an almost continuous basis, steel ladle refractories exposed to more severe conditions during ladle treatment of steels, or the increased use of stream shrouding devices during casting to prevent oxidation.

The consumption of refractories (in terms of lbs of refractories per ton of steel produced) continues to decrease both from improved life and because modern processes are ever more productive. At the same time, the cost of refractories per ton is increased as refractories are improved to meet the new process conditions. Table 3.39 shows the current refractory consumption in two United States steelplants. As previously stated, these figures vary widely and comparisons between shops should not be attempted without a full knowledge of process conditions and parameters. The cost of refractories produced will also vary widely from $0.20 to over $2.50 per lb depending on composition and complexity. The cost of refractories per ton of steel produced for an integrated plant of the type shown in Table 3.39 would range from $9.00 to $12.00 per ton of steel produced.

It should also be noted that refractory costs are often secondary to other parameters including steel quality, performance predictability and safety, energy considerations, and environmental criteria.

References

1. W.T. Lankford *et al.*, *The Making, Shaping and Treating of Steel,* 10th ed. (Pittsburgh: Association of Iron and Steel Engineers, 1985), 37–96.

2. M.A. Rigaud and R.A. Landy, eds., *Pneumatic Steelmaking, Volume Three, Refractories* (Warrendale, PA: Iron and Steel Society, 1996).

3. *Modern Refractory Practice*, 5th ed., (Pittsburgh: Harbison-Walker Refractories, 1992).

4. C.A. Schacht, "Improved Mechanical Material Property Definition for Predicting the Thermomechanical Behavior of Refractory Limings of Teeming Ladles", *J. Am. Ceram. Soc.*, 76 (1), 202-206.

5. J.A. Boggs, et al, "Thermal Mechanical Effects on the Structural Integrity and Service Life of Ladle Refractory, in Secondary Metallurgical Process", *Iron and Steelmaker*, June 1992, 33.

Chapter 4

Steelmaking Refractories

D. H. Hubble, Chief Refractory Engineer, U.S. Steel Corp. (Retired)
R. O. Russell, Manager, Refractories, LTV Steel Co.
H. L. Vernon, Applications Manager, Harbison-Walker Refractories Co.
R. J. Marr, Manager, Application Technology, J. E. Baker Co.

4.1 Refractories for Oxygen Steelmaking Furnaces[†]

4.1.1 Introduction

The essential goal in the development of refractory practices for basic oxygen furnaces is to obtain a useful lining life that will provide maximum furnace availability for the operators to meet production requirements at the lowest possible refractories cost per ton of steel produced. To this end, the operators and refractories engineers seek to optimize their lining design, maintenance practices, and control of operating practices that are known to affect lining life. In most shops, longer lining life can result in lower refractories cost, but, in high production shops, longer life achieved with minimal downtime for maintenance will also enable increased productivity through increased furnace availability.

To optimize the lining design, most operators try to develop a balanced lining, that is, a lining in which different refractory qualities and thicknesses are assigned to various areas of the furnace lining on the basis of a careful study of the wear patterns. In a balanced lining, the refractories are zoned such that a given segment of lining known to receive less wear is assigned a lower quality or less thickness of refractory, whereas refractories of greater wear resistance and generally of higher costs are reserved for those segments of the furnace that will be subjected to the most severe wear.

The refractory qualities available to use in BOF linings range from pitch-bonded magnesia or dolomitic types to the advanced refractories that are made with resin bonds, metallics, graphites, and sintered and/or fused magnesia that can be 99% pure. Bricks are designed with a combination of critical physical properties to withstand the high temperatures and rapidly changing conditions/environment throughout the BOF heat cycle. A balance of properties such as hot strength, oxidation resistance, and slag resistance are necessary for good performance.

With the wide variety of available brick qualities, there is a wide range of prices; the more expensive brick can cost as much as six times that of a conventional pitch-bonded brick of the type used in many furnace bottoms. As lining designs are upgraded and more of the higher priced products

---

[†] Excerpted from *Pneumatic Steelmaking, Vol. 3, Refractories* with permission. Please see the Acknowledgment at the end of this chapter.

are used in a lining, determining if the changes are cost-effective is important. For example, when the cost of a lining is increased by 25% in a shop that is averaging 2000 heats, the lining life will need to increase to 2500 heats for the refractories costs to be maintained. However, in shops where furnace availability is needed for productivity, a lesser increase in lining life and a higher refractory cost may be justified if the furnace availability is greater during periods of high production needs.

As lining designs are upgraded to optimize performance and costs, the effects of operating variables on lining wear are important to know. With this information, the possibility of controlling those parameters that affect lining wear adversely and the economic tradeoffs of increasing operating costs to extend lining life can be better evaluated. In general, the practices that improve process control, such as sub-lances, will benefit lining life. In addition, lining life is helped by charging dolomitic lime to provide slag MgO, minimizing the charge levels of fluorspar, controlling flux additions and blowing practices to yield low FeO levels in the slags. These practices need to be optimized to yield the most cost-effective lining performance.

Even when many operating conditions are improved, lining designs are optimized for balanced wear, and the best brick technology is used, wear does not occur uniformly, and, generally, maintenance practices that involve gunning of refractories and coating with slag are used to extend the life of a lining. See Section 4.2.

The above discussion illustrates some of the many factors that need to be considered in a strategy for BOF lining performance. Some details that are needed for developing the optimum refractories practices are provided in the following sections.

4.1.2 Balancing Lining Wear

4.1.2.1 General Considerations

In theory, linings for oxygen furnaces should be designed by refractory type and/or thickness so that no materials are wasted at the end of a furnace campaign; that is, so that all areas of the furnace are worn to a stopping point such as the safety lining at the same time. In the real case, however, some areas will show higher wear despite the latest brick technology and efforts to use internal maintenance techniques. A continuing and dynamic effort is always in progress in any shop to minimize wear and to provide longer life in these severe wear areas.

4.1.2.2 Areas of Severe Wear

The areas of severe wear are dictated to a large extent by the type of oxygen process involved (top-blown, bottom-blown, combination-blown, bottom-stirred, etc.). In all oxygen practices, the area where scrap and hot metal are charged into the furnace (charge pad) is subject to impact and abrasion. More uniform wear will occur during the oxygen blowing period on top-blown vessels, but bottom-blown vessels will be subject to accelerated bottom wear during the blowing period. On turndown for sampling and tapping, the furnaces (normally to opposite and generally horizontal positions), localized contact with slag will also produce localized wear. Accelerated wear is also experienced in the trunnion areas of any oxygen furnace, mainly because this area is the most difficult to protectively coat with slags or gunning. Other unusual wear areas may result from unique features of a particular oxygen process; for example, cone wear from post-combustion lances or the damage inflicted by deskulling.

Fig. 4.1 shows the different areas of a typical oxygen steelmaking vessel, where different types or thicknesses of refractory may be used to obtain balanced wear.

4.1.2.3 Wear-Rate Measurements

Consistent and predictable lining life is very important to avoid production delays in the steelmaking facility and related operations such as ironmaking, casting, or finishing; predictable life is

Steelmaking Refractories

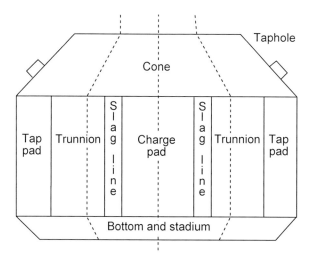

Fig. 4.1 Areas of a basic oxygen furnace.

also a vital part of a safe and stable operation. For these reasons, a variety of tools have been used to determine the condition of the lining at any time and to dictate the maintenance required to balance wear.

The most widely used method currently is a laser-measuring device as illustrated in Fig. 4.2. In this practice, a laser beam is rebounded off calibrated points on the furnace proper and compared to points in the worn lining. A computer analysis is then used to plot the remaining lining thickness. While this information is invaluable in comparing wear rates for different refractories and avoiding shell damage or breakouts, its primary usefulness is in determining and controlling furnace maintenance by gunning. Using the laser as a guide, the areas actually requiring gunning maintenance can be isolated, and the amount of gunning material required can be controlled. Fig. 4.3 illustrates one furnace campaign in which rapid trunnion wear was experienced in the first 500 heats.

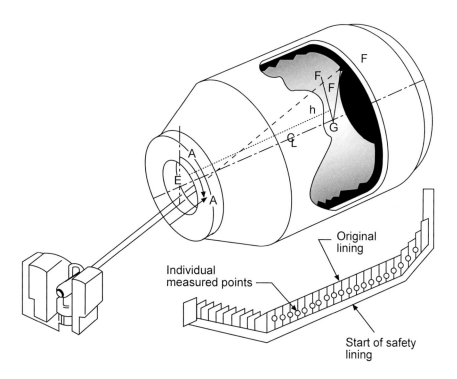

Fig. 4.2 Use of a laser to measure refractory wear.

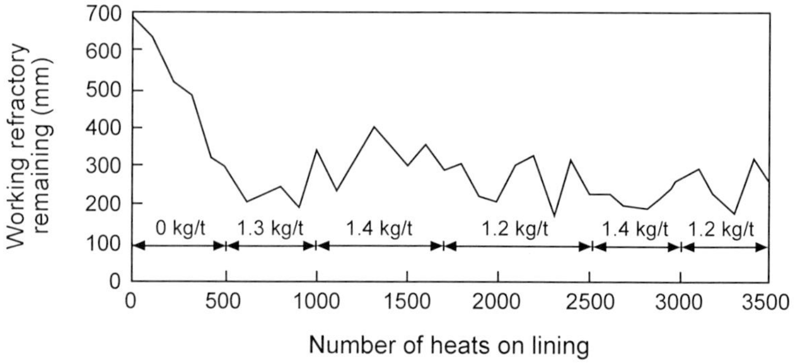

Fig. 4.3 Wear and gunning rate in the trunnion area.

Gunning was initiated at the indicated approximate rate at that time and continued throughout the balance of the campaign.

4.1.3 Zoned Linings by Brick Type and Thickness

The previous chapter on refractories described the types of carbon-bearing basic brick available for use in modern oxygen steelmaking furnaces. The engineers who are responsible for designing today's sophisticated vessel linings now may choose from greater than 30 proven compositions to construct a working lining to meet the service conditions found in a particular vessel. Normally, five to ten compositions have been found to cover most current operating practices and associated wear mechanisms.

Fig. 4.4 and Fig. 4.5 describe the zoning used in two types of operations to provide the optimum refractory behavior. The lining zones may also vary in working lining thickness from some 18 to 30 in. as an additional zoning method.

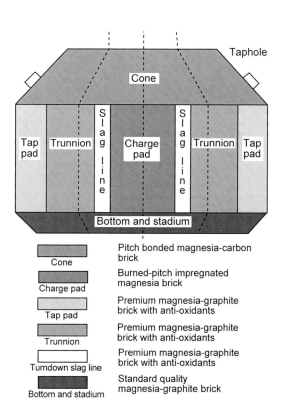

Fig. 4.4 Typical LD-BOF lining.

Steelmaking Refractories

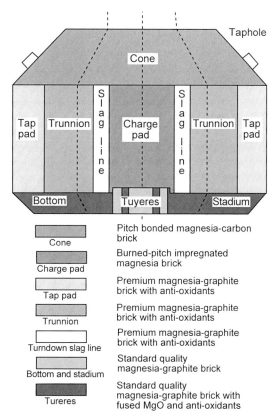

Fig. 4.5 Typical bottom-stirring LBE-BOF lining.

The zoned lining for a particular shop evolves as the operating conditions change and as new refractory types and/or maintenance practices are developed. Table 4.1 is an idealized description of the known wear condition(s) for the several furnace areas along with the best current refractories for each environment.

4.1.4 Refractory Construction

The refractory constructions in oxygen steelmaking furnaces are relatively simple in comparison to other refractory applications in that:

a. Tear-out of the spent working lining can be readily accomplished by mechanical removal of the top brick courses followed by rotation of the furnace to an inverted position.

b. Only two refractory layers are normally used—a thicker working lining (18–30 in.) and a thinner safety lining (6–9 in.) for shell protection. (A large part of the safety lining will last multiple working lining campaigns.)

c. The working lining is installed without mortar while the safety lining is held in place with steel retainer rings and mortar.

d. In principle, the working lining is installed with minimum brick cutting in a ringed-keyed construction where the brick are held in place by the brick taper (smaller hot-face than cold-face).

In reality, the brick construction in actual vessels is more complex as illustrated in Figs. 4.6 through 4.11. Fig. 4.6 and Fig. 4.7 show cross-sections of the refractory construction in typical top

Table 4.1 Furnace Area Wear Conditions and Recommended Materials

Furnace Area	Wear Conditions	Recommended Materials
Cone	Oxidizing atmosphere Mechanical abuse Thermo-mechanical stress High temperature	Standard-quality magnesia–graphite refractories containing anti-oxidants Pitch-bonded magnesia brick Resin-bonded low-carbon refractories with with anti-oxidants
Trunnions	Oxidizing atmosphere Slag corrosion Slag and metal erosion	Premium-quality magnesia–graphite refractories containing anti-oxidants Premium-quality magnesia–graphite refractories containing fused MgO and anti-oxidants High-strength premium-quality magnesia–graphite refractories
Charge Pad	Mechanical impact Abrasion from scrap and hot metal	Pitch-impregnated burned magnesia brick Standard-quality high-strength magnesia–graphite refractories containing anti-oxidants High-strength low-carbon magnesia brick containing anti-oxidants
Tap Pad	Slag erosion High temperature Mechanical erosion	Premium-quality magnesia–graphite brick containing anti-oxidants High-strength low-graphite magnesia brick with metallic additives Standard-quality magnesia–graphite refractories containing anti-oxidants
Turndown Slaglines	Severe slag corrosion High temperature	Premium-quality magnesia–graphite brick containing anti-oxidants Premium-quality magnesia–graphite brick containing fused magnesia and anti-oxidants
Bottom and Stadium (bottom-stirred vessels)	Erosion by moving metal, slag and gases Thermo-mechanical stresses as a result of expansion Internal stresses as a result of thermal gradients between the gas-cooled tuyeres and the surrounding lining	High-strength standard-quality magnesia–graphite refractories containing anti-oxidants Magnesia–graphite refractories without metallic additives characterized by low thermal expansion and good thermal conductivity Pitch-impregnated burned magnesia refractories

and bottom-blown furnaces, respectively. Note that the refractory construction is simple keyed rings for the cone, stadium and barrel sections, but involves other shapes in the furnace bottom areas. These more complex shapes are, however, still layered entirely dry (no mortar) and close dimensional requirements are necessary for all such shapes. Fig. 4.8 shows alternate methods of construction for the area surrounding the removable bottoms of bottom-blown furnaces. The inner rings of these furnace linings are often replaced when bottoms are installed.

Steelmaking Refractories

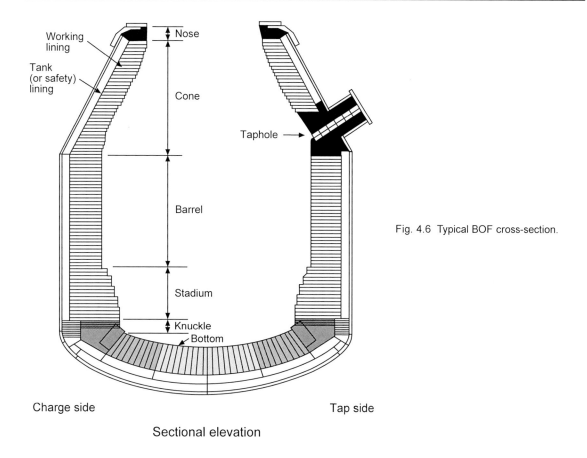

Fig. 4.6 Typical BOF cross-section.

Fig. 4.9 shows the construction of a removable bottom where brick are layered around bottom tuyeres in a manner that only one tuyere is contained in a given brick row to avoid stress buildup in the refractories. As shown in Fig. 4.10, two special shapes are used to obtain a tight fit around the tuyere itself. The construction and operation of tuyeres have been discussed in prior chapters. Similar complex constructions may be required when bottom stirring by inert gas injection is used.

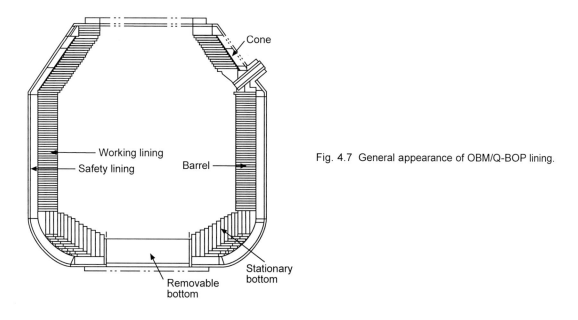

Fig. 4.7 General appearance of OBM/Q-BOP lining.

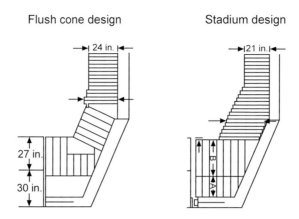

Fig. 4.8 Stationary bottom designs.

Fig. 4.11 shows construction when such inert gas devices (canned bricks, porous bricks, annulus tuyeres, etc.) are employed.

The previous illustrations were presented to give a flavor of the variety of refractory constructions possible in oxygen steelmaking furnaces.

As refractory linings become more complex, the increased thermal rigidity of the newer types of materials resulted in problems with overstressing the refractory and/or steel shell or trunnion support ring. Many modern linings now incorporate thermal expansion allowance on the brick using compressible materials such as special tapes between the brick in a given ring (horizontal expansion) and using compressible fibers between rings (vertical expansion). The amounts of such expansion may be estimated using stress-strain measurements on refractories and computer finite element analysis.

Relines on oxygen steelmaking furnaces are highly organized and planned to achieve furnace turnaround times of four to ten days, depending on other work done on the furnace auxiliaries at the same time.

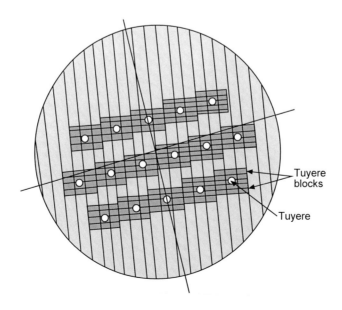

Fig. 4.9 OBM/Q-BOP bottom using bias brick lay-up.

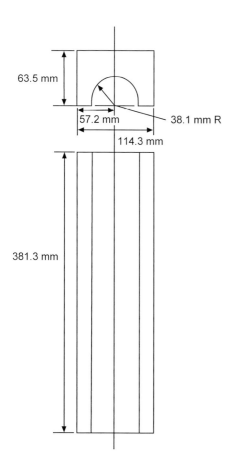

Fig. 4.10 Special shapes used around tuyeres.

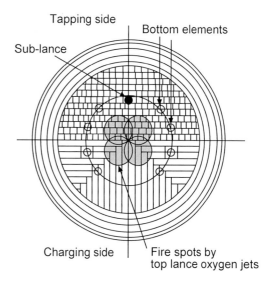

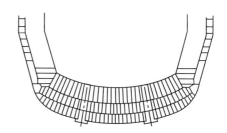

Fig. 4.11 Top and side cross-sectional views of one possible element pattern relative to the top lance oxygen jets and sub-lance sampling point.

4.1.5 Furnace Burn-In

The refractories in all types of oxygen steelmaking furnaces are heated rapidly to avoid unnecessary carbon oxidation from the brick. Fig. 4.12 shows a typical burn-in of four hours followed by immediate charging of the first heat. As described in Chapter 8, fuel for the burn-in is provided by coke charged on the furnace bottom and burned by the oxygen lance and/or by fuels through the tuyeres in a bottom-blown furnace.

4.1.6 Wear of the Lining

Wear of the lining is a complex process as illustrated in Fig. 4.13. As shown, wear may occur by a combination of various physical and chemical parameters and can be expected to differ not only by inherent differences in the process, (for example, top vs. bottom-blown) but also from site-specific parameters such as vessel shape or hot metal composition. The following is a brief discussion of some of the parameters which may affect lining wear.

4.1.6.1 Considerations of Slag Formation

Slags are a necessary part of steelmaking in that they remove impurities, such as sulfur and phosphorus, from hot metal. However, slag attack is the main mechanism of wear to a BOF lining. The

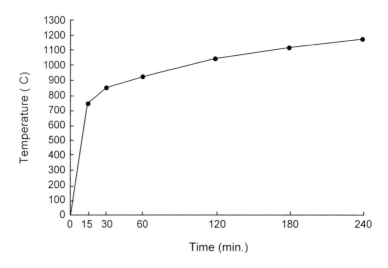

Fig. 4.12 Typical time-temperature curve for burn-in.

considerations of slag formation begin with the scrap and hot metal composition. In the following section, only the hot metal composition, the primary source of many of the oxide constituents in the slag, will be addressed. The components of most concern in the hot metal are: sulfur, phosphorus, silicon, manganese, and titanium. During the blow, liquid slags are formed by the oxidation of the metallic components of the molten metal (including iron), coupled with the addition of fluxes, such as CaO (burned lime), MgO (burned dolomitic lime), and CaF_2 (fluorspar or spar). The fluxes serve two purposes—they aid in achieving the proper chemical composition to remove sulfur and phosphorus from the metal bath, and, except for fluorspar, they protect the lining from attack by FeO and SiO_2.

The main constituents of slag are: SiO_2, CaO, MgO, Fe_xO_y, TiO_2, Mn_xO_y, P_2O_5, and CaF_2. The composition of the slag is shown in Fig. 4.14. The oxidation of silicon to form silica occurs early in the process and is very rapid. The FeO and manganese oxide are also formed early in the blow.

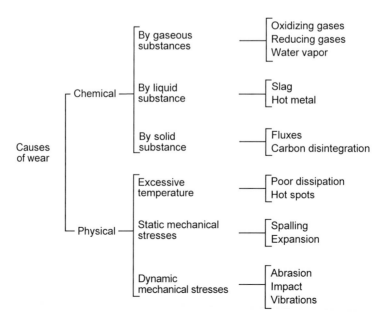

Fig. 4.13 Causes of wear in BOF linings.

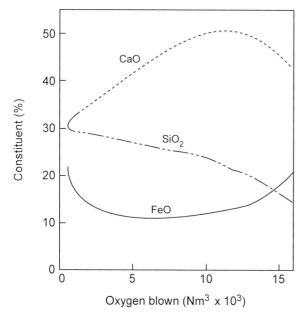

Fig. 4.14 Evolution of slag composition during the blow.

The percentage of FeO decreases as the blow progresses; however, toward the end, when the carbon levels are low, FeO will again increase. Silica reaches a maximum approximately 20–40% into the blow and then decreases while CaO continues to dissolve in the liquid slag. In the final stages, as FeO increases rapidly, the percentages of both CaO and SiO_2 are reduced.

The FeO that is formed both oxidizes the carbon in the carbon-containing refractories and forms liquids with the calcium silicate bonding phases found in the magnesite portion of the brick. The oxidation of carbon is of concern as many BOF refractories are carbon-bonded, and thus become quite weak when the carbon is removed. Also, carbon acts to prevent slag intrusion into the brick structure, and once the bricks are decarburized, the slag resistance of the refractory is diminished significantly.

The silica that is formed early in the blow is very corrosive to the refractory lining and reacts with the MgO to form low-melting compounds. Furthermore, as lime is dissolved into the liquid slag, the early slags, which have a lime-to-silica ratio close to 1.0, are extremely corrosive to the lining.

CaO must dissolve into the slag quickly to protect the lining for the reasons discussed above and to aid in removal of impurities from the molten bath. A limiting aspect of the dissolution of lime is that SiO_2 reacts with the CaO to form a dicalcium silicate shell around the lime particle. This dicalcium silicate layer is very refractory and dissolves slowly.

MgO additions to the slag are well known in reducing slag attack on a BOF lining. Two approaches are used. The first is to add sufficient MgO via dolomitic lime to guarantee adequate saturation of the slag throughout the blow. The second approach, which prevails today, is to add more than enough MgO to saturate the slag. The latter approach results in a buildup on the vessel walls which can greatly extend lining life when properly used.

Very thick layers of MgO-enriched slag may also be frozen on a specific worn area of the furnace such as the charge and tap pads. This practice requires some amount of furnace delay and must be accomplished while the other furnaces in the shop are available.

The reactivity of the lime has an influence on the resulting dicalcium silicate layer formed. A low-reactivity lime forms heavy, coarsely crystalline and adherent rims while a reactive lime produces the opposite results. The amount of fluorspar and other fluxes that must be added to dissolve this rim depends on the reactivity of the lime.

4.1.6.2 Other Factors Affecting Lining Life

A multitude of other factors which may be site or process specific can affect lining life and a brief review of these parameters may be helpful in understanding differences in lining behavior.

4.1.6.2.1 Hot Metal Differences Hot metal with higher silicon or phosphorus content will require the use of different slag practices and will cause more severe lining wear. Metal pre-treatments to reduce sulfur or other impurities greatly simplify slag processing and produce less severe conditions.

4.1.6.2.2 Scrap Variations Processing larger amounts of scrap can complicate slag development and reduce lining life. Some shops have increased damage from scrap charging because of limitations in charging equipment design.

4.1.6.2.3 Tap Temperature Requirements Tap temperature requirements depend on the subsequent processing to be involved before continuous casting (degassing, refining) and the ability to reheat steel in ladles. Higher tap temperatures will significantly lower lining life. Some processes also remove carbon during degassing which can further reduce tap temperatures.

4.1.6.2.4 Reblows and Heat Times An increased number of reblows and/or longer heat times will also have significant effects on lowering lining life. Improvements in processing which reduce these variables (dynamic controls, etc.) will improve lining life.

4.1.6.2.5 Lance/Tuyere Design and Practices Changes in lance practice and design or improved tuyere designs can cause substantial changes in refractory behavior.

4.1.6.2.6 Production Rates Shops operating in a manner to fully utilize furnaces minimize furnace cycling and improve lining life.

4.1.7 Lining Life and Costs

The current life of oxygen steelmaking furnaces varies widely from 1500 to 15,000 heats. The longer lining lives are achieved by extensive use of slag splashing to protect the brick lining as will be described in Section 4.2. Lives of 1500 to approximately 5000 heats can be achieved using brick linings balanced by refractory gunning and slag patching on a periodic basis. In the gunning process, refractory materials ranging in MgO content, from 40% to above 90%, are applied pneumatically to selected areas of the furnace. The so-called "programmed" gunning is designed to extend lining life without increasing overall costs or causing delays which would reduce steel production. Fig. 4.15 shows a typical campaign where gunning was used starting after approximately 1400 heats to extend the overall life to approximately 3500 heats. Note that the lining was removed when overall costs started to slightly increase.

The use of improved brick linings, dolomitic lime to promote slag buildup, programmed gunning and slag splashing or periodic patching have all extended lining lives. These techniques have not

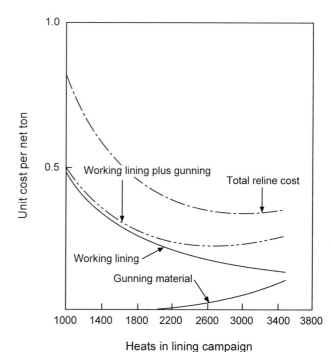

Fig. 4.15 Cost curve showing how gunning material can increase costs in longer campaigns.

proven as successful to-date in shops using the bottom-blown processes. Significant cost reductions have also been achieved so that the current refractory costs of oxygen processes are now only 8–15% of the total refractory costs in the modern steelplant.

4.2 BOF Slag Coating and Slag Splashing

4.2.1 Introduction

For the steel industry, as occurs in other industries, about the time a process appears to plateau in development, a new infusion of technology takes off and another round of experiments and associated technologies are applied to the process. In the past, several areas have been going through these stages simultaneously—bottom stirring in steelmaking, switching from ingot to continuous caster production, and resin-bonded magnesia–carbon brick in the refractory sector. While the emphasis of Section 4.2 is on slag coating and splashing of BOF refractories, the importance of operations and operational changes on refractory performance cannot be stressed strongly enough and have been described in recent publications.[1,2]

4.2.2 Slag Coating Philosophy

Slag coating and splashing substantially contribute to maintaining furnace lining profiles for safety and performance. Slag coating is an art form that requires considerable attention if it is to be done most effectively. Actions that make coating practices successful include: selecting the right slag, making the right and proper amount of additions, rocking the vessels correctly, disposing of the slag when necessary, and coating when it is the best time. These items need to be thought out and well executed.

As in the case of gunning, time to slag coat can often be found, even in most two-vessel operations. A successful slagging program utilizes established rules, and strict adherence to those rules is a key. It should be kept in mind that the vessel does not have to look pretty to accomplish long lining life, Fig. 4.16.

4.2.3 Magnesia Levels and Influences

Magnesia in the slag works against the steelmaking demands such as yield, phosphorus removal, and bottom stirring. The trick, therefore, is to maintain the lowest magnesia level required to accomplish the metallurgical and operational goals and still protect the refractory lining.

The magnesia in the slag is also a very important factor in slag coating technology (with and without slag splashing). It is not desirable to have low magnesia slag on long life vessels. The objec-

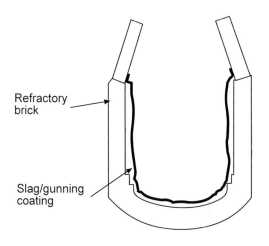

Fig. 4.16 Slag coating techniques by rocking the vessel create a working lining of slag that may be constantly replenished.

tive is to charge more magnesia than the saturation level of the slag to make the slag more refractory. This results in improved coating characteristics of the slag, so when it is thrown on the vessel walls, as with slag splashing, it will resist removal by the following heat. Alternate sources of sized magnesia of +90% are potential materials for slagging pads in combination with the vessel slag. These magnesia sources are typically lower cost magnesia units and possibly solve basic brick recycle problems.

Thus far, magnesia levels (MgO) in the 8–11% range have not adversely affected phosphorus removal in North American melt shops.

4.2.4 Material Additions

Alongside the normal flux charge, there are a number of potential material additions that can be made at different times in a heat, from beginning to end. A cost effective approach is to use low cost materials such as limestone or dolomitic stone, slag, ground brick, etc. Also, BOF slag can be added with the initial flux charge for early slag development. Another early charge technique, called stone-to-foam, is done for trunnion protection. This utilizes a carbonate limestone or dolomite to create CO_2 gas that aids in the foaming of the slag. When using carbonates, care should be taken regarding the effect on the offgas system.

Towards the end of a heat, dolomitic lime, dolomite stone, ground-up recycled brick, and magnesia are used to cool the slag and condition the slag regarding refractoriness and coating characteristics. Other materials (magnesia and carbon) are added at the end of the heat to foam the slag, similar to EAF slagging techniques. A magnesia (MgO) addition may be charged to the slag on extra low and ultra low carbon steels for additional lining protection.

4.2.5 Equilibrium Operating Lining Thickness

The ideal operating mode of a long life vessel is one where a lining wear equilibrium is reached and maintained. From experience, thermal equilibrium appears to be established at a lining thickness of 380 mm (15 in.) from a starting thickness of 762 mm (30 in.). The lining settles in between 127 mm (5 in.) and 381 mm (15 in.) of remaining working lining thickness, Fig. 4.17. In most shops, special attention is paid to the tap pad and the trunnion areas, which may have as little as 25 mm (1 in.) and 76–102 mm (3–4 in.) brick remaining, respectively. These thin areas demand that the lining be laser-measured a minimum of once a day and as often as every turn, with special attention being paid to the worn areas. From an operating perspective, a worn vessel produces heats with increased yield when compared to a new vessel.

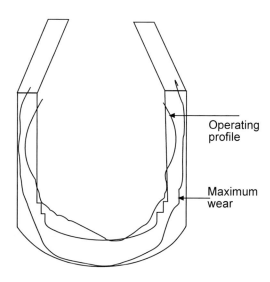

Fig. 4.17 Operating and maximum wear profiles on a extended campaign life of approximately 15,000 heats at a gunning rate of 0.37 kg/tonne.

4.2.6 Other Refractory Maintenance Practices

Those steps, independent of slag splashing, that will aid the maintenance of BOF linings are presented in Table 4.2. Note that these are recommended practices that are beneficial with or without slag splashing as they will extend refractory lining life.

Table 4.2 Additional Steps for Refractory Maintenance

Slag furnace as often as possible
   Selectively coat worn areas
   Space high FeO-slag heats if permissible
   O_2 wash bottom when needed
   Use crushed brick, especially for tap and charge pads
Gun furnace to fill holes in slagged or refractory lining
Use shooter-type gun
Keep cone skulls from building up
   Use proper lance height settings
   Deskull often, do not let skulls get too large
   Apply parting agent after deskulling
Laser measure vessel daily

4.2.7 Laser Measuring

It is best to laser-measure the lining thickness once per day to establish the lining status (profile) and to distribute color coded copies of the lining profile to the furnace crew, vessel pulpit, and operations management. Areas of the vessel that should be measured daily are the tap pad, the bottom, and the trunnions. When areas of less than 127 mm (5 in.) thickness are noted, an action plan should be implemented. This plan should include gunning as required, the use of a slag conditioner, slag coating the lining after every heat in the vessel from nose to nose, and taking a follow-up laser measurement reading eight hours after the low thickness reading.

Early in the campaign, wear rates on brick can be measured and the influences of the operation can be determined due to the presence of little or no gunning. Once gunning and slag splashing commences, laser readings are best used to maintain lining/slag thickness.

4.2.8 Slag Splashing

The technique of slag splashing has reduced refractory costs and increased productivity by increasing vessel campaign life to years instead of months. This technology was maximized in the early 1990s and has proven to be cost effective.[2] It entails the use of the oxygen lance to blow nitrogen on the residual slag, after tapping a heat, to coat the walls and cone of the converter with slag, Fig. 4.18. The process has greatly reduced the need for gunning of the lining by more than half. Two slag splashing practices are known: (1) with the vessel empty of steel and all the slag remaining in the vessel, as shown in Fig. 4.18, and (2) with the both the molten steel bath and slag in the vessel.[3] The latter method is specific in coating the trunnions and the upper reaches of the furnace. The blowing practices are different for these two techniques.

Slag splashing utilizes the technique of working off a slag coating instead of the working lining brick or its gunned coating. Slag splashing requires only a minute or two to perform and is done while the vessel is in a vertical position after a heat is tapped. The nitrogen flows are often automatic and are based on lance height at the start of a splash. Currently slag splashing refinements are continuing to be optimized, i.e., tuyere cleaning, placing the slag in a specific area, and/or best

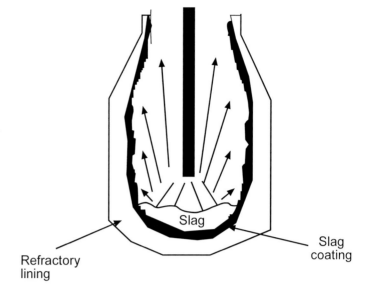

Fig. 4.18 Schematic presenting how nitrogen blowing onto a slag bath produces a slag coating that protects the working lining brick.

slag chemistry for coating. The emissions during the slag splashing period have caused no pollution problems.

4.2.8.1 Lance Buildup

Because slag splashing tosses the molten slag in all directions, the lance is also coated with slag. It is quite important that no steel in the vessel is present as it adds to the difficulty of the lance skull removal. Lance slag buildup removal can be aided by a low pressure water spray that eliminates the slag. Also, alternating lances has been found to be beneficial in slag removal from the lance.

4.2.8.2 Slag Splashing Augmented by Gunning

To attain extended campaign life on a lining, it is necessary to gun areas that encounter severe wear such as the trunnions and slag lines. The best refractory costs, as expressed in cost/ton, Fig. 4.19, are achieved by the selective use of gunning materials.[2] In the past, on long campaigns, gunning material costs could exceed the cost of the original magnesia–carbon lining. Therefore, the use of

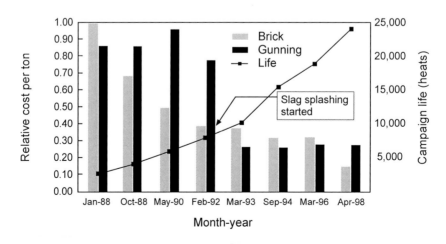

Fig. 4.19 Cost reductions have been considerable with the institution of slag splashing; especially note the reduction of the gunning cost/ton.

the exact amounts of gunning material in the right areas of the vessel, as directed by the laser readings, will yield the overall lowest refractory cost per ton.

Gunning has the potential of being a major improvement area in refractories. It hold great promise as considerable research is being applied to gunning materials. The shooter-type gun has made the application process what it should be under the hostile environment of the hot vessel, and with new gunning materials, it is anticipated that more effective gunning will occur in the future.

4.3 Refractories for Electric Furnace Steelmaking

Refractories for electric arc furnace (EAF) steelmaking are selected based on operating conditions and furnace design features that impact refractory performance. Steelmakers use electric arc furnaces to produce steel from scrap and/or alternative iron units under similar operating conditions. The charge is melted, using fluxes to maintain a basic slag chemistry, tapped at roughly 3000°F, and the furnace is again charged for another heat cycle. These operating conditions require chemically basic refractory products with excellent resistance to high temperature and thermal cycling. Furnace design features of present day electric melting technology require specialized refractory linings. There are various design features, but they are broadly grouped into three areas: tapping design, side tap vs. bottom tap; power source, AC vs. DC power; and the use of supplemental oxygen to increase the melting rate.

4.3.1 Electric Furnace Design Features

Side tapping electric furnaces have a spout extending from their furnace sidewall to transfer the heat to the ladle. The spout is refractory lined, and a taphole through the furnace sidewall connects the furnace interior to the spout. Side tap furnaces tilt approximately 45°, requiring a higher refractory sidewall lining on the tap side of the furnace to contain the molten steel. See Fig. 4.20 for a typical side-tapping electric arc furnace. Bottom tapping furnaces have their taphole through the bottom hearth section of the furnace and require special taphole refractory products. Bottom tapping also enables reduced height refractory sidewalls due to the reduced tapping tilt angle of approximately 15–20°. See Fig. 4.21 for typical bottom-tapping furnace.

Alternating current (AC) power sources require three electrode columns within the furnace for the three electrical phases. These three electrodes have increased arc flare during operation that can

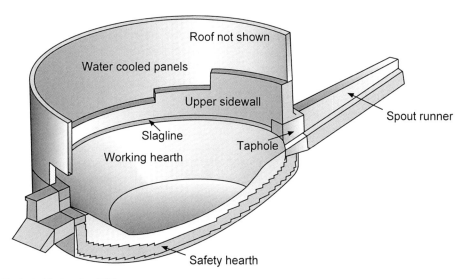

Fig. 4.20 Typical side-tapping EAF.

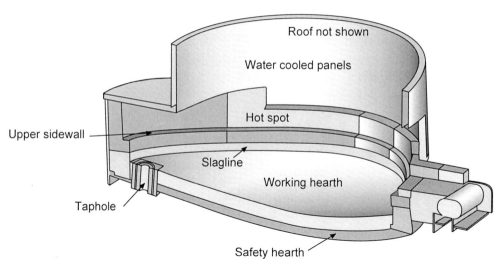

Fig. 4.21 Typical bottom-tapping EAF.

impinge on the refractory sidewalls, resulting in hot spots which must be addressed by refractory design. AC furnaces also require three holes through the refractory roof and the center section of the refractory roof between the electrodes is often an area that limits furnace performance. Direct current (DC) furnaces utilize a single electrode through the roof with the electric arc passing directly to the steel bath that contacts the bottom anode electrode to complete the electrical circuit. DC furnaces have less arc flare to the refractory sidewalls and therefore no hot spots. Roof design is also less complicated with less difficult operating conditions. However, the furnace hearth must contain the bottom electrode, which complicates refractory design there.

The use of supplemental oxygen lances and burners to increase melt rates impacts refractory design and performance. Oxygen directed from lances or burners can be deflected by scrap or charge materials and impinge on the refractory lining, resulting in localized overheating and rapid refractory wear. Localized oxidizing conditions can also occur which cause rapid refractory lining erosion.

4.3.2 Electric Furnace Zone Patterns

Even with the individual differences of operating conditions and furnace features, EAF steelmaking can be broken down into specific zones that have differing refractory requirements. These zones within the furnace, the specific operating conditions for that zone, and the analysis for refractory lining selection are discussed below. Fig. 4.20 shows the zones for side tapping electric furnaces; Fig. 4.21 for bottom tapping furnaces. Key zones of EAF steelmaking furnace are first, the hearth, which contains the molten steel as well as the initial charge materials. The furnace slagline is the transitional area between the hearth and the furnace sidewall. The upper sidewall is the refractory portion of the furnace walls above the slagline and below the water-cooled panels. The taphole is the opening that permits the molten steel to exit from the furnace. The roof has a refractory portion, occasionally referred to as the delta, which provides openings for the electrode(s) to enter the furnace, and an opening for furnace exhaust fume to exit the furnace to a bag house. Relative refractory consumptions by zone are presented in Fig. 4.22.

4.3.2.1 Hearth

The hearth zone of an EAF must contain molten steel at high temperature and resist the impact of heavy charge materials. In addition the hearth must withstand corrosion by molten slag as the furnace is drained. A typical EAF uses a two component refractory hearth of approximately 9 in. of

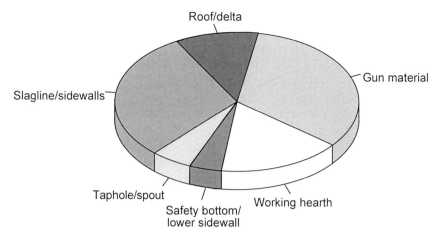

Fig. 4.22 Relative EAF refractory costs by zone, including hearth gunning material.

brick as a safety hearth against the bottom steel shell and 12 to 24 in. of monolithic magnesite as a working hearth. The safety lining brick should be 90–97% MgO content. These brick have the strength and slag resistance capability to contain the molten bath in the unlikely event steel or slag penetrates the monolithic working hearth. The safety hearth brick function as a permanent lining and would be changed infrequently, possibly every year or two. The monolithic hearth material is also a high magnesia product (60–95% MgO) and is a dry granular material. The grain sizing of the hearth material is formulated to compact easily by using vibration equipment during installation. High temperatures from the initial heat of steel causes sintering—bonding, densification and strengthening—in the monolithic hearth product, and the hearth becomes quite strong and penetration resistant. The hearth lining is designed so that approximately the top third of the thickness of the monolithic material is fully sintered, while the middle third of the lining is only partially sintered and the bottom third of the monolithic material against the safety lining brick is unsintered. This layering effect facilitates patching the hearth refractories when steel or slags damage the hearth through penetration or corrosion. The damaged area can be cleaned out by removing the penetrated, sintered magnesite and repaired with new monolithic material, which sinters in place forming an effective patch.

In DC electric arc furnaces the hearth refractory design must incorporate a bottom electrode. Operating conditions for the bottom electrode refractories are severe. Localized high temperatures and extreme turbulence are common at the surface of the bottom electrode. These conditions require refractories with high temperature stability and strength. For DC furnaces using pin or fin bottom electrodes, the same dry vibratable magnesite product can be used; although some steelmakers prefer a more temperature resistant, higher MgO product for the electrode. DC furnaces with bottom electrodes using larger diameter pins or billets often use special brick shapes surrounding each pin or billet to form the bottom electrode. Another design utilizes conductive refractories, either brick or monolith, which are a combination of magnesite and carbon, to carry the electrical current from the bath to the copper electrical connections at the bottom of the furnace. Conductive refractories are a complicated blend of high purity magnesia, graphite, and powdered metals to achieve the required combination of high temperature refractoriness and electrical conductivity to contain the molten steel as well as conducting current. (Bottom tapping electric furnaces also have a taphole through the hearth, but their refractory considerations will be covered in Section 4.3.2.4 on tapholes.)

4.3.2.2 Slagline

The slagline of electric arc furnaces is the transitional area between the hearth and upper sidewall. This area is subject to high temperatures from exposure to the electric arc, oxidation and flame impingement from supplemental oxygen injection, and most importantly, slag attack from high temperature slags containing FeO, SiO_2, and MnO. Slagline refractory design in the electric furnace is a combination of brick and monolithic products. Most common is a slagline 12 to 18 in. in thickness using magnesia–carbon brick with 10–20% carbon content. The carbon phase of the brick is composed of graphite and a carbonaceous resin bond. These carbon materials have excellent resistance to attack by steelmaking slags and have excellent high temperature resistance. However, the carbon is susceptible to oxidation. Often powdered metallic aluminum, silicon, or magnesium are added to the slagline magnesia–carbon brick to protect the carbon from oxidation. These metallics combine with carbon to form carbides which are more oxidation resistant and strengthen the refractory brick as well. Strength is important to resist the erosive action of molten slag and steel washing against the slagline zone in the furnace. Magnesia–carbon brick in the slagline are additionally protected by monolithic refractories. Initial installation of the monolithic hearth usually covers all or part of the slagline brick. However, this hearth material corrodes or erodes rapidly due to the difficult operating conditions in this zone of the furnace. Additional protection for the slagline is afforded by injecting a magnesite gunning mix into the furnace and building up a protective layer on the slagline. Again, this protective layer does not last long and must be replaced at regular intervals depending on the severity of the operating conditions.

4.3.2.3 Upper Sidewall

The upper sidewall zone of electric arc furnaces is lined with magnesia–graphite brick of a similar quality to the brick used in the slagline. The upper sidewall is subject to arc flare (very high temperatures) and impingement by heavy scrap during the charging process. During the tapping process, as the furnace tilts, molten steel and slag contact the upper sidewall brick on the tap side of the furnace. Lastly, upper sidewall brick must be able to withstand corrosion by steelmaking slags and flame impingement from oxygen lances and oxy-fuel burners. Magnesia–carbon brick of 5–20% carbon give cost effective service in upper sidewall linings. Various qualities and purities of magnesite, graphite, and powdered metals are utilized. In AC electric arc furnaces, which have hot spots in the upper sidewall, higher quality brick are used in the hot spots. These higher quality products are based on fused magnesia grain compared to the sintered magnesite grain in standard quality brick products. The fused magnesia brick have improved high temperature resistance but are considerably more costly.

4.3.2.4 Taphole Refractories

Taphole refractories are required for both side tapping and bottom tapping electric furnaces. Operating conditions in both types of furnaces are similar; hot molten steel, and to a lesser degree slag, flowing through a 5 to 8 in. diameter hole at considerable velocity erodes the refractories in the taphole. In the side-tapping furnace the taphole refractories have several alternative designs. The first and simplest design is to leave an opening in the sidewall brickwork while constructing the refractory lining. Then when all brickwork is completed, either a refractory taphole sleeve or a steel pipe is positioned in the taphole opening. Then MgO-based gunning mix is used to fill in the void between the sleeve or pipe and the adjacent brickwork. A second alternative is to use a large taphole assembly with a pre-formed taphole. This assembly is set in place in the furnace prior to installing the sidewall brickwork. Once the taphole assembly is properly positioned, the adjacent brickwork is completed, creating a tight fit between sidewall brickwork and the taphole assembly. Refractory products are always high quality in the taphole. If the taphole is constructed with gunning mix sprayed around a steel pipe mandrel, the highest quality magnesia gunning mix is used. This product has maximum strength and erosion resistance to minimize the erosive action of the molten steel. If a taphole assembly is used or a refractory sleeve, these are also of high quality magnesia brick with carbon and metals. Metallic additions to magnesia–carbon taphole brick assemblies provide added strength and oxidation resistance in this critical application. The side-tapping furnace has a spout extending from the taphole to enable the molten steel to flow into the ladle. The spout is a precast monolithic runner shape placed in the runner steel shell during lining construction. This precast runner shape is made from a magnesia based castable if furnace operating practices result in slag entering the taphole and the furnace runner; the slag resistance of MgO is required to counteract slag attack. On the other hand, if furnace operating practices limit the introduction of slag into the taphole and runner, then high strength, high alumina castables are used for the precast runner. These high alumina runners generally last longer due to improved thermal shock resistance and higher strength as compared to magnesite precast runners.

Bottom-tapping electric arc furnaces require specially designed taphole sleeves and an end block to comprise the taphole design. The taphole sleeves sit within the hearth in a taphole seating assembly. The assembly can be constructed from either brick shapes or precast shapes which result in a roughly 18 in. diameter hole through the furnace hearth refractories. The taphole sleeves are centered within the taphole seating assembly and a basic castable or ramming mix is tamped in the annular opening between the sleeves and the seating blocks. Taphole sleeves are magnesia–carbon shapes made from high purity magnesia or fused magnesia grain and with 10–15% carbon content. Powdered metals are used as a strengthening agent to maximize erosion and oxidation resistance. The bottom of the taphole extends beyond the furnace shell utilizing a shape called an end block. This end block is a similar magnesia–carbon brick that is held in place by an end block casting attached to the furnace. The end block is exposed to the atmosphere outside the furnace and must have excellent oxidation resistance as well as maximum erosion resistance to withstand the erosive action of the taphole stream. The end block is most often the limiting factor in taphole performance. As refractory erosion occurs the tap stream begins to flare, increasing reoxidation of the steel. It is necessary to then do a hot repair to replace the end block and taphole sleeves. Typical taphole life in bottom tapping furnaces is 5–10 days of operation.

4.3.2.5 Roof

Electric arc furnace roof refractories for both AC and DC furnaces are generally high strength, high alumina (70–90% Al_2O_3) precast shapes. Because the roof lifts and swings away from the furnace body during the charging process, refractories in the roof experience excessive thermal shock. The lesser thermal expansion of high alumina castables compared to basic castables offer an advantage in withstanding thermal shock. In addition, high alumina castables are much stronger than basic castables; therefore high alumina roofs are better able to resist the stresses developed as the roof is lifted and moved during furnace operations. Electric furnace roof refractories last anywhere from less than a week to up to ten weeks in some steelmaking operations. The roof also enables furnace exhaust fumes to exit the furnace and be transported to a baghouse for dust control. While the immediate exit from the furnace is usually water cooled, there is a refractory lined zone in the duct system. Refractories in the exhaust ductwork must be capable of withstanding slag carryover and slag abrasion from particulate-laden gases moving at high velocity. Refractories in the ductwork are 50–70% alumina brick or fireclay/high alumina gunning mix (40–60% Al_2O_3). Both of these materials have the required combination of thermal shock resistance and slag resistance to withstand the operating conditions.

4.3.3 Electric Furnace Refractory Wear Mechanisms

4.3.3.1 Corrosion

Electric furnace refractories are subject to a variety of wear mechanisms which must be understood to properly design and manage electric furnace refractory systems. The most important wear mechanism is corrosion. This is the chemical reaction of metallic oxides in the slag, iron oxide (FeO), silica (SiO_2), or manganese oxide (MnO), with the refractory products. Magnesia from the refractory lining is soluble in the steelmaking slag, with saturation levels varying from 6 to 14%, depending on temperature and FeO content. These chemical corrosion reactions result in wearing away the furnace refractory lining; the products of reaction become part of the slag. Corrosion reactions can be minimized by neutralizing FeO with fluxes and controlling the oxygen content of the slag. Another way to control corrosion is to use refractory brick that contain carbon. This carbon in the refractory lining deoxidizes corrosive slag at the refractory/slag interface, minimizing lining corrosion.

4.3.3.2 Oxidation

A second critical wear mechanism in electric arc furnace linings is oxidation. In this process, carbon in the refractory lining is oxidized by reacting with oxygen or FeO in the slag. As the carbon in the refractory lining reacts, the brick loses its strength and is washed away. This carbon oxidation mechanism also occurs at the cold face of the brick if there are holes in the steel shell. Oxygen from the air reacts with the brickís carbon, and the back part of the brick lining turns to powder.

4.3.3.3 Erosion

Erosion is another prevalent refractory wear mechanism. This is the physical wearing away of the refractory due to molten steel or slag moving over the face of the refractory lining and physically abrading or eroding the lining. Erosion is most common in electric arc furnace tapholes, slaglines, roof electrode openings, and in the offtake ductwork.

4.3.3.4 Melting

Melting is also a common wear mechanism in the electric arc furnace. The unshielded electric arc generates temperatures that are well beyond the melting point of all commercial refractories. Melting is the simple phase change of the refractory from solid to liquid, and the liquid phase is easily washed away. Melting is a serious problem in electric arc furnace linings if not detected and corrected immediately.

4.3.3.5 Hydration

Because water cooling is used extensively in modern electric arc furnaces, there are occasional water leaks. Refractories are easily damaged by water or steam due to hydration of the magnesia or lime phases in the refractory product. Hydration results in expansion of the individual grains comprising the refractory lining. These grains grow and burst, disrupting the lining.

4.3.3.6 Spalling

A more subtle refractory wear mechanism is known as spalling. In this type of wear, rapid heating or cooling of the refractory lining cause stresses in the refractory lining. These stresses often exceed the inherent strength of the refractory material, resulting in cracking. As these cracks intersect, chunks of refractory fall out of the lining. This condition is most common on furnace roofs that are opened and closed to expose the refractory portion of the roof to cold air.

4.3.4 Conclusion

Refractory lining systems for EAF steelmaking are selected using a thorough knowledge of both operating conditions in the furnace and the impact of electric arc furnace design features. Refractory wear mechanisms can be minimized by a similar understanding of furnace operating conditions and their interaction with the available refractory products. As electric melting technology advances, refractory systems must be improved or developed which can counteract the increasingly difficult operating conditions.

4.4 Refractories for AOD and VOD Applications

4.4.1 Background

In the past 30 years, the AOD and similar type processes (CLU, ASM, K-OBM-S etc.) and the VOD, have become the dominant methods for the production of stainless steels throughout the world. In 1995, the total worldwide production of stainless steels was reported to be 14.9 million metric tons.[4] Over 85% of the stainless steel produced was through the AOD vessel or related processes. The balance was produced via various VOD ladle processes.

Dolomite and dolomite–magnesia brick are the most common (> 75% worldwide) refractory lining for the AOD vessel, but with significant magnesia–chrome and small amounts (< 2%) of magnesia–carbon brick also being used. In VOD ladle applications the same types and grades of refractories are used, but with more equal amounts of magnesia–chrome and dolomite–magnesia based brick, with the worldwide consumption estimated at 45% magnesia–chrome, 40% dolomite–magnesia and 15% magnesia–carbon brick. Monolithic refractories are not currently used as primary linings in AOD or VOD applications.[5]

4.4.2 AOD Refractories

4.4.2.1 Introduction

The types of refractories used in the AOD have evolved since the first AOD was put into commercial operation at Joslyn Stainless Steel, Ft. Wayne, Indiana in 1968. Magnesia–chrome refractories were initially used in the AOD, ranging from 60% MgO direct-bonded magnesia–chrome for use in the barrel to rebonded fused grain 60% MgO tuyere pads. In the early 1970s, the first dolomite based bricks were used in Europe.[6] The first dolomite lining used in the U.S. was in 1976. Since those initial trials, dolomite refractories have become predominant in AOD applications for three main reasons: economics; longer lining life—experience with dolomite based linings has shown that most vessels achieve or exceed lining life obtained with other basic refractories; and metallurgical benefits—the very basic nature of dolomite refractories allows the steelmaker to operate with higher basicity ratio slags, $(CaO + MgO)/(SiO_2 + Al_2O_3) > 1.4$, which improves chrome recovery, and desulfurization.[7,8] Dolomite bricks are chrome-free, enabling the steel producer to make both stainless and low alloy or chrome-free steel in a single AOD.

For those AOD applications which run at lower basicities, magnesia–chrome refractories may give lower overall costs. The factors governing the choice of magnesia–chrome over dolomite are generally lower slag basicities (< 1.4), extensive use of alternative alloy sources (recycled high-alloy slag, or direct reduction of chrome or manganese ore in the vessel), high process temperatures (> 1800°C), and local availability of competitively-priced magnesia–chrome refractories.

4.4.2.2 Life and Wear Rates

Each area of the vessel is subjected to different types of wear mechanisms. The primary wear mechanisms for AOD refractories can be classified as follows: corrosion by decarburization and reduction slags that can vary widely in basicity; erosion as a result of turbulence created in the tuyere region; thermal shock due to temperature variations during and between heats; and metal attack resulting from chemical reactions with strong reductants and fuels (Si, Al).

Life in the AOD vessel is partially a function of vessel size, due to turbulence effects, and ranges from 30–50 heats in vessels under 50 tons in size to greater than 120 heats in 90 ton and larger vessels. Consumptions range from 5–15 kg of refractory/ton of steel. Typically the highest wear/failure area is in the AOD sidewall tuyeres, or bottom tuyeres for a CLU type vessel. Side-blown vessels experience tuyere area wear rates of 2–8 mm/heat, with 5 mm/heat being typical. Bottom-blown vessel tuyere wear rates vary from 8–12 mm/heat. In the balance of both vessels, wear rates are typically 1–4 mm/heat.[5]

4.4.2.3 AOD Lining Construction and Zoning

4.4.2.3.1 Safety Lining   AOD safety linings are constructed using standard grade, see Fig. 4.23, Table 4.3 and Table 4.4, fired magnesia–chrome or rarely, dolomite brick. Magnesia–chrome is used in 76–100 mm thick, fully mortared arch or bevel construction laid against the shell without insulation. When dolomite brick are used, special non-aqueous mortars must be employed. The cone usually is constructed without a safety lining in vessels less than 90 tons. Alumina brick or monolithics are not used for AOD safety linings because they lack sufficient resistance to the operating conditions to allow a heat to be completed on the safety lining without a breakout.

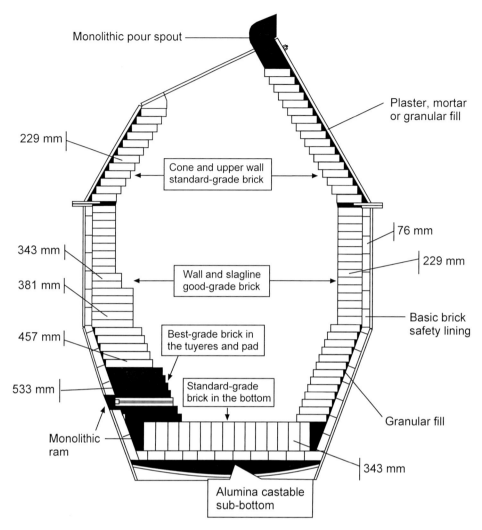

Fig. 4.23 Typical 45-ton flat bottom, side tuyere AOD vessel, showing quality zoning and relative brick thicknesses.

4.4.2.3.2 Working Lining One key to balancing AOD refractory wear patterns and reducing refractory cost per ton is careful attention to zoning and tight construction of the working lining. Zoning options include zoning the refractory lining by brick thickness, composition, or a combination of both thickness and composition. Fig. 4.23 illustrates the typical zoning of an AOD vessel. Dolomite working linings are typically constructed of straights and keys, laid dry (without mortar) and tight against the safety lining without expansion allowance. Backfill is only used where small gaps exist between the working and safety linings, especially in the stadium area. Sizing tolerances for dry dolomite linings must be tighter than ±1 mm, and are commonly ±0.8 mm. Magnesia–chrome linings frequently use mortared construction. If mortared construction is not used with magnesia–chrome, expansion allowances may be needed.

4.4.2.3.3 Bottoms Two types of bottom refractory construction are used in AOD vessels, flat and dished. Flat bottoms are simple to install, usually requiring only one size of straight sided brick. Dished bottoms require two to three keys and/or straight shapes, see Fig. 4.24. The main advantage of using the dished bottom design is that several tons of increased volume can usually be gained. The keyed construction also helps to retain the bottom refractories in position as the brick wears. Dolomite bottoms are laid dry, and dusted with dolomite fines. Magnesia–chrome bottoms are usually laid with mortar. The perimeter of the bottom (under the sidewalls) is rammed with monolithic

Table 4.3 Simplified Chemical and Physical Properties of Typical Dolomite AOD and VOD Bricks

	Grade						
	Standard	Standard	Standard	Good	Good	Best	Best
AOD Applications	Cone, barrel and bottom	Barrel and bottom	Barrel and bottom	Intermediate wear areas	Intermediate wear areas	Tuyere and tuyere pad	Tuyere and tuyere pad
VOD Applications	Freeboard and bottom	Wall, bottom and impact pad	Freeboard, wall and bottom	Upper and lower slagline	Slagline	Slagline and stir pad	Slagline and stir pad
Brick Type	Resin-bonded dolomite	Resin-bonded dolomite	Fired dolomite	Fired MgO-enriched dolomite	Fired MgO-enriched dolomite	Fired MgO-enriched dolomite	Fired MgO-enriched dolomite
Chemical Analysis (%)							
MgO		>38		~50	~60	~60	65–70
CaO		55–60		~45	~40	~40	~30
SiO_2				Total typically 2.3–2.8			
Al_2O_3				Typically <5 ppm			
Fe_2O_3 + MnO				2–10 ppm			
Cr		—		—	—	<1	<1
B		—		—	—	—	—
ZrO_2							
C	2–3	4–6					
Physical Properties							
Bulk Density (g/cm³)			>2.95		~60	2.98	>3.0
Porosity (%) MOR	<13*	<15*	>13	11–15		>9	<11
As-received (MPa) MOR	>2						
As-coked (MPa)		—	—			0	—
Permanent Linear Change (after 1600°C)	—			−0.1 to −0.2			
Hot MOR at 1371°C (MPa)	>3				3–4		>5

Table 4.4 Simplified Chemical and Physical Properties of Typical Magnesia–Chrome AOD and VOD Bricks

	Grade				
	Standard	Good	Good	Best	Best
AOD Application					
VOD Application					
Brick Type					
Chemical Analysis (%)					
MgO	~50	~60	60–70	~60	~60
CaO			<1		
SiO_2	<2.5	<2	<2	<2.5	<2.5
Al_2O_3			7–14		
Fe_2O_3			6–11		
Cr_2O_3	>20	>16	>11	>19	>19
Physical Properties					
Bulk Density (g/cm^3)		>3.1		>3.2	>3.25
Porosity (%)	18	17	16	15	13
MOR As-received (MPa)		>5		>10	>14
Permanent Linear Change (after 1700°C)			+0.1 to +0.3		
Hot MOR at 1482° (MPa)	1	>2	>2	>4	>5

material. A typical non-aqueous basic monolithic used with dolomite linings is shown in Table 4.5. Alumina or chrome-based monolithics are used with magnesia–chrome linings, and should have a continuous service rating of >1700°C.

4.4.2.3.4 Tuyere Zone—Knapsacks Because of the high wear rate in the tuyere area, increasingly longer refractories (up to 1000 mm) are being used to extend vessel campaign life. However, longer tuyere pads in standard AOD shell designs can result in unexpected side effects, such as a reduction in vessel volume, increased slag splashing and changes in lining wear patterns. One method to increase the tuyere length without sacrificing vessel volume is the use of the knapsack. Also referred to as a doghouse or pod, an example is shown in Fig. 4.9. In the stadium section, notice the use of two bricks, one in front of the other. Using this construction technique allows a greater total tuyere pad length.

The tuyere brick and pad experience thermal shock from gas cooling, erosion from turbulence, and extremes in temperature from oxidation reactions. Corrosion from the reaction products may also contribute to wear. In this area the best grade of available refractory is used. In dolomite linings, the tuyere brick are usually high-density, low permeability with increased MgO levels and additions of ZrO_2 for thermal shock resistance. The one or two guard bricks immediately surrounding the tuyere are frequently of the same composition. The balance of the tuyere pad is usually constructed using slightly less premium brick to save costs. The tuyere pad typically extends up to the vessel slagline and encompasses 160–180 degrees of circumference. Magnesia–chrome tuyere/pad compositions employ either fused-grain or chrome oxide enriched bricks fired to temperatures >1750°C.

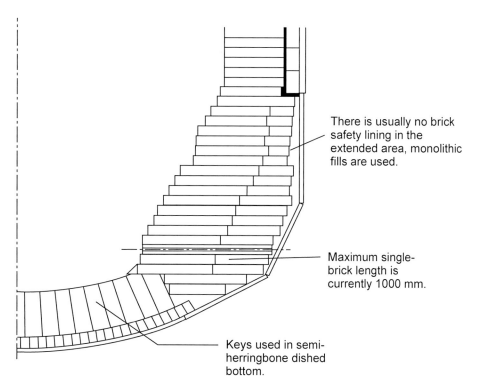

Fig. 4.24 Close-up view of a knapsack-extended tuyere pad design, and also of a dished bottom.

Table 4.5 Simplified Chemical and Physical Properties of Typical Non-Aqueous Basic Monolithic AOD and VOD Refractories

Product Type	Resin-Bonded Dolomite–Magnesia	Magnesia–Carbon	MgO-Enriched Dolomite–Carbon
Use	Monolithics for AOD cone seal, behind and around tuyere, VOD/AOD bottoms well-stir blocks	Pre-formed AOD lip rings VOD well and stir blocks	
Chemical Analysis (%)			
MgO	38–95	88.0	55.0
CaO	balance	2.1	43.2
SiO_2		<1	
Al_2O_3	<3 Total	8.8	0.4
Fe_2O_3 + MnO		<0.7	
C	—	~8	~3
Physical Properties			
Bulk Density (g/cm^3)	>2.7	2.75	2.80
Coked Porosity (%)	<18	17	17
MOR As-cured (MPa)	>9	>15	>15
MOR As-coked (MPa)	>3	>4	>4

4.4.2.3.5 Walls and Slagline
The second highest wear area of the AOD vessel is the slagline, especially in the trunnion area. This area is under constant contact with slag in all vessel positions. If the slag chemistry is less than optimum, wear will occur from chemical corrosion combined with some erosion. Dolomite linings use a good grade of MgO-enriched brick in this area. The balance of the wall is constructed with standard grade brick.

4.4.2.3.6 Vessel/Cone Flange and Seal
Several techniques are used for sealing the joint between the vessel and cone flange, see Fig. 4.25. The method used depends on flange design, the height of the last barrel course and whether the cone is bricked separately. If a basic monolithic seal material, see Table 4.5, is used between bricks, as shown in Fig. 4.25, the seal thickness at the brick hot face should be 25–75 mm. Fig. 4.25 also shows the use of a cut flange brick. Magnesia–chrome linings may use either the dolomite material from Table 4.5 or a 90% alumina plaster.

4.4.2.3.7 Cone
The cone is the lowest wear area and frequently may be reused on a second vessel with minor repairs to the tapping side. The cone is constructed with standard grade keys using corbelled construction, being dry for dolomite and mortared for magnesia–chrome. Large dolomite-lined vessels often use a double-chamfered (or parallelogram key) lining in the cone. This construction is faster and more stable than standard keys. Vessels <15 tons in size may use cast magnesia or alumina monolithic cones.

4.4.2.3.8 Pour Spouts
Fig. 4.23 shows an example of a monolithic pour spout. Pour spouts help to concentrate the metal stream and help to reduce nitrogen pickup in the steel during tapping. Other benefits of monolithic pour spouts include increased metal yield, optimized slag-off, easy slag deskulling and reduced cone steelwork maintenance. Table 4.5 lists the properties of some basic monolithic spout compositions.

4.4.2.4 Preheating of Linings

Fig. 4.26 shows the recommended preheat curve for all AOD linings. The preheat begins with a temperature rise of 60°C per hour up to 650°C. At this point the lining is soaked for four hours, after which heating should continue at 65°C per hour up to 1100°C. If time does not permit a soak period, the cycle can go to 65°C per hour after 650°C has been reached. Linings should be held at 1100°C for >4 hours prior to being placed in service. Reheating of a used lining can be at a rate of 65° to 95°C per hour up to 1100°C.

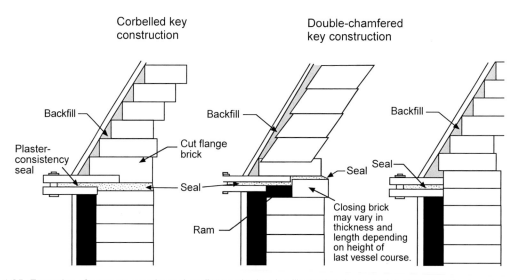

Fig. 4.25 Examples of common cone/vessel sealing methods, also illustrating the use of double-chamfer cone keys.

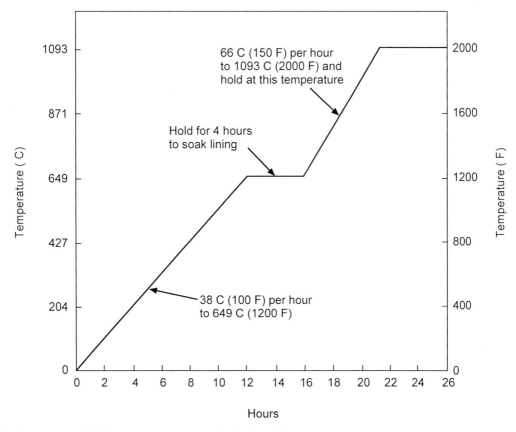

Fig. 4.26 Recommended AOD preheat schedule for all refractory linings.

4.4.2.5 Process Variables Significantly Affecting Life

It is necessary to optimize certain process variables in order to maximize the performance of AOD refractories. Slag chemistry and fluidity control in the AOD during all refining steps is important to maintain good refractory life. This requires that sufficient levels of lime and magnesia are present in the slag. The optimum slag conditions are similar for all three major refractory types, although dolomite and magnesia–carbon are more tolerant of high basicities. Slags with V-ratios >1.4 are compatible with basic AOD refractories and can be important with regards to metallurgical considerations. Other operational parameters that need to be controlled in order to improve refractory life include process temperatures <1700°C, tuyere and knurdle control, and short average heat times. With improper control of tuyere and slag conditions, tuyere area wear rates can exceed 20 mm/heat.

4.4.2.5.1 Slag Control The control and predictability of the slag chemistry throughout the AOD process is critical for optimizing refractory wear and metallurgical control.

The type of refractory used in the AOD will influence the slag practice. If a magnesia–chrome refractory is used, good refractory life is generally obtained if V-ratios are in the range of 1.2 to 1.5. Operating with higher slag basicities will have a detrimental affect on these refractories due to chemical corrosion of the chrome component of the brick.

In comparison, a dolomite-lined AOD vessel requires slags having higher basicities (V-ratios 1.4 to 2.0). One of the reasons for the extensive use of dolomite linings in the AOD is that several metallurgical benefits can result by operating with higher slag basicities. These benefits include better desulfurization potential and alloy recovery. In Fig. 4.27, the effect of higher reduction slag basicities on chrome recovery is shown.

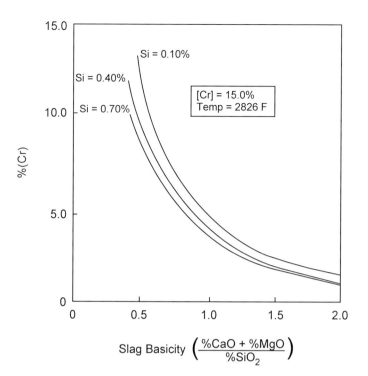

Fig. 4.27 Estimated equilibrium chromium content in slag versus basicity.

The following is a synopsis of AOD slag practices with dolomite linings.

4.4.2.5.1.1 Transfer Slag Control of transfer slag chemistry and volume brought to the AOD is important if proper slag control is to be maintained. Highly variable transfer slags make it difficult to calculate the proper amount of flux additions necessary in the vessel. If possible, the transfer slag should be CaO and MgO-saturated, so that the initial slag in the vessel is basic and can neutralize the acidic oxides generated during the oxygen blow. If the transfer slag is acidic and the amount of slag is variable, erosion of the AOD refractory lining will start at the beginning of the process.

4.4.2.5.1.2 Decarburization Slag Slag control during the decarburization blow is critical if maximum refractory life is to be obtained. At the beginning of this stage, corrosive slags with high levels of SiO_2, Fe_2O_3 and Cr_2O_3 are generated. In order to neutralize the acidic oxides, CaO and MgO should be added as a pre-charge material, or very early in the blow. The role of MgO is to react with the metallic oxides of Cr, Mn, Fe and Al. Their reaction with MgO forms solid or near-solid complex spinels during decarburization and limits corrosion of the lining. It is recommended that the CaO source be in the form of dolomitic lime. Part or all of the MgO can come from dolomitic lime, however this can result in additional slag volume. Decarburization slags that are fluid have insufficient CaO and MgO levels and can be corrosive to dolomite linings. Typical MgO content of stainless decarburization slags is in the 10 to 20% range.

4.4.2.5.1.3 Reduction Slag During the reduction stage, the AOD slags change considerably. The solid MgO spinel phases disappear as the metal oxides of Cr, Mn, etc. are reduced by silicon or aluminum back into the bath. The SiO_2 and Al_2O_3 generated during reduction must be neutralized with CaO to minimize corrosion of the dolomite lining. The lime also serves to tie up the SiO_2 and effectively reduce any reaction with the transition metals. Reduction slags for silicon-killed practices should target V-ratios of 1.6 to 1.8, with an MgO content of 8 to 12%. Better refractory life, desulfurization and alloy recovery can be achieved with V-ratios >2.0.

If aluminum is used during reduction or as a fuel, maintaining CaO saturation is critical if Dolomite refractory wear is to be minimized. Slags with high Al_2O_3 content can dissolve more lime and will do so from the dolomite refractory lining unless enough lime is added. Table 4.6 shows the effect of Al_2O_3 content in the slag on lime demand. If CaO saturation is not possible, higher MgO levels in the slag are required for refractory protection.

Table 4.6 Aim V-Ratios used with Dolomite Linings as a Function of Al_2O_3

Al_2O_3 Content	V-Ratio*
0–10%	1.55
15%	1.65
20%	1.78
>25%	1.82

*V-ratio = (CaO + MgO)/(SiO_2 + Al_2O_3)

4.4.2.5.2 Tuyere Knurdle Control Consistent control of knurdle length and shape is essential for preventing premature failure of the refractory tuyere zone. Knurdle size is controlled by making adjustments to shroud gas pressure and/or flow. A large knurdle, indicating excessive shroud pressure, can become plugged and misdirect the gas stream toward the refractory lining. A small knurdle or no knurdle at all, resulting from low shroud pressure, will not protect the tuyere and can result in tunneling of the tuyere brick. Optimum knurdle size depends on the individual shop conditions, but generally, 25 to 75 mm is best.

If as little as 2% oxygen is present in the shroud gas, the cooling effect can be reduced and tuyere brick tunneling may begin to occur.

4.4.2.5.3 Temperature Temperature can have a significant impact on the amount of CaO and MgO needed to saturate slags during reduction. Higher CaO and MgO levels are needed if processing temperatures are higher. If adjustments are not made, dissolution of the refractory lining by the slag will occur. In Table 4.7, the affect of temperature on lime and MgO requirements for saturation is shown.

Table 4.7 Effect of Temperature on CaO and MgO Demand for Slag Saturation

Temp (°C)	CaO (%)	MgO (%)	SiO_2 (%)	CaO/SiO_2
1600	44.3	17.3	38.4	1.15
1700	48.5	16.1	35.4	1.37
1750	51.6	15.1	33.3	1.55
1800	55.1	14.0	30.9	1.78

The bath temperature becomes elevated under normal conditions by metallic oxidation of the bath as the carbon content reduces to lower levels. The temperature of the bath can also be increased by oxidation of fuels (aluminum or silicon) during a reblow. A reblow to achieve temperature increases of 20°C or higher, however, is equivalent in refractory wear to making another heat on the lining. Temperature control is best accomplished by keeping transfer metal temperature and chemistry levels consistent and controlling the peak temperature at the end of the oxygen blow by varying the O_2/inert gas ratios.

An excessive inert gas stir for temperature reduction can also adversely affect refractory performance. Extended stirs can form long, pipe-like knurdles that will sag from their own weight. When these long knurdles bend over, the injected gases are directed downward during the early stage of the next heat, and tuyere pad or bottom erosion may occur.

4.4.2.5.4 Back Tilt Gas injected through the tuyeres tend to roll back and up along the sidewall, carrying the products of oxidation with it. The gas stream creates localized high temperature and

4.4.3 VOD Refractories

4.4.3.1 Introduction

As mentioned in the Section 4.4.1, the VOD process uses more linings of fired magnesia–chrome brick than of fired or resin-bonded dolomite–magnesia primarily because of the use of highly variable multiple slag processes, and secondarily because of much more thermal cycling than in the AOD. Some of these slags are very fluid or insufficiently basic, $(CaO + MgO)/(SiO_2 + Al_2O_3) \ll 1.4$, for optimum life with dolomite refractories.

The choice of brick composition, especially in the slagline, is controlled by the acceptable carbon pickup during processing, the acceptable bath chrome pickup, and slag chemistries. Where carbon pickup from the brick in the range of ~0.2 ppm per minute (under vacuum) is tolerable, magnesia–carbon slaglines are preferred for their wide range of slag compatibility, and have been used to produce steel as low as 15–30 ppm carbon.[9] If carbon pickup is not acceptable, then fired brick must be used, and for chrome-free alloys, fired magnesia–dolomite brick is the only choice. Where chrome pickup is allowable, the choice of fired magnesia–chrome or magnesia–dolomite is determined primarily by slag chemistry.

VOD slags have the requirement of high fluidity, because of the relatively small slag volume, and the limited stirring energy usable in the ladle. This fluidity may be obtained by having slags with V-ratios $\ll 1.4$, or with higher V-ratio slags by using fluorspar or alumina as fluidizers. Magnesia–chrome brick generally give better slagline performance against slags having V-ratios <1.2 or having Al_2O_3 contents >15%, while magnesia–dolomite is preferred for V-ratios >1.4 and for slags containing >10% fluorspar.

The effect of slag on the wall refractories is less than in the main slagline, and with the increasing use of magnesia–carbon slaglines, compatible with a wide range of slags, and tighter control over the process, and especially thermal cycling, the choice of magnesia–chrome or dolomite-based wall and bottom linings is becoming primarily economic.

4.4.3.2 Lives and Wear Rates

Life is typically equally short for either type of lining, ranging from 15–25 heats. The primary wear/failure areas are the slagline, the stirring pad and the stir plug/block itself. Refractory consumption is similar for all refractories, is highly variable between shops and may be as low as 10 kg/ton or over 30 kg/ton of steel. In the severe areas, slagline and stir pad, wear rates are 7–15 mm/heat. In the lower wear areas of the walls and bottom, typical rates are 3–7 mm/heat.

4.4.3.3 VOD Lining Construction and Zoning

The selection of refractories for VOD applications should consider the conditions of aggressive fluid slags, high temperatures (>1750°C), long process and stirring times (>3 hr), and the effects of vacuum on reducing heat transfer and increasing shell temperatures where the VOD ladle is in a vacuum tank. Because of the very wide range of processing conditions encountered in VODs, there is considerable variation in safety and working lining construction. The following typical design, Fig. 4.28, should be considered the minimum requirement in the absence of a shop refractory history.

4.4.3.3.1 Safety Lining The safety lining in a VOD ladle is intended to prevent a steel breakout if there is a failure of the working lining during the processing, and the quality and construction must be sufficient to allow one complete heat to be made directly on the safety lining. A VOD ladle safety lining is generally similar to the safety linings used for ladles used for severe ladle furnace or tank degasser applications, see Section 4.5.3.

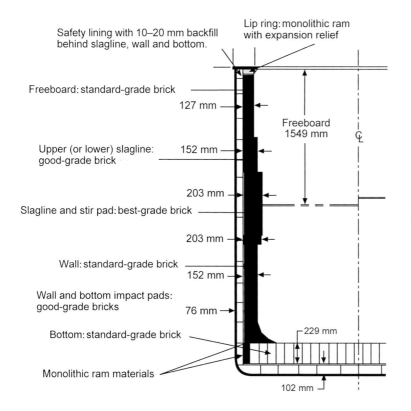

Fig. 4.28 Typical 45-ton VOD lining showing quality zoning, relative brick thickness, and large freeboard.

Total thicknesses of 75–150 mm are used consisting of primarily of 70–85% alumina fired brick in the walls and bottom, with fired 95% magnesia or direct-bonded magnesia–chrome brick commonly used in the slagline area for their better slag resistance. The entire brick safety lining should be installed fully mortared, using arch brick or bevels with a maximum joint thickness of 2 mm. Monolithic safety linings are also used based on both castable and dry-vibratable alumina or magnesia materials of similar quality to bricks. All safety lining materials for VOD ladles should be volume stable, be resistant to cyclic oxidation-reduction, and have a continuous service rating >1650°C.

Where insulation is required to hold shell temperatures below 400°C, 5–8 mm of non-compressible insulating panels are most frequently used between the shell and the main safety lining. Where less insulation is needed, 20–30 mm of clay or forsterite brick has been used in place of the insulating panels. This requires a corresponding reduction in the thickness of the main safety lining, but is less expensive. The maximum permissible in-service compression or shrinkage of insulating materials for VOD ladle applications is ~2 mm. A continuous service rating of >800°C is suggested for VOD insulating materials.

4.4.3.3.2 Backfill Backfills are sacrificial granular monolithic materials used between the safety and working linings to absorb working lining thermal expansion, capture metal penetration, and reduce reactions between the working lining, slag, and safety linings. They generally protect the safety lining and make removal of the working lining easier. More than 90% of dolomite–magnesia lined VOD ladles use backfill. Magnesia–chrome lining customarily have no backfill.

Backfill thicknesses of from 10–20 mm are used by placing the working lining brick a desired distance from the safety lining, and pouring and tamping the backfill into the gap every second course. Compositions are most commonly 80–85% alumina, with >90% magnesia sometimes used in the slagline areas. Dolomite-based backfills are also used. Backfills should have a continuous service rating of 1650°C.

4.4.3.3.3 Working Lining The primary zones of a VOD working lining, Fig. 4.28, are: the bottom, including well blocks, stir plug, and impact pad; walls; sidewall impact and stirring pads; slagline and hotspots; and the freeboard. These zones do not vary significantly between magnesia–chrome or dolomite linings. Because of the fluid slags and metal, and extensive stirring, tight sizing tolerances are needed for VOD brick. Sizing ranges for resin-bonded compositions are typically less than ±0.5 mm, and for fired brick are typically less than ±1.0 mm.

> Note: Magnesia–chrome, alumina-based and dolomite-based refractories are normally not mixed in a single VOD working lining. Under vacuum, destructive interactions may occur between the CaO in dolomite and the spinels in magnesia–chrome or alumina–silica phases in alumina refractories. Where the two types of brick are used together, an interface layer of magnesia–carbon brick should be used. An interface may also be required if the stir plug block or well block, and their surrounding monolithics, are not of a refractory type (basic vs. acid) compatible with the adjacent brick. Table 4.5 shows monolithics suitable for use with dolomite linings. Because VOD conditions greatly increase the potential for incompatibility reactions, refractory suppliers should be consulted prior to mixing compositions.

4.4.3.3.4 Bottoms Standard grades of magnesia–chrome and dolomite straight bricks, see Tables 4.3 and 4.4, are used in thicknesses of 187–250 mm. Recently, limited amounts of alumina–magnesia–carbon (AMC) brick have been used in bottoms. Any of the common brick patterns, (rowlock, soldier, semi-herringbone, herringbone), may be used with magnesia–chrome, AMC, or resin-bonded dolomite for the convenience of fitting around the stir plug and well block. Fired dolomite brick are usually laid in a semi-herringbone pattern. Bottom brick may be installed under the sidewalls or as a plug bottom. Plug bottoms are easier to repair and/or replace, but expose the monolithics used around the bottom perimeter. Fig. 4.28 shows the most common construction, with the bottom installed under the sidewalls. The bricks are typically laid dry, with finish grouting or dusting to fill any minor open joints.

4.4.3.3.5 Bottom/Wall Impact Pads Usually the good-grade of brick is employed in the bottom/wall impact pad, frequently slightly thicker than the main lining. For dolomite linings, the impact pads are often made of resin-bonded brick. VOD ladles duplexed with BOFs or EBTs may require either the best-grade brick or metals-containing magnesia–carbon brick in the bottom pad, because of more concentrated tap streams.

4.4.3.3.6 Stir Plug/Blocks and Well Blocks VOD ladles use plug and block compositions similar to those required in severe LMF and degasser operations, see Section 4.5.5. The conditions of very fluid slags, high temperatures, and long stirring times cause these areas to be one of the main limitations on VOD life. A very wide range of compositions are being used, often requiring magnesia–carbon brick as an interface layer or window to maintain compatibility with the main lining. The current trend (1998) is to increase the use of basic compositions (magnesia–carbon, magnesia–spinel, magnesia–zirconia, and magnesia–dolomite) for the blocks, and to use slotted or directional porosity plugs.

4.4.3.3.7 Walls Standard-grade bricks from 114–250 mm thick are used in the walls. Key or arch construction is preferred over semi-universals for greater stability when bricks wear thin. Magnesia–chrome is frequently mortared, while dolomite is laid dry. A majority of dolomite linings use 10–20 mm of backfill behind the walls.

4.4.3.3.8 Main Slagline plus Sidewall and Slagline Stir Pads The best grade of brick is used in these critical wear areas. Selection of the refractory type is governed by the metallurgical/slag conditions as discussed above. The use of metals-containing, large crystal size and fused grain magnesia–carbon brick in slaglines is increasing, even in otherwise carbon-free linings. For VOD applications, carbon levels are commonly 7–10%, to minimize oxidation and carbon pickup, and anti-oxidants must be carefully chosen for slag compatibility. Magnesia purity and large crystal size are more critical in VOD ladles than in other ladle applications.

4.4.3.3.9 Upper Slaglines As a result of foaming under vacuum, VOD ladles must have an extended slagline zone. Where a typical LMF ladle may have a 700–1000 mm high slagline, a VOD will use

1000–1500 mm of slag zone. Most of the additional height is in the upper slagline, which is in intermittent contact with the low-density foamed vacuum slag. This area typically uses either 20–50 mm thinner brick and/or one grade lower brick than the main slagline. Even when the main slagline is magnesia–carbon, the upper slagline frequently is fired brick because of the greater oxidation potential in this zone.

4.4.3.3.10 Freeboard Fired standard-grade brick are used, and are chosen for their oxidation and thermal cycling resistance. In this low wear area, thickness may be as little as 114 mm. Construction is usually keys or arches, laid dry and tight against the safety lining without backfill. In ladles with <800 mm of freeboard, this zone will see some minor slag attack and is usually of a composition similar to the wall brick. In VOD ladles with >1000 mm freeboard, the upper portion sees little slag and may even be of 85% alumina brick. The freeboard is an area where dolomite brick may be used in a magnesia–chrome lining and vice versa.

4.4.3.3.11 Lip Rings For VOD ladles, the lip ring design is determined by whether the vacuum seal is on the lip or the ladle is tank degassed. In most VODs, the large freeboard allows the use of alumina brick or castables and pre-formed shapes with alumina-based monolithics. These lip materials should be rated for continuous service at >1700°C. With mortared magnesia–chrome linings, vertical thermal expansion allowances are not usually needed, but with dolomite–magnesia linings, vertical expansion allowances of approximately 0.5% should be incorporated in the lip construction. Many dolomite linings use a dolomite or magnesia-based resin-bonded ram, Table 4.5, which is temporarily thermoplastic on heatup to provide this relief.

4.4.3.4 Preheating of Linings

Fig. 4.29 presents a schedule that is suitable for all lining types. When there are no carbon-containing brick in the lining, slower preheat rates and longer soaks may be used. Preheat and/or soak temperatures above 1100°C are not recommended because of brick reactions with the relatively low melting point VOD slags.

4.4.4 Acknowledgments

The author acknowledges the significant contributions of Lowell Johnson, Nobu Mimura, and Eugene Pretorius, all of Baker Refractories, and Don Moritz, retired from Baker Refractories and Eastern Stainless Steel. And also wishes to thank Victor Ardito, retired from Allegheny Ludlum Corp., and Richard Choulet, of Praxair, Inc. for their helpful suggestions.

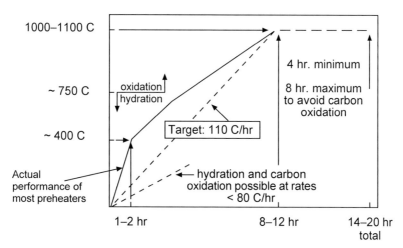

Fig. 4.29 Preheat schedule for fired and resin-bonded VOD brick.

4.5 Refractories for Ladles

4.5.1 Function of Modern Steel Ladle

The modern steel ladle is used in a significantly more complex manner than the older ingot ladles used simply to transport steel from a furnace to ingot molds. Table 4.8 lists typical functions of the modern steel ladle.

Table 4.8 Functions of Modern Ladles

Metal transport from furnace to caster
Metal processing/treatments—minimum wear
Minimize O_2 pickup from refractory
Minimize heat loss
Safety/reproducibility
Environmental effects minimized—in use and disposal
Low cost/ton

While the overall purpose of the ladle is still to deliver steel for a caster, steel processing between the furnace and caster may occur by a complex combination of steel and slag treatments. These includes furnace slag skimming and the introduction of a new artificial slag, steel alloying and stirring from porous plug and lances, steel reheating using electric arcs, and degassing by various procedures. These steps must be accomplished without significant refractory wear or influence on any of the refining processes. Fig. 4.30 illustrates the process steps in one shop where both bloom and slab heats are produced in degassed or regular grades. The increased number of process steps in ladles means that, in essence, they function as traveling components of skimming, stirring, reheating, and degassing processes. Also the exposure time for a given heat has expanded from two to five times of that for ingot teeming.

The ladle lining can also influence the quality of the steel produced in the lining if oxygen is picked up from the lining during any stage of processing. Linings containing SiO_2 in uncombined forms

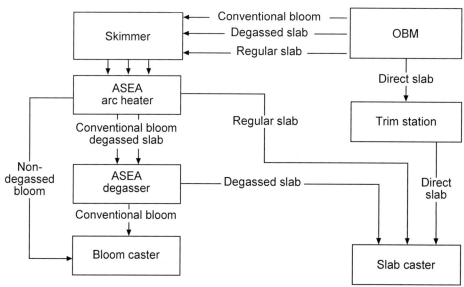

Fig. 4.30 Process routes.

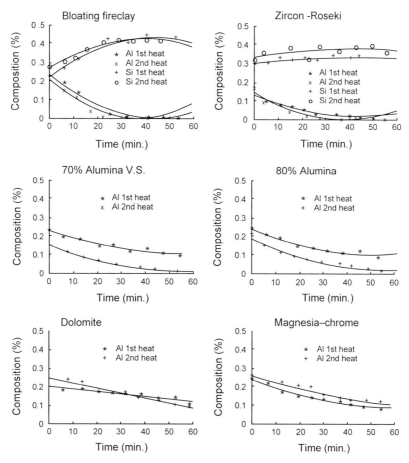

Fig. 4.31 Aluminum loss and silicon pickup on holding steels in a laboratory furnace.

can cause problems in modern ladles as illustrated in Fig. 4.31. In those laboratory tests where Al-killed steels were held in various lining materials, fire clay or zircon-Roseki linings gave rapid Al loss (and Si pickup) in comparison to significantly less Al loss with high alumina refractory linings where the lower levels of SiO_2 were largely combined as mullite. Refractories with very low levels of SiO_2 (dolomite, magnesia–chrome) obviously perform well in such tests. In actual service, the ladle surface is clean only on the first heat, and oxygen pickup is influenced by the amount and oxidation state of the coating remaining on the lining from the prior heats.

Ladles must also conserve heat by minimizing heat loss during transport and the various process steps. In this regard, significant developments have been made to properly preheat ladles prior to the first heat, and to cycle ladles on subsequent heats in a manner to minimize heat losses. Fig. 4.32 shows a ladle cycling procedure for one plant. Fig. 4.33 shows some ladle temperatures for this plant for the first and subsequent heats. In this case, the first heat was tapped into a ladle preheated to 1900°F, and later heats were tapped in a cycle of approximately three hours. While every effort is made to reuse ladles without delay and to provide covers on ladles between heats, processing of ladles between heats is necessary to remove ladle slag and provide for ladle well cleanout, slide-gate inspection and repairs and ladle well sanding.

Ladle preheat and cover devices have improved significantly in recent years, and many types of preheaters (ladle horizontal or vertical) are available, Fig. 4.34. Table 4.9 lists some of the benefits of proper ladle preheating and cycling, including longer refractory life by minimizing thermal shock. The actual ability to rapidly cycle ladles and make the most efficient use of preheaters can vary significantly between operations, depending on shop layout and ladle transfer ability.

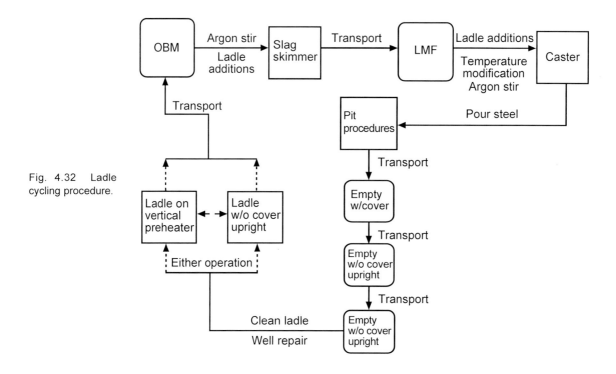

Fig. 4.32 Ladle cycling procedure.

Ladle life must be very predictable and reproducible for safety reasons and to avoid process delays. As a result, ladles are often removed from service to provide an adequate supply of ladles. The cost of ladle refractories can be significant ($1–4 per ton of steel) and disposal costs of spent linings must also be considered.

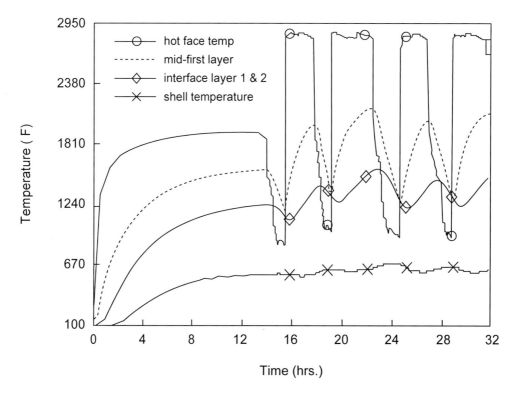

Fig. 4.33 Transient phenomena in ladle cycling.

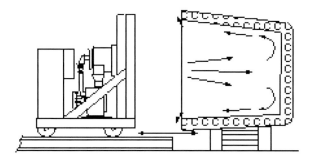
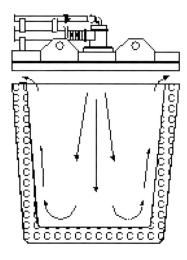

Fig. 4.34 Horizontal and vertical preheat systems for ladles.

4.5.2 Ladle Design

Steel ladle shells must be constructed to allow safe transport of very heavy loads by transfer cars, cranes, and other devices. For example, a typical full ladle could involve 200 tons of molten steel in a lining and steelwork weighing 100 tons. The design of such a structure is highly complex as

Table 4.9 Ladle Preheating: Goals and Benefits

Goals

Provide ladles with consistently high refractory heat content to the BOF for tapping heat after heat
    a. less tap temperature biasing for cold/hot ladles
    b. consistent first heat practice

Benefits

Reduction in average steel temperature loss in the ladle
    a. lower tap temperatures
    b. less arcing required at the LMF
Reduction in refractory costs
    a. reduced thermal shock to ladle refractories
    b. increased steelmaking furnace refractory life
    c. reduced ladle slagline wear through less reheating at the LMF
Provide more consistent steel temperature to the caster
    a. less temperature related terminations
    b. increased productivity and quality by maintaining optimum tundish temperatures

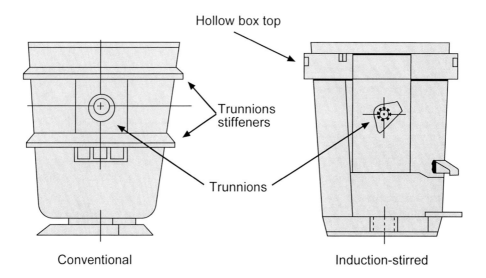

Fig. 4.35 Appearance of steel ladles.

shown by the design criteria of *AISE Technical Report No. 9, Specifications for Design and Use of Ladles*.[10] Fig. 4.35 shows the general appearance of some steel ladles indicating the trunnion supports for ladle lifting and rotation and stiffeners to provide lining structural support. In reality, ladle structures are often even more complex because of shop specific criteria. For example, Fig. 4.36 shows an obround ladle in which a flat section was added to a round ladle to increase ladle capacity while using existing crane facilities.

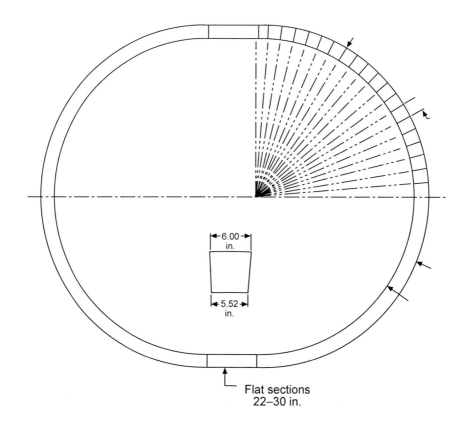

Fig. 4.36 Obround ladle section.

Steelmaking Refractories

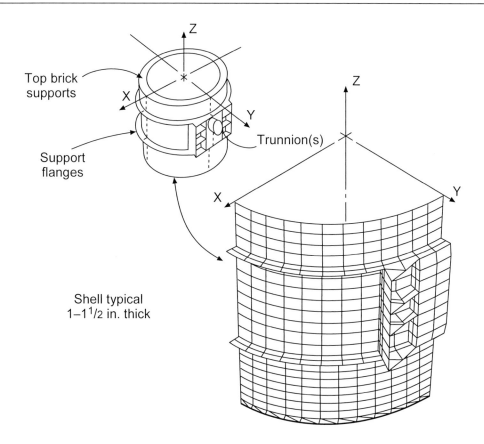

Fig. 4.37 Shell model of ladle for finite element analysis.

In recent years, finite element analysis using non-linear refractory properties and transient temperature regimes are being employed to study the behavior of refractories in steel ladles, Fig. 4.37. While highly complex, the studies essentially endeavor to maintain the proper degree of compression on the ladle refractory during all phases of their use in modern ladles. Excessive compressive forces can result in refractory cracking and/or buckling in areas such as the ladle flat section. Lower than desirable compressive stresses can cause joints or gaps to form, which may permit metal or slag penetration.

The properties of refractories may be adjusted to provide proper behavior in ladles. For example, Fig. 4.38 and Fig. 4.39 show the use of specific refractories to either increase or decrease refractory expansion to more desirable levels. In Fig. 4.38 a refractory with higher reheat expansion is

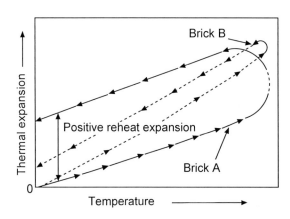

Fig. 4.38 Thermal expansion characteristics of high alumina brick. Brick A, with a positive reheat expansion, will keep joints tighter during cooling than the Brick B, with the higher thermal expansion.

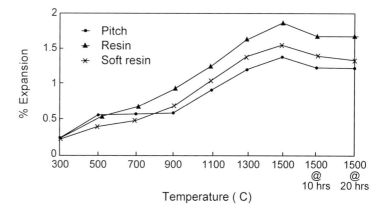

Fig. 4.39 Thermal expansion under load. In oval ladles, pitch or soft resin bonded magnesia–carbon brick are desirable over resin due to their thermal plastic behavior that relieves expansion stresses. From 500 to 900°C expansion forces in pitch and soft resin brick are relieved.

used to prevent joint openings in the high alumina section of a ladle. The same ladle might use a pitch or soft resin brick in the slagline section to reduce expansion and prevent brick cracking or movement. Ladle finite element analysis has provided valuable guidance for refractory service trials to improve behavior.

4.5.3 Ladle Refractory Design and Use

The refractories in steel ladles are zoned in type and thickness to provide maximum service minimum cost. Fig. 4.40 shows a typical ladle to illustrate the overall size and style of a ladle and some ideas of refractory thickness. Table 4.10 shows ranges of refractory thickness used in ladles. Linings are relatively thin to meet ladle capacity and weight requirements.

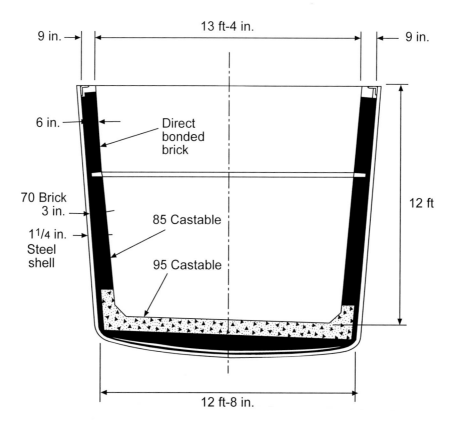

Fig. 4.40 Lining in a caster ladle using a castable barrel and bottom.

Table 4.10 Refractory Thickness by Area

Area	Thickness range, in.
Working Lining	
Slagline	5–7
Upper Barrel	6–7
Lower Barrel	6–9
Bottom	9–12
Safety Lining	2–6

The types of refractory construction vary widely depending on the caster operating conditions and on the ability to cycle ladles rapidly. Table 4.11 presents some of the more widely used working lining combinations.

Table 4.11 Working Lining Combinations

Construction	Slagline	Balance of Working Lining
A	Magnesia–carbon brick	High-aluminum brick
B	Magnesia–carbon brick	High-aluminum castable
C	Dolomite brick	Dolomite brick
D	Magnesia–chrome	High-aluminum brick
E	Magnesia–carbon brick	Aluminum carbon brick

Safety linings function to hold steel or slag for limited time periods, but essentially provide shell insulation. Multiple component safety linings may be used to further lower shell temperatures as illustrated in Fig. 4.41. In this illustration, the 4 in. thick safety lining has now changed from all

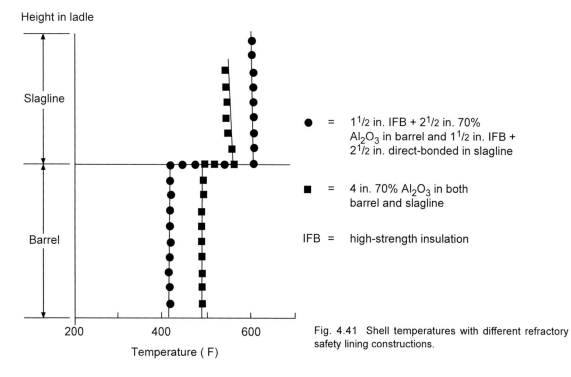

Fig. 4.41 Shell temperatures with different refractory safety lining constructions.

high-alumina brick to a composite with a high-strength insulation brick to lower the shell temperature. The higher shell temperatures in the slagline are caused primarily by the higher thermal conductivity of the slagline working lining brick. In general, few ladles use true insulating materials as part of safety linings because of reduced safety lining life and/or increased danger of steel penetration and possible breakouts.

The particular refractory constructions used are under constant change in each operating shop. The following will comment briefly on the factors and refractory properties important in each area of the ladle.

4.5.3.1 Stream Impact Pad

As illustrated in Fig. 4.42, wear in this zone occurs as the high-momentum steel stream strikes the ladle bottom (and in some cases the lower sidewall) during the initial moment of tap. The severity of this wear is quite shop-specific and requires that additional thickness or quality of refractory be used. In general, refractories for the stream impact are selected to have maximum erosion resistance based on hot-strength. Fig. 4.43 shows the temperature–hot strength relation for several refractories used in impact pads where the 96 Al_2O_3 castable provides improved performance.

4.5.3.2 Bottom and Lower Barrel Refractories

As shown in Fig. 4.44, wear occurs in this area from erosion during stirring or reheating and from physical damage during deskulling between heats. In some cases, slag remaining in this area in the

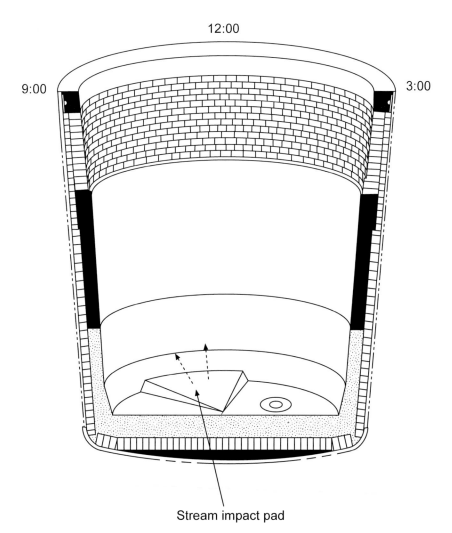

Fig. 4.42 Primary causes of wear in stream impact area of ladle are: 1. metal (and slag) erosion, 2. thermal cycling.

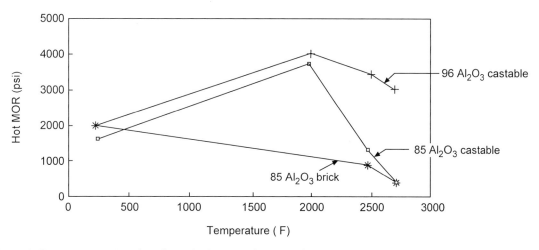

Fig. 4.43 Property comparison for refractories in stream impact pad.

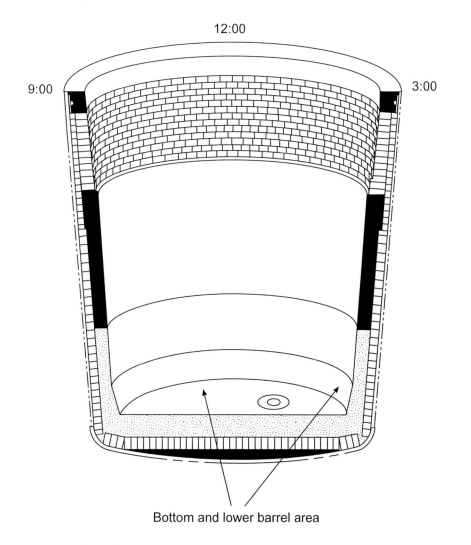

Bottom and lower barrel area

Fig. 4.44 Primary causes of wear in the bottom and lower barrel area of the ladle include: 1. erosion during stirring and other processing, 2. damage during deskulling, 3. slag attack during transport back from caster, and 4. thermal cycling.

time between steel shutoff at the caster and slag dumping can cause slag erosion problems. In general, the slag erosion in this area is not sufficient to zone for, except to provide for additional refractory thickness. Damage from skull removal can occasionally be sufficiently enough severe to require bottom repairs.

4.5.3.3 Barrel

The barrel normally has the least severe wear in the ladle and can be zoned for quality and/or thickness. Fig. 4.45 shows a ladle in which the barrel castable lining is zoned with a lower quality (lower Al_2O_3) castable.

4.5.3.4 Slagline

The most severe wear area of most modern ladles occurs in the ladle slagline where the refractory is subject to severe corrosion, Fig. 4.46. The slags encountered vary widely, and include high iron oxide slag carried over from the furnace, artificial slags introduced after partial slag skimming, slags added or formed during specific metallurgical purposes such as stirring or injection, and slags formed or circulated during degassing. As the slags are generally basic in nature, basic refractories are required in ladle slaglines. Fig. 4.47 shows the range of erosion resistance between high-alumina and various basic refractories. The high-alumina refractories are suitable for most areas of the ladle other than the slagline proper. Table 4.12 shows the wide range of slag compositions experienced in one shop producing a wide range of Al-killed caster products.

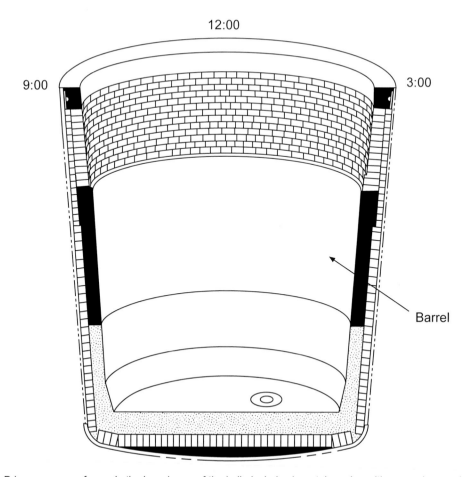

Fig. 4.45 Primary causes of wear in the barrel area of the ladle include: 1. metal erosion with some slag erosion, 2. thermal cycling.

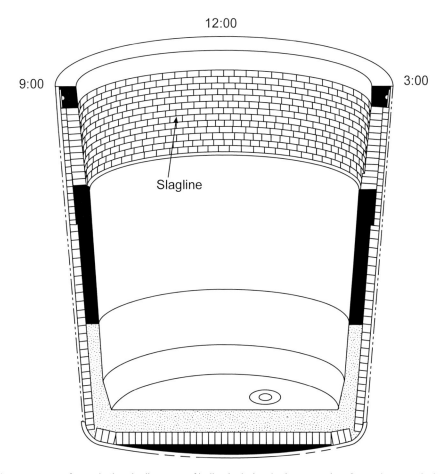

Fig. 4.46 Primary causes of wear in the slagline area of ladles include: 1. slag corrosion, 2. arc damage, 3. thermal cycling, and 4. ladle flexing.

The corrosive effect of ladle slags is particularly severe when arc reheating is used to control and add steel temperature by superheating the ladle slag. Fig. 4.48 shows the degree of superheat added to the slag for various slag thicknesses and stirring conditions. As shown, the temperature of the slag can be expected to be 100–300°F above the steel temperature. Laboratory tests show that slag erosion rates can increase two to five times for such changes in temperature. Field tests have shown that slag erosion can be reduced by control of slag basicity, Al_2O_3 content, and additions of MgO to the slag as illustrated in Fig. 4.49. Significant control over the amount of erosion during arc reheating can therefore be obtained using controlled slag obtained using compositions with added

Table 4.12 Range of Slag Compositions for a Specific Shop

	Range	Median
CaO	20–55	42
SiO_2	5–18	10
Al_2O_3	12–50	26
MgO	6–12	8
MnO	1–10	5
FeO	1–15	5

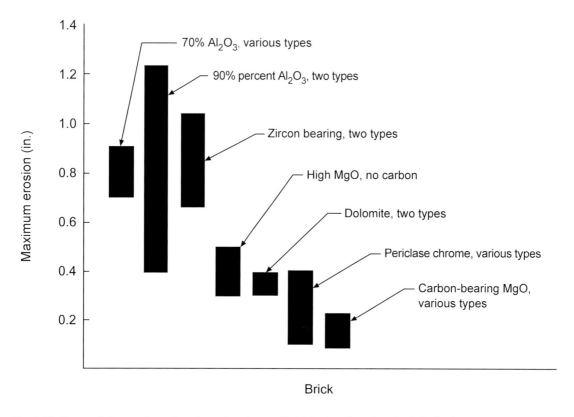

Fig. 4.47 Range of slag erosion values for various types of brick tests with various basic ladle slags.

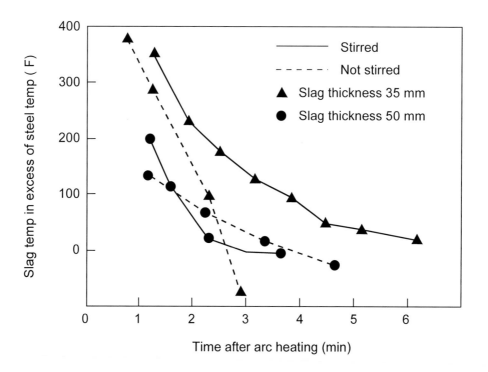

Fig. 4.48 Effect of slag thickness and stirring on slag superheat.

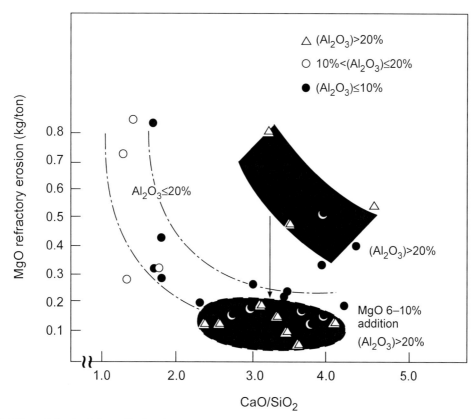

Fig. 4.49 Effect of slag basicity on MgO refractory erosion.

MgO and the use of consistent slag stirring to control slag superheat. Reference should also be made to sections on electric furnace operations which clearly illustrate the importance of arc heating conditions such as arc length and slag submergence on refractory wear.

Table 4.13 shows the properties of various MgO–C brick for slaglines and how specific properties may be improved. Higher purity raw material (sintered or fused MgO and graphite) and metal additions can be used to enhance the properties of magnesia–carbon brick. Table 4.14 shows some of the properties of these bricks for ladle slaglines with bricks ranging from 5–15% carbon.

Table 4.13 Properties of Various MgO–C Brick for Ladle Slaglines

Property	Better Resistance Obtained by
Arc Resistance	More graphite
	Higher purity graphite
Oxidation Resistance	Metal
	Metals
Hot Strength	Metal
	Metals
	Metals and higher purity graphite
Corrosion Resistance	High purity sintered grain
	High purity graphite
	Fused grain
	Combinations of above

Table 4.14 Properties of Some MgO–C Brick for Ladle Slaglines

%C	5	5	10	10	10	10
Density (lb/ft³)						
as received	188	187	183	182	186	179
% Porosity						
as received	4.5	5.0	6.0	6.0	4.5	6.5
coked	11	11	11	11	9.5	12
ignited	18	19	24	22	21	28
MOR (lb/in²)						
70°F	4000	3500	500	1900	2500	1800
2200°F	2800	2700	1200	2700	1800	1500
2820°F	2000	2000	—	2200	1600	1300
Relative oxidation						
2200°F	0.41	0.63	0.91	0.62	0.74	0.97
2700°F	0.39	0.45	0.79	0.43	0.51	1.26
%MgO	85	85	84	79	79	76

Direct-bonded magnesite–chrome or dolomite refractories may also be used in steel ladle slaglines. Dolomite refractories provide lower initial cost and excellent performance if ladle cycles can avoid long delays. Magnesite–chrome refractories can provide lower cost operations if arc reheating conditions are not severe.

It should be noted that thermal cycling damage was listed as a wear mechanism in all areas of the ladle. The extent of such damage has been greatly minimized in recent years with the proper use of preheating and more extensive use of ladle covers. Experience in each shop to use the minimum possible number of ladles at any time and to cycle those ladles as rapidly as possible has also minimized the extent of thermal cycle damage.

Ladle flexing on lifting and during other parts of the ladle cycle is known to influence ladle life. Efforts to combat this effect with improved design in the ladle and lining are continuing.

4.5.4 Ladle Refractory Construction

The majority of ladles in North America are lined with brick construction using semi-universal shapes as illustrated in Fig. 4.50. This construction allows the use of an upward spiral of brick

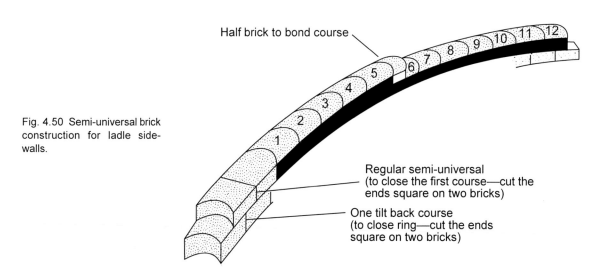

Fig. 4.50 Semi-universal brick construction for ladle sidewalls.

Table 4.15 Cast Steel Ladle Advantages and Disadvantages

Advantages	Disadvantages
Labor	**Costs**
Reduced labor hours per ladle	Total costs of labor, material, ladle availability difficult to quantify
	High startup costs due to equipment
	Changes in mandrel design can be costly
Technical	**Technical**
Joint-free lining, less steel penetration	Low-moisture castables require controlled installation
Physical properties equal to brick	
No void between working and safety lining	Vertical dryers with controls are required
Can zone castables in high wear areas with no joints	Technology for slagline castable does not exist
Portions (25%) of castable is reused	

against the sloping sides of a ladle. Brick locking is produced by the curved mating surfaces. Other types of brick construction (arch-wedge or keyed) are more widely used in other parts of the world. In all cases, tight construction with very thin (or no) mortar joints are necessary to keep the lining under compression and prevent joint penetration. Certain operations have converted to cast ladles in the barrel and bottom sections, Fig. 4.40, but efforts to cast basic slaglines have not proven successful. Table 4.15 summarizes the advantages and disadvantages of cast ladles under present conditions. In general, castables provide an excellent joint-free construction and many offer cost advantages where a portion of the spent lining can be reused. Castable ladles do require special equipment, including space, and must be very carefully installed and dried. Future changes in material and labor costs or environmental changes may require reexamination of castable ladles for specific shops. At present, ladles are being used where combinations of brick and castable are employed to obtain the best technical and economic combination of castable and brick approaches to ladle lining. For example, Fig. 4.51 shows construction in a ladle bottom using a precast impact pad with the balance of the bottom cast. This construction improves life due to the jointless construction, but avoids most of the special equipment and space requirements required for a fully cast ladle.

4.5.5 Refractory Stirring Plugs

Refractory stirring cones or plugs are used in ladle bottoms to introduce gas, mainly argon, for ladle rinsing or stirring in the various metallurgical processes. Fig. 4.52 shows cross sections of three types of plugs using different directional mechanisms to provide controlled argon flow.

Argon flow may be controlled by the space between a solid refractory cone and its metal casing, through a permeable refractory cone, or through and solid refractory cone with pre-formed holes or other channels. The amount of flow required will vary widely from rinsing (low flow) to mixing during arc reheating or other processes. Fig. 4.53 shows the relation between flow and pressure for several plug types.

The reliable performance and life of plugs is very important in producing consistent steel product quality. To insure proper flow, it is often necessary to clean the plug surface after a given heat by oxygen burning or mechanical cleaning. Fig. 4.54 shows the sequence of wear of the plug when a penetrated plug surface is cleaned to restore flow. The refractories for plugs are high-Al_2O_3 or burned MgO materials, designed specifically for this application, and are installed from outside the ladle by mechanical or manual devices such as the bayonet system shown in Fig. 4.55. This system

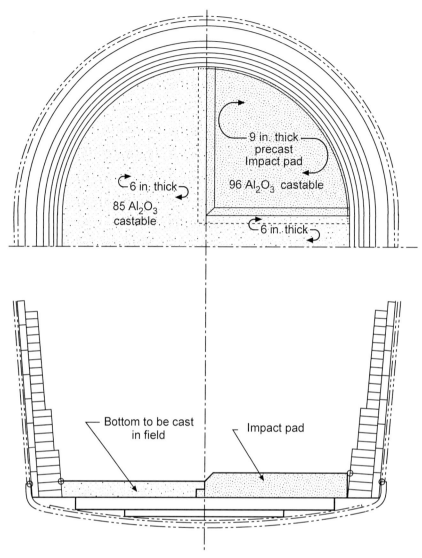

Fig. 4.51 Castable refractory in ladle bottoms.

permits a quick plug exchange in a hot ladle. Plugs are removed from service after a predetermined number of minutes of use or when visual wear indicators built into the plug are overused. Because of wear in the refractory seating block around the plug, hot repairs of the area around the plug may be required using pneumatic refractory placement or from inside the ladle using a diving bell.

The control of steel flow from ladles to caster molds is accomplished by one of a variety of sliding gate systems. Fig. 4.56 shows the concept of steel flow by a sliding gate, where refractory plates held under pressure by springs or other devices are moved to control flow. The design and construction of the various slidegate systems vary widely according to the steel pouring demands of the particular caster. For example, spring location and method of cooling differ between the various gate systems and the movement of plates may be accomplished by hydraulic or other mechanisms. All the slidegate systems provide a rapid means of removing pressure from the plates between heats to allow inspection of the refractories and to permit rapid replacement of plates or the lower nozzle.

The refractory construction of a typical gate system is illustrated in Fig. 4.57. Included are refractories in the seating block and upper nozzle in the ladle bottom, the fixed and sliding plates (in this

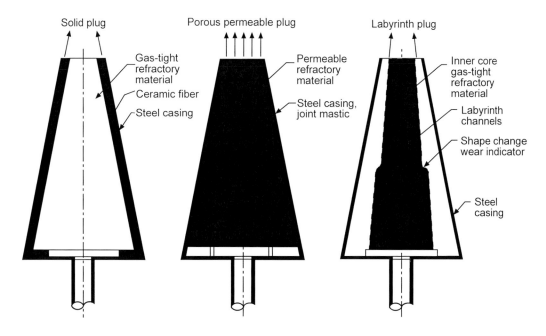

Fig. 4.52 Standard gas stirring plugs.

case a three plate system) and a lower nozzle connection for a tube or shroud into a caster tundish. Table 4.16 and Table 4.17 give refractory properties of the upper and lower nozzles and seating blocks, respectively, which are selected to balance the overall life of the gate system.

The sliding and fixed plates are among the most unique and durable refractories used in any steelplant application. These plates must withstand severe thermal shock and steel erosion for long

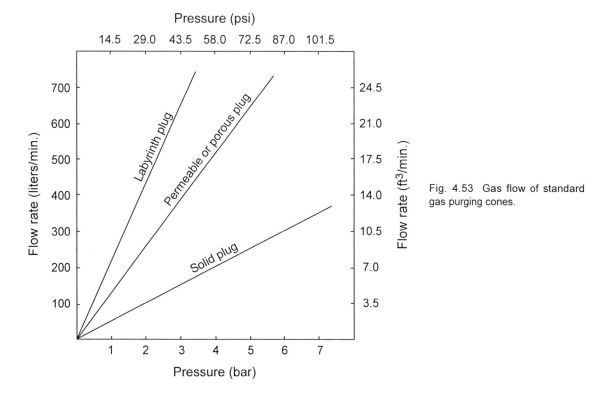

Fig. 4.53 Gas flow of standard gas purging cones.

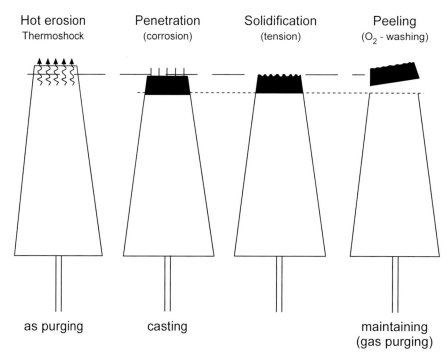

Fig. 4.54 Wear behavior of a porous purging plug.

periods of operation. The composition of these plates may vary from simple alumina to zirconia in the oxide system, Table 4.18, to complex oxide–carbon systems, Table 4.19. The exact plates used depend largely on the steel compositions to be cast and the frequency of plate replacement. Such plate replacement can be accompanied after inspection of the plates. Such inspection is performed after each cast or less frequently based on experience. Fig. 4.58 shows the appearance of plates after use with either erosion or erosion/cracking requiring the plates to be replaced. Plates may be changed after only one heat or may have lives of 10 to 20 heats, depending on steel grades and/or the refractory quality used.

Table 4.16 Typical Properties and Refractory Compositions for Upper and Lower Nozzles

Brand	A	B	C	D	E
Physical Properties					
Bulk density (g/cm^3)	2.25	2.57	2.77	2.86	3.01
Apparent porosity (%)	13.5	8.5	7.0	7.5	7.0
Cold crushing strength (psi)	6500	9000	13,500	11,500	11,500
Chemical Composition (wt.%)					
Al_2O_3	44	60	74	82	90
SiO_2	45	31	16	2	4
ZrO_2	—	—	—	4	—
C	4	4	4	7	4
Corrosion Resistance	⎯⎯⎯⎯⎯⎯⎯⎯⎯⎯⎯⎯ Increasing ⎯⎯⎯⎯⎯⎯⎯⎯⎯⎯⎯⎯→				
Thermal Shock Resistance	←⎯⎯⎯⎯⎯⎯⎯⎯⎯⎯⎯⎯ Increasing ⎯⎯⎯⎯⎯⎯⎯⎯⎯⎯⎯⎯				
Main Application	Erodable lower nozzle	Lower nozzle	Lower nozzle	Upper and lower nozzle	Upper nozzle

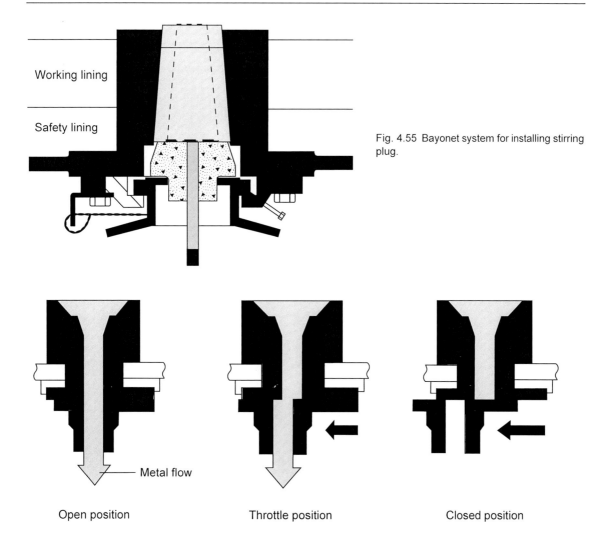

Fig. 4.55 Bayonet system for installing stirring plug.

Fig. 4.56 Slidegate operation concept. In a typical installation, the upper nozzle and plate are stationary in relation to the ladle lining and ladle bottom. Coil springs, used to apply face pressure, prevent molten metal leakage without restricting movement of the lower plate. In a sliding motion, nozzle openings in the upper and lower plates are brought into or out of alignment to initiate or terminate metal flow. Intermediate positions, with partial alignment of nozzle openings, provide a throttling action and a range of reduced controlled flow rates.

After each heat, the entire gate system must be cleaned of residual metal and slag by oxygen lancing and a granular refractory filler installed before the next heat. As illustrated in Fig. 4.59, this filler (ladle sand) prevents molten metal from entering the gate system before the gate is opened at the proper time at the caster. The ladle sands may be silica, zircon, or other refractory combinations, Table 4.20, which will flow freely from the slidegate when opened without requiring mechanical probing or lancing.

4.5.6 Refractory Life and Costs

The life of ladle refractories will obviously vary widely between shops, but historically show significant improvement during the first one to three years of operation. These initial improvements occur from the development of standardized operating conditions and the increased ability to control ladle cycling. After this initial period, life will improve at a moderate rate from changes in refractory type and construction. In most cases, slagline life will be less than that for the remaining ladle, with slaglines being replaced one or more times during a ladle campaign. Safety lining

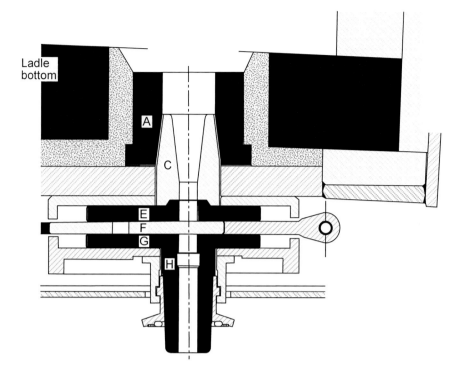

Fig. 4.57 Typical refractories in a slidegate system.

A = Seating block
C = Upper nozzle
E = Upper plate
F = Middle plate (sliding plate)
G = Lower plate
H = Lower nozzle

Table 4.17 Typical Properties and Refractory Compositions for Seating Blocks

Type Material	Cast High Alumina		Pressed Alumina–Carbon			
Physical Properties						
Bulk density (g/cm³)	2.64	2.98	>2.15	2.60	2.80	3.00
Apparent porosity (%)	23	16.5	<16.0	9.50	6.00	8.00
Cold crushing strength (psi)	—	—	>4000	9000	14,500	14,000
Chemical Composition (wt.%)						
Al_2O_3	82	94	>40	50	80	90
SiO_2	11	—	<50	31	13	4
ZrO_2	—	—	—	—	—	—
C	—	—	4	4	4	4
Corrosion Resistance	⎯⎯⎯⎯⎯⎯⎯⎯⎯⎯⎯⎯⎯⎯⎯⎯ Increasing ⎯⎯⎯⎯⎯⎯⎯⎯⎯⎯⎯⎯⎯⎯⎯⎯→					
Main Use	Large one-piece blocks for thick ladle bottoms		Lower blocks	Upper, lower and one-piece blocks	Upper and one-piece blocks	Upper and one-piece blocks

Table 4.18 Oxide Refractories for Slidegate Plates

Properties	Standard Alumina	High Purity and Density Alumina	Magnesia	Zirconia
Chemical Composition (%)				
Al_2O_3	89.3	96.0	—	0.4
SiO_2	9.7	1.0	0.3	0.7
ZrO_2	—	3.0	—	94.8
MgO	—	—	98.0	0.1
Apparent Porosity (%)	15.2	2.1	16.0	14.6
Bulk Density (g/cm^3)	2.98	3.98	2.92	4.80
MOR (lb/in^2)				
at room temperature	2900	13,340	2980	6000
at 1480°C	1680	11,280	1320	—
Thermal Shock Resistance	Good	Very Poor	Very Poor	Good
Steel Erosion Resistance	Poor	Excellent	Excellent	Excellent
Performance	1.0	3.0	2.0	3.0

Table 4.19 Carbon Bonded Refractories for Slidegate Plates

Properties	Brand A	Brand B	Brand C	Brand D	Brand E
Chemical Composition (%)					
Al_2O_3	75.8	83.3	72.6	71.6	0.4
SiO_2	8.2	3.3	4.5	6.0	0.6
ZrO_2	—	4.8	8.0	7.0	83.4
Carbon	12.7	8.6	13.6	11.8	12.0
Apparent Porosity (%)	6.5	1.7	6.0	4.0	5.0
Bulk Density (g/cm^3)	2.92	3.26	3.07	3.20	4.40
MOR (lb/in^2)					
at room temperature	4700	4470	4830	4550	4000
at 1480°C	1010	2410	3270	2540	2000
Thermal Shock Resistance	Poor	Fair	Fair	Fair	Fair
Steel Erosion Resistance	Fair	Good	Good	Good	Very Good
Performance	1.5	2.0	3.0	3.5	4.0

will last multiple working campaigns with some localized repairs. Fig. 4.60 shows the ladle life history of a caster during the first two years of its operation, where ladle life improved from 50 to 86 heats. This shop used two slaglines per ladle campaign. At present, ladle lives from 50 to 200 heats are being experienced with slagline lives from 30 to 200 heats. For a given shop, analysis of process data can usually be used to explain differences in the life for a given ladle or a given time period. For example, one could predict or explain the life of individual ladles by the frequency of heats with abnormal wear parameters such as long ladle arc times, long steel holding times during caster delays, and unusual situations where ladles were taken out of service of repairs to slidegates or plugs.

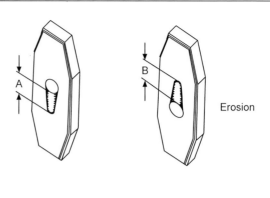

Fig. 4.58 Appearance of slidegate plates after use.

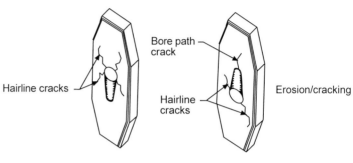

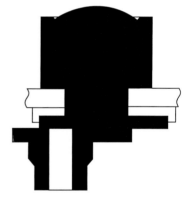

Fig. 4.59 Nozzle plugging agents prevent molten metal from entering slidegate before it is opened and assure free opening of the slidegate after long ladle hold times.

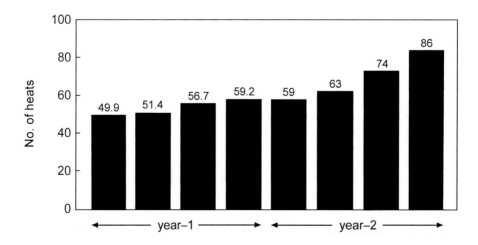

Fig. 4.60 Ladle life during startup of a caster operation.

Table 4.20 Characteristics of Nozzle Plugging Agents

Properties	Silica	Zirconia
Chemical Composition (%)		
SiO_2	95.3	0.2 max
Al_2O_3	2.4	2.0 max
R_2O_3	1.7	—
C	—	—
TiO_2	—	0.35 max
ZrO_2	—	65.0
Fe_2O_3	—	0.05 max
Grain Size (%)		
>2.00	1.0	—
2.00–1.00	47.0	—
1.00–0.50	51.0	—
0.50–0.21	1.0	<1.0
0.21–0.15	—	13.0
0.15–0.10	—	48.0
0.10–0.08	—	36.0
<0.08	—	<3.0
Bulk Density (g/cm^3)	1.4	2.8
Refractoriness (SK)	33	>34
Sinterability		
at 1200°C	Slight	None
at 1300°C	Moderate	None
at 1400°C	Significant	Slight
Features	Large grains, low alkali content to prevent excessive sintering	Small round grains, high melting point, low alkali content to prevent excessive sintering

4.6 Refractories for Degassers

The wear of refractories in degassers can present limitations to equipment availability and cost and affect steel quality in producing low carbon, hydrogen, and nitrogen steel for modern requirements. The refractory use and maintenance technologies vary widely with location in a given degassing unit and between operations.

Fig. 4.61 through Fig. 4.63 illustrate a typical vessel degasser (RH or RH-OB) and refractory construction. Table 4.21 lists refractories used for each area. The RH degasser is a recirculating type degasser. The snorkels are immersed into a ladle held steel bath. Evacuation of the degasser causes metal to rise through the legs into the lower vessel. Circulation of the liquid steel results from injecting argon into the steel in the snorkel of one of the legs (called the lifting leg or up leg) thus lowering the steel density and causing flow upward. The degassed steel in the lower vessel, now considerably denser, returns to the ladle through the other leg (appropriately call the return leg or down leg), descending in fact to the ladle bottom. Steel flow rates, or circulation rates, can be quite high (e.g., 80,000 kg/min) and depend on argon flow rates and snorkel inside diameter. RH degasser treatment times generally range between 15 and 30 minutes, the longer times for attaining low hydrogen levels. While originally used primarily for hydrogen removal, alloying, and

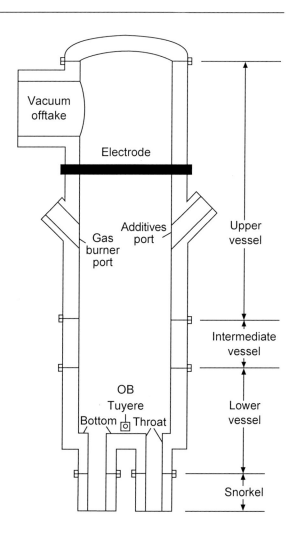

Fig. 4.61 Typical configuration of RH/RH-OB degassers.

inclusion flotation, the RH degasser is now also used for decarburization, nitrogen removal, and desulfurization.

RH-OB adds oxygen blowing capability to the RH process. With oxygen blowing, heats with higher furnace tap carbons can be decarburized, and some reheating can be done by aluminum burning.

Table 4.21 Vessel Degasser Refractory Use by Area (Refer to Fig. 4.62 and Fig. 4.63.)

Area	Refractory Description
Bottom Brick	Magnesia–chrome brick containing fused grain
Offtake Brick	60% Al_2O_3 brick
Alloy Chute Brick	90% Al_2O_3
Barrel Brick	Magnesia–chrome brick
Leveling Castable	Magnesia–chrome
Port Facing Brick	Magnesia–chrome brick with fused grain
Dome Brick	50–60% Al_2O_3 brick
Snorkel Castable	Tabular alumina or spinel castable
Insulating Board	High-strength insulting board

Steelmaking Refractories

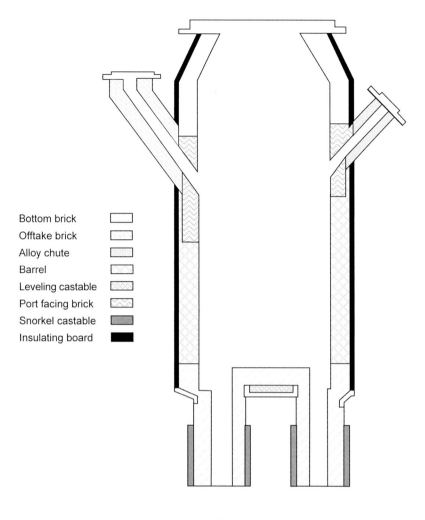

* Drawing not to scale

Fig. 4.62 Refractory areas of a vessel degasser.

Working lining refractories are exposed to a variety of conditions depending on vessel location. All locations are exposed to thermal and atmospheric cycling. Use of auxiliary heating between treatments can significantly reduce the severity of thermal cycling, but rapid temperature changes can range from 200 to 600°C depending on the adequacy of that heating. Atmospheric cycling can range from ambient down as to low as 0.5 Torr. Besides being exposed to liquid steel at temperatures as high as 1650°C, snorkel and lower vessel (particularly throat and bottom) refractories are subjected to the erosive action of high velocity, turbulent steel flow. Refractories in these locations are also exposed to contact with slags, which include ladle carryover slags (despite precautions to exclude them) and those generated during treatment (especially if the treatment includes desulfurization). These refractories may also suffer iron oxide alteration from oxidation of skulls left on their hot faces and from skull oxide melt runoff from higher vessel locations. Tuyere area refractories of RH-OB degassers experience greater erosive bath action than do lower vessel sidewall refractories in RH degassers. Aluminum burning may also cause localized heating. Upper vessel refractories, spared from erosive bath action, are coated with metal and slag splash accompanying the violent bath agitation and gas evolution. Subsequent oxidation of the adherent metal results in major alteration.

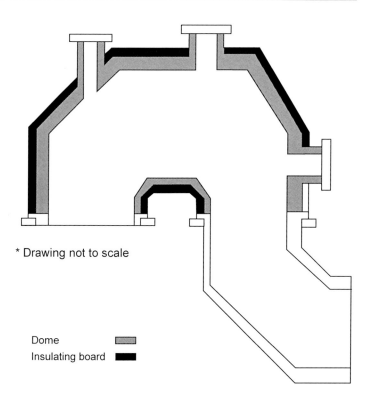

Fig. 4.63 Refractory areas of a degasser offtake.

Knowledge of the service conditions and nature of the refractory alterations as described above provide insight into wear mechanisms. In those areas experiencing the greatest steel bath circulation rates, such as snorkel, throat, bottom, and OB tuyere, the likely primary mode of wear is erosion of the disrupted hot face refractory structure. Structural disruption or weakening occurs by a variety of processes, such as: thermal shock damage resulting from rapid heating or cooling of the hot face between preheating and steel processing (the less the difference between the temperatures involved, the less the damage); Fe^{2+}/Fe^{3+} oxide cycling because of oxygen pressure and/or temperature changes (both refractory chrome ore and chromite spinel phases and absorbed iron oxide zones will be affected); and slag infiltration with dissolution of the normal refractory bonding (silicate bonds are particularly susceptible). Also lost to erosion in these lining areas will be the iron-oxide-rich immediate hot face portion of the iron oxide absorption zone. Considering the temperatures involved, wherever the iron oxide content exceeds roughly 80%, appreciable amounts of liquid are formed. Even in those lining areas subjected to less aggressive contact with the circulating steel (e.g., upper portion of lower vessel sidewall exclusive of OB tuyere areas), the partially liquefied hot face zone will be washed away.

Finally, some wear is likely the result of loss of partial hot face from spalls, as the hot face wear front approaches internally formed cracks. For a given brick, these losses are probably relatively discontinuous (as opposed to an essentially continuous erosion–corrosion loss of material). The loss rate of spalls would be greater in high erosion areas due to more rapid movement of the wear front and the stripping power of the flowing steel.

RH and RH-OB degasser working linings are commonly direct-bonded magnesia–chromite brick. To counter the different exposure conditions, linings are often zoned by product quality.

Table 4.22 shows some properties of various types of magnesia–chrome brick. The more corrosion-resistant products would obviously be selected for the snorkel and lower vessel areas where more severe conditions are encountered. Similarly, castables for the outside of the snorkel must be selected for maximum hot-strength and high-temperature stability so that only the most high purity materials prove successful.

Table 4.22 Properties of Various Magnesia–Chrome Brick used in Vessel Degassers

	Regular	Rebonded Fused-Grain	
		A	B
Composition (wt%)			
MgO	60	60	60
Cr_2O_3	15	19	24
CaO	6.7	6.6	0.5
SiO_2	1.4	1.6	1.5
Porosity (%)	16–18	13–14	12–14
Modulus of rupture			
at 1300°C, MPa	12–13	16–17	15–18
at 1500°C, MPa	3	4–6	6
Corrosion index (Low = less erosion)	100	50	45–50

Table 4.23 Refractory Life for Areas of Vessel Degasser

Area	Life (number of heats)
Snorkel	150–200
Throat/bottom	150–200
Sidewalls	400–600
Balance of degasser	About one year

Refractory life obviously varies with the location in the degasser as illustrated below in Table 4.23 for one typical degasser. The degasser lining must be designed to facilitate rapid replacement of the higher wear sections.

Acknowledgment

The information on oxygen steelmaking refractories presented in Section 4.1 is excerpted, with permission, from the Iron & Steel Society publication titled *Pneumatic Steelmaking, Vol. 3, Refractories*.

This comprehensive reference text, consisting of six chapters, provides an understanding of the underlying principles of refractories for oxygen steelmaking and presents information on improving, both technically and economically, the steelmaking process.

To obtain more information about this publication, contact the Iron & Steel Society by writing to 410 Commonwealth Drive, Warrendale, PA 15086.

References

1. M. A. Rigaud and R. A. Landy, eds., *Pneumatic Steelmaking, Vol. 3, Refractories*, (Warrendale, PA: Iron and Steel Society, 1996).

2. K. M. Goodson, N. Donaghy, and R. O. Russell, "Furnace Refractory Maintenance and Slag Splashing," *Proceedings of the 78th Steelmaking Conference* (Warrendale, PA: Iron and Steel Society, 1995), 481–5.
3. J. V. Spruell and J. B. Lewis, "Method for Increasing Vessel Life for Basic Oxygen Furnaces," U.S. Patent No. 4,373,949, February 15, 1983.
4. A. Jones, A. Hanson, and N. Kam, "Stainless Steels Monthly," *Metal Bulletin Research* (August 1996), 1–12.
5. Refractory Consumption and Wear Rates: Internal data, Baker Refractories, York, PA.
6. D. A. Brosnan and R. J. Marr, "The Use of Direct Bonded Burned Dolomite Brick in the AOD Vessel," *Electric Furnace Conference Proceedings* (Warrendale, PA: Iron and Steel Society, 1977), 179–83.
7. C. W. McCoy, A. F. Kolek, and F. C. Langenberg, "Application of Thermochemical Data During Stainless Steel Melting," *Proceedings of the First Conference on the Thermodynamic Properties of Materials* (1967), 262–70.
8. E. B. Pretorius and R. J. Marr, "The Effect of Slag Modeling to Improve Steelmaking Processes," *Electric Furnace Conference Proceedings* (Warrendale, PA: Iron and Steel Society, 1995), 407–15.
9. T. Kishida et al., "Secondary Metallurgy to Meet Demands from both Continuous Casters and Customers for the Optimization of Plant Economy," *SCANINJECT 5*, June 1989.
10. *AISE Technical Report No. 9, Specifications for Design and Use of Ladles* (Pittsburgh: Association of Iron and Steel Engineers, 1991).

Chapter 5

Production and Use of Industrial Gases for Iron and Steelmaking

R.F. Drnevich, Manager of Hydrogen Technology and Process Integration, Praxair Inc.
C.J. Messina, Director Primary Metals, Praxair Inc.
R.J. Selines, Corporate Fellow, Praxair Inc.

5.1 Industrial Gas Uses

5.1.1 Introduction

The iron and steel industry, one of the largest users of industrial gases, consumes substantial quantities of oxygen, nitrogen, argon, and hydrogen, and a small but growing amount of carbon dioxide (CO_2). Table 5.1 shows estimates of total consumption by the US industry for the most commonly used gases.

Table 5.1 Consumption of Industrial Gases by the United States Steel Industry

(Millions of Cubic Feet in Gaseous Form)		
	1990	1995
Oxygen	199,420	288,105
Nitrogen	133,250	135,665
Argon	2,480	2,513
Hydrogen	830	940

Oxygen and nitrogen represent the largest volumes by far. Oxygen consumption has grown substantially in recent years while use of the other gases has remained relatively constant.

In order to provide a framework for reviewing the various uses of industrial gases, Table 5.2 summarizes the generic process functions they perform in iron and steelmaking operations.

Table 5.2 Industrial Gas Functions

Gas	Heating	Oxidation	Reduction	Stirring	Inerting	Injection
Oxygen	X	X				
Nitrogen	X (plasma)			X	X	X
Argon	X (plasma)			X	X	X
Hydrogen			X			
CO_2				X	X	X

Oxygen is used to provide process heat through various exothermic reactions which occur upon contact with hot metal and liquid steel and through combustion of fossil fuels. Injected oxygen also participates in the formation of carbon monoxide for the reduction of iron ore in the blast furnace and in some natural gas-based processes for producing direct reduced iron. Nitrogen, argon, and CO_2 are used for inerting vessels, equipment, and metal transfer streams in tapping and casting operations; to eliminate the formation of explosive mixtures in enclosed spaces; and to prevent undesirable reactions between iron and steel with oxygen and hydrogen in the surrounding atmosphere. Nitrogen and argon are also injected into molten iron and steel to provide metal stirring and slag/metal mixing, and as a carrier gas for powder injection. Additional uses for these gases include providing conditions which favor the oxidation of carbon instead of chromium in the argon oxygen decarburization (AOD) process for producing stainless and high alloy steels and as a coolant to protect oxygen injection tuyeres in various oxygen converters. The choice of gas for such inerting, stirring, injection, and steelmaking converter operations is dictated by overall cost and effect on metal chemistry and quality. Nitrogen is usually the gas of choice so long as its reactivity with molten metal and the associated increase in dissolved nitrogen or nitride formation is acceptable. Argon is used whenever a totally inert gas that does not affect metal chemistry is required. CO_2 does not result in increased nitrogen levels but can result in increased carbon, oxygen, and oxide inclusion content depending on the molten metal chemistry and extent of contact. A more detailed discussion of the use of each gas follows.

5.1.2 Oxygen Uses

The introduction of the basic oxygen furnace (BOF) for steelmaking in the 1950s also marked the genesis for tonnage supply of oxygen from on-site cryogenic air separation plants. Today, oxygen is the most widely used industrial gas due to its continued use in the BOF and its growing uses for enrichment in the blast furnace and providing supplemental chemical heat in electric arc furnace based steelmaking. Over 90% of this oxygen comes from plants that are owned and operated by industrial gas suppliers. Table 5.3 provides a summary of the more significant oxygen uses for the integrated and electric arc furnace industry segments.

Table 5.3 Oxygen Applications

	O_2 Consumption (millions of cu. ft.)		Raw Steel Production (thousands of net tons)		Specific O_2 Consumption (cf/ton* of steel)	
	1995	1990	1995	1990	1995	1990
Blast Furnace Enrichment	90,700	37,750	62,525	58,470	1450	645
Basic Oxygen Process	124,500	112,720	62,525	58,470	1990	1930
Electric Arc Furnaces	55,400	27,235	42,410	36,940	1305	735
Cutting, burning, etc.	17,520	15,890	104,930	95,410	165	165

*One ton equals 2000 pounds
Source: American Iron and Steel Institute, 1995 Annual Statistical Report

5.1.2.1 Blast Furnace

The use of significant quantities of oxygen for enrichment of the blast has become widespread. Table 5.3 shows that specific consumption rates have more than doubled over the last five years, and 40% of the total oxygen requirement of integrated steel mills is consumed in the blast furnace. This increase is a result of the desire for higher furnace productivity and the growth in powdered coal and natural gas injection as a means of reducing coke consumption and lowering production costs. Typical levels of enrichment are in the range of 3–8%, but enrichment levels as high as 12% have been used in furnaces using large amounts of coal or natural gas. Oxygen enrichment increases furnace productivity by reducing the blast volume and the associated pressure drop which allows a higher total oxygen throughput rate. It is also used to compensate for the reduction in raceway adiabatic flame temperature (RAFT) that results from the injection of powdered coal and natural gas into blast furnace tuyeres, thereby maintaining smooth furnace operation and the required hot metal temperature.

Oxygen is injected through spargers into the cold blast line between the blower and stoves. While process considerations would allow the use of relatively low purity oxygen for enrichment, the need to supply high purity oxygen for the BOF and to produce nitrogen and argon usually results in the use of standard 99.5% purity plants. However, the trend to very high rates of powdered coal and natural gas injection can easily result in the use of 300 to 1000 tons of oxygen per day in a single furnace. There have been a few recent installations of on-site cryogenic plants that are dedicated to the blast furnace thereby allowing optimization concerning product purity as well as plant integration and cogeneration schemes. The pressure required depends on the operating pressure of the blast furnaces and is usually about 60 psig.

5.1.2.2 Basic Oxygen Furnace (BOF)

Oxygen is used primarily for decarburization and conversion of blast furnace hot metal to liquid steel in the BOF. This accounts for 55% of the total oxygen consumption in integrated steel mills. The heat which results from the exothermic reactions of oxygen with silicon and carbon in the hot metal and the post combustion of a portion (about 10–15%) of the carbon monoxide (CO) which is generated in the converter is sufficient to melt scrap in quantities that amount to about 25% of the total vessel charge weight. A supply system capable of providing high flow rates of up to 30,000 scfm at nominal pressures of around 250 psig for about 20 minutes is required. High purity (99.5% oxygen content) is required to achieve required low steel nitrogen contents and to maximize vessel productivity and scrap melting capability. In most vessels, all of the oxygen is injected through a water cooled top lance with tips that have four or five nozzles which generate supersonic jets that impinge on the molten bath surface. Recently, lances which also have a number of secondary subsonic oxygen nozzles have been introduced to increase the degree of CO post combustion in order to control skull formation and increase scrap melting capability. There are also several modifications to the BOF such as the K-BOP and OBM for which some or nearly all of the required oxygen is injected through hydrocarbon shrouded tuyeres located in the bottom of the converter.

5.1.2.3 Electric Arc Furnace (EAF)

Table 5.3 also shows that the total consumption of oxygen in electric arc furnaces has also more than doubled in the last five years, and that much of this increase is due to a 75% increase in specific oxygen consumption during this time period. Typical uses for oxygen in the EAF include oxy-fuel burners for scrap heating and melting; high velocity lancing for localized scrap melting, steel decarburization, slag foaming; and sub-sonic injection for post combustion of carbon monoxide. Recent trends to use more supplemental chemical energy to increase furnace productivity and reduce melting time and electric power consumption, scrap pre-heating, and the growing use of high carbon content materials such as direct reduced iron (DRI), pig iron, iron carbide, and hot metal have all contributed to this significant increase in specific oxygen consumption over the past five years. Oxygen derived chemical energy can provide 30% or more of the total energy required to make steel in a modern high productivity EAF.

The equipment used to deliver the oxygen includes water cooled burners located on furnace walls, slag doors, and pre-heat shafts to provide direct flame impingement on cold scrap; water cooled or consumable lances which are positioned through the slag door or sidewall for scrap cutting, bath decarburization, and slag foaming; and door and sidewall lances or wall mounted nozzles for post combustion.

Growth in the rate of steel production and in specific oxygen consumption at many EAF based mills have combined to increase oxygen use to levels that make an on-site plant an increasingly typical mode of supply. Cryogenic, vacuum pressure swing adsorption (VPSA), and pressure swing adsorption (PSA) type plants are all used depending on the total industrial gas requirement of the plant and surrounding local market, purity requirement, use pattern, power rate, etc. Oxygen purity is usually in the range of 90–99.5%, and supply pressure is usually in the range of 150–250 psig.

5.1.2.4 Cutting and Burning

High purity (above 98%) oxygen is also used extensively throughout integrated and EAF based mills for steel cutting and burning as well as general lancing requirements. Significant uses in this category include automatic cut-off torches on continuous casters, periodic lancing to remove skulls from the mouths of vessels and ladles, and cutting of crops, skulls, and other forms of mill scrap into pieces that can be readily fed to the BOF or EAF.

5.1.2.5 Steel Reheating

Oxygen is also sometimes used for enrichment or with oxy-fuel burners in steel reheat furnaces. Benefits associated with the use of oxy-fuel burners include a 25–60% reduction in fuel consumption and associated sulfur dioxide (SO_2) and CO_2 emissions, increased furnace productivity, up to 90% reduction in nitrous oxide (NOx) emissions, and elimination of recuperators.

5.1.3 Nitrogen Uses

Table 5.4 identifies the common uses for nitrogen in both integrated and EAF plants.

Table 5.4 Nitrogen Applications

Integrated Plants[1]	EAF Plants[2]
Inerting of coal grinding and storage equipment	EAF stirring
Inerting of blast furnace charging equipment	Injecting powder for steel desulfurization
Injecting powder for hot metal desulfurization	Ladle stirring
BOF slag splashing	AOD refining
BOF stirring	Caster inerting
Annealing atmospheres	Tundish stirring
Controlling Zn thickness on hot dipped galvanizing lines	Annealing atmospheres
Instrumentation and control equipment	Instrumentation and control equipment
[1]In 1995, US plants consumed 130 billion cubic feet to produce 62.5 million tons of raw steel which is an average consumption of 2,085 scf/ton.	[2]In 1995, US plants consumed 5.2 billion cubic feet to produce 42.4 million tons of raw steel which is an average consumption of 122 scf/ton.

As noted, both overall consumption and intensity of use are considerably higher at integrated plants. Nitrogen purity for most applications is typically 99.999%. Occasionally, nitrogen with a nominal purity of around 97–99% may be used for inerting to prevent explosive mixtures in confined spaces. High purity nitrogen containing 5–10% hydrogen is used to provide a protective reducing atmosphere in batch and continuous type bright annealing furnaces for carbon steel grades.

5.1.3.1 Integrated Plants

Traditional uses of nitrogen in integrated plants include purging and inerting of blast furnace raw material charging equipment to prevent reactions of air with furnace offgas; injecting lime, lime-magnesium mixtures, or other reagents in powder form through refractory coated lances to desulfurize hot metal in torpedo cars and ladles; injection through bottom tuyeres to provide stirring and slag/metal mixing during the initial period of refining in the BOF which improves yield, reduces slag iron oxide (FeO) content, and lowers metal oxygen content; as the inert component with

5–15% hydrogen for bright annealing in continuous or batch furnaces; and as a clean and dry gas for various types of pneumatic instrumentation and process control equipment.

The adoption of powdered coal injection (PCI) in blast furnaces usually creates a new requirement for nitrogen to inert and prevent explosions in coal grinding, storage, and handling equipment. A process called slag splashing, which uses high pressure and flows of nitrogen through the top lance to coat the inside of BOF vessels and extend its refractory life, has also been recently adopted by most plants in North America. Finally, nitrogen can be used instead of air in the pneumatic knives used to control coating thickness in hot dip galvanizing lines. The benefits of nitrogen wiping include less dross formation, improved coating thickness control, and fewer coating defects such as edge build-up or entrapped oxides.

5.1.3.2 EAF Plants

Traditional uses for nitrogen in EAF plants include injection through top lances or porous plugs in ladles to provide stirring for steel temperature and composition control and to remove oxide inclusions; injecting powdered reagents containing one or more constituents such as calcium, silicon, and aluminum to desulfurize steel in ladles and ladle furnaces; preventing reaction with surrounding air of open teem streams between the tundish and molds in continuous casters; and as a clean and dry gas for pneumatic instrumentation and process control equipment.

Uses that are more site specific include injection through multi-hole elements to provide stirring in the EAF for faster melting, power savings, and improved temperature and composition control; injection through porous elements of various designs to modify flow patterns and remove inclusions in tundishes; to protect the tuyeres and promote carbon removal with minimum oxidation of chromium in the AOD process for producing stainless and other high alloy grades; and as the inert component with hydrogen for bright annealing in continuous or batch furnaces.

5.1.4 Argon Uses

Table 5.5 identifies the common uses for argon in both integrated and EAF plants. Argon purity is typically 99.998%, and it is usually supplied from liquid storage tanks at pressures up to 200 psig.

Table 5.5 Argon Applications

Integrated Plants[1]	EAF Plants[2]
BOF stirring	AOD refining
Ladle and ladle furnace stirring	Ladle and ladle furnace stirring
Injecting powder for steel desulfurization	Injecting powder for steel desulfurization
Vacuum degasser recirculation and stirring	Vacuum degasser stirring
Inerting submerged entry nozzles on casters	Inerting submerged entry nozzles on casters
Tundish inerting	Tundish inerting

[1] In 1995, US plants consumed 1.54 billion cubic feet to produce 62.5 million tons of crude steel which is an average consumption of 24.6 scf/ton.

[2] In 1995, US plants consumed 920 million cubic feet to produce 42.4 million tons of crude steel which is an average consumption of 21.7 scf/ton.

5.1.4.1 Integrated Plants

Traditional uses for argon in integrated plants include injection through bottom tuyeres to provide stirring and slag/metal mixing during the final stages of refining in the BOF which improves yield, reduces slag FeO content, and lowers metal oxygen content; injection through porous plugs to provide stirring for improved temperature and chemistry control and to promote desulfurization in ladles and ladle furnaces; injection of various powdered reagents in ladles and ladle furnaces for desulfurization; injection in vacuum degassers to facilitate carbon, hydrogen, and nitrogen removal and to provide circulating metal flow in RH type degassers; and inerting submerged entry nozzles to prevent clogging and re-oxidation or nitrogen pick-up by the steel during ladle to tundish and tundish to mold transfers in continuous casters. Recently, argon has also been used to inert the tundish headspace during transient periods of operation to improve slab surface quality.

5.1.4.2 EAF Plants

Argon is primarily used in plants that produce stainless and high alloy grades, flat carbon steel products, and high quality long products. Traditional uses include many of those described above for the integrated plants such as ladle, ladle furnace, and vacuum degasser stirring, powder injection, and inerting of submerged entry nozzles and tundishes in continuous casters. In addition, argon is also used extensively to facilitate carbon, nitrogen, and hydrogen removal in the AOD process for producing a wide variety of stainless, tool, heat resisting, and other high alloy content grades.

5.1.5 Hydrogen Uses

Hydrogen is used in significant quantities for natural gas based production of direct reduced iron (DRI). In the Midrex process, the CO_2 and H_2O which is present in the offgas from the direct reduction shaft reactor is used to reform natural gas and produce the high temperature H_2 and CO containing reducing gas required to achieve the desired level of iron ore conversion to iron. In Hylsa's HYL III process, a conventional steam reformer is used to produce the reducing gas, and the CO_2 and H_2O in the reactor offgas is removed to allow recycle of unreacted H_2 and CO back into the reactor. The natural gas consumption is approximately 11,000 scf per ton of DRI, and the reducing gas generators are part of the overall plant installation.

The other common use for hydrogen is for annealing atmospheres. For carbon steel annealing, atmosphere compositions of 5–10% hydrogen with the balance nitrogen are commonly used in both continuous and batch furnace types. For stainless and other high alloy products, the hydrogen content is typically in the range of 75–100%. Typical hydrogen purity is 99.995%. These atmospheres serve to transfer heat, remove residual rolling oils, prevent surface oxidation, and reduce any surface oxides that may be present resulting in a bright annealed surface free from oxides, deposits, and discoloration.

A recent trend is to anneal coils in batch bell-type furnaces that have been designed to use a 100% hydrogen atmosphere. The main benefit is a shorter cycle time and increased productivity due to the higher diffusivity and thermal conductivity of pure hydrogen and improved uniformity of mechanical properties as a result of more uniform temperature distribution throughout the coil.

5.1.6 Carbon Dioxide Uses

CO_2 is sometimes used as a stirring gas in BOF vessels in place of nitrogen and argon. This option provides the ability to achieve lower steel nitrogen contents and lowers stirring gas costs by reducing or eliminating the use of argon which is more expensive per unit volume. However, since CO_2 is a reactive gas when exposed to molten steel under the conditions associated with submerged injection, increased wear of the bottom injection tuyeres and surrounding refractory materials and an inability to achieve low aim carbon contents may result. CO_2 is occasionally used for ladle stirring. As for BOF stirring, it offers a combination of lower nitrogen content and lower gas cost versus the alternatives of nitrogen or argon. However, the same potential for increased porous plug and refractory wear exists, and there is also the risk of higher oxide inclusion contents due to the reaction of the CO_2 with deoxidants such as silicon or aluminum. CO_2 has also been used in continuous billet casters for protecting the tundish to mold stream from contact with the surrounding atmosphere.

A more recent application for CO_2 is to suppress fugitive fume emissions during electric arc furnace charging and tapping operations and to lower nitrogen pick-up during EAF or BOF tapping operations. Typical practice is to inject CO_2 in snow form to eliminate air from the furnace during charging and from around the tap stream and its impact zone in the ladle. Industry experience with this type of CO_2 application is extremely limited, and more work needs to be done to optimize practices and quantify potential benefits.

5.2 Industrial Gas Production

5.2.1 Introduction

The industrial gases that are used in the steel industry are atmospheric gases (oxygen, nitrogen, and argon) or gases that are produced from hydrocarbon processing either directly or as byproducts (hydrogen and carbon dioxide.) Hydrogen is also obtained as a byproduct from a number of processes including the cracking of ethane to produce ethylene and the electrochemical process used to produce chlorine (the chlor-alkali process). The atmospheric gases are separated from air using cryogenic processes for large quantity5, pressure swing adsorption (PSA), vacuum swing adsorption (VSA), or vacuum pressure swing adsorption (VPSA) for moderate quantities of oxygen, or membranes when a moderate quantity of nitrogen is the only product desired. Hydrogen is typically produced from natural gas using steam reforming or separated as byproduct streams from hydrocarbon containing streams. The carbon dioxide that is supplied to industrial users is most often obtained as a byproduct from ammonia and hydrogen plants using an acid gas (CO_2, hydrogen sulfide H_2S, etc.) removal system. CO_2 can also be economically recovered from fermentation systems, natural gas wells, and, in some cases, from combustion systems.

Industrial gases are supplied from on-site plants that are dedicated to the needs of one steel mill, from industrial gases complexes by pipelines that serve multiple steel mills and/or other users, or as cryogenic liquid products that are trucked in from a production facility that can supply hundreds of small volume users. For atmospheric gases the process used to produce the gas will depend on which gases are needed and the quantity and quality requirements. Fig. 5.1 shows the approximate flow and purity boundaries that define the low cost supply options for oxygen. Similarly, Fig. 5.2 shows the flow/purity regions where the alternative nitrogen production processes are most economical. Argon is almost always obtained from cryogenic air separation units and supplied as a liquid.

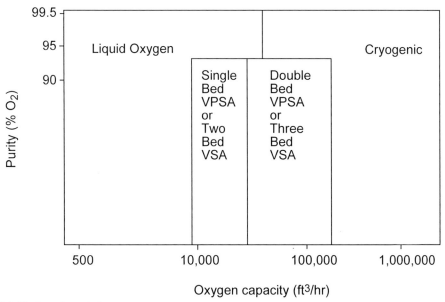

Fig. 5.1 Modes of supply for oxygen.

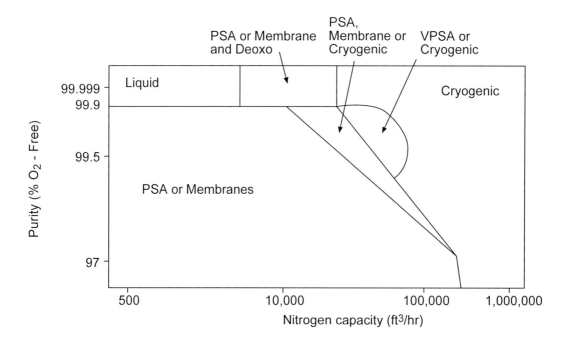

Fig. 5.2 Modes of supply for nitrogen.

The industrial supply options for hydrogen are presented in Table 5.6.

Table 5.6 Modes Of Supply For Hydrogen

Supply Option	Average monthly use rate, scfh
Tube trailers	>135
Liquid	120-30,000
Modular Reformers (on-site)	6,000-40,000
Custom Designed Reformers (on-site)	>40,000

The hydrogen supplied by tube trailer or as liquid is normally produced from steam reforming of methane or obtained as a byproduct, most often from chlorine production. The tradeoff between tube trailer and liquid is also affected by the location of the user relative to the production site. Tube trailer supply is favored when the distances are short (<200 miles). Modular on-site reformers are relatively new to the supply picture and are filling a void between the liquid and the custom designed reformer option.

Except for rare cases where the user is near a high quality source of CO_2, carbon dioxide is supplied as a liquid.

5.2.2 Atmospheric Gases Produced by Cryogenic Processes

Cryogenic air separation plants can be designed to produce oxygen, nitrogen, argon, or any combination of the three. Since oxygen typically is the dominant product for steel industry applications, this discussion will focus on it with nitrogen and argon treated as optional products that can be made available with system modifications.

The commercial production of oxygen in cryogenic plants is well established with hundreds of plants operating worldwide and individual capacities ranging from 20 to over 3000 tons per day. Present-day processes are basically the same as those used in the early 1900s. Although plant

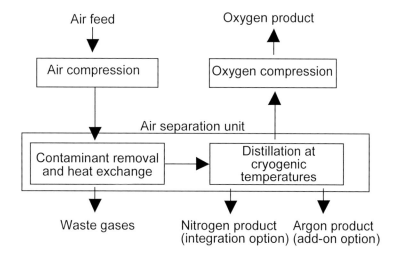

Fig. 5.3 Basic steps in cryogenic air separation plants.

capacities and efficiencies have increased dramatically and many equipment variations have been developed, the principle of distilling air to separate the oxygen from the nitrogen remains the heart of the process. This distillation technology is like that used in the petroleum and chemical industries except that air distillation requires operation at cryogenic temperatures instead of the higher temperatures used in most other applications. The four main process steps and sequence necessary to produce gaseous oxygen from air are shown in Fig. 5.3. These steps are supported by subsystems such as expansion turbines to provide refrigeration, and superheaters to recover refrigeration.

The composition of air and the characteristics of all of its constituents must be addressed in designing cryogenic separation systems. Components considered as contaminants in air separation include carbon dioxide, water, and hydrocarbons in various concentrations. The four process steps and supporting subsystems are designed to effectively remove the contaminants and separate the air into its various products.

5.2.2.1 Air Compression

Air compression, the simplest step, uses standard commercial equipment. As atmospheric air, drawn through a filter, is compressed to a pressure of about 90 psia, the heat of compression is removed by coolers and entrained water is removed by water separators. The compressed air supplied to the air separation unit contains the energy required to separate the air into oxygen and nitrogen. Power is the primary operating cost element in the production of oxygen. Since most of the power is used for compression, design of the entire air separation process must be keyed to obtaining maximum efficiency in the compression step.

5.2.2.2 Contaminant Removal and Heat Exchange

The air stream that leaves the compressor aftercooler at near-ambient temperature contains carbon dioxide, water vapor, and hydrocarbons. These contaminants must be removed, and the stream must be cooled to the liquefaction temperature before the distillation process can be started. Carbon dioxide and water freeze before the air liquefaction temperature (–278°F at 90 psia) is reached; therefore, the removal of these contaminants is critical to prevent heat exchanger plugging.

The primary method for performing contaminant removal is the pre-purifier system. Pre-purifiers use adsorbents such as molecular sieves to remove bulk and trace contaminants at ambient temperatures. The filtered air from the compression system normally is cooled to about 40°F using a refrigerant chiller to condense and remove water vapor so that smaller adsorption beds can be used. Typically, there are two beds with one on-stream while the other is being regenerated. They are sized to reduce the water content to 0.1 ppm and the carbon dioxide content to 0.25 ppm. These beds also reduce hydrocarbon concentrations to acceptable levels. The clean, dry process air from the pre-purifier is then cooled to the liquefaction temperature through heat exchange with the cold product and waste streams. This is accomplished in a primary heat exchanger (PHX) that is of plate-and-fin construction.

5.2.2.3 Cryogenic Distillation

The distillation process is used for separating air into its desired components of oxygen, nitrogen, etc. If the maximum amount of oxygen is recovered from the feed air stream, the feed air quantity required to produce a given amount of oxygen is minimized. Thus, the energy required for air compression (the power required for the process) is also minimized.

The distillation process is based on allowing vapor to rise through a descending flow of liquids. When a liquid mixture of two components with different boiling points is thoroughly mixed with vapors of a similar composition, the liquids and vapors reform. After the first contact, the vapor contains more of the component with the lower boiling point (nitrogen), and the liquid contains more of the component with the higher boiling point (oxygen). Thus, after one contact stage, the two components are partially separated with the vapor enriched with one component and the liquid enriched with the other. The separation becomes more complete—higher component purities are obtained—if successive vapor-liquid contacting stages are provided.

The distillation column provides many contact stages in a single vessel. Traditionally, each contact stage consisted of a single tray—a perforated plate that promotes intimate mixing of the liquids and vapors. The descending liquid spreads across the plate and contacts the vapor rising through the perforations. Part of the vapor condenses and joins the liquid which becomes progressively richer in oxygen, and some of the liquid boils off and joins the vapor which becomes progressively richer in nitrogen. Part of the fluid at the bottom of the column is withdrawn as product oxygen and the rest is heated in a reboiler to form the vapor that rises through the column. Similarly, at the top of the column, part of the fluid is withdrawn as waste-nitrogen vapor and the rest is cooled in a condenser to form the liquid that descends through the column. The components of the air are used to heat the reboiler and cool the condenser. Over the past few years packing has been used instead of trays for many air separation applications. In either case, column operation is essentially the same.

The double-column concept has been used by the air separation industry for most of this century. It enables oxygen recovery of over 95%. The key is the relationship of the boiling points of oxygen and nitrogen at different pressures. The boiling points become higher as the pressure increases. The boiling temperature of nitrogen at 90 psia is higher than the boiling temperature of oxygen at 22 psia. Under these conditions, nitrogen can be condensed by transferring heat to the boiling oxygen. The boiling of oxygen by condensing nitrogen in the double column provides the driving force for this distillation process. The double column has a combination reboiler/condenser unit (main condenser) located between the two sections of the column. Each column operates at a different pressure to exploit the boiling point variations at different pressures. By condensing nitrogen to provide reflux liquid for both the upper and lower columns, the main condenser provides vapor boil-up (gaseous oxygen) to the upper column. The lower column provides a crude product split and the upper column provides a finer separation.

The air stream that feeds the distillation column is primarily in the gaseous state. Refrigeration must be provided to cool a portion of the stream to temperatures low enough to form the liquid necessary for the distillation process. Modern plants use an expansion turbine to produce the low temperatures; the compressed air feed is expanded through a turbine and energy is recovered externally. These turbo-expanders recover energy and produce refrigeration efficiently.

5.2.2.4 Oxygen Compression

The fourth process step, oxygen compression, boosts the pressure of the product oxygen leaving the air separation unit from near-atmospheric to the desired end-use pressure. Most oxygen users in the steel industry require pressures of 250 psig or higher. Compression of large quantities of oxygen has traditionally been accomplished by specially designed centrifugal machines that are commercially available from several suppliers.

5.2.2.5 Cryogenic Process Description

The four major process steps of a double-column air separation system integrated into a typical process arrangement for producing gaseous oxygen are shown in Fig. 5.4.

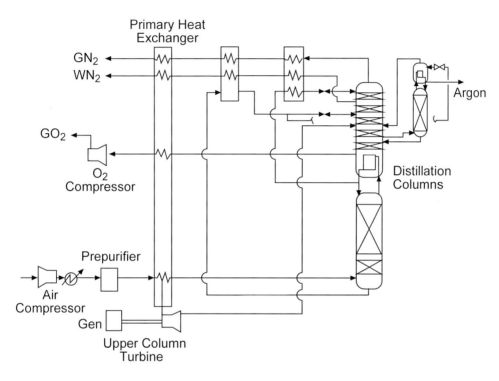

Fig. 5.4 Basic double-column plant for gaseous oxygen.

The feed air is drawn in through an inlet air filter, compressed to the head pressure of the plant, and then cooled in the aftercooler. Condensed moisture is removed in the moisture separator which also serves to smooth the flow of air to the air separation unit.

The air then passes through the switching valves into the prepurifier where the contaminants are removed. The clean air stream leaving the prepurifier is fed to the brazed aluminum heat exchanger (primary heat exchanger, PHX). The vast majority of the air (normally greater than 90%) is cooled to near its liquefaction temperature by the returning product and waste streams and flows from the PHX to the bottom of the lower column. The remaining air is removed from the primary heat exchanger at about the midpoint and fed to the air-expansion turbine which supplies the refrigeration for the process. As the vapor rises through the lower column, essentially pure nitrogen is formed. This vapor is liquefied in the main condenser and split into two streams, lower column reflux and upper column reflux. The upper column reflux is cooled against waste nitrogen gas in the nitrogen superheater. The bottoms product of the lower column, a liquid rich in oxygen, is also cooled in the nitrogen superheater before being fed to the middle of the upper column.

The bottoms product of the upper column is pure oxygen. Although a small fraction of this oxygen can be withdrawn as a liquid, the bulk of it is reboiled in the main condenser to form both the vapor that rises in the upper column and the gaseous oxygen product. The product oxygen is warmed by a heat exchange with feed air in the oxygen superheater, and then further heated to ambient temperature as it flows through the primary heat exchanger. This product oxygen leaves the air separation unit at a pressure of about 2 psig and is compressed to the desired end-use pressure.

The overhead vapor of the upper column, the waste nitrogen stream, has the lowest temperature of any stream in the plant. The low temperature refrigeration available from this stream is recovered in the nitrogen superheater that cools the feed streams to the columns. The nitrogen is warmed to ambient temperature through heat exchange with the air stream in the primary heat exchanger before being vented to the atmosphere.

5.2.2.6 Product Options

The basic plant configuration presented in Fig. 5.4 is very versatile and can be used to provide many product variations. The product options of interest for steelmaking include gaseous oxygen, gaseous nitrogen, and argon.

5.2.2.6.1 Gaseous Oxygen Gaseous oxygen can be produced in purities varying from enriched air, slightly greater than 21 mol% O_2, to beyond standard high purity (99.5 mol%). A basic plant designed for 95 mol% purity will produce oxygen at a lower cost than a system designed for 99.5 mol% oxygen. The lower purity requires a less complete separation and, consequently, fewer trays in the upper column. Since the pressure drop caused by the trays in the upper column has a significant impact on the required compressor discharge pressure, low purity plants consume less power.

5.2.2.6.2 Gaseous Nitrogen Low purity nitrogen (~99 mol% N_2) can be obtained from the top of either the upper or lower column. Small quantities of high-purity nitrogen (99.999 mol% N_2) can be produced by increasing the number of trays in the lower column and drawing off nitrogen gas from the top of this high-pressure unit. This method of high purity nitrogen recovery has only a limited effect on the overall power requirement for oxygen production.

If substantial quantities of high purity nitrogen are required, provisions must be integrated into the original process and equipment design for the plant. Basically, this involves adding a "top hat" section on the upper column to provide the necessary separation and purification capacity. The top hat section provides additional trays which increase the pressure drop and add to the power for the air separation system. The ratio of high purity nitrogen to oxygen product can be as high as 3:1.

5.2.2.6.3 Argon All oxygen or oxygen-nitrogen plants can be designed to accommodate the add-on side column and the refinery required for argon recovery and purification. With the inclusion of the proper number of contacting surfaces in the upper and lower columns and provision for sufficient refrigeration, a relatively high percentage of the argon available in the feed air can be produced. The add-on system shown in Fig. 5.4 recovers, refines, and stores liquid argon. Argon production can range between 3–4% of the oxygen product rate.

5.2.3 Atmospheric Gases Produced by PSA/VSA/VPSA/Membranes

Cryogenic technology has been the dominant supply option for the steel industry since oxygen was first used in steel production. The development of separation systems that operate at near atmospheric temperatures and pressures have begun to supply oxygen to smaller volume users such as mini-mills (VPSA Systems). Membranes have not proven to be a cost effective method for oxygen production but can competitively produce nitrogen depending on the quantity and quality required.

5.2.3.1 PSA/VSA/VPSA Systems

Pressure swing adsorption for oxygen production involves the use of synthetic zeolites (molecular sieves) to adsorb nitrogen from a pressurized air stream in a fixed bed system. During the adsorption portion of the cycle 90–93% purity oxygen is produced while the adsorbent becomes saturated with nitrogen. When nearly all the adsorbent is at the nitrogen saturation point, the bed is removed from production service and is regenerated. Regeneration is mainly accomplished by lowering the pressure of the bed which allows most of the nitrogen to be desorbed and released from the bed followed by purging the low pressure bed with a small portion of the product oxygen. After the bed has been regenerated, it is repressurized and placed back into service. The total cycle time (the time associated with the adsorption period through regeneration to the point in time when the bed is again making product) ranges from 1–3 minutes. PSA systems operate at feed pressures ranging from 40–80 psia and regeneration pressures near atmospheric. VSA systems use a feed pressure of one atmosphere and regeneration at 3–5 psia. VPSA systems operate with feed pressures in the 20–30 psia region with regeneration occurring at 3–10 psia. The vacuum regeneration allows for greater effective nitrogen adsorption capacity for a specific molecular sieve. VPSA systems have become the technology of choice because of their lower capital and operating costs.

5.2.3.1.1 Process Description The number of adsorption beds used in PSA/VSA/VPSA systems have ranged from one to four depending on the choice of molecular sieve and the approach to regeneration. Fig. 5.5 shows a schematic of a two bed VPSA system.

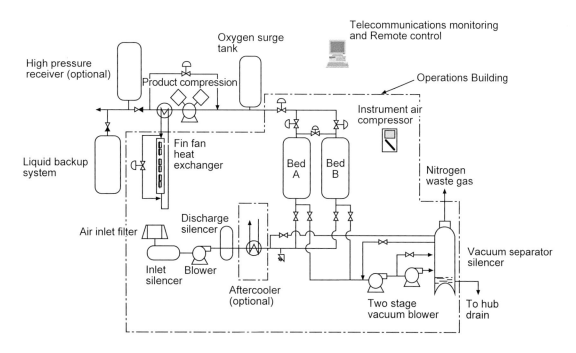

Fig. 5.5 VPSA process.

The system consists of an inlet air filter, an air blower, a feed discharge silencer, an aftercooler, switching valves, two adsorption beds, a vacuum blower, a vacuum separator/silencer, product surge tanks, an oxygen compressor, and high pressure receivers.

Air is compressed to 20–30 psia by the feed blower and cooled to near ambient temperature in the aftercooler. The air then passes through switching valves to one of the two adsorption beds, Bed A, where water vapor, carbon dioxide, hydrocarbons, and nitrogen are adsorbed. Oxygen product at 90–93% purity leaves the adsorption bed through a second series of switching valves and is fed to the product compressor. The oxygen surge tank is used to dampen pressure fluctuations at the product compressor inlet that result from the switching of the adsorption beds. The product compressor delivers oxygen at the desired pressure for use in the electric arc furnace. The high pressure receiver is used to accommodate the flow fluctuations that result from the intermittent use patterns of most EAFs.

After the sorbent in Bed A has been nearly saturated with nitrogen the feed air is switched to Bed B which had been previously regenerated. Bed A undergoes regeneration by being pumped to 5–10 psia using the vacuum blower. The nitrogen, hydrocarbons, carbon dioxide, and water vapor desorb from the sieve and are evacuated from the bed. At the end of the evacuation period a small quantity of product oxygen is used to purge the remaining desorbed material from the void spaces around the adsorbent particles prior to repressurizing the bed for use in the adsorption step.

5.2.3.1.2 Product Options PSA/VPSA systems that use synthetic zeolites as the sieve have generally been successful for the production of oxygen. The waste nitrogen from the process can contain as much as 10% oxygen as well as carbon dioxide, water vapor, and small quantities of hydrocarbons. PSA systems based on carbon molecular sieves are used to produce nitrogen (97–99.5%) in moderate quantities. The waste oxygen product from nitrogen PSAs has too much nitrogen to be of use in steel applications.

5.2.3.2 Membranes

Membranes for atmospheric gases are used exclusively for the production of nitrogen. Hollow fiber polymeric membranes use differences in permeation rates to separate nitrogen from air. Generally oxygen, carbon dioxide, and water vapor permeate faster than nitrogen so nitrogen at purities up to about 99.9% can be economically obtained at flow rates up to 10,000 scfh. The two essential characteristics of a membrane are its permeability and selectivity. Permeability defines the rate at which constituents (oxygen, nitrogen, etc.) will pass through a membrane for a given pressure drop while selectivity defines the rate that one component will flow through the membrane relative to the rate of a second component. For membranes used for nitrogen production, selectivity is the ratio of oxygen permeability to nitrogen permeability. The most cost effective membrane systems would have high permeability and high selectivity. However, membranes with high permeability usually have low selectivity and membranes that are highly selective normally have low permeability.

5.2.3.2.1 Process Description As shown in Fig. 5.6 membrane systems are relatively simple consisting of an air compressor, a membrane unit which contains bundles of hollow fiber membranes, and a control unit.

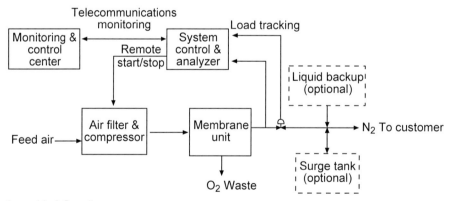

Fig. 5.6 Membrane block flow diagram.

The air compressor is used to pressurize the air to 100–150 psig. The air is then cooled and compressor oil and water are removed. The air enters the membrane unit which is composed of bundles containing more than a million membrane fibers. Each fiber is thinner than a human hair. The hollow fiber membranes are arranged in the bundles so that the unit's flow configuration is similar to that of a shell and tube heat exchanger. Bundles normally contain a fixed number of fibers having a fairly constant surface area. The number of bundles used is proportional to the desired product rate from the unit. The bundles all operate in parallel if low purity nitrogen (97%) is acceptable. If higher purities are required, then a second set or even a third set of bundles can be operated in series with the first set.

5.2.4 Hydrogen Production

Hydrogen production is dominated by the steam reforming of methane. Fig. 5.7 presents the basic schematic for a steam reformer system used to produce hydrogen. In a typical steam reformer the natural gas provides both feedstock and fuel. The feedstock portion is compressed to about 400 psia, preheated to about 750°F, and desulfurized in a zinc oxide bed. The sulfur free natural gas is mixed with steam at approximately a 3:1 steam to carbon ratio and further heated to about 925°F prior to entering the reformer. The reformer reactions proceed to near equilibrium in catalyst filled tubes that are heated from the outside by the combustion of natural gas and/or PSA tail gas. The reformed gas leaves the furnace at about 1450°F and is cooled slightly prior to entering the high temperature shift reactor (HTS). The feed gas entering the HTS has a hydrogen to carbon ratio of about 5:1. The catalyst in the HTS increases this ratio to more than 30:1. The hydrogen rich gas leaving the HTS unit is cooled to near ambient temperature and fed to a Pressure Swing Adsorption (PSA) unit. Between 80–90% of the hydrogen is recovered as product with purities of 99.999%. The unconverted CO and the unrecovered hydrogen (PSA Tail Gas) is used to fuel the reformer.

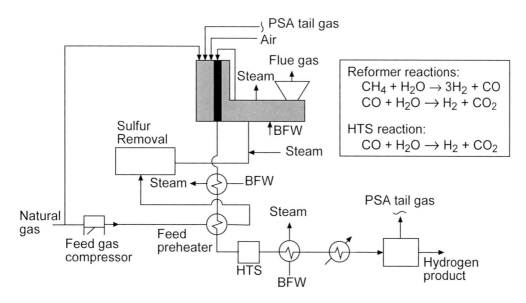

Fig. 5.7 Steam reformer for hydrogen.

Most hydrogen needs in the steel industry are relatively small, less than 1 million scfd. Consequently, hydrogen is often supplied as liquid because the usage rate is too low to justify an on-site plant. Recently, modular reformer systems have been developed that are cost effective at flow rates as low as 150,000 scfd.

Hydrogen has also been recovered from coke oven gas which contains 40–50% hydrogen. This has been accomplished by using a PSA to adsorb all of the coke oven gas constituents except hydrogen. For this system to be effective the coke oven gas has to be compressed and nearly all of the heavy hydrocarbons have to be removed to prevent PSA fouling. This approach has been used when other sources of hydrogen have not been readily available.

5.2.5 Carbon Dioxide Production

Except for some carbon dioxide that is recovered from gas wells, most sources of carbon dioxide are dilute streams, less than about 30% CO_2, at low pressure that are very expensive to recover. Most of the carbon dioxide that is recovered for commercial use is a byproduct of another process such as ammonia production. CO_2 removal is a requirement to make the chemical synthesis process economical and is paid for by the chemical manufacturer. Ammonia plants use either chemical absorbents (amines or hot potassium carbonate) or physical absorbents (methanol or Selexol solvent) to remove carbon dioxide from the process stream. The vent gas from these absorbent systems contains more than 90% CO_2 by volume. Normally only a portion of the CO_2 produced by these plants is converted into liquid CO_2 because more is produced than can be sold within a reasonable distribution area. Fig. 5.8 shows a common CO_2 liquefaction process.

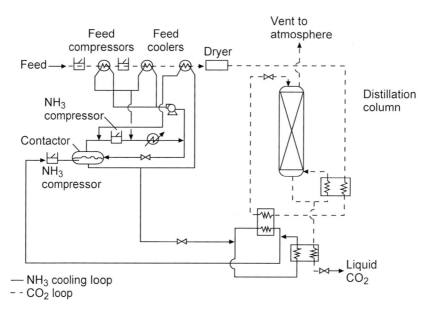

Fig. 5.8 Carbon dioxide liquefaction cycle.

A CO_2 rich stream that has been purified to remove water and other impurities is compressed to about 300 psia and dried prior to entering the heat exchangers that cool the stream to its liquefaction temperature. Part of the cooling is accomplished by boiling pure CO_2 at the bottom of the distillation column. The remainder of the refrigeration is obtained from an ammonia refrigeration loop. The distillation column provides the final purification and delivers a product that is 99.99% CO_2.

5.3 Industrial Gas Supply System Options and Considerations

5.3.1 Introduction

The choice of supply system to provide industrial gases to the steel industry depends upon several variables. The decision to supply gases by either pipeline, on-site plant or merchant liquid is determined by the number of gases required, the purity of the gases, the volume of gas, the use pressure, the use pattern, the cost of power, backup requirements and integration with other utilities.

5.3.2 Number of Gases

If oxygen, nitrogen and argon are required and the volume requirement is modest, the supply can be merchant (liquid) product delivered into tanks at the mill. If the volume of gases is large (more than 100,000 cfh oxygen) the supply will most likely be a cryogenic on-site plant. Finally, if the requirement is large for oxygen, but small for argon or nitrogen, the supply can be a combination of VPSA/VSA/PSA for oxygen and merchant for argon and nitrogen.

5.3.3 Purity of Gases

If high purity is not required for oxygen or nitrogen, the supply mode can be non-cryogenic on-site (VPSA/VSA/PSA or membrane) if the required volume is in the range of about 10,000 to 120,000 standard cubic feet per hour. These non-cryogenic supply systems deliver product purity between 90–99%. Because of separation process requirements, argon is always delivered as high purity product.

5.3.4 Volume of Gases

The amount of gas required will determine supply mode and plant size. Small users, merchant; larger users, VPSA/VSA/PSA; largest users, on-site cryogenic plants. See Fig. 5.1 and Fig. 5.2.

5.3.5 Use Pressure

The pressure requirement of the gas at the use point is no longer a variable that determines supply mode. However, it does determine requirements for compression equipment and high pressure gas storage receivers. In the case of merchant supply, high pressure tanks are used to supply high pressure product to the use point.

5.3.6 Use Pattern

The use pattern of the gases will determine the number of high pressure storage receivers required to deliver the gas when and in the volumes that are needed.

5.3.7 Cost of Power

Merchant supply requires no power to vaporize or compress the gas to use pressure. On-site supply modes require power to separate and compress the gases. The on-site supply modes have different power requirements that are dependent upon the volume and pressure requirements.

5.3.8 Backup Requirements

On-site supply systems are usually backed up with some quantity of merchant product. This quantity is determined by the process requirements of the steel mill and cost.

5.3.9 Integration

Because on-site supply systems are a steady draw on the power grid, they are good customers for the electrical company. As a major user of electrical power, the on-site supply plant should be included in an overall energy/utility strategy for the steel mill.

5.4 Industrial Gas Safety

Table 5.7 shows the boiling points and densities of the industrial gases commonly used for steelmaking. These physical properties as well as chemical reactivity are among the most important attributes that determine the potential for risk and safe handling practices for these gases.

Table 5.7 Physical Properties

	O_2	N_2	Ar	H_2	CO_2
Atomic Number	8	7	18	1	NA
Atomic or Molecular Wt.	32	28	40	2	44
Boiling Point, °F	-297	-320	-303	-423	-109**
Density NPT*, lb/ft³	0.08281	0.07245	0.1034	0.00521	0.1144

*NPT = 70°F and 14.7 psi
**Sublimation

5.4.1 Oxygen

At ambient temperature and atmospheric pressure, oxygen is a colorless, odorless and tasteless gas. For all ironmaking and steelmaking applications, oxygen is used in its gaseous form. Approximately 21% of the air we breathe is oxygen.

5.4.1.1 Chemical Properties

Ironmaking and steelmaking are processes of controlled reduction and oxidation. Oxygen in ironmaking is used to reduce iron ore by reacting with coke to form CO and to create heat by taking advantage of the chemical reaction between carbon and oxygen in the blast furnace. In steelmaking, oxygen is used to selectively oxidize carbon, silicon, manganese and phosphorus from the blast furnace iron, steel scrap and other metallic additions to the furnace. Oxygen is also combined with many different fuels to create heat which is used to increase the temperature of refractories in ladles and steel in reheat furnaces.

5.4.1.2 Safe Practices

Oxygen is not flammable, but vigorously accelerates the burning of combustible materials. Therefore, it is important to keep combustibles away from oxygen and eliminate ignition sources. Oxygen concentrations exceeding 23% by volume in air significantly increase the risk of ignition and fire. Do not smoke or use an open flame where oxygen is stored, handled or used. Do not use cleaning agents that leave organic deposits on the cleaned surfaces. Do not place oxygen equipment on asphalt surfaces or allow equipment to come into contact with grease or oil. Equipment to be used in contact with oxygen should be handled only with clean gloves or hands washed clean of oil. Any equipment that comes into contact with oxygen should be cleaned for oxygen service using the guidelines published by the Compressed Gas Association, Inc. using Pamphlet G–4.1 Cleaning Equipment for Oxygen Service and CGA Pamphlet G–4.4 Industrial Practices for Gaseous Oxygen Transmission and Distribution Piping Systems. Read and understand the supplier Material Safety Data Sheet (MSDS) for oxygen.

5.4.2 Nitrogen

Nitrogen is colorless, odorless and tasteless. It is only slightly soluble in water and is a poor conductor of electricity. The volume percentage of nitrogen in the air we breathe is approximately 78%.

5.4.2.1 Chemical Properties

Nitrogen neither burns nor supports combustion. Although it does not react readily with most materials, at high temperatures nitrogen combines with hydrogen to from ammonia; with oxygen to form nitrogen oxides; and with some of the more active metals, such as calcium, sodium, chromium, magnesium and silicon to form nitrides. At room temperature, nitrogen reacts with lithium.

5.4.2.2 Safe Practices

The displacement of breathing air is a potential hazard when using nitrogen to purge or inert a confined space. Its density at ambient temperature is about 3% lighter than air. Nitrogen can displace normal breathing air and cause rapid asphyxiation. Therefore, the oxygen level in any enclosed space that may be entered by personnel must be maintained above 19.5% using forced-air convection, if necessary. If this is not practical, then access to the enclosed area should be locked and entry allowed only after the area has been purged with air and tested for a safe oxygen level. Read and understand the supplier MSDS for nitrogen.

5.4.3 Argon

Argon is colorless, odorless and tasteless Approximately 1% of the air we breathe is argon.

5.4.3.1 Chemical Properties

Argon is non-toxic and chemically inert. It is non-flammable, does not present a fire hazard, and in fact, inhibits combustion. Because of its properties, argon is used to provide an inert atmosphere for molten steel to prevent the reoxidation of deoxidants such as aluminum, silicon and calcium. Argon is also used as a purge gas by injecting it subsurface into liquid steel to remove dissolved oxygen, nitrogen and hydrogen.

5.4.3.2 Safe Practices

The displacement of breathing air is a potential hazard when using argon to purge or inert a confined space. As a gas at ambient temperature, argon is about 38% heavier than air. This means that in confined spaces or large containers with closed bottoms, argon will accumulate in the bottom of the space or container. Argon can displace normal breathing air and cause rapid asphyxiation. Therefore, the oxygen level in any enclosed space that may be entered by personnel must be maintained above 19.5% using forced-air convection, if necessary. If this is not practical, then access to the enclosed area should be locked and entry allowed only after the area has been purged with air and tested for a safe oxygen level. Read and understand the supplier MSDS for argon.

5.4.4 Hydrogen

Hydrogen is an element which at atmospheric temperatures and pressures exists as a colorless, odorless, tasteless, flammable non-toxic gas. It is the lightest gas with a specific gravity of only 0.0695. Its concentration in the atmosphere is only 0.0001–0.0002% by volume at sea level. Hydrogen diffuses rapidly through porous materials and through some metals at red heat. It may leak out of a system which is gas tight with respect to air or other common gases at equivalent pressure.

5.4.4.1 Chemical Properties

Hydrogen burns in air with a pale blue, almost invisible flame. The ignition temperatures of hydrogen-air and hydrogen-oxygen mixtures vary depending on composition, pressure, water vapor content, and initial temperature. At atmospheric pressure, the ignition temperature of either hydrogen-air or hydrogen-oxygen mixtures will not vary greatly from the range 1050–1074°F. The flammability limits of hydrogen-air and hydrogen-oxygen mixtures depend on initial pressure, temperature, and water vapor content. In dry air at atmospheric pressure the lower limit is 4.1% hydrogen, and the upper limit is 74.2% hydrogen by volume. In dry oxygen at atmospheric pressure the lower limit is 4.7% hydrogen and the upper limit is 93.9% hydrogen by volume. Hydrogen is fundamentally a reducing agent. Either used alone or mixed with other gases, it is used to provide a non-oxidizing atmosphere in annealing and heat treating.

5.4.4.2 Safe Practices

Hydrogen is non toxic, but it can act as a simple asphyxiant by displacing the oxygen in the air. Unconsciousness from inhaling air which contains a sufficiently large amount of hydrogen can occur without any warning symptoms such as dizziness. Still lower concentrations than those which could lead to unconsciousness would be flammable, since the lower flammable limit of hydrogen in air is only 4% by volume. All the precautions necessary for the safe handling of any flammable gas must be observed with hydrogen. Read and understand the supplier MSDS for hydrogen.

5.4.5 Carbon Dioxide

Carbon dioxide in its most common form is a gas, but depending on the temperature and pressure it may exist as a gas, liquid or solid. Carbon dioxide comprises approximately 0.038% by volume of the earth's atmosphere at sea level. It is a product of human and animal metabolism and is important to the life cycle of all types of vegetation. Gaseous carbon dioxide is about 1.5 times as heavy as air.

5.4.5.1 Chemical Properties

Carbon dioxide is nontoxic, soluble in water, colorless and normally odorless. In high concentrations, above 5,000 ppm, carbon dioxide can be detected as it dissolves in the moisture of the mucous membranes in the nasal passages. The sensation produced has been described as a pungent or irritating odor. Carbon dioxide is cooled and compressed to form a colorless liquid with approximately the same density as water. At temperatures above 87.8°F (31°C), carbon dioxide cannot be liquefied regardless of the pressure applied. Carbon dioxide is used to create an inert atmosphere for various metal working applications and for some molten metal subsurface stirring applications.

5.4.5.2 Safe Practices

Carbon dioxide can cause rapid asphyxiation if released in large quantities into a confined area lacking adequate ventilation. Carbon dioxide will replace normal air, giving no warning that a non-life-supporting atmosphere is developing. Any atmosphere that does not contain enough oxygen (19.5% by volume) can cause dizziness, unconsciousness and death. An oxygen-deficient atmosphere is created when carbon dioxide gas displaces air, causing the oxygen content to decrease. Because carbon dioxide is 1.5 times more dense than air, it accumulates in low (below grade) or confined areas. Breathable air is forced upward as the carbon dioxide concentration in a confined area increases. Continued increases in carbon dioxide concentration could force all breathable air out of an area.

Carbon dioxide is absorbed by the blood via the lungs and, to some extent, the skin. The symptoms associated with inhalation of carbon dioxide are not normally noticed until the concentration of carbon dioxide in the air is so high that there is insufficient oxygen in the atmosphere to support life. The sign and symptoms of carbon dioxide over-exposure are the same as those preceding asphyxia, namely, headache, dizziness, shortness of breath, muscular weakness, drowsiness and ringing in the ears. Read and understand the supplier MSDS for carbon dioxide.

Chapter 6

Steel Plant Fuels and Water Requirements

A. Lehrman, Development Engineer, LTV Steel Co.
C. D. Blumenschein, Senior Vice President, Chester Engineers
D. J. Doran, Manager of Market Development, Metals, Nalco Chemical Co.
S. E. Stewart, District Account Manager, Nalco Chemical Co.

6.1 Fuels, Combustion and Heat Flow

Any substance capable of producing heat by combustion may be termed a fuel. However, it is customary to rank as fuels only those which include carbon and hydrogen and their compounds. Wood was the earliest fuel used by man. Coal was known to exist in the fourth century B.C., and petroleum was used by the Persians in the days of Alexander. Prehistoric records of China and Japan are said to contain references to the use of natural gas for lighting and heating.

Heat generated by the combustion of fuel is utilized in industry directly as heat or is converted into mechanical or electrical energy. Fuel has become the major source of energy for manufacturing enterprises.

Fuel enters significantly into manufacturing costs, and in some industries represents one of the largest items of expense. The steel industry is one of the major consumers of metallurgical coal and also consumes large quantities of electricity, natural gas and petroleum.

Energy conservation efforts and technological improvements have combined to decrease domestic steel industry energy consumption from 34.40 gigajoules per net tonne (29.58 million Btu/ton) of shipments in 1980 to 24.44 gigajoules per net tonne (21.02 million Btu/ton) of shipments in 1995 per AISI survey data. The actual total steel industry average has dropped to an even lower value because most of the non-surveyed companies are electric arc furnace based, which inherently consume fewer gigajoules per net tonne (Btu/ton).

6.1.1 Classification of Fuels

There are four general classes of fuels; namely, fossil, byproduct, chemical and nuclear. Of these classes, the first three listed achieve energy release by combustion of carbon and/or hydrogen with an oxidant, usually oxygen; the process involves electron exchange to form products of a lower energy state and results in an energy release in an exothermic reaction. The fourth class liberates energy by fission of the nucleus of the atom and converting mass into energy.

Fossil fuels are hydrocarbon or polynuclear aromatic compounds composed principally of carbon and hydrogen and are derived from fossil remains of plant and animal life. These fossil remains have been transformed by biochemical and geological metamorphoses into such fuels as coal, natural gas, petroleum, etc.

Byproduct and waste fuels are derived from a main product and are of a secondary nature. Examples of these fuels are coke breeze, coke-oven gas, blast-furnace gas, wood wastes, etc.

Chemical fuels are primarily of an exotic nature and normally are not used in conventional processes. Examples of these fuels are ammonium nitrate and fluorine.

Nuclear fuels are obtained from fissionable materials. The three basic fissionable materials are uranium–235, uranium–233 and plutonium–239.

Fossil and byproduct fuels currently used in the steel industry are classified further into three general divisions; namely, solid, liquid and gaseous fuels. Fuels in each general division can be classified further as natural, manufactured or byproduct. Fuels found in nature sometimes are called primary fuels; those manufactured for a specific purpose or market, together with those that are the unavoidable byproduct of some regular manufacturing process, are called secondary fuels. The primary fuels serve as the principal raw materials for the secondary fuels. Table 6.1 gives a classified list of the important fossil fuels. It also lists some interesting byproduct fuels, many of which have been utilized by industry to conserve primary fuel.

6.1.1.1 Importance of Each Class

Coal is the major fuel of public utilities for the generation of power and is essential to the steel industry for the manufacture of coke.

Coal has been supplanted almost entirely by liquid fuels for the generation of motive power by railroads in North America. However, coal continues as a major raw material for many chemical plants as a source of carbon, hydrogen, and their compounds.

The growth of petroleum consumption has resumed after the price shocks of the past two decades due to the increasing demand for its distillation products. Gasoline, the most important product, is used as a motor fuel. Diesel engine fuel is a distillate of crude oil. Distillate and residual fuel oils, and some crude petroleums of too low commercial value for distillation are used for industrial and domestic heating. Crude and refined petroleum of various grades are used for lubrication of all types of machinery and prime movers. Petroleum and natural gas are raw materials for the petrochemical industry.

Natural gas has replaced coal to a considerable extent for domestic and industrial heating due to the installation of very large pipelines from producing to consuming centers, the relative level in the price of natural gas over the intervening time, and its convenience, cleanliness, controllability and versatility as a fuel. The byproduct gaseous fuels—coke-oven gas and blast furnace gas—are major integrated steel industry fuels.

The nuclear energy industry has fallen on hard times. The development of a practical method for fission of the atom and the release of nuclear energy in controlled chain reactions had given rise to a different type of power generation system. Many large reactors were built throughout the country and are still in operation. However, no new units are under construction or are being designed. Nuclear power will contribute an ever decreasing share of power to the electric grid as units are taken out of service unless some major breakthrough in design and operation occurs.

6.1.2 Principles of Combustion

Fossil and byproduct fuels consist essentially of one, or a mixture of two or more, or of four combustible constituents: (1) solid carbon, (2) hydrocarbons, (3) carbon monoxide, and (4) hydrogen. In addition to these combustible constituents, nearly all commercial fuels contain inert material, such as ash, nitrogen, carbon dioxide, and water. Bituminous coal is an example of a fuel which contains all four of the combustible constituents named above, and coke is an example of a fuel containing only one (solid carbon). The constituents which make up liquid fuels and many coals are quite complex, but because these complex constituents decompose or volatilize into the four

Table 6.1 Classification of Fuels [a]

	Primary Fuels	Secondary Fuels	
	Natural	**Manufactured**	**Byproduct**
Solid	Anthracite coal	Semi-coke (low-temperature carbonization residue)	Charcoal—low-temperature distillation of wood
	Bituminous coal		
	Lignite	Coke	Wood refuse—chips, shavings, trimmings, tan bark, sawdust, etc.
	Peat	Charcoal	Bagasse—refuse sugar cane
	Wood		Anthracite culm—silt refuse of anthracite screening
		Briquettes { Coal slack and culm, Lignite, Peat, Sawdust, Petroleum-refining residue }	Coke breeze { Byproduct coke—screenings, Petroleum coke—petroleum-refining residue }
		Pulverized coal	Waste materials from grain { Corn, Barley, Wheat, Buckwheat, Sorghum }
Liquid	Petroleum	Gasoline	Coal distillates { Tar, Napthalene, Pitch, Benzol } —coke manufacture
		Kerosene	
		Alcohol	
		Colloidal fuels	
		Fuel oil { Residual oils, Distillate oils, Crude petroleum }	Acid sludge—petroleum-defining residue
		Naphtha	Pulp-mill waste
		Vegetable oils { Palm, Cottonseed }	
Gaseous	Natural gas	Producer gas	Blast-furnace gas—pig-iron manufacture
		Water gas	Coke-oven gas[c]—coke manufacture
		Carburetted water gas	Oil-refinery gas
		Coal gas	Sewage gas—sewage sludge
		Oil gas	Basic oxygen furnace gas—steel manufacture
		Reformed natural gas	
		Butane[b]	
		Propane[b]	
		Acetylene	
		Hydrogen	

[a] Excluding chemical and nuclear fuels.
[b] Liquefiable heavier constituents of natural gas.
[c] Considered byproduct of coke manufacture in steel industry but a manufactured fuel in the gas industry.

simpler constituents named above before actual combustion takes place, a knowledge of the combustion characteristics of these constituents is sufficient for nearly all practical applications. All of these four constituents of fuels except carbon are gases at the temperatures where combustion occurs. Combustion takes place by combining oxygen, a gas present in air, with the combustible constituents of a fuel. The complete combustion of all fuels generates gases. It is apparent, therefore, that a review of the properties, thermal values and chemical reactions of gases is necessary for an understanding of any class of fuel.

Because fuels are used to develop heat, a knowledge of heat terms and the principles of heat flow are also essential for the efficient utilization of this heat. The combustion of fuels involves, besides combustion reactions, the factors and principles which influence speed of combustion, ignition temperature, flame luminosity, flame development, flame temperature and limits of flammability. The ensuing divisions of this section deal generally with these subjects. Sections 6.2, 6.3 and 6.4, respectively, deal specifically with the combustion of solid, liquid and gaseous fuels.

6.1.2.1 Units for Measuring Heat

Heat is a form of energy and is measured in absolute joules in SI units.

In the centimetre-gram-second (cgs) the unit for measuring heat was the calorie (abbreviated cal), defined as the amount of heat required to raise the temperature of one gram of pure, air-free water 1°C in the temperature interval of 3.5° to 4.5°C at normal atmospheric pressure: this unit was the gram-calorie or small calorie, identified in the Table 6.2 as $cal_{4°C}$. The temperature interval chosen for this definition was selected because the density and, therefore, the heat capacity of water varies slightly with temperature and the temperature of maximum density of water is very nearly 4°C. A larger heat unit in the cgs system was the kilocalorie (kilogram-calorie or large calorie), equal to 1000 gram-calories and abbreviated kcal.

Other values for the calorie were obtained by selecting other temperature intervals, resulting, for example, in the $cal_{15°C}$ and the $cal_{20°C}$ listed in Table 6.2. Yet another variation was the mean calorie (abbreviated cal_{mean}), defined as 1/100 of the amount of heat required to raise the temperature of one gram of water from 0°C (the ice point) to 100°C (the boiling point).

None of the foregoing definitions of the calorie were completely satisfactory because of the variation of the heat capacity of water with temperature. Consequently, on the recommendation of the Ninth International Conference of Weights and Measures (Paris, 1948), the calorie came to be defined in energy units in ways that made its value independent of temperature. The thermochemical calorie (abbreviated $cal_{thermochem}$) was defined first in international electrical-energy units and later (1948) in terms of mechanical-energy units. The calorie used in the present International Tables, identified as cal_{IT} was adopted in 1956 at the International Conference on Properties of Steam in Paris, and is expressed in mechanical-energy units.

As stated above, the SI unit used to define the calorie in terms of mechanical-energy units is the absolute joule: the word "absolute" differentiates the SI joule based on mechanical-energy units from the international joule formerly used which was based on international electrical units.

The presently accepted values in absolute joules of the various calories discussed above are presented in Table 6.2 here.

Table 6.2 cgs/SI Equivalent Values for Measuring Heat

1 $cal_{4°C}$	=	4.2045 joules
1 $cal_{15°C}$	=	4.18190 joules
1 cal_{mean}	=	4.19002 joules
1 cal_{IT}	=	4.1868 joules (exactly)
1 $cal_{thermochem}$	=	4.184 joules (exactly)

Steel Plant Fuels and Water Requirements

In the foot-pound-second (fps) system, the principal unit adopted for measuring heat was the British thermal unit (Btu). Defined as the amount of heat required to raise by 1°F the temperature of one pound of pure, air-free water, its value depended upon the temperature interval chosen for its complete definition. As in the case of the calorie, several values came into use, notably the $Btu_{39°F}$ based on a 1°F rise in temperature at or near the temperature of maximum density of water (39.2°F), the $Btu_{59°F}$ based on the temperature interval of 58.5°F to 59.5°F corresponding nearly to the 14.5°C to 15.5°C interval of the $cal_{15°C}$, the $Btu_{60°F}$ based on the temperature interval from 60°F to 61°F, and the mean Btu (Btu_{mean}) that represented 1/180 of the heat required to raise the temperature of a pound of water from 32°F (the freezing point) to 212°F (the boiling point). Other values for the Btu were adopted, based on definitions that made the unit independent of the properties of water: these included the Btu of the International Tables referred to above (abbreviated Btu_{IT}, and the thermochemical Btu (designated $Btu_{thermochem}$). Following the recommendations of the 1948 International Conference on Weights and Measures, all of the foregoing values for the Btu came eventually to be expressed in SI mechanical-energy units, and now have been assigned the values presented in Table 6.3 in absolute joules.

Table 6.3 fps/SI Equivalent Values for Measuring Heat

$Btu_{39°F}$	=	1059.67 joules
$Btu_{59°F}$	=	1054.80 joules
$Btu_{60°F}$	=	1054.68 joules
Btu_{mean}	=	1055.87 joules
Btu_{IT}	=	1055.056 joules
$Btu_{thermochem}$	=	1054.350 joules

When both are determined for the same temperature interval, 1 Btu equals very nearly 252 calories, and 1 kilocalorie (1000 cal) very nearly equals 3.9683 Btu.

6.1.2.2 Calorific Value of Fuel

The heat given up or absorbed by a body between two temperatures, provided no change of state or of allotropic form is involved, is known as sensible heat. The heat given up or absorbed by a body when a change of state or of allotropic form takes place and no temperature change is involved is known as latent heat. For example, 1 kilogram of water absorbs 418.68 kilojoules of sensible heat when being heated from 0°C to 100°C, and absorbs 2257.1 kilojoules of latent heat when converted to steam at 100°C. Likewise, 1 pound of water absorbs 189.9 kilojoules (180 Btu) of sensible heat on being heated from 32°F to 212°F, and absorbs 1023.8 kilojoules (970.4 Btu) of latent heat when converted to steam at 212°F.

Sensible heat and latent heat are used frequently in combustion calculations, particularly in problems dealing with heat losses in flue gases. Their significance is indicated in describing gross and net heating values.

The gross heating value of a fuel is the total heat developed by the combustion of a fuel at constant pressure after the products of combustion are cooled back to the starting temperature, assuming that all of the water vapor produced is condensed; that is, the gross heating value includes both sensible and latent heat. The net heating value of a fuel is defined as the heat developed by the combustion of a fuel at constant pressure after the products of combustion are cooled back to the starting temperature, assuming that all of the water vapor remains uncondensed. Accordingly, the net heating value includes only the sensible heat.

Where combustion calculations in this chapter are in SI the starting point is 273.15K (0°C) at 101.325 kPa (760 mm Hg) absolute pressure. Where the calculations are in cgs units, the starting point is 0°C at 760 mm absolute pressure. The starting point for calculations in the fps system in this chapter has been taken as 60°F at 30 in. of Hg absolute pressure; this has generally been the base for combustion calculations in the American steel industry.

When a fuel contains neither hydrogen nor hydrocarbons, no water vapor is produced by combustion and the gross and net heating value will be the same, as in the case of burning carbon or carbon monoxide. The heating value or calorific value of a fuel may be determined on a dry or wet basis. The determination may be made by laboratory tests employing calorimeters, or by calculation. The process of determining the calorific value of solid and liquid fuels by a calorimeter consists in completely oxidizing the fuel in a space enclosed by a metal jacket (called the bomb) so immersed that the heat evolved is absorbed by a weighed portion of water contained in an insulated vessel. From the rise in temperature of the water, the heat liberated by one gram of the fuel is calculated. The best types of calorimeters for solid and liquid fuels are those called oxygen-bomb calorimeters in which the fuel is burned in the presence of compressed oxygen. Gas calorimeters are of different construction to permit volumetric measurement of the gas and its complete combustion under non-explosive conditions, as well as absorption of the heat produced in a water jacket.

A saturated gas is one which contains the maximum amount of water vapor it can hold without any condensation of water taking place. The usual basis for reporting the calorific value of a saturated fuel gas in SI units is in gross kilojoules per cubic metre measured at 273K and 101 kPa absolute pressure. In the cgs system, calorific value usually has been reported in gross kilocalories per cubic metre measured at 0°C and 760 mm Hg absolute pressure. In fps units in the American steel industry, calorific value usually has been reported in gross Btu per cubic foot of saturated gas measured at 60°F and 30 in. Hg absolute pressure.

The heating value of a given fuel can be obtained by multiplying the calorific value of each gas by its percentage of the total fuel volume, and then totaling the individual values of the separate constituents. The heat of combustion for various dry elementary gases may be found in Table 6.4. For instance, the gross heating value of dry blast-furnace gas is 3633 kilojoules per cubic metre (92.5 Btu per cubic foot) for the composition used in the following calculations.

In the calculation of the heating value of gases saturated with water vapor, the volume of water vapor must be deducted from the unit volume of the gas. For instance, a cubic metre of dry carbon monoxide gas has a heating value of 12,623 kilojoules, but when saturated with water vapor at 273K (0°C) and 101 kPa (760 mm Hg), a cubic metre has a heating value of only 12,405 kilojoules (see Table 6.4). Likewise, a cubic foot of dry carbon monoxide gas has a heating value of 321.4 Btu, but when saturated with water vapor at 60°F and 30 in. Hg absolute pressure, a cubic foot has a heating value of only 315.8 Btu. The amount of water vapor present in saturated mixtures can be calculated from data in Table 6.5, as discussed in Section 6.1.2.4.

6.1.2.3 Thermal Capacity, Heat Capacity and Specific Heat

The thermal capacity or heat capacity of a substance is expressed as the amount of heat required to raise the temperature of a unit weight of the substance one degree in temperature. In SI, it is expressed in joules per kilogram Kelvin (J/kg K). The fps system has used Btu per pound per degree Fahrenheit (Btu/lbm °F), while the cgs system has used calories per gram per degree Celsius (cal/g °C). The specific heat is always a ratio, expressed as a number; for example, the specific heat of wrought iron is 0.115. There is no further designation, as this means that if it takes a certain number of joules to heat a certain number of kilograms of water a certain number of Kelvins, it will take only 0.115 times as many joules to heat the same number of kilograms of wrought iron the same number of Kelvins, and the same figure, 0.115, obviously applies if the centimetre-gram-second or foot-pound-second system were used.

The amount of heat required to raise the temperature of equal masses of different substances to the same temperature level varies greatly; that is to say, the specific heat varies greatly; also the specific heat of the same substance varies at different temperatures. Usually, it is necessary to know

Table 6.4 Essential Gas Combustion Constants [a]

Gas	Formula	Molecular Weight	Specific Gravity (Air = 1)	Heat of Combustion [b]								Unit Volumes per Unit Volume of Dry Combustible (m³ or ft³)						
				Btu per ft³ [c]		Btu per lb [c]		kJ per m³ [d]		kJ per kg [d]		Required for Combustion				Flue Products		
				Gross	Net	Gross	Net	Gross	Net	Gross	Net	O_2	N_2 [e]	Air	CO_2	H_2O	N_2
Carbon (Graphite)	C	12.01	—	—	—	14,093	14,093	—	—	32,780	32,780	—	—	—	—	—	—
Hydrogen	H_2	2.016	0.06959	325.02	274.58	60,991	51,605	12,767	10,786	141,865	120,033	0.5	1.882	2.382	—	1.0	1.882
Oxygen	O_2	32.00	1.1053	—	—	—	—	—	—	—	—	—	—	—	—	—	—
Nitrogen	N_2	28.016	0.9718	—	—	—	—	—	—	—	—	—	—	—	—	—	—
Carbon monoxide	CO	28.01	0.9672	321.37	321.37	4347	4347	12,623	12,623	10,111	10,111	0.5	1.882	2.382	1.0	—	1.882
Carbon dioxide	CO_2	44.01	1.5282	—	—	—	—	—	—	—	—	—	—	—	—	—	—
Methane	CH_4	16.042	0.5543	1012.32	911.45	23,875	21,495	39,764	35,802	55,533	49,997	2.0	7.528	9.528	1.0	2.0	7.528
Ethane	C_2H_6	30.068	1.0488	1773.42	1622.10	22,323	20,418	69,660	63,716	51,923	47,492	3.5	13.175	16.175	2.0	3.0	13.175
Ethylene	C_2H_4	28.052	0.9740	1603.75	1502.87	21,636	20,275	62,995	59,033	50,325	47,160	3.0	11.293	14.293	2.0	2.0	11.293
Propylene	C_3H_6	42.078	1.4504	2339.70	2188.40	21,048	19,687	91,903	85,960	48,958	45,792	4.5	16.939	21.439	3.0	3.0	16.939
Acetylene	C_2H_2	26.036	0.9107	1476.55	1426.17	21,502	20,769	57,999	56,020	50,014	48,309	2.4	9.411	11.911	2.0	1.0	9.411
Benzene	C_6H_6	78.108	2.6920	3751.68	3600.52	18,184	17,451	147,366	141,428	42,296	40,591	7.5	28.232	35.732	6.0	3.0	28.232
Hydrogen sulfide	H_2S	34.076	1.1898	646	595	7097	6537	25,375	23,372	16,508	15,205	1.5	5.646	7.146	SO_2 = 1.0	1.0	5.646
Sulfur dioxide	SO_2	64.06	2.264	—	—	—	—	—	—	—	—	—	—	—	—	—	—

[a] Adapted from "Gas Engineers Handbook" (See Segeler listing in bibliography at end of chapter).
[b] Based on perfect combustion.
[c] Based on dry gases at 60°F and 30 in. Hg. For gases saturated with water, 1.74% of the heating value must be deducted.
[d] Based on dry gases at 273K (0°C) and approx. 101 kPa (760 mm Hg). For gases saturated with water at 273K (0°C), 0.60% of the heating value must be deducted. (To convert kJ to kcal, multiply by 0.239.)
[e] N_2 in air accompanies O_2 (N_2 not required for combustion).

Note: Conversion Factors: Btu/ft³ to kJ/m³ = 39.28; Btu/lb to kJ/kg = 2.326.

Table 6.5 Water Vapor Pressure [a]

Temp. (°C)	Pressure mm Hg	Pressure Pa[b]	Temp. (°F)	Pressure (in. Hg)	Pressure Pa[c]	Temp. (°C)	Pressure mm Hg	Pressure Pa[b]	Temp. (°F)	Pressure (in. Hg)	Pressure Pa[c]
0	4.579	610.5	32	0.1803	608.8	50	92.51	12,334	125	3.956	13,360
2	5.294	705.8	35	0.2035	687.2	52	102.09	13,611			
4	6.101	813.4	40	0.2478	836.8	54	112.51	15,000	130	4.525	15,280
6	7.013	935.0				56	123.80	16,505			
8	8.045	1072.6	45	0.3004	1014	58	136.08	18,143	135	5.165	17,440
10	9.209	1227.8				60	149.38	19,916	140	5.881	19,860
12	10.518	1402.3	50	0.3626	1224	62	163.77	21,834	145	6.682	22,560
14	11.987	1598.1	55	0.4359	1472	64	179.31	23,906	150	7.572	25,570
16	13.634	1817.7	60	0.5218	1762	66	196.09	26,143			
18	15.477	2063.4	65	0.6222	2101	68	214.17	28,554	155	8.556	28,890
20	17.535	2337.8	70	0.7393	2497	70	233.7	31,160	160	9.649	32,580
22	19.827	2643.4				72	254.6	33,940			
24	22.377	2983.4	75	0.8750	2955	74	277.2	39,960	165	10.86	36,670
26	25.209	3360.9	80	1.032	3485	76	301.4	40,180	170	12.20	41,200
28	28.349	3779.6				78	327.3	43,640			
30	31.824	4242.9	85	1.213	4096	80	355.1	47,340	175	13.67	46,160
32	35.663	4754.7	90	1.422	4802	82	384.9	51,320	180	15.29	51,630
34	39.898	5319.3	95	1.660	5606	84	416.8	55,570	185	17.07	57,640
36	44.563	5941.2				86	450.9	60,120			
38	49.692	6625.1	100	1.933	6527	88	487.1	64,940	190	19.02	64,230
40	55.324	7375.9	105	2.243	7574	90	525.76	70,100	195	21.15	71,420
42	61.50	8199	110	2.596	8766	92	566.99	75,590	200	23.46	79,220
44	68.26	9101				94	610.90	81,450			
46	75.65	10,086	115	2.995	10,110	96	657.62	87,680	205	25.99	87,760
48	83.71	11,160	120	3.446	11,640	98	707.27	94,290	210	28.75	97,080
						100	760.00	101,325	212	29.92	101,035

[a] Values for °C and mm Hg from "International Critical Tables" (E. W. Washburn, ed. in chief), published for National Research Council by McGraw-Hill Book Company, New York, 1928. Values for °F and in. Hg from "Gas Engineers Handbook (see Segeler listing in bibliography at end of chapter).
[b] Calculated, using factor 133.3224 to convert mm Hg (0°C) to Pa.
[c] Calculated, using factor 3376.85 to convert in. Hg (60°F) to Pa. Factor for converting in. Hg (32°F) to Pa is 3386.38.

the amount of heat required to raise the temperature of a substance some appreciable amount. For that purpose, formulae and tables are usually accessible in handbooks for supplying the mean specific heat between various temperature levels. Two values of specific heat for gases are usually given: (1) specific heat at constant pressure, and (2) specific heat at constant volume. The difference is due to the heat equivalent of the work of expansion caused by an increase of volume resulting from a temperature rise. Normal combustion practice with gases in steel plants deals with a constant pressure condition (or nearly so), and for this reason specific heat at constant pressure is used. The mean specific heat is the average value of the specific heat between two temperature levels. It is obtained by integrating the equations for instantaneous specific heat over the temperature limits desired, and dividing this quantity by the difference between the temperature limits.

The heat content is the heat contained at a specified temperature above some fixed temperature. It is calculated by multiplying the weight of a substance by the mean specific heat times the temperature difference, or H_t = weight × mean specific heat × $(T_2 - T_1)$. For convenience in calculations with gases, the unit weight of the volume of a cubic metre or a cubic foot of gas is often used.

6.1.2.4 Gas Laws

Calculations based on the gas laws to be discussed involve the concepts of absolute zero and absolute temperature. Absolute zero in SI is 0K, in the cgs system it is –273.15°C and in the fps system it is –459.67°F: for practical purposes to facilitate calculations, –273°C and –460°F can be taken as absolute zero in the cgs and fps systems, respectively.

Absolute temperature in SI is the temperature expressed in Kelvins above 0K at an absolute pressure of 101.325 kPa (one standard atmosphere). In the cgs system, absolute temperature (°C_{abs}) has been the temperature in degrees Celsius (formerly called degrees centigrade) above –273.15°C at 760 mm Hg (millimetres of mercury) at one standard atmosphere absolute pressure. Absolute temperature in the cgs system also has been expressed according to the Kelvin scale of temperature, using K instead of °C, (1°C = 1K). Absolute zero on the Kelvin scale is 0K, and on the Celsius scale absolute zero, as stated above, is –273.15°C. Temperatures on the Kelvin and Celsius scales have the following relation:

$$T_K = T_{°C} + 273.15$$

or, to facilitate calculations,

$$T_K = T_{°C} + 273 \tag{6.1.1}$$

Absolute temperature in the fps system is the temperature in degrees Fahrenheit (°F_{abs}) above 459.67°F at 29.921 inches of mercury (one standard atmosphere) absolute pressure. Also in the fps system, absolute temperatures have been expressed according to the Rankine temperature scale, using degrees Rankine (°R) instead of degrees Fahrenheit (°F), (1°R = 1°F). Absolute zero on the Rankine scale is 0°R, and on the Fahrenheit scale absolute zero, as stated above, is –459.67°F. Temperatures on these two scales have the following relation:

$$T_{°R} = T_{°F} + 459.67$$

or, to facilitate calculations,

$$T_{°R} = T_{°F} + 460 \tag{6.1.2}$$

Again to facilitate calculations, rounded values for an absolute pressure of one atmosphere (101 kPA in SI and 30 in. Hg in the fps system) can be used instead of the more precise values in the definitions.

The volume of an ideal gas varies in direct proportion to its absolute temperature (Charles' Law) and inversely as its absolute pressure (Boyle's Law).

For example, in SI units, the volume of 1000 m³ of a gas measured at 288K (15°C) and 101 kPa absolute pressure, when heated to 1253K (980°C) and 101 kPa absolute pressure, is equal to:

$$1000 \times 1253/288 = 4351 \text{ m}^3$$

and the volume of 1000 m³ of fuel gas measured at 288K(15°C) and 101 kPa absolute pressure is equal to 802 m³ when compressed to 25 kPa gauge pressure at 288K, calculated as follows:

$$1000 \times 101/(101+25) = 802 \text{ m}^3$$

Similarly, in the fps system, the volume of 40,000 ft³ of gas measured at 60°F and 30 in. Hg absolute pressure, when heated to 1800°F and 30 in. Hg absolute pressure, is equal to:

$$40,000 \times (460 + 1800)/(460+520) = 174,000 \text{ ft}^3$$

and the volume of 40,000 ft³ of fuel gas measured at 60°F and 30 in. Hg (standard conditions) is equal to 31,579 ft³ when compressed to 8 in. Hg gauge pressure at 60°F, calculated as follows:

$$40{,}000 \times 30/(30+8) = 31{,}579 \text{ ft}^3$$

The total pressure of any gas mixture is equal to the sum of the pressures of each component. Each component produces a partial pressure proportional to its concentration in the mixture. Therefore, in a mixture of water vapor and any other gas, each exerts a pressure proportional to its percentage by volume, and since water has a definite vapor pressure at various temperatures, as shown in Table 6.5, the concentration of water vapor in a gas is limited. When this limit of water vapor is reached, the gas is said to be saturated. Any drop in temperature or increase in pressure from that point will cause condensation of some of the water vapor; for instance, the water vapor in 1000 m³ of saturated fuel gas at 293K (20°C) and 101 kPa would equal:

$$1000 \times 2.3378/100 = 23.1 \text{ m}^3$$

(2.3378 kPa is the partial pressure of water vapor in a saturated mixture at 293K (20°C) and 101 kPa as shown in Table 6.5.) In fps units, water vapor in 1000 ft³ of saturated fuel gas measured at 60°F and 30 in. Hg is calculated as follows:

$$1000 \times 0.522/30 = 17.40 \text{ ft}^3$$

(0.522 is the partial pressure of water vapor in a saturated mixture at 60°F and 30 in. Hg, from Table 6.5.) The amount of water vapor which will condense at various temperatures may be ascertained by the use of Table 6.6.

In some combustion calculations, it is necessary to convert volumes to weights and vice versa. Such conversions can be made very conveniently by using molar units; namely, the mole (abbreviated mol and expressed in kilograms) in SI; the gram-mole (abbreviated g-mol and expressed in grams) in the cgs system; and the pound-mole (abbreviated lb-mol and expressed in pounds) in the fps system. A mol, g-mol or lb-mol of a substance is that quantity whose mass expressed in the proper units stated above is the same number as the number of the molecular weight. Thus, the molecular weight of oxygen is 32, so that the mol in SI is 32 kg of oxygen, the g-mol in the cgs system is 32 grams of oxygen, and the lb-mol in the fps system is 32 lb of oxygen.

In SI, a mol of any gas (its molecular weight in kilograms) theoretically occupies 22.414 m³ at 273.15K and 101.325 kPa absolute pressure. (Values of 22.4 m³, 273K and 101 kPa are close enough for most calculations.) In the cgs system, a g-mol of any gas (its molecular weight in grams) theoretically occupies 22.414 dm³ at 0°C and 760 mm Hg absolute pressure. In the fps system, a lb-mol of any gas (its molecular weight in pounds) theoretically occupies 359 ft³ at 32°F and 29.921 in. Hg absolute pressure; or, at 60°F and 30 in. Hg absolute pressure (the usual reference points for combustion problems in the steel industry) a lb-mol occupies 378.4 ft³. The simplicity of using molar units in combustion calculations is shown by the following examples: The weight of a cubic metre of dry air is calculated in SI as follows:

$$0.21 \text{ (\% vol. of O}_2 \text{ in air)} \times 32 \text{ (mol. wt. of O}_2) = 6.72$$
$$0.79 \text{ (\% vol. of N}_2 \text{ in air)} \times 28 \text{ (mol. wt. of N}_2) = 22.12$$
$$\text{Weight in kg of a mol of dry air} = 28.84.$$

$28.84/22.4 = 1.29$ kg (weight per m³ of dry air at 273K and 101 kPa absolute pressure)

The volume of 1 kg of dry air at 273K and 101 kPa absolute pressure is equal to:

$$22.4/28.84 = 0.78 \text{ m}^3$$

In the cgs system, the calculations would be similar to those in SI, except that the weight of a gram-mol of dry air would be determined to be 28.84 grams, and the weight of a cubic decimetre (liter) of dry air would be found to be 1.29 grams: the volume of 1 gram of dry air at 0°C and 760 mm Hg would be found to be 0.78 dm³.

Table 6.6 Properties of Dry Air [a]

	SI Units[b]				fps Units[d]		
Temp. (°C)	Volume of 1 kg (m³)	Mass of 1 m³ (kg)	Mass of Water Vapor to Saturate 100 kg of Dry Air at 100% Humidity[c] (kg)	Temp. (°F)	Volume of 1 lbm (ft³)	Mass of 1 ft³ (lbm)	Mass of Water Vapor to Saturate 100 lbm of Dry Air at 100% Humidity[c] (lbm)
0	0.7735	1.2928	0.3774	32	12.360	0.080906	0.3767
5	0.7874	1.2700	0.5403	40	12.561	0.079612	0.5155
10	0.8019	1.2471	0.7638	50	12.812	0.078050	0.7613
15	0.8160	1.2255	1.0649	60	13.063	0.076550	1.1022
20	0.8302	1.2046	1.4702	70	13.315	0.075103	1.5726
25	0.8443	1.1844	2.0082	80	13.567	0.073710	2.2184
30	0.8584	1.1649	2.7194	90	13.818	0.072370	3.0998
35	0.8726	1.1460	3.6586	100	14.069	0.071077	4.2979
40	0.8868	1.1277	4.8872	110	14.321	0.069829	5.0913
45	0.9010	1.1099	6.5279	120	14.571	0.068627	8.0981
50	0.9151	1.0928	8.6686	130	14.823	0.067463	11.0935
55	0.9293	1.0761	11.5070	140	15.074	0.066338	15.2441
60	0.9434	1.0600	15.3233	150	15.327	0.065244	21.1155
65	0.9576	1.0443	20.5315	175	15.954	0.062679	52.6416
70	0.9717	1.0291	27.8395	200	16.584	0.060298	226.384
75	0.9859	1.0143	38.5230				
80	1.0001	0.9999	55.1492				
85	1.0142	0.9860	83.6207				
90	1.0284	0.9724	141.5052				
95	1.0425	0.9592	312.9940				

[a]At an absolute pressure of 101.325 kPa (760 mm Hg) in SI units, and 30 in. of Hg in fps units.
[b]Calculated on basis of density of dry air at 273.15K (0°C) and 101.325 kPa (760 mm Hg) equal to 1.2928 kg per m³.
[c]Mass of water vapor at lower humidities is approximately proportional to the humidity; e.g., at 50% humidity, the mass will be one-half that at 100% humidity for a given temperature.
[d]From "Gas Engineers Handbook" (see Segeler listing in bibliography at end of chapter).

In the fps system, the weight of a lb-mol of dry air would be determined to be 28.84 lbm, (lbm denotes pounds-mass and replaces the previous designation lb) and the weight of 1 ft³ of dry air at 60°F and 30 in Hg would be:

$$28.84/378.4 = 0.076 \text{ lbm (weight per ft}^3 \text{ of dry air at 60°F and 30 in. Hg absolute pressure)}$$

Also, the volume of 1 lbm of dry air at 60°F and 30 in. Hg would be

$$378.4/28.84 = 13.1 \text{ ft}^3$$

The relation of an ideal gas to its volume and pressure is expressed by the formula:

$$PV = nRT \quad (6.1.3)$$

where:

R = gas constant
P = absolute pressure
V = volume
n = number of mols
T = absolute temperature of gas

The numerical value of R in the above equation depends upon what units (SI, cgs or fps) are used to measure P, V, n and T. Values of R for various combinations of units for measuring the other quantities are as follows:

SI units   R = 8.3144 kJ/mol K
CGS units   R = 8.3144 J/g-mol K
                 R = 62.37 mm Hg-dm^3/g-mol K
                 R = .08206 dm^3-atm/g-mol K
FPS units   R = 10.703 lbf ft^3 in^2/lb-mol °R
                 R = 21.83 in. Hg-ft^3/lb-mol °R

In SI units, an example of the use of the foregoing formula would be to calculate the volume occupied by 100 kg of natural gas having a composition of 80% CH_4, 18% C_2H_6 and 2% N_2 by volume at a gauge pressure of 27 kPa and a temperature of 38°C (using the data from Table 6.4).

The weight of a mol of the gas is:

CH_4 = 0.80 × 16 = 12.8
C_2H_6 = 0.18 × 30 = 5.4
N_2 = 0.02 × 28 = 0.56 kg
                       18.76 kg

P   = 27 + 101   = 128 kPa absolute
n   = 100/18.76 = 5.33
R   = 8.3144
T   = 273 + 38   = 311

Substituting these values into the equation 6.1.3 for a perfect gas:

$$128V = 5.33 \times 8.3144 \times 311$$
$$V = 107.7 \text{ m}^3$$

In fps units, a similar application of the formula would be to calculate the volume occupied by 100 lbm of natural gas of the same composition as above at 8 in. Hg gauge pressure and a temperature of 100°F.

P = 30 + 8   = 38 in. Hg absolute
n = 100/18.75 = 5.33 lb-mols
R = 21.83
T = 460 + 100 = 560°R

$$38V = 5.33 \times 21.83 \times 560$$
$$V = 1715 \text{ ft}^3$$

6.1.2.5 Combustion Calculations

The combustion of fuels is carried out by chemical reaction with air, and occasionally with air enriched with oxygen, or with pure oxygen. Dry air is a mixture of the following gas volumes under average conditions:

N_2 = 78.03%
O_2 = 20.99%
Ar   = 0.94%
CO_2 = 0.03%
H_2 = 0.01%
Total = 100.00%

In combustion calculations it is customary to include all elements in dry air (other than oxygen) with the nitrogen, as shown below:

	% by Volume	% by Weight
Oxygen	20.99	23.11
Nitrogen	79.01	76.89

Only the oxygen in the air reacts with a fuel in combustion processes. The nitrogen acts as a diluent which must be heated up by the heat of the reaction between the oxygen and the fuel. It, therefore, reduces, the temperature of the flame and reduces the velocity of combustion.

Water vapor which is present in air also acts as a diluent. The amount of moisture present in air is generally stated in terms of humidity. Air is capable of being saturated with water vapor the same as other gases as described in Section 6.1.2.4. Air which is saturated completely with water vapor has a humidity of 100%; if only 50% saturated, it has a humidity of 50% (Table 6.6).

The principal combustion reactions are:

$$C + O_2 = CO_2$$
$$2CO + O_2 = 2CO_2$$
$$2H_2 + O_2 = 2H_2O$$
$$CH_4 + 2O_2 = CO_2 + 2H_2O$$
$$2C_2H_6 + 7O_2 = 4CO_2 + 6H_2O$$
$$C_2H_4 + 3O_2 = 2CO_2 + 2H_2O$$
$$2C_3H_6 + 9O_2 = 6CO_2 + 6H_2O$$
$$2C_2H_2 + 5O_2 = 4CO_2 + 2H_2O$$
$$2C_6H_6 + 15O_2 = 12CO_2 + 6H_2O$$
$$2H_2S + 3O_2 = 2SO_2 + 2H_2O$$

The amount of oxygen required and consequently air, together with the amount of the resultant products of combustion, may be calculated in SI by the use of mols and the proper chemical reaction. For instance, it will require $(32 \div 12)$ or 2.667 kg of O_2 to burn 1 kg of C, and as dry air contains 23.11% by weight O_2, the weight of dry air required to burn one kilogram of carbon will be (2.667×0.2311) or 11.540 kg. The product of combustion, CO_2, will amount to $[(12 + 32) \div 12]$ = 3.667 kg.

Combustion calculations using gases are more conveniently made in volumetric units. For instance, to burn a cubic metre of CO completely to CO_2 requires $\frac{1}{2}m^3$ of O_2 in accordance with the molecular relationship in the reaction. The dry air required would be $(0.5 \div 0.209)$ or $2.382\ m^3$. For burning a cubic metre of methane, CH_4, to CO_2 and H_2O, the air required would be $(2.0 \div 0.209)$ or $9.528\ m^3$.

The foregoing calculations may be performed with fps units by substituting pounds-mass (lbm) for kilograms and cubic feet for cubic metres.

Combustion calculations are necessary to determine the air requirements and the products of combustion for burning fuels of various compositions. The percent of air used above theoretical requirements is called percent excess air; the percent below, the percent deficiency of air. Typical combustion data on a dry basis for burning gaseous fuels of the compositions stated are shown in Table 6.7. In making calculations to include the water vapor which may be present in a saturated or partially saturated gas and in air, the same general method may be used by adding water vapor to the fuel gas composition, and by adding the volume of water vapor which is introduced through air in the products of combustion column, headed H_2O.

In order to maintain combustion, a fuel must, after it has been ignited, be able to impart sufficient heat to its air-gas mixture so that it will not drop below ignition temperature, the minimum point of self-ignition. Too lean or too rich a mixture of a fuel with air is unable to support combustion. An upper and lower limit of flammability exists for all gases. The limits of flammability, as well as ignition temperatures, for a number of gases are shown in Table 6.8.

In the design of burners or in the selection of fuel for a specific purpose, consideration of velocity of combustion is of major importance. Because gaseous fuels are composed usually of a mixture of combustible gases, a knowledge of the relative combustion speed of each elementary gas will provide means for evaluating this factor in any gaseous mixture. The velocity of combustion, or rate of flame propagation, of a given fuel, is influenced by three factors: (1) degree to which the air and gas are mixed, (2) temperature of the air-gas mixture, and (3) contact of the air- gas mixture with a hot surface (catalyst). By intimately mixing air and gas, combustion may be accelerated

Table 6.7 Combustion Data[a] for Blast Furnace, Coke Oven and Natural Gas

Blast Furnace Gas (all volumes at 0°C and 101.325 kPa).

Gas Comp.	% by Volume	m³ of Air Required For Combustion Per m³	m³ of Air Required For Combustion Each Component	m³ Products of Combustion per m³ of Fuel No Excess Air CO₂	H₂O	SO₂	O₂	N₂	10% Excess Air CO₂	H₂O	SO₂	O₂	N₂	50% Excess Air CO₂	H₂O	SO₂	O₂	N₂
CO₂	11.5	. . .	. . .	.115	. . .	. . .	. . .	. . .	.115	. . .	. . .	. . .	.054[b]	.115	. . .	. . .	.014[b]	.269[b]
N₂	60.0	. . .	. . .	. . .	. . .	. . .	. . .	.600	. . .	. . .	. . .	. . .	.600	. . .	. . .	. . .	. . .	.600
CO	27.5	2.382	.655	.275	. . .	. . .	. . .	.517	.275	. . .	. . .	. . .	.517	.275	. . .	. . .	. . .	.517
H₂	1.0	2.382	.0238	. . .	.010	. . .	. . .	.019	. . .	.010	. . .	. . .	.019	. . .	.010	. . .	. . .	.019
Total	100.0	. . .	.679	.390	.010	. . .	. . .	1.136	.390	.010	. . .	.014	1.190	.390	.101	. . .	.071	1.405

Coke Oven Gas (all volumes at 0°C and 101.325 kPa).

Gas Comp.	% by Volume	Per m³	Each Component	CO₂	H₂O	SO₂	O₂	N₂	CO₂	H₂O	SO₂	O₂	N₂	CO₂	H₂O	SO₂	O₂	N₂
CO₂	1.4	. . .	. . .	.014	. . .	. . .	. . .	. . .	.014	. . .	. . .	. . .	. . .	.014	. . .	. . .	. . .	. . .
H₂S	0.6	7.146	.0429	. . .	.006	.006	. . .	.034	. . .	.006	.006	. . .	.034	. . .	.006	.006	. . .	.034
O₂	0.4	. . .	−.0190	. . .	. . .	. . .	. . .	−.015	. . .	. . .	. . .	. . .	−.015	. . .	. . .	. . .	. . .	−.015
N₂	4.3	. . .	. . .	. . .	. . .	. . .	. . .	.043	. . .	. . .	. . .	. . .	.043	. . .	. . .	. . .	. . .	.043
CO	5.6	2.382	.1334	.056	. . .	. . .	. . .	.105	.056	. . .	. . .	. . .	.105	.056	. . .	. . .	. . .	.105
H₂	55.4	2.382	1.3196	. . .	.554	. . .	. . .	1.042	. . .	.554	. . .	. . .	1.042	. . .	.554	. . .	. . .	1.042
CH₄	28.4	9.528	2.7060	.284	.568	. . .	. . .	2.138	.284	.568	. . .	. . .	2.138	.284	.568	. . .	. . .	2.138
C₂H₄	2.5	14.293	.3573	.050	.050	. . .	. . .	.282	.050	.050	. . .	. . .	.282	.050	.050	. . .	. . .	.282
C₂H₆	0.8	16.675	.1334	.016	.024	. . .	. . .	.105	.016	.024	. . .	. . .	.105	.016	.024	. . .	. . .	.105
Illuminants	0.6	26.208	.1572	.018	.018	. . .	. . .	.124	.018	.018	. . .	. . .	.124	.018	.018	. . .	. . .	.124
Total	100.0	. . .	4.831	.438	1.220	.006	. . .	3.858	.438	1.220	.006	.101	4.240	.438	1.220	.006	.507	5.767

Natural Gas (all volumes at 0°C and 101.325 kPa).

Gas Comp.	% by Volume	Per m³	Each Component	CO₂	H₂O	SO₂	O₂	N₂	CO₂	H₂O	SO₂	O₂	N₂	CO₂	H₂O	SO₂	O₂	N₂
CO₂	0.08	. . .	. . .	.001	. . .	. . .	. . .	. . .	.001	. . .	. . .	. . .	. . .	.001	. . .	. . .	. . .	. . .
O₂	0.17	. . .	−.002	. . .	. . .	. . .	. . .	−.002	. . .	. . .	. . .	. . .	−.002	. . .	. . .	. . .	. . .	−.002
N₂	1.02	. . .	. . .	. . .	. . .	. . .	. . .	.010	. . .	. . .	. . .	. . .	.010	. . .	. . .	. . .	. . .	.010
CH₄	81.88	9.528	7.802	.819	1.638	. . .	. . .	6.164	.819	1.638	. . .	. . .	6.164	.819	1.638	. . .	. . .	6.164
C₂H₆	16.85	16.675	2.810	.337	.506	. . .	. . .	2.220	.337	.506	. . .	. . .	2.220	.337	.506	. . .	. . .	2.220
Total	100.00	. . .	10.610	1.157	2.144	. . .	. . .	8.392	1.157	2.144	. . .	.223	9.230	1.157	2.144	. . .	1.114	12.583

[a] The same numerical values apply if all volumes are expressed in cubic feet at 60°F and 30 in. Hg absolute pressure.
[b] From excess air.

Table 6.8 Limits of Flammability and Ignition Temperature for Simple Gases and Compounds [a]

Simple Gases and Compounds	Limits of Flammability		Ignition Temperature (In Air)	
	Lower % by Volume Gas in Air	Upper % by Volume Gas in Air	(°C)	(°F)
H	4.0	75	520	968
CO	12.5	74	644–658	1191–1216
CH_4	5.0	15.0	705	1301
C_2H_6	3.0	12.5	520–630	968–1166
C_3H_8	2.1	10.1	466	871
C_2H_4	2.75	28.6	542–548	1008–1018
C_3H_6	2.00	11.1	458	856
C_4H_8	1.98	9.65	443	829
C_2H_2	2.50	81	406–440	763–824
C_6H_6	1.35	6.75	562	1044
C_7H_8	1.27	6.75	536	997

[a] From U.S. Bureau of Mines Bulletin 503 (1952); see also U.S. Bureau of Mines Bulletin 627 (1965); also "Gas Engineers Handbook" under Segeler listing at end of chapter.

and a shorter, sharper flame developed. In the case of a gas containing large amounts of hydrogen, intimate mixing will provide a combustion reaction of explosive velocity relative to that of a gas containing large amounts of methane. Inert gases, such as carbon dioxide and nitrogen, present in fuel gases or in a gas-air mixture, reduce combustion velocity. The proportion of nitrogen in a fuel gas-air mixture may be reduced by oxygen enrichment of air for combustion, and combustion speed may, by this means, be accelerated many fold. Such measures also will raise the flame temperature. The use of preheated air for combustion also accelerates combustion of gases. In order to burn large volumes of fuel in a small space, a mixture of air and gas is sometimes directed against a hot, incandescent surface. By increasing the velocity of combustion, higher temperatures are localized close to the burner point. This condition is desirable for some processes and highly undesirable for others. For instance, the scarfing process requires a highly intensive localized heat, while the heating of steel for rolling requires a lower intensity distribution of heat over the full surface of the pieces being heated. In order to reduce combustion speed of a gaseous fuel, the air and gas streams may be stratified to produce slow mixing. Such a method creates a diffusion flame, a long flame of relatively uniform temperature with a relative higher degree of cracking of the hydrocarbon components.

Theoretical flame temperature is the temperature which would be attained by the products of combustion if the combustion of a fuel took place instantaneously, and there were no loss of heat to the surroundings. Such a condition never exists, but theoretical flame temperature represents another measure for comparing fuels. Fuels which develop a high flame temperature by combustion are more capable of producing a higher thermal efficiency in practice than those which develop low flame temperatures. The efficiency of heat utilization is the relation of the total heat absorbed by a substance to the heat supplied. Because the temperature level at which waste gases leave a furnace is usually fixed within a relatively narrow range, the higher the flame temperature the higher the potentiality for heat absorption by the substance to be heated. The theoretical flame temperature of a fuel may be calculated by balancing the sum of the net heating value of a given quantity of fuel and the sensible heat of the air-gas mixture against the heat content of the products of combustion.

Theoretical flame temperature so calculated should be corrected for dissociation of CO_2 and H_2O at temperatures in excess of 1650°C (3000°F). The theoretical flame temperatures for a number of important gaseous fuels are given in Table 6.23 in Section 6.4.5. The reader is referred to the *Gas Engineers Handbook* and the books by Lewis and von Elbe, by Hougen et al., and Trinks and Mawhinney and others listed in the bibliography at the end of this chapter for a full explanation of combustion stoichiometry, fuel economy calculations and the calculation of theoretical flame temperatures, and the dissociation of gases at elevated temperatures.

There are a number of factors which determine the character, size and shape of a gas flame. Gases burned at very high combustion velocity will produce very little or no luminosity regardless of the kind of gas. The velocity and volume with which the air-gas stream leaves a burner or furnace port, the fuel-air ratio, and the amount of non-combustible material in the fuel will influence the length and shape of a flame. The kind of gas to be burned has a very great effect upon the character of the flame. Carbon monoxide and hydrogen burn with an invisible to a clear blue flame, while the hydrocarbon gases, methane, ethane, etc. are capable of developing highly luminous flames. The principal reason that these gases burn with a luminous flame is due to the thermal breakdown of the hydrocarbons into carbon and hydrogen, and under combustion conditions which permit this, the carbon particles are heated to incandescence thereby giving the flame its luminous appearance. The luminosity of a flame may be decreased or increased by varying the supply of air. A deficiency of air below theoretical requirements will increase luminosity and it also usually will lengthen the flame. An excess of air will decrease luminosity and shorten the flame with most burners or furnace ports. Increasing the temperature of preheat of the air for combustion will reduce luminosity, as is also the case when water vapor (steam), which may be introduced with the gas, air, or for atomization of liquid fuels, is increased. A luminous flame has a number of desirable qualities, the principal one being its greater ability to transfer heat by radiation from a fixed temperature level. However, it should be noted that a luminous flame is obtained usually at a lower temperature level than when the same fuel is burned with a lower degree of luminosity.

6.1.3 Heat Flow

Heat flow is caused by a difference in temperature, and heat is transmitted in three ways, namely, by conduction, by convection, and by radiation

6.1.3.1 Conduction

Conduction is the transmission of heat through a solid body without visible motion of the body, as through a steel bar. The amount of heat transferred through a homogeneous solid by conduction is expressed by the formula:

$$q = (k\ A\ \Delta T)/ x \qquad (6.1.4)$$

where, in SI, the quantities are expressed in the following units:

q = watts transmitted (1 W = 1 J/s)
k = conductivity factor in W/m K
A = area in m^2
ΔT = temperature difference in K
x = length of heat-transfer path in metres

and, in fps units:

q = Btu transmitted per hour
k = conductivity factor in Btu in. / ft^2 h °F
A = area in square feet
ΔT = temperature difference in °F
x = length of heat-transfer path in inches

The flow of heat through a non-homogeneous solid body by conduction is expressed by the formula:

$$q = \frac{\Delta T}{\left(\dfrac{x_1}{k_1 A_1}\right) + \left(\dfrac{x_2}{k_2 A_2}\right) + \cdots \left(\dfrac{x_n}{k_n A_n}\right)} \qquad (6.1.5)$$

where, in SI units,

x_1, x_2 and x_n = the respective lengths of heat-transfer path through the various resistances in metres

k_1, k_2 and k_n = the corresponding conductivity factors of the various resistances expressed in W/m K

A_1, A_2 and A_n = the corresponding areas expressed in m^2

6.1.3.2 Convection

When heat is transmitted by the mechanical motion of gas or water currents in contact with a solid, or by gas currents in contact with a liquid, the transfer of heat is by convection. In the transfer of heat by convection, it is necessary to conduct heat through the relatively stationary film between the moving and stationary bodies. This film becomes thinner as the velocity of the currents parallel to its surface increases. The transfer of heat by convection is expressed by the formula:

$$q = U A \Delta t \qquad (6.1.6)$$

where, in SI, the quantities are expressed in the following units:

q = watts transmitted
U = film coefficient, expressed in W/m^2 K, dependent upon the velocity, specific gravity and viscosity of the moving fluid and the conductivity of the film
A = area in m^2
Δt = temperature difference in K

and, in fps units,

q = Btu transmitted per hour
U = film coefficient (Btu per ft^2 per °F per h) dependent upon the velocity, specific gravity and viscosity of the moving fluid and the conductivity of the film
A = area in ft^2
Δt = temperature difference in °F

6.1.3.3 Radiation

Radiation refers to the transmission of heat through space without the help or intervention of matter. This is the means by which the heat of the sun reaches the earth, and by which much of the heat of combustion of fuels is utilized in high-temperature processes in the steel industry. When radiant energy strikes any body a certain proportion of the total is reflected, while that absorbed is reconverted to heat energy. A perfect black body is one that will not reflect radiation falling upon it but absorbs all of it. The coefficient of reflectivity of a body receiving radiation is equal to one minus its black body coefficient. Emissivity refers to the rate at which a body radiates heat in relation to a black body of equal area, and this rate depends upon the temperature of the body and the nature of its surface. Kirchoff's Law shows that the absorptivity and emissivity of a given surface are numerically equal at the same temperature. The Stefan-Boltzmann Law states that the total energy of a black body is proportional to the fourth power of its absolute temperature, that is:

$$W = \sigma T^4 \qquad (6.1.7)$$

where, in SI units, W equals the total emissive power of a black body, expressed in watts per square metre (W/m²), σ is the Stefan-Boltzmann constant equal to 5.71 × 10⁻⁵ ergs/cm² s K⁴ or 5.71 × 10⁻⁸ watts per m² K⁴ and T is the absolute temperature in Kelvins (K).

In fps units, W is expressed in Btu/ft² h °R⁴ and σ equals 0.173 × 10⁻⁸ Btu/ft² h °R⁴ with T representing the absolute temperature in °R.

The net effect of heat transfer between two bodies, neither of which can be considered a black body, must take into account the emissivity factor ε, which is the ratio of the emissive power of an actual surface to that of a black body; this results in the following equation in SI units:

$$q = 5.71\varepsilon A\left[\left(\frac{T_1}{100}\right)^4 - \left(\frac{T_2}{100}\right)^4\right] \tag{6.1.8}$$

where

- q = watts transmitted
- 5.71 = Stefan-Boltzmann constant expressed in W/m²K⁴
- ε = emissivity factor
- A = surface area in m²
- T_1 = absolute temperature of body giving off heat, in Kelvins (K)
- T_2 = absolute temperature of body receiving heat, in Kelvins (K)

In fps units, equation 6.1.8 becomes:

$$q = 0.173\varepsilon A\left[\left(\frac{T_1}{100}\right)^4 - \left(\frac{T_2}{100}\right)^4\right] \tag{6.1.9}$$

where

- q = Btu transmitted per hour
- 0.173 = Stefan-Boltzmann constant expressed in Btu/ft² °R⁴
- ε = emissivity factor
- A = surface area in ft²
- T1 = absolute temperature of body giving off heat, in °R
- T2 = absolute temperature of body receiving heat, in °R

The emissivity factors for various materials at specified temperatures are presented in Table 6.9. Emissivities vary from almost zero to slightly less than one, depending on the nature of the mate-

Table 6.9 Emissivity Factors (a perfect absorber or radiator = 1)

Material	ε
Polished aluminum at 230°C (445°F)	0.039
Polished aluminum at 580°C (1075°F)	0.056
Polished brass at 300°C (570°F)	0.031
Polished nickel at 380°C (715°F)	0.086
Polished nickel-plated steel at 22°C (72°F)	0.052
Bright tinned steel plate at 24°C (75°F)	0.071
Polished mild steel	0.288
Cast iron—machined—at 22°C (72°F)	0.437
Cast iron—liquid—at 1330°C (2425°F)	0.282
Cast iron—rough oxidized	0.97
Mild steel—dull oxidized—from 26° to 355°C (79° to 672°F)	0.96
Firebrick glazed through use at 1000°C (1830°F)	0.75
Silica brick (rough)	0.81

rial, its surfaces face finish, and its temperature. Polished metal surfaces have low emissivities, whereas those of oxidized surfaces and non-metals generally approach a value of one. In the generation of heat from fuels, the character of the flame and its proximity to the receptor of heat is particularly significant in the transfer of heat by radiation. The amount of heat transferred from a flame varies widely and in proportion to its degree of luminosity. The transfer of heat by radiation varies inversely with the square of the distance between the transmitter and receptor of radiant energy. For that reason, flames should be kept close to the substance to be heated where high heat transfer rates are desirable.

6.2 Solid Fuels and Their Utilization

Solid fuels have played a significant role in the evolution of our modern, industrial civilization. Coal in particular has been of far-reaching importance in that it has provided the prodigious amount of energy essential to the development of the iron and steel industries. Vast quantities of this energy source remain to be exploited, but the rate of utilization far exceeds the rate at which coal is being formed. It follows that the efficient use of the remaining supply is desirable. Toward this end, modern coal research is directed.

Geologically, the earliest-formed coal thus far encountered occurs in the Silurian strata of Bohemia. It is not until Lower Carboniferous time, see Table 6.10, however, that the source materials

Table 6.10 Geologic Time Divisions

Era	Period		Epoch	Millions of Years
Cenozoic	Quaternary		Recent / Pleistocene	70
	Tertiary		Pliocene / Miocene / Oligocene / Eocene	200
Mesozoic	Cretaceous / Jurassic / Triassic			
Paleozoic		Permian		500
	Carboniferous	Pennsylvanian (Upper Carboniferous)		
		Mississippian (Lower Carboniferous)		3000 +
	Devonian / Silurian / Ordovician / Cambrian			
Proterozoic	Algonkian	Keweenawan / Huronian		
Archeozoic	Archean	Timiskamian / Keewatin		

of coal began to accumulate in significant quantities. Every continent, including Antarctica, contains some coal and no system of rocks younger than the Silurian is devoid of this important substance. In North America major concentrations of source materials were accumulated during the Carboniferous, Cretaceous and Tertiary periods. A similar statement can be made for Europe but, in contrast, some of the most important Asiatic coals occur in Triassic and Jurassic rocks of the Mesozoic Era.

6.2.1 Coal Resources

The known coal deposits in the U.S. are greater than those of any other country. Based on 1995 International Energy Annual data, the U.S. identified recoverable coal reserves of all ranks were estimated to be 246 billion metric tons (272 billion short tons) which is approximately 24% of the known world reserves. This figure represents coal of all ranks in the ground at depths and bed thicknesses generally considered mineable under current economic conditions. This would be enough to supply requirements for a long period in the future if all present coal reserves were available economically and of acceptable quality. A considerable quantity of the reserves of better quality coking coal has been utilized in the past and it is apparent that in the future it will be necessary to use coals requiring efficient extraction, cleaning, and other processing to assure proper utilization.

For obvious reasons, the steel industry has been striving to use coals which would produce metallurgical coke of optimum quality with a minimum of processing. Concentrations of coals of this class are found chiefly in the Appalachian area, although isolated deposits also exist in some Central and Western states. The preponderance of total coal reserve in the U.S. is in the form of lower-rank coals in the Great Plains, the Rocky Mountains, the Pacific Coast states and the Gulf region, see Fig. 6.1. These western coals are taking a larger share of the electric utility market due to lower sulfur content and lower mining costs. As a result, the eastern share of national production fell to 53% in 1995 from the 93% level in 1970. Total domestic coal production is still increasing and reached 937 million metric tons (1,033 million short tons) in 1995.

6.2.1.1 Origin and Composition of Coal

Coal is known to be a complex mixture of plant substances which have been altered in varying degrees by physical and chemical processes. Ordinarily, plant material, upon death, completely decomposes because of the action of microorganisms. Under certain circumstances, notably those associated with forested fresh-water swamps, this action is inhibited by antibiotic solutions which are common in this type of environment. As a result, the rate of accumulation of the plant material exceeds that of its decomposition and dispersion. Under such conditions a brown fibrous deposit known as peat is formed. Peat is the first step in the formation of coal.

Peat deposits, formed millions of years ago, subsequently were submerged through vertical movements of the earth's crust, in which position they became covered by deposits of sedimentary rocks. Later movements of the earth's crust raised many of these deposits to various heights above sea level. In the meantime, the peat had been changed, through agencies of biological action, pressure, and heat, into coal. The better ranks of coal in this country were formed during the Carboniferous period, the geologic period when conditions were most favorable for plant accumulation and decomposition. Included in the present deposits that originated in that period are the coal fields of the Appalachian and Central states.

The rate at which peat forms depends upon the rapidity of plant growth and the manner in which tissue increment is related to the rate of decomposition. It has been estimated that approximately one century is required to form a deposit of mature, compacted peat about one-third metre (one foot) in thickness. Certain studies of volatile matter relationships suggest that about a one-metre thick (a three-foot-thick) deposit of mature peat is required to produce a one-third-metre thick (one-foot-thick) layer of bituminous coal. These and other data indicate that a coal seam which is

Steel Plant Fuels and Water Requirements

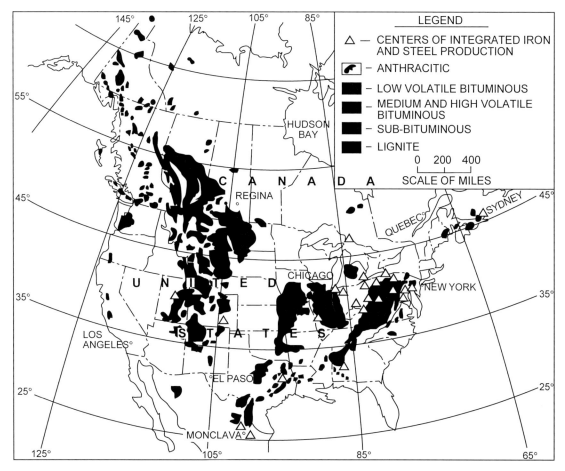

Fig. 6.1 Map showing general location and extent of the important coal fields of North America and centers of integrated iron and steel production. (Map prepared by the Canadian Department of Energy, Mines and Resources, Ottawa, Canada.)

a metre or more (several feet) thick may require a time span of thousands of years for its formation. If, in the course of time, the peat is subjected to the necessary conditions it becomes modified to brown coal and, when adequately consolidated, to lignite. From the lignitic stage, the material passes progressively through the sub-bituminous, bituminous, semi-anthracite and anthracite stages with a gradual change in the composition of the individual components of the complex mass. The proximate and ultimate compositions of coal, defined later in Section 6.2.3 and shown in Tables 6.11, 6.12 and 6.13 illustrate the gradual concentration of carbon and loss of oxygen in the various stages of coal formation.

Peat varies in appearance from a light, brown-colored, fibrous material to a very black and dense, muck-like sediment. Lignite is usually brown in color and commonly shows a woody texture. It contains a large amount of moisture and usually disintegrates, or slacks, into small pieces as it dries on exposure to air. Sub-bituminous coal varies in color from very dark brown to black and fractures irregularly. Bituminous coal is black in color and usually exhibits a banded structure due to the alternate dull and vitreous layers of varying thickness. Coals of the high-volatile bituminous rank commonly burn with a smoky, yellow flame. Anthracite is black, hard and brittle and has a high luster. It ignites less easily than bituminous coal and burns with a short, bluish, yellow-tipped flame producing very little or no smoke. The characteristics of semi-anthracite coal are intermediate between those of bituminous coal and anthracite.

All of the solid natural fuels contain both combustible and non-combustible materials. The combustible material is composed mainly of carbon, hydrogen and, to a lesser extent, sulfur. The non-

Table 6.11 Typical Moisture and Ash Content of Raw Solid Fuels [a]

Fuel	Moisture Content (%)	Ash Content (%)
Peat	65–90	[b]
Lignite (North Dakota)	35–40	7.6
Sub-Bituminous (Wyoming)	15–25	3.3
Bituminous (Low-Volatile B)	2.5	11.4
Anthracite (Northeastern Pa.)	5.5	9.6

[a] For additional coal analyses, see Bureau of Mines R.I. 7104 (1968), "Analyses of Tipple and Delivered Samples of Coal" and previous reports in the same series; also "Combustion Engineering" and "Steam, Its Generation And Use" published by Combustion Engineering, Inc. and the Babcock & Wilcox Co., respectively, cited in the references.

[b] Highly variable, from 2–15% or higher.

combustible constituents are water, nitrogen and oxygen, and a variety of mineral materials usually referred to as ash.

The bituminous coals are of greatest interest to the steel industry in view of the fact that essentially all coking coals fall in this category. The lustrous black bands which are conspicuous in a lump of bituminous coal are generally referred to as vitrain although some American coal petrographers employ the term anthraxylon in preference. Following U.S. Bureau of Mines terminology, the anthraxylon is derived from woody plant tissues and is surrounded by a dull ground mass made up of translucent attritus, opaque attritus and fusain. The attrital portion is composed of finely comminuted fragments of altered plant materials. Fusain is a friable, charcoal-like substance derived from woody tissues and is a term used universally without modification.

In addition to the readily recognizable bands of vitrain and fusain, European and Asiatic coal investigators have found it useful to identify silky, minutely striated layers within a coal as clarain. Layers of dull, compact coal are called durain. Thus, coal seams can be thought of as being composed, usually, of various mixtures of vitrain, fusain, clarain and durain, each occurring in the form of layers which are visually observable. Coals made up largely of vitrain and clarain are spoken of as bright coals whereas coals containing a high percentage of durain are called splint coals. Bright coals are generally better coking coals than splint coals, vitrain apparently playing an important part in the carbonization process. Fusain will not coke, but in small percentages it may actually increase coke strength provided the particle size is fine enough. The fixed carbon content is higher and the volatile matter content is lower in fusain than in the other banded ingredients.

Microscopic study has shown the banded components to be composed of identifiable plant entities called phyterals, but of greater significance is the fact that the vitrain, fusain, clarain and durain are made up of numerous components or macerals which can be defined by their physical and chemical properties. Durain, for example, may include several macerals (vitrinite, semi-fusinite, micrinite, cutinite, etc.) which are easily distinguished by their differing optical properties. Additional information regarding the nature and variability of these individual coal components and their contribution to the effective and efficient utilization of all types of coal appears in references on applied coal petrography at the end of this chapter.

Table 6.12 Typical Compositions of Peat and Coals of Different Ranks (Dry Basis)

Group (ASTM Designation D 388)	Proximate Analyses (%)			Ultimate Analyses (%)					Gross Heating Value	
	Volatile Matter	Fixed Carbon	Ash	Carbon	Hydrogen	Nitrogen	Sulfur	Oxygen	kJ/kg[a] (dry)	BTU/lbm[b] (dry)
Meta anthracite	1.2	90.7	8.1	86.8	1.6	0.6	0.9	2.0	31,797	13,682
Anthracite	3.4	87.2	9.4	84.2	2.8	0.8	0.6	2.2	32,094	13,810
Semianthracite	13.0	74.6	12.4	78.3	3.6	1.4	2.0	2.3	31,560	13,580
Bituminous										
Low-Volatile	16.0	79.1	4.9	85.4	4.8	1.5	0.8	2.6	34,860	15,000
Medium-Volatile	22.2	74.9	2.9	86.4	4.9	1.6	0.6	3.6	35,274	15,178
High-Volatile A	34.3	59.2	6.5	79.5	5.2	1.4	1.3	6.1	33,456	14,396
High-Volatile B	39.2	55.4	5.4	78.3	5.2	1.5	1.4	8.2	32,787	14,108
High-Volatile C	36.4	54.5	9.1	73.1	4.8	1.5	2.6	8.9	31,302	13,469
Subbituminous										
A	38.9	56.4	4.7	75.1	5.0	1.4	1.0	12.8	31,595	13,595
B	42.8	54.4	2.8	75.0	4.9	1.3	0.5	15.5	30,788	13,248
C	39.4	47.4	13.2	64.2	4.4	1.2	0.4	16.6	25,796	11,100
Lignite A and B	41.8	49.4	8.8	64.4	4.2	1.1	0.8	20.7	25,643	11,034
Peat	67.3	22.7	10.0	52.2	5.3	1.8	0.4	30.3	21,048	9,057

(a) To convert to kilocalories per kilogram, multiply by 0.2390.
(b) To convert to kJ/kg from Btu/lbm, multiply by 2.326.

Table 6.13 Approximate Range of Moisture Contents for Peat and for Coals of Different Ranks (ASTM Designation D 388)

Fuel	Moisture Content (%)
Meta anthracite	3–10
Anthracite	1–8
Semianthracite	1–10
Low-Volatile Bituminous	2–4
Medium-Volatile Bituminous	1–4
High-Volatile A Bituminous	2–11
High-Volatile B Bituminous	4–15
High-Volatile C Bituminous	7–17
Subbituminous A	10–20
Subbituminous B	14–25
Subbituminous C	16–34
Lignite A and B	23–60
Peat	55–90

6.2.1.2 Chemical Composition and Coal Classification

There are two methods commonly employed to determine the chemical composition of coal; namely ultimate analysis and proximate analysis (Table 6.12). An ultimate analysis determines the quantities of carbon, hydrogen, oxygen, nitrogen, sulfur, chlorine and ash in dry coal; a proximate analysis determines the fixed carbon, volatile matter, moisture and ash contents. The proximate analysis is used most commonly because it furnishes most of the data required for normal commercial evaluations.

The analysis of coal can be made in the laboratory on an 'air-dry' basis. The coal sample is delivered to the laboratory in sealed containers. In the laboratory, the coal is weighed and then exposed to the air of the laboratory for a period of time and then weighed again. The percent loss in weight is the 'air-dry' loss. However, since the air-dry analysis is of little value to the user, the analysis is converted to the 'as-received' basis by combining air-dry loss and final moisture content.

$$1.0 - \frac{\% \text{ air-dry loss}}{100} \qquad (6.2.1)$$

The heating value of coal is reported on as-received basis and dry basis. To convert the analysis to dry basis from the as-received basis, each constituent value (except the moisture because that is being eliminated) is divided by the factor

$$1.0 - \frac{\% \text{ moisture in as-received analysis}}{100} \qquad (6.2.2)$$

Typical ranges of moisture are listed for various coals in Table 6.11 and Table 6.13.

Using data provided by chemical, physical or petrographic analyses, coals are classified according to rank, grade, and type. Classification according to rank is based upon the degree of metamorphism within the coal series from the level of lignite to that of anthracite coal. The American Society for Testing and Materials ranks coals according to their fixed-carbon content on a dry basis, and the lower rank coals according to Btu content on a moist basis. The classification of coals by rank adopted by the American Society for Testing and Materials (ASTM Specification D388), is shown in Table 6.14.

Steel Plant Fuels and Water Requirements

Table 6.14 Classification of Coals by Rank [a],[f]

Class	Group	Fixed Carbon Limits, per cent (Dry, Mineral-Matter-Free Basis)		Volatile Matter Limits, per cent (Dry, Mineral-Matter-Free Basis)		Calorific Value Limits (Moist,[b] Mineral-Matter-Free Basis)[g]				Agglomerating Character
		Equal or Greater Than	Less Than	Greater Than	Equal or Less Than	Equal or Greater Than		Less Than		
						Btu/lb	kJ/kg	Btu/lb	kJ/kg	
I. Anthracitic	1. Meta-anthracite	98	...	...	2	...	...	...	...	Nonagglomerating
	2. Anthracite	92	98	2	8	...	...	...	...	
	3. Semianthracite[c]	86	92	8	14	...	...	...	...	
II. Bituminous	1. Low volatile bituminous coal	78	86	14	22	...	...	...	...	Commonly agglomerating[e]
	2. Medium volatile bituminous coal	69	78	22	31	...	...	...	...	
	3. High volatile A bituminous coal	...	69	31	...	14,000[d]	32,500[d]	...	...	
	4. High volatile B bituminous coal	...	...	...	...	13,000[d]	30,200	14,000	32,500	Agglomerating
	5. High volatile C bituminous coal	...	...	...	...	11,500	26,700	13,000	30,200	
						10,500	24,400	11,500	26,700	
III. Subbituminous	1. Subbituminous A coal	...	...	...	...	10,500	24,400	11,500	26,700	Nonagglomerating
	2. Subbituminous B coal	...	...	...	...	9,500	22,100	10,500	24,400	
	3. Subbituminous C coal	...	...	...	...	8,300	19,300	9,500	22,100	
IV. Lignitic	1. Lignite A	...	...	...	...	6,300	14,600	8,300	19,300	
	2. Lignite B	...	...	...	...	...	...	6,300	14,600	

(a) This classification does not include a few coals, principally nonbanded varieties, which have unusual physical and chemical properties and which come within the limits of fixed carbon or calorific value of the high-volatile bituminous and subbituminous ranks. All of these coals either contain less than 48% dry, mineral-matter-free fixed carbon or have more than 15,500 moist, mineral-matter-free British thermal units per pound (36,100 kJ per kg).
(b) Moist refers to coal containing its natural inherent moisture but not including visible water on the surface of the coal.
(c) If agglomerating, classify in low-volatile group of the bituminous class.
(d) Coals having 69% or more fixed carbon on the dry, mineral-matter-free basis shall be classified according to fixed carbon, regardless of calorific value.
(e) It is recognized that there may be nonagglomerating varieties in these groups of the bituminous class, and there are notable exceptions in high volatile C bituminous group.
(f) From ASTM Designation D-388-66 in "ASTM Standards 1975," Part 26, page 215, to which reference may be made for method of calculation to mineral-matter-free basis and other information. Reproduced by permission of the American Society for Testing and Materials. The complete specification is obtainable from the society.
(g) Rounded values for kJ/kg, obtained by calculation, are not part of the original specification.

In the U.S., coals are also classified into types and such terms as bright, semi-splint, splint, cannel and boghead coal are applied. The data required are obtained from microscopic studies. The U.S. Bureau of Mines defines bright coal as containing less than 20% opaque matter; semi-splint must have between 20 and 30%, and splint coal must be made up of more than 30% of this ingredient. Cannel and boghead coals are non-banded and are characterized by a small percentage of anthraxylon (vitrain). Boghead possesses a high percentage of volatile oils and gases, and contains an abundance of algal material. Cannel, or candle, coal is so named because it can be ignited with a match or a candle flame and it burns with unusual brilliance. Cannel coal is non-coking, often contains large quantities of spore and pollen materials, and like boghead, has a high content of volatile oil and gas.

Coals are classified to grade by their ash and sulfur contents. The mineral constituents of the ash are also important because they influence fouling and slagging in the furnace.

6.2.2 Mining of Coal

It is found that seams of coals vary in thickness throughout the world from a few millimetres to over 75 metres (a fraction of an inch to over 250 feet). In this country the thickest seams are found in the sub-bituminous coals of the West, one of which approaches about 30 metres (100 feet). In the East, the Mammoth bed in the anthracite fields of Pennsylvania attains a thickness of 15 to 18 metres (50 to 60 feet) but is found to be quite variable when traced laterally. The Pittsburgh seam at the base of the Monongahela series in the Appalachian area is noteworthy because of its exceptionally uniform thickness (approximately 2 metres or 7 feet) over thousands of square miles. Fig. 6.2 shows the western portion of Pennsylvania in such a manner as to make clear the areal extent as well as the sub-surface relations of the coal-bearing formations of this region. Data are provided in Table 6.15 as to thickness of seams and distance between coals.

Coal seams may dip gently as shown in Fig. 6.2, or they may be horizontal, or they may exist almost vertical with respect to the Earth's surface. Mining problems are often complicated by the fact that seams seldom remain in the same plane throughout their extent. Under present conditions, a coal bed must be at least 0.75 to 0.90 metres (30 inches to 36 inches) thick to be profitable for

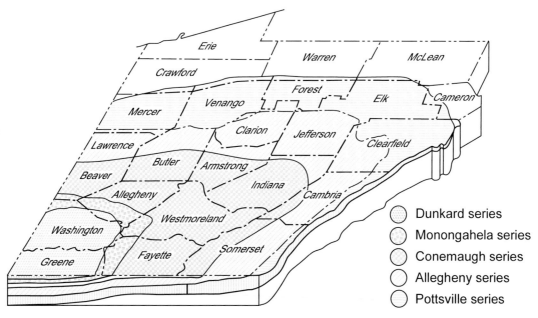

Fig. 6.2 Distribution of coal-bearing strata in Western Pennsylvania.

Table 6.15 Important Coals of the Northern Portion of the Appalachian Coal Region

Series	Strata	Thickness		Totals From Top	
		m	ft	m	ft
Dunkard	Proctor Sandstone	12.2–12.2	40–40	12.2	40
	Windy Gap Coal	0.15–0.30	$1/_2$–1	54.9	180
	Gilmore Coal	0.15–0.30	$1/_2$–1	88.4	290
	Nineveh "A" Coal	0.076–0.30	$1/_4$–1	138.7	455
	Nineveh Coal	0.15–0.30	$1/_2$–1	153.9	505
	Hostetter Coal	0.15–0.30	$1/_2$–1	176.8	580
	Fish Creek Coal	0.15–0.30	$1/_2$–1	205.7	675
	Dunkard Coal	01.5–03.0	$1/_2$–1	225.6	740
	Jollytown Coal	0.15–0.30	$1/_2$–1	240.8	790
	Hundred Coal	0.076–0.30	$1/_4$–1	254.5	835
	Washington "A" Coal	0.15–0.30	$1/_2$–1	280.4	920
	Washington Coal	0.61–1.52	2–5	313.9	1030
	Little Washington Coal	0.38–0.61	$1^1/_4$–2	321.3	1054
Monongahela	Waynesburg Coal	0.91–1.52	3–5	374.6	1229
	Uniontown Coal	0.30–0.91	1–3	391.7	1285
	Lower Uniontown Coal	0–0.30	0–1	413.0	1355
	Sewickley Coal	0.91–1.52	3–5	452.6	1485
	Redstone Coal	0.91–1.52	3–5	470.9	1545
Conemaugh	Pittsburgh Coal	1.52–2.43	5–8	484.6	1590
	Morgantown Coal	0.30–1.52	1–5	490.7	1610
	Little Pittsburgh Coal	0.30–1.83	1–6	507.5	1665
	Little Clarksburg Coal	0.61–2.13	2–7	530.4	1740
	Normantown Coal	0–0.30	0–1	545.6	1790
	Clarysville Coal	0–0.30	0–1	560.8	1840
	Elk Lick Coal	0.61–1.22	2–4	591.3	1940
	Harlem Coal	0.15–0.61	$1/_2$–2	621.8	2040
	Upper Bakerstown Coal	0.15–0.30	$1/_2$–1	650.7	2135
	Bakerstown Coal	0.61–1.52	2–5	652.3	2140
	Brush Creek Coal	0–1.83	0–6	714.8	2345
	Mahoning Coal	0.30–1.83	1–6	722.1	2369
Allegheny	Upper Freeport Coal	0.61–1.52	2–5	730.0	2395
	Lower Freeport Coal	0.61–0.91	1–3	751.3	2465
	Upper Kittanning Coal	0–0.30	0–1	768.1	2520
	Middle Kittanning Coal	0.91–1.52	3–5	775.7	2545
	Lower Kittanning Coal	0–2.74	0–9	797.1	2615
	Clarion Coal	0.30–1.22	1–4	829.1	2720
Pottsville	Brookville Coal	0.30–0.91	1–3	836.7	2745
	Mercer Coal	0–0.61	0–2	853.4	2800

mining. Table 6.15 also shows that coal seams vary in their distance below the Earth's surface. U.S. Geological Survey estimates of coal reserves do not include coal seams deeper than about 910 metres (3000 feet) from the surface, although in Great Britain and Europe coal seams at greater depth are being mined.

The mining of coal is performed by either one of two methods: (1) open pit or stripping, also called contour mining, or (2) underground or deep mine. The first method involves removing the formation (over-burden) above the seam by stripping with scrapers, bulldozers, or mechanically operated shovels, followed by removing the exposed coal. Stripping is applied to coal seams which are relatively close to the surface, particularly to thick seams underlying overburden about 25 to 50 metres (80 to 150 feet) deep, although the development or larger equipment and improved techniques in recent years has justified removal of layers of overburden thicker than this. Auger mining is being used extensively to recover coal where the overburden is too great for strip-mining practices to be employed. A large-diameter auger or drill with cutting bits on its end is propelled into the exposed edge of a coal seam. As the auger progresses into and along the seam, the broken coal is conveyed away from the face through the tube to the outside for transport away from the auger. Production by strip mining has increased greatly since World War I due to reduced labor and material costs and a quicker return on capital investment compared to underground mining. In the U.S., surface mining accounted for slightly over 25% of the coal produced in 1957. By 1995, surface mining has climbed to about 62% of the total coal mined.

6.2.2.1 Underground Mining

Underground mining is performed by either the room-and-pillar or the longwall method. The room-and-pillar method is in more common use in the U.S. The longwall method is particularly adaptable to mining seams up to about 4.6 metres (15 feet) thick under conditions where the roof may be permitted to cave. It is used more extensively in the mines of the eastern U.S. There are a number of modifications applicable to each method. The room-and-pillar system consists essentially of working out rooms, chambers, or breasts in the coal seam from passages (entries) driven from the mine entrance. Entrance to an underground mine is by drift, shaft or slope. The rooms vary in width from about 3.5 to 12 metres (12 feet to 40 feet), 6.1 metres (20 ft.) being the most common, and from about 45 to 90 metres (150 feet to 300 feet) in length, depending on such factors as weight and character of the overlying and underlying structure and thickness of seam. Pillars separating the rooms vary in width from about 2 to 30 metres (6 feet to 100 feet), depending on conditions and mining practice. These pillars are sometimes removed by retreat mining and the coal recovered.

In the longwall method, a continuous mining face is maintained in the coal seam. After mining, the roof is permitted to cave, about 5 to 9 metres (15 or 30 feet) from the mine working face.

Prior to the advent of mechanical mining, undercutting of the coal seam preparatory to blasting was done manually. Production per man was low by this method and required a number of working faces in the mine to produce high mine tonnage. Hand loading of coal into mule-drawn cars was the prevailing practice for many years until development of machinery for cutting, loading and haulage. Electric trolley-type locomotives capable of hauling longer underground trains of cars of increased capacity displaced mule-drawn trains as mine capacity increased.

In some modern underground mines, the coal is carried out of the mine by a system of conveyor belts to a shipping station or cleaning plant. In other mines, the coal is carried in mine cars to a rotary dumper. From the dumper the coal is fed by way of a conveyor or elevator to a shipping station or cleaning plant.

6.2.2.2 Continuous Mining

The cutting machines and loading machines characteristic of mechanical mining in the past were single-purpose units, and each performed essentially a single function of mining at the working face. After either unit completed its work it was withdrawn from the face to allow other units of the production setup to move up to the face to carry out succeeding functions. To keep all operating units working at full efficiency, it was necessary to have additional working places near at hand so that the single-purpose machines could enter the places in rotation and carry out their functions without interference.

To eliminate some of the difficulties attendant upon the addition of extra working places, multi-purpose machines known as continuous miners have been developed and the operation carried out by such machines has been given the name continuous mining. Continuous miners combine in a single unit the actions of dislodging the coal from the solid seam and loading it into some unit of a transportation system. Such machines, therefore, combine in one operation the separate steps of cutting, drilling, blasting and loading common to earlier mechanical mining methods. Coal planers and shearers with long-wall mining achieve these combined objectives.

There are several types of continuous miners in operation, one of which is a ripper-type miner that has cutting bits mounted in the rims of multiple wheels that are rotated to rip the coal out of the seam while the ripper wheels are propelled into and up or down in the coal seam. The coal that is ripped loose from the seam falls into the gathering head of the loader, which has dual gathering arms that sweep the broken coal into the conveyor section of the machine for loading into shuttle cars or other suitable conveying equipment.

Continuous miners of some other types employ auger-type cutters that bore into the face, the cut coal in both cases being carried by a conveyor on the machine to a shuttle car or other means of transportation.

6.2.3 Coal Preparation

As one phase of coal preparation, the objective of coal cleaning (often called washing) is removal of solid foreign matter, such as rock and slate, from the coal prior to its use. Reduction of ash and sulfur contents; control of ash fusibility; increase of heating (calorific) value; and improvement of coking properties of the coal can be achieved by this practice. From a coal-cleaning standpoint, the impurities in coal are of two types; namely, those which cannot be separated from the coal, usually called fixed or inherent impurities; and those which can be removed, herein referred to as free impurities. Altogether, these impurities are of eight types, named as follows: (1) residual inorganic matter of the coal-forming plants from which the coal was derived; (2) mineral matter washed or blown into the coal-forming mass during the periods of its formation; (3) pyrites (FeS_2) formed by bacterial reaction of iron and sulfur in the coal-forming matter; (4) sedimentary deposits during the coal-forming periods which appear as partings, sometimes called "bone," that usually must be mined with the coal; (5) massive deposits formed through deposition on bedding planes; (6) saline deposits, somewhat rare in coal beds of the U.S.; (7) slate, shale, clay, etc. from the underlying and overlying strata accidentally included in mining; and (8) water or moisture, which includes that naturally carried by the coal in air-dry condition, and excess moisture producing a condition of wetness. Items (1) (2) and, for the most part (3) form fixed ash, while (4), (5), (6), and (7) are partly free ash-forming materials that can be removed by hand-picking and suitable mechanical cleaning treatments. Item (8) involves drying operations differing from those required to separate mineral impurities, which is the primary objective of cleaning. Mechanical cleaning is possible because of the difference in specific gravity between the major impurities and the coal, the density of the former being 1.7 to 4.9, while pure coal has a density of about 1.3. Sulfur is present as pyrites, organic compounds, and sulfates, and only part of the pyrites can be removed by cleaning. Phosphorus is usually associated more with bony and impure coal than with clean coal and is, therefore, reduced by washing. Salts, particularly the alkali chlorides, lower the fusion point of the ash, affect coke-oven linings and are troublesome in waste liquors from coking operations.

The advent of full-seam mechanical coal mining and the increasing need for metallurgical coke of low and uniform ash and sulfur contents has focused attention on the needs for the most efficient types of washers.

The preparation of coal starts at the production face in the mine. If loading is done by hand, the miner is required to discard all rock and slate over 75 mm (3 in.) size. As practically all loading is done mechanically in the U.S. little attempt is made to prepare the coal at the face other than to control the tonnage from various sections of the mine if sulfur content of the coal is high or variable.

The cleaning qualities of a particular coal are determined by the float and sink test, commonly referred to as a washability test. Fundamentally, this test effects a fractionation of the coal by size and specific gravity. This test consists in crushing coal to proper size and floating individual sizes of it on liquids having specific gravities of 1.30, 1.40, 1.50, 1.60, etc., to determine the weight and character of the material that floats and sinks in each liquid. The proportion of coal, and the ash and sulfur content of the different fractions, provides reasonably complete data on the washability characteristics of a tested coal. Extreme fines may be evaluated by froth flotation.

Coal preparation is accomplished by a combination of crushing, sizing, cleaning, and dewatering operations:

1. Crushing
    a) Mine breakers
    b) Bradford breakers
    c) Roll crushers
    d) Impact crushers

2. Sizing
    a) Grizzlies
    b) Vibrating screens
    c) Classifiers
    d) Cyclones

3. Cleaning
    a) Jigs
    b) Dense media processes
    c) Cyclone processes
    d) Tables
    e) Froth flotation

4. Dewatering
    a) Screens
    b) Centrifuges
    c) Vacuum filters
    d) Thermal dryers

By far the largest percentage of coal is cleaned by wet methods.

A complete description of each of the foregoing processes would be too lengthy for inclusion herein; hence only a brief review will be given of the principles of some of the more important types of cleaning processes in use at present. A reference list for further study of this subject is appended to this chapter.

Jigs were probably the earliest type of machine used in the mineral industry to separate materials of different densities. They consist essentially of a box with a perforated base into which the material is placed, and by alternate surges of water upward and downward through the perforations, materials of different specific gravities stratify. Materials having the highest specific gravities remain at the bottom while the lighter material rises. With proper mechanical facilities, a continuous separation is achieved. While jigs are not very efficient in cleaning a mixture of various sizes, they are capable of satisfying some market requirements, and capacities up to about 450 metric tons (500 net tons) per hour have been obtained.

In dense-media processes, only a part of the power for separating coal and refuse is supplied by an upward flow of liquid, this separating power being supplemented by using a liquid medium which is heavier than water. The medium employed is a mixture of water and some finely divided solid material, such as sand, magnetite, or barite, which can be separated readily from the washed coal and reused. In the high-density suspension process, the upward flow is discarded entirely, the liquid medium consisting of a mixture which is just dense enough so that the coal floats in it, and the impurities sink. The size of coal has less significance in the efficiency of this process than of those previously described, and material ranging from 1.6 to 254 mm (1/16 inch to 10 inches) can be cleaned in one operation. However, difficulty is encountered in separating the solid material from coal of fine size. Capacities up to about 545 metric tons (600 net tons) per hour have been obtained with bituminous coal. The Chance cone method, which uses a mixture of sand and water, is also widely used in the U.S. The Tromp and Barvoys processes, using magnetite and barite respectively as the solid material in the mixture, are used extensively in Europe and in the U.S. In these processes, the specific gravity of the mixture of solid material and water can be varied by changing their proportions to suit the optimum conditions in cleaning. Agitation in the separating cone is supplied by an upward current of water and by mechanical stirring.

Cyclones, which have an inverted conical shape and are fed tangentially at the widest part of the cone, are used with the dense medium process to increase the separating rate of particles in the medium. This device is particularly effective on sizes between $3/4$ inch (17 mm) and 100–mesh (0.15 mm). Cyclones can also be used in the absence of a specific medium, but the efficiency is lower and they must often be combined in stages. In this practice they are referred to as hydroclones or water-only cyclones.

Coal is also cleaned on table concentrators. Essentially these tables consist of a slightly inclined rectangular surface having a series of parallel grooves or cleats. The tables are mechanically agitated to permit stratification of the light and heavy material and to cause the heavy material to move with the long axis of the table. A current of water is introduced at the top edge of the table to wash the coal which has settled above the refuse to the discharge edge of the table. The refuse which settles underneath the coal moves longitudinally down the table and is discharged at the end. Tables have been used principally for cleaning coal of the smaller sizes, from about 0.3 to 12.7 mm (48–mesh up to about $1/2$ in.).

Froth flotation of coal involves agitating fine coal with a mixture of water and a relatively small quantity of some frothing agent. In this process, coal is buoyed to the surface by the froth and removed while refuse settles. It is widely used to recover coal finer than 100–mesh (150 mm).

With practically all wet-washing systems the water is recirculated. When the water passes through the dewatering screens it contains a considerable amount of small-size coal solids which must be recovered for efficiency reasons. Also, the circulating water and effluent from flotation must be clarified before it is returned to the cleaning unit. This clarification is accomplished in various ways, the most important being by the use of hydraulic cyclones followed by a Dorr-type thickener. Settling cones and settling tanks are also used for this purpose. Where the Dorr-type thickener and settling tanks are used, it is customary to draw off the settlings in the form of a slurry containing 40% to 60% solids and to further separate the slurry in a vacuum-type filter. The filters deliver a product with approximately 25% moisture. For a more complete discussion of dewatering and waste disposal, the reader is referred to a standard book on coal preparation.

6.2.4 Carbonization of Coal

The most important use of coal in the modern steel industry is in the manufacture of metallurgical coke, which is discussed in detail in Chapter 7 in the Ironmaking Volume.

The carbonization of coal in byproduct ovens entails the production of large amounts of coke-oven gas and tar, important fuels in the steel industry, as well as light oils and various coal chemicals. The yields of gas and tar are largely a matter of the type of coal used and the temperatures employed in coke manufacture.

6.2.5 Combustion of Solid Fuels

The principal combustion reactions of solid fuels have been given in Section 6.1.2, and this present discussion will deal with operating factors pertinent to the combustion of solid fuels in steel plants.

The combustion of coke in blast furnaces has been studied by a number of investigators, each of whom has found that combustion takes place in a relatively small space directly in front of each tuyere.

Coke breeze, produced by screening coke at both the coke plant and blast furnaces, is utilized as a fuel in steel plant boiler houses to generate steam and in ore-agglomerating plants. When used as boiler fuel, coke breeze is burned on chain-grate stokers. Of importance in the combustion of coke breeze on chain-grate stokers is the maintenance of a relatively uniform fuel bed on the grate, approximately 200 to 300 mm (8 to 12 in.) thick, to prevent blowholes, and a balanced or slight positive pressure in the furnace at fuel-bed level. The operation of the grate should permit the normal combustion of approximately 146 kilograms per square metre (30 pounds of coke breeze per square foot) of effective grate area per hour. Chain-grate stokers are particularly adaptable to solid fuels with an ash of low fusion point. The design of front and back arches must take into consideration the fuel to be burned on chain-grate stokers. The arches are utilized to reflect heat and thereby aid ignition on the fuel bed.

Stokers for firing coal were generally used in steel plant boilers on units whose capacity is under approximately 45,000 kilograms (100,000 pounds) of steam per hour and for units using exclusively a solid fuel. They were often used on boilers to provide flexibility for the adjustment of boiler output to the steam load in plants where there is an insufficient or fluctuating supply of gaseous byproduct fuels. When coal is used as a boiler fuel today, pulverized boilers dominate the field. The advantage of stokers lies in their ability to control easily the rate of combustion of a solid fuel with efficient use of air. The combustion process on stoker-fired boilers consists essentially in first driving the volatile matter from a continuous supply of fuel, and then oxidizing the carbon in the residue on the stoker. The combustion of the coke-like residue on the grate produces CO_2 and CO. The CO and volatile matter are burned over the grate by secondary air admitted over the fuel bed. The temperature of the fuel bed is affected by the rate of firing and, at the top or hottest part of the bed, varies from about 1230°C (2250°F) at low to 1510°C (2750°F) at high rates. The amount of primary air supplied determines the capacity of stoker-fired furnaces and the effective use of secondary air determines the efficiency of combustion. In well operated and carefully sealed boilers, approximately 20 to 30% excess air will permit combustion of the gases within seven or eight feet above the grate.

Stokers are classified in general according to the travel of the fuel. In an overfired stoker the fuel is fed on top of the bed, and in an underfired or a retort stoker the fuel is fed at the bottom or side of the bed. A traveling-grate or chain-grate stoker carries the bed horizontally on the flat upper surface of a conveyor as in a chain-grate stoker. There are a number of modifications of these stoker types. The spreader stoker projects the coal into the furnace above the fuel bed and the fuel is burned both in suspension and on the fuel bed. While the fuel bed of a stoker-fired boiler is relatively thin, usually from 100 to 300 mm (4 to 12 in.), compared to a gas-producer bed, similar zones of reaction occur. In overfired stokers the ash zone is immediately above the grate, followed by the oxidation, reduction and distillation zones. In underfired or retort stokers the distillation of the volatile matter takes place in an oxidizing atmosphere and the volatile products pass through the incandescent residue from combustion rather than through green coal, as in the case of overfired stokers. The normal combustion rates on coal-fired stokers amounts to approximately 150 to 300 kilograms of coal per square metre (30 to 60 pounds per square foot) of effective grate area per hour.

6.2.5.1 Pulverized Coal

The cement industry was the first to use pulverized (powdered) coal extensively as a fuel. Public utilities and the steel industry began applying pulverized coal on an experimental basis as a boiler fuel around 1917, and by 1935 practically all large boilers (above about 45,000 kilograms or 100,000 pounds of steam per hour) in public utility power stations used this fuel, except for those stations located in the vicinity of oil and natural gas fields where local fuels were more competitive than coal. Large modern boiler installations in integrated steel plants generally use pulverized coal if coal is utilized, either as a standby or as an auxiliary fuel, in conjunction with blast-furnace gas for steam or power generation. Although pulverized coal has been used as a fuel for metallurgical purposes in steel plants, such as in open-hearth, reheating, forge and annealing furnaces, today it is employed as an injectant in the blast-furnace tuyere to increase furnace performance and efficiency.

Pulverized-coal firing offers important combustion advantages over grate firing and an economic advantage over gaseous and liquid fuels in most sections of the country. Boiler capacities are not limited as is the case with boilers equipped with stokers. Fine particles of coal burned in suspension are capable of developing a highly luminous high-temperature flame. Coal in this form may be burned normally with less excess of air above theoretical requirements than with a solid fuel, and the rate of heat release from the combustion of pulverized coal is greater than that accomplished with the solid fuel. Coal, when pulverized to the degree common for boiler uses (70% through a 0.074 mm or 200–mesh screen), has the control flexibility of gaseous and liquid fuels. Practically all ranks of coal, from anthracite to lignite, can be pulverized for combustion and each possesses specific combustion characteristics which largely influence the extent of pulverization. Pulverized-coal firing in modern boilers has certain inherent problems. Excessive fly-ash discharge from the stack, high operating power consumption rates, excessive pulverizer maintenance cost, erosion of induced-draft fan blades and other boiler components, and requirements for large furnace volumes impose practical limitations in selecting this type of firing for low-capacity boilers. Dust collectors are required to control stack particulate emissions, while scrubbers are required to remove the sulfur from the waste gas stream.

The ash-disposal problem has been one of the principal deterrents to a more extended use of pulverized coal. In the cement kiln, coal ash is no problem as it is absorbed by the cement in the kiln without adverse effect on the final product. In boilers, the principal difficulty of clogged boiler tubes and deterioration of furnace walls has been overcome by the use of slagging-type furnaces in which the ash in molten form is granulated by water jets at the bottom of the furnace well. The introduction of the cyclone furnace, which offered the removal of the ash as liquid slag, further increased the application of pulverized-coal firing for steam and power generation. This equipment was developed to solve two major problems that beset the power engineer: (1) the increasing necessity to use low quality, high-ash fuels for steam generation; and (2) the requirement that as much of the coal ash as possible be kept in the furnace and not permitted to go through the furnace and out of the stack. However, the removal of ash as a liquid slag requires the use of coal having ash of low fusion point and such special coal is sometimes difficult to obtain.

The problem of ash contamination resulting from burning fine particles of coal in suspension above a metallic liquid bath or mass of hot steel, damage to refractories from the chemical or physical action of ash, and the clogging of furnace checkers or recuperators from ash accumulation, as well as the normal availability of other fuels, has prevented widespread use of the fuel for metallurgical purposes in steel plants.

Pulverized coal offers high boiler efficiency, and means for quick regulation of boiler load. The rank of coal pulverized and the extent of pulverization particularly determine the speed of combustion. A high-volatile coal will burn faster than anthracite coal, also one with a lower ash content will burn faster. The process of combustion with pulverized coal is similar to that of lump coal but is of much higher velocity due to the introduction of the particle in suspension in a high-temperature chamber, and the greater surface exposure relative to weight. The release of volatile matter in pulverized coal is practically instantaneous when blown into the furnace, and the speed of combustion of the resulting carbonized particle and volatile gases depends upon the thoroughness with which the pulverized coal has been mixed with air. High combustion temperatures, low ash losses, and low excess air needs (10–20%), with resultant high boiler efficiencies (85– 90% with good practice), make pulverized coal an ideal boiler fuel.

Air for combustion of pulverized fuel is generally preheated, with 10–50% of that required introduced ahead of the pulverizer and the balance made up at a point near the burner. This method of introducing the air helps dry the coal and maintains a nonexplosive mix in the pulverized coal transmission system.

The combustion chamber size for pulverized coal is generally proportioned for a heat-release range of from approximately 207,000–1,035,000 watts per cubic metre or 745,000–3,725,000 kilojoules per cubic metre per hour (20,000–100,000 Btu per cubic foot per hour) of combustion space. However, the cyclone furnace has heat-release rates of approximately 5,175,000–9,315,000 watts per cubic metre or 18,625,000–33,525,000 kilojoules per cubic metre per hour (500,000–900,000 Btu per cubic foot per hour) within the cyclone chamber and the boiler furnace is used only for extracting heat from the flue gases. The difference in requirements is dependent upon whether pulverized coal is the sole fuel to be used in the chamber, the size of the coal particles, the rank of coal to be pulverized, the ash-slagging temperature of the coal, and the desired temperature for the combustion chamber. Spreader-stoker installations offer low first cost for smaller-size boilers, and the fly-ash emission from the boiler is not as severe as with the pulverized-fuel boilers.

Another technology is fluidized bed combustion which has been used to provide emission control for high sulfur coals. In this type of firing, the fuel limestone sorbent particles are kept suspended and bubbling or fluidized in the lower section of the furnace through the action of air under pressure through a series of orifices in a lower distribution plate. The fluidization promotes the turbulent mixing required for good combustion, which in turn promotes the three required parameters for efficient combustion; time, turbulence, and temperature. The limestone captures the freed sulfur from the coal products of combustion to form calcium sulfate. The mixture of ash and sulfated limestone sorbent discharged is a relatively inert material, disposal of which presents little hazard. This type firing also achieves a measure of NO_x control in that it burns at a lower temperature, somewhat below maximum NO_x formation.

6.3 Liquid Fuels and Their Utilization

Liquid fuels are essential to practically all parts of the American transportation system. The movement of passengers and freight by highway and air is dependent upon gasoline and other products of petroleum. The railways of the country have nearly all been equipped with diesel locomotives powered by fuel oil. Nearly all ocean-going ships are driven by oil, as are the majority of lake and river craft. Liquid fuels have also become of major importance as a source of heat and power in manufacturing plants. The particular advantages of petroleum as a source of energy and the available supplies have brought about a phenomenal growth in the petroleum industry.

The oil industry has recovered from the traumatic price increases of the 1970s and early 1980s which temporarily reduced the demand for oil products. The U.S. domestic demand for all oil products reached its low point in 1983 at 2.42 million cubic metres (15.23 million barrels) per day and returned to its highest level since 1979 in 1994 at 2.82 million cubic metres (17.72 million barrels) per day. During this period, the share of crude oil produced by the domestic oil industry has continued a steady decline to 1.06 million cubic metres (6.66 million barrels) per day in 1994 since reaching a peak of 1.53 million cubic metres (9.64 million barrels) per day in 1970. Unless a major new field is discovered, this decline in production and reserves will continue indefinitely.

Worldwide crude oil production in 1994 was also back to its highest levels since 1979 with a production rate of 9.63 million cubic metres (60.58 million barrels) per day with OPEC countries accounting for 41.1% of the total. This is still well below the 48.9% share OPEC held in 1979. Fortunately, as consumption of oil products has risen over the last decade, so has the known worldwide reserves. These reserves have stood at nearly 159 billion cubic metres (1,000 billion barrels) since the beginning of the 1990s. This is over 30% above the levels of the 1970s with the OPEC countries accounting for the majority of the increase.

Fuel oil is the principal liquid fuel used in the steel industry. Tar and pitch consumption has over the years decreased considerably. Fuel oil is utilized predominantly in the blast furnace as an injectant for coke replacement, in reheating furnaces as a substitute for gaseous fuels and in boilers as a backup fuel for other byproduct fuels.

Tar and pitch are byproducts of the manufacture of coke. The virgin tar as it comes from the ovens contains valuable tar-liquor oils which can be extracted and the residue pitch used as a fuel. It is customary to mix virgin tar with this highly viscous residue to provide fluidity for facilitating handling and burning, or to utilize tar in which only the lighter products have been removed by a topping process by which sufficient fluidity is retained. Pitch-tar mixtures and topped tar make available for use as a fuel 78 to 83% of the heat in the crude tar recovered in the distillation process.

6.3.1 Origin, Composition and Distribution of Petroleum

Classified according to their origins, three main types of rocks make up the outer crust of the earth: igneous, sedimentary and metamorphic rocks. Igneous rocks are formed from magma, a molten (liquid or pasty) rock material originating at high pressures and temperatures within the earth. Lava is magma that reaches the surface in the liquid or pasty state. Very commonly, the magma cools and solidifies before reaching the surface. In any case, when the molten material cools sufficiently to become solid, igneous rocks are the result. If cooling is slow, the rocks will be crystalline (granite, for example); but if the cooling is rapid, the rocks will not be crystalline but glassy in nature (obsidian, for example). Because of the nature of their origin and their usually dense, nonporous structures, igneous rocks are never hosts to petroleum deposits.

Sedimentary rocks are formed from eroded particles of rocks and soil, carried away by wind or water (and sometimes glacial action) and deposited in seas, lakes, valleys and deltas in relatively even, sometimes very thick, beds or strata (sandstones and shales are formed from deposits of this type). Other types of stratified deposits may be formed by evaporation of land-locked seas (beds of rock salt), by accumulation of the mineral remains of animals (composed chiefly of calcium carbonate, which is the principal constituent of limestone), or by chemical precipitation (gypsum and some limestones originate in this manner). The beds of sand, silt, clay, calcium carbonate or whatever eventually are covered by other sedimentary deposits, sometimes to very great depths. With the passage of long periods of time, pressure of the overlying strata, heat, cementation by chemical means, earth movements, or a combination of these or other agencies, the strata are consolidated into sedimentary rocks, typified by the few mentioned parenthetically earlier. Petroleum occurs almost entirely in sedimentary rock formations, principally sandstones and limestones, under certain ideal conditions to be described later.

Metamorphic rocks originally were sedimentary or igneous rocks. Their composition, constitution or structure have been changed through the single or combined action of natural forces such as heat, pressure, or other agencies. Marble, for example, is metamorphosed limestone.

The organic theory of the origin of petroleum, generally accepted by geologists, is that petroleum has been derived from either animal or vegetable matter, or both, by a process of slow distillation, after its burial under beds of sediments. There is evidence to indicate that the animal and vegetable matter was of marine origin; such evidence includes the association of brines with oil, the visible oily coating on seaweed found in certain localities, and the optical phenomenon of light polarization by oils similar to that of substances found in certain plants and animals and which is not shown by inorganically synthesized petroleum. The accumulation of the matter from which petroleum has been derived, its burial by sedimentary material, and the action of pressure and heat to cause distillation, has resulted in petroleum formation in many parts of the world. Geological studies indicate that petroleum was not formed in the pools in which it is found, but that the action of water pressing against oil formations caused the petroleum to flow, over a period of many years, through porous beds or strata to points of accumulation. Pools of oil occur in traps in sedimentary rocks such as sandstone or limestone. Essentially, such traps are formed by an impervious layer which prevents upward migration of the petroleum to any further extent. The oil is obtained by drilling wells into these zones of accumulation. The well is encased in a steel pipe through which it is often customary to run a number of smaller pipes to bring the product to the surface. These traps may be

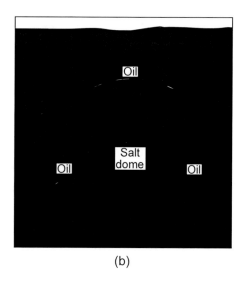
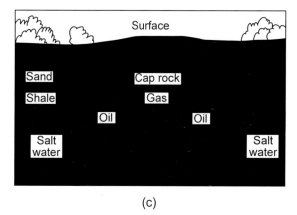
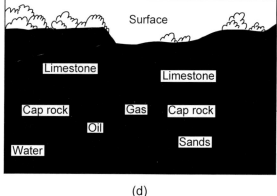

Fig. 6.3 Schematic representation of four geologic structures associated with the underground accumulation of petroleum through natural agencies: (a) stratigraphic trap, (b) salt domes, (c) anticlines, (d) faults. *From Fundamentals of Petroleum, NAVPERS 10883, Superintendent of Documents, U.S. Government Printing Office, Washington, DC.*

formed in various ways, a few of which are illustrated schematically in Fig. 6.3. In the stratigraphic trap, the producing formation gradually pinches out and disappears up the structure. An impervious layer is deposited on top of the sand, thus forming a cap rock. The solid black section in Fig. 6.3(a) represents petroleum accumulated below the cap rock. The salt dome is believed to be the result of intrusion of large masses of salt into the sediments where they are found. This intrusion creates an upward pressure and results in the doming of the overlying sedimentary rocks. In this type of structure, petroleum accumulates within the upturned porous beds about the summit and flanks of the salt core, as indicated by the solid black sections in Fig. 6.3(b). In an anticlinal structure, Fig. 6.3(c), the rocks comprising the crust of the earth are folded upward. The oil and gas are usually found on the crest of an anticlinal structure. An impervious cap rock must be present to seal the reservoir and prevent the escape of the gas and oil into higher layers. This cap rock, in one form or another, must be present in all reservoirs to contain the oil and gas within the structure. A fault, Fig. 6.3(d) is a structural closure caused by the fracturing of the crustal rocks during earth movements. In the process of folding, a reservoir for oil may be formed when a porous rock is brought into contact with an impervious layer, thus forming a trap.

Crude petroleum is a liquid containing a complex mixture of solid, liquid and gaseous hydrocarbons. The solid hydrocarbons are in solution and the liquid is at least partly saturated with gases (methane, ethane, etc.). The elementary composition of American crude oils from representative fields covers the following ranges: carbon 84–87%, hydrogen 11.5–14.0%, sulfur 0.05–3.0%, and nitrogen 0.01–1.70%

Ordinary crude petroleum is brownish-green to black in color with a specific gravity from approximately 0.810 to 0.981, and an ash content of 0.01 to 0.05%.

The principal constituents in crude oil are the paraffin (C_nH_{2n+2}), naphthene (C_nH_{2n}), and aromatic (C_nH_{2n-x}) series of hydrocarbons, and asphaltic compounds. In paraffin-base crudes, such as found in Pennsylvania, the asphaltic content is low, only traces of sulfur and nitrogen are found, and the specific gravity averages approximately 0.810. In mixed-base crudes which have a lower content of paraffins and a higher content of naphthenes than the paraffin-base crudes, the content of asphaltic compounds is higher, the sulfur content usually is under 0.4% and the paraffin-wax content is generally high. Mixed-base crudes occur in the mid-continent region. The naphthene-base crudes contain a high percentage of naphthenes and very little paraffin wax. They occur in the central, south-central and south-western areas of the U.S. Light naphthene-base crudes contain a low proportion of asphalt, compared to reverse proportions in heavy naphthene-base crudes. The sulfur content varies widely. The aromatic crudes, which occur chiefly in California, generally have a high asphaltic-compound content, sulfur content varying from 0.1 to 4.13% and a relatively high nitrogen content. The presence of wax is often widespread, although some crudes of this class are free of wax.

Crude oil is delivered by rail, ocean tankers, inland and intercoastal waterways, in specially constructed tanks, and by pipelines, including the Alaskan pipeline.

6.3.2 Grades of Petroleum Used as Fuels

Fuel oils may be classified generally as: (1) raw or natural crude petroleums, (2) distillate fuel oils, (3) residual fuel oils, and (4) blended oils. By the older methods of refining, the products from many of the oil refineries west of the Mississippi River were gasoline, naphtha, kerosene and fuel oil, while eastern refineries usually carried the fractionation of oil much further, their output being such products as gasoline, benzene, naphtha, kerosene, light machine oil, automobile oils, cylinder oils, paraffin wax and tar, pitch, or coke. Recent improvements in thermal cracking at both high and low pressure and the use of catalytic conversion processes have enabled refiners to convert more of the petroleum to gasoline and to produce lubricants from western petroleum relatively high in asphalt.

Distillate fuel oils consist of the fractions distilled intermediate between kerosene and lubricating oils. Residual fuel oils are the viscous residual products remaining after the more volatile hydrocarbons have been driven off in the refining process. Blended oils are mixtures of any or all of the three classes of fuel oils. The distribution of products obtained from crude oil in a sample of eight refineries is listed in Table 6.16.

Table 6.16 Percentages of Yield of Refined Petroleum Products from Crude Oil [a]

Fuel Gas	2%
Propane + Butane	5%
Gasoline	55%
Low Sulfur Diesel/#2 Oil	7%
Jet Fuel/Kerosene	28%
Bunker/Asphalt	3%
Total	**100%**

[a] Based on the average yield from eight different refineries.

6.3.3 Properties and Specifications of Liquid Fuels

Before discussing the more important properties and specifications of fuel oil, some of the common terms will be reviewed. Further details may be found in the ASTM Standards for Petroleum, Section 5, cited in the references at the end of this chapter.

Specific gravity is the ratio of the weight of a volume of a body to the weight of an equal volume of some standard substance. In the case of liquids, the standard is water. Baumé gravity is an arbitrary scale for measuring the density of a liquid, the unit being called Baumé degree. Its relation to specific gravity is shown by the formula:

$$\text{Be}° = \frac{140}{\text{Sp. Gr.}} - 130 \text{ (for liquids lighter than water)} \tag{6.3.1}$$

For example, the Baumé hydrometer will read 10° Bé in pure water, when the specific gravity scale reads 1.00.

The American Petroleum Institute (API) Gravity is a modification of the Baumé scale for light liquids. API gravities are always reported at 15°C (60°F). The relation between API gravity and specific gravity is:

$$°\text{API} = \frac{141.5}{\text{Sp. Gr.}} - 131.5 \tag{6.3.2}$$

The greater the degrees Baumé or API, the lighter or lower in density the fluid. There are roughly 90 API degrees between the heaviest and lightest oils which, therefore, make this scale valuable for determining differences between the density of various oils.

Flash point is the lowest temperature at which, under specified conditions, a liquid fuel vaporizes rapidly enough to form above its surface an air and vapor mixture which gives a flash or slight explosion when ignited by a small flame. It is an indication of the ease of combustion or of the fire hazard in handling or using oil.

Pour point is the lowest temperature at which oil will pour or flow when chilled without disturbance under specified conditions.

Viscosity is the property of liquids that causes them to resist instantaneous change of shape or rearrangement of their parts due to internal friction. Since this property has a direct relation to resistance of flow in fuel-oil pipe systems and to atomization, it is an important specification.

Absolute viscosity is a measure of internal fluid friction. It is defined as the tangential force on unit area of either of two parallel planes a unit distance apart when the space between the two planes is filled with liquid and one of the planes moves relative to the other with unit velocity in its own plane, and is also referred to as dynamic viscosity.

The unit for expressing absolute viscosity in SI is the Pascal-second (Pa s). By definition the pascal equals one Newton per square metre (N/m^2) and the Newton has the formula $kg\ m/s^2$, thus the dimensions for the Pascal-second are kilograms per second per metre (kg/s/m). One Pascal-second is equal to 0.1 poise or 10 centipoises (the poise and centipoise are defined below).

The cgs unit of absolute viscosity is the poise, which has the dimensions, grams per centimetre per second (g/s/cm). The centipoise is 1/100 of a poise and is the unit of absolute viscosity most commonly used in cgs. One poise equals 10 Pa s (the SI unit) and one centipoise equals 0.1 Pa s.

Absolute viscosity in the fps system is expressed in pounds force per second per foot (lbf/s/ft) The absolute viscosity of water at 20°C (68°F) in SI equals 0.1 Pa s; in the cgs system it equals 1 centipoise; and in the fps system it equals 0.00209 lbf/s/ft.

Relative viscosity of a fluid is defined as the ratio of the absolute viscosity of the fluid to the absolute viscosity of water, with both the fluid and water at the same temperature and with their respective viscosity measured in the same units.

Kinematic viscosity relates to the time for a fixed amount of a fluid to flow through a capillary tube under the force of gravity; it may be defined as the quotient of the absolute viscosity in centipoises divided by the specific gravity of a fluid, both at the same temperature. The unit of kinematic viscosity is the stoke or centistoke (0.01 stoke), derived as follows:

$$\frac{\text{centipoises}}{\text{specific gravity}} = \text{centistokes} \qquad (6.3.3)$$

The viscosity of all liquids decreases with increasing temperature and ASTM viscosity determinations are made at oil temperatures of 37.8°, 50.0°, 54.4° and 99.2°C (100°, 122°, 130° and 210°F, respectively), and are often expressed as Saybolt Universal at 37.8°C (100°F) or Saybolt Furol at 50°C (122°F). The terms "Saybolt Universal" and "Saybolt Furol" represent the type of instrument used in making the viscosity determinations. Viscosity measurements made by either method may be converted by the use of tables.

Reid Vapor Pressure is a test for the vapor pressure of gasoline at 37.8°C (100°F). It shows the tendency of gasoline to generate vapor bubbles and is expressed in SI in kiloPascals, in the cgs system in kilograms per square centimetre, and in the fps system in pounds per square inch absolute.

Octane number is the anti-knock rating of gasoline. The rating is made by matching the fuel in a test engine with a mixture of normal heptane, which detonates very easily and has an octane rating of zero, and iso-octane, which has exceptionally high anti-knock characteristics and is rated at 100. A fuel knock that matches a mixture of say 60% octane and 40% heptane would have an octane rating or number of 60. Cetane number is used to show the ignition quality of diesel oils. The rating is based on a scale resembling those of octane numbers by matching the ignition delay of the fuel against blends of cetane, a fast-burning paraffin, and methyl naphthalene, a slow-burning aromatic material.

The ASTM has developed a table for grading fuel oils, consisting of six grades. According to this classification, heating oils generally used for domestic and small industrial heating furnaces comprise Grades 1, 2 and 3. Grades 5 and 6, formerly known as Bunker B and Bunker C fuel oils, respectively, are used extensively in the steel industry. Grade 5 fuel oil is usually cracking-still tar and Grade 6 fuel oil, a straight-run or cracked residual, or a mixture of residual and cracking-still tar blended to reduce the viscosity as required by the consumer.

All grades of fuel oil are normally sold to meet specifications mutually satisfactory to buyer and seller. A typical specification of Grade 6 fuel oil is given in Table 6.17. Some purchasers may also specify the desired calorific value of Grade 6 fuel oil (for example, 41,805 kilojoules per cubic decimetre or 150,000 Btu per gallon, minimum) and the fire point of the oil (for example, 205°C (400°F) minimum and 232°C (450°F) maximum).

Table 6.17 Typical Specification for Grade 6 Fuel Oil

	Minimum	Desired	Maximum
Gravity — API at 15.6°C (60°F)	10	12	14
Viscosity — SSF at 50°C (122°F)	280	300	310
Pour Point	—	0°C (32°F)	10°C (50°F)
Flash Point	107°C (225°F)	121°C (250°F)	149°C (300°F)
Sulfur (%)	0	0.6	1.0[a]
Sodium Chloride	0	0	0.719 kg/m³ (0.006 lb/gal)
Ash (%)	0	0	0.15
Bottom Sediment and Water (%)	0	0	2.0

[a] For blast-furnace use, 1.5% maximum.

The yield of tar produced in byproduct coke ovens by high-temperature distillation between 1000° and 1100°C (1832° and 2012°F) differs within very wide limits according to the kind of bituminous coal coked, and to the temperature, coking time, and design of oven employed in the process. Virgin tar as produced in the byproduct ovens consists essentially of tar acids, neutral oils which are principally aromatic hydrocarbons, and a residue pitch.

The residue pitch from the distillation of tar is highly viscous or brittle. Pitch contains a substantial percentage of free carbon and some high-boiling and complex organic chemicals. The composition and properties of a typical pitch-tar mix and Grade 6 fuel oil are shown in Table 6.18.

Table 6.18 Composition and Properties of Typical Liquid Fuels

Fuel	Ultimate Analysis of Fuel (%)						
	H_2O	C	H_2	N_2	O_2	S	Ash
Pitch-Tar (Dry)	. . .	90.78	5.35	1.39	1.65	0.61	0.22
Pitch-Tar (Natural Basis)	1.33	89.57	5.28	1.37	1.63	0.60	0.22
Grade 6 Fuel Oil (Dry)[a]	. . .	88.60	10.50	0.30	0.00	0.55	0.05

Fuel	Specific Gravity at 15.6°C (60°F)	Mass		Dry Air Required for Combustion		Theoretical Flame Temperature	
		kg/dm³	lb/gal	m³/kg	ft³/lbm	°C	°F
Pitch-Tar (Dry)	—	—	—	—	—	—	—
Pitch-Tar (Natural Basis)	1.199	1.196	9.9855	9.41	158.13	1924	3495
Grade 6 Fuel Oil (Dry)	0.9529	0.951	7.935	10.64	180	2093	3800

Fuel	Calorific Value			
	Gross		Net	
	kJ/kg	Btu/lbm	kJ/kg	Btu/lbm
Pitch-Tar (Dry)	—	—	—	—
Pitch-Tar (Natural Basis)	37,577	16,155	36,458	15,674
Grade 6 Fuel Oil (Dry)	43,938	18,890	41,449	17,820

[a] Courtesy of Sun Oil Co. — Typical composition.

The viscosity of liquid fuels such as virgin tar, pitch-tar mixtures and topped tar decreases with temperature increase as shown in Table 6.19.

Table 6.19 Effect of Temperature on Viscosity of Various Tars and Tar Mixtures

Fuel	Test Temperature		Viscosity in Saybolt Universal Seconds		
	(°C)	(°F)	Max.	Min.	Avg.
Virgin Tar	79.4	175	189.4	73.3	109.4
Pitch-Tar Mix	79.4	175	1940	181	946.1
Pitch-Tar Mix	98.9	210	687	97	561.7
Topped Tar	93.3	200	700	550	600

6.3.4 Combustion of Liquid Fuels

The combustion of liquid fuel usually is obtained by atomizing the fuel. Atomization breaks up the fuel into fine, mist-like globules, thus permitting an increased area for intimate contact between the air supplied for combustion and the fuel. The chemistry of combustion of liquid fuels is complex. The small particles of fuel either vaporize to form gaseous hydrocarbons which burn to CO_2 and H_2O through a chain of reactions, or the fuel cracks to form carbon (soot) and hydrogen which also burn with complete combustion to CO_2 and H_2O. Both of these conditions normally occur in the combustion of liquid fuels. The first condition predominates with good atomization and proper mixing with sufficient air. A deficiency of air or poor atomization will cause smoke. For large furnaces the atomizing agent is usually steam at a pressure anywhere between 415 and 860 kPa (60 and 125 pounds per square inch) gauge. The steam consumed in atomization varies from 0.3 to 0.7 kilograms per kilogram of fuel. When liquid fuels are used in smaller furnaces, atomization usually is achieved by compressed air or by mechanical action. The character of a liquid-fuel flame, that is, its shape, size and luminosity, may be altered with a fixed burner design by changing the degree of atomization which is controlled by the steam pressure. Liquid fuels normally are burned in steel plants to produce a highly luminous flame at an intensity of flame propagation intermediate between that generally secured with coke-oven gas and that with natural gas.

Liquid fuels are often preferred because they permit better control of flame direction and, because of their high calorific value, control of flame temperature and luminosity.

The amount of air required to burn liquid fuels depends upon the chemical composition of the particular fuel. Grade 6 fuel oil requires approximately 10.64 m^3 per kilogram (180 ft^3 per lbm) of dry air for perfect combustion, and tar-pitch approximately 9.34 m^3 (158 ft^3). From the ultimate analysis of a liquid fuel, the theoretical air requirements and products of combustion may be calculated, as explained in Section 6.1.2.5.

6.3.5 Liquid-Fuel Burners

There are many different designs of burners for liquid fuels. Burners designed for atomization by steam or air may be classified into two general types, the inside mixing and the outside mixing. In the inside-mixing type the fuel and atomizing agent are mixed inside the burner or burner system, while in the outside-mixing type the two fluids meet immediately outside the burner tip. In large reheating furnaces the inside-mixing type is used. The inside-mixing type is sometimes classified as an emulsion type or a nozzle-mix type of burner. In the emulsion type the mixing is performed at a point several feet from the burner tip, while in the nozzle-mix type the two fluids meet inside the burner but very close to the burner tip. In the latter type, mixing probably takes place both inside the burner and as the stream enters the furnace. Liquid-fuel burners used in reheating, forge and annealing furnaces seldom require water cooling.

The handling of liquid fuels at consuming plants requires a system for their transportation, storage and conditioning. Where liquid fuels are received by tank car, a system of receiving basins, unloading pumps, strainers and storage tanks generally is required. The storage tanks must be of ample size to meet fuel demands between deliveries and should be provided with heaters to maintain proper fluidity for flow through pipelines to the system pressure pumps. Pressure pumps are used to deliver the liquid fuel through a pipe system to the point of consumption. Where there are a number of consuming units being served from a common fuel-storage system, the pipe feeder line is designed in the form of a loop through which the fuel flows at constant pressure and temperature. The various units tap into this loop. The fuel-oil lines are lagged and provided with tracer steam lines to maintain uniform fluidity throughout the system and to provide fuel at the burners at the proper viscosity for atomization. The temperature at which liquid fuel is delivered to the burners varies with the character of the fuel and burner design. Where pitch is used, a temperature as high as 150°C (approximately 300°F) in the lines is sometimes required. A temperature level usually somewhere between 95° and 120°C (approximately 200° and 250°F) is maintained for pitch-tar mixtures, and 65° and 95°C (approximately 150° and 200°F) for Grade 6 fuel oil. Additional details on burners and firing practices for liquid fuels appear in several references cited at the end of this chapter.

6.4 Gaseous Fuels and Their Utilization

The availability of natural gas in so many sections of this country has had a profound influence upon our industrial progress. It was first used as an illuminating gas at Fredonia, New York, in 1821. The discovery of new fields and the installation of pipelines to consuming centers led to increasing demands, as the convenience, cleanliness, and general utility of this form of fuel became better known. The initial use of natural gas for steel manufacture was at a rolling mill plant at Leechburg, Pa., in 1874. A well in this area permitted exclusive use of natural gas for puddling, heating, and steam generation for a period of six months. Since 1932 there has been an accelerated demand for natural gas, which peaked in 1972 at 0.626 trillion cubic metres (22.10 trillion cubic feet). The rapid price escalation which followed the oil shocks cut demand by 1986 to less than 75% of the previous peaks. As supplies have increased and prices have fallen, the consumption of natural gas has returned to 0.587 trillion cubic metres (20.73 trillion cubic feet) in 1994. Deregulation of the gas industry has been the main contributing factor in the resurgence of gas as the fuel of choice in the U.S.

Many industrial and some commercial gas users purchase their own gas from suppliers in the Gulf of Mexico and Southwestern states, arrange for transportation on the interstate pipeline system and if necessary pay the local utility to move the gas into the plant. In some cases, industrial users have bypassed the local utility completely and have built short pipelines from the interstate pipeline to their plant. End users have many options available to them with regards to the quality of service required. For example, a company could request firm or interruptible transportation services from the interstate and/or sign up for storage capacity with the local distribution company to hedge against winter curtailments on the interstate system.

As with oil, the U.S. natural gas reserves are shrinking while the worldwide reserves have risen substantially. For example, over the past two decades, U.S. natural reserves have fallen 30% to 4,599 billion cubic metres (162,415 billion cubic feet) and the world natural gas reserves have nearly doubled to 141,024 billion cubic metres (4,980,278 billion cubic feet). The majority of the world increase has come from the Middle East and the CIS (former Soviet Union Republics). The predominant domestic supply sources are the Gulf of Mexico, Southwestern states and Western Canada.

Producer gas was the first gaseous fuel successfully utilized by the iron and steel industry. This gas permitted the early experimentation in regeneration, and the utilization of this principle started a new era of steel manufacturing. The advantages of preheated gas and air were so clearly indicated in 1861 that producer gas rapidly became the major fuel utilized by open-hearth furnaces and maintained its position for almost sixty years, or until about 1920, when byproduct coke plants, supplying coke-oven gas and tar, began to challenge this leadership.

Blast-furnace gas utilization by the iron and steel industry probably should rank first historically, although its adoption by the industry was slower than in the case of producer gas. The sensible heat in the blast-furnace top gases was first utilized in 1832 to transfer heat to the cold blast. Originally, this heat exchanger was mounted on the furnace top. In 1845, the first attempts were made to make use of its heat of combustion, but history indicates that the burning of blast-furnace gas was not successful until 1857. It is probable that progress in the utilization of blast-furnace gas was delayed by its dust content, the problems of cleaning and handling, and the low cost of solid fuel. Increasing cost of other fuels and competition forced its use, and by the turn of this century, blast-furnace gas had become one of the major fuels of the iron and steel industry. In 1995, the steel industry used approximately 18 billion cubic metres (651 billion cubic feet) of blast-furnace gas (based on 3,535 kJ/m^3 or 90 Btu/ft^3) for blast-furnace stove heating, coke-oven underfiring, the raising of steam in boilers and heating steel in various types of furnaces. This quantity of blast-furnace gas would, in joules (Btu's), equal 1.6 billion cubic metres (57 billion cubic feet) of natural gas.

Steel Plant Fuels and Water Requirements

The initial use of byproduct coke-oven gas in the iron and steel industry was at the Cambria Steel Company, Johnstown, Pa., in 1894. This installation was followed by only a few byproduct coke-plant additions until a shortage of transportation facilities and the rising price of coal and natural gas during the first World War accelerated installations throughout the steel industry. The utilization of coke-oven gas has been very profitable as it reduced the purchase of outside fuels. It is estimated that plants operating steelmaking furnaces in the U.S. used approximately 11,794 million cubic metres (28,052 million cubic feet) of coke-oven gas as fuel in 1995 (based on 19,640 kJ/m^3 or 500 Btu/ft^3). This is only a small fraction of the quantities used as recently as 1982.

The gas produced by the basic oxygen process has not been utilized by the U.S. steel industry due to the cost of the equipment required to gather and clean the gas versus the cost of alternative fuels. In Europe and Japan, where substitute fuels are more expensive, BOF gas has been collected and utilized by many of the steel works for a number of years. Approximately 0.70 GJ/tonne (0.60 million Btu/ton) of offgas can be collected from each heat in a BOF with a typical size of 181 tonne (200 ton). The recovered gas has a calorific value of 8837 kJ/m^3 (225 Btu/ft^3).

6.4.1 Natural Gas

Natural gas and petroleum are related closely to each other in their chemical composition and in geographical distribution. Both are made up predominantly of hydrocarbons. Petroleum rarely is free of natural gas, and the same fields usually produce both fuels. When natural gas exists indigenous to an oil stratum and its production is incidental to that of oil, it is called casinghead gas. Gas found in a field is usually under pressure which diminishes with extended use or, sometimes, from the presence of too many other wells. The life of a well varies from a few months to twenty years. Rocks bearing gas are sandstones, limestones, conglomerates, and shales, never igneous rocks. Natural gas is derived from the remains of marine animal and plant life—in theory, the same as described previously for petroleum.

Natural gas, as found, is usually of singular purity and is composed principally of the lower gaseous hydrocarbons of the paraffin series, methane and ethane, some of the heavier liquefiable hydrocarbons (which are recovered as casinghead gasoline or sold in bottled form as butane, propane, pentane, etc.) and a small amount of nitrogen or carbon dioxide. Some natural gases contain small quantities of helium. Occasionally, wells are found in which the gas contains hydrogen sulfide and organic sulfur vapors. Sour gas is defined as a natural gas which contains in excess of 0.0343 grams of hydrogen sulfide or 0.686 grams of total sulfur per cubic metre, equivalent to $1^1/_2$ grains of hydrogen sulfide or 30 grains of total sulfur per 100 cubic feet. It is fortunate, however, that by far the greater part of natural gas available in this country is practically sulfur-free.

The principal constituent of natural gas is methane, CH_4. Because natural gas contains from 60 to 100% CH_4 by volume, the characteristics of methane gas, which were shown in Section 6.1, largely dominate the parent gas. Comparing methane with the other principal combustible gases, it will be noted that it has a low rate of flame propagation, a high ignition temperature, and a narrow explosive range. Methane, as well as all other hydrocarbons (of which it is the lowest member), burns with a luminous flame. Typical compositions of natural gas are presented in Table 6.20.

The iron and steel industry consumed a total of 7,892 million cubic metres (278,711 million cubic feet) of natural gas in 1995 in blast furnaces and other uses in the blast-furnace area, steel-melting furnaces, heating and annealing furnaces, heating ovens for wire rods, and other uses.

6.4.2 Manufactured Gases

The four most important of the commercially used manufactured gases are producer gas, water gas, oil gas, and liquefied petroleum gases. Because none of these gases except liquefied petroleum gases are used presently in steel manufacturing or processing in the U.S., only a brief description of their manufacture and characteristics will be given here.

Table 6.20 Typical Composition of Natural Gas in Various States in the United States (Based on GRI Survey for 1992) [a]

Constituents (% by Volume)	Louisiana[b]	Illinois[c]	Ohio[d]	Pennsylvania[d]	California[e]
Methane	93.7	92.6	89.7	96.1	91.2
Ethane	2.3	3.6	4.5	2.2	4.1
Propane	0.6	0.6	1.3	0.3	1.1
Butanes	0.3	0.2	0.4	0.2	0.4
Pentanes	0.1	0.1	0.1	0.1	0.1
Hexane +	0.1	0.1	0.1	0.1	0.1
Oxygen	0	0	0	0	0
Inerts (CO_2 & N_2)	2.9	2.9	3.9	1.1	3.0
Gross Heating Value					
Btu/Scf [g]	1023.1	1031.3	1044.9	1029.8	1048.4
MJ/m^3 [h]	40.19	40.51	41.04	40.45	41.18
Specific Gravity	0.597	0.601	0.619	0.581	0.615

[a] Bulletin GR92-0123
[b] State Wide Summary Data
[c] Gate Station B
[d] Company #1 Station A
[e] City #1 Gate Station E
[f] At 30″ Hg, 60°F, Dry
[g] At 760 mm Hg, 0°C, Dry

6.4.2.1 Manufacture of Producer Gas

Producer gas is manufactured by blowing an insufficient supply of air for complete combustion, with or without the admixture of steam, through a thick, hot, solid-fuel bed. A large proportion of the original heating value of the solid fuel is recovered in the potential heat of carbon monoxide, hydrogen, tarry vapors, and some hydrocarbons, and in the sensible heat of the composite gas which also contains carbon dioxide and nitrogen. When the gas is cleaned, the sensible heat of the gases and the potential heat of the tar vapors is lost.

Table 6.21 gives the composition of clean producer gas made from various fuels in a well-operated updraft producer.

The gross heating value of raw producer gas, including tar, made from a high-volatile coal, 8% ash, is about 6678 to 7463 kilojoules per cubic metre (170 to 190 Btu per cubic foot).

Producer gas has a very low rate of flame propagation due to the relatively large amount of inert gases, N_2 and CO_2, it contains. The hot gas, containing tar, burns with a luminous flame; the cold gas is only slightly luminous, while it is non-luminous if made from anthracite coal or coke. Producer gas is a relatively heavy gas and has a wide explosive range. The theoretical flame temperature is low, approximately 1750°C (3180°F), and the gas generally was preheated when utilized in steel plant processes.

6.4.2.2 Manufacture of Water Gas

Water gas or blue gas is generated by blowing steam through an incandescent bed of carbon. The gas-forming reactions are primarily:

$$C + H_2O = CO + H_2 \qquad (6.4.1)$$

$$C + 2H_2O = CO_2 + 2H_2 \qquad (6.4.2)$$

Table 6.21 Composition of Clean Producer Gas [a]

		Solid Fuel Feed			
		Coke		Bituminous Coal	
Constituent	Anthracite Coal	100 to 125 mm (4 to 5") Lump	Breeze	A	B
CO_2	6.3%	9.2%	8.7%	3.4%	9.2%
Illuminants	0.0	0.1	0.0	0.8	0.4
O_2	0.0	0.0	0.0	0.0	0.0
CO	25.0	21.9	23.3	25.3	20.9
H_2	14.2	11.1	12.8	9.2	15.6
CH_4	0.5	0.2	0.4	3.1	1.9
N_2	54.0	57.5	54.8	58.2	52.0
Total	100.0%	100.0%	100.0%	100.0%	100.0%
Gross Heating Value,					
kJ/m^3 [b]	5185	4753	5146	6088	6128
Btu/ft^3 [c]	132	121	131	155	156

[a] U.S. Bureau of Mines, Bulletin 301.
[b] At 760 mm Hg, 0°C, dry.
[c] At 30 in. Hg, 60°F, dry.

Fuel oil may be cracked in a separate heating chamber to form gases that are added to water gas to enrich it when carburetted water gas is made.

While coke generally is used as the fuel in the production of water gas because of its high carbon content and cleanliness, anthracite and bituminous coal and mixtures of coal and coke also have been used successfully, but with some sacrifice in overall operating efficiency.

Water gas burns with a clear blue flame; hence, the name blue gas. It is used in a number of chemical processes to supply a basic gas for synthetic processes, but it is not suitable for distribution as a domestic fuel unless it has been enriched with cracked fuel oil, when it is called carburetted water gas.

Water gas made from coke burns with a nonluminous flame. Carburetted water gas burns with a highly luminous flame. Both gases have a high rate of flame propagation. The speed of combustion for water gas exceeds that of any other extensively used fuel gas; that for carburetted water gas is practically the same as for coke-oven gas. Water gas has a slightly lower specific gravity than natural gas, but is somewhat heavier than coke-oven gas. Carburetted water gas is heavier than natural gas but lighter than producer gas. The theoretical flame temperature of both blue and carburetted water gas is very high, approximately 2020°C and 2050°C (3670° and 3725°F), respectively, exceeding that of all other industrial fuel gases commonly used. Both gases have a relatively wide explosive range.

6.4.2.3 Oil Gas

Oil gas is a combination of cracked petroleum and water gas made by passing oil and steam through hot refractory checker work. Oil gas is commercially important in localities where coal or coke is expensive and oil is cheap.

6.4.2.4 Liquefied Petroleum Gas

Liquefied petroleum gases (LPG), sometimes referred to as bottled gases, have become commercially important because of the concentration of fuel energy in liquid form which may be converted easily into a gas. They are distributed for household use in steel cylinders, called bottles, and in tank cars or trucks for industrial purposes. They are sometimes sold under various trade names but are composed mainly of butane, propane, and pentane. A steel cylinder of propane as sold for domestic purposes contains approximately 50,300 kilojoules per kilogram (21,640 Btu per pound) of liquid gas. Larger versions of these containers have been used by the steel industry as back up supply for natural gas systems.

6.4.2.5 Alternative Fuel Sources

Drastic escalation of oil prices by the Organization of Petroleum Exporting Countries (OPEC) in the 1970s stimulated interest in alternative sources of fuels to supplement existing supplies. A number of projects began operation during the early 1980s. Tennessee Eastman built a coal gasification plant to provide synthetic gas for conversion to chemicals. The Cool Water Project in California integrated coal gasification with the generation of electricity by a combination of gas and steam turbines. Union Oil Company began operation of a shale oil facility at Parachute, Colorado, and a consortium built the Great Plains Coal gasification plant to produce synthetic natural gas. Companies also started to recover gas from sanitary landfills.

Once oil and gas prices stabilized at more normal levels, many of these projects became uneconomic. Over time, these types of alternative fuel sources may once again become economically attractive.

6.4.3 Byproduct Gaseous Fuels

The two major byproduct gaseous fuels recovered by the steel works are blast-furnace and coke-oven gases. A number of other unavoidable gaseous fuels are created by regular manufacturing processes. Some of these are of minor economic consequence, but the majority are useful and generally utilized at the plant where they are produced. An exception is oil-refinery gas which is sometimes piped and marketed to industries adjacent to refineries. The calorific value and flame characteristics of byproduct gases have wide ranges. Blast-furnace gas has probably the lowest heat content of any, and oil refinery gas the highest, respectively about 3535 and 72,668 kilojoules per cubic metre (90 and 1850 Btu per cubic foot), although both vary from these values.

6.4.3.1 Blast-Furnace Gas

Blast-furnace gas is a byproduct of the iron blast furnace. The paramount objective in blast-furnace operation is to produce iron of a specified quality, economically; the fact that usable gas issues from the top of the furnace is merely a fortunate attendant circumstance. When air enters the tuyeres, its oxygen reacts with the coke. The resulting gas passes up through the shaft of the furnace which has been charged with coke, ore, and limestone, and after a number of chemical reactions and a travel of some 25 metres (80 feet), issues as a heated, dust-laden, lean, combustible gas. The annual volume production of this gas is greater than that of any other gaseous fuel. Two and one-half to three and one-half tons of blast-furnace gas are generated per ton of pig iron produced. While the purpose of the gases generated by the partial combustion of carbon is to reduce iron ore, the value of a blast furnace as a gas producer is evident from the relation just noted.

The percentage of CO and CO_2 in blast-furnace gas is directly related to the amount of carbon in the coke and the amount of CO_2 in the limestone charged per ton of iron produced. The rate of carbon consumption depends principally upon the kind of iron to be made, the physical and chemical characteristics of the charged material, the distribution of the material in the furnace stack, the furnace lines, and the temperature of the hot blast. The total $CO+CO_2$ content of the top gas is approximately 40% by volume, and when producing ordinary grades of iron the ratio of CO to CO_2

will vary from 1.25:1 to 2.5:1. The hydrogen content of the gas varies from 3 to 5% depending on the type and amount of tuyere-injected fuels. The remaining percentage is made up of nitrogen, except for approximately 0.2% CH_4.

Blast-furnace gas leaves the furnace at a temperature of approximately 120° to 370°C (approximately 250° to 700°F), and at a pressure of 345 to 1380 mm Hg gauge pressure (15 inches w.g. to 14.5 psig), carrying with it 22 to 114 grams of water vapor per cubic metre (10 to 50 grains per cubic foot) and 18 to 34 grams of dust per cubic metre (8 to 15 grains per cubic foot). The particles of dust vary from 6.4 to 0.000254 mm in diameter. In early days of blast-furnace operation, the gas was used as it came from the furnace without cleaning, causing a great deal of trouble with flues, combustion chambers, and stoves due to clogging. The gas now is cleaned almost universally, the degree depending upon the use.

The outstanding characteristics of blast-furnace gas as a fuel are: (1) very low calorific value—approximately 2946 to 3535 kilojoules per cubic metre (75 to 90 Btu per cubic foot) depending on blast-furnace coke rate, (2) low theoretical flame temperature—approximately 1455°C (2650°F), (3) low rate of flame propagation—relatively lower than any other common gaseous fuel, (4) high specific gravity—highest of all common gaseous fuels, and (5) burns with a non-luminous flame.

6.4.3.2 Coke-Oven Gas

The steel industry, which uses the majority of the total coke-oven gas generated in the U.S., generally classifies coke-oven gas as a byproduct of coke manufacture. This undoubtedly is due to the former waste of coke-oven gas and other coal products for so many years in the beehive coke process. Actually, the production of coke-oven gas and other coal chemicals is a part of an important manufacturing process, in which large sums have been expended for their recovery, as they have a value almost equal to that of the coke. Coke-oven gas is produced during the carbonization or destructive distillation of bituminous coal in the absence of air. Approximately 310 cubic metres of 19,640 kJ/m^3 gas are produced per metric ton of coal coked (about 11,000 cubic feet of 500-Btu gas per net ton of coal coked) in conventional high-temperature coking processes.

The composition of coke-oven gas varies in accordance with the grade and density of coal and operating practices. Typical percentage ranges for constituents of dry coke-oven gas by volume are presented in Table 6.22.

Table 6.22 Typical Properties of Coke-Oven Gas

CO_2 (includes H_2S)	1.3–2.4%
O_2	0.2–0.9%
N_2	2.0–9.6%
CO	4.5–6.9%
H_2	46.5–57.9%
CH_4	26.7–32.1%
Illuminates	3.1–4.0%
Specific Gravity	0.36–0.44
Heating Value (Gross)	
kJ/m^3	21,093–22,782
Btu/ft^3	537–580
Heating Value (Net)	
kJ/m^3	18,854–20,543
Btu/ft^3	480–523

Coke-oven gas contains hydrogen sulfide, H_2S. Approximately 40% of the sulfur in coal, not removed in the washing process, is evolved with the distillation products. Much of this remains in the gas. Carbonization of coals containing 1.20% sulfur evolves a gas containing approximately 9.7 grams of sulfur per cubic metre (424 grains per 100 cubic feet), and those containing 1.60% sulfur approximately 14 grams of sulfur per cubic metre (600 grains per 100 cubic feet). Commercial coals in the eastern part of the U.S. usually run from 0.5 to 1.5% sulfur. Gases high in sulfur content are very undesirable for metallurgical purposes.

Coke-oven gas normally is saturated with water vapor. In distribution systems, means must be provided for draining off the condensation due to any temperature change.

Coke-oven gas burns with a non-luminous to semi-luminous flame, depending upon the degree of mixing air and gas. Its rate of flame propagation is high, considerably higher than natural, producer, or blast-furnace gas. It has a low specific gravity, in fact the lowest of any of the gaseous fuels commonly utilized by the steel industry. It has a high theoretical flame, approximately 1980°C (3600°F), which is a little higher than that of natural gas. The explosive range is roughly twice that of natural gas.

6.4.3.3 Basic Oxygen Furnace Gas

One fuel that has not been economically recovered by the domestic steel industry is basic oxygen furnace gas. The utilization of this fuel requires special collection and cleaning equipment which would allow for its use in the steel plant. The special equipment includes such items as an adjustable skirt on the vessel hood, a flare, a gas holder, by-pass valving and a wet electrical precipitator. BOF gas of sufficient quality to be used is collected only during part of the blowing cycle. For example, the start of gas recovery begins when the CO content is above 30% and the O_2 content falls below 2%. The stoppage of gas recovery occurs when the O_2 content is above 2% or the CO content falls below 30%. These parameters may vary from plant to plant. The composition of the gas varies throughout the blow but a typical analysis (by volume) is as follows: 57% CO, 26% CO_2, 2% H_2O, 15% N_2, and O_2 <1% (most of the cycle).

In steel plants in other countries which utilize BOF gas, the most predominate use is in the boiler plant, either directly or blended with blast-furnace gas. More recently, BOF gas has also been utilized along with blast-furnace gas in gas turbine combined cycle units, which are much more efficient in producing power than a conventional boiler and steam turbine generator set.

6.4.4 Uses for Various Gaseous Fuels in the Steel Industry

Gaseous fuels are ideal for many steel plant applications. Below are the more important applications where gaseous fuels either must be used because of the nature of the work or facility, or where they are preferred over liquid or solid fuel:

- Coke-Oven Heating
- Blast-Furnace Stoves
- Gas Turbines for Power Generation
- Boilers
- Soaking Pits
- Reheating Furnaces
- Forge and Blacksmith Furnaces
- Normalizing and Annealing Furnaces
- Controlled-Cooling Pits
- Foundry Core Ovens
- Blast Furnace and Steel Ladle Drying
- Drying of Blast-Furnace Runners
- Hot-Top Drying
- Ladle Preheating
- Oxy-Fuel Burners

The choice of the most desirable fuel for each of the many facilities in a steel plant is not always possible, but by judicious planning the most efficient fuel or combination can be selected from those available. The general characteristics of each gas govern, wherever possible, its selection for a specific purpose in a steel plant. An outline of the important applications of the major gaseous fuels follows.

6.4.4.1 Use of Blast-Furnace Gas

For many years the use of blast-furnace gas for purposes other than for the firing of stoves and boilers was not economical. A number of factors have contributed, however, to the enlarged use of blast-furnace gas, the more important of which are: (1) rising cost of purchased fuel; (2) technical progress in gas cleaning, in the use of regeneration and recuperation, and in the mixing of gaseous fuels; (3) the economic advantage of using pulverized coal in boiler houses to substitute for blast-furnace gas, thereby permitting its substitution elsewhere for the more expensive liquid and gaseous fuels; and (4) seasonal shortages in the availability of purchased liquid and gaseous fuels.

In certain applications, in addition to preheating the air, the gas itself may be preheated to provide higher temperature potential. For the facilities listed, blast-furnace gas may be utilized successfully without preheat: blast furnace stoves, soaking pits, normalizing and annealing furnaces, foundry core ovens, gas engines for blowing or power generation, gas turbines for power generation, and boilers.

The thermal advantage of using blast-furnace gas in gas engines for blowing and for electric-power generation must overcome the heavy investment and maintenance expense of this equipment. The modern boiler house utilizing high steam pressure and temperature with efficient turbo-blowers and generators has sufficiently reduced the thermal advantage of gas engines, such that their use is difficult to justify. Some steel plants in Asia and Europe have been successful in the use of direct-connected gas turbines for driving generators.

Preheated blast-furnace gas burned with preheated air has been used successfully in coke-oven heating, soaking pits, and reheating furnaces.

When blast-furnace gas is preheated, it should have a minimum cleanliness of 0.023 grams per cubic metre (0.01 grains per cubic foot); and in all cases where this gas is used, extra precautions must be taken to prevent the escape of fuel or unburned gas into attendant surroundings because it contains a large percentage of toxic CO gas. Blast-furnace gas is used for many applications in the steel plant and, in addition, is used frequently for heating coke ovens and sometimes is mixed with other gases as a fuel. In blast furnace operations, where the blast-furnace gas has a heating value approaching a low value of 2946 kilojoules per cubic metre (75 Btu per cubic foot), it is necessary to switch the gas with other fuels to obtain very high hot-blast temperature from the stove.

6.4.4.2 Use of Coke-Oven Gas

Coke-oven gas has had a more extended use than blast-furnace gas because of: (1) relatively low distribution costs due to its low specific gravity and high calorific value; (2) its ability to develop extremely high temperatures by combustion; and (3) the high rate at which it can release heat, thereby eliminating excessively large combustion chambers. The low specific gravity of coke-oven gas is a disadvantage, and for this reason, it is supplemented wherever possible with a driven liquid fuel. In addition, the sulfur (in the form of H_2S) present in raw (not desulfurized) coke-oven gas is a distinct disadvantage, particularly in heating certain grades of alloy steel for rolling. Its presence also requires the use of materials resistant to sulfur attack in pipelines, valves, and burners.

There are a number of fuel applications in a steel plant where neither blast-furnace gas nor coke-oven gas, when burned alone, develop the desired flame characteristics or temperature level for optimum results. By mixing two fuels of such great variance in characteristics, a more ideal fuel can be obtained for specific applications.

The speed of combustion is very high for coke-oven gas and very low for blast-furnace gas. The desired speed can be attained through the proper proportioning of the two fuels. The speed also can be modified to a limited extent when necessary by suitable combustion technique. Mixed blast-furnace and coke-oven gas is particularly suitable for application to soaking pits and reheating furnaces.

6.4.4.3 Use of Natural Gas

Due to plant balances requiring the purchase of outside gaseous fuels, mixtures of coke-oven gas and natural gas are often utilized. While the temperature-developing characteristics of these two gases are nearly identical, they have differences in other characteristics, notably in the rate of flame propagation and in luminosity. By proper proportioning, the advantage of a short, intensive cutting flame or a long, luminous, soft flame may be had to suit the applications. Use of natural gas for flame cutting and scarfing has been increasing steadily.

6.4.4.4 Use of Producer Gas

Raw, hot, producer gas was used extensively in the past in steel plant operations for open-hearth furnaces, soaking pits, and reheating furnaces. It was customary to preheat this gas regeneratively when used in batch-type reheating furnaces. In continuous-type reheating furnaces, the fuel seldom was preheated. With good gas making, producer gas develops a soft, heavy, long, luminous flame desirable for reheating steel. The use of this gas has been superseded in all plants by natural gas and byproduct gaseous and liquid fuels.

6.4.5 Combustion of Various Gaseous Fuels

The major combustion reactions of the components of gaseous fuels with air and a table of essential gas combustion constants were given in Section 6.1.2.5. From chemical equations, the quantity of air required to provide perfect combustion and the resultant products may be calculated for any given gaseous fuel. Table 6.23 shows the air requirements, products of combustion, and pertinent characteristics of several gaseous fuels. The degree of mixing of air with a gaseous fuel, and the degree of excess or deficiency of air to the theoretical requirements are pertinent combustion problems. The degree of mixing is controlled by burner design. Burners have been developed to produce short, intense flames or long, slow-burning flames. The short, intense flame is usually nonluminous or semi-luminous, while the long flame is luminous. This relation is not always the case, however, because a gas must contain hydrocarbons to develop luminosity. Burners capable of producing short, intense flames will liberate a large amount of heat in a small space. Some gases, due primarily to the constituents of which they are composed, are capable of a high rate of heat release; others, of a very low rate of heat release. The two extremes are evident in two common steel-plant fuels, coke-oven gas and blast-furnace gas, which give high and low rates of heat release, respectively. There is also a limit to the length of flame which can be produced. It is determined by the ability of the flame to provide enough heat to propagate itself. If the short, intense-flame type burner is used with coke-oven or natural gas, combustion will be so intense that no flame will be visible, and heat can be liberated at rates over 40 million watts per cubic metre per hour (up to several million Btu per cubic foot per hour) of combustion space; while the long, slow-burning-flame burner firing the same gases is capable of developing a visible flame 6 to 9 metres (20 to 30 feet) long with a heat liberation of 155,000 to 210,000 watts per cubic metre (15,500 to 21,000 Btu per cubic foot per hour). Both types of flames are desirable for specific steel plant applications. It is obvious that burner selection based on degree of mixing is important. Carrying an excess or deficiency of air for combustion is practiced usually to control scale formation, but this is done sometimes to control flame characteristics. An excess of air tends to shorten, while a deficiency lengthens, a flame. An excess of air above theoretical requirements causes higher heat losses as any extra air absorbs its share of the heat of combustion. Fuels containing hydrogen must have the exit stack temperature high enough to avoid condensation of water within the furnace system and subsequent corrosion of furnace parts. In some cases, it is necessary to burn the fuel with

Table 6.23 Properties of Typical Gaseous Fuels

Fuel Gas	Constituents of Fuel Gas % by Volume (Dry Basis)						Illuminants		Specific Gravity	Unit Vols. of Air Required for Combustion of Unit Vol. of Gas (m^3 or ft^3)	Heating Value per Unit Volume of Gas [a]			
											kJ/m^3		Btu/ft^3	
	CO_2	O_2	N_2	CO	H_2	CH_4	C_2H_6	C_2H_4	C_3H_6			Gross	Net	Gross	Net
Natural Gas (Pittsburgh)	—	—	0.8	—	—	83.4	15.8	—	—	0.61	10.58	44,347	40,105	1129	1021
Reformed Natural Gas	1.4	0.2	2.9	9.7	46.6	37.1	—	1.3	0.8	0.41	5.22	23,529	21,054	599	536
Coke-Oven Gas	2.2	0.8	8.1	6.3	46.5	32.1	—	3.5	0.5	0.44	4.99	22,547	20,190	574	514
Water Gas (Coke)	5.4	0.7	8.3	37.0	47.3	1.3	—	—	—	0.57	2.10	11,273	10,291	287	262
Carburetted Water Gas	3.0	0.5	2.9	34.0	40.5	10.2	—	6.1	2.8	0.63	4.60	21,604	19,954	550	508
Oil Gas (Pacific Coast)	4.7	0.3	3.6	12.7	48.6	26.3	—	2.7	1.1	0.47	4.73	21,643	19,483	551	496
Producer Gas (Bituminous Coal)	4.5	0.6	50.9	27.0	14.0	3.0	—	—	—	0.86	1.23	6403	6010	163	153
Blast Furnace Gas	11.5	—	60.0	27.5	1.0	—	—	—	—	1.02	0.68	3614	3614	92	92
Butane (Commercial) 93% C_4H_{10}—7% C_3H_8	—	—	—	—	—	—	—	—	—	1.95	30.47	126,678	116,937	3225	2977
Propane (Commercial) 100% C_3H_8	—	—	—	—	—	—	—	—	—	1.52	23.82	101,028	93,133	2572	2371

Table 6.23 (continued)

Fuel Gas	Products of Combustion in Unit Vol. per Unit Vol. of Fuel (m³ or ft³)				Ultimate % CO_2	Net Heat Content per Unit Volume of Products of Combustion[b]		Theoretical Flame Temperature, No Excess Air	
	H_2O	CO_2	N_2	Total		kJ/m³	Btu/ft³	°C	°F
Natural Gas (Pittsburgh)	2.22	1.15	8.37	11.73	12.1	3417	87.0	1961	3562
Reformed Natural Gas	1.30	0.53	4.16	5.99	11.3	3519	89.6	1991	3615
Coke-Oven Gas	1.25	0.51	4.02	5.78	11.2	3417	87.0	1988	3610
Water Gas (Coke)	0.53	0.44	1.74	2.71	20.1	3794	96.6	2021	3670
Carburetted Water Gas	0.87	0.76	3.66	5.29	17.2	3779	96.2	2038	3700
Oil Gas (Pacific Coast)	1.15	0.56	3.77	5.48	12.9	3555	90.5	1999	3630
Producer Gas (Bituminous Coal)	0.23	0.35	1.48	2.06	18.9	2930	74.6	1746	3175
Blast Furnace Gas	0.02	0.39	1.14	1.54	25.5	2337	59.5	1454	2650
Butane (Commercial)	4.93	3.93	24.07	32.93	14.0	3555	90.5	2004	3640
Propane (Commercial)	4.0	3.0	18.82	25.82	13.75	3582	91.2	1967	3573

(a) From: "Combustion," American Gas Association (Third Edition); "Gaseous Fuels," American Gas Association (1948); and "Gas Engineers Handbook" (see bibliography at end of chapter).
(b) kJ/m³ at 0°C, 760 mm Hg (dry); Btu/ft³ at 60°F, 30 in. Hg (dry).

excess air to maintain the flue gases above their dewpoint. When there is a deficiency of air, potential heat is lost. In problems of design and fuel conservation, the air requirements and volume and constituents of the products of combustion must be known to effect a practical solution.

6.5 Fuel Economy

Because fuel represents the largest single item of raw material expense for the manufacture of iron and steel, the subject of fuel economy is of consequence to both the producer and consumer of steel products. The world steel industry accounts for nearly 20% of the worldwide total industrial energy consumption. The efficient utilization of this large quantity of fuel is also pertinent to the conservation of our fuel resources. The history of the steel industry shows great progress has been made in reducing the amount of fuel required to produce a ton of steel. During the Revolutionary War, ironmaking required large quantities of charcoal, as the source of carbon, to reduce the ore. If a substitute had not been found for charcoal, our forests would have disappeared many years ago and our industrial progress arrested. In the past one hundred years, which really represents the modern era of steelmaking, a number of important developments have taken place to reduce the fuel requirements in producing steel. Some of these developments could be listed by historical sequence, while others are of such a nature that they cannot be designated by any period of time.

The major contributions to fuel economy in ironmaking and steelmaking plants have been:

1. Development of the Bessemer converter.
2. Development of the Siemens-Martin regenerator.
3. Development of the hot blast.
4. Utilization of blast-furnace gas.
5. Installation of byproduct coke plants and utilization of byproduct fuels.
6. Integration of steel plants.
7. Electric drives for rolling mills.
8. Improved efficiency of steam-generating equipment and steam prime movers.
9. Large producing units.
10. Balancing of producing units.
11. Use of raw materials with improved chemical and physical quality.
12. Recovery of waste heat by recuperators, boilers and other forms of heat exchangers.
13. Development and utilization of instruments and control equipment.
14. Insulation of high-temperature facilities.
15. Utilization of the optimum fuel for specific facilities.
16. Improvements in manufacturing technique and production control.
17. More highly skilled operators.
18. Development of oxygen-blown steelmaking processes.
19. Development of continuous casting of steel and, more recently, thin slab casting.
20. Recovery of sensible heat by hot charging and direct rolling.
21. Gas turbine combined-cycle facilities.

The results of the above contributions now have made it possible to produce a ton of raw steel utilizing far less energy than was ever thought possible. The consumption of primary fuels in the iron and steel industry for 1995 is shown in Table 6.24.

Table 6.24 Primary Fuels Consumed by the Steel Industry in 1995 [a]

Fuel	Gigajoules[b]	Million BTUs[b]
Coal and Coke	688,302,130	652,510,419
Natural Gas	294,056,779	278,765,826
Electricity[c]	91,068,079	86,332,539
Liquid Fuels	25,496,395	24,170,582
Oxygen[d]	77,337,694	73,316,134
Steam	3,088,522	2,927,919
Net Energy Consumed	**1,179,349,599**	**1,118,023,419**
Gross Steel Shipments	48,248,612 (metric tons)	53,194,095 (short tons)
Fuel Consumption/Ton	24.44	21.02

[a] From: "Annual Statistical Reports" (1995), American Iron and Steel Institute
[b] Higher Heating Valve
[c] 3600 Kilojoules/kW (3413 Btu/kW)
[d] 37.2 Kilojoules/m^3 (175 Btu/ft^3)

In addition to these outstanding contributions to fuel economy in steel mills, the importance of the effect that the rate of operations has on fuel economy should be stressed. Historically, the iron and steel industry follows the general business level maintained in the country, but its rate of operations often fluctuates more than that of many other industries. During peak production, optimum fuel economy is the natural result of operating the facilities which require fuel under the conditions for which they were designed to operate most economically. During periods of low production, fuel consumption undergoes a severe increase per unit of output; careful scheduling of production and facilities are required during this period to maintain minimum fuel losses.

The effectiveness with which byproduct fuels are used in steel plants is of major significance in reducing the quantity of primary or purchased fuel required to produce a ton of steel. The efficiency of heat utilization by steelmaking furnaces is discussed in the chapters dealing with the design and operation of these units in Chapters 8 and 10.

The heat lost from the combustion of fuel in steel plant metallurgical and service facilities represents an appreciable part of the total supply. The amount lost differs among the various processes. In general, those processes having the higher temperature levels have the greatest thermal losses and, therefore, offer the best opportunity for heat recovery. For a specific facility, the amount and causes for these losses can be determined quantitatively by conducting a heat balance, and the results of this heat balance can then be used for planning a program aimed at heat conservation. Usually, the largest losses are contained in the waste flue gases and in radiation from the furnace walls. Additional losses are associated with combustion control (providing insufficient air for combustion at the burners and/or inadequate mixing of fuel and air), heating practice and air infiltration. Most important common denominators underlying all successful programs of heat conservation are proper maintenance of the facility and its instrumentation and proper scheduling of operations so that the facility is, as closely as practical, fully utilized for its intended purpose.

6.5.1 Recovery of Waste Heat

The recovery of heat from waste flue gases of high-temperature processes has been practiced for nearly 100 years. These high-temperature flue gases contain both sensible heat and the latent heat of vaporization of water and sometimes potential heat (unburned flue gases). The recovery of the heat of vaporization of water is not practical; however, the recovery of the sensible heat is accomplished by one or a combination of several methods, including the use of regenerators, recuperators, waste-heat boilers, or utilization of a furnace design in which the waste heat is used to preheat the product.

Regenerators are used alternately to absorb heat from one fluid and then transfer it to another fluid; recuperators are used to transfer heat continuously from one fluid to another. The fluids referred to in these two definitions are: (1) hot gaseous products of combustion which give up heat during passage through the regenerator or recuperator and (2) fuel gases or air for combustion which undergo heating while passing through the regenerator or recuperator. The importance of recuperators or regenerators for preheating the air to be used for combustion is shown in Fig. 6.4. In this illustration, the amount of fuel saving (in percent) is plotted vertically, and the temperature of the flue gas at exit is plotted horizontally. The temperature of the preheated air is shown on the curves, and from these an estimate can be obtained of the amount of fuel saving to be gained from using preheated air.

Regenerators are applied usually to furnaces which can be fired alternately from the ends, the flow of gases through the furnace and regenerators being reversed according to predetermined time and/or temperature cycles. Coke ovens, and many batch-type reheating furnaces and soaking pits, are equipped with regenerators. Blast-furnace stoves also use the regenerative principle but operate over a much longer cycle and in a somewhat different manner than that practiced in other installations. In a blast-furnace stove, checker brick of the regenerator are heated by burning of fuel exclusively for the purpose of heating the regenerator brick, while in other furnace installations the checker brick are heated by waste gases. In both cases, the heat stored in the regenerators is used to preheat air for the combustion of fuel in the furnace they serve.

Recuperators have been applied to many modern batch and continuous-type reheating furnaces, and to steam boilers. When applied to steam boilers, they commonly are called air preheaters.

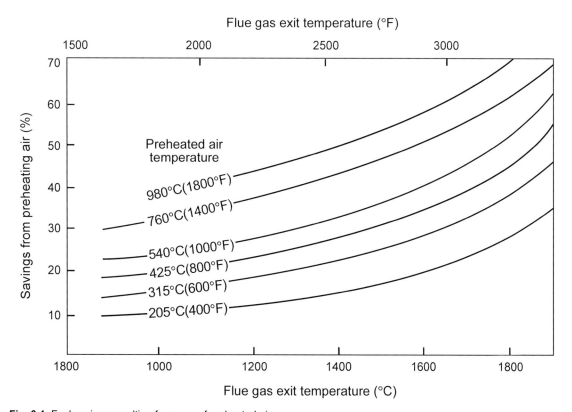

Fig. 6.4 Fuel savings resulting from use of preheated air.

Recuperators are of three general types, classified according to the direction of flow of the waste gases and air, as follows: counter-flow, parallel or co-current flow, and cross-flow.

Counter-flow is used to attain maximum air-preheat temperatures, and cross-current flow is used to secure optimum heat-transfer rates (kilojoules per square metre of recuperator surface per degree Celsius temperature difference per hour, or Btu per square foot of recuperator surface per degree Fahrenheit temperature difference per hour). Parallel flow is used where it is necessary, such as in metallic recuperators, to maintain the temperature difference of the division wall between the two fluids as uniform as possible throughout its length and to keep the temperature of the hot end below a maximum so that the metallic elements will not be overheated. Generally, a combination of counter-flow and cross-flow is applied to many steel-mill furnace applications where a refractory material is used to divide the two fluids. A combination of the two types is accomplished by baffling the flow of one of the fluids. In such designs, the general direction of flow of fluids exchanging heat is counter-current and the flow in each baffle is cross-current. When the temperature of waste gas from which heat is to be extracted is relatively low, under 980°C (approximately 1800°F), metallic tubes (e.g., stainless steel) are generally used because they possess an advantage against leakage. Higher-temperature recuperators are generally constructed of clay or silicon-carbide materials; often, however, these recuperators suffer the disadvantage of air leakage.

Waste-heat boilers are used to obtain heat recovery when a practical limit of recovery has been obtained by regenerators or recuperators and there is still sufficient heat left in the waste gases to justify expenditures for the waste-heat boilers. Boilers sometimes are used in place of regenerators or recuperators, depending upon conditions such as where preheated air is undesirable or where the generation of steam solves the problem of fuel conservation more satisfactorily. Waste-heat boilers are most applicable to high-temperature, continuous processes and have been used principally in the steel plant in conjunction with basic oxygen vessels, gas turbines and, to a lesser degree, with reheating furnaces and soaking pits. Fire-tube and water-tube boiler types have been installed, the former being the preferred type and generally of horizontal single-pass design. Waste-heat boilers usually are provided with superheaters and sometimes with economizers.

6.5.2 Minimizing Radiation Losses

Radiation losses from the walls and roofs of the furnaces can be minimized by the selection and use of an insulating system. Thermal insulating materials have been used in steel plants for a great many years. There are many different kinds of insulating materials, each being most suitable for a specific temperature level and for the degree of insulation desired and improvements in insulating materials are continuing. A detailed discussion of insulating systems for steel-mill furnaces is given in Section 3.6.3.

6.5.3 Combustion Control

As mentioned previously, one of the large losses in a heating process is the heat contained in the waste flue gases. The amount of sensible heat in the waste flue gases is the product of the heat content per unit volume (m^3 or ft^3) or unit weight (kg or lbm) of gas multiplied by the total volume or total weight of the gas respectively, in the appropriate units. The total loss of heat in waste flue gases can be minimized by providing the proper amount of air for combustion. If too much air is provided, the volume of the waste gases is increased. In addition, an excess of air over that required for combustion lowers the flame temperature and increases the time necessary to heat the product. In a similar manner, an excess of fuel (or a deficiency of air) also decreases the flame temperature and prolongs the heating time and, in addition, results in unburned fuel being carried into the waste flue gases. This results in unnecessary usage of fuel and can be dangerous because this unburned fuel can burn in the flue and cause damage to the stack and/or waste heat recovery systems. In other words, fuel economy is optimized by the use of proper combustion conditions.

In modern steel plant furnaces, and in boilers, the amount of air supplied for combustion is maintained only a little above theoretical requirements by special pressure regulators and valves which accurately proportion the amounts of fuel and air fed to the burners. For most burners, an amount of air approximately 10% above theoretical is considered to be optimum. The use of instruments to continuously monitor the products of combustion and automatically control the air-fuel ratio to achieve this desired level of excess air has led to improved fuel economy and heating efficiency.

The amount of waste flue gases can be minimized, and the heating rate of the unit can be increased, by the oxygen enrichment of combustion air. Air is composed of only approximately 20.9% oxygen (by volume), with the remainder consisting of inert nitrogen plus a small amount of several other inert gases. When combustion takes place, the oxygen combines with the carbon and hydrogen of the fuel and liberates heat. The inert gases of the air absorb heat from the combustion and carry it out of the furnace, and it is lost so far as the furnace process is concerned. These gases reduce flame temperature by absorbing heat, thus reducing the rate of transfer of heat to the work.

Obviously, if the inert content of the air could be diminished, more efficient combustion could be obtained. Recent technical developments that have lowered the cost for producing oxygen of commercial purity have made large-scale use of this gas economical for some industrial processes. Consequently, many plants have experimented with the addition of oxygen to ordinary air used for combustion, with generally good results. In effect, increasing the oxygen content lowers the inert content of the air; consequently, when a given amount of fuel is burned with oxygen-enriched air, the volume of the waste gas is less than if ordinary air were used. If the temperature of the waste gas is not increased, the sensible-heat loss in the flue gas will be decreased, due to the smaller heat capacity of the smaller volume. In furnaces operated at high thermal head, a decrease in the amount of the inert gases usually results in a decrease in the waste-gas temperature. With the same fuel input, enriched air for combustion raises the flame temperature of a given fuel, thereby improving heat-transfer rate and increasing production; alternatively, the fuel input may be decreased when enriched air is used to maintain the same production rate as obtained with fuel using ordinary air. Increased production rates almost always reduce the heat losses per ton of product in any furnace employing a high thermal head.

6.5.4 Air Infiltration

If the pressure of the gases in the heating chamber of the furnace is below atmospheric, cold outside air will be drawn into the furnace through any openings that exist. If the interior pressure is above that of the outside air, the hot gases will be forced out of the furnace through these same openings, and if too much higher will, in addition, tend to penetrate the refractories and overheat the furnace bindings with, in some cases, damaging effect. Generally, it is desirable to operate a furnace with a slight positive pressure in the heating chamber (i.e., furnace pressure slightly higher than atmospheric). It should be noted that the pressure from top to bottom of the heating chamber is not uniform, due to the stack effect of the hot gases. Control, therefore, is aimed at maintaining the desired pressure at the hearth level. Air drawn into a furnace operating under negative pressure upsets the fuel-air ratio which is controlled automatically or by valve settings. In some units such as reheating or heat-treating furnaces, this air aggravates the problem of oxidation (or scaling) of the work because of the ingressed oxygen.

If the pressure in the furnace at the hearth level is equal to atmospheric or slightly positive, better heating conditions are obtained. This is especially so in furnaces where most of the heating of the work takes place through heat transfer by radiation from the flame to the bath or work. The positive pressure must be controlled to prevent excessive sting-out of flame from furnace openings (a small pressure imparts a high velocity to hot gases), as well as to avoid the buildup of excessive back pressure that would interfere with proper flow of fuel (if gaseous) and combustion air. Positive pressures maintained at hearth level in practical work are quite low, ranging only up to approximately one millimetre (a few hundredths of an inch) of water. Furnace pressure is controlled by adjusting the opening in the stack damper.

Positioning of the damper can be done manually, using the flame sting-out as an indication of the existence of positive pressure, but it is difficult to adjust the opening for the frequent changes in furnace conditions. The development, about the year 1928, of industrial-type instruments with sufficient sensitivity to measure differential gas pressures with an accuracy of ±0.06 mm (0.0025 inch) of water made possible the use of automatic control of furnace pressure.

Automatic furnace-pressure control has been provided for a majority of steel plant furnaces and has been a principal factor in the improvements in fuel economy and efficiency of melting and reheating furnaces during the years following its adoption.

6.5.5 Heating Practice

The primary objective of any furnace operation is to heat the product (steel, as in the case of most furnaces, or air as in the case of blast-furnace stoves) to the desired temperature at the desired heating rate. The actual practice used to achieve the desired temperature and the desired rate will vary with each specific installation, and in each installation will vary with the level of operation. Consequently, it is beyond the scope of this discussion to provide such specific details. However, several general principles apply to all facilities.

The use of excessive amounts of fuel (high heating rate) is wasteful of the fuel itself and results in high furnace exit-gas temperatures and damage to refractories. In some processes, high fuel rates not only do not hasten transfer of heat to the material being processed, but also may cause it actual damage. On the other hand, the use of insufficient fuel reduces the rate of heat transfer and prolongs process time, thereby increasing thermal losses. The optimum rate for protection either of the material being heated or the furnace refractories, and often for control of heating or production rate, is maintained in most furnaces by automatic temperature-measuring instruments which control the fuel rate through a system of electrical relays or other units which control the operation of motors, hydraulic systems or other means for regulating fuel valves. Many heating installations have adopted computerized control for the heating operations.

A basic requirement for all heating operations is good temperature measurement. One problem involved in temperature measurement is the difficulty and almost impossibility of measuring the temperature inside a solid piece of steel, and in most cases the difficulty is encountered in measuring the surface temperature in one spot. Because of this inability to measure the temperature of solid steel directly, measurement is made of some other temperature that is closely related to the steel temperature. Depending upon the specific furnace installation, this involves measuring the temperature of the roof or wall of the furnace or a measurement of the temperature of the gases in the furnace. These measurements are achieved in the steel industry by instruments operating on four main principles: (1) by measuring the intensity of radiation emitted by the hot furnace or object being heated, (2) by measuring the minute electrical current generated in a circuit composed of two wires of dissimilar metals, joined end to end, when one of the joints is heated (this is the principle of the thermocouple), (3) by measuring the change in the electrical resistance of conductors when heated to the temperature in question, and (4) by measuring the change in the ratio of two separate wavelengths of radiation emitted from the hot object or furnace.

6.6 Water Requirements for Steelmaking

6.6.1 General Uses of Water in Steelmaking

Water is such a common substance and generally has been so abundant that its importance to the iron and steelmaking processes is seldom emphasized in the discussion of operations and metallurgical problems. Without water, steel could not be made. Because early steel mills were built adjacent to ample sources of fresh water, the availability and quality of water was taken for granted. Today, greater attention is given to the management of available water resources in the steel mill environment, particularly in terms of water quality, quantity, and how it is used.

Water is used for direct contact cooling and cleaning of the steel in process, for cooling the process offgases, for product rinsing, and for process solution makeup; however, the vast majority of the water used in steel making is for non-contact cooling of associated processing equipment. Water is also used for steam and power generation, potable uses, and dust and moisture control.

Several factors make water a versatile material. It is normally plentiful, readily available, and inexpensive. It is easily handled. It can carry large amounts of heat per unit volume (high specific heat). It neither expands nor compresses significantly within ambient temperature ranges. It does not decompose. It can dissolve, entrain, suspend, and subsequently transport other material.

The overall use of water in the steel industry has been reduced since the 1970s as a result of modernization, more continuous processes, and greater reuse (recycle) of the available water supplies. This reduction has been the result of the more efficient use of existing water supplies and compliance with environmental regulations governing the quantity and quality of water discharged from the plants.

6.6.1.1 Raw Water Sources

Water can be generally classified as: seawater, with a total dissolved solids (TDS) or salt content of more than 3.5% (35,000 ppm); brackish water, with dissolved solids between 1500 and 35,000 ppm; and fresh water, with a dissolved solids content less than 1500 ppm. While some facilities are located adjacent to oceans, bays or other sources of high salinity waters, fresh water is the overwhelming preference and source of water consumed and utilized in steel mills.

Fresh water can be further classified into surface water and ground water. All fresh water originates as rainfall in the continuous cycle of atmospheric evaporation and condensation. Once fallen, the water either collects on the surface in streams, lakes, and rivers, or seeps into the earth's crust. Ground water is 'old' water originated from wells and springs. Depending on the size of the aquifer being used, ground water will most likely be the most consistent in terms of temperature and chemical content. Water chemistry can also vary relative to the depth and extraction point of the well. As compared to surface waters, good clarity, higher TDS, and low total suspended solids (TSS) levels would be characteristics of ground water because it has flowed through miles of porous rock strata prior to extraction.

Ground water quality is a function of the geological characteristics of the area. The chemical content will vary, depending on the rock strata through which the water has percolated. Water flow through the strata is usually slow, measured in feet per year. Seasonal changes will typically be minimal. Over time, the chemical content may change, particularly if large quantities are extracted, reflecting ground water migration in the aquifer.

Surface water is typically obtained from shallow wells, rivers, and large lakes or reservoirs. Compared to ground water, surface water in general will have higher suspended solids and lower dissolved solids, depending on rainfall in the area being drained. Quality may vary seasonally, depending on local rainfall patterns, as dilution affects constituent concentrations. In addition, water quality can be affected by upstream use and sources, including farm runoff, mine drainage, treatment plants, and practices such as road salt application. In larger water bodies, such as the Great Lakes, observable change may be minimal because of the volume of water contained and its turnover rate. Table 6.25 lists typical makeup water chemistries from various sources, by geography.

6.6.1.2 Cooling Water and Other Applications

Water usage in the steelmaking process can be generally classified as either non-contact or contact water. Non-contact application is the use of water for general cooling purposes where only equipment is being cooled and the water does not contact the steel in process, offgases or fluids. Contact water includes all applications of water directly on the steel in process, process gas cleaning applications, process fluid applications (oils, cleaners, etc.) and rinse applications. Contact waters are subject to treatment, as discussed in Section 6.6.4.

Other applications include steam generation, process solutions makeup, dust control, and potable systems. Each of these uses is discussed below.

Table 6.25 Typical Water Chemistry

Constituent	Units	Lake Michigan	Well South East	Well Mid-West	Ohio River
Calcium	ppm as CaCO$_3$	80	17	282	70
Magnesium	ppm as CaCO$_3$	41	6	86	30
Sodium	ppm as Na	9	5	39	9
Alkalinity – Bicarb	ppm as CaCO$_3$	113	20	352	30
– Carb	ppm as CaCO$_3$	0	0	0	0
Sulfate	ppm as SO$_4$	17	2	52	76
Chloride	ppm as Cl	7	6	18	13
Nitrate	ppm as NO$_3$	not detected	1	1	1
Fluoride	ppm as F	trace	not detected	1	trace
pH	S.U.	8.2	6.4	7.6	7.3
Silica	ppm as SiO$_2$	2.3	5	31	5.6
Iron	ppm as Fe	0.1	.2	6.5	.3
Turbidity	NTU	12	<5	<5	88
TDS	ppm	171	46	488	143

6.6.1.2.1 Steam Generation Prior to steam generation in a boiler, most raw waters must be pretreated to meet the water quality requirements set by the boiler manufacturer. Such treatment typically involves filtration, hardness removal (softening) and/or silica reduction, deaeration, and pH/alkalinity adjustment. If improperly conditioned, water-related problems, including mineral precipitation and corrosion, can result in permanent damage to the steam generating unit along with energy inefficient operation.

Higher-pressure power boilers are used in most integrated mills and coke plants to provide the energy source for blowers to the ironmaking facilities, and to generate electrical power. Waste heat recovery boilers (steaming BOF hoods, slab reheat furnaces, etc.) are sometimes utilized to provide additional steam to plant systems, while doubling as the heat removal equipment for the specific process. Smaller capacity and lower pressure satellite or package boilers will typically generate steam for specific user requirements (vacuum degassing, acid heating, etc.). In all cases, recovery of uncontaminated condensed steam can improve operational economics and reduce makeup water requirements.

6.6.1.2.2 Process Solutions Pickling, cleaning, and coating operations typically performed in the finishing mills utilize water as the diluent for inhibitors, lubricants or plating solutions prior to their application to the steel in-process. Once applied, excess material is recaptured, or transferred to a suitable waste treatment facility prior to discharge, or sent offsite (spent acids) for disposal or reuse. Water that meets the necessary quality requirements for rinse applications is used to remove the residue of process chemicals from the surface of the steel. These rinse waters are normally treated in a wastewater treatment facility.

6.6.1.2.3 Raw Materials Preparation The principal raw materials are coal (most of which is converted to coke), iron-bearing materials (natural lump ore, iron pellets, recycled agglomerates, and more recently iron carbide and DRI), and fluxes (lime, limestone, and dolomite). In coal preparation, considerable quantities of water are used for dust control in the mines and in coal preparation (washing) facilities for foreign matter removal. Similarly, large amounts of water are used in the beneficiation of iron ores to remove foreign matter and concentrate the iron content. Water may also be added in the forming operation (balling mill) to facilitate agglomeration.

In primary facilities, water may be used for dust suppression to meet plant air emissions requirements with or without the use of dust suppressants for roads, conveyor transfer points, and storage piles. In pulverized coal injection (PCI) for blast furnaces, moisture content is monitored and additions may be made.

6.6.1.2.4 Potable Systems These systems provide the drinking and convenience water for human consumption. Primary concerns are for heavy metal content, water clarity, selected organic compounds, and the absence of biological activity. Except in remote plant sites, both potable supplies and sanitary wastewater treatment are normally provided by municipal systems.

6.6.2 Water-Related Problems

Some of the properties of water that make it useful for so many purposes can also contribute to the creation of problems in its use. Water's ability to dissolve, entrain, and suspend solids, liquids, or gases can affect the purity of the water being used and can subsequently affect piping, equipment, and the process. Contamination of the water can subsequently limit its use in many instances until suitable treatment can remove or modify the concentration of offending constituents.

Water-related problems can be generally classified in four different categories: corrosion, scaling, fouling, or biological activity.

6.6.2.1 Corrosion

Corrosion is an electrochemical process by which a base metal reverts to its oxide form. Corrosion can be of a general nature, localized (pitting), or galvanic. For corrosion to occur, a corrosion cell, consisting of an anode, a cathode, and an electrolyte, must be present. Water is an excellent electrolyte. The most common evidence of corrosion is the visible loss of base metal in piping and wetted equipment, and the presence of the resultant oxide (rust). Impurities in the water can affect the rate of metal loss.

Specific mechanisms can be responsible for the metal loss or failure. Galvanic corrosion occurs when two dissimilar metals are in contact in an aqueous environment, resulting in loss of the anodic metal; this can occur on a macro or micro level. Beneath deposition, concentration cell corrosion can be established via various mechanisms, resulting in potentially aggressive localized attack (pitting). Stress corrosion cracking can occur when metal is subjected to tensile stress in a corrosive environment. Fatigue corrosion occurs in an acidic environment when the metal is subjected to cyclic application of stress. Impingement or erosion/corrosion attack occurs when entrained solids or gases repeatedly wear the affected metal due to physical/chemical mechanisms.

6.6.2.2 Scaling

Scaling is a chemical process by which certain minerals become insoluble, precipitate from the solution, and deposit onto equipment surfaces. Scaling is most commonly associated with salts of calcium and magnesium. Common to both of these elements is their inverse solubility in water relative to temperature; as the localized water temperature rises, the solubility of these species drops. The solubility of these minerals is greatly dependent on their relative concentrations, water temperature, pH, alkalinity, types of anions present, and other water parameters. Should localized boiling to dryness occur, elements including normally soluble ions such as sodium may also precipitate out. The most common evidence of scale formation is the growth of mineral deposition in piping and heat transfer surfaces.

6.6.2.3 Fouling

Fouling is the deposition of suspended material in the water, originating from internal or external sources. The most common evidence of fouling is the accumulation of foulants in low-flow areas and in flow channel restrictions. Foulants can include items such as insects, mollusks, silt, rust, oils and greases, scale formed in other areas, and other miscellaneous debris. The presence of soft films (i.e., oils/greases/biological masses) can facilitate the growth of deposition by capturing normally suspended material and forming a fouling matrix of material.

6.6.2.4 Biological Activity

The aqueous nature of a cooling water system can provide an extremely favorable environment for microorganisms to grow and rapidly reproduce. These organisms include algae, fungi, and different classifications of bacteria: aerobic, anaerobic, and iron depositing. The most common evidence is the accumulation of slimy, possibly odorous, biomass in or on wetted equipment. Specific anaerobic bacteria found beneath deposits can accelerate localized corrosion via the metabolic acids generated by these species. Other species, namely fungi, can deteriorate wooden structures (cooling towers).

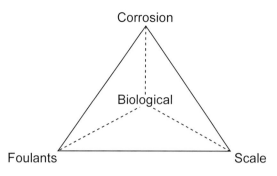

Fig. 6.5 Interrelationship of water-related problems.

Fig. 6.5 illustrates the interrelationship of the water-related problems. For example, when a mild steel pipe corrodes, a significant amount of soluble and insoluble iron is released, which can lead to fouling. Should the pipe or water passage be sufficiently restricted, the lower flow may result in localized overheating, resulting in scale formation. The presence of biological mass would only compound the problem because it has high volume but relatively light mass. Anaerobic bacteria beneath the deposition could result in a localized pitting attack.

The various undesirable properties of water can be combatted by the consideration of the following: materials of construction; steelmaking equipment design with sufficient water flows and velocities; water treatment equipment designed to meet the intended use; operating modes; maintenance procedures; control of additives to the water, both intentional and incidental; and proper chemical treatment.

Because there are usually several possible alternatives or combinations of measures which can be applied, and because two or more problems can co-exist, selection of the proper methods for water use and treatment is a highly specialized activity.

Table 6.26 briefly summarizes the most common water treatment techniques for various constituents.

6.6.3 Water Use by Steelmaking Processes

Because of the great variety of steelmaking facilities, their layout, discharge limitations, and available water supplies, water usage varies widely in both integrated and mini-mill facilities. In the following paragraphs, some specific water uses are discussed.

6.6.3.1 Cooling Water

By far, the largest application and usage of water in iron and steelmaking is for cooling purposes. In addition to categorization by contact and non-contact criteria, cooling water systems use can be generalized in the following categories.

6.6.3.1.1 Once-through Systems Water is used once prior to discharge, as in the cooling of condensers, furnace components, and other miscellaneous applications. This mode of cooling is typically in non-contact cooling applications where the water is intentionally restricted from coming into contact with the steel in process or resultant process streams originating from the facility, such as exhaust gas, steelmaking fluids, etc. This approach is more prevalent in mills located adjacent to large water bodies, rivers, lakes, and at times, oceans.

6.6.3.1.2 Evaporative Recirculating Systems Both contact and non-contact applications utilize evaporative systems. When water absorbs heat from a process, this heat can be easily transferred to the environment (air) by exposing the water to the atmosphere. This evaporation reduces the bulk water

Table 6.26 Common Water Treatment Methods

Substance in Water	Removal Process	Chemicals Used
Hardness	Precipitation	Lime, soda ash, caustic, phosphate
	Ion exchange	Salt, acid
Alkalinity	Precipitation	Lime, gypsum
	Ion exchange	Acid, salt
	Neutralization	Acid
CO_2	Precipitation	Lime, gypsum
	Ion exchange	Acid, salt
	Neutralization	Lime
	Degasification	None
Dissolved Solids	Reverse osmosis	Anti-scalants, biocides
	Demineralization	Acid, caustic, salt
	Specific ion removal	Ion specific resin
Particulate	Screening, Straining	None
	Settling basins	None
Suspended Solids	Coagulation	Alum, aluminate, ferric chloride
	Flocculation	Coagulant, flocculants
	Sedimentation	None
	Filtration	None
Fe/Mn	Oxidation	Chlorine, air
	Precipitation	Lime, caustic
	Filtration	None
	Ion exchange	Salt, acid
SiO_2	Precipitation	Iron salts, magnesium oxide
	Ion exchange	Caustic
Organic Matter	Clarification	Alum, aluminate
	Bio-oxidation	None
	Oxidation	Chlorine, chlorine dioxide, ozone
	Adsorption	Activated carbon
O_2	Degasification	None
	Reduction	Sulfite, O_2 scavengers
Macro Aquatic Organisms (Zebra Mussels, Asiatic Clams, Fish)	Heating	None
	Straining	None
	Sterilization	Chlorine, bromine
	Organism specific toxin	Proprietary biocides/ biocide neutralization
Micro-organisms	Clarification	Coagulants, flocculants
	Filtration	None
	Heating	None
	Sterilization	Chlorine, ozone, sterilants, ethylene oxide

temperature, enabling the water to be reused. This cooling method is commonly used in evaporative cooling towers, and to a lesser extent, in lagoons and spray ponds.

In non-contact systems, the limit to which water can be recycled in this manner is dependent on the water quality requirements of the process being cooled, the raw water quality, and ambient air temperature. As the water is evaporated, the minerals and other constituents for the most part remain in the water and become concentrated. This concentration mechanism increases the TDS in the system and, if not compensated, may lead to water-related problems, as discussed in Section 6.6.2. Normally, the TDS concentration in the system is controlled by bleeding off some of the water from the system (blowdown).

The same is true for contact cooling water systems, and consideration must be given to the constituents originating from the process. These systems often require some treatment for the removal of solids, oil, etc., prior to cooling and recycle.

In both contact and non-contact cooling systems, the amount of recycle that is obtainable is highly dependent on the water quality requirements and constraints of the systems being cooled. Because of this, some discharge blowdown is required from these systems to maintain the desired water quality. The balance for an evaporative water system is dictated by the following interrelationship between makeup, evaporation, and blowdown. (Note that blowdown in this equation includes uncontrolled losses such as tower drift, windage losses, exhaust entrainment, spillage, and leakage.)

$$\text{Makeup (MU)} = \text{Evaporation (EV)} + \text{Blowdown (BD)} \tag{6.6.1}$$

Cycles of concentration and concentration ratio are terms used to indicate the degree of concentration of the recirculating water as compared to that of the makeup water. The degree of concentration is dictated by the relationship of makeup to blowdown, given that evaporation (heat load) is constant over time. The evaporating water is pure water vapor, leaving behind soluble water impurities that can be measured. Analytically, measurement of these dissolved minerals in the makeup and blowdown can also indicate this concentration phenomenon as shown below:

$$\text{Concentration Ratio (CR)} = \frac{\text{MU (gpm)}}{\text{BD (gpm)}} = \frac{\text{Concentration (BD)}}{\text{Concentration (MU)}} \tag{6.6.2}$$

An accepted rule of thumb to estimate evaporation in a recirculating system is for every 6°C (10°F) temperature drop across the evaporative process, 1% of the recirculating water is lost to evaporation. This 1% rule of thumb may vary. In areas of high humidity (Gulf Coast), evaporative loss may be as low as 0.75%, while in arid areas, as high as 1.2%. A second method to determine makeup (MU) is as follows:

$$MU = EV\left(\frac{CR}{CR-1}\right) \tag{6.6.3}$$

In direct contact systems, when water is applied to hot steel in process, analytical measurement of dissolved salts may be skewed. Some of the mineral content may precipitate out on the hot surface of steel in process as the water evaporates. Conversely, soluble elements from the process may also dissolve in the water, increasing their level, as in the gas cleaning applications of water.

6.6.3.1.3 Non-evaporative (Closed) Recirculation Systems Incorporating the use of a heat exchanger mechanism, the recirculated water system heat is transferred to a second water system or to the air through conductive cooling. This transfer is accomplished through water/water heat exchangers of plate and frame or shell and tube design, or through air coolers of wet surface or dry design. The same cooling water is used in a continuous cycle with little or no water loss. Usually associated with higher water quality applications, these systems are typically not exposed directly to the environment and do not have a concentration mechanism. In everyday life, the automobile radiator cooling system is a common example of a closed loop cooling system.

6.6.3.1.4 Application of Cooling Water Systems Most once-through systems are limited to non-contact or indirect cooling applications. Previously, contact or direct cooling applications were of the once-through design, but these generally have been converted to recirculating systems, incorporating primarily evaporative cooling towers to meet discharge requirements. Most closed non-evaporative systems are used to cool process equipment requiring higher quality water in terms of hardness, suspended solids, and dissolved solids. Often these requirements necessitate pretreatment of the water to meet the desired characteristics.

In terms of water resource management, many mills utilize cascading water systems, whereby water is intentionally transferred from one system to another. A common example would be to use the blowdown from a non-contact system as makeup to a contact system. Rarely is the reverse employed without additional treatment because the contact waters may contain potentially troublesome constituents originated from the steelmaking process. Because of constituent levels in contact water, the use of non-contact water is preferred in indirect heat exchange equipment, where practical and feasible. In some applications, boiler feed water is required.

6.6.3.2 Water Use and Wastewater Generation by Ironmaking and Steelmaking Operation

The following is a brief outline of water use and wastewater generation for each major operation in steel making. Table 6.27, summarizes the water use and wastewater generation of these operations.

Table 6.27 Water Usage and Wastewater Generation at Steel Making Operations

Operations	Water Use	Wastewater Generation
Coke plant	Gas and product cooling Process water Steam heating Coke quenching	Excess ammonia liquor Final coolers Light oil recovery Desulfurization processes Condensates
Sinter plant	Process water Dust control Sinter cooling Flue gas cooling/cleaning	Flue gas cleaning blowdown
Blast furnace	Process cooling Flue gas cooling/cleaning Slag granulation	Flue gas cleaning blowdown
Steelmaking (EAF, BOF and open hearth furnaces)	Process cooling Flue gas cooling/cleaning Cooling system blowdown	Flue gas cleaning blowdown
Vacuum degassing	Gas cleaning system Process cooling	Gas cleaning blowdown
Continuous casting	Mold cooling Machine cooling Product/Process cooling Scale removal	Cooling system blowdown Contact cooling waters
Rolling mills (hot and cold rolling)	Reheat furnace cooling Product/process cooling Scale removal Equipment lubrication Flume finishing	Contact cooling waters Run out table Scale pit
Finishing mills	Pickling Cleaning Process solutions Process water Rinse water	Spent cleaning solutions Spent pickle acids Spent rinse water Spent process solutions

6.6.3.2.1 Coke and Coal Chemical Plants
The largest volume of water used in these plants is for non-contact cooling in a variety of cooling and condensing operations. Moisture from the coal being coked and from process steam condensation makes the byproduct operation a net generator of process water. Process wastewater sources include: excess ammonia liquor from the primary cooler tar decanter, and barometric condenser wastewater from the crystallizer, the final coolers, light oil recovery operations, desulfurization processes, and air pollution control operations. Most of the wastewater generated at the coke plant results from moisture in the coal and steam condensate. Additional sources can include condensates from drip legs and gas lines.

The largest consumption of water in coke plants is for the quenching of coke. The amount of water required for coke quenching can vary and has been reported to be from 400 to 3000 litres/tonne (120 to 900 gallons per ton). A sufficient quantity of water is needed to cool the coke, yet leave enough heat in the coke to evaporate any entrained water. Some facilities utilize untreated process wastewater for coke quenching, while others use treated process waters or service waters. In some areas, the level of TDS in the quench water must be controlled due to potential air pollution concerns. This requirement has impacted the use of untreated process wastewater in coke quenching.

6.6.3.2.2 Sinter Plants
The principal uses of water in a sintering plant are for controlling the moisture content of the pre-sinter mix, for dust control, and for sinter product cooling. Some indirect cooling of equipment is involved. For emissions control in sinter plants, either electrostatic precipitator or wet venturi-type scrubber technology is typically employed for dust control. In precipitator applications, some water may be added to control exhaust temperature and to condition the particulate prior to capture.

In scrubber systems, water is sprayed into the gas stream to capture the emissions from the sinter operation to meet stack emissions requirements. The associated water system will typically include clarifying thickeners, a cooling tower, and pH Adjustment.

6.6.3.2.3 Blast Furnace
The blast furnace is one of the largest water users in an integrated steel mill operation. The primary water use is for non-contact cooling of various parts of the furnace and auxiliaries, including the tuyeres, hearth staves, bosh, cooling plates and staves, cinder notch, and stove valves. The historical approach has been to use once-through cooling water, but recent construction and modernizations have incorporated evaporative cooling systems along with closed-loop water systems for hot blast valves, tuyeres, and staves. Additional water is used for furnace moisture injection (steam), dust control, and slag granulation.

Contact water use is primarily associated with blast-furnace gas cleaning operations necessary to recover the fuel value of the offgas for use in stove and boiler operations. In venturi-type scrubbers and spray-cooled gas coolers, water contacts, cleans, and cools the gas for reuse. This water system usually consists of thickeners, dewatering devices, and evaporative cooling towers. Chemical treatment may consist of settling aids, pH adjustment, and inhibitors, depending on the nature of the particulate and gas exiting the furnace. As the amount of recycle is increased in these systems, the control of total dissolved solids (TDS) and the increased concentration of specifically monitored elements or compounds become concerns.

Treatment for the reduction of TDS would be dependent upon the parameters of concern and could involve more sophisticated treatment systems. Typically, the blast-furnace process waters contain ammonia, phenols, cyanide, lead, and zinc. Additional treatment for these parameters may be required prior to discharge of any blowdown from the recycle system. In many facilities, a significant portion of the recirculated water can be consumed rather than discharged by utilizing it for slag quenching.

6.6.3.2.4 Alternative Iron Technologies
With the increased tonnage produced via the electric arc furnace, the direct reduced iron, iron carbide, and Corex processes play a meaningful role in the steel industry. Likewise, they also have water requirements. Because these technologies produce a source of iron units in varying forms, their process requirements can be similar to those of a blast furnace. However, their gas handling system is much more complex due to the nature of the technologies. In general, each of these processes has non-contact cooling requirements for compres-

sors, coolers, and various furnace shell components. Closed non-evaporative systems are prevalent for this application.

The gas exiting the shaft furnace in these processes is utilized extensively in the process and requires cleaning prior to reuse. Wet scrubbers are installed in the dirty system to perform this function. The water system usually consists of thickeners, dewatering devices, and evaporative cooling towers. Chemical treatment may consist of settling aids, pH adjustment, and inhibitors because of the nature of the particulate and gas composition in the dirty gas exiting the furnace.

A clean gas contact system, using water mainly as a process gas coolant, is also used to maintain process gas temperature requirements. This system is cooled via an open evaporative cooling tower.

Water treatment to meet the process requirements for each of the systems can be extensive. The contact systems in particular may require extensive water treatment and quality adjustment in process to maintain plant operations.

6.6.3.2.5 Electric Arc Furnaces The EAF can also be an extensive user of water, particularly for non-contact cooling functions. In the melt shop, emissions in the exhaust gas stream are typically handled using baghouse technology, requiring no contact water. Non-contact water applications include water-cooled ductwork, roof, sidewalls, doors, injection lances, panels, electrode clamps, cables, and arms. These systems usually incorporate evaporative cooling towers, while a number of applications incorporate closed cooling loops. Some applications utilize closed-loop high-purity systems for electrical apparatus cooling requiring low conductivity (dissolved solids) water. As with other primary operations, the control of TDS in evaporative cooling systems for electric arc furnaces is a significant concern.

6.6.3.2.6 Basic Oxygen Furnace The BOF and its derivatives are extensive users of water. Non-contact cooling is used for the vessel hood, ductwork, trunnion, and oxygen lance. Both closed-loop and evaporative systems are prevalent for handling the cooling requirements for these components, based on manufacturer recommendations. The trend for hood cooling in recent years has been toward the closed-loop cooling approach to extend hood life. Some BOF hoods are steam generating, with water requirements similar to those of high pressure boiler applications.

Gas handling for the various basic oxygen processes can be categorized as either full combustion (open) or suppressed combustion (closed) hood systems. In full combustion systems, excess air is allowed to enter the exhaust stream in the hood. This enables the oxygen blow to accomplish a complete reaction, converting CO to CO_2. In suppressed combustion systems, the hood is skirted to prevent excess air from entering the exhaust. Typically, the amount of particulate emissions generated by either approach equals 1–3% of the steel tonnage produced. The suppressed combustion approach has gained favor because the particulate generated is generally larger (except in the OBM or Q-BOP process), the temperatures generated in the hood are generally lower, and the volume of gas is lower, thereby reducing the size of the equipment required to convey and clean the gas.

In either full or suppressed combustion gas cleaning systems, water is used for initial cooling of the gas in the ductwork and/or in a quencher to reduce gas temperature and volume. Full combustion gas cleaning systems can incorporate either electrostatic precipitator (semi-wet) or venturi scrubber (wet) technology. After initial conditioning in a precipitator-based system, little or no water is used because the captured dust is handled on a dry basis. Precipitators are generally not used on suppressed combustion systems because the potential for explosion exists due to the presence of CO gas; the few facilities with suppressed combustion and precipitators have incorporated design considerations accommodating the potential explosion hazards.

The scrubber-based system typically incorporates separate quencher and venturi-type scrubbers for gas cooling and cleaning purposes. The auxiliary water systems typically consist of a clarifying thickener and dewatering devices for the solids captured. Upon exiting the thickener, the cleaned water is returned to the venturi, typically without cooling. After the venturi scrubber, the venturi effluent is typically reused as supply to the quencher. In some facilities, a separate gas cooler with a dedicated cooling tower is in line, after the scrubber but prior to the exhaust fan and stack.

Because of the lime carryover in the exhaust gas, the high gas temperature, and the fine particulate nature of the dust evolved, water-related problems, including mineral precipitation and fouling, can interfere with scrubber operations. Problems include plugging of the venturi nozzles and fouling in the venturi/quencher throats. Various water quality treatment approaches are incorporated, including pH/alkalinity control, solids management, and various inhibitor programs.

6.6.3.2.7 Ladle Metallurgy Furnace Water usage for the LMF is similar to the EAF, although its function and design are different. Non-contact water is used to cool the roof, clamps, arms, cables, and associated equipment. Cooling is typically via an open recirculating system and is combined with the EAF system if plant layout allows.

6.6.3.2.8 Vacuum Degasser Water usage is primarily associated with the vacuum generating/gas cleaning system. The vacuum is generated by injecting steam through multi-point eductors leading to a barometric condenser. As the steel exhaust emissions are drawn in contact with the steam, the particulate is wetted and collects in the condenser water. The water system may consist of an evaporative cooling tower, solids settling, and solids removal equipment.

Depending on the facility design, non-contact water may be used to meet equipment cooling requirements. Typically, these cooling requirements will be met by evaporative towers associated with the EAF/LMF facilities.

6.6.3.2.9 Continuous Casting Water use and quality are critical to the success of continuous casting. Water use in the caster is categorized by function in the casting process: primary (mold); secondary (spray); and auxiliary (equipment). When continuous casting was initially adopted by the steel industry, little attention was given to water quality and one evaporatively cooled system was installed to provide the cooling requirements.

As with the casting process itself, water management practices evolved as demands for equipment reliability and quality grew. A significant improvement was made when the mold circuit was isolated as a non-contact (closed) system, virtually eliminating process contamination and mineral scaling as the cause for mold related failures. Further improvements were made when contact cooling and non-contact requirements were separated into separate water systems.

The primary cooling process is the non-contact cooling of the molten steel shell in the mold to its semi-finished form by passing high quality water through a highly conductive copper mold. Closed-loop non-evaporative cooling is primarily employed when high surface and strand quality are required. Because of the high heat flux encountered in the mold, pretreatment is normally specified for removal of scale-forming constituents. Open recirculating evaporative systems are sometimes used where makeup water quality and caster quality requirements allow.

Secondary or spray cooling occurs as the strand exits the mold, with contact water sprays covering the surface of the strand. Spray cooling allows for the extraction of heat as it migrates from the molten core to the surface until solidification is complete. This contact water system will typically incorporate settling basins (scale pits), oil skimmers, straining devices, and deep bed filtration equipment when low suspended solids levels are to be maintained. Water treatment should include consideration of contamination from grease, hydraulic fluids, and mold lubricants that will be collected and concentrated in this system. As with other contact systems, this system will typically utilize evaporative tower systems for cooling.

Auxiliary cooling is non-contact or internal cooling of the casting equipment. Applications typically include heat exchangers, internal roll cooling, frames, bearings, compressors, exposed instrumentation, and other miscellaneous components. Evaporative cooling systems are prevalent for these applications. Electromagnetic stirring devices and high temperature exposed components such as center bearings may require the installation of closed-loop cooling systems, which may require high-quality water.

Cascading water systems are not unusual in caster systems as water quality requirements are typically less stringent for contact systems than for non-contact systems. Cascading often occurs unintentionally because of equipment leaks during normal operations.

6.6.3.2.10 Hot Rolling Mills After casting, the semi-finished steel is transferred to a reheat furnace to bring the metal to a uniform temperature. Whether the furnace is of skid, walking beam, or tunnel design, non-contact cooling water is typically utilized to provide the cooling for doors, internally cooled rolls, skid pipes, beams, bearings, and miscellaneous equipment. Though once-through water can be used, an evaporative cooling tower is generally employed to allow water recycle.

In a thin slab caster complex, the strip enters the finishing stands directly, after passing through a continuous in-line reheating furnace. Direct rolling of thin cast slabs without reheating is more energy efficient and thus requires less indirect non-contact cooling water.

In the hot mill, the mill stand work rolls are cooled by a contact water system to maintain roll contour, to prevent surface cracking of the steel rolls due to sudden temperature changes, to minimize fire checking, and to generally extend roll life. Water is used in the form of high pressure jets to remove scale from the steel (descale) before rolling to maintain surface quality. It is also used between certain roll stands to maintain surface cleanliness of the steel in process. In addition, water is used as flume flushing to transport the scale to the scale pits for removal. When the product is coiled, cooling sprays or laminar cooling are employed to cool the strip to enable the coiling operation.

The rolling mill water system will typically incorporate settling basins (scale pits), oil skimmers, straining devices, solids removal with clarifiers and/or deep bed filtration equipment (when low suspended solids levels are to be maintained), and a cooling tower. Wastewater treatment should include consideration of contamination from fine mill scale, grease, hydraulic fluids, and rolling oils which will be collected in this system. The laminar system used on the runout tables may be separated from the rolling mill, with separate settling, cooling, straining, and water filtration facilities.

6.6.3.2.11 Finishing Mills In finishing operations including pickling, cold reduction, annealing, temper, cleaner, and coating lines (tin, galvanized, terne, etc.), water is used primarily as non-contact cooling water, solution makeup, and rinse water. Non-contact cooling typically incorporates evaporative cooling towers, while some equipment may include closed-loop systems to meet specific high purity water quality requirements. Typical process wastewaters from these operations include rinses and spent concentrates from alkaline cleaners, pickling solutions, plating solutions, and electrochemical treating solutions. Many technologies are being utilized to recycle and/or reuse the concentrated solutions; however, the rinse waters require treatment to meet discharge requirements. In treatment of these wastewaters, consideration should be given to the type of plating solutions, cleaners, and pickling solutions utilized; oil and grease; dissolved metals; organic compounds; and lubricants that may be present in the wastewater.

6.6.4 Treatment of Effluent Water

In the iron and steelmaking industry, as in any major industry, the large volumes of process water that come into direct contact with the raw materials, products, and offgases must be treated for removal of regulated parameters prior to discharge or reuse of water. As discussed previously, the main operations in an integrated steel plant that require wastewater treatment include cokemaking; ironmaking; steelmaking; hot and cold rolling; and finishing operations such as pickling, electrolytic tinning, and other coating processes.

For the iron and steel industry, the parameters of most significance, which are generally regulated by the terms of discharge permits, are suspended solids, oil and grease, phenol, cyanide, ammonia, and heavy metals such as lead, zinc, chromium, and nickel. In addition, several organic compounds that are on the priority pollutant list compiled by the U.S. Environmental Protection Agency are regulated for cokemaking and cold rolling operations. The following discussion describes the conventional wastewater treatment technologies employed for effective treatment of steel industry wastewaters. Table 6.28 presents a summary of where these technologies would be applicable in steel mill operations.

Table 6.28 Steel Processing Wastewater Control Parameters

Mill Operation	Dissolved Organics	Dissolved Inorganics[a]	Oils	Suspended Solids	Metals	pH
Coke Plant	X	X	X	X		X
Sinter Plant		X		X	X	X
Blast Furnace	X	X	X	X	X	X
Electric Arc Furnace				X	X	X
Basic Oxygen Steelmaking				X	X	X
Vacuum Degassing				X	X	X
Continuous Casting			X	X	X	X
Hot Forming			X	X	X	X
Finishing Mills:						
Cold Forming	X		X	X	X	X
Acid Pickling			X	X	X	X
Alkaline Cleaning			X	X		X
Hot Coating			X	X	X	X

[a] Ammonia, cyanide, sulfide, thiocyanate, etc.

6.6.4.1 Control of Suspended Solids

Removal of suspended solids is necessary for the wastewaters from practically all of the production steps in the iron and steel industry, from cokemaking to product finishing. Solid particulates become suspended in process water streams during cleaning and cooling of flue gases, descaling, roll and product cooling, and flume flushing in rolling mills, and during product rinsing in finishing operations. The three general methods for removing suspended solids are sedimentation, centrifugal separation, and filtration. Sedimentation or clarification, which is settling by gravity, can be accomplished in a clarifier or inclined plate separator specially designed for a given application. Clarifiers are generally circular but also may be constructed in a rectangular shape. An advantage of inclined plate separators over clarifiers is that inclined plate separators occupy much less ground space; however, care should be exercised in their use when the wastewater contains a high oil and grease concentration. The disadvantage is the small storage volume for sludge at the bottom of inclined plate separators. Both clarifiers and inclined plate separators are designed for continuous removal of the collected sludge from the bottom of the unit. The underflow sludge may be gravity thickened before being further dewatered in one of several types of sludge dewatering units such as a filter press, a belt press, or a centrifuge, to reduce the volume for ease in handling and economy of disposal.

Coagulant aids, such as alum, ferric chloride, ferric sulfate, ferrous sulfate, ferrous chloride, and commercial organic polyelectrolytes, are often added to the wastewater prior to clarification to promote flocculation of the solid particles, which increases their effective size and thus increases their settling rate.

Centrifugal separation is a technique to remove suspended materials from the water column via centrifugal forces, sometimes termed cyclone separation. The process is highly dependent upon particle size and specific gravity. Larger particles and higher specific gravity enhances performance.

Multi-media or single media filtration, by either pressure or gravity, is another method for removal of fine suspended particulates which is commonly applied to steel industry wastewaters. The water is passed through a filter media contained in a vessel. The system usually is comprised of a number of individual filtration units in parallel. Often side-stream filtration can be utilized to treat a portion of the wastewater that is then blended with the unfiltered portion. It is desirable to design a filter system with the highest feasible flow rate through the filter media to minimize the required size and cost.

In a typical multi-media system, the water first passes through a relatively coarse layer of media such as anthracite coal, and then through a layer of fine sand. Most of the particulates are removed by the coarse layer, while the fine layer does the final polishing. Multi-media filters are often used when oil and grease concentrations are elevated in the wastewaters. High oil and grease concentrations can result in fouling and/or plugging of the media in both single and multi-media filters. Periodically, the collected particulate must be removed from the filter media by backwashing. In this operation, the influent flow of wastewater is shut off and a stream of treated water, and occasionally air, is passed through the filter in the opposite direction to flush out the collected solids. By having a number of filter units installed in parallel, one unit can be put through the backwash cycle without interrupting the continuous treatment of the wastewater stream. The backwash stream is usually settled in a backwash holding tank, and the solids are processed through a thickener and sludge dewatering equipment. Both single media and multi-media filters can produce a high degree of clarity in effluent streams. However, clarifiers should be considered for pretreatment of wastewater heavily laden with solids to remove the majority of the particulates prior to filtration. Filters can be used alone without prior clarification for waste streams with less solids.

The quantity of suspended solids and other particulates discharged to a receiving stream can usually be greatly reduced by recirculating the water back to the process. However, the degree of recirculation that is feasible is limited by the amount of suspended solids present in the wastewater and the buildup in the concentration of dissolved salts in the system, which can eventually lead to deposition and plugging in equipment and piping. Therefore, a certain portion of the circulating water volume must be released as blowdown to control the concentration of dissolved salts to a tolerable level.

6.6.4.2 Control of Oil

Oil and grease are commonly found in wastewaters from continuous casting, hot and cold rolling, pickling, electroplating, and coating operations. The oils originate from machinery and product lubricants and coolants; hydraulic systems; and preservative coatings applied during certain phases of the production operations. Oil and grease can be removed from process wastewaters by several methods including skimming, filtration, gravity separation, air flotation, and ultrafiltration. If the oils are insoluble in water, they can be controlled by gravity separation and skimming. Gravity oil separators are usually rectangular chambers in which the velocity of the water stream is slowed down sufficiently to allow time for the oil to float to the surface, from where it is removed by one of several types of skimming devices. Such devices include the rotary drum, rope and belt type skimmers, and scraper blades, which are also used to scrape the heavier solids that have settled to the bottom. Insoluble oils also can be removed along with suspended solids in the multi-media filters previously described. If the oils are emulsified or water soluble, such as those found in waste cold rolling solutions or rinse waters, they can be treated by acid or commercial emulsion breakers to break the emulsion, followed by gravity sedimentation and skimming, or by air flotation and/or membrane separation techniques.

Skimming may be used on any wastewater containing constituents that float to the surface and is commonly used to remove free oil, grease, and soaps. Skimming is often used with air flotation or clarification to improve removal of both settling and floating materials. The removal efficiency of a skimmer is a function of the density of the material to be floated and the retention time of the wastewater in the tank. API or other gravity-type separators tend to be more suitable for use where the amount of surface oil flowing through the system is fairly high and consistent.

Air flotation is a process that is used to separate floatable materials having a specific gravity close to that of water, which therefore cannot be effectively separated by gravity alone. In a flotation system, gas bubbles, usually air, are released in the wastewater and attach to the oil and fine solid particles, causing them to float more rapidly to the surface where they are skimmed off as a froth.

Foam, dispersed air, dissolved air, gravity, and vacuum flotation are the most commonly used techniques. Chemical additives are often used to enhance the performance of the flotation process.

Ultrafiltration (UF) includes the use of pressure and semi-permeable polymeric or ceramic membranes to separate emulsified or colloidal materials suspended in a liquid phase. The membrane of an ultrafiltration unit forms a molecular screen which retains molecular particles based upon their differences in size, shape, and chemical structure. The membrane permits passage of solvents and lower molecular weight molecules.

In the ultrafiltration process, the wastewater is pumped through a tubular membrane unit. Water and some low molecular weight materials pass through the membrane under the applied pressure of 10 to 100 psig. Emulsified oil droplets and suspended particles are retained, concentrated, and removed continuously.

6.6.4.3 Control of Heavy Metals

Limitations on the discharge of heavy metals have been established by the U.S. Environmental Protection Agency (EPA) for steel industry process waters from blast furnaces; steelmaking furnaces; and pickling, cold rolling, electroplating, and hot coating operations. The conventional method used for removal of these trace metals is chemical precipitation followed by clarification or filtration. As shown in Fig. 6.6, the solubility of heavy metals in water is a function of pH. Generally, metals become less soluble as the pH increases; therefore, to remove dissolved metals, a wastewater stream is treated with an alkaline material in a mixing tank with a pH controller.

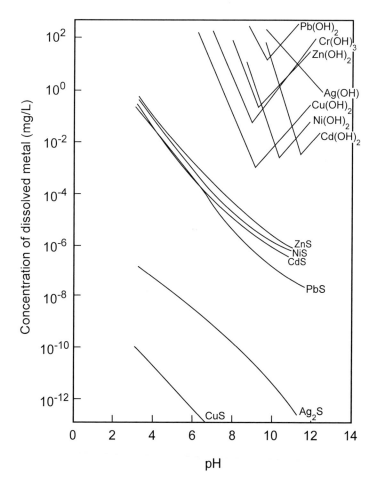

Fig. 6.6 Solubility of metal hydroxides and sulfides as a function of pH. *From Ref. 77.*

Common chemical precipitation processes include hydroxide and sulfide precipitation. In hydroxide precipitation, lime, which is the least expensive reagent, is most often used, although caustic soda, magnesium hydroxide, or other alkali may also be employed for this purpose. After the pH is raised to a level where the dissolved metals will precipitate as hydroxides, the water passes to either a clarifier and/or a filter for removal of the precipitated solids. The addition of a coagulant aid is usually required. The use of a coagulating agent such as ferric chloride at an alkaline pH results in the formation of an oxy-hydroxide surface, which enhances additional removal of metals by adsorption. Other coagulants such as alum, ferrous sulfate, and polymeric flocculants also can be used to enhance particle formation.

If chromium is present in the hexavalent form, it must first be chemically reduced to the trivalent form before it will precipitate. The rate of this reduction reaction is a function of the pH condition of the system. For example, the pH of the system must be adjusted to between 2.0 and 3.0 if sulfur dioxide, sodium bisulfite, or spent pickle liquor is used as the reducing agent. Hexavalent chromium can also be reduced to trivalent chromium with sodium hydrosulfite at a relatively higher pH (between 8.5 and 9.5). The reduced trivalent chromium ion is then converted to insoluble chromium hydroxide and is removed by sedimentation.

As indicated above, dissolved metal ions and certain anions may be chemically precipitated and removed by physical means such as sedimentation or filtration. In addition to the use of alkaline compounds, the following other reagents can be used:

1. Metal sulfides—Except for chromium sulfide, the solubility of metal sulfides is lower than that of metal hydroxides (see Fig. 6.6). Therefore, the removal of dissolved metals can be enhanced using the sulfide precipitation process. Both soluble sulfides, such as hydrogen sulfide or sodium sulfide, and insoluble sulfides, such as ferrous sulfide, may be used to precipitate many heavy metal ions as insoluble metal sulfides. Currently, the use of organic sulfide has become popular in the wastewater treatment industry. Typically, the sulfide precipitation process includes clarification and filtration.

2. Carbonates—Carbonate precipitates may be used to remove metals either by direct precipitation using a carbonate reagent such as calcium carbonate or by converting hydroxides into carbonates using carbon dioxide.

Chemical precipitation as a mechanism for removing metals from wastewater is a complex process consisting of at least two steps: precipitation of the unwanted metals and removal of the precipitate. A small amount of metal will remain dissolved in the wastewater after complete precipitation. The amount of residual dissolved metal depends on the treatment chemicals used, the solubility of the metal, and co-precipitation effects. The effectiveness of this method of removing any specific metal depends on the fraction of the specific metal in the raw waste (and hence in the precipitate) and the effectiveness of suspended solids removal.

6.6.4.4 Biological Treatment

Biological oxidation is the most commonly applied technology for final treatment of coke plant wastewaters. These waters contain significant levels of phenol, cyanide, thiocyanate, and ammonia, plus lesser concentrations of other organic compounds, primarily as a result of condensation from coke-oven gases that contain these substances. A conventional system for treatment of coke plant waste would include ammonia distillation followed by biological oxidation. The ammonia still system may have a free leg, where dissolved gaseous ammonia is removed by steam distillation, and a fixed leg, where ionized ammonia is converted to free ammonia by alkali addition, and then is removed in the free leg. While this distillation process will remove a large percentage of the ammonia and some of the free cyanide from the water, enough of these constituents will remain in the ammonia still effluent, in addition to phenol, thiocyanate, and other organics to require further treatment, such as biological oxidation, prior to disposal.

Because biological oxidation is highly sensitive to fluctuations in constituent loadings and pH, the effluent from the ammonia still is first passed through an equalization basin to level out the concentrations, temperature, and flow volume. Conventional systems consist of either a single-stage or two-stage system. In a single-stage system, the process is designed to reduce the organic compounds as well as ammonia. In a typical two-stage system, the first stage is designed to reduce the organic compounds and the second stage is utilized for nitrification (ammonia removal). Sometimes, with careful control, both carbonaceous and nitrogen treatment can occur in the same aeration basin. It should be noted that the biological treatment of thiocyanates results in an increase of ammonia in the wastewater. This must be accounted for in the design of the treatment system. Both aeration systems typically utilize an activated sludge process, followed by a clarifier. The activated sludge process is a suspended growth process similar to that applied in sewage treatment plants. In the aeration system, a mass of microorganisms or biomass in the form of suspended solids, called an activated sludge, is supplied with oxygen, which enables it to reduce the biologically degradable constituents in the wastewater. Populations of microorganisms can be developed that can effectively degrade phenol and other organics, thiocyanate, free cyanide, and ammonia. The required oxygen is supplied either by mechanical surface aerators or by diffusion of air bubbles through the basin, with or without the use of submerged turbine mixers. The treated water overflows the basin to a clarifier, where the activated sludge is settled out to be recycled to the aeration basin. The overflow water from the clarifier is discharged. Other biological treatment processes also can be used, including fixed film, packed towers, fluidized beds, and suspended growth processes with integral clarification.

6.6.4.5 Terminal Treatment

A common practice in wastewater treatment within the steel industry is to combine wastes from several different types of operations for treatment in a so-called terminal treatment plant. This practice has been particularly successful in the handling of wastes from the various finishing operations. These wastes typically might contain suspended solids, free and emulsified oils from cold rolling, acids from pickling rinse waters, and heavy metals from pickling and coating processes. In a typical system, acid streams are mixed with the emulsified oil streams to break the emulsions. The combined wastes are then passed through a gravity-oil separator, neutralized with lime to remove acids and precipitate heavy metals, and treated for removal of solids and any remaining oils in a clarifier or filter. This general type of terminal treatment system can be applied for environmental control of a number of different steel industry processes more economically than the alternative of providing a separate treatment system for each process.

6.6.4.6 Breakpoint Chlorination

Chlorination is one of the technologies recognized by U.S. EPA for the treatment of ammonia, phenols, and free cyanide. Chlorine has long been utilized as a biocide in municipal drinking water facilities and is known for its strong oxidation potential. When chlorine is added to water, hydrolysis of the chlorine molecule occurs, creating hypochlorous acid and hypochlorite ions which together comprise the free available chlorine. Alkaline chlorination (at pH > 9.5 in the presence of excess chlorine) is required for the destruction of free cyanide.

The term breakpoint chlorination comes from the observation of the point of the maximum reduction of chlorine residual while subjecting a sample to increasing chlorine dosages. The theoretical chlorine dosage for the treatment of ammonia is 7.6 parts of chlorine to one part of ammonia. In practice, dosages of 8:1 to 10:1 may be required. The optimum pH is usually in the range of 6.0 to 7.0. Care must be taken to provide sufficient chlorine to complete the reaction to avoid the formation of chloroamines. Additionally, the competing demand from phenols, nitrite, ferrous iron, sulfites, hydrogen sulfide, free cyanide, and other organics must be accounted for in the total chlorine demand. Alkalinity may also need to be added to the wastewater to maintain the desired pH. Approximately 14.3 mg/litre of alkalinity (as $CaCO_3$) is consumed for each 1.0 mg/litre of ammonia nitrogen oxidized. In some cases, dechlorination of the final discharge will be required. This can be accomplished by adding sulfur dioxide, sodium bisulfite, or activated carbon.

The advantages of the process are that it has rather consistent process performance, has low space requirements, and can reduce the ammonia concentrations in one step. The disadvantages of this treatment method are the potential formation of trihalomethanes (THM), an increase in total dissolved solids (TDS), and the relatively high operating costs. This is why the technology is usually only applied to treat small concentrations of contaminants, or as a polishing treatment.

6.6.5 Effluent Limitations

The Federal Clean Water Act has established effluent limitations for any discharge to a river or lake (public waterway) or to the local publicly owned treatment works (POTW). A direct discharge to a river or lake is regulated by a permit under the National Pollutant Discharge Elimination System (NPDES). The terms of each individual permit can vary from state to state and watershed to watershed. However, at a minimum, the discharge from an existing facility must meet the Effluent Limitation Guidelines (ELG) for the discharge to a public waterway or to a POTW. These guidelines were developed by the U.S. Environmental Protection Agency (EPA) and are established as regulations in 40 CFR 420.

U.S. EPA reviewed the parameters of concern for each category of iron and steel manufacturing operation and established effluent limitations accordingly. This information is presented in the *Development Document for Effluent Limitations Guidelines and Standards for the Iron and Steel Industry* published by the U.S. EPA. These guidelines establish specific categorical limitations for discharges to public waterways and to POTWs. The limitations were developed based on the production operations in industrial facilities as well as the number of tons per day of product produced. The limitations are intended to be uniform treatment standards, such that the location of the facility is not meaningful. The U.S. EPA established limitations for existing sources as well as new sources.

The Agency evaluated the best practicable control technology currently available (BPT) and the best available technology economically achievable (BAT) for each steelmaking process. In some cases, the U.S. EPA indicated that BPT is equivalent to BAT and established BPT as BAT. For our purposes, these cases are referred to as BAT in the following text. All existing sources discharging to a public waterway have to meet the best available technology economically achievable (BAT). All existing sources discharging to a POTW have to meet Pretreatment Standards for Existing Sources (PSES) limitations. Similarly, all new sources have to meet New Source Performance Standards (NSPS) or Pretreatment Standards for New Sources (PSNS).

In addition to the federally-mandated effluent limitations, the states have the authority to impose more restrictive conditions in order to comply with local water quality standards. The more restrictive of the federal or state water quality limitations will be imposed in the NPDES permit. Local municipalities can also impose more restrictive effluent limitations for discharge into their sewer system, based on their need to meet state water quality standards or the ability of the POTW to handle the parameters of concern.

It should be noted that the achievable limitations proposed by U.S. EPA, and summarized below, were used to establish effluent limitation guidelines as published in Federal Regulations 40 CFR 420. These regulations establish limitations on a production basis and are presented as lb/1000 lb of a product or kg/1000 kg. These limitations are typically presented as a maximum for any one day (daily maximum) and an average of daily values for 30 consecutive days (monthly average). Historically, the U.S. EPA has utilized production data from the previous five-year period and has calculated the limitations based on the highest actual monthly production converted to a daily value. This is done by dividing the monthly production by the number of operating shifts in that month and then multiplying by three to determine a daily production rate. The production value is then multiplied by the factors in 40 CFR 420 to calculate the effluent limits. The achievable limitations presented below and the associated treatment methodologies are all taken from the U.S. EPA development document. This information is presented as a guideline; it is not intended to represent an absolute condition. There are many ways of achieving the required effluent limitations, including alternate treatment methods and/or alternate flow reduction methods. Additionally, the model flows presented in this text are not intended to be requirements to achieve BPT or BAT; rather, these flows are simply guidelines that U.S. EPA believes could be achieved. Table 6.29 presents a general summary of the treatment technologies utilized for the parameters of concern in steel making wastewaters.

Table 6.29 Treatment Technologies for Parameters of Concern in Steelmaking Wastewaters

PARAMETER	Gravity Separation	Filtration	Mechanical Separation	Chemical Precipitation	Neutralization	Carbon Adsorption	Chemical Oxidation	Biological
Dissolved Organics						X	X	X
Dissolved Inorganics[a]							X	X
Metals	X	X	X	X	X			
Suspended Solids	X	X	X					
Oil and Grease	X	X	X					
pH					X			

[a] Cyanide, ammonia, sulfide, thiocyanate, etc.

The expected effluent quality of the treated wastewaters from each of the steelmaking operations is shown in daily maximum and monthly average values in Table 6.30. Note that these values are based on BAT treatment systems. New facilities would have more restrictive effluent limitations as defined under the NSPS. Following Table 6.30 is a discussion of the recommended treatment system for each of the major steel making operations.

6.6.5.1 Cokemaking

The BAT treatment system for cokemaking consists of: 1. the recycle of crystallized wastewaters, if any, to minimize the flow to be treated; 2. the treatment of the remaining wastewater flow in a two-stage or extended biological system. Fig. 6.7 presents the treatment system while Table 6.31 provides the model flows.

6.6.5.2 Sintering

The BAT treatment system for the sintering process involves recycling process wastewaters and filtering a blowdown flow of 120 gal/ton to reduce the levels of toxic materials and suspended solids. The pH of the effluent is adjusted using acid. The BAT system assumes a recycle rate of 92%. The system is illustrated in the Fig. 6.8.

6.6.5.3 Ironmaking

The BAT treatment system for ironmaking consists of recycling process wastewaters and treating the blowdown, as shown in Fig. 6.9. The applied flow rate for BAT is 13,344 litres/tonne (3200 gal/ton). After recycling, the blowdown (70 gal/ton) may be treated with two-stage alkaline chlorination, if required. Lime is added to the blowdown to raise the pH to 10.5 or greater. The toxic

Steel Plant Fuels and Water Requirements

Table 6.30 Summary of Expected Effluent Quality

Subcategory		Coke-making	Sintering	Iron-making	Steel-making	Vacuum Degassing	Continuous Casting	Hot Forming	Salt Bath Descaling	Acid Pickling	Cold Forming	Alkaline Cleaning	Hot Coating
pH (SU)		6–9	6–9	6–9	6–9	6–9	6–9	6–9	6–9	6–9	6–9	6–9	6–9
TSS (mg/l)	Max.	270	150	150	150	150	70	40	70	70	60	70	70
	Avg.	140	50	50	50	50	25	15	30	30	30	30	30
Oil & Grease (mg/l)	Max.	10	30	—	—	—	30	10	—	30(c)	25	30	30
	Avg.	—	10	—	—	—	10	—	—	10	10	10	10
Ammonia-N (mg/l)	Max.	85	30(a)	30	—	—	—	—	—	—	—	—	—
	Avg.	25	10	10	—	—	—	—	—	—	—	—	—
Benzene (mg/l)	Max.	0.05	—	—	—	—	—	—	—	—	—	—	—
	Avg.	—	—	—	—	—	—	—	—	—	—	—	—
Benzo-pyrene (mg/l)	Max.	0.05	—	—	—	—	—	—	—	—	—	—	—
	Avg.	—	—	—	—	—	—	—	—	—	—	—	—
Cyanide (mg/l)	Max.	10	6(a)	6	—	—	—	—	0.75(b)	—	—	—	—
	Avg.	5.5	3	3	—	—	—	—	0.25	—	—	—	—
Chromium (mg/l)	Max.	—	—	—	—	—	—	0.3	1	1.0(d)	1.0(e)	—	0.06(f)
	Avg.	—	—	—	—	—	—	0.1	0.4	0.4	0.4	—	0.02
Lead (mg/l)	Max.	—	0.9	0.9	0.9	0.9	0.9	0.3	—	0.45	0.45	—	0.45
	Avg.	—	0.3	0.3	0.3	0.3	0.3	0.1	—	0.15	0.15	—	0.15
Naphthalene (mg/l)	Max.	0.05	—	—	—	—	—	—	—	—	0.1	—	—
	Avg.	—	—	—	—	—	—	—	—	—	—	—	—
Nickel (mg/l)	Max.	—	—	—	—	—	—	—	0.9	0.9(d)	0.9(e)	—	—
	Avg.	—	—	—	—	—	—	—	0.3	0.3	0.3	—	—
Phenols (mg/l)	Max.	0.1	0.2(a)	—	—	—	—	—	—	—	—	—	—
	Avg.	0.05	0.1	0.1	—	—	—	—	—	—	—	—	—
Residual Chlorine (mg/l)	Max.	—	0.5(a)	0.5	—	—	—	—	—	—	—	—	—
	Avg.	—	—	—	—	—	—	—	—	—	—	—	—
Tetrachloroethylene (mg/l)	Max.	—	—	—	—	—	—	—	—	—	0.15	—	—
	Avg.	—	—	—	—	—	—	—	—	—	—	—	—
Zinc (mg/l)	Max.	—	1.35	1.35	1.35	1.35	1.35	0.45	—	0.3	0.3	—	0.3
	Avg.	—	0.45	0.45	0.45	0.45	0.45	0.15	—	0.1	0.1	—	0.1

(a) Applicable only when sintering wastewater is treated with ironmaking wastewater.
(b) Applicable at reducing operations only.
(c) Applicable only when pickling wastewater is treated with cold rolling wastewater.
(d) Applicable at combination acid pickling only.
(e) Applicable only when cold rolling wastewaters are treated with descaling or combination acid.
(f) Hexavalent chromium.

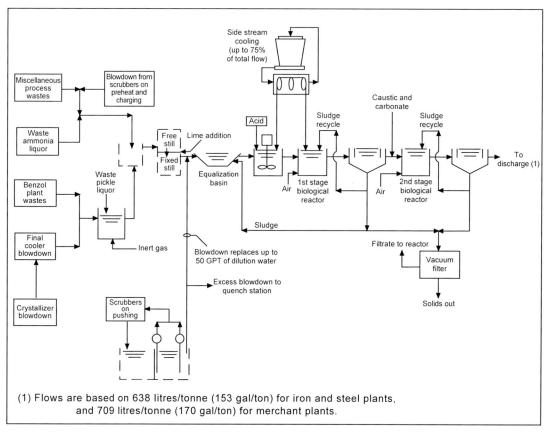

Fig. 6.7 Byproduct cokemaking subcategory, biological treatment system; BAT model.

Table 6.31 BAT Flow Summary for Byproduct Cokemaking Subcategory

(All flows in gal/ton of coke.)

Wastewater Source	Flow Basis BAT Effluent	
	Integrated Producers	Merchant Producers
Waste Ammonia Liquor	32	36
Final Cooler Blowdown	10	12
Barometric Condenser Blowdown	3	5
Benzol Plant Wastewater	25	28
Steam & Lime Slurry	13	15
Miscellaneous Sources (leaks, seals, test taps, drains)	20	24
Subtotal—Process Wastewaters	103	120
Dilution to optimize bio-oxidation	50[a]	50[a]
BASIC TOTAL FLOW	**153**	**170**
Additional Flow Allowances Provided in the Regulation:		
For Qualified Desulfurizers (Wet), up to:	25	25
For Indirect Ammonia Recovery, up to:	60	60
No Additional Allowances For: Air Pollution Control Scrubbers:		
Coal Drying or Preheating – up to 15 gal/ton blowdown*	0	0
Charging/ Larry Car – up to 5 gal/ton blowdown*	0	0
Pushing Side Scrubber – up to 100 gal/ton blowdown*	0	0
MAXIMUM TOTAL FLOW	**238**	**255**

[a] Up to 50 gal/ton of dilution water is replaced by blowdown from air pollution control scrubbers. Any excess blowdown (from pushing only) is disposed via quenching operations, or treated and reused in the scrubber system.

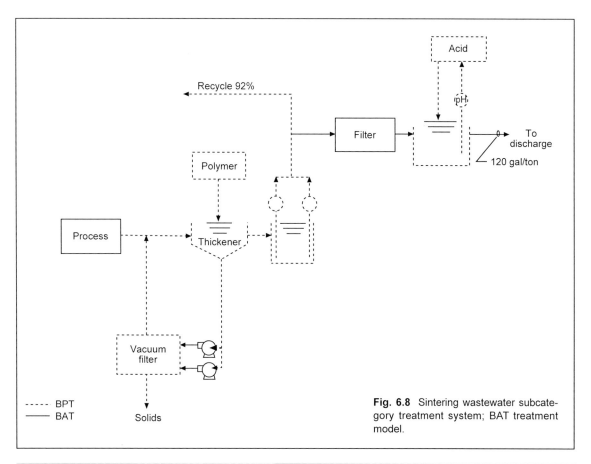

Fig. 6.8 Sintering wastewater subcategory treatment system; BAT treatment model.

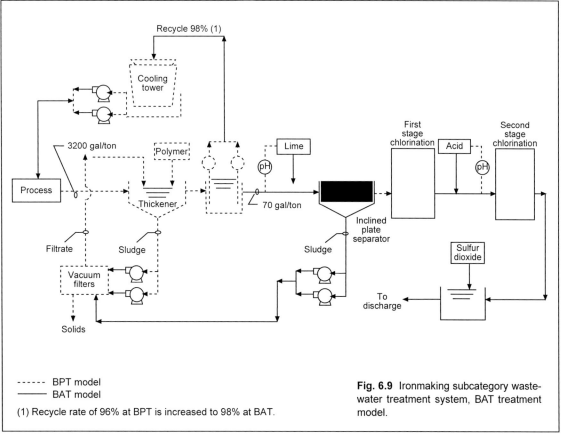

Fig. 6.9 Ironmaking subcategory wastewater treatment system, BAT treatment model.

(1) Recycle rate of 96% at BPT is increased to 98% at BAT.

metal precipitates and other suspended solids formed by lime addition are separated prior to alkaline chlorination. Chlorine is added in a two-stage system. In the first reactor, chlorine converts the cyanides to cyanates and oxidizes ammonia–N and phenolic compounds. As the wastewaters leave the first reactor, acid is added to reduce the pH to 8.5. Additional chlorine is added in the second reactor to complete the oxidation of the cyanides as well as residual ammonia–N and phenolic compounds. The effluent is then dechlorinated with appropriate reducing agents prior to discharge.

6.6.5.4 Steelmaking

The applied and effluent flows included in the BAT treatment system for each segment are shown in Table 6.32.

Table 6.32 Flow Rates for Steelmaking Subcategory

Operation	Applied Flow (gal/ton)	BAT Effluent Flow (gal/ton)
BOF—Wet-Suppressed Combustion	1000	50
BOF—Wet-Open Combustion	1100	110
Open Hearth—Wet	1700	110
Electric Arc—Wet	2100	110

The BAT treatment system for the wet subdivision consists of pH adjustment and solids separation, as shown in Fig. 6.10. Lime is added to increase the pH for the purpose of providing both dissolved and particulate toxic metals removal. The suspended solids generated in this process are then removed. Final pH adjustment of the treated effluent is also required.

6.6.5.5 Vacuum Degassing

The BAT treatment system for the vacuum degassing process includes lime precipitation, sedimentation, and pH control to remove dissolved and particulate toxic metals, as presented in Fig. 6.11. The applied and discharge flows are 1400 gal/ton and 25 gal/ton, respectively.

6.6.5.6 Continuous Casting

The model flow for continuous casting is 25 gal/ton. The BAT system includes recycling and treatment of the blowdown with lime precipitation and sedimentation to remove both particulate and dissolved toxic metals, as shown in Fig. 6.12.

6.6.5.7 Hot Forming

In the BAT treatment system for hot forming waste streams, the water discharged from the mill is collected in a scale pit where large particles settle out and surface skimmers remove floating oils. The BAT treatment system includes recycle of the process wastewaters followed by a roughing clarifier and then a filter for additional suspended solids and oil removal, as presented in Fig. 6.13. Depending on the hot forming subdivision, anywhere from 58 to 77% of the primary scale pit effluent is recirculated to the mill. A vacuum filter is used to dewater the underflow from the clarifier. Although EPA has suggested model flows for the hot forming processes, as presented in Table 6.33, these recycle rates are not mandatory.

6.6.5.6 Salt Bath Descaling

The BAT system for oxidizing operations includes acidification followed by chemical reduction of hexavalent chromium with sulfur dioxide, sodium metabisulfite, or sodium hydrosulfite. Chemical

Steel Plant Fuels and Water Requirements

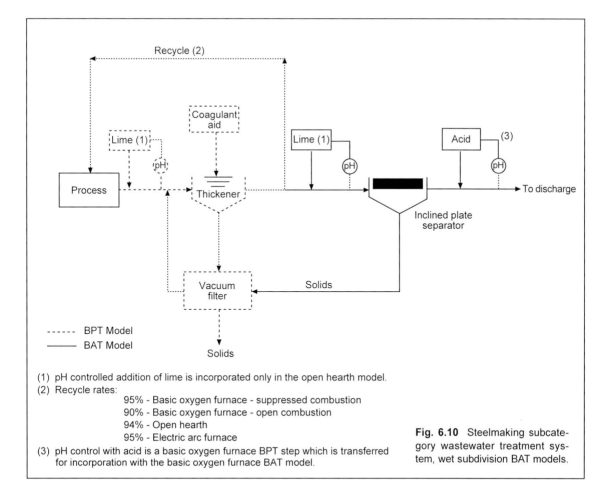

(1) pH controlled addition of lime is incorporated only in the open hearth model.
(2) Recycle rates:
   95% - Basic oxygen furnace - suppressed combustion
   90% - Basic oxygen furnace - open combustion
   94% - Open hearth
   95% - Electric arc furnace
(3) pH control with acid is a basic oxygen furnace BPT step which is transferred for incorporation with the basic oxygen furnace BAT model.

Fig. 6.10 Steelmaking subcategory wastewater treatment system, wet subdivision BAT models.

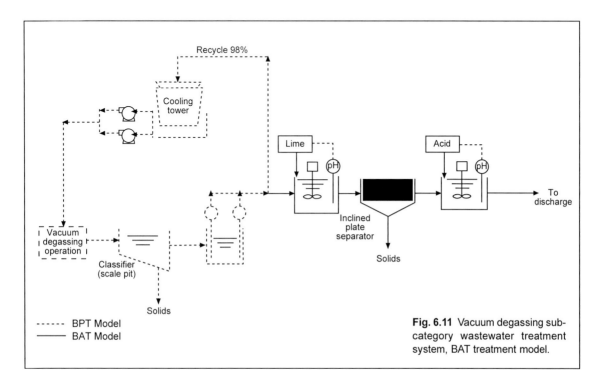

Fig. 6.11 Vacuum degassing subcategory wastewater treatment system, BAT treatment model.

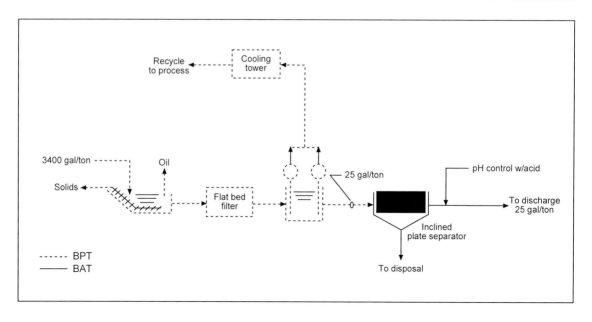

Fig. 6.12 Continuous casting subcategory wastewater treatment system, BAT treatment model.

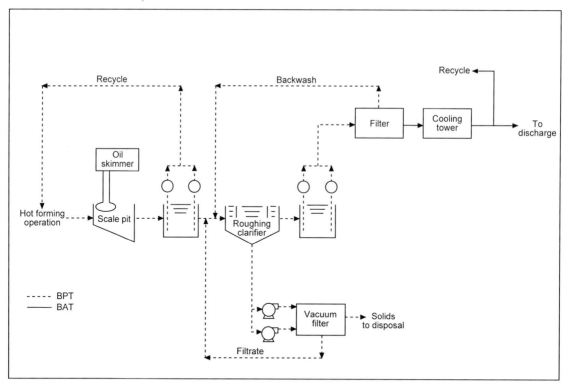

Fig. 6.13 Hot forming subcategory wastewater treatment system, BAT models.

reduction is followed by oil separation or skimming, neutralization/precipitation with lime, chemical coagulation with polymer, and settling in a clarifier. Sludges are dewatered in vacuum filters. In reducing operations, chemical oxidation with chlorine for cyanide destruction is followed by neutralization/precipitation with acid, chemical coagulation with polymer, and settling in a clarifier. Sludges are dewatered. The model flows are shown in Table 6.34.

Table 6.33 Flow Rates for Hot Forming Subcategory

Subdivision	Applied Flow (gal/ton)	Primary Scale Pit Recycle (%)	BPT Discharge Flow (gal/ton)	BAT Recycle Increase[a] (%)	BAT Discharge Flow (gal/ton)
Primary:					
w/o Scarfer	2300	61	897	35	90
w/ Scarfer	3400	61	1326	35	140
Section:					
Carbon	5100	58	2142	38	200
Specialty	3200	58	1344	38	130
Flat:					
Hot Strip	6400	60	2560	36	260
Carbon Plate	3400	60	1360	36	140
Specialty Plate	1500	60	600	36	60
Pipe and Tube	5520	77	1270	19	220

[a] Increase over Primary Scale Pit recycle. Total overall BPT/BAT system recycle = 96%.

Table 6.34 Flow Rates for Salt Bath Descaling Subcategory

Oxidizing Operations	**Flow** (gal/ton)
Batch	
Sheet, Plate	700
Rod, Wire, Bar	420
Pipe, Tube	1700
Continuous	330
Reducing Operations	**Flow** (gal/ton)
Batch	325
Continuous	1820

6.6.5.7 Acid Pickling

The BAT flows for acid pickling are shown in Table 6.35. The BAT system includes equalization of spent pickle liquor, fume scrubber recycle, equalization of spent pickle liquor, rinse water, fume scrubber blowdown, and absorber vent scrubber wastewaters, where applicable. Following equalization, treatment consists of lime neutralization/precipitation, polymer addition, aeration, clarification, and vacuum filtration, as shown in Fig. 6.14. The BAT flow rates are presented in Table 6.35.

6.6.5.8 Cold Forming

The BAT flows are 90 gal/ton, 400 gal/ton, and 300 gal/ton for single-stand, multiple-stand and combination-mill operations, respectively. The BAT treatment system for the cold rolling subdivision includes oil separation and equalization, chemical addition (alum and acid) to break any oil emulsions, neutralization, flocculation with polymer, and dissolved air flotation, as shown in Fig. 6.15.

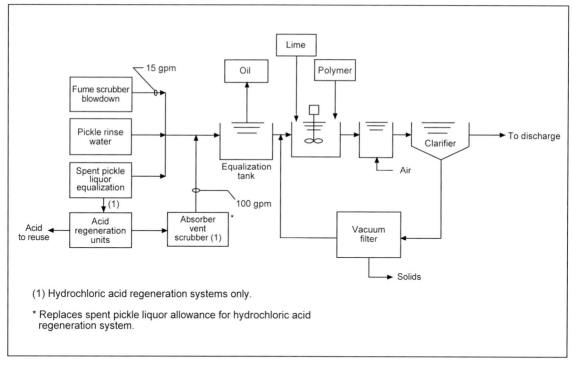

Fig. 6.14 Acid pickling wastewater treatment system; BPT/BAT model.

Table 6.35 Development of Applied Flows Acid Pickling Subcategory

	Discharge Flow (gal/ton)
Sulfuric Acid	
Strip/Sheet/Plate	180
Rod/Wire/Coil	280
Bar/Billet/Bloom	90
Pipe/Tube/Other	500
Fume Scrubber[a]	15 gpm
Hydrochloric Acid	
Strip/Sheet/Plate	280
Rod/Wire/Coil	490
Pipe/Tube	1020
Absorber Vent Scrubber[b]	100 gpm
Fume Scrubber[a]	15 gpm
Combination Acid	
Batch – Strip/Sheet/Plate	460
Cont. – Strip/Sheet/Plate	1500
Rod/Wire/Coil	510
Bar/Billet/Bloom	230
Pipe/Tube	770
Fume Scrubber[a]	15 gpm

[a] The fume scrubber limitations which is given in Kg/day is in addition to the kg/tonne limitations shown for other acid pickling segments.

[b] The absorber vent scrubber limitation, which is given in Kg/day, is in addition to the kg/Kkg limitation for other pickling segments and the Kg/day fume scrubber limitation for the hydrochloric acid subdivision.

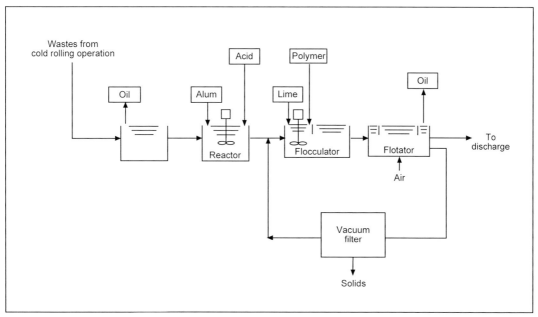

Fig. 6.15 Cold forming subcategory wastewater treatment system; cold rolling BPT/BAT model.

The BAT treatment systems for the cold worked pipe and tube subdivision is divided into two further categories: plants using water and plants using soluble oil solutions. For the plants using water, the BAT treatment system includes settling of the raw wastewater in a primary scale pit equipped with oil skimming equipment. All of the treated wastewater is then recycled to the process. The BAT treatment system for the plants using soluble oil includes settling of the raw wastewater in a primary scale pit equipped with oil skimming equipment that removes tramp oils. Nearly all of the solution is then recycled to the process. The spent solution is periodically removed by a contract hauler so that there is no discharge to navigable waters.

6.6.5.9 Alkaline Cleaning

The model flows in the alkaline cleaning subcategory are 250 gal/ton and 350 gal/ton for batch and continuous operations, respectively. The BAT treatment system includes the following wastewater treatment steps: equalization, oil skimming, neutralization with acid, and addition of a polymer followed by sedimentation in a flocculation-clarifier, as shown in Fig. 6.16.

6.6.5.10 Hot Coating

The BAT flows in the hot coating subcategory are 600 gal/ton and 2400 gal/ton for strip, sheet, and miscellaneous (SSM) and wire products and fasteners (WPF), respectively. The treatment system alternative selected for hot coating operations relies on flow reduction by recycling fume scrubber wastewaters and limiting blowdown streams from the scrubber system to 15 gpm. This reduced scrubber discharge is combined with the rinse water and treated in the BAT treatment system. The BAT system for galvanizing consists of chrome reduction, equalization, pH adjustment, polymer addition, and solids separation, as shown in Fig. 6.17. The BAT system for terne and other metals is similar, but does not include a chromium reduction step.

6.6.6 Boiler Water Treatment

Of the many uses for energy in the U.S. today, in industry, in transportation, in homes and commercial buildings, the largest portion of total use is directed toward producing steam through the

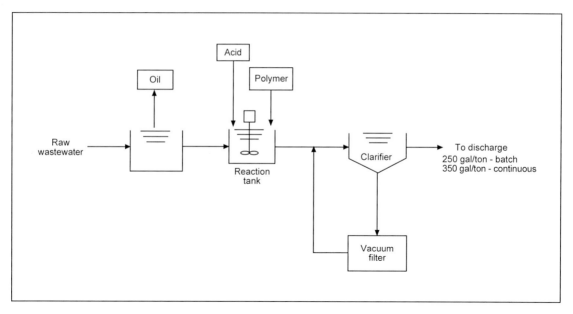

Fig. 6.16 Alkaline cleaning wastewater treatment system; BPT/BAT model.

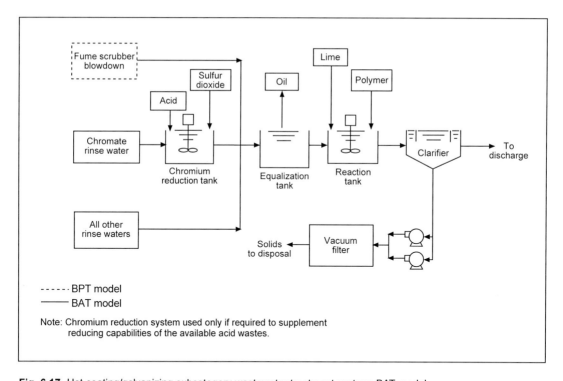

Fig. 6.17 Hot coating/galvanizing subcategory wastewater treatment system; BAT model.

combustion of fossil fuels. Utilities account for the greatest share of this, but industrial plants also produce enormous quantities of steam for process uses, often generating electric power through turbines of a byproduct fuel (co-generation).

The treatment of water for steam generation is one of the most sophisticated branches of water chemistry. An understanding of the fundamentals of boiler water chemistry is essential to the power engineer who continually strives to increase the efficiency of the boilers and steam-using equipment.

The pressure and design of a boiler determine the quality of water it requires for steam generation. Municipal or plant water of good quality for domestic use is seldom good enough for boiler feed water. These sources of makeup are nearly always treated to reduce contaminants to acceptable levels; in addition, corrective chemicals are added to the treated water to counteract any adverse effects of the remaining trace contaminants. The sequence of treatment depends on the type and concentration of contaminants found in the water supply and the desired quality of the finished water to avoid the three major boiler system problems: deposits, corrosion, and carryover.

6.6.6.1 Deposits

Deposits, particularly scale, can form on any water-washed equipment surface, especially on boiler tubes, as the equilibrium conditions in the water contacting these surfaces are upset by an external force, such as heat. Each contaminant has an established solubility in water and will precipitate when it has been exceeded. If the water is in contact with a hot surface and solubility of the contaminant is lower at higher temperatures, the precipitate will form on the surface, causing scale. The most common components of boiler water deposits are calcium phosphate, silicate, various forms of iron oxide, silica adsorbed on the previously mentioned precipitates, and alumina; see Table 6.36. If phosphate salts are used to treat the boiler water, calcium will preferentially precipitate as the phosphate before precipitating as the carbonate, and calcium phosphate becomes the most prominent feature of the deposit.

Table 6.36 Expected Composition of Boiler Sludge

Constituent	Coagulation-Type Treatment	PO_4 Residual Treatment
Calcium carbonate	High	Usually less than 5%
Calcium phosphate	Usually less than 15%	High
Calcium silicate	Usually less than 3%	Trace or none
Calcium sulfate	None	None
Calcium hydroxide	None	None
Loss on Ignition	Usually less than 5%	Usually 8–12% except higher in very pure feed waters
Magnesium phosphate	None	Usually less than 5% except in some high-pressure boilers
Magnesium hydroxide	Moderate	Moderate
Magnesium silicate	Moderate	Moderate
Silica	Usually less than 10%	Usually less than 10%
Alumina	Less than 10%	None
Oil	None	None
Iron oxide	Usually less than 5%	Usually less than 5% except in high-purity feed waters
Sodium salts	Usually less than 1.5%	Usually less than 1.5%
Copper	Trace	Usually low
Other Metals	Trace	Low

At the high temperatures found in a boiler, deposits are a serious problem, causing poor heat transfer and a potential for boiler tube failure. In low-pressure boilers with low heat transfer rates, deposits may build up to a point where they completely occlude the boiler tube.

In modern intermediate and higher pressure boilers with heat transfer rates in excess of 5000 cal/m^2/hr (200,000 Btu/ft^2/h), the presence of even extremely thin deposits will cause a serious elevation in the temperature of tube metal. The deposit coating retards the flow of heat from the fur-

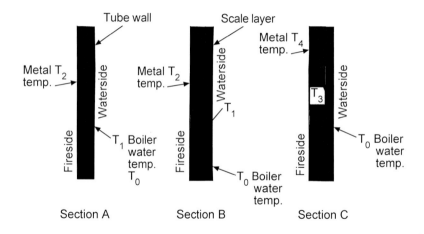

Fig. 6.18 Temperature profile across a clean tube, Section A, and tubes having a water-side deposit, Section B and Section C.

nace gases into the boiler water. This heat resistance results in a rapid rise in metal temperature to the point at which failure can occur. The action that takes place in the blistering of a tube by deposit buildup is illustrated by Fig. 6.18. For simplification, no temperature drops through gas or water films have been shown. Section A shows a cross section of the tube metal with a completely deposit-free heating surface. There is a temperature drop across the tube metal from the outside metal (T_2) to the metal in contact with boiler water (T_1). Section B illustrates this same tube after the development of a heat-insulating deposit layer. In addition to the temperature drop from T_2 to T_1, there would be an additional temperature drop through the deposit layer from T_1 to T_0. This condition would, of course, result in a lower boiler water temperature T0. However, boiler water temperature is fixed by the operating pressure, and operating conditions require that the same boiler water temperature be maintained as before the development of the deposit layer. Section C illustrates the condition that actually develops. Starting at the base boiler water temperature of T0, the increase through the scale layer is represented by the line from T_0 to T_3. The further temperature increase through the tube wall is represented by the line from T_3 to T_4. The outside metal temperature T_4 is now considerably higher than the temperature T_2, which was the outside metal temperature prior to the formation of deposits on the tube surfaces.

If continued deposition takes place, increasing the thickness of the heat-insulating deposits, further increases will take place in the tube metal temperature until the safe maximum temperature of the tube metal is exceeded. Usually this maximum temperature is 480 to 540°C (900 to 1000°F). At higher heat transfer rates, and in high-pressure boilers, the problem is more severe. At temperatures in the 482 to 732°C (900 to 1350°F) range microstructural transformations take place in carbon steel. At temperatures above 427°C (800°F) spheroidization of carbon begins to take place, with a resultant reduction in strength properties. Temperatures within the boiler furnace are considerably above this critical temperature range.

Water circulating through the tubes normally conducts heat away from the metal, preventing the tube from reaching this range. Deposits insulate the tube, reducing the rate at which this heat can be removed, as shown in Fig. 6.19; this leads to overheating and eventual tube failure. If the deposit is not thick enough to cause such a failure, it can still cause a substantial loss in efficiency and disruption of the heat transfer load in other sections of the boiler.

Deposits may be scale, precipitated in situ on a heated surface, or previously precipitated chemicals, often in the form of sludge. These drop out of water in low-velocity areas, compacting to form a dense agglomerate similar to scale, but retaining the features of the original precipitates. In the operation of most industrial boilers, it is seldom possible to avoid formation of some type of precipitate at some time. There are almost always some particulates in the circulating boiler water which can deposit in low-velocity sections, such as the mud drum. The exception would be high-purity systems, such as utility boilers, which remain relatively free of particulates except under conditions where the system may become temporarily upset.

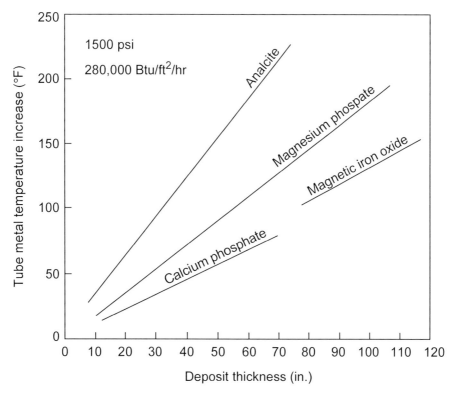

Fig. 6.19 Tube metal temperature increase vs. deposit thickness.

6.6.6.2 Corrosion

The second major water-related boiler problem is corrosion, the most common example being the attack of steel by oxygen. This occurs in water supply systems, preboiler systems, boilers, condensate return lines, and in virtually any portion of the steam cycle where oxygen is present. Oxygen attack is accelerated by high temperature and by low pH. A less prevalent type of corrosion is alkali attack, which may occur in high pressure boilers where caustic can concentrate in a local area of steam bubble formation because of the presence of porous deposits.

Some feed water treatment chemicals, such as chelants, if not properly applied can corrode feed water piping, control valves, and even the boiler internals.

While the elimination of oxygen from boiler feed water is the major step in controlling boiler corrosion, corrosion can still occur. An example is the direct attack by steam of the boiler steel surface at elevated temperatures, according to the following reaction:

$$4H_2O + 3Fe \rightarrow Fe_3O_4 + 4H_2 \qquad (6.6.4)$$

This attack can occur at steam-blanketed boiler surfaces where restricted boiler water flow causes overheating. It may also occur in superheater tubes subjected to overheating. Because this corrosion reaction produces hydrogen, a device for analyzing hydrogen in steam is useful as a corrosion monitor.

6.6.6.3 Carryover

The third major problem related to boiler operations is carryover from the boiler into the steam system. This may be a mechanical effect, such as boiler water spraying around a broken baffle; it may be caused by the volatility of certain boiler water scales, such as silica and sodium compounds; or it may be caused by foaming. Carryover is most often a mechanical problem, and the chemicals found in the steam are those originally present in the boiler water, plus the volatile components that distill from the boiler even in the absence of spray.

There are three basic means for keeping these major problems under control. External treatment of makeup water, condensate, or both, is performed before their entry into the boiler to reduce or eliminate chemicals (such as hardness or silica), gases or solids. Internal treatment of the boiler feed water, boiler water, steam, or condensate with corrective chemicals may be performed. Control of the concentration of chemicals in the boiler water may be performed by blowdown of the water from the boiler.

6.6.6.4 External Treatment

Most of the unit operations of water treatment can be used alone or in combination with others to adapt any water supply to any boiler system. The suitability of the processes available is judged by the results they produce and the costs involved. Water treatment processes include: sodium zeolite softening, lime softening, lime softening followed by sodium zeolite softening, split stream dealkalization, demineralization, and reverse osmosis.

The boiler treatment program aims at control of seven broad classifications of impurities: suspended solids, hardness, alkalinity, silica, total dissolved solids (TDS), organic matter, and gases.

6.6.6.4.1 Suspended Solids The removal of suspended solids is accomplished by coagulation/flocculation, filtration, or precipitation. Just about all unit processes require prior removal of solids. For example, water to be processed by ion exchange should contain less that 10 mg/litre suspended solids to avoid fouling of the exchanger and operating problems.

6.6.6.4.2 Hardness A number of unit operations can remove calcium and magnesium from water, as well as other impurities. Partial cold lime softening and complete hot lime softening are used in most heavy industrial plants. These processes remove hardness, alkalinity, silica, and suspended solids. Sodium exchange can be used alone or in conjunction with lime softening.

6.6.6.4.3 Alkalinity It is desirable to have some alkalinity in boiler water, so complete removal of alkalinity from boiler makeup is seldom practiced except in demineralization. Some alkalinity is also needed to provide optimum pH in the feed water to prevent corrosion if piping and equipment.

The makeup alkalinity may be present as HCO_3^-, CO_3^{2-}, or OH^-. If the makeup is city water that has been zeolite softened, alkalinity is usually in the bicarbonate (HCO_3^-) form; if lime softened, it is mostly carbonate (CO_3^{2-}), but the water may also contain some hydroxide (OH^-). When bicarbonates and carbonates are exposed to boiler temperatures, they break down to release CO_2:

$$2NaHCO_3 \rightarrow Na_2CO_3 + H_2O + CO_2 \quad (6.6.5)$$

The sodium carbonate then breaks down further to caustic:

$$Na_2CO_3 + H_2O \rightarrow 2NaOH + CO_2 \quad (6.6.6)$$

The carbon dioxide gas redissolves when the steam condenses, producing corrosive carbonic acid:

$$CO_2 + H_2O \rightarrow H^+ + HCO_3^- \quad (6.6.7)$$

The amount of CO_2 generated is proportional to alkalinity. For a given alkalinity twice as much CO_2 is formed from HCO_3^- as from CO_3^{2-} because the bicarbonate breakdown is the sum of both reactions 6.6.5 and 6.6.7 above. The carbonic acid is usually neutralized by chemical treatment of the steam, either directly or indirectly through the boiler, to produce a condensate pH in the range of 8.5 to 9.0. Reduction of feed water alkalinity is desirable, then, to minimize CO_2 formation and reduce chemical treatment costs.

The hydroxide produced by the breakdown of HCO_3^- and CO_3^{2-} is beneficial for precipitation of magnesium, to provide a good environment for sludge conditioning, and to minimize SiO_2 carryover. However, too high an excess of caustic can be corrosive, particularly if localized concentration can occur. The breakdown of HCO_3^- is complete, but not all the CO_3^{2-} converts to caustic. The conversion varies from one boiler to another and increases with temperature. As a general rule, at 600 lb/in² 65 to 85% of boiler water alkalinity is NaOH, the remainder being Na_2CO_3. (This is based on the equilibrium in the cooled sample of boiler water).

The degree of alkalinity reduction is therefore dictated by boiler water control limits and steam quality goals. The best unit process for alkalinity reduction may be chosen for the other benefits it provides as well as its efficiency in alkalinity reduction

6.6.6.4.4 Silica The permissible concentrations of silica in boiler water at various operating pressures are given in Table 6.37. Silica reduction is not always necessary, especially in the absence of a condensing turbine. Low concentrations of silica can be sometimes produce sticky sludge in low-pressure boilers treated with phosphate. A makeup treatment process may be selected to provide just the proper degree of silica reduction required by the steam system.

Table 6.37 Permissible Concentrations of Silica in Boiler Water at Various Operating Pressures

Drum pressure, (lb/in²)	Silica concentration, (mg/litres)	
	Recommended	To produce 0.02 mg/litre
0–300	150	150
301–450	90	90
451–600	40	55
601–750	30	35
751–900	20	20
900–1000	8	15

6.6.6.4.5 Total Dissolved Solids Some treatment processes increase dissolved solids by adding soluble byproducts to water; sodium zeolite softening increases solids by adding an ion (sodium) having a higher equivalent weight (23) than calcium (20) or magnesium (12.2) removed from the raw water. Processes to reduce dissolved solids achieve various degrees of success. Usually, reduction of dissolved solids is accomplished by a reduction of several individual contaminants.

6.6.6.4.6 Organic Matter Organic matter as a general classification is only a qualitative term. It includes a wide variety of compounds that are seldom analyzed as specific materials. Problems in boiler systems attributed to organic matter have often been traced to organic materials from plant processes in returned condensate, rather than makeup water contaminants. However, in higher pressure utility systems, organic matter is the major impurity in makeup and can result in formation of organic acids.

6.6.6.4.7 Dissolved Gases Degasifiers are commonly used to remove gas mechanically rather than chemically. Blower types are used for CO_2 removal at ambient temperatures following acid or hydrogen-exchange units. Vacuum degasifiers provide the same extent of CO_2 removal, but also reduce O_2 to less than 0.5 to 1.0 mg/litre, offering corrosion protection, especially if the vacuum degasifier is part of a demineralizing system. Steam-scrubbing degasifiers, call deaerating heaters, usually produce an effluent free of CO_2 with O_2 concentrations in the range of 0.005 to 0.01 mg/litre. Direct reaction of this low residual with catalyzed sulfite, hydrazine, or hydrazine substitutes (all-volatile oxygen-reducing compounds) eliminates O_2 completely to prevent preboiler corrosion.

6.6.6.5 Condensate Returns

In addition to makeup treatment, acceptable feed water quality may require cleanup of condensate to protect the boiler system, particularly if there is process condensate containing oil. Boilers requiring high-quality demineralized water also demand high-quality condensate. Some plants operate both high- and low-pressure boilers; high-quality feed water for the high pressure boilers may be provided entirely by a demineralizer, with lower quality condensate segregated for return to the low-pressure boilers.

Septum filters are usually selected for oily condensate treatment. A cellulose type filter aid (processed wood pulp) is applied both as a precoat and a body feed. The temperature should be less than 93°C (200°F) to avoid degradation of the filter aid. Anthracite filters precoated with a floc produced from alum and sodium aluminate are also effective. However, the pH of the condensate must be controlled in the range of 7.0 to 8.0 to avoid solubilizing the alumina floc. Condensate contaminated with corrosion products and inleakage of hard water is cleaned up through specially designed, high flow rate sodium exchangers. They have been used to process condensate at temperatures up to 149°C (300°F). One serious limitation of the simple sodium exchanger is its ability to pick up neutralizing amines such as morpholine (present in the condensate as morpholine bicarbonate) and exchange this for sodium. This causes excessive use of amines, but a more serious problem arises if the condensate is returned to a high-pressure boiler where the presence of sodium may be objectionable in deterioration of steam quality. Special regeneration procedures would then be needed.

6.6.6.6 Internal Treatment

Scale formation within a boiler is controlled by one of four chemical programs: coagulation (carbonate), phosphate residual, chelation, or coordinated phosphate.

6.6.6.6.1 Coagulation Program In this process, sodium carbonate, sodium hydroxide, or both are added to the boiler water to supplement the alkalinity supplied by the makeup, which is not softened. The carbonate causes deliberate precipitation of calcium carbonate under favorable, controlled conditions, preventing deposition at some subsequent point as scale. Under alkaline conditions, magnesium and silica are also precipitated as magnesium hydroxide and magnesium silicate. There is usually a fairly high concentration of suspended solids in the boiler water, and the precipitation occurs on these solids. This method of treatment is used only with boilers (usually firetube design) using high-hardness feed water and operating below 250 lb/in^2 (17 bars). This type of treatment must be supplemented by some form of sludge conditioner. Even with a supplemental sludge conditioner, heat transfer is hindered by deposit formation, and blowdown rates are excessive because of high suspended solids. Coagulation programs are becoming obsolete as pretreatment systems become more common and competitive with the high internal treatment cost.

6.6.6.6.2 Phosphate Program Where the boiler pressure is above 250 lb/in^2, high concentrations of sludge are undesirable. In these boilers, feed water hardness should be limited to 60 mg/litre, and phosphate programs are preferred. Phosphate is also a common treatment below 250 lb/in^2 with soft makeup.

A sodium phosphate compound is fed either to the boiler feed water or to the boiler drug, depending on water analysis and the preboiler auxiliaries, to form an insoluble precipitate, principally hydroxyapatite, $Ca_{10}(PO_4)_6(OH)_2$. Magnesium and silica are precipitated as magnesium hydroxide, magnesium silicate (often combined as $3MgO \cdot 2SiO_2 \cdot 2H_2O$), or calcium silicate. The alkalinity of the makeup is usually adequate to produce the OH^- for the magnesium precipitation. Phosphate residual programs which produce high suspended solids require the addition of a sludge conditioner/dispersant. Because these programs restrict heat transfer, owing to the deposition of calcium and magnesium salts, precipitation programs of this type are often replaced with solubilizing treatments such as chelants and polymer/dispersants.

6.6.6.6.3 Chelant Programs A chelate is a molecule similar to an ion exchanger; it is low in molecular weight and soluble in water. The sodium salts of ethylene diamine tetraacetic acid (EDTA) and nitrilotriacetic acid (NTA) are the chelating agents most commonly used for internal boiler treatment. These chelate (form complex ions with) calcium and magnesium. Because the resulting complex is soluble, this treatment is advantageous in minimizing blowdown. The higher cost compared to phosphate usually limits the use of chelates to feed waters having low hardness. There is the risk that breakdown of the organic molecule at higher temperatures could create a potential problem of control that could result in corrosion, so chelate programs are usually limited to boilers operating below 1500 lb/in^2 (100 bars). The addition of polymers as scale control agents increases the effectiveness of chelate programs. It also reduces the corrosion potential by reducing the chelant dosage below theoretical requirements, so that there is no chelant residual in the boiler water.

Chelates can react with oxygen under boiler water conditions, which can increase the cost of a chelate program substantially. Overfeed of chelates and concentration mechanisms in the boiler can lead to severe localized corrosion and subsequent unit failure.

6.6.6.6.4 Coordinated Phosphate Program In high-pressure, high heat transfer rate boilers, the internal treatment program must contribute little or no solids. The potential for caustic attack of boiler metal increases with increasing pressure, so free caustic alkalinity must be minimized. The coordinated phosphate program is chosen for these conditions. This differs from the standard program in that the phosphate is added to provide a controlled pH range in the boiler water as well as to react with calcium if hardness should enter the boiler. Trisodium phosphate hydrolyzes to produce hydroxide ions:

$$Na_3PO_4 + H_2O \rightarrow 3Na^+ + OH^- + HPO_4^{2-} \tag{6.6.8}$$

This cannot occur with the ionization of disodium and monosodium phosphate:

$$Na_2HPO_4 \rightarrow 2Na^+ + HPO_4^{2-} \tag{6.6.9}$$

$$NaH_2PO_4 \rightarrow Na^+ + H^+ + HPO_4^{2-} \tag{6.6.10}$$

The program is controlled by feeding combinations of disodium phosphate with trisodium or monosodium phosphate to produce pH without the presence of free OH^-. To successfully control a coordinated phosphate program, the feed water must be extremely pure and consistent quality. Coordinated phosphate programs do not reduce precipitation; they simply cause precipitation of less adherent calcium phosphate in the absence of caustic. A dispersant must be added to condition deposits that would otherwise reduce the heat transfer rate. The coordinated phosphate program was first developed for high-pressure utility boilers, and most experience with this program has been gained in this field.

6.6.6.6.5 Complexation and Dispersion The newest addition to internal treatment technology is the use of synthetic organic polymers for complexation and dispersion. This type of program can be used to 1500 lb/in^2 (100 bars) and is economical in all low-hardness feed water systems typical of those produced by ion exchange. Heat transfer rates are maximized because these polymers produce the cleanest tube surfaces of any of the available internal treatment programs. This treatment solubilizes calcium, magnesium, and aluminum, and maintains silica in solution while avoiding corrosion potential side effects as determined by hydrogen levels in the steam. Iron particulates returned from the condensate system are likewise dispersed for removal via blowdown. A simple measure of ion transport is used to demonstrate on-line performance of this program.

6.6.6.6.6 Program Supplements In addition to controlling scale and deposits, internal treatment must also control carryover, defined as entrainment of boiler water into the steam. Boiler salts carried as a mist may subsequently deposit in the superheater, causing tube failures or deposit on the blades of a turbine. They may also contaminate a process in which the steam is used. Because a high percentage of carryover is caused by foaming, this program is usually solved by the addition of an antifoam agent to the boiler feed water.

Sludge in boiler water may settle to form deposits, which are as serious a problem as scale. Chemicals are used to condition boiler water particulates so that they do not form large crystalline precipitates; smaller particles will remain dispersed at the velocities encountered in the boiler circuit. At lower pressures both the coagulation and phosphate residual programs incorporate sludge conditioning agents for this purpose. A variety of natural organic materials are used, including starches, tannins, and lignins.

At intermediate pressures, chemically reacted lignins have been widely used, though synthetic polymers are replacing them. At pressures up to 1800 lb/in^2 (120 bars), heat-stable polymers such as anionic carboxylates and their derivatives are used as effective dispersants. An alkaline environment generally increases their effectiveness. Lignin-type dispersants and other natural organic derivatives are being replaced by these more effective synthetic organic polymers. These dispersants have been designed for specific dispersion problems, with tailored molecules for magnesium silicate, calcium phosphate, and iron particulates being available.

Somewhat related to carryover, in that steam quality is affected, is the discharge of contaminants that volatilize under boiler operating conditions. The major volatiles are CO_2, created by the breakdown of carbonate and bicarbonate mentioned earlier, and SiO_2. Although the CO_2 can be neutralized, it is prudent to reduce feed water alkalinity to minimize its formation. For all practical purposes, external treatment for silica reduction and blowdown are the only means to avoid excessive SiO_2 discharges for protection of turbine blades. Hydroxyl alkalinity helps reduce silica volatility.

Oxygen is the chief culprit in boiler systems corrosion. Deaeration reduces the oxygen to a low concentration in the preboiler system, but does not completely eliminate it. Application of sulfite, hydrazine, or hydrazine-like (all-volatile) compounds after deaeration scavenges the remaining O_2 and maintains a reducing condition in the boiler water. An advantage of hydrazine is that it is discharged into the steam to become available in the condensate as protection against oxygen corrosion in the return system. If oxygen is present, ammonia can attack copper alloys in condensers and stage heaters. The removal of NH_3 by external treatment may be necessary. The corrosive aspects of CO_2 have already been mentioned in relation to condensate systems. The beneficial and detrimental aspects of NaOH in the boiler circuit in relation to corrosion control have also been discussed earlier.

6.6.6.7 Blowdown

Boiler feed water, regardless of the type of treatment used to process the makeup, still contains measurable concentrations of impurities. In some plants, contaminated condensate contributes to feed water impurities. Internal boiler water treatment chemicals also add to the level of solids in the boiler water.

When steam is generated, essentially pure H_2O vapor is discharged from the boiler, leaving the solids introduced in the feed water to remain in the boiler circuits. The net result of impurities being continuously added and pure water vapor being withdrawn is a steady increase in the level of dissolved solids in the boiler water. There is a limit to the concentration of each component of the boiler water. To prevent exceeding these concentration limits, boiler water is withdrawn as blowdown and discharged to water. Blowdown must be adjusted so that the solids leaving the boiler equal those entering and their concentration is maintained at the predetermined limits. Of course it is apparent that the substantial heat energy in the blowdown represents a major factor detracting from the thermal efficiency of the boiler, so minimizing blowdown is a goal in every steam plant. One way of looking at boiler blowdown is to consider it a process of diluting boiler water solids by withdrawing boiler water from the system at a rate that induces a flow of feed water into the boiler in excess of steam demand.

There are two separate blowdown points in every boiler system. One accommodates the blowdown flow that is controlled to regulate the dissolved solids or other factors in the boiler water. The other is an intermittent or mass blowdown, usually from the mud drum or waterwall headers, which is operated intermittently at reduced boiler load to rid the boiler of accumulated settled solids in relatively stagnant areas. The following discussion of blowdown will be confined only to that used for adjusting boiler water dissolved solids concentrations.

Blowdown may be either intermittent or continuous. If intermittent, the boiler is allowed to concentrate to a level acceptable for the particular boiler design and pressure. When this concentration level is reached, the blowdown valve is opened for a short period of time to reduce the concentration of impurities, and the boiler is then allowed to reconcentrate until the control limits are again reached. In continuous blowdown on the other hand, the blowdown valve is kept open at a fixed setting to remove water at a steady rate, maintaining a relatively constant boiler water concentration. Because the average concentration level in a boiler blown down intermittently is substantially less than that maintained by continuous blowdown, intermittent blowdown is less efficient, more costly, than continuous blowdown.

It is common to express blowdown as a percentage of feed water. However, this may give the utilities engineer a false sense of security. If the plant has 80% condensate return and 20% makeup, a 5% blowdown would appear satisfactory, but it indicates that the makeup is being concentrated only four times; of the four units of makeup entering the boiler, one unit is being thrown away. Perhaps that is as much usage as can be made of that particular quality makeup, but the operator should be aware of it.

Because the main purpose of blowdown control is to reach the maximum permissible concentrations for best boiler efficiency without exceeding concentrations that would harm the system, the first step in developing a blowdown control program is to establish allowable limits. The conventional limits recommended to provide boiler cleanliness and adequate steam quality should be followed. These limits cover most situations encountered in industrial boiler operations, but not the coagulation treatment used in low-pressure boilers. With the coagulation treatment, total dissolved solids are usually limited to 3500 mg/litre, and adequate alkalinity is maintained to provide the carbonate for calcium precipitation and the hydroxide for magnesium precipitation. These levels can be established only after the nature of the makeup treatment system has been considered.

The limits on such things as total dissolved solids, silica, and alkalinity are basically related to the amounts of these materials entering the makeup water; these concentrations can be adjusted by blowdown and also by some adjustment in the makeup treatment system if that flexibility is provided. On the other hand, such constituents as phosphate, organics, and sulfite are introduced as internal treatment chemicals, and their concentration can be adjusted both by blowdown and by rate of application.

6.6.6.8 Contaminants in Returned Condensate With increasing use of demineralized water, even for intermediate pressure boilers (600 to 900 lb/in^2), the major impurities in the feed water are no longer introduced by the makeup water but rather by the returned condensate, principally as corrosion products.

These corrosion products are not solubilized by chelates and are difficult to disperse. If they deposit, boiler salts may concentrate under them because they are relatively porous and permit boiler water to enter with only steam escaping. This may lead to caustic attack. To prevent this, boilers should be cleaned at a set frequency.

The likelihood of alkalinity attack is increased by variable loads and firing conditions, which cause flexing and cracking of the normal dense magnetite film on the boiler metal, exposing fresh metal to attack.

In forced circulation boilers, there is a tendency for deposits to form on the downstream side of flow control orifices. To avoid this, feed water should be free of deposit-forming material, especially corrosion products, so that a chemical program can be applied successfully. Condensate polishing may be required since the condensate is the source of the corrosion products. Steam separators are essential to production of acceptable steam, and their deterioration will depreciate steam quality.

Boilers are often damaged by corrosion during out-of-service periods. Idle boilers are very vulnerable to attack when air contacts moist metal surfaces. To prevent this corrosion, the boiler metal must be protected by either keeping the surfaces completely dry or by excluding all forms of air from the boiler by completely filling it with properly treated water. Because of variations in boiler design, there is no single, detailed procedure that covers all steps in boiler lay-up, including both chemical and mechanical aspects. The basic principles in protecting boilers against corrosion are simple.

There are two basic ways of laying up boilers: wet and dry. In storing a boiler dry, trays of moisture absorbing chemicals, such as quicklime, are distributed on trays in the boiler drum (or drums) and the boiler is sealed. The alternate method, wet storage, involves forcing air out of the boiler by completely filling to overflow with water which has been specially treated. Nitrogen gas under slight pressure can also be used to displace air and blanket the boiler surfaces. Special consideration must be given to protecting superheaters during lay-up, particularly, the non-drainable type.

The choice between the wet and dry methods of lay-up depends to a great extent on how long the boiler is to be out of service. Dry lay-up is preferable for long outages; the wet method has the advantage of permitting the boiler to be returned to service on reasonably short notice. It is a good idea to drain, flush, and inspect a boiler prior to any lay-up. When times does not permit this, the boiler may be stored wet without first draining it. In this case, the chemical treatment for lay-up, including catalyzed sulfite, caustic, and organic dispersants, is injected into the boiler just before it comes off-line.

6.6.6.9 Condensate Return Systems

Condensate produced when steam is used in any kind of process is seldom cooled measurably below the steam temperature. Because it is hot, close to the steam temperature, the collection system and piping used to handle it must be carefully selected. The piping is generally larger than that used for cold water because pressure drop will cause steam to flash from the flowing condensate, choking the pipe and restricting flow. Condensate may be picked up by a pump for boosting to the best point of return, or it may be delivered to a point of lower pressure simply by the pressure gradient.

Turbine condensate is normally collected in a hotwell, and a level is maintained in the hotwell so that the pump transferring the turbine condenser to the deaerator will have adequate net positive suction head. The level in the hotwell is maintained by returning some of the pump discharge, depending on load fluctuations.

Process condensate is also usually returned by pumps to the deaerating heater. Condensate receivers collect the process condensate and maintain level control so that the pump handling the hot condensate will have adequate suction head. Even so, special designs of centrifugal pumps are usually required for handling hot condensate. Stage heater condensates are normally handled by gravity, returning to a lower pressure stage heater, the deaerator, or even the condenser hotwell.

Once the condensate has been collected, the proper point of return must be decided upon. In the utility station, the entire flow of condensate may be polished through some type of ion exchange system before being returned to the deaerating heater. In industrial plants, if the condensate is contaminated it will be sent to a treatment plant before returning to the deaerator. For the most part, condensates are returned to the deaerating element itself, as they may contain dissolved oxygen and other gases; however, high-pressure returns free of dissolved oxygen may be sent directly to the storage section of the deaerating heater to flash and supply steam for the deaerating operation.

For the most part, the condensate handling system is of ordinary carbon steel construction, although pump impellers, valve trim, and heat exchange tubes are usually of copper alloys. Because the condensate is usually hot, if corrosive agents are present the rate of corrosion will usually be greater than what would be expected in cold water. The principal agents of corrosion are carbon dioxide and oxygen. The CO_2 is normally produced by the breakdown of alkalinity in the boiler and the oxygen may be drawn into the system by inleakage of air or of water containing dissolved oxygen (pump sealing water, for example). Inspection of the condensate piping provides a good clue as to the cause of corrosion. The principal cause of corrosion of copper alloys is ammonia in systems containing O_2.

Without adequate treatment of the steam, the pH of condensate would normally be low because of the presence of carbon dioxide. The application of volatile alkaline amines will control attack by neutralizing carbonic acid, thus raising the pH value of the system. In a tight system, the neutralizing amine is often adequate for the complete corrosion control program. However, many systems are operated intermittently or under throttling (flow-restricting) conditions where oxygen inleakage can occur. At these higher levels of oxygen, neutralization is inadequate as the sole protective measure against corrosion of steel piping. In such cases, volatile filming amines are added to the stream, which upon condensation produce a waxy substance on the metal and provide a barrier between the flowing condensate and the pipe wall so that corrosion cannot occur.

Hydrazine and other all-volatile oxygen scavengers may be used both for pH correction and oxygen scavenging, but it becomes uneconomical when high levels of carbon dioxide and oxygen occur, a common condition in most industrial operations.

In plants where gross contamination occurs, the source and cause should be located and corrected. An example of such an occurrence is the use of steam for producing hot water through a heat exchanger. The industrial operation may require hot water at a specific temperature, 150°F for example, and a thermostatic element is installed in the water line to regulate the steam flow into the heat exchanger according to water flow and exit temperature. At low water flows where the steam demand is low, the steam admission valve may be so fully throttled that there is actually a vacuum in the vapor space. Most of these systems are designed for pressure operation, and under vacuum, air leakage is common. If a neutralizing amine is used in a system of this kind, it may be easy to locate inleakage of this sort because the air in the industrial atmosphere will contain enough CO_2 to drop the pH of the condensate at that particular point. Thus, a way to find the source of air inleakage in a complex industrial plant is to sample condensate at all sources and compare the pH of the sample condensate with the pH of a condensed steam sample or a sample of condensate known to be free of atmospheric contamination.

Where the attack of copper alloy has been found to be caused by ammonia, filming inhibitors will usually prevent further attack by preventing O_2 from reaching the surface. If the ammonia concentration is high, reduction of ammonia in the pretreatment system should be considered.

6.6.6.9.1 Neutralizing Amines The most commonly used neutralizing inhibitors are amines such as morpholine, cyclohexylamine, and diethylaminoethanol. The ability of each product mentioned to enter the condensate or water phase is indicated by its vapor-to-liquid distribution ratio, Table 6.38. This ratio compares the concentration of amine in the vapor phase to the concentration in the water phase.

Table 6.38 Amine Vapor-to-Liquid Distribution Ratios

Product	Vapor-to-liquid distribution ratio[a]
Morpholine	0.4
Cyclohexylamine	4.0
Diethylaminoethanol	1.7

[a] At atmospheric pressure.

In order to neutralize carbonic acid, the amine must be present in the water phase. The distribution ratio indicates the preference of an amine for the water phase or the vapor phase. An amine such as morpholine, preferring the water phase, will be present in the initially formed condensate at high temperatures. On the other hand cyclohexylamine tends to remain with the steam to enter the condensate as the temperatures decrease.

Because of their differing vapor-to-liquid distribution ratios, two or more such amines may be used together to provide the effective neutralization programs for complex systems.

Neutralizing amines are fed to the feed water after deaeration, boiler steam drum, or steam header. They are controlled by monitoring the returned condensate pH from samples taken at the beginning, middle, and end of a condensate system.

6.6.6.9.2 Filming Inhibitors Inhibitors used to film condensate systems are amines with chain-like molecules. One end of each molecule is hydrophilic (loves water), and the other end hydrophobic (hates water). The hydrophilic end attaches to the metal, leaving the other end to repel water. As the molecules accumulate, the surface becomes non-wettable. The film, therefore, provides a barrier against metal attack by water containing carbon dioxide, oxygen, or ammonia. Because the molecules also repel each other, they do not tend to build up layers or thick films. Instead, they remain a monomolecular protective film.

A film one molecule thick actually improves heat transfer in condensers, dryers, and other heat exchange equipment. By promoting dropwise condensation, an insulating water film between the water and metal surface is prevented.

Good distribution of filming inhibitors is of prime importance in preventing condensate corrosion. Protection depends on the maintenance of a continuous film. Because steam and condensate can wash away the film, it must be constantly repaired by continuous feed of the inhibitor.

Octadecylamine and certain of its salts were the first chemicals to be used as filming inhibitors in steam-condensate systems. However, because of their wax-like nature (whether supplied in flake or emulsion form), it was difficult to put these chemicals into uniform solutions for feeding. A relatively narrow condensate pH range of 6.5 to 8.0 is required for the octadecylamine to form a film and to remain on the metal surface. To overcome these limitations molecules formulated specifically for boiler plant conditions have been developed as alternatives to octadecylamine.

Most filming inhibitors are normally fed to the steam header, but may be fed to the feed water or boiler drum; the latter feed points result in some loss of product to blowdown. If only the process equipment needs protection, the inhibitor may be conveniently fed into the desuperheating water at the process steam header. Regardless of the feed point, however, the inhibitor should be fed continuously for best results. Dosages are not based on oxygen or carbon dioxide content of the steam. The amount of inhibitor required is set according to the system surface area. Creation of an effective film is a physical process, highly dependent on flow rates, feeding, and testing techniques.

6.6.6.10 Evaluating Results

There are several good ways of determining how much corrosion is occurring in a system and how effective a prevention program is. Because this requires extensive monitoring, it is essential that sampling be done at the significant points of the system and with adequate facilities. There must be a quill installed in the line that projects into the flowing stream; a sample taken along the pipe wall is meaningless. Sample lines must be of stainless-steel tubing.

6.6.6.10.1 pH Monitoring This involves checking the pH throughout a system, necessary to make sure that: sufficient amine is being fed to neutralize carbon dioxide, the proper amine is used to give total system protection, and process contamination or air inleakage is not occurring.

When taking condensate samples, to avoid carbon dioxide flashing off and giving false results, the sample must be cooled prior to contact with the atmosphere. This means using a sample cooler attached to the condensate line. The sample may be throttled at the outlet but not at the inlet. This is done to prevent a vacuum from occurring in the coil and drawing in air, which will give false results. Failure to set up a cooler properly will give inaccurate and misleading results.

6.6.6.10.2 Conductivity Monitoring The conductivity of the returning condensate can be an indicator of process contaminants and corrosion.

6.6.6.10.3 Carbon Dioxide Testing By actually measuring the carbon dioxide content of the condensate, the problem of corrosion from carbon dioxide can be directly monitored.

Steel Plant Fuels and Water Requirements

6.6.6.10.4 Hardness Similar to conductivity measurement, this can be an indicator of process contaminants, particularly from cooling water leakage.

6.6.6.10.5 Test Nipples The use of test nipples, installed in steam or condensate lines, permits both visual inspection of system conditions and a measure of corrosion.

6.6.6.10.6 Test Coupons These have also been used to evaluate condensate corrosion conditions. Preweighed coupons in holders are inserted into condensate lines. After an arbitrary time interval (usually at least 30 days), the coupons are removed, cleaned, and reweighed. The difference in original and final weight, when coupon surface area and exposure time are known, gives an indication of rate of metal loss owing to corrosion. It should be noted, however, that such test coupons usually cannot identify bottom grooving and threaded-joint corrosion as they occur in actual piping.

6.6.6.10.7 Iron and Copper Testing for corrosion products (iron and copper) in condensate is a preferred and widely used means of observing corrosion trends. Because metal corrosion products in condensate are mainly present as insoluble particulates rather than in dissolved form, methods of sampling that insure representative and proportional collection of particulates are important.

A general level of corrosion products in the 0–50 μg/litre range indicates the system is in control. Because of the problem of sampling error when looking for this small quantity of particulates, a relatively large amount of sample needs to be run. Two accepted methods used are visual estimation and quantitative measurement.

Visual estimation of iron concentration is based on the relative degree of discoloration of a membrane filter through which a known volume of condensate is passed. Estimation is made by comparing the membrane pad, after filtration of the sample, with prepared standard pads having color equivalents for specific iron values.

Very precise quantitative determinations of total iron and copper values are easily made by the Analex method. In this procedure, a sample stream of condensate is allowed to flow through a small plastic cartridge containing a high-purity filter of ion exchange materials for a period of 7 to 30 days. Particulates are captured by filtration, while dissolved solids are retained by ion exchange. By a unique laboratory process, the total weight of each metal is calculated. A series of successive analyses gives a complete and accurate picture of condensate corrosion trends. Normally, when inhibitors are used in a system previously untreated, the corrosion products observed in the condensate will increase temporarily. Most inhibitors have a detergent effect and tend to slough off old oxides. This must be taken into consideration when evaluating test results.

Normally a plant uses more than one of the above monitoring methods (e.g. pH monitoring and corrosion coupons). Although it may seem inconvenient when initially setting up a program to monitor the condensate, the results are more than worth the time invested in avoiding downtime and gaining energy savings in the high-heat value and high-purity of the returned condensate.

6.6.6.11 Steam Sampling

The most difficult sampling is of the steam itself as a representative sample is hard to obtain without making special provisions for it. The ASTM recommended procedure for sampling steam and condensing this for analysis should be followed. The analysis may only require a determination of pH to establish the level produced by the application of a neutralizing amine as a benchmark for interpreting the pH of samples of condensate. On the other hand, steam sampling is extremely valuable as a means of locating sources of trouble in a system.

Hydrogen analysis of steam will help determine boiler corrosion rate with respect to steaming load. Hydrogen evolution results from active corrosion cells or chemical decomposition. Experienced consultants should conduct these tests so that correct interpretations can be made. The corrosion action within the boiler can be easily observed on the hydrogen analyzer chart; destruction of the protective film of magnetite produces hydrogen directly from the attack of the boiler steel.

A common method for determining steam quality is by use of the specific ion electrode for sodium. As the salts dissolved in the boiler water are sodium salts, the presence of sodium in the steam sample is a direct indication of carryover. A trained steam purity consultant should be contracted.

References

General

1. *Steel at the Crossroads*, (Washington D.C.: American Iron and Steel Institute).
2. T. Baumeister, ed., *Marks Standard Handbook for Mechanical Engineers*, 8th ed. (New York: McGraw-Hill, 1978).
3. J. S. S. Brame and J. G. King, *Fuel—Solid, Liquid and Gaseous* (London: Edward Arnold, 1967).
4. A. J. Chapman, *Heat Transfer*, 4th ed. (New York: MacMillan, 1984).
5. D. M. Considine, ed., *Energy Technology Handbook* (New York: McGraw-Hill, 1977).
6. R. M. Fristrom and A. A. Westenberg, *Flame Structure*, (New York: McGraw-Hill, 1965).
7. B. Gebhart, *Heat Transfer* (New York: McGraw-Hill, 1961).
8. W. H. Giedt, *Principles of Engineering Heat Transfer* (New York: Van Nostrand, 1957).
9. J. Griswold, *Fuels, Combustion and Furnaces* (New York: McGraw-Hill, 1946).
10. *Hauck Industrial Combustion Data Handbook*, 3rd ed. (Lebanon, PA: Hauck Manufacturing Company, 1953).
11. R. F. Hill, ed., *A Decade of Progress, Energy Technology X*, (Government Institutes, Inc., annual publication, June, 1983).
12. H. C. Hottel and A. F. Sarofim, *Radiative Transfer* (New York: McGraw-Hill, 1967).
13. O. A. Hougen et al., *Chemical Process Principles: Part I*, 2nd ed. (New York: John Wiley and Sons, 1954).
14. O. A. Hougen et al., *Material and Energy Balances: Part II*, 2nd ed. (New York: John Wiley and Sons, 1959).
15. O. A. Hougen et al., *Thermodynamics: Part III* (New York: John Wiley and Sons, 1947).
16. O. A. Hougen et al., *Kinetics: Charts*, 3rd ed. (New York: John Wiley and Sons, 1964).
17. "Waste-Heat Recovery," in *Proceedings of 1961 Institute of Fuel Conference* (London: Chapman and Hall, Ltd., 1963).
18. M. Jakob, *Elements of Heat Transfer*, 3rd ed. (New York: John Wiley and Sons, 1957).
19. A. J. Johnston and G. H. Auth, *Fuels and Combustion Handbook* (New York: McGraw-Hill, 1951).
20. B. Lewis and R. N. Pease, *Combustion Processes* (Princeton University Press, 1956).
21. B. Lewis and G. von Elbe, *Combustion, Flames and Explosions of Gases* (New York: Academic Press, 1951).
22. G. N. Lewis et al., *Thermodynamics*, 2nd ed. (New York: McGraw-Hill, 1961).
23. C. O. Mackey et al., *Engineering Thermodynamics* (New York: John Wiley and Sons, 1957).
24. W. H. McAdams, *Heat Transmission*, 3rd ed. (New York: McGraw-Hill, 1954).
25. *North American Combustion Handbook*, 2nd ed. (The North American Manufacturing Company Company, 1978).
26. J. R. O'Loughlin, "Generalized Equations for Furnace Combustion Calculations," *Combustion* 34:5 (1963): 23–4.
27. R. H. Perry et al., eds. *Chemical Engineers' Handbook*, 6th ed. (New York: McGraw-Hill, 1984).
28. A. Schack, *Industrial Heat Transfer*, 6th ed. (New York: John Wiley and Sons, 1965) (translated from German).
29. M. L. Smith and K. W. Stinson, *Fuels and Combustion* (New York: McGraw-Hill, 1952).
30. R. A. Strehlow, *Fundamentals of Combustion* (Scranton, PA: International Textbook Company, 1967).

31. "Temperature—Its Measurement and Control in Science and Industry," *American Institute of Physics* 5 (1982).
32. W. Trinks and M. H. Mawhinney, *Industrial Furnaces, Vol. I*, 5th ed. (New York: John Wiley and Sons, 1961)
33. W. Trinks and M. H. Mawhinney, *Industrial Furnaces, Vol. II*, 4th ed. (New York: John Wiley and Sons, 1967).
34. U.S. Department of Energy Research, Development and Demonstration for Energy Conservation: Preliminary Identification of Opportunities in Iron and Steelmaking, Arthur D. Little, 1978.

Coal and Coke—General
35. F. J. Donnelley and L. T. Barbour, "Delayed Coke—A Valuable Fuel," *Hydrocarbon Processing* 45:11 (1966): 221–4.
36. M. A. Elliot, ed., *Chemistry of Coal Utilization, 2nd Supplementary Volume* (New York: John Wiley and Sons, 1981).
37. R. A. Glenn and H. J. Rose, *The Metallurgical, Chemical and Other Process Uses of Coal*, (Pittsburgh: Bituminous Coal Research, Inc., 1958).
38. R. E. Machin, *Science in a Coalfield*, 2nd ed. (New York: Pitman, 1952).
39. J. W. Leonard, ed., *Coal Preparation*, 4th ed. (New York: American Institute of Mining, Metallurgical and Petroleum Engineers, 1979).
40. A. Raistrick and C. E. Marshall, *Nature and Origin of Coal and Coal Seams*, (New York: British Book Centre, 1952).
41. D. W. Van Krevelen, *Coal* (New York: Elsevier, 1961).
42. D. W. Van Krevelen and J. Schuyer, *Coal Science* (New York: Elsevier, 1957).
43. P. J. Wilson and J. H. Wells, *Coal, Coke and Coal Chemicals* (New York: McGraw-Hill, 1950).

Coal and Coke—Testing
44. *ASTM Standards, Volume 06.05, Gaseous Fuels, Coal and Coke* (Philadelphia: American Society for Testing and Materials, 1983).
45. O. W. Rees, *Chemistry, Uses and Limitations of Coal Analyses* (Illinois State Geological Survey, R. I. 220, 1966).
46. *Bulletin 638, Methods of Analyzing and Testing Coal and Coke* (Washington, D.C.: U.S. Bureau of Mines, 1967).

Coal Petrography
47. R. J. Gray et al., "Distribution and Forms of Sulfur in a High-Volatile Pittsburgh Seam Coal," *Transactions, American Institute of Mining, Metallurgical and Petroleum Engineers*, Vol. 226 (1963): 113–21.
48. J. A. Harrison, "Coal Petrography Applied to Coking Problems," *Proceedings, Illinois Mining Institute* (1961): 17–43.
49. J. A. Harrison, "Application of Coal Petrography to Coal Preparation," *Transactions of SME* 226 (1963): 346–57.
50. N. Kaye, "The Application of Coal Petrography in the Production of Metallurgical Coke," *The Coke Oven Manager's Yearbook* (London: The Coke Oven Managers' Association, 1967).
51. M. Th. Mackowsky, "Practical Possibilities of Coal Petrography," *Compterendu 31e Congress Intern. de Chemie Indust.*, (Liege, Sept. 1958), Special Libraries Association translation 59-17595.
52. C. E. Marshall, *Coal Petrology*, Economic Geology, 50th Anniversary Volume, 1955.
53. D. M. Mason and F. C. Schora, Jr., "Coal and Char Transformation in Hydrogasification," *Fuel Gasification, Advances in Chemistry Series No. 69*, American Chemical Society, 1967, 18–30.
54. B. C. Parks and H. J. O'Donnell, *Petrography of American Coals*, (Washington, D.C.: U.S. Bureau of Mines Bulletin 550, 1957).

55. N. Schapiro and R. J. Gray, "The Use of Coal Petrography in Coke Making," *Journal Institute of Fuel* 37:6 (1964): 234–42.
56. C. M. Thomas, "Coal Petrology and Its Application to Coal Preparation," *Coal Preparation*, March–April (1968): 50–9.

Gaseous and Liquid Fuels
57. ASTM Standards, Petroleum Products, etc., Section 5, Volumes 05.01 to 05.04 (Philadelphia: American Society for Testing and Materials, 1983).
58. S. G. Brush, ed., *Kinetic Theory, Vol. I* (Pergamon, 1965).
59. S. G. Brush, ed., *Kinetic Theory, Vol. II* (Pergamon, 1966).
60. S. G. Brush, ed., *Kinetic Theory, Vol. III* (Pergamon, 1972).
61. S. Chapman and T. G. Cowling, *Mathematical Theory of Non-Uniform Gases; An Account of the Kinetic Theory of Viscosity, Thermal Conduction, and Diffusion in Gases*, 2nd ed. (London: Cambridge, 1952).
62. M. G. Eilers et al., "Producer Gas," *Gas Engineers Handbook*, (New York: The Industrial Press, 1977).
63. C. B. Glover, "Blue Gas and Carburetted Water Gas," *Gas Engineers Handbook* (New York: The Industrial Press, 1977).
64. W. Gumz, *Gas Producers and Blast Furnaces*, (New York: John Wiley and Sons, 1950).
65. W. Jost, *Diffusion in Solids, Liquids, Gases* (Academic Press, 1952).
66. B. J. Moore, *Analyses of Natural Gases, 1917–1980* (Washington, D.C.: U.S. Bureau of Mines I. C. 8870, 1982).
67. J. J. Morgan, "Water Gas," in *Chemistry of Coal Utilization, Vol. II*, H. H. Lowry, ed. (New York: John Wiley and Sons, 1945).
68. C. G. Segeler, ed., *Gas Engineers Handbook*, (New York: The Industrial Press, 1977).
69. The Gas Research Institute (GRI), 8600 West Bryn Mawr Avenue, Chicago, Illinois 60631.
70. B. J. C. van der Hoeven, "Producers and Producer Gas," in *Chemistry of Coal Utilization, Vol. II*, H. H. Lowry, ed. (New York: John Wiley and Sons, 1945).
71. S. A. Weil et al., *Fundamentals of Combustion of Gaseous Fuels. Research Bulletin 15*, (Chicago: Institute of Gas Technology, 1957).

Power and Steam Generation
72. *Steam, Its Generation and Use*, 39th ed. (New York: Babcock and Wilcox Company, 1978).
73. R. J. Bender, "Low Excess Air and Sonic Atomization Team Up," *Power* 110:7 (1966): 71.
74. L. Cizmadia and F. J. Fendler, "Design and Operation of Fairless Package Boiler," *Iron and Steel Engineer* 45:4 (1968): 109–14.
75. G. R. Freyling, ed., *Combustion Engineering,* revised ed. (New York: Combustion Engineering, Inc., 1967).
76. R. S. Rochford, "Considerations in Converting Multiple Fuel Fired Industrial Boilers," *Iron and Steel Engineer* 42:12 (1965): 169–70, 173.

Water Requirements
77. R. Nilsson, "Removal of Metals by Chemical Treatment of Municipal Wastewater," *Water Res.* 5 (1971): 5.
78. F. N. Kemmer, ed., *The Nalco Water Handbook,* 2nd ed. (New York: McGraw-Hill, 1988).

Chapter 7

Pre-Treatment of Hot Metal

P.J. Koros, Senior Research Consultant, LTV Steel Co.

7.1 Introduction

Pre-treatment of hot metal is the adjustment of the composition and temperature of blast furnace produced hot metal for optimal operation of the oxygen converter process; as such, it is one of the interdependent chain of processes that constitute modern steelmaking[1], Fig. 7.1. When taken to the extreme case, the converter process function is reduced to scrap melting and carbon reduction subsequent to the prior removal of silicon, phosphorus and sulfur in preparatory steps under thermodynamically favorable conditions. An important benefit of removing phosphorus and sulfur from the hot metal prior to the oxygen converter process is the ability to produce steels with phosphorus and sulfur contents lower than otherwise achievable without severe penalty to the converter process. Silicon removal is beneficial to the converter to reduce the chemical attack on the basic refractory lining and to allow the use of only minimal amounts of slag-making fluxes, thereby maximizing process yield.

Hot metal pre-treatment by North American and European steel producers presently is focused on desulfurization due to the common use of relatively low phosphorus containing iron ores.

In a unique approach to pre-treatment, ISCOR, in South Africa, installed a hot metal mixer equipped with channel inductors to provide electrical energy to heat the liquid, and thereby raise the scrap melting capability of the steel plant.

Details of the process steps introduced above are provided in the following sections.

7.2 Desiliconization and Dephosphorization Technologies

The introduction of oxygen converter technology in Japan occurred at a time of limited availability of high quality scrap, and, as a result, the desire was to minimize the use of this expensive resource. Steel production was focused on the use of controlled, prepared raw materials. The technologies developed for the efficient removal of silicon and phosphorus from the hot metal, both fundamentally endothermic when carried out using the customary oxide reagents, provided an economic benefit by consuming chemical energy otherwise available for melting scrap in the converter. By 1983, a large number of pre-treatment facilities were in use, Table 7.1.[2,3]

Initially, these pre-treatment processes were performed by adding iron ores or sinter to the hot metal[4] during its flow in the blast furnace runner. Further improvements and control over chemical

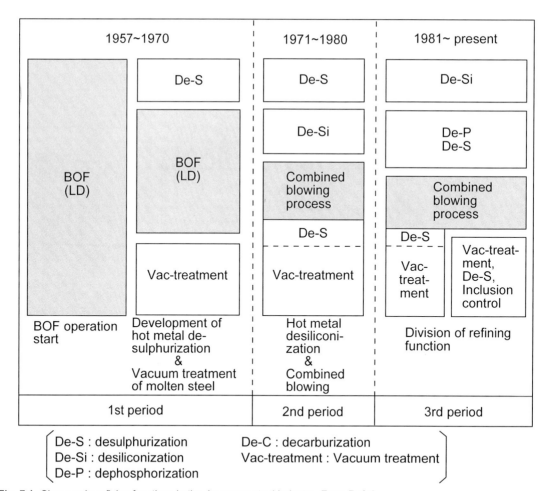

Fig. 7.1 Changes in refining functions in the Japanese steel industry. *From Ref. 2.*

results were attained[4-8] by the addition via subsurface injection of the reagents in dedicated vessels, such as oversized torpedo or submarine cars. This brought on the use of a variety of chemical reagents, including soda ash (sodium carbonate), which also provides for significant removal of sulfur. When using iron oxides for desiliconization, it is essential to separate, i.e., remove, the process slag before the hot metal is desulfurized as this operation requires low oxygen potential for efficient performance. It is important to recognize that phosphorous removal occurs only in hot metal containing less than 0.15% Si, additionally, phosphorus held in the slag could be subject to reduction, i.e., reversal, into the hot metal if it were present during desulfurization. An interesting technical development was the combination of dephosphorization and desulfurization in a single vessel whereby[5] phosphorous is reacted with the oxidizing reagents as they rise in the liquid and sulfur is removed by the top slag in the vessel, Fig. 7.2.

Desiliconization and dephosphorization are accompanied by losses of carbon from the hot metal and evolution of CO_2 from carbonate reagents.[4] Thus, control strategies such as addition of coke breeze[9] or equipment accommodations must be made in the reaction vessel and gas capture systems to contain foaming and flame evolution. In the recent timeframe, environmental considerations over disposal of sodium-containing slags has forced the use of limestone based reagents, often mixed with iron ore or sinter fines and delivered with oxygen, the latter used to diminish the thermal penalty from the pre-treatment process. Oxygen consumption in these process steps is illustrated in Fig. 7.3.

Table 7.1 Hot Metal Pre-treatment Facilities in Japan (1983). From Ref. 3.

		Desiliconization equipment		Dephosphorization equipment		
		Desiliconization in blast furnace runner	Desiliconization transport vessel	Dephosphorization in transport vessel (soda ash)	Dephosphorization in transport vessel (lime-based flux)	Dephosphorization in furnace for exclusive use (converter)
In operation	NSC	Kimitsu (No.2, 4 BF) Yawata (No. 4 BF)	Muroran Yawata Sakai Nagoya (No. 1 LD plant)	Yawata (No. 1 LD plant)	Kimitsu Nagoya (No. 1 LD plant) Muroran Yawata	
	NKK	Fukuyama (No. 4 BF)	Fukuyama	Fukuyama		
	KSC	Chiba (No. 6 BF)			Chiba (No. 1 LD plant)	Chiba (No. 2 LD plant) Mizushima
	SMI	Kashima (No. 3 BF) Wakayama (No. 4 BF)	Kashima	Kashima		
	KSL	Kobe (No. 3 BF)				Kobe
Planned or under construction	NSC	Oita (No. 2 BF)	Nagoya (No. 2 LD plant)		Oita Nagoya (No.2 LD plant)	
	NKK	Fukuyama (No.2 BF) Keihin (No. 1 BF)	Keihin		Keihin	
	SMI	Kokura (No. 2 BF)	Wakayama Kakogawa		Chiba (No. 3 LD plant) Wakayama Kokura	
	KSC					
	KSL				Kakogawa	
	NISSHIN	Kure (No. 2 BF)		Kure		

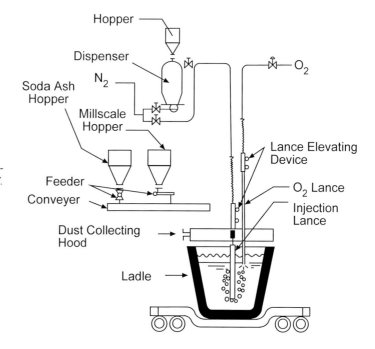

Fig. 7.2 Equipment for concurrent dephosphorization and desulfurization. *From Ref. 7.*

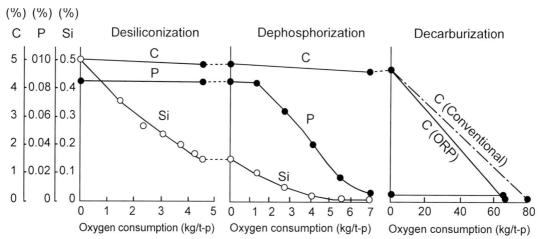

Fig. 7.3 Changes in silicon, phosphorus and carbon contents of iron in each stage of steelmaking. *From Ref. 4.*

In some plants, the silicon and phosphorus removal steps occur in full size oxygen converter vessels and the resulting carbon containing liquid is transferred, after separation of the low basicity primary process slag, into a second converter, Fig. 7.4, for carbon removal by oxygen top blowing.[4,7] In this sequence, the slag from the second vessel is used as a starter slag for the first step. In a way, this is todayís equivalent of the former open hearth process, which provided for flushing of the initial silica and phosphorus rich slag and thus allowed the use of hot metal made from phosphorus bearing ores for production of what was then considered low phosphorus steels.

7.3 Desulfurization Technology

7.3.1 Introduction

"…sulfur is frequently found in metallic ores, and, generally speaking, is more harmful to the metals, except gold, than other things. It is most harmful of all to iron…", so wrote Agricola four and one half centuries ago.[10] From ancient times, through puddling furnaces and into blast furnaces,

Pre-Treatment of Hot Metal

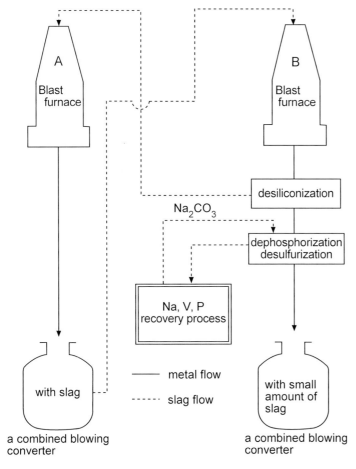

Fig. 7.4 Refining process based on soda ash treatment. *From Ref. 8.*

the control and elimination of sulfur has been a major task for the steelmaker. The cost of sulfur is enormous. In its simplest form, a modern coal to steel flow sheet involves separating more than 99% of the sulfur dug out of the ground at the coal pits. To control the 11 kg/ton of sulfur contributed by the coal and other feedstocks, the typical steel plant spends over $5/ton of steel in addition to capital charges for equipment and exclusive of processes for sulfide shape control in the steel product.

In the oxygen driven converter operation, the heavily oxidizing environment of the metal and slag and the inability to attain the equilibrium sulfur partition ratio between slag and metal limit the sulfur removal capability of the process. Thus, to bring the sulfur content of the steel to within the range manageable by the far more costly steel desulfurization, the lower cost hot metal treatment technologies have been developed to remove sulfur prior to the oxygen steelmaking step.

Initially these technologies were used to help the steelmaker, but, in time, it was recognized[11] that significant cost savings and production increases in ironmaking would result if sulfur limits formerly imposed on the blast furnace operation were lifted. In most North American steel plants, the hot metal leaves the blast furnace containing 0.040%–0.070%S, while the oxygen converters are charged with hot metal containing as little as 0.010%–0.001%S, to conform to limits on steel composition set by caster operations and final product quality requirements.

The importance of sulfur management and the huge costs involved have led to worldwide efforts to develop and implement an array of different desulfurization technologies. The different reagent and delivery systems in use are the result of local economic and environmental factors and the preferences of technical and operating management at the individual plant sites.

The following sections address the main chemical reactions for sulfur removal from hot metal, the range of process permutations, the specifics of reagent delivery systems, the importance of reaction vessel selection and of slag management issues as these topics bear on a well-functioning system.

7.3.2 Process Chemistry

The variety of process permutations adopted worldwide depend on one or a combination of the following reactions:

$$Na_2CO_3(s) + \underline{S} + \underline{C} \rightarrow Na_2S(l) + CO_2(g) + CO(g) \qquad (7.3.1)$$

$$Mg(s) + \underline{S} \rightarrow MgS(s) \qquad (7.3.2)$$

$$\text{CaC}_2 + \underline{\text{S}} \rightarrow \text{CaS(s)} + 2\underline{\text{C}} \qquad (7.3.3)$$

$$\text{CaO} + \underline{\text{S}} + \underline{\text{C}} \rightarrow \text{CaS(s)} + \text{CO(g)} \qquad (7.3.4)$$

$$\text{Mg} + \text{CaO} + \underline{\text{S}} \rightarrow \text{CaS(s)} + \text{MgO(s)} \qquad (7.3.5)$$

$$\text{CaO} + 2\underline{\text{Al}} + \underline{\text{S}} + 3\underline{\text{O}} \rightarrow (\text{CaO} \cdot \text{Al}_2\text{O}_3)\,(\text{S}) \qquad (7.3.6)$$

$$(\text{CaO} \cdot \text{Al}_2\text{O}_3)(\text{s}) + \underline{\text{S}} \rightarrow (\text{CaO} \cdot \text{Al}_2\text{O}_3)\,(\text{S}) \qquad (7.3.7)$$

Initially, most plants relied on reaction 7.3.1, that is, the addition of soda ash (Na_2CO_3) at the blast furnace cast house or at the steelworks while filling iron transfer ladles. This approach was abandoned as process control and environmental management were very difficult. In Europe and Japan, mechanical stirrers (KR) were introduced into the blast furnace runners[12], Fig. 7.5, and later for use in hot metal transfer ladles[13], Fig. 7.6. In the U.S. and Canada, the next step was dependent on reaction 7.3.2 with the use of Magcoke (a product made by filling the pores of coke with magnesium and submerging this material into the hot metal in a sequence of multiple dunks). Results were reproducible[11], but, attainment of sulfur contents of less than 0.020% was costly in reagents and process time. Capture of the copious magnesium fumes was nearly impossible without total building evacuation.

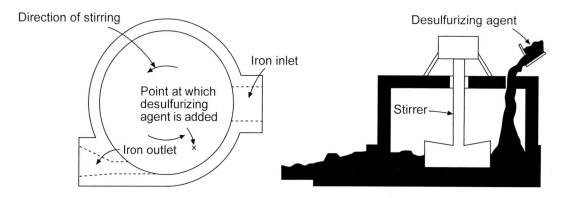

Fig. 7.5 Plan (left) and elevation (right) views of a unit for continuous desulfurization of iron. *From Ref. 12.*

The chemical behavior of magnesium in hot metal has been the subject of extensive study[14], Fig. 7.7. It is important to realize that the solubility product of Mg and S is strongly dependent on temperature (Fig. 7.8) and silicon and carbon content of the iron. This results in improved sulfur removal, or the reduced need for magnesium for colder iron. The practical effect is to lessen the cost penalty of having to load relatively cold hot metal with magnesium for attainment of sulfur levels lower than 0.002%S. An interesting technical side effect of the increase in solubility of magnesium in hot metal at low sulfur levels, e.g. near 10 ppm S, is the observation that some magnesium appears to be oxidized from the iron as soon as the raker blade clears the surface of slag. The effect is noticeable as a light white fume even after emptying the transfer ladle into the steelmaking vessel. Sampling has shown the plume material to consist mostly of MgO.

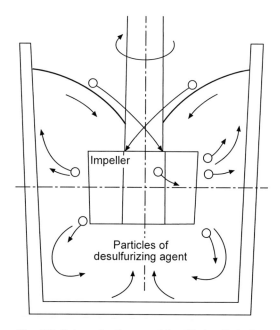

Fig. 7.6 Schematic diagram of the KR desulfurization method. *From Ref. 13.*

$[Mg][S] = -1.4 \times 10^{-4}$

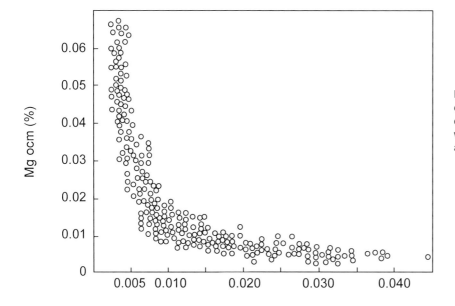

Fig. 7.7 Experimental data on magnesium and sulfur content in hot metal treated with granulated magnesium at 1400°C. *From Ref. 14.*

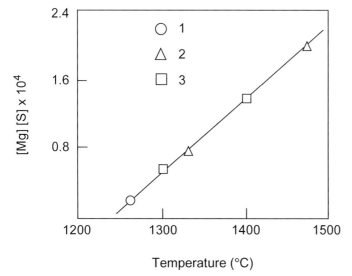

Fig. 7.8 Dependence of Mg·S product on temperature. *From Ref. 14.*

A critical step in process development was adoption of subsurface pneumatic injection of calcium carbide powder[15] (reaction 7.3.3) and of pulverized lime[16] (reaction 7.3.4) or combinations, i.e., mixtures of pulverized magnesium and lime (reaction 7.3.5). Because calcium carbide is inert, it is difficult to distribute it throughout the liquid; to improve on this, one of two reagents is added (~15–20%) to create surface and stirring: limestone, which cools the liquid or diamid lime, which is less endothermic. The latter version, known as CaD, was developed by SKW in Germany.

An important improvement came in the development of co-injection technology[17]: the controlled mixing in the transport line of reagents supplied separately, Fig. 7.9. This technique, now in universal use, allows for a wide array of reagent combinations and permits independent adjustment of the rates of the delivery of the reagents during the process, Fig. 7.10. This is most useful for magnesium based systems wherein splash and fuming during lance insertion and removal can be kept to a minimum by starting and stopping magnesium reagent flow with the lance tip at the deepest immersion in the hot metal. Another advantage is that the rate of delivery of the magnesium can

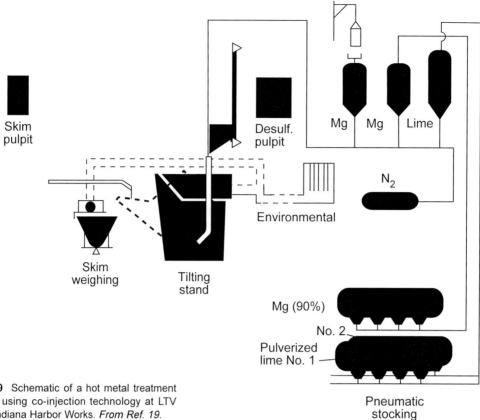

Fig. 7.9 Schematic of a hot metal treatment station using co-injection technology at LTV Steel Indiana Harbor Works. *From Ref. 19.*

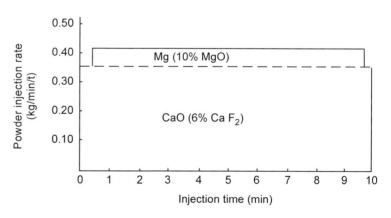

Fig. 7.10 Reagent injection patterns by co-injection. *From Ref. 19 and Ref. 21.*

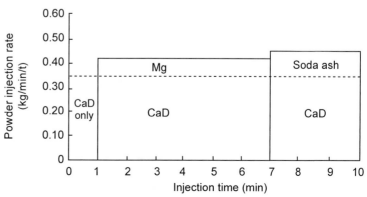

Pre-Treatment of Hot Metal

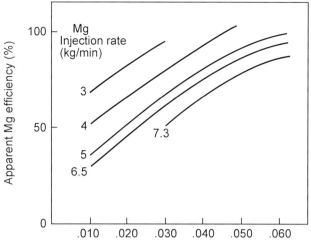

Fig. 7.11 Influence of the rate of injection of magnesium on desulfurization apparent efficiency. *From Ref. 18.*

be reduced at low sulfur levels when low magnesium solubility limits its dissolution in the iron[18], Fig. 7.11. Co-injection affords a cost benefit by allowing the user to purchase the individual components from the least costly material supplier rather than being solely reliant on a supplier for a proprietary mixture. A further benefit over blended reagent mixtures is elimination of segregation (i.e., separation) of individual reagent components14 with differing size or density while the mixture is in transit or in storage.

Although in Europe and Canada carbide based systems had been favored for a number of years, several European shops also have adopted co-injection for lime and magnesium. In another variant, carbide and magnesium are used in combination by co-injection.[16] For a fixed magnesium feed rate (splash limit) carbide + Mg is faster than lime + Mg. In the CIS countries[14], magnesium granules, coated with passivating layers of salts, have been in use with delivery by subsurface injection. Several North American plants[22] have also used this reagent in the past, but environmental concerns and limited supply have eliminated it from current use. In Japan, with environmental constraints on the disposal of slags containing residual amounts of calcium carbide, reagent systems based on vigorous stirring of lime and soda based reagents (KR process, Fig. 7.6) have been used successfully. The benefit of intense hot metal reagent mixing is demonstrated in one plant where the spent desulfurization slag from a shop equipped for conventional injection treatment is used as the main reagent in a shop using the KR process.[23] Recently, environmental concerns (in Japan) on disposal of soda slags has brought on adoption of lime plus magnesium systems.[24–26]

A variant to the use of reactive agents like calcium carbide and magnesium is the use of lime powder either preceded by addition of aluminum[27–29] to the hot metal (reaction 7.3.6), or lime delivered with organic stirring agents such as natural gas[16, 30] and/or solid hydrocarbons.[31] The latter, used as ground solid hydrocarbons, has been adopted to improve mixing even for magnesium based reagents. With the use of aluminum, $CaO \cdot Al_2O_3$ globules form which have a large solubility for sulfur.[32] Recently, in the U.S., desulfurization by injection of prefluxed $2CaO \cdot Al_2O_3$ has been introduced with some commercial success (reaction 7.3.7). Although these reagents have lower unit cost that carbide or magnesium, there is a limitation shared with lime systems; the greater mass of reagent needed increases the time required for treatment and for the follow on raking step.

Two other methods for delivery of desulfurization agents into hot metal transfer ladles are worthy of note. One approach, paralleling the use of cored wires for steel ladle treatment, is to feed magnesium cored wire[33] at high rates to reach the ladle bottom for release of the reagent at maximum depth. Magnesium in this form is far costlier than as an injectable solid, albeit the delivery system is simpler to operate and maintain. Another approach, with somewhat larger implementation[34,35], known by the commercial name ISID, consists of feeding the reagents through a rotatable bayonet system installed low in the wall of the hot metal transfer ladle, Fig. 7.12. Maintenance concerns and cost have limited broad implementation of this technology.

7.3.3 Transport Systems

Delivery and use of pulverized desulfurization reagents, e.g. magnesium, lime, calcium carbide, entail distinct technological requirements, principally the avoidance of contact with air. Magnesium powder, produced by atomization or grinding, must be transported in sealed, air-tight

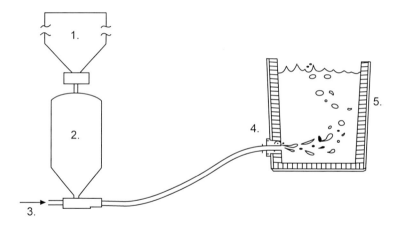

1. Storage silo. 2. Powder feed vessel. 3. Transport gas. 4. Side injection device. 5. Ladle.

(a) Equipment schematic.

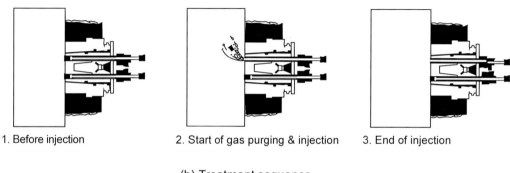

1. Before injection   2. Start of gas purging & injection   3. End of injection

(b) Treatment sequence.

Fig. 7.12 The ISID powder injection process. *From Ref. 34.*

containers of limited capacity (20,000 kg each). Thus, to provide the capability to move this material in bulk, the industry developed a 90% Mg–10% lime product that is flowable and can be delivered and stored in bulk trailers. Calcium carbide also must be kept from the moisture in air and is transported and stored in bulk, sealed, pneumatic system equipped trailers. Salt-coated magnesium is relatively impervious to moisture and generally is stored in bulk trailers as well.

An important adjunct to facilitate the use of pulverized materials such as lime and calcium carbide has been the development of a technology for improvement of the flowability of these materials by application of silicone oil based flow aids during pulverization.[36] Powders prepared in this manner can be delivered by dense phase injection techniques, which minimize the amounts of reagent and iron droplets carried out of the liquid by the transport gas (e.g., for lime, transport line loading of 2 kg/l of gas). This allows delivery of injection reagents at rates of 50 kg/min through an 18mm transport line.

7.3.4 Process Venue

When the pneumatic delivery of reagents was first introduced in the 1970s, these processes were relatively time consuming, i.e. 20 to 30 minutes, and the task was relegated to be carried out in torpedo cars. This resulted in interference from slag reactions as the ever present high sulfur blast

Pre-Treatment of Hot Metal

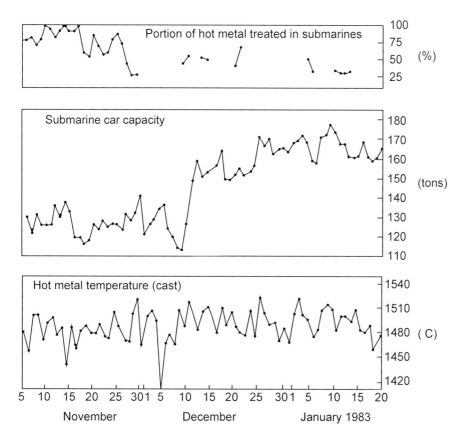

Fig. 7.13 Illustration of gain in submarine (torpedo) car capacity upon cessation of desulfurization treatment (lime-Mg). From Ref. 37.

furnace slag can cause sulfur reversals in the case of carbide based reagents. Furthermore, the post-reaction slags are viscous and stick to the roof and sides in the submarine fleet, reducing holding capacity.[37] Over-treatment was the rule as the submarine ladles were treated some time prior to their arrival at the melt shop and the coordination of their use for an intended product order.

Most shops changed to transfer ladle treatment to resolve these issues. The hot metal carrying capacity of the submarine fleet increases due to elimination of slag build ups on the sub roofs[37] and refractory wear is reduced, Fig. 7.13. Other problems associated with submarine ladle treatment include relatively poor mixing caused by the shape of the elongated bottle configuration and the relatively shallow liquid depth. In the BOF transfer ladle, reagent efficiency of magnesium is increased by the greater depth of lance immersion, which provides longer residence time for the magnesium bubbles to travel to the surface. The major benefit of treatment in the transfer ladle rather than in the submarine ladle is that it provides the opportunity to treat individual hot metal charges to specific sulfur levels set by the requirements of the intended steel grade.

7.3.5 Slag Management

As in all metallurgical processes, management of the slag produced during hot metal desulfurization is critical to success. After conclusion of treatment, the slag usually is removed with a raking device, which typically is an articulated arm and paddle assembly. The raking process requires some time which may become a production penalty in some operations. Process yield suffers as some hot metal is lost from the ladle with each stroke of the paddle. In some shops, the iron transfer ladle has a retention dam across the mouth with hole(s) for the metal to pour out. The slag is retained in the ladle as the hot metal is charged into the furnace. While effective at separating the slag and minimizing yield loss, this may slow the rate of charging the vessel and, therefore, extend the converter heat cycle. Additionally, the slag retained in the ladle must be dumped after each use; as this is done by reversing direction (to keep the pour holes open), this step may create difficulties in some shops.

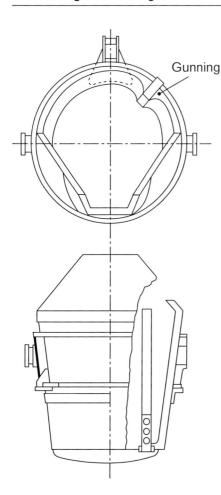
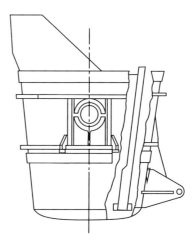

Fig. 7.14 Illustration of a transfer ladle slag stirring lance. From Ref. 39.

During raking the post-treatment slag will take with it a significant metallic content (approximately 40% by weight)[38], this can represent a yield loss of nearly 1%. Methods to minimize this loss include the use of dense phase injection (to minimize the volume of gas for delivery of the injected powders), and the addition of a fluxing agent, ~5% CaF_2 or Na_2CO_3 to the desulfurizing reagents, to produce a less viscous liquid slag for release of the iron globules. A further help is to provide a small amount of gas bubbling[39] (Fig. 7.14) during the raking process; this promotes flotation of the slag towards the lip of the ladle and thereby reduces the number of strokes required for slag removal.

Typical figures for slag removal are in the range of 15—25 kg/ton hot metal for most U.S. shops. Two viscosity related factors combine to make the quantity of slag raked from transfer ladles dependent on hot metal temperature[16], Fig. 7.15: colder hot metal appears to "hold" more entrained blast furnace slag and the retention of iron droplets in slags increases as temperature drops. The trend shown in Fig. 7.15 is typical for most BOF shops. Disposal of the spent slag usually is by mixing it into the blast furnace slag management system despite the remaining unused sulfur holding capacity.[23] In some plants special controls are in place to cope with the effect of remaining unreacted reagents such as carbide or soda ash. This is not an issue with magnesium or lime.

7.3.6 Lance Systems

Fig. 7.16 illustrates the commonly used injection lance designs. Nozzles may be directed vertically or exit the side of the lance at various angles. Typical life experience is 80 treatments or up to 1200 minutes for the hockey stick design, which prevents the reactive gases from attacking the refractory coating. A lower figure, 70 treatments, is typical for the simpler lazy L lances.

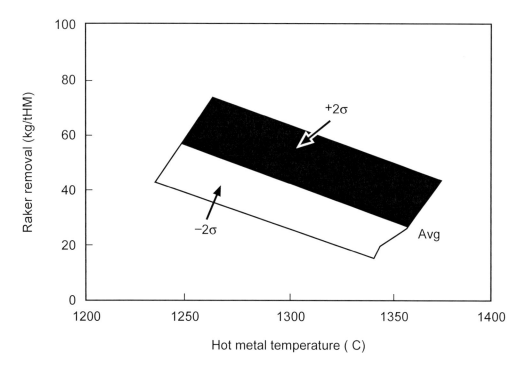

Fig. 7.15 Effect of hot metal temperature on raking loss after desulfurization with lime-Mg. From Ref. 19.

Lances typically are constructed of lengths of square steel tubing which contain the transport pipe(s). These assemblies are then cast within a refractory mold and cured in temperature controlled drying ovens. The cross sectional shape may be square or round with an area of 40–50 cm^2 (6–8 in^2). The refractory typically is high alumina.

A recent development directed at increasing the injection rate of magnesium containing combinations, without causing otherwise intolerable violence at the surface of the metal in the transfer ladle, is to use two transport lines contained within a single lance each fed by a separate injection materials source. The outlet nozzles are positioned to deliver the reagent at 30° from the vertical on opposite sides of the lance. It is also possible to use two separate lances immersed into the metal

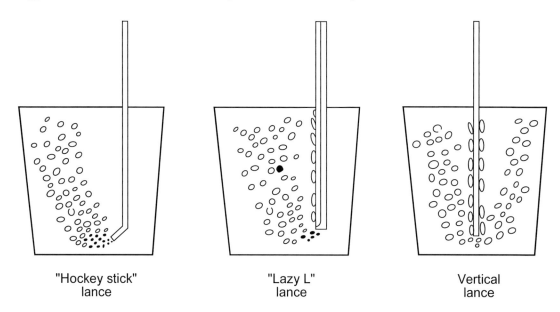

Fig. 7.16 Commonly used injection lance configurations.

at the same time. These approaches have reduced injection time by 40~50% without increasing iron losses to the ladle slag or reduction in reagent efficiencies.

7.3.7 Cycle Time

In a typical BOF shop, the hot metal treatment operation can become a limit on productivity. Pour-out and movement of the submarine ladles may require five to ten minutes. Injection time will consume seven to twenty minutes, depending upon the amount of sulfur to be removed, followed by raking time of five to ten minutes.

7.3.8 Hot Metal Sampling and Analysis

Hot metal sampling and analysis is a critical issue, particularly for magnesium reliant systems, because time must be provided for removal by flotation of the MgS reaction product. Best reproducibility in results is obtained when the post-treatment sample is obtained after the raking step is completed[19], Fig. 7.17. Similarly, gas stirring systems installed to aid the raking operation[30], Fig. 7.14, can result in improved sampling accuracy as both the pre-and post-treatment samples are made more representative.

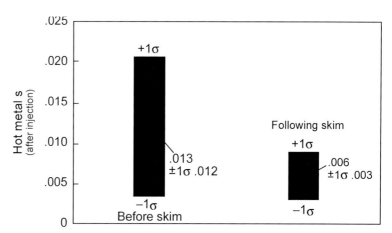

Fig. 7.17 Decrease in variability of after-treatment analyses (lime-Mg process). From Ref. 19.

The method for determination of endpoint sulfur is, in part, a function of the precision required at the individual melt shop and other operating considerations. Gas combustion methods (Leco) and optical emission spectrometers (OES.) are used commonly. Many shops have these instruments at the treatment sites for rapid determination of results.

Optical emission spectrometers have demonstrated sufficient precision for determination of ppm concentrations of sulfur in hot metal provided the disk samples obtained for analysis are sound and prepared properly. Pits, holes and cracks will affect the results, as will surfaces with improper texture and/or flatness. OES determination has the advantage of providing silicon and manganese analysis for use in the furnace charge model—a significant benefit when erratic blast furnace operations result in varied hot metal chemistry.

The gas combustion method is recognized as probably the most accurate for determination of sulfur (and carbon) due to ease in achieving appropriate sample integrity. Prevention of process slag entrapment in the pin sample is the greatest concern. Properly designed immersion samplers inserted quickly to sufficient depth will provide slag-free samples. The latest Leco C–S determinator is advertised to have a precision of 0.15 ppm.

7.3.9 Reagent Consumption

Reagent consumptions for lime plus magnesium and carbide plus magnesium process systems are illustrated in Fig. 7.18. Control of powder flowability and rates is critical for achievement of these consumption levels. It is essential to avoid excess transport gas and/or visible magnesium flares or powder plumes.

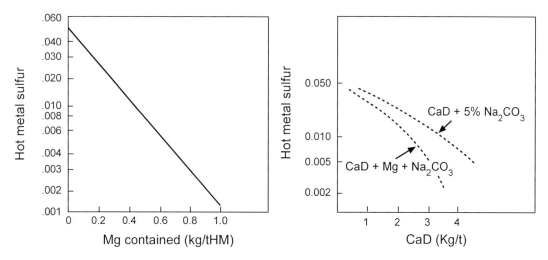

Fig. 7.18 Typical reagent consumptions for desulfurization by co-injection. Left: lime-Mg, (lime containing 6% spar). *From Ref. 19*. Right: carbide-Mg. *From Ref. 21*. Refer to Fig. 7.10 for blow patterns

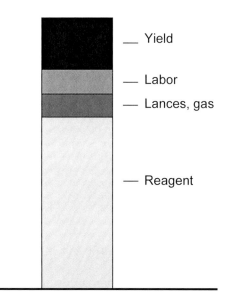

Fig. 7.19 Relative cost components for desulfurization of hot metal. Yield refers to loss of hot metal during slag raking.

7.3.10 Economics

Fig. 7.19 presents a breakdown of the cost components for a typical transfer ladle hot metal desulfurization operation.

7.3.11 Process Control

Modern facilities are generally fully automated to control the operation from the start of treatment to finish. Powder initiation and lance insertion are automatically controlled along with lance withdrawal after the predetermined amount of reagent has been injected. Reagent rates may be adjusted during treatment by controlling injection tank pressure or transport line throat diameter. In addition to the automatic functioning, the controlling computer provides for process data capture and storage. Real-time graphics provide information on process efficiency and dispenser performance to aim standards. Typical, for a well-controlled operation, is to achieve within 0.002%S of aim at end points ordered <0.003%S.

7.4 Hot Metal Thermal Adjustment

A one-of-a-kind facility to heat hot metal has been installed by ISCOR in South Africa.[40,41] The 1500 ton hot metal mixer, Fig. 7.20, is equipped with channel inductors which provide induction heating and stirring of the liquid. Options for use are (a) to melt scrap in the unit by energy addition to the hot metal or (b) to raise the hot metal temperature and thereby improve the scrap melting capability of the converter. Superheat of 225°C (440°F) for a 40% scrap rise increase is achievable. Electrical energy conversion for temperature gain has been reported[41] as 85%. Whereas technically successful in performing its functions, the capital cost of such a system, in contrast to the scrap melting benefit obtainable with chemical energy sources, has made this approach unattractive for other steel plants.

7.5 Acknowledgments

The guidance and help of many friends and colleagues are much appreciated, in particular Professor Toshishiko Emi, Dr. Paul Nilles, and Robert G. Petrushka and C. J. Bingel of LTV Steel.

7.6 Other Reading

Excellent general technical information on pre-treatment of hot metal is provided in Ref. 1 and Ref. 42.

Fig. 7.20 1500 tonne superheater channel furnace, rated at 15,000 kW. *From Ref. 41.*

References

1. T. Emi, Proceedings, "The Howard Warner International Symposium on Injection in Pyrometallurgy,"(1996) TMS, 19.
2. T. Shima, *Proceedings of The Sixth International Iron and Steel Congress*, (Nagoya: 1990), *ISIJ,* Vol. 3:1.
3. S. Ishihara, Nippon Steel, address given at 1984 IISI Technology Committee Meeting.
4. K. Sasaki et. al., *Steelmaking Proceedings*, Iron and Steel Society 66 (1983): 285.
5. M. Wada et. al., *Proceedings,* Scaninject IV Conference (1986): 19:1.
6. S. Yamato et. al., McMaster University Symposium on Hot Metal Preparation (1983): 218.
7. K. Yamada et. al., *Proceedings,* McMaster University Symposium on Hot Metal Preparation (1983): 245.
8. T. Ueda et. al., *Proceedings,* McMaster University Symposium on Hot Metal Preparation (1983): 289.
9. K. Tomita, *CAMP-ISIJ* 8 (1995): 936.
10. G. Agricola, *De Re Metallica*, 1556 (from 1912 translation, *The Mining Magazine*, London): 273.
11. R. F. Potocic and K. G. Lewis, *Proceedings,* McMaster Symposium on External Desulfurization (1975): paper no. 2.
12. K. D. Haverkamp, *Iron and Steel Engineer*, (1972): 49; and *Proceedings,* McMaster University Symposium (1975), paper no. 14.
13. H. Kajioka, *Proceedings,* McMaster University Symposium (1975), paper no. 15.
14. N. A. Voronova, "Desulfurization of Hot Metal by Magnesium," International Magnesium Association and Iron and Steel Society (1983).
15. W. Mieschner et. al., in *Journal of Metals*, The Metallurgical Society, 26 (1974): 55; and H. P. Haastert et. al., McMaster Symposium on External Desulfurization (1975): paper no. 6.
16. U. Nölle et. al., *Stahl u. Eisen* 92 (1972): 1085; and V. Puckoff and H. Kister, *Proceedings,* McMaster Symposium on External Desulfurization (1975): paper no. 8.
17. P. J. Koros, U.S. Patent 3,998,625 (1976)
18. P. J. Koros et. al. *Iron and Steelmaker* 4 (1977): 34.
19. J. H. Kaminski et. al., *Steelmaking Proceedings,* Iron and Steel Society 68 (1985): 333.
20. T. H. Bieniosek, *Steelmaking Proceedings,* Iron and Steel Society 69 (1986): 349; and G. E. DeRusha et. al., *Steelmaking Proceedings,* Iron and Steel Society 73 (1990): 351.
21. M. Bramming et. al., *Proceedings,* Scaninject VI Conference, Part II, MEFOS (1992): 91.

22. V. Kendrick et. al., *Proceedings,* McMaster University Symposium on Hot Metal Preparation (1983): 13.
23. N. Kurakawa et. al., Technical report, Sumitomo Metal Industries, 45 (1993): 52.
24. A. Aoygi, *CAMP-ISIJ* 7 (1994): 1080.
25. S. Sasakawa, *CAMP-ISIJ* 9 (1996): 223.
26. M. Iguchi, *CAMP-ISIJ* 8 (1995): 106.
27. T. Mitsuo et. al., *Trans. Japan Institute of Metals*, 23 (1982): 768.
28. H. T. Kossler and P.J. Koros, *Steelmaking Proceedings,* Iron & Steel Society 67 (1984): 341.
29. G. Carlsson et. al., *Scandinavian Journal of Metallurgy*, 16 (1987): 50.
30. S. T. Pliskanovskii et. al., *Stal* 6 (June 1967): 449 (in English).
31. P. J. Koros, US Patents No. 4,266,969 (1981) and No. 4,345,940 (1982); and Kossler et. al., *Proceedings,* McMaster Symposium on Developments in Hot Metal Preparation (1983): 310.
32. P. J. Koros, *Proceedings,* Scaninject III Conference (1983): paper no. 24; and *Iron and Steelmaker*, 23 (1983): 23.
33. A. F. Kablukouskij et. al., *Metallurg* (April 1995): 23.
34. J. Kolsi, *Scandinavian J. of Metallurgy* 19 (1990): 110.
35. S. K. Srivastava et. al., *Proceedings,* Scaninject VI Conference, MEFOS (1992): Part II, 73.
36. P. J. Koros et. al., *Proceedings,* Scaninject V Conference (1989): 183; and International Symposium on Injection in Process Metallurgy, The Metallurgical Society (1991): 81.
37. P. J. Koros, *Proceedings,* Scaninject III Conference (1983): paper no. 24; and *Iron and Steelmaker*, 23 (1983): 23.
38. R .G. Petrushka, *Steelmaking Proceedings,* Iron and Steel Society 72 (1989): 167.
39. R. G. Petrushka and D. H. Winters, Jr., *Steelmaking Proceedings,* Iron and Steel Society 70 (1987): 325.
40. P. R. Morrow, *Proceedings,* McMaster Symposium on Hot Metal Preparation (1983): 345.
41. J. A. Dalessandro et. al., *Iron and Steel Engineer*, 66 (1985): 29.
42. Gmelin, *Handbook of Inorganic and Organometallic Chemistry*, 8th ed. (New York: Springer-Verlag, 1992): 135a.

Chapter 8

Oxygen Steelmaking Furnace Mechanical Description and Maintenance Considerations

K. J. Barker, Manager of Technology—Steelmaking and Continuous Casting, USX Engineers and Consultants, Inc.
J. R. Paules, Manager, Technical Services, Berry Metal Co.
N. Rymarchyk, Jr., Vice President, Sales and Engineering, Berry Metal Co.
R. M. Jancosko, Exec. Vice President, Vulcan Engineering Co.

8.1 Introduction

The intent of this chapter is to provide a mechanical description of the basic oxygen furnace (BOF), as well as the maintenance of certain BOF components. The components covered in this report include basically all components of the BOF vessel and the trunnion ring up to the trunnion pins. Excluded areas are probes, couplings, bearings, foundations and the various drive units. The BOF components covered are: top ring and lip ring; cone, barrel, bottom shells and transition knuckle sections or flanges; brick retainer rings, slag shields and taphole assembly; working and safety refractory linings; vessel support system; trunnion ring, trunnion blocks and trunnion pins; cooling system for the vessel or trunnion ring; and oxygen lances.

This chapter is an abridged version of *AISE Technical Report No. 32, Design and Maintenance of Basic Oxygen Furnaces,*[1] with the addition of Section 8.5 which addresses oxygen lance design, and Section 8.6 which addresses sub-lance design.

8.2 Furnace Description

8.2.1 Introduction

This chapter is established to provide a description and preliminary design considerations for the manufacture and supply of BOFs. Basic oxygen furnaces are so called by virtue of the refractory and the additives used in their steelmaking processes. The processes referred to in this chapter are those which involve the treatment of a mixture of steel scrap and molten iron, generally transported from the blast furnace to the BOF. The steel scrap and hot iron are charged into the BOF vessel and oxygen is injected in one of many different methods into the furnace for purposes of producing a steel melt of specific chemical and physical properties.

The processes involved share one common operating factor; the injection of oxygen into the furnace is the agent for decarburizing the molten hot iron and generating the reaction heat required to melt the scrap.

For purposes of this chapter, the terms used will be consistent with the following descriptive definitions. A BOF installation consists of the basic oxygen furnace, furnace support foundation, furnace tilt drive and controls, furnace water cooling system, fume exhaust and cleaning system, oxygen injection system, auxiliary furnace bottom stirring system, process additives system, scrap and hot metal charging system, molten steel delivery and slag disposal system, furnace deskulling system, and other auxiliary steelmaking requirements such as sampling, refractory inspection and relining systems, process computers, etc.

An operating BOF, Fig. 8.1, consists of the vessel and its refractory lining, vessel protective slag shields, the trunnion ring, a vessel suspension system supporting the vessel within the trunnion ring, trunnion pins and support bearings, and the oxygen lance.

The BOF vessel consists of the vessel shell, made of a bottom, a cylindrical center shell (barrel), and a top cone; reinforcing components to the cone, such as a lip ring and top ring; auxiliary center shell and top cone flanges for bolted-on top cones; auxiliary removable bottoms for bottom reline access, or for individual bottom reline of bottom-blown vessels; and a taphole. This list is not intended to be either restrictive or comprehensive, e.g., top cone flanges are not universal.

BOF vessels can be one of the general classifications presented in Fig. 8.2. These are top-blown vessels, in which the oxygen is injected above the hot metal bath by means of a retractable lance; top-blown vessels, in combination with bottom stirring, the latter usually by introducing metered amounts of inert gas at specific locations under the hot metal bath—the introduction of the inert gas is either through porous plugs or tuyeres; bottom-blown vessels, in which the oxygen is injected under the molten metal bath through tuyeres arranged in the bottom of the vessel, and usually carrying pulverized additives; bottom-blown vessels utilizing a calculated source of heat energy provided by hydrocarbon fuel, in a very similar arrangement as the bottom blown vessel; and combination-blown vessels, in which the oxygen is introduced under the bath through tuyeres

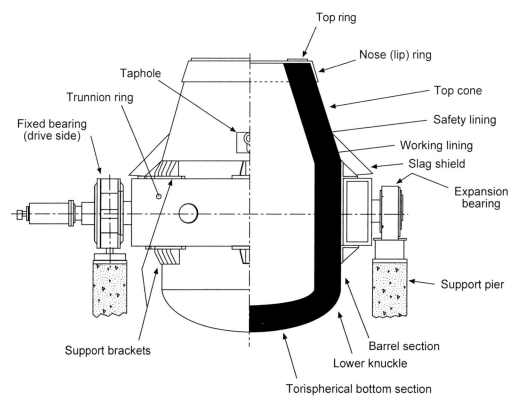

Fig. 8.1 BOF configuration.

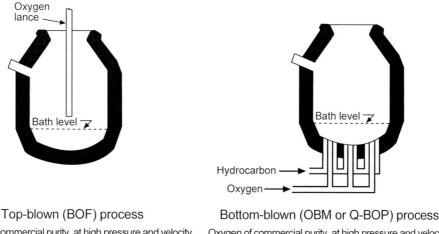

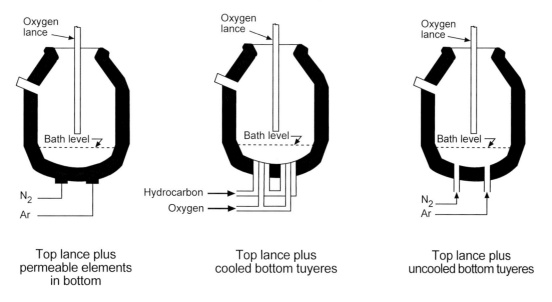

Combination-blown processes

Oxygen is blown downward into the bath, and oxygen and/or other gases are blown upward through permeable elements or tuyeres.

Fig. 8.2 General BOF vessel classifications.

in the bottom of the vessel, as well as above the bath through a lance—the oxygen blown through the bottom usually carries pulverized additives. Fig. 8.3 presents commercially available bottom stirring processes.

8.2.2 Vessel Shape

The shape of the vessel has an influence on the efficiency of the steelmaking process inside. This is particularly the case when the oxygen to the steel bath is supplied only from the top (top blowing). Fig. 8.4 shows a variety of shapes and sizes which were used in North America during the

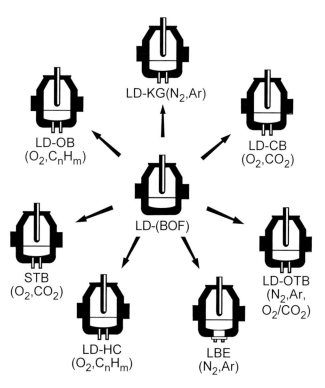

Fig. 8.3 Commercial variations of bottom stirring practices.

period of 1954–1973. Although many factors would influence the shape of the vessel, an approximate rule of thumb which has yielded favorable designs relates to the specific volume of the vessel. In conjunction with the rate of oxygen blowing, Fig. 8.5, and with hot metal composition as controlling factors, operating experience has shown that in general when the specific volume is in excess of 26 cubic feet per net ton of processed steel, Fig 8.6, the yield loss due to slopping is highly reduced.

Except for the case of a greenfield site installation, the optimum vessel shape for the particular application is usually determined by factors other than the oxygen blowing efficiency. The height clearance and the distance between trunnion pin bearing centers are two common factors which limit vessel volume increase in a vessel replacement project.

8.2.3 Top Cone-to-Barrel Attachment

There are generally two methods for the attachment of the top cone of a vessel to its cylindrical section, namely welding or bolting. Welded top cones can be either corner welded or welded with a rounded knuckle. In each case, the inside surface of the shell must be free from offsets because of stress concentrations. Corner welded transitions are more susceptible to cracking than rounded knuckle transitions.

Bolted top cones, on the other hand, necessitate outfitting both the cone and the top of the cylindrical center section with adequately heavy flanges and elevated temperature-resistant bolts and nuts.

The taphole on BOFs built in the U.S. has been placed mostly in the cone section. The centerline of the taphole usually falls on the intersection of the refractory linings of the top cone and the cylinder. In vessels with a top cone knuckle, most of the tapholes fall partly within the knuckle area. In other countries, there are numerous vessels where the tap nozzle is actually located in the top cylinder of the center shell. This is done to utilize the properties of the molten steel flow which are associated with the angle of the tapped molten steel jet, the ferrostatic head, and the control of the vessel movement during the tapping sequence. This results in improved ladle yield through reduced entrainment of slag in the steel tapped into the ladle.

8.2.4 Methods of Top Cone Cooling

Two components in the top cone of a basic oxygen furnace vessel can benefit from water cooling as a means to maintain their low operating temperature. These are the conical shell itself and the lip ring at the top corner of the cone. At full combustion, these components are exposed to radiant heat reflected back from the BOF exhaust fume hood while the furnace is in the oxygen blow mode. Additionally, they are subjected to convective heat from the vessel interior and radiant heat during tapping and slagging off into the steel ladle and the slag pot, respectively. Experience in a majority of steel mills in the U.S. indicates that the water cooling of these two components has a significantly advantageous effect in prolonging their life.

There are different designs for achieving the water cooling of the conical shell, all of which utilize closure channels for forced circulation of given amounts of cooling water. The design differences are in the pattern of flow based on the arrangement of the channels, and their cross-section. The wetted area, as a percentage of the total conical shell area, plays a significant role in the amount of heat removed. It is possible to create a cooler than desirable cone refractory, which could contribute to the generation of large skull buildups and kidneys inside the cone.

One design type consists of horizontally oriented channels, Fig. 8.7, mostly half-pipes, feeding from and returning into adjacent vertical inlet and outlet headers, respectively. Another design type consists of vertically oriented channels, Fig. 8.8, fed by a series of circumferentially placed inlet and outlet headers in a pattern of upward and downward flows covering the entire surface of the top cone shell. The channels in this type would be either half pipes or equal-legged angles. Water circulation is maintained by forcing the cooling water to the lip ring, either through series or a parallel connections. Most often series connections are used. More flexibility in shutting off individual systems is provided where both the cone and the trunnion ring have their own individual cooling water circuit, controlled from outside the vessel.

In rare occasions, a combination between the two systems has been used, whereby the orientation of the cooling channels was helical. In each case, the

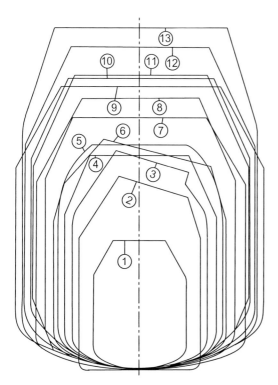

	Company	Estimated Capacity, ton	Operation Date
1	Allegheny	18	1964
2	McLouth	60	1954
3	Acme	75	1957
4	Interlake	75	1972
5	Kaiser	110	1956
6	Dofasco	105	1966
7	Armco	150	1963
8	Wisconsin	120	1964
9	Great Lakes	300	1962
10	Inland	210	1973
11	United States	220	1970
12	Algoma	250	1973
13	Bethlehem	300	1973

Fig. 8.4 Evolution of the size and shape of the BOF vessel.

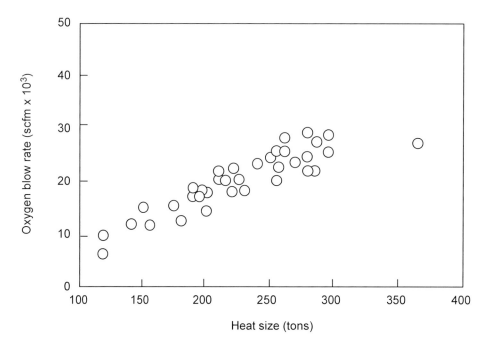

Fig. 8.5 BOF oxygen blowing rate in a sampling of North American shops.

flow has to be carefully calculated to result in a sufficient water velocity, such that the temperature of the water in the individual channels does not exceed the boiling point. Drains must be provided to enable emptying the system of the contained water, and to relieve the water pressure in case of emergency. Additionally, provisions should be made to allow for venting of the system.

A development which has been recently marketed in the industry is the air-mist cooling process which incorporates the use of a system of sprays, Fig. 8.9. Atomized water droplets are applied to

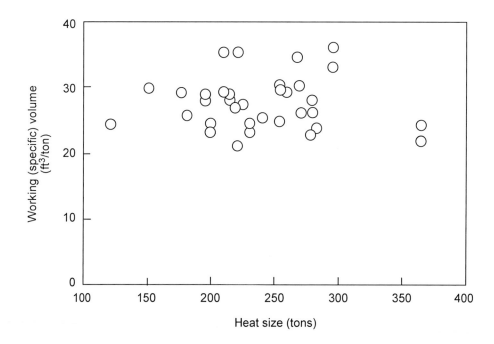

Fig. 8.6 BOF working volume in a sampling of North American shops.

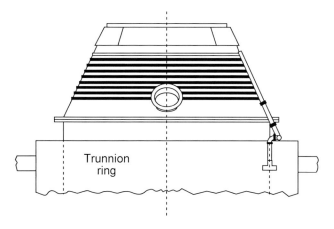

Fig. 8.7 Method of water cooling the nose cone using horizontal channels.

Channel configurations

the surface of the top cone to provide necessary cooling to reduce its temperature. The rate of flow in the sprays is controlled by thermocouples which are applied to the top cone shell.

Lastly, with the application of high conductivity magnesia–carbon bricks, the necessity of cooling the barrel section also becomes evident. To meet this need some plants operate air cooling systems. In these systems compressed air is routed through the trunnion pin to panels located between the

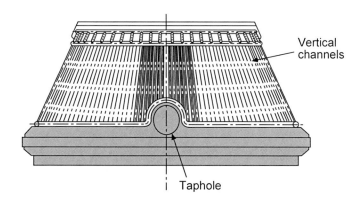

Fig. 8.8 Method of water cooling the nose cone using vertical channels.

Channel configurations

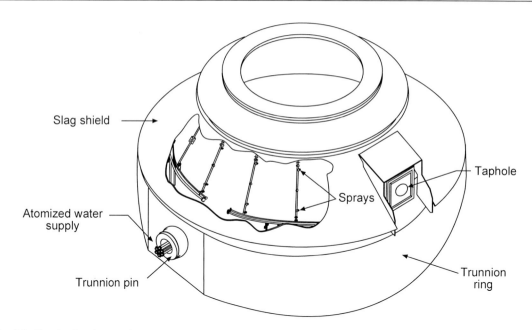

Fig. 8.9 Atomized water supply.

vessel shell and the trunnion ring. Air nozzles on the panels distribute the air uniformly over the barrel surface.

8.2.5 Vessel Bottom

The vessel bottom design is influenced by the process and the weight balance required to optimize the tilt drive system. The common shape is torispherical. For processes requiring the introduction of gases from the bottom of the vessel (through tuyeres), the shape of the bottom tends to be flatter than those which have only top blowing. Also, because some bottom stirring/blowing processes pose more of a burden on the bottom refractory, the bottom is designed to be interchangeable to enhance relining. For example, in the OBM (Q-BOP) process, the refractory lining in the bottom of the vessel wears more than twice as fast as that in the rest of the vessel. Therefore, the bottom is replaced at mid-campaign. This also allows for the maintenance of the tuyeres.

Interchangeable bottoms also allow for relining of the whole vessel from underneath through the bottom hole.

8.2.6 Types of Trunnion Ring Designs

For the basic intended function, the design shape of a trunnion ring normally is circular, Fig. 8.10. In such a design, the vessel and trunnion ring are assembled once for the duration of the useful life of either or both. In certain melt shops, the logistics of operation and maintenance have demanded the assembly and disassembly of the vessel to and from the trunnion ring for refractory reline purposes. The design of the trunnion ring to satisfy this type of service is usually in the form of a horseshoe, that is open-ended, to allow the ease of vessel removal and reassembly. However, circular rings have also been designed for exchange of vessels.

Structurally, the trunnion ring is designed to accommodate the types of load to which it is subjected by the specific method of vessel suspension. In any given design, certain sections of the trunnion ring are subject to shearing forces, bending moments, internal pressure, and/or torsion and to thermal stresses due to non-uniform temperature distribution between the ring inside and outside diameter and the top and bottom flange. Thermal loads must be considered for trunnion ring design,

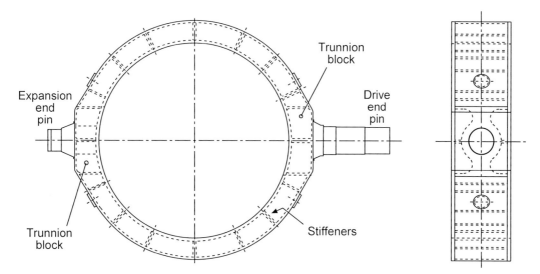

Fig. 8.10 Trunnion ring design.

even in the case of water-cooled rings. The design, therefore, should provide the necessary sizing of the components and manner of assembly (welding, etc.) which can accomplish the required integrity.

The cross-section of the trunnion ring is usually rectangular in shape, made up of top and bottom flanges, inner and outer wrapper plates, and inner baffles and diaphragm plates suitable to the design. In water-cooled and non-water-cooled trunnion ring designs, accesses are normally provided, and are located on the outside wrapper plate.

The bearing shaft or pins are normally connected to the trunnion ring via cast or fabricated blocks which are integral parts of the ring. These pins are secured to the ring either by shrinking fitting, welding, or bolting. Other design specifics in the trunnion blocks stem from process needs such as bottom blowing or stirring, slag detecting and stopping, cooling, etc.

8.2.7 Methods of Vessel Suspension

Virtually all commercially available BOF designs utilize the basic concept of supporting the vessel by the trunnion ring. The function of the trunnion ring is to support the vessel at all phases of operation, while allowing it to undergo the thermal expansion necessitated by the process within. As will be discussed in more detail, there is a multiplicity of technically workable vessel suspension system designs, and several are commercially available and industrially tested.

With the exception of a greenfield site installation, not all designs can befit the application. Usually, the limitations arise in the confined space within which the vessel system has to be installed. If the replacement vessel system requirement also carries with it the need for an increase in the heat size, the problem is further magnified. Fortunately, with the advent of better design methods (e.g., finite element analysis) and materials, these particular design needs can be met.

There are basically two methods to support the vessel by the trunnion ring. In some designs, the vessel rests on the trunnion ring, and in others, it hangs from it. Until the late 1970s, the most popular vessel suspension design in North America was that involving support brackets. The number of brackets used to support the vessel load was a function of the design concept used. Because of the operating and maintenance limitations which the industry has come to realize with bracket support systems, there is now some inclination toward the concept of hanging the vessel from the trunnion ring. To better identify the features of the commercially available and industrially proven systems, the following is a description of such, without any intent to commercially or technically discredit any design. It should further be emphasized here that these suspension systems are covered by various patents and should not be regarded as public domain.

8.2.7.1 Bracket Suspension Systems

During the 1960s and 1970s, most designs in the U.S. employed the multi-bracket system which incorporated a certain degree of redundancy in load support. When the design involves more than three support brackets, the load distribution among the brackets cannot be accurately predicted. Examples of multi-bracket support systems are shown in Fig. 8.11 and Fig. 8.12. In all bracket design systems, the vessel rests on the trunnion ring. Tilting of the vessel is affected by some form of connection between the vessel and the trunnion ring, some using the same securing mechanism between the bracket-to-trunnion ring while others would have a separate tilt bracket.

In general, bracket suspension systems include cast or fabricated brackets attached by bolting or welding to the vessel shell and supported at the top and bottom surfaces of the trunnion ring. Additionally, the brackets are permitted to move radially, but restricted from lateral (tangential) movement with respect to the supporting trunnion ring. The restriction of the tangential movement is sometimes provided by means of stop blocks, engaged to the brackets by various systems of keys, grooves, tapered wedges and spherical seats.

There are many designs to compensate for the inevitable longitudinal thermal expansion of the vessel. Inclined planes may be located at the bottom and sometimes the top surfaces of the trunnion rings, in the location of the brackets, such that the radial and longitudinal displacement components of the vessel shell's thermal expansion describe the combined inclination of the top and bottom planes. Another method is to restrict the vessel top brackets from axial movement within the trunnion ring, while allowing the bottom brackets to thermally move in the longitudinal direction in equal magnitude to the longitudinal thermal displacement of the vessel shell between the two brackets. In this design it becomes necessary to tie down the top brackets to the top of the trunnion ring by such means as a gib, Fig. 8.12, which enables radial bracket displacement, while preventing the latter's movement away from the trunnion ring.

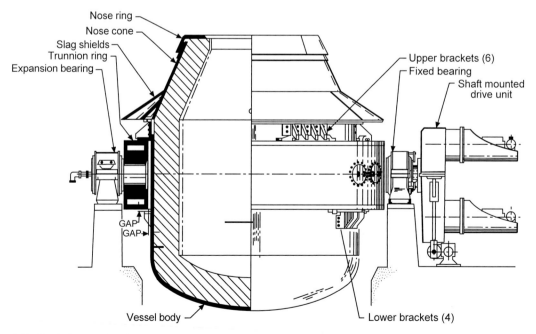

Fig. 8.11 Bracket support system.

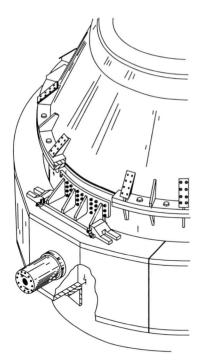

Fig. 8.12 Bracket support system showing detail of gibs.

Vessel suspension brackets, in general, induce a large magnitude of a combination of localized circumferential, longitudinal and twisting moments on the shell, together with membrane and radial forces. If the vessel shell material, geometry, plate thickness and operating temperature are such that the applied bracket stresses could cause plastic deformation and/or creep, the bracket system would then suffer from misalignment symptoms and unequal stress distribution and would become vulnerable to high maintenance requirements.

Geometrically stable vessel shells are the best guarantee for the success of the bracket suspension system. Creep resisting materials for the shell have been used successfully with brackets. Brackets, in general, are located in areas vulnerable to vessel breakouts in cases of burn-throughs. Distorted shells or repaired brackets adversely affect the performance of the bracket suspension system.

Bolted brackets disturb the clean surface inside the vessel and render the bricking of the furnace difficult and time consuming, compared to a smooth internal vessel shell surface. There are, however, solutions to internal surface bolt head protrusion into the shell, which were successfully applied, i.e., bolt heads let into a thickened strake so they are flush on the inside.

To alleviate the load redundancy issue associated with statically undetermined multi-bracketed support systems, some designers chose to reduce the number of supporting members to only two main brackets, one at each trunnion pin location, to support the weight of the vessel and its contents, and a third, smaller bracket for purposes of only tilting the vessel. Thus a statically determined three-point suspension system is provided.

8.2.7.2 Support Disc Suspension System

This system relies on two main circular discs protruding from the trunnion ring center of rotation engaging into two large circular rings attached by welds and radially braced by heavy gussets to the shell, Fig. 8.13. These guarantee the support of the vertical gravitational loads of the vessel and its auxiliary equipment while enabling the vessel to radially and longitudinally expand without restriction. To enable tilting the vessel without relative motion between the shell and the trunnion ring, a third bracket called a tilting claw or toggle is attached to the vessel at a point at the cross-axis of the vessel, supported on the trunnion ring, while a fourth member, called a guide claw, prevents lateral shifting of the vessel in the trunnion ring. This member does not take any gravitational or tilting load. The disc and ring are the supporting members, and are in permanent engagement in all vessel tilting positions, thus avoiding shocks during tilting.

This system creates a smooth surface inside the vessel shell enhancing good and easy bricking of the BOF. It is statically determinate, and enables calculating the precise loads and stresses applied to both the trunnion ring and the shell. Due to the proximity of the disc to the shell, it is expected that the bending and twisting moments can be managed from the design point of view, considering that the load is carried at only two points, compared to at least four in the case of other suspension systems.

The system does not require any adjustment of components and is therefore virtually free of maintenance.

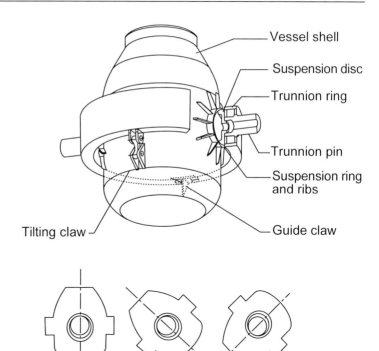

Fig. 8.13 Support disc suspension system.

However, this suspension method requires a larger space between the vessel and the trunnion ring than usually required when utilizing other suspension systems. So, it would be difficult or even impossible to use this system in the case of a replacement BOF vessel, where the available space is already dictated.

The cooling effect on the vessel shell by natural convection is significantly improved as compared to other designs. Furthermore, this large space enables a future installation of an air cooling system for the barrel section.

The disc suspension system is more vulnerable in the case of a breakout due to a burn-through, by virtue of the large shell area required by the ring and its bracing gusset. Repairs to a damaged shell and/or suspension system components in this area become more difficult, as the majority of that area is inaccessible behind the trunnion ring.

8.2.7.3 Tendon Suspension System

This system relies on tying the vessel to the bottom of the trunnion ring from four single brackets distributed circumferentially along the lower shell underneath the trunnion ring. With quick change converters these brackets can be combined into two bracket segments. Pre-stressed (pre-tensioned) tendons penetrate the trunnion ring and are tightened to its top to retain the vessel lower brackets tightly in contact with the bottom surface of the trunnion ring at the point of suspension, while allowing radial displacement of the vessel with respect to the trunnion ring, Fig. 8.14. For absorption of the vessel loads during tilting of the converter, transversal tendons are arranged underneath (with larger vessels above and beneath) the trunnion ring in the region of the trunnion pins as shown in Fig. 8.15. Horizontal stabilizer brackets, similar to the one shown in Fig. 8.16, are sometimes used instead of transversal tendons.

Radial expansion of the vessel shell has to overcome the friction at the supporting surfaces because of the pre-stressed tendons.

This system also provides a clean smooth inside shell surface for easy dependable bricking of the BOF. The largest load carrying member of the lower shell bracket is provided in generally one of

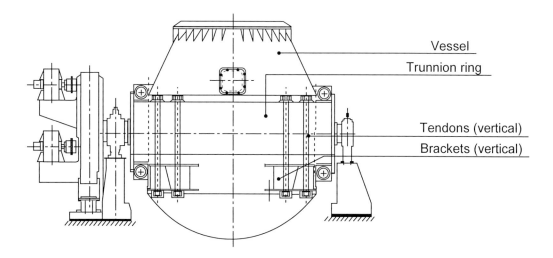

Fig. 8.14 Tendon suspension system, front view.

the lowest temperature shell regions. The stability of the shell, therefore, is expected to be favorable. The tensioning of the tendons is easy to perform with the aid of a tensioning device with direct indication of the tensioning forces. A breakout due to a burn-through could cause difficulties in repairing or replacing the lower brackets.

Experience to date shows that the system appears to require very little maintenance. Also, this system does not occupy any additional room between the vessel and the trunnion ring than that required by a bracket suspension system.

8.2.7.4 Lamella Suspension System

This system relies on suspending the BOF vessel from the lower surface of the trunnion ring by means of a series of two flexible plates oriented in an inclined tangential plane to the shell,

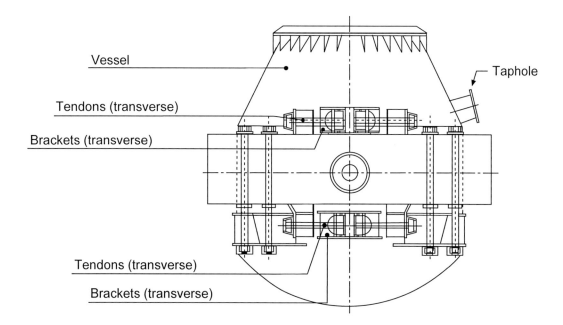

Fig. 8.15 Tendon suspension system, side view.

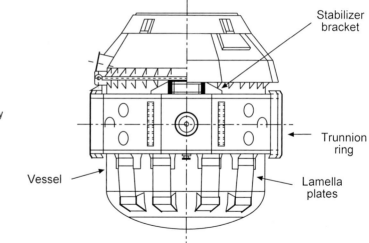

Fig. 8.16 BOF vessel suspended by lamella plates.

Fig. 8.16 and Fig. 8.17. These plates, called lamellae, are separated from each other by means of a spacer, and enable radial movement of the lower vessel shell against flexural deflection of the two plates about their weak axes. In the direction of their own plane, they possess large strength and stiffness. Therefore, with the aid of a stabilizer bracket between the top shell at the trunnion ring, the system can sustain loads resulting from a tilted or inverted furnace, in addition to an upright vessel.

This system also provides a clean smooth inside shell surface for easy dependable bricking of the BOF. One of the positive features of this design is that the lamella load carrying bracket attached to the lower shell is provided in the shell region of lowest temperature.

The stability of the shell, therefore, is expected to be favorable. A breakout due to a burn-through will not cause difficulties in repairing or replacing the lower brackets of the lamellae.

Experience to date shows that the system requires very little maintenance. Also, this system does not occupy any additional room between the vessel and the trunnion ring than that required by a bracket suspension system.

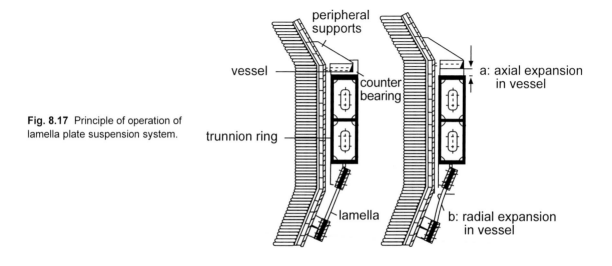

Fig. 8.17 Principle of operation of lamella plate suspension system.

8.2.8 Vessel Imbalance

The operation of a BOF vessel includes tilting it to either side for purposes of charging, sampling and tapping. The drive train required for this function has to overcome the imbalance, which is created initially by design, of the vessel about the center of rotation. Additionally, the drive mechanism with associated components has to either be able to absorb or attenuate the dynamic or transient type loads which primarily arise during tilting operations.

Two schools of thought exist in the industry regarding the nature of the imbalance in the vessel. One concept is to design the vessel system such that at any time during its operation the imbalance forces are self-righting. In other words, should the drive system fail to hold the vessel, it would return to an upright position. The other concept, which is influenced by purely economic considerations, is one where the vessel system at some stage in its operation becomes overturning if the drive control is lost. The drive system in the latter concept is smaller and therefore lower in cost. However, most of the BOF vessels have adopted the self-righting concept.

The unbalanced torque at the drive side trunnion pin results from the eccentricity of the center of gravity of the rotating masses from the center of rotation of the furnace. The center of gravity of the solid rotating masses, namely the steelwork, refractory and contained water, remains constant with respect to the center of rotation, and the static torque resulting from a 360° rotation of the furnace is a true sine wave. The molten masses on the other hand assume the inside configuration of the refractory lining and under a given angle of tilt will result in a hot metal torque corresponding to the weight of the molten masses and the location of their center of gravity. This latter property is a function of the geometry of the molten mass. By adding the two torques, the static torques of the solid and the molten masses, the total static torque is obtained. This is important in determining the degree of equilibrium of the vessel, and its tendency to movement from a given parked condition. For that reason, the effect of both new and worn lining configurations must be considered.

Coupled with the static unbalanced torque are magnitudes of transient or dynamic torque which can be higher than the static values, depending on the acceleration and deceleration settings in the drive control circuitry, Fig. 8.18. While the operator would prefer a quick responding vessel for

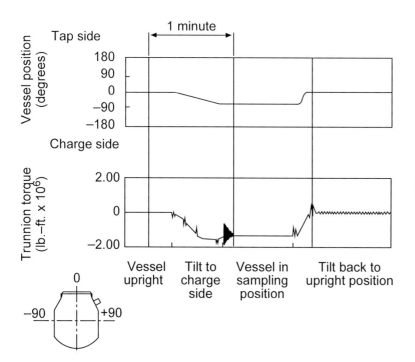

Fig. 8.18 Vessel position and tilt drive torque for a 200 ton BOF.

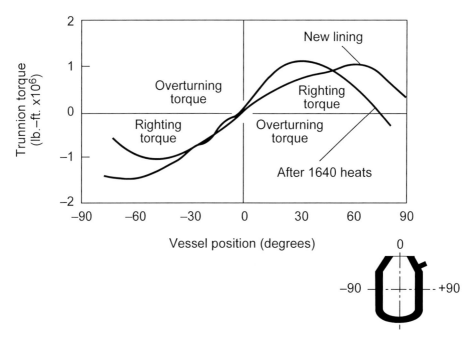

Fig. 8.19 Effect of vessel lining wear on tilt drive torque.

control during the tapping process, such responsiveness usually is detrimental to the integrity of the drive components. An acceleration/deceleration ramp time of four seconds has proven to be an acceptable threshold for imposing the least damage on the drive components.

Uneven and/or non-symmetric refractory lining wear, coupled with skull buildups, play a significant role in changing the vessel unbalance torque. As shown in Fig. 8.19, in one specific installation a 50% increase in static torque was experienced after the refractory lining had produced 1640 heats.

8.2.9 Refractory Lining Design

Fig. 8.20 shows the areas of the refractory lining. The BOF refractory systems are discussed in greater detail in Chapters 3 and 4. A brief description of the requirements of each areas is presented here.

8.2.9.1 Safety Lining

The safety lining is to be made up of burned and/or burned pitch impregnated magnesite refractories. Typical thickness of the safety lining is nine inches. Some companies prefer to use 18 inches on the bottom of the vessel. Refractory retaining bars and rings are installed on the inside of the shell to segment the safety lining so that, during repair, only the worn sections of refractory will be removed and good refractory will be retained. The size and location of the refractory retainers shall be mutually determined jointly by the user, the designer and the refractory supplier.

8.2.9.2 Working Lining

The working lining can vary in thickness depending upon the type of operation and upon the wear rate generally experienced. Higher wear areas should have greater thickness or higher quality materials. The refractories for each area should be selected to have properties that reflect the wear mechanisms of the area where placed. The highest wear areas of the lining should contain brick with properties that reflect the mechanism of wear in that area. Less severe wear areas may contain refractories that are less resistant to the wear mechanisms of the BOF process.

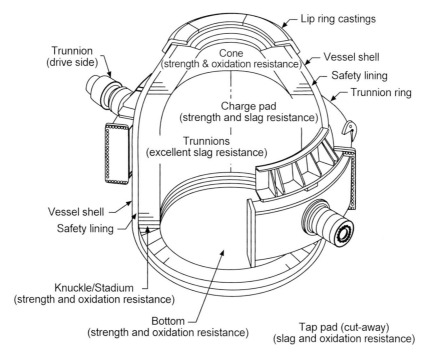

Fig. 8.20 Areas of the refractory lining.

8.2.9.2.1 Cone The cone should contain refractories having good oxidation resistance and good strength. Generally, slag resistance is not a serious consideration.

8.2.9.2.2 Charge Pad The charge pad represents the most important mechanical impact zone. This area is subject to impact during the charging of scrap, and it is subject to erosion during the charging of liquid hot metal. The charge pad should contain refractories with high hot strength to withstand the charging of scrap and hot metal. Slag resistance and oxidation resistance are also desirable properties for charge pad brick.

8.2.9.2.3 Tap Pad The tap pad should contain refractories that are resistant to molten steel and to BOF slag. The oxidation resistance of these materials should also be good.

8.2.9.2.4 Taphole Taphole refractories must have good resistance to molten hot metal erosion and to slags. Tapholes are replaced frequently and a magnesite-based material is gunned around the taphole brick to fill any holes left by the removal of the old taphole refractory. This gunning material must also be resistant to slag attack and molten metal erosion.

8.2.9.2.5 Trunnions Trunnion brick should possess exceptional resistance to BOF slags. Good high temperature strength and oxidation resistance are also desirable.

8.2.9.2.6 Knuckle/Stadium Knuckle/stadium brick should have good resistance to molten metal erosion and thermal shock.

8.2.9.2.7 Bottom Bottom brick should have good resistance to molten metal erosion, oxidation and thermal shock, especially for bottom blown converters.

8.2.9.3 Refractory Shapes

Refractory shapes for BOF linings are generally key shapes. Certain areas of the vessel may contain special shapes. The cone may contain a parallelogram shape that permits easy deskulling of the cone early in the campaign. Also, the bottom may contain BOF key-wedge-arch shapes to accommodate the concentric ring bottom design.

8.2.9.4 Burn-in Practice

Before a newly lined vessel system can be rotated for operation, a burn-in procedure is necessary to heat the brick for expansion to secure it in place. The burn-in procedure consists of loading coke and other combustible materials into the vessel, igniting them, and starting the heating procedure by lowering the oxygen lance into the vessel. The proper burn-in procedure requires the following.

8.2.9.4.1 Time A controlled burn-in requires 3.5 to 4 hours at a heatup rate of 500°F per hour to 2000°F to allow for a more gradual temperature adjustment of the refractory. The vessel can be immediately charged once burn-in is complete.

Some shops still use tar-bonded magnesite refractories in the cones. These materials are normally manufactured to withstand a slow heatup and can be burned in as described above. However, some materials may still be available that have not been so manufactured, and in this case, they may exhibit slumping as they are heated slowly through the temperature range of 400–600°F. The refractory supplier should be consulted on this matter. For this latter material, a rapid heatup avoids the slumping, and burn-in should consist of heatup to 2000°F over a time period of 1.5 hours.

8.2.9.4.2 Temperature Monitoring Temperature during burn in should be monitored and controlled by thermocouple. The thermocouple wire can be inserted through the taphole and should be of sufficient length to reach well into the vessel. Although it will be difficult to achieve a steady state, the rate of temperature increase should follow the desired rate as closely as possible. The initial flare-up upon ignition will exceed the desired rate, but will fall within 30 minutes to the expected curve. It is at this time that the blowing of the oxygen should commence and be increased to allow the temperature to rise at the desired rate.

8.2.10 Design Temperatures

8.2.10.1 Vessel Shell Temperature

A general idea of the temperature levels which prevail in the shell are shown in Fig. 8.21. These levels vary around the periphery of the vessel, and are influenced by the variation in refractory lining thickness.

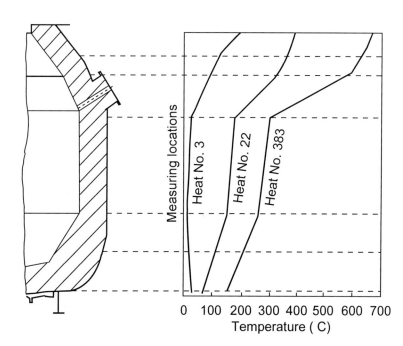

Fig. 8.21 Shell temperature measurements on a 220 ton BOF.

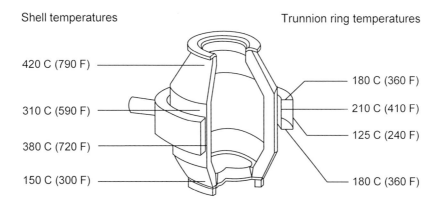

Fig. 8.22 Temperature measurements on a converter shell and trunnion ring.

At certain locations in the shell, the temperature level can become high enough to affect the useful life of the vessel by lowering the material resistance to deformation under load. In the barrel section, temperature levels at mid-height are usually highest, Fig. 8.22. The cooling effect from a water-cooled trunnion ring plays a role in lowering the temperature levels in this portion of the vessel. An example of such is shown in Fig. 8.23.

Designs have been worked out to provide external cooling for the shell around the trunnions. An example of one is shown in Fig. 8.24.

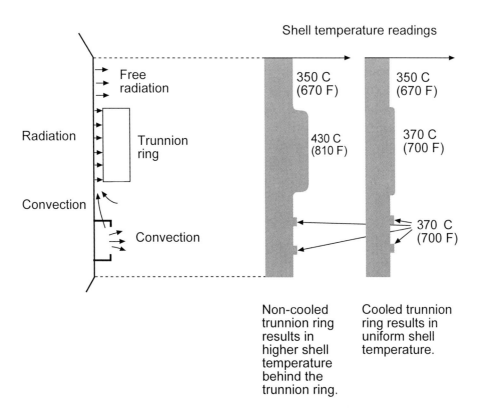

Fig. 8.23 Influence of trunnion ring cooling on BOF shell temperature.

The most significant benefit of external cooling on the vessel has been on the nose cone. In addition to maintaining or prolonging the structural integrity of the cone, there are basically two methods of closed system of water cooling utilized. One method is to apply the cooling tubes in the horizontal direction (circumferential flow) and the other is to apply them along the cone generator. The life of a non-water-cooled nose cone is usually curtailed by the severe distortion which then precludes the ability to maintain the refractory lining. While the use of material which has higher resistance to deformation at the elevated temperatures in the cone can help prolong its life, the concept of water cooling has far more advantages. Without water cooling, the cone shell can reach temperature levels in excess of 1000°F.

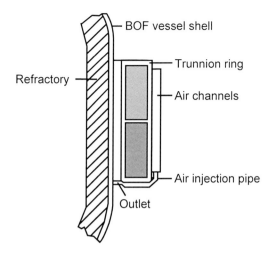

Fig. 8.24 Method of cooling vessel shell with air jets, U.S. Patent No. 5,039,067.

8.2.10.2 Effect of Shell Temperature on Vessel Suspension System

The temperature distribution along and around the vessel is not uniform. This non-uniformity in temperature, especially around the shell periphery, affects certain suspension designs more than others. In multi-bracket suspension designs, where axi-symmetric radial expansion growth is essential to the proper functioning of the bracket securing system, overstress problems occur in both the suspension components and the vessel shell. Even in bracket designs where axi-symmetric radial expansion is not essential, the existence of a temperature distribution other than that assumed by design can affect the proper functioning of the support system. Yet there are suspension systems, the performance of which is much less dependent on the temperature variation along and around the vessel.

8.2.10.3 Trunnion Ring Temperature

In the early to mid 1960s, virtually all BOF systems installed in the U.S. had dry or non-water-cooled trunnion rings. After an average of five years most of these systems were replaced because of severe distortion in both the vessel shell and the trunnion ring. In the early 1970s almost every system was replaced with a water-cooled trunnion ring, allowing the useful life to be extended threefold. While water cooling maintains the structural integrity of the trunnion ring, it also serves to lower the temperature level of the portion of the vessel which is in close proximity to it. As is shown in Fig. 8.23, a temperature reduction of 100°F or higher can be achieved.

The dimensional stability in the trunnion ring, which is affected by water cooling, has many merits, among which are: alignment of the trunnion pins is preserved; structural integrity of the suspension system, especially in the case of a multi-bracket support system, is better preserved; and a cooled trunnion ring can be used as a reference in monitoring the permanent growth (due to creep) of the vessel shell.

While many BOF installations utilize water-cooled trunnion rings, there are several which are still operated dry (without water cooling). The concern and therefore the refrain from using water cooling stems from the potential hazard associated with the large volume of water present in the event of a vessel breakout.

8.2.10.4 Refractory Design Temperatures

The temperature-related destructive mechanism of refractories is slag attack—the higher the temperature, the greater the slag attack. The refractory lining must be designed to withstand slag attack at the maximum operating temperature of the furnace.

8.2.11 Design Pressures and Loading

8.2.11.1 Shell

During the operation of a BOF, the vessel shell is subjected to a complex system of loads and stresses, some uniform and some localized, depending on the design of the trunnion ring, the suspension system, and the geometrical configuration of the vessel.

In general, the loads acting on a BOF vessel shell can be categorized as: gravitational loads, including the weight of the steelwork, refractory lining, charge materials and residual skull buildup; impact loads due to charging of scrap into the furnace; impact loads due to deskulling the furnace mouth or cones, by either a floor deskuller, or mechanical means such as a pneumatic chisel or a swung-loaded chisel; and dynamic loads resulting from vessel rotation for turndown and tapping.

The above loads are usually transmitted to the trunnion ring and the support or suspension system will interact, to a larger or a smaller degree, depending on the system design, with the vessel shell at the attachment areas. Such loads are ultimately transmitted as localized bending, twisting or membrane stresses, to the shell plate.

The following loads, on the other hand, generate vessel shell stresses usually unsensed by the outside support system or the trunnion ring, if the design and function of the suspension system are not flawed. Internal pressures result from the thermal expansion of the hot refractory brick, acting in the radial direction and the axial vessel direction—these pressures are reflected as membrane stresses on all shell components, except at corners of cones and bottoms, where bending stresses will result. Ferrostatic pressures are generated from the molten masses in the furnace—the effect of the ferrostatic head is similar to that in a tilted charging ladle. Non-uniform temperature distributions in the shell produce non-uniform stress loading of the shell.

A type of load which affects both the shell and the support system arises from large variations in temperature in the shell. These large differences in temperature result from both process variations within the vessel and non-uniform refractory lining wear. Yet another condition of significant temperature variations develops in both the vessel shell and trunnion ring during sustained vessel tapping.

The performance of the vessel shell under the applied loads and stresses is a function of many variables, namely: the magnitude of the stresses and the location, area and configuration of the shell at the attachment of the suspension system; the design, and the effect of load transmittal from the shell to the trunnion ring; and the material of the shell, its thickness and its behavior at elevated operating temperatures.

In North America, the experience has been that shell temperatures reach levels which affect the long term stability under the computed acceptable levels of stresses at room temperatures. For the same wall thickness, and correspondingly the same operating shell stresses, utilizing alternate materials of the creep resistant variety has proved successful.

8.2.11.2 Minimum Thickness Requirement

Selection of the suitable shell thickness in a BOF vessel shell is influenced by many factors, among which are: tonnage rating of the vessel; actual geometrical size of the vessel; particulars of design of certain areas of the vessel shell, such as water cooling of the conical shell; physical and creep properties of the material of the shell; nature and magnitude of loads applied from the suspension system to the shell; thickness and type of refractory used; operating conditions and degree of wear of refractory during the campaign; temperature of operation within the furnace; and methods of deskulling and applied mechanical loads.

Due to the number of uncontrolled factors in operating a BOF, it will be a difficult, if not an impossible, task to derive formulae to compute the above effects on the magnitude of the loads applied

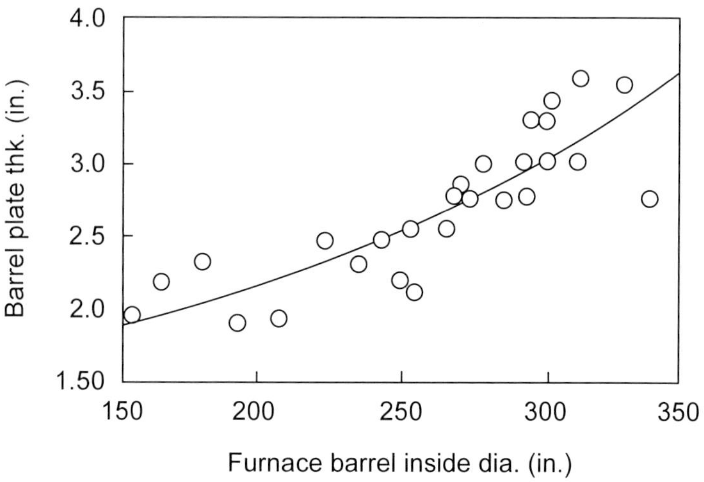

Fig. 8.25 Historical design data for BOF barrel plate thickness.

to the shell, whether from the interaction with the suspension system, or from within the vessel interior. It is anticipated that the recent investigative work, which was done and is being continued on the analysis of the structural behavior of the BOF vessel, that the load mechanisms involved can be better defined.[2]

It is for this reason that quantification of the shell thickness relies mostly on experience of the designers and the BOF manufacturers with their particular designs and suspension systems. A survey of BOF installations shown in Fig. 8.25 provides some idea of the complexity involved in selecting the suitable shell thickness. A similar collection of historically established dimensional information on the design of trunnion rings is shown in Fig. 8.26.

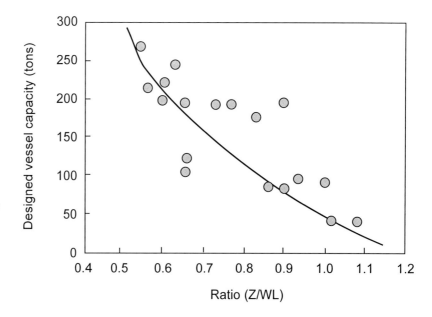

Fig. 8.26 Historical design data for BOF trunnion rings.

Z = Section modulus of trunnion ring (in.^{3})
W = Total load on trunnion ring (tons) including vessel, lining and heat weight
L = Distance between bearings (ft.)

8.2.11.3 Bottom Stirring and Bottom Blowing

The term bottom stirring means the injection of essentially inert gases into the bottom of the BOF vessel, penetrating the bottom shell and the bottom refractory lining, under the molten bath, to agitate the molten masses for such purposes as homogeneity of the melt after introducing additions in the furnace and improving the interaction between the steel bath and the slag. There is usually no direct chemical reaction associated with bottom stirring. The gas injection is either by means of refractory material porous plugs embedded in the bottom lining, or by means of tuyeres, penetrating the bottom lining.

On the other hand, the term bottom blowing means injection of oxygen, singularly, with additives (such as pulverized lime) or in addition to hydrocarbon fuels (such as natural gas, pulverized coke or fuel oil), all routed in the same manner to initiate reactions in the bath. The injection of the oxygen under the molten bath is usually done by means of tuyeres, mostly of annular design, in which the fuel shrouds the oxygen nozzle. The preservation of the tuyere and the bottom refractory are major objectives of the annular design.

In each case the gas lines and the fuel and additive lines (if any) are piped through one or both hollowed trunnion pins of the BOF and down to the bottom shell of the vessel.

As recommended in pressure vessel design, cutting a hole in a shell requires structural reinforcement. Additionally, special care shall be taken in separating the oxygen and the fuel lines. They are usually introduced through opposite trunnion pins. The piping must take into consideration the relative displacement of the vessel with respect to the trunnion ring. Therefore, a series of expansion joints must be designed into the piping. External protection of the piping from heat and falling debris is required. Oxygen piping must be properly cleaned for oxygen service and must be of stainless steel construction. If additives are injected with the oxygen, spark and abrasion resistant materials will be required in key areas of the pipe line.

In association with bottom stirring and, to a more pronounced degree, bottom blowing, it is important to expect generation of additional low frequency vibration in the vessel. These vibrations result partially from the thermodynamic changes and chemical reactions encountered during the passage of either gas in the molten bath, and the propagation of hot metal waves generated in the surface of the bath due to the induced agitation. These oscillations will encumber (and increase the stresses in) the shell, suspension system, trunnion ring, bearings and support pedestals and foundations. Therefore, it is important to carefully plan for possible future process requirements when considering new BOF installations.

In top blown vessel systems, the bearing support is normally not subject to forces along the trunnion axis. However, in systems where gases are introduced in the vessel from the bottom or sides, forces along the trunnion axis are developed, and if the supporting bearing pedestals and piers are not sized properly, undesirable vibration problems result. The expansion side trunnion pin support is normally not subject to this axial load, because the ladder bearing there allows axial movement, without significantly resisting it. The drive side bearing support, however, has to resist these forces, not only in terms of strength, but also dynamic compliance. Both bearing supports, however, have to resist horizontal vibration loads acting perpendicularly to the axis of rotation.

The excitation forces in a bottom-blown or bottom-stirred system are primarily a result of the mass motion of the molten steel inside the vessel. An example of such lateral vibration is shown in Fig. 8.27, where actual field measurements were acquired on the support system of an OBM (Q-BOP) installation. Combined with the forces developed due to chemical reactions within, the overall frequency spectrum contains many components. Coincidence of the excitation frequency of force with the natural frequency of the vessel support system will create resonant vibration problems not only in the support structure, but also in the vessel components.

In past installations, where vibration measurements were acquired, the frequency of axial vibration from melt activity was in the range of 0.3–0.4 Hz. The expansion side bearing support does not participate in resisting this horizontal force developed as a result of the melt activity in the vessel.

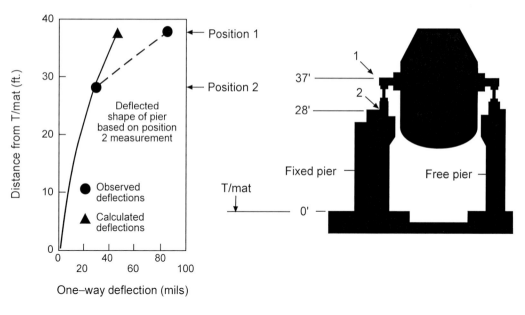

Fig. 8.27 OBM (Q-BOP) support vibration.

As a result, the magnitude of the horizontal force resulting to the drive side pier from melt activity (in the OBM or Q-BOP process) in the vessel is roughly 15% of the total weight supported by both piers.

8.2.11.4 Creep

The useful life of a BOF vessel is normally determined by the change in the space or air gap between the vessel and the supporting trunnion ring. Examples are shown in Fig. 8.28, Fig. 8.29 and Fig. 8.30. The permanent deformation in the vessel due to the mechanism of material creep causes an outward growth, which reduces the air gap.

When this gap is or nearly exhausted, the vessel system is taken out of service. The stresses in the shell which are largely responsible for this growth are due to the thermal expansion of the refractory within.

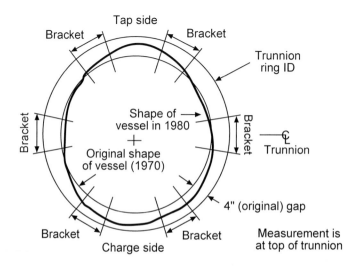

Fig. 8.28 BOF vessel distortion.

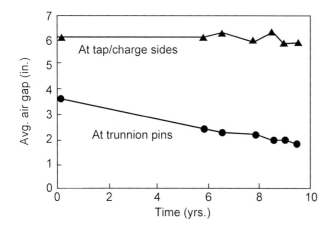

Fig. 8.29 Change in vessel-to-trunnion ring air gap in a water-cooled trunnion ring system.

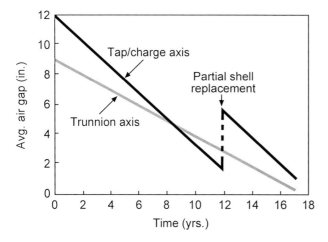

Fig. 8.30 Monitoring of vessel-to-trunnion ring air gap in a non-water-cooled trunnion ring system.

Material creep is a time related phenomenon. In steel, the mechanism is most pronounced at temperature levels above 800°F. As shown in Fig. 8.31, the complete creep life cycle is composed of three stages.

In the primary stage, large deformation or strain occurs in a relatively short time. Depending on the material, temperature and level of stress, this period could be a matter of a few days or even hours. In the secondary stage, the rate of strain is significantly lower, and this stage essentially represents

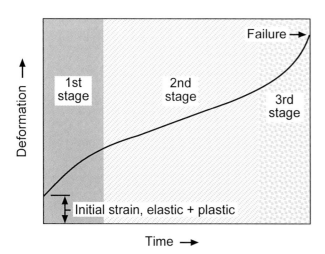

Fig. 8.31 Classical creep life cycle.

the useful life span of the structure. In well designed structures, this period normally represents many years of service. The tertiary stage represents an unstable region, in which the point of rupture is not very predictable. As can be seen, the strain rate is very high here.

The factors which, therefore, influence the rate of the BOF shell growth are stress level, shell temperature level (i.e., refractory practice), and the shell material resistance to deformation at elevated temperature.

Proper refractory lining design and operating practice influence both the stress and temperature levels developed in the shell. Material selection for the application can serve a very significant role here. The use of chrome-molybdenum steel grades has brought about a significant improvement in BOF vessels in North America. Fig. 8.32 shows the change in strength in various BOF vessel steels at elevated temperature levels.

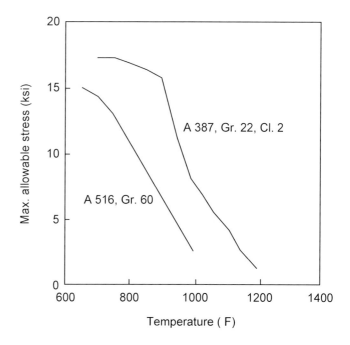

Fig. 8.32 Effect of service temperature on strength of BOF vessel steels.

8.2.11.5 Suspension System

The nature of loading on the vessel shell and the trunnion ring is a function of the specific type of suspension mechanism incorporated.

8.2.11.6 Trunnion Ring and Pin

Again, the nature of vessel suspension influences the type of loading on the trunnion ring and in some cases the trunnion pin.

8.2.11.7 Refractory Lining Design

BOF working and safety linings are made from standard refractory shapes. The lining has to be strong enough to resist severe operating conditions associated with relining, burn-in and with high temperature operation of the vessel.

8.2.11.7.1 Loading on Refractories Loading on refractories will develop as a result of the weight of brick courses, the weight of the cold scrap and hot metal during charging of the vessel, the weight of liquid metal and slag during operation, and the contained pressure from the vessel shell during refractory lining expansions.

8.2.11.7.2 Stresses in the Lining Stresses in the lining will develop during operation and therefore refractories must resist thermomechanical stresses caused by thermal and permanent expansion of the lining in the horizontal directions and by the restriction imposed on the brick by the steel shell. Linings also must withstand mechanical abuse caused by deskulling of the lip ring and the cone and stresses caused by sudden temperature increases during burn-in of the lining and as a result of thermal cycling between heats.

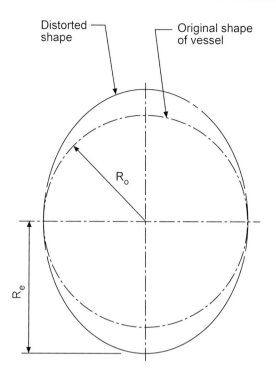

Fig. 8.33 Simulated distortion in a BOF vessel due to creep.

8.2.12 Method of Predicting Vessel Life

The absolute useful life of the vessel material is limited by the time it takes to rupture under a given stress and temperature. In the practical sense, the useful life is limited only by the primary and secondary stages of creep, as shown in Fig. 8.31. In the case of a BOF vessel, the practical and useful life is determined or measured by the depletion of the air gap between the vessel and the trunnion ring. One method, which has been commonly used in the industry, is to periodically monitor this air gap change and predict the end of the vessel life using a straight line projection.

Rarely do BOF vessels deform symmetrically with respect to the central or vertical axis. In some systems, the deformation or distortion occurs along the charge to tap axis and in others along the trunnion axis.

Schematically, therefore, the common distortion in the BOF vessel is in the form shown in Fig. 8.33. Using the anticipated or measured changes in the vessel, the total creep strain which the material at the tap and charge side would or has undergone can be evaluated in one of many approximate methods.

Documented creep data is normally provided in the form of strain rate at the test temperature and stress levels. As indicated earlier, for steel, the information is usually developed at temperature levels above 800°F. Test data also include the total strain at which rupture occurs, again for the specific temperature and stress levels.

For an estimated circumferential (hoop) stress level, coupled sometimes with the mechanical stresses resulting from suspension loads, and the average operating temperature level of the shell in the trunnion ring area, a strain rate can be determined from the data established for the shell material. An example of such data is shown in Fig. 8.34 and Fig. 8.35. If the projection of the

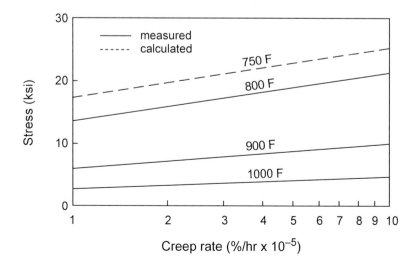

Fig. 8.34 Creep rate curves for medium carbon steel.

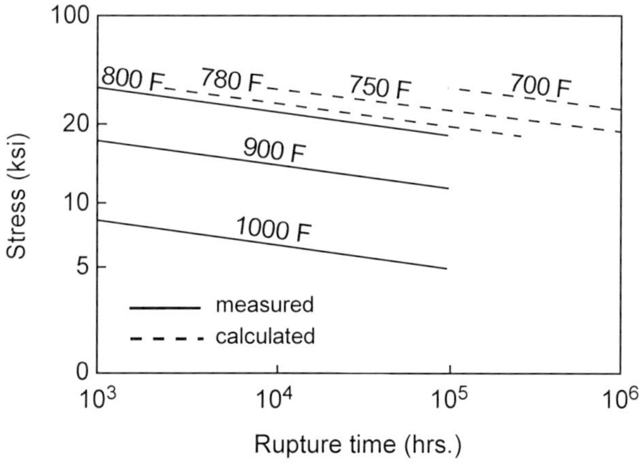

Fig. 8.35 Stress rupture curves for medium carbon steel.

remaining useful life is being made for a vessel system which has already been put in service, a better method of establishing the strain rate would be to use the recorded permanent shell growth.

8.2.13 Special Design and Operating Considerations

8.2.13.1 Protective or Slag Shields

Certain components of the vessel–trunnion ring system require protection from the relatively hostile environment resulting from the oxygen blowing process. In general, the suspension system components require protection. The air gap provided between the vessel and the trunnion ring is protected with slag shields. These are segments of steel, usually bolted to supports on the cone, and serve the function of covering the top of the trunnion ring including the mentioned air gap. There are other forms of protective shields designed to suit the particular concept of suspension.

8.2.13.2 Skull Buildup and System Stability

The effects of the refractory lining wear and skull buildup in altering the unbalance torque in the tilt drive are discussed in Section 8.2.8. What is to be emphasized here are the necessary measures which should also be taken to provide for the unusual loading on the vessel suspension, which may not have been anticipated, such as if the user decides to use a skull cleaning system which puts undue loads on the vessel system parts. Deskulling systems which incorporate stationary blades mounted on the charge floor can induce unusually high loads on the vessel system if not used with discretion.

8.2.13.3 Taphole Slag Detection and Retention

Many attempts have been made over the past decade or so to retain the slag in the BOF vessel during tapping. The concept of a floating ceramic ball or a pyramid, Fig. 8.36, has worked well in some

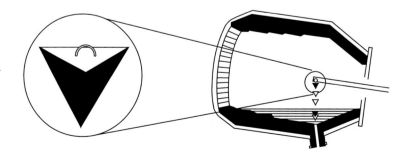

Fig. 8.36 Refractory plug used for slag retention in the BOF vessel.

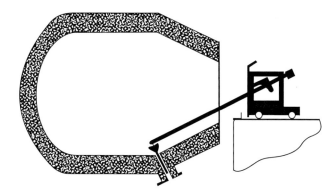

Fig. 8.37 Method of positioning the dart-type slag stopper in the taphole.

shops but not others. The basic principle involved is to drop a floating stopper in the bath prior to tapping and let the stopper plug the taphole from inside the vessel after all the steel is tapped out, while retaining all the slag. The specific density of the stopper material is made such that it floats on the molten steel but not the slag. A modified version of such floating stoppers shaped in the form of a dart has found better performance in shops where a special machine was designed to position the stopper at the taphole from inside the vessel, Fig. 8.37. Usually, BOF shops with adequate floor space on the tap side appear to have the most success with such systems, as this design then allows the installation of the proper machinery and, thus, the precise placement of the floating stopper.

Another concept of slag retention which has come into use over the last fifteen years is a pneumatic device which is installed at the taphole on the outside of the vessel. An electromagnetic slag detector actuates a stopper which closes the taphole after all the steel is tapped. An example is shown in Fig. 8.38. With this system the stream emerging from the taphole is cut by the dynamic impact of the gas blown into the taphole. Sealing performance does not depend on taphole wear.

8.2.13.4 Bottom Blowing

Bottom blowing injection devices, practices and gases differ from one shop to another. The refractories in the area where bottom stirring devices are located should be very high quality magnesia–graphite brick with graphite contents of 10 to 20%. These refractories may contain fused grain. In some cases, burned, impregnated brick designed to have excellent thermal shock resistance is used.

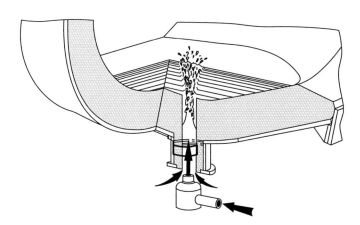

Fig. 8.38 Pneumatic slag stopping device.

8.3 Materials

Materials of construction and welding consumables must be selected with due regard for the stresses, the service temperatures, the interaction with lining materials, and the thermal and mechanical cycling of the BOF process.

A detailed description of the specific type of steel materials, welding materials, materials testing and refractory selections commonly used are discussed in detail in Reference 1.

8.4 Service Inspection, Repair, Alteration and Maintenance

8.4.1 BOF Inspection

Proper inspection and maintenance of a BOF is necessary to produce long life of the equipment and a minimum of maintenance cost throughout its life.

The recommended areas for inspection do not rule out the need for inspecting other components which may be unique to the user's BOF design, and as stated in the Manufacturer's Maintenance Manual.

8.4.1.1 Frequency of Inspection

At a minimum, visual inspection of an active BOF is recommended during any operational cycle. Physical inspection of certain areas and components is recommended during scheduled and non-scheduled downturns. Physical inspection of all areas is recommended after each campaign or sooner if duration of campaign exceeds regular intervals of inspection for certain areas.

The recommended frequencies are stated in the following critical areas.

8.4.1.2 Critical Areas for Inspection

The following components are recommended for inspection. The components may not be representative of all BOFs but are intended to provide guidance in identifying the more critical areas for inspection.

8.4.1.2.1 Lip Ring and Top Ring The buildup of skull on the top of the vessel is to be completely removed and the lip ring and top ring are to be checked for excessive warpage, cracks and/or loose bolts. Any sections that are warped or cracked and can hinder the retention of the refractory in the vessel are to be repaired and/or replaced. Any loose bolts and/or broken bolts are to be tightened or replaced.

Skull removal and inspection or repair of these components are to be accomplished during the campaign on scheduled and non-scheduled downturns as required to maintain their integrity. Complete inspection and repair and/or replacement are to be made after each campaign.

8.4.1.2.2 Slag Shields The slag shields are to be inspected for warpage due to heat. If this warpage is excessive and the shields no longer protect the suspension system, nose cone cooling member, or trunnion ring, these sections should be replaced.

Inspection, repair and/or replacement are to be accomplished during the campaign on scheduled and non-scheduled downturns as required to maintain their integrity. Complete removal of slag shields is to be accomplished after each campaign to facilitate inspection of the slag shield attachment brackets and other components under the slag shields.

8.4.1.2.3 Tapping Nozzle Welded shell nozzles are to be inspected after each campaign for cracked welds and for burn-throughs. These should be repaired before placing the furnace back into operation.

Removable nozzles are to be inspected during and after a campaign to check for cracked welds, burn-throughs and damage to the retaining lugs. All of these conditions should be repaired before placing the furnace back in operation.

8.4.1.2.4 Bolting Flanges The bolting flanges of a removable nose cone are to be inspected after each campaign for cracked welds, loose or broken bolts and any warpage causing excessive gaps.

All cracked welds should be repaired. Loose bolts should be tightened and broken bolts replaced. Excessively warped areas of the flange should be cut out and replaced.

8.4.1.2.5 Water-Cooled Nose Cone and/or Lip Ring Any leakage that would occur during operation of the furnace should be eliminated by either repairing or isolating the leak by means of diverting the water through valving or through a complete shutoff.

After each campaign all water cooling components should be inspected and repaired as required.

8.4.1.2.6 Removable Bottoms The removable bottoms on bottom-blown vessels are to be inspected throughout the campaign for leaks in the tuyere supply piping. Any leaks are to be isolated immediately and repaired.

During bottom changes the removable bottom proper and the attachment bolts or clamp-link assemblies are to be inspected for any damage. Repairs should be made as required.

8.4.1.2.7 Refractory Retainers Horizontal and vertical refractory retainers are to be inspected during each safety lining replacement. Cracked welds should be repaired and any distorted or missing sections should be replaced.

8.4.1.2.8 Vessel Suspensions System Components Every BOF is designed and furnished with a suspension system to support the vessel in the trunnion ring. Many are furnished with vessel connections that are either welded or bolted to the vessel shell and are restrained to the trunnion ring with welded blocks and tapered wedges.

During the campaign any noticed movement of the vessel in the trunnion ring or abnormal noises observed during rotation should be investigated immediately.

After each campaign the following vessel suspension system components are to be inspected.

8.4.1.2.8.1 Vessel Shell Connections All welds are to be inspected for cracking.

8.4.1.2.8.2 Vessel Bracket Connections Connection bolts are to be inspected and any loose bolts are to be tightened. Any broken bolts are to be replaced. The fit of back plates to the shell and of bottom plates to the trunnion rings is to be inspected for distortion.

8.4.1.2.8.3 Retention Hardware Retention blocks are to be inspected for attachment and for distortion of machined surfaces. Adjustable retaining wedges are to be inspected for looseness and distortion. All necessary repairs and adjustments should be made in order to maintain the integrity of the system and to prevent any damage to the shell and the trunnion ring.

8.4.1.2.8.4 Trunnion Ring After each campaign the trunnion ring is to be checked for deformation. The pin alignment in the main trunnion bearings is to be checked and recorded, as well as the axial displacement of the trunnion pin in relation to the center of free bearing. Also, the position of the trunnion pin in each trunnion casting is to be checked. Once per year, the access covers to the outer wrappers are to be removed (after removal of water lines if the trunnion rings are water cooled) and all internal compartments and diaphragms inspected for any cracked welds or members; repairs are to be made as required. Before access covers are replaced, water cooling passages should be cleaned and freed of all debris so as not to cause clogging.

8.4.1.3 Refractory Service Inspection

Service inspections of the refractory lining are necessary to insure that there are no uncontrolled problems with the lining. The areas to be inspected include the cone, charge pad, tap pad, trunnions, knuckle/stadium and bottom. These inspections are normally carried out visually; however, many companies also rely on a laser lining thickness measuring instrument.

Visual inspection of the lining shall be made by the operator after each heat.

Laser readings should be obtained on a new vessel every other day. As the vessel wears, the frequency of laser readings should be increased to daily. If heats known to be detrimental to linings are made, laser readings should be made after these heats.

In addition to being useful for judging when the reline of the furnace shall take place, these measurements can also provide an indication of the effectiveness of zoning practices (i.e., the practice of installing different quality refractories in various areas of the furnace in an attempt to achieve uniform wear rates). Among the other uses for these readings, two significant uses are to identify operating practices that accelerate wear rates, and to help in scheduling routine maintenance of the working lining of the vessel.

8.4.2 BOF Repair and Alteration Procedures

8.4.2.1 General

During the life of the BOF, repair and alteration by the user may be required to maintain the structural integrity of the BOF. It is important to distinguish between a BOF repair and a BOF alteration.

A repair constitutes a rework of the BOF in order to maintain the structural integrity of the various BOF components. A repair is defined as a rework of one or more components of the BOF without deviation from the original BOF design.

An alteration can be either a form of rework in which a structural component of the BOF is changed from the original specified geometry, or a change in the BOF operating conditions, such as deviation from specified deskulling practices, increasing heat size, adding bottom gas blowing, addition or deletion of cooling systems and/or slag shields or any other such process changes.

8.4.2.2 Procedure for BOF Repair

A BOF repair may be made in the field by the BOF user or his agent. To make a satisfactory repair the following procedure is recommended:

1. The BOF user shall prepare a list of problem areas and operating restrictions.

2. The BOF manufacturer or qualified agent shall submit a rework plan to the BOF user for review and comment.

3. If the BOF manufacturer or qualified agent agrees the user has adopted an appropriate repair procedure, then the user can proceed with the proposed repair.

4. Repair welding of the integral vessel and trunnion ring components shall meet the requirements of ANSI/NB–23 Chapter III for welding, preheat and postweld heat treatment.

5. An inspection of the complete repair shall be made by the user or a qualified agent selected by the user. If the repair was made by a party selected by the user, the complete repair inspection results shall be made available to the repair party.

6. All work on the BOF shall be done in accordance with OSHA safety requirements.

8.4.2.3 Procedure for BOF Alteration

A BOF alteration may be made in the field by the BOF user or his agent. To make a BOF alteration the following procedure is recommended:

1. The BOF user shall prepare an outline of the proposed alteration.

2. The BOF manufacturer or qualified agent shall perform a structural analysis of the alteration consistent with the requirements of these specifications. Complete structural details and requirements shall be determined by the analysis and made available to the user.

3. Following the analysis, the BOF manufacturer or qualified agent shall prepare a proposal which includes details of the alteration and follow the procedures recommended for BOF repair.

4. In the case where the alteration is a change in the operating condition such as a significant increase in tap temperatures or a major change in refractory types, and not a structural alteration, the BOF user shall review the change in operating conditions with a BOF manufacturer or qualified agent to determine the impact of the operating conditions on the structural integrity of the BOF. If necessary the recommended alteration procedure shall be followed.

8.4.3 Repair Requirements of Structural Components

8.4.3.1 General

The BOF shall be maintained such that the structural components can function as intended in the original design. The refractory linings shall be maintained such that the design temperatures for the BOF shell and structural components are not exceeded during typical BOF operations.

8.4.3.2 Crack Repair

Temporary crack repair shall be made by drilling holes at the ends of crack tips. The end of cracks shall be detected by appropriate NDE methods.

The hole drilling technique is a temporary repair used only to curtail the crack growth. When the holes are drilled, immediate plans should be made to make a permanent repair of the crack. Permanent crack repair shall be made by either removing the cracked section and replacing with patch plates, or by complete gouging and grinding of the cracked section and replacement of the removed material with weld deposits.

8.4.3.3 Vessel Shell or Trunnion Ring Wrapper Patch

The shell patch shall be made to replace severely distorted or buckled sections of the shell, as well as any portion of the shell in which the thickness is less than 70% of the original design thickness due to a washout (partial burn-through).

In the case of a molten metal breakout, the opening should be enlarged to a distance of three times the plate thickness beyond the edge of the breakout. The patch and matching cut out portion of the shell plate or wrapper plate must have rounded corners with a radius equal to or greater than three times the plate thickness. The recommended maximum misalignment tolerance should be 25% of plate thickness. The patch plate must be welded with butt joints attained by double bevel welding or other means which will result in the same quality of deposited weld on the inside and outside weld surface. To minimize weld shrinkage, the root opening between the patch plate and the cutout in the shell plate should be minimized in accordance with sound welding practice.

The patch plate material must have a material specification number and grade equal to the original plate material. Alternate grades can be used with guidance from the designer/builder.

Temporary or permanent patching by the use of overlapping patch plates attached with fillet welds is not recommended.

8.4.3.4 Repair of Welds

The repair of welds is necessary when the weld strength has deteriorated due to wear, corrosion or other causes, resulting in loss of weld dimension or due to weld cracks and other signs of weld strength loss.

The existing weld should be removed, preferably by grinding, and replaced with a weld defined by the BOF design.

Preheat temperatures and postweld heat treatment must be imposed based on the criticality of the weld, the thickness of material being welded and other factors of the repair.

8.4.3.5 Refractory Maintenance

The practice of gunning, spraying, slagging, and slag splashing the refractories in a BOF is very important to maintaining the structural integrity of the BOF components. Such maintenance helps to protect a lining from excessive wear caused by either mechanical damage and/or corrosive mechanisms. Maintenance also helps to effectively utilize downtime to extend campaign life.

The refractory at the mouth of the vessel needs to be maintained by spraying in order to maintain the proper mouth opening and to protect the top ring and lip ring from being subjected to molten metal contact during tapping and slagging operations. The taphole area must be repaired and/or sprayed regularly in order to maintain proper thickness in this area so as not to experience burn-throughs during tapping. Trunnions are a weak area in the vessel and spraying may be required in these areas. In the case of a bottom blown or bottom stirred vessel, the tuyere areas must be maintained to prevent damage to the removable bottom and tuyeres. Any location in the refractory lining that has deteriorated either by normal wear or premature damage should be sprayed and built up in order to alleviate vessel shell temperatures higher than designed for and possible burn-throughs. Spraying should be in thin coats of approximately two inches, or as specified by those qualified (such as the material manufacturer or other responsible expert). Heavy coats are likely to fall off and not adhere to the refractory lining.

Slagging maintenance is possible on the charge pad and the tap pad. This is carried out by allowing a small portion of the slag to remain in the bottom of the vessel after a heat. The vessel is then rotated from one pad to another to place thin coatings of this slag on the pads. A shallow pool of slag may also be allowed to remain on one or the other of the pads while the furnace lays on its side between heats. Slag that is used for maintenance may also be treated by adding dolomite to make it more refractory. For maximum effectiveness, slagging should be done frequently. Techniques to slag the trunnion areas, such as nitrogen blow slag splashing, are recommended when available.

8.4.4 Deskulling

8.4.4.1 General

External and internal deskulling is a very important operation of maintaining the BOF. Failure to remove skull buildup can adversely affect the center of gravity of the vessel, causing unsafe operating conditions.

8.4.4.2 Gradall-Mounted Pneumatic Deskullers

This is the most effective method used as it is possible to remove the external and internal skull buildup and, with proper operation, it does not impose extreme impact loads on the BOF components.

8.4.4.3 Floor-Mounted Deskullers

Various types of deskulling devices mounted into the floor structure on the charging side of the vessel are used to continuously scrape the skull buildup from the nose of the vessel. These are effective, but must be maintained so as not to cause damage to the BOF, should the scraping members come into direct contact with the vessel during rotation. This method can cause damage to the tilt drive gearing, trunnion bearings and suspension system from the impact loads transmitted during deskulling.

8.4.4.4 Crane-Suspended Deskullers

Various items such as the scrap charging box, purposely designed battering rams, etc. can be used to impact the nose of the vessel as they are suspended from the charging crane hook(s). Such methods are not recommended due to the significant damage that can occur to the tilt drive gearing, trunnion bearings and suspension system from the impact loading.

8.4.4.5 Post-combustion

This technique can be used to control skull buildup in the vessel, particularly in the upper cone area. Consideration must be given to the potential for excessive refractory wear and damage to the vessel lip structure and brick retaining top plates.

8.5 Oxygen Lance Technology

8.5.1 Introduction

In modern steelmaking production, a water-cooled lance is used as the refining tool by injecting a high velocity stream of oxygen onto a molten bath. The velocity or momentum of the oxygen jet results in the penetration of the slag and metal to promote oxidation reactions over a relatively small area. The jet velocity and penetration characteristics are functions of the nozzle design. This section will discuss the design and operation of water-cooled oxygen lances as they apply to modern steelmaking practices in the BOF.

8.5.2 Oxidation Reactions

The primary reason for blowing oxygen into steel is to remove carbon to endpoint specifications. The principle reaction which results from the oxygen lancing is the removal of carbon from the bath as CO. This is an exothermic reaction which adds heat to the system. A small amount of CO_2 is also produced, but 90% or more is usually CO. As will be discussed later, the burning of this CO inside the furnace by reacting with oxygen is called post-combustion.

Other elements such as Si, Mn, and P are also oxidized and are absorbed in the slag layer. These reactions are also exothermic, further contributing to the required heat to melt scrap and raise the steel bath to the necessary temperature. The oxidation of the silicon is particularly important because it occurs early in the oxygen blow and the resultant silica combines with the added lime to form the molten slag. Table 8.1 presents the oxidation reactions during the steelmaking process.

Table 8.1 Oxidation Reactions in Steelmaking

Reaction	Change in Free Energy at 1600°C (kcal/mole)
$C + \frac{1}{2}O_2 = CO$	−66
$2CO + O_2 = 2CO_2$	−57.4
$Si + O_2 = SiO_2$	−137.5
$Mn + \frac{1}{2}O_2 = MnO_2$	−58.5
$2P + \frac{5}{2}O_2 = P_2O_5$	−148.5

The oxidation reactions occur in the jet impact zone, called a dimple, which is created by the impingement of the oxygen. The depression in the liquid bath is a function of the oxygen jet momentum or the thrust, which is calculated by the following equation:

$$F = W \frac{V_e}{g}$$

(8.5.1)

where,

F = force
W = mass flow rate
V_e = exit velocity
g = the acceleration of gravity

The jet thrust and impact angle are optimized to achieve the desired chemical reactions and bath agitation through the design of the nozzle, Fig. 8.39.

8.5.3 Supersonic Jet Theory

Nozzles are designed for a certain oxygen flow rate, usually measured in scfm (Nm^3/min), resulting in a certain exit velocity (Mach number), with the required jet profile and force to penetrate the slag layer and react with the steel bath in the dimple area.

Supersonic jets are produced with convergent/divergent nozzles, Fig. 8.40. A reservoir of stagnant oxygen is maintained at pressure, P_o. The oxygen accelerates in the converging section up to sonic velocity, Mach = 1, in the cylindrical throat zone. The oxygen then expands in the diverging section. The expansion decreases the temperature, density, and pressure of the oxygen and the velocity increases to supersonic levels, Mach > 1.

As the oxygen jet exits into the furnace, at a pressure P_e, it spreads and decays. A supersonic core remains for a certain distance from the nozzle. Supersonic jets spread at an angle of approximately 12°.

Proper nozzle design and operation are necessary both to efficiently produce the desired steelmaking reactions and to maximize lance life. If a nozzle is overblown, which means that the oxygen jet is not fully expanded at the time it exits the nozzle, shock waves will develop as the jet

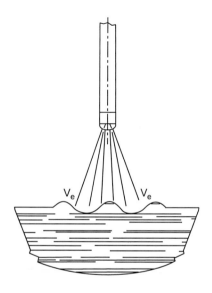

Fig. 8.39 Effect of nozzle design on impact angle and jet thrust.

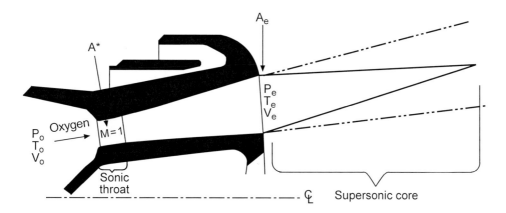

Fig. 8.40 Mechanics of supersonic jet formation.

expands outside of the nozzle. Useful energy is lost in these shock waves, and an overblown jet will impact the steel bath with less force than an ideally expanded jet.

Nozzles are underblown when the jet expands to a pressure equal to the surrounding pressure and then stops expanding before it exists the nozzle. In this case, the oxygen flow separates from the internal nozzle surface. Hot gases from the steel vessel then burn back or erode the nozzle exit area. This erosion not only decreases the lance life, but also results in a loss of jet force, leading to a soft blowing condition. Overblowing and underblowing conditions are demonstrated in Fig. 8.41.

Fig. 8.42 displays the major components of the BOF oxygen lance. These include oxygen inlet fittings, the oxygen outlet (lance tip), which is made of a high thermal conductivity cast copper design with precisely machined nozzles to achieve the desired flow rate and jet parameters. Cooling water is essential in these lances to keep them from burning up in the vessel. The lance

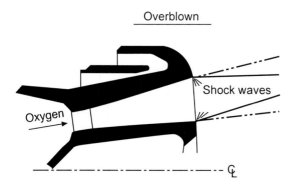

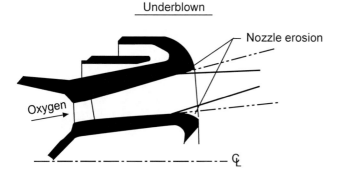

Fig. 8.41 Overblown and underblown conditions at the nozzle tip.

barrel is a series of concentric pipes, an outer pipe, an intermediate pipe and the central pipe for the oxygen. Lances must be designed to compensate for thermal expansion and contraction. The outer barrel/pipe of the lance is exposed to the high temperatures in the furnace. As its temperature increase it expands and the overall lance construction internally is constructed with O-ring seals and various joints, but can accommodate the thermal expansion and contraction while in service. The lance also has a stress-free design and it must be built with mill duty construction quality to be able to withstand the normal steel mill operating conditions. Fig. 8.43 presents various types of BOF lance tips.

Fig. 8.44 is a schematic cutaway of a lance tip displaying the oxygen nozzle and the water channels. This figure shows a typical 5-hole BOF lance tip. Its important components are the water cooling channels where the cooling water flows through the center of the tip and exits through the outer pipe of the lance. It is designed to get maximum velocity of cooling water in the tip area, which is exposed to the highest temperatures.

8.5.4 Factors Affecting BOF Lance Performance

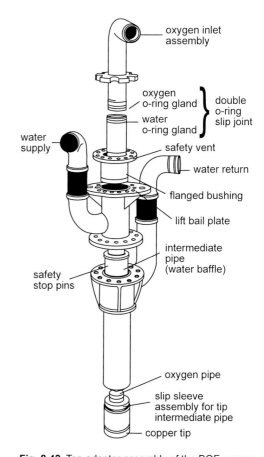

Fig. 8.42 Top adapter assembly of the BOF oxygen lance.

There are a number of factors affecting the performance and the efficiency of the lance. The performance of the lance depends on the conditions in the furnace. The hot metal silicon content is very important—this effects the amount of slag that forms, the amount of slag that has to be penetrated by the jet, and also controls the amount of steel slopping in the furnace. The lance operating height is also very important and must be included in the nozzle design calculations. If the lance is too

Fig. 8.43 Various types of BOF lance tips.

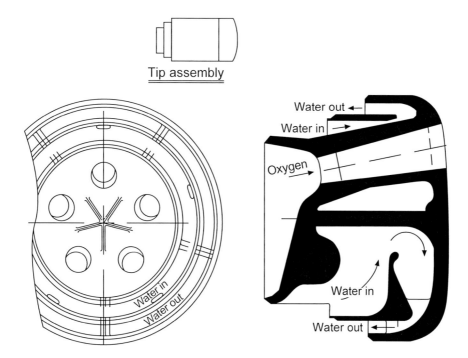

Fig. 8.44 Schematic of a typical 5-hole BOF lance tip.

low in the furnace it is exposed to extremely high temperatures and the heat transfer from the cooling water may not be sufficient to keep the face of the lance from melting or being burned away prematurely. If the lance is too high, the thrust of the jet becomes less efficient and the refining time will take longer, and more oxygen will be required to achieve the necessary decarburization and steel temperature. The oxygen flow rate is a design parameter that is sometimes limited by the oxygen supply system, and/or emissions concerns. The exit velocity of the Mach number is also a factor that is used in the design of the lances. Generally the higher the Mach number the more forceful the jet is.

Also, important considerations are the number of nozzles and the nozzle hole angle. When BOFs were originally developed, the first ones were operated with a single nozzle blowing directly down unto the bath. This caused a lot of slopping and molten material was ejected straight up the mouth of the vessel. Three hole nozzles slightly angled were developed to minimize slopping, resulting in a high process yield. Currently many BOFs are operating with 4, 5, or 6-nozzle configurations. Fig. 8.45 shows the effect of increasing the number of nozzles and nozzle exit angle on the impact area in the bath. As the nozzle angle is increased more of the lateral force component, rather than a vertical force component develops, contributing to more stirring and agitation in the bath. However, if the lateral component of the jet becomes excessive, higher refractory wear will occur.

8.5.5 Factors Affecting BOF Lance Life

Lance life varies from shop to shop, depending on the various operating practices. A typical lance life may be 200 heats, although there are some shops which can achieve up to 400 heats per lance and others may not be able to achieve 150 heats. Cooling water is critical to maintain high lance life. The flow rate must be maintained at the design rate. The cooling water outlet temperature should not exceed 140° to 150°F. Water quality is also important. If the water is contaminated with oxides or dirt, deposits will form inside the lance pipes and tip, resulting in a negative effect on heat transfer and ultimately reducing lance life. Operating height is critical for achieving the proper jet penetration of the bath. However if the lance is too low, the tip face may erode or melt.

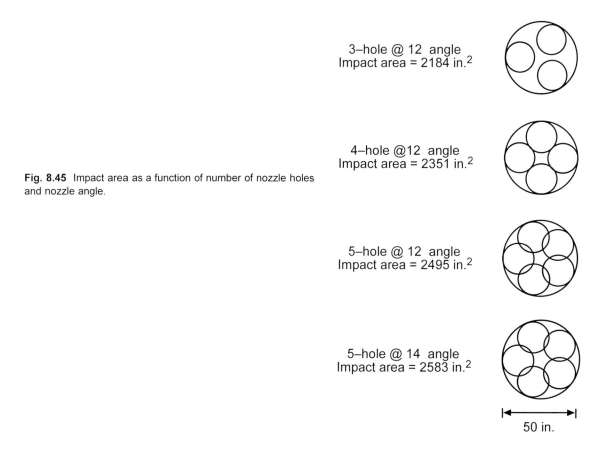

Fig. 8.45 Impact area as a function of number of nozzle holes and nozzle angle.

Underblowing causes nozzle exit erosion and lance tip failure. Excessive skull buildup on the lance tip must be mechanically removed or burned off—both of these practices could cause damage to the lance.

8.5.6 New Developments in BOF Lances

The first recent development was the post-combustion lance. Because 90% of the gas evolving from the oxidation reactions of the bath is carbon monoxide, it is desirable to further combust this carbon monoxide to form carbon dioxide. This reaction is highly exothermic, resulting in additional heat for the steelmaking process. This is a practice that is being done in several BOF shops in North America. This practice requires a dual-flow oxygen lance, which has two oxygen outlets. The main supply of oxygen is distributed through the lance tip similarly to a conventional lance, Fig. 8.46. The auxiliary oxygen is controlled separately and is blown at a higher elevation in the vessel. The function of the auxiliary oxygen is to react with the carbon monoxide coming off the bath, thus creating additional thermal energy which could be used to melt additional scrap, and help to control skull buildup in the mouth of the vessel.

The second recent development for oxygen lances is its use to splash a protective coating of slag containing high levels of MgO onto the walls of the BOF, Fig. 8.47. This is done after the steel has been tapped out of the furnace with some residual slag remaining. The oxygen supply is switched

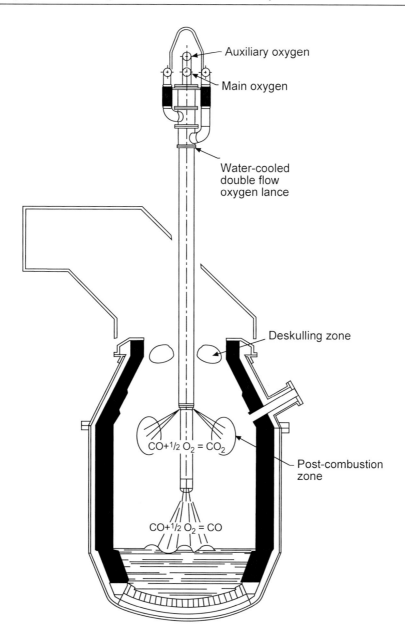

Fig. 8.46 Schematic of a post-combustion oxygen lance.

off and the nitrogen supply is switched on. The lance is lowered to 2–4 feet above the vessel bottom. The nitrogen is then turned on, splashing the molten slag onto the walls of the vessel and creating a protective slag coating over the refractories. This slag coating has successfully increased the typical refractory life from 3000 heats to over 20,000 heats per campaign. Furthermore the gunning requirements have also been decreased to less than one pound per ton of steel produced.

8.6 Sub-Lance Equipment

The purpose of a sub-lance is to measure carbon content, bath temperature, soluble oxygen, and bath level, and to secure a steel sample in an oxygen steelmaking vessel while in the upright position. Such sub-lance systems employ mechanical and electrical devices to perform measuring and

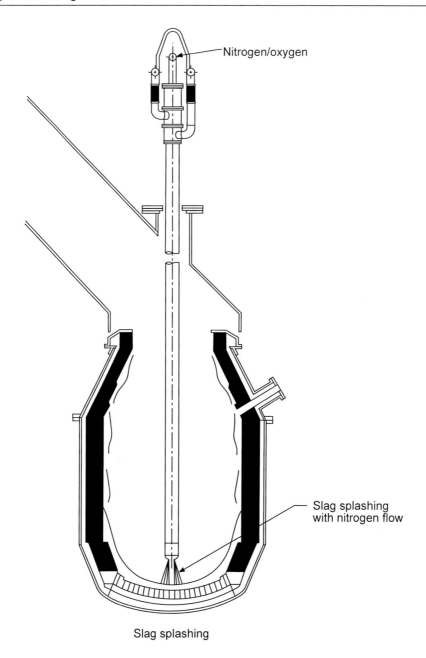

Fig. 8.47 Schematic of a slag splashing oxygen lance.

monitoring functions, and to provide a steel sample for spectrographic analysis. Fig. 8.48 provides a schematic of the process control outputs generated from sub-lance measurements.

During the heat cycle, after a predetermined amount of oxygen has been blown, the oxygen blow rate is reduced and an in-blow test is made by lowering the sub-lance at a controlled speed through the hood opening. The sub-lance is stopped at the test elevation 20 in. below the bath surface. This test elevation is fixed and is regulated from heat to heat by the process computer to compensate for lining wear throughout the campaign. All pertinent data gathered is transmitted to process computers for adjustments to the oxygen blow cycle and coolant additions to finish the heat at the tap carbon and temperature aims. After the blow cycle is completed, the sub-lance is lowered to the test elevation to measure the finished tap carbon and temperature and to retrieve a metal sample for

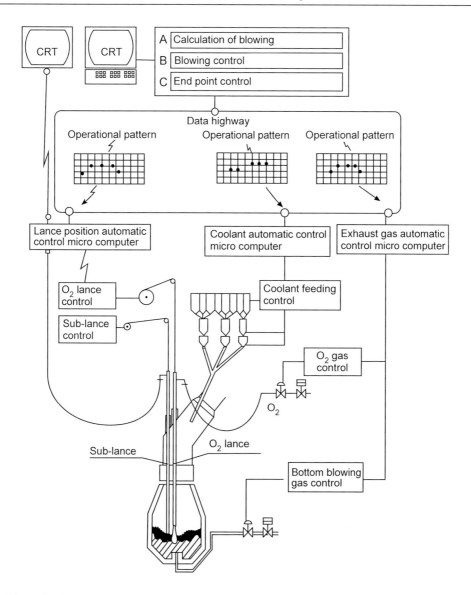

Fig. 8.48 Schematic of process control outputs generated from sub-lance measurements.

spectrographic analysis. Fig. 8.49 illustrates the heat cycle and process control models with in-blow and end-blow sub-lance measurements.

The sub-lance is raised and lowered within a guide frame by means of a hoist drive incorporating a drum, wire rope and a series of sheaves. An encoder mounted on the hoist drive indicates position of the sub-lance and provides data for a programmable limit switch to control sub-lance travel and positioning. A cylinder-actuated slidegate is located on the BOF hood to provide a porthole for sub-lance access to the vessel. The slidegate is continuously cooled by a nitrogen purge arrangement. Cooling water for the sub-lance is continuously supplied from the return system on the oxygen lance. Should the system pressure be insufficient, booster pumps may be utilized.

Consumables for measurement of carbon content, bath temperature, and soluble oxygen, and for securing a steel sample are pre-loaded into a cassette and are conveyed mechanically to the probe tip. Prior to connection of the probe to the sub-lance, and also during the measurement cycle, nitrogen purging is performed to provide good electrical contact between the probe and the probe holder. Following capture of the metal sample from the bath, a probe retrieval system is used to

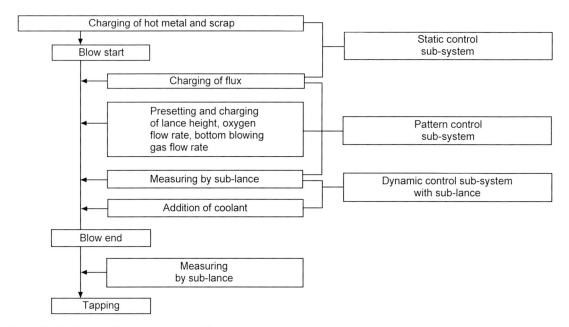

Fig. 8.49 Heat cycle with in-blow and end-blow sub-lance measurements.

detach the sample from the sub-lance and deposit it in a vertical transport tube for delivery to the charging floor. The metal sample may then be sent via pneumatic tube transport to the laboratory for spectrographic analysis.

References

1. "Design and Maintenance of Basic Oxygen Furnaces," *AISE Technical Report No. 32* (Pittsburgh, PA: Association of Iron and Steel Engineers, 1996).
2. E. S. Chen et al., "Thermomechanical Analysis of a 225 Ton BOF Vessel," *Iron and Steel Engineer* 70:11 (1993): 43.

Chapter 9

Oxygen Steelmaking Processes

T.W. Miller, Supervisor, Bethlehem Steel Corp.
J. Jimenez, Associate Research Consultant, U. S. Steel Corp.
A. Sharan, Engineer, Bethlehem Steel Corp.
D.A. Goldstein, Engineer, Bethlehem Steel Corp.

9.1 Introduction

9.1.1 Process Description and Events

The oxygen steelmaking process rapidly refines a charge of molten pig iron and ambient scrap into steel of a desired carbon and temperature using high purity oxygen. Steel is made in discrete batches called heats. The furnace or converter is a barrel shaped, open topped, refractory lined vessel that can rotate on a horizontal trunnion axis. The basic operational steps of the process (BOF) are shown schematically in Fig. 9.1.

The overall purpose of this process is to reduce the carbon from about 4% to less than 1% (usually less than 0.1%), to reduce or control the sulfur and phosphorus, and finally, to raise the temperature of the liquid steel made from scrap and liquid hot metal to approximately 1635°C (2975°F). A typical configuration is to produce a 250 ton (220 metric ton) heat about every 45 minutes, the range is approximately 30 to 65 minutes. The major event times for the process are summarized below in Table 9.1.

These event times, temperatures, and chemistries vary considerably by both chance and intent. The required quantities of hot metal, scrap, oxygen, and fluxes vary according to their chemical compositions and temperatures, and to the desired chemistry and temperature of the steel to be tapped. Fluxes are minerals added early in the oxygen blow, to

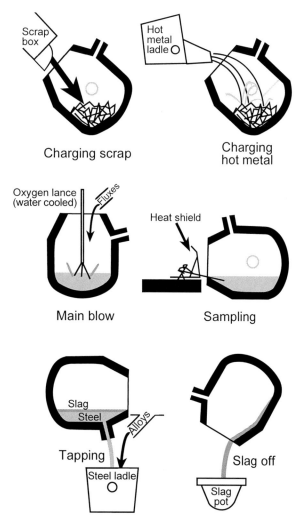

Fig. 9.1 Schematic of operational steps in oxygen steelmaking process (BOF).

control sulfur and phosphorous and to control erosion of the furnace refractory lining. Input process variations such as analytical (hot metal, scrap, flux and alloy) and measurement (weighing and temperature) errors contribute to the chemical, thermal and time variations of the process.

Table 9.1 Basic Oxygen Steelmaking Event Times

Event	Min.	Comments
Charging scrap and hot metal	5–10	Scrap at ambient temperature, hot metal at 1340°C (2450°F)
Refining–*blowing* oxygen	14–23	Oxygen reacts with elements, Si, C, Fe, Mn, P in scrap and hot metal and flux additions to form a slag
Sampling–chemical testing	4–15	Steel at 1650°C (3000°F), chemistry and temperature
Tapping	4–8	Steel is poured from furnace into a ladle, typical size = 250 tons
Pouring *slag off* at furnace	3–9	Most slag is removed from furnace, in some shops slag is used to coat furnace walls

The energy required to raise the fluxes, scrap and hot metal to steelmaking temperatures is provided by oxidation of various elements in the charge materials. The principal elements are iron, silicon, carbon, manganese and phosphorous. The liquid pig iron or hot metal provides almost all of the silicon, carbon, manganese and phosphorous, with lesser amounts coming from the scrap. Both the high temperatures of the liquid pig iron and the intense stirring provided when the oxygen jet is introduced, contribute to the fast oxidation (burning or combustion) of these elements and a resultant rapid, large energy release. Silicon, manganese, iron and phosphorous form oxides which in combination with the fluxes, create a liquid slag. The vigorous stirring fosters a speedy reaction and enables the transfer of energy to the slag and steel bath. Carbon, when oxidized, leaves the process in gaseous form, principally as carbon monoxide. During the blow, the slag, reaction gases and steel (as tiny droplets) make up a foamy emulsion. The large surface area of the steel droplets, in contact with the slag, at high temperatures and vigorous stirring, allow quick reactions and rapid mass transfer of elements from metal and gas phases to the slag. When the blow is finished the slag floats on top of the steel bath.

Controlling sulfur is an important goal of the steelmaking process. This is accomplished by first removing most of it from the liquid hot metal before charging and later, inside the furnace, by controlling the chemical composition of the slag with flux additions.

9.1.2 Types of Oxygen Steelmaking Processes

There are basically three variations of introducing oxygen gas into the liquid bath. These are shown schematically in Fig. 9.2. Each of these configurations has certain pros and cons. The most common configuration is the **top-blown** converter (BOF), where all of the oxygen is introduced via a water-cooled lance. The blowing end of this lance features three to five special nozzles that deliver the gas jets at supersonic velocities. In top blowing, the stirring created by these focused, supersonic jets cause the necessary slag emulsion to form and keeps vigorous bath flows to sustain the rapid reactions. The lance is suspended above the furnace and lowered into it. Oxygen is turned on as the lance moves into the furnace. Slag forming fluxes are added from above the furnace via a chute in the waste gas hood. A process description is in Section 9.4 of this chapter.

In the **bottom-blown** converters (OBM or Q-BOP), oxygen is introduced via several tuyeres installed in the bottom of the vessel, Fig. 9.2. Each tuyere consists of two concentric pipes with the oxygen passing through the center pipe and a coolant hydrocarbon passing through the annulus between the pipes. The coolant is usually methane (natural gas) or propane although some shops have used fuel oil. The coolant chemically decomposes when introduced at high temperatures and absorbs heat in the vicinity, thus protecting the tuyere from overheating. In bottom blowing, all of the oxygen is introduced through the bottom, and passes through the bath and slag thus creating vigorous bath stirring and formation of a slag emulsion. Powdered fluxes are introduced into the bath through the tuyeres located in the bottom of the furnace. The first part of Section 9.5 is a process description of the OBM (Q-BOP).

The **combination blowing** or **top and bottom blowing**, or **mixed blowing** process (Fig. 9.2 shows these variants) is characterized by both a top blowing lance and a method of achieving stirring from the bottom. The configurational differences in mixed blowing lie principally in the bottom tuyeres or elements. These range from fully cooled tuyeres, to uncooled tuyeres, to permeable elements. Section 9.5 summarizes further details about combination blowing processes.

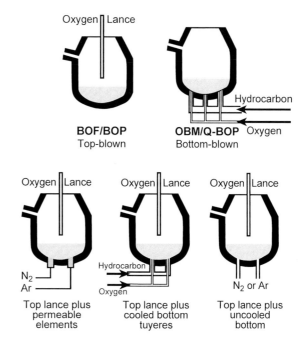

Fig. 9.2 Methods of introducing oxygen and other gases into the steelmaking converter.

9.1.3 Environmental Issues

The oxygen steelmaking process is characterized by several pollution sources and most require emission control equipment. These sources are: hot metal transfer, hot metal desulfurization and skimming of slag, charging of hot metal, melting and refining (blowing), BOF tapping, handling of dumped BOF slag, handling of fluxes and alloys, and maintenance (burning of skulls, ladle dumping. etc). Thus, compliance to emission standards is an important design and operating cost factor for the operation.

9.1.4 How to Use This Chapter

The process falls into several basic component parts which determine important control and economic parameters. Accordingly, this chapter is organized into the process component headings of Sequence of Operations, Raw Materials, Process Reactions and Energy Balance, Variations of the Process, Process Control Strategies, and Environmental Issues.

The use of this chapter depends on your level of knowledge and interest in the details. It is geared to bring together several more detailed chapters, such as furnace design, refractory practices, oxygen lance design and physiochemical principles, and to describe how these underlying factors contribute to the oxygen steelmaking process. If you are a steelmaking novice, reading this chapter first is a good way to get a brief, yet coherent description of the process. Once you have the big picture, then it is easier to focus on the chapters on detailed design and first principles. If you are more experienced and want to review or deal with a specific issue, you may turn directly to one of the component sections or to one of the underlying principle chapters.

A note on References: The material for this chapter is an assimilation of many different sources. For continuity, rather than specifically cite every source, the references are listed at the end of the chapter both as a source acknowledgment and as a supplemental reading list. This list is organized in the same chapter section format. The references are numbered for citing figures and tables.

9.2 Sequence of Operations—Top Blown

9.2.1 Plant Layout

To understand the sequence of the oxygen steelmaking process, one must examine the design, layout and materials flow of the facilities. Figs. 9.3, 9.4 and 9.5 show a 275-ton BOF that illustrates the process. Shops vary considerably in basic layout. Reasons for these layout differences are: type of product (ingots, cast product or both), the parent company's operating and engineering culture, the relationship of the infrastructure and material flows to the rest of the plant, and age of the facility. (Is the plant an updated older facility or a new greenfield site?) Flow of materials plays a key role in the design of the shop. Handling of raw materials (scrap, hot metal, fluxes, alloys, refractories), oxygen lances in-and-out, slag handling, gas cleaning, and transport of steel product must be accomplished smoothly with minimum delays and interference.

Fig. 9.3 is a plan view of a two-furnace shop and Fig. 9.4 is an elevation of the same plant but looking to the west. Figure 9.5 is an elevation looking to the north. BOFs, OBMs (Q-BOPs) and other variants can have similar layouts except for oxygen conveying and flux handling details. All shops feature transportation systems for hot metal and scrap.

9.2.2 Sequence of Operations

9.2.2.1 Scrap Handling

Scrap for a heat is ordered and prepared well in advance of actually charging the furnace. It is selected according to size and quality and then is brought into the plant via railroad cars, usually gondolas. It is loaded and mixed into an open ended scrap box which sits on a transfer car. Loading the box is usually done by magnet or grapple crane in a remote area from the shop. The box/car is frequently weighed during loading. Some shops use a crane scale to weigh and accumulate each magnet load. Weights are entered into the shop computer when the loading is completed. The transfer of scrap from rail cars to charging box is done in an attached bay to the BOF shop which is large enough to handle eight to 24 hours of scrap supply. The scrap box is then conveyed by rail to the charging aisle. A few shops use rubber tired platform carriers rather than rail cars to move the scrap box into the shop.

9.2.2.2 Hot Metal Pouring

The hot metal system consists of a track(s) and one to three pouring stations. The liquid pig iron arrives from the blast furnace in a train of torpedo shaped, refractory lined railroad cars called submarines (subs) or torpedoes. Each car is positioned over a track scale and weighed prior to pouring. There is a trunnion at each end of the car which allows the operator to rotate the open top toward a transfer ladle located in an adjacent pit. Generally, it takes one or two subs to fill the hot metal transfer ladle. The control room or operator's pulpit is equipped with controls for rotating the sub, operating the ladle transfer car, reading scales, taking temperatures, desulfurization equipment, and sending samples to a chemistry lab. The pouring operation, which generates considerable dust emissions, is accomplished under an enclosed hood equipped with an evacuation system and a baghouse. The dust generated at pouring, called kish, is mainly fine flaked graphite which precipitates from the carbon saturated metal as its temperature drops during pouring. The poured weights and measured temperatures are entered into the shop process control computer.

9.2.2.3 Hot Metal Treatment

The hot metal transfer ladle sits on a transfer car at the outside wall of the charging aisle, usually out of the crane's reach. Here many shops treat the hot metal by injecting a mixture of lime and magnesium to remove sulfur. This process is called hot metal desulfurization. During hot metal treatment, sulfur is removed from approximately 0.025 wt% to as low as 0.002 wt% and the time of injection will range from five to twenty minutes. A gas collecting and filtration system collects the fumes from

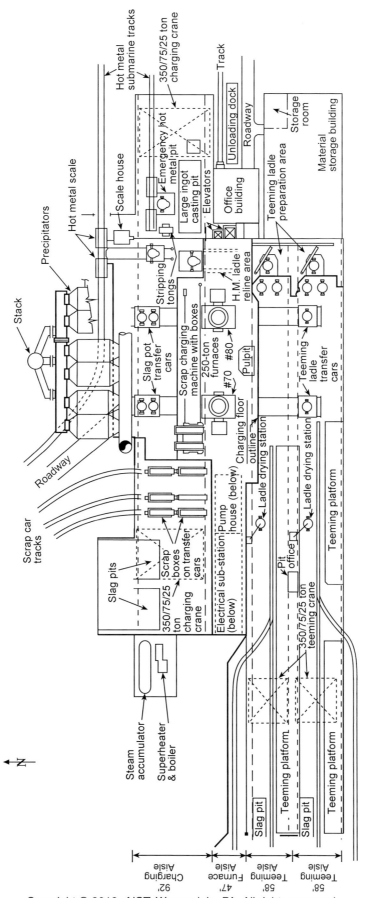

Fig. 9.3 Plan view of 275 ton BOF shop.

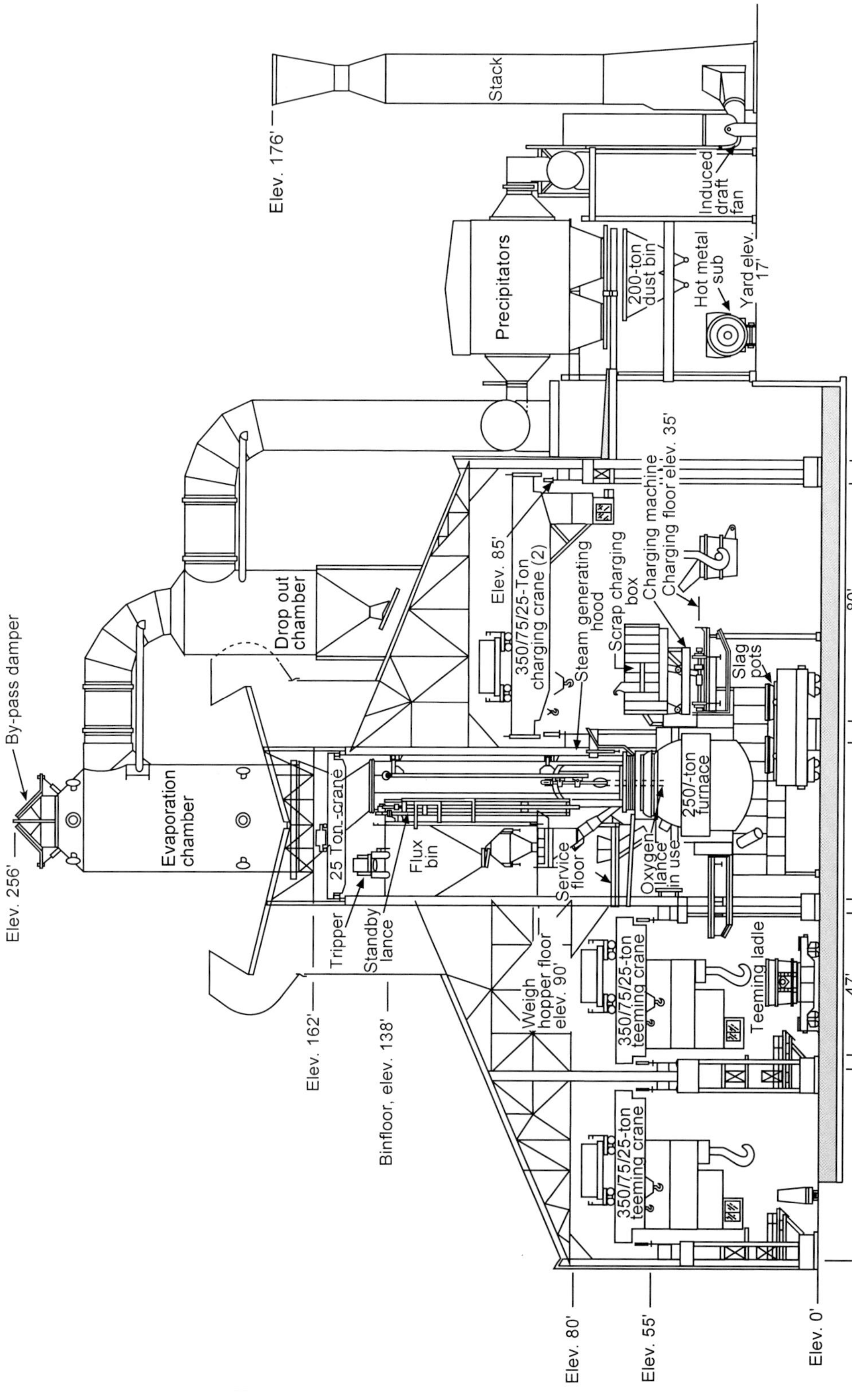

Fig. 9.4 Elevation of 275 ton BOF shop—looking west.

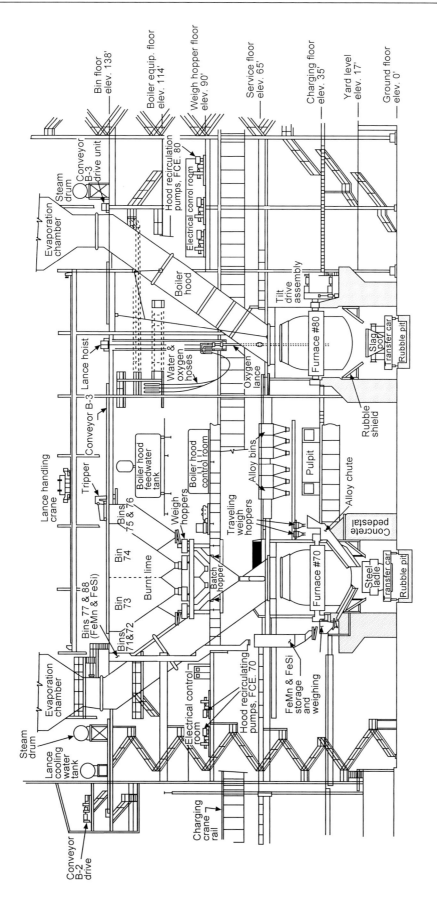

Fig. 9.5 Elevation of 275 ton BOF shop—looking north.

the desulfurization process as well as collecting the pouring fumes. After desulfurization, the ladle is tilted by the crane or in a special cradle just to the point of pouring. In this position, the sulfur containing slag floating on the iron is scraped off into a collection pot using a hydraulic manipulating arm. This slag removal process is called skimming. Often, much metal is scraped out of the ladle along with the sulfur containing slag. Thus there is an iron yield loss due to deslagging that ranges from 0.5 to 1.5% depending on equipment design and operator skill. Hot metal temperature is measured using disposable thermocouples mounted on a mechanical arm called a pantograph. When the hot metal pouring and treatment are finished, the ladle car moves into the charging aisle and the ladle becomes available for pickup and charging by the charging crane. Details of hot metal treatment are found in Chapter 7.

9.2.2.4 Charging the Furnace

The BOF furnaces are open-topped, refractory-lined vessels located in the adjacent aisle called the furnace aisle. The furnaces rotate on trunnions so they can be tilted both toward the charging and tapping aisles. The furnace refines the steel in an upright position although it is capable of rotating 360°. The furnace aisle contains the furnaces, flux conveying system, furnace alloy chute and oxygen lances in a top-blown shop. Here, the furnace aisle is very high to accommodate loading and operation of 60–70 foot long oxygen lances. There is a space and capital cost advantage of an OBM (Q-BOP) that does not require the high elevation for lances.

In nearly all North American shops, scrap is charged first. Many shops lift and tilt the box emptying the scrap into the furnace with the charging crane. Charging scrap before hot metal is considered a safer practice that avoids splashing. The crane method usually has faster scrap charging times. However, many shops load the scrap boxes onto special charging machines that can move on rails in front of the furnace. This scrap charging machine has a hydraulic tilting mechanism that raises the scrap box to 45° and charges them one at a time. It usually holds two scrap boxes. While the scrap charging machine is often slower than the crane, it frees up the charging crane for other duties and more quickly handles two box charges per heat when required.

After scrap is charged, liquid hot metal is charged into the furnace using the charging crane in the charging aisle. The ladle is tilted and the liquid hot metal is poured into the furnace. This process takes one to five minutes depending on the design of the furnace hood and shop fugitive emission systems. Some shops can charge quickly because the fume from pouring into the furnace is effectively collected by the hood and a closed roof monitor collection system. Other shops with less advanced fume collection systems, must pour rather slowly to minimize the heat and fume cloud, thus taking a lengthy three to five minutes to charge hot metal.

9.2.2.5 Computer Calculations

Prior to pouring hot metal and charging scrap, a computer calculation is initiated by the pulpit operator to determine the charge recipe. The grade and temperature and chemistry aims are loaded into the shop computer beforehand when a heat is put on the schedule line-up. The temperature and chemical content of the hot metal can vary significantly. The hot metal is sampled and analyzed at the chemistry lab—a process that takes from three to ten minutes—and the results and ID are transmitted to the computer. The temperature of the hot metal is measured in the ladle after it is poured and that result is transmitted to the computer. The chemistry of the scrap is calculated from the known mixture of the scrap; its temperature is assumed to be ambient. Thus, all of the characteristics of the charged materials and the heat aims are available for the charge recipe calculation. The principal aim parameters for the furnace heat are carbon and temperature. Other specific aims are sulfur, phosphorous and slag composition (%FeO level). Missing any of these aims can be very costly later and can require time consuming corrective actions or create significant quality problems. Often, there are several update calculations as items are weighed and charged to correct for minor weighing irregularities and mistakes. The results of these calculations are principally the amounts of coolants, fluxes and oxygen. Most shops' calculations have features that allow early calculation of hot metal and scrap weights as well. All shops use the charge recipe calculation.

Some shops have sensors (sub-lance, light meter, bomb-drop thermocouples, etc.) that measure carbon and/or temperature near the end of the blow. The sensor's measurements are used to make late oxygen and coolant corrections to bring the carbon and temperature to aims. This saves time and money by reducing the frequency of corrective actions.

9.2.2.6 Oxygen Blow

After scrap and hot metal are charged, the furnace is set upright and the oxygen is supplied through a water-cooled lance. There are two lance lift carriages above each furnace but only one lance is used at a time; the other is a spare. The oxygen blow times typically range from 13 to 25 minutes from one shop to another with an average of about 20 minutes. The oxygen is added in several batches. Each batch is characterized by a different lance height above the static steel bath and sometimes by an oxygen rate change. These blowing rates and lance heights vary considerably from shop to shop and depend on the pressure and quality of the oxygen supply. The oxygen blow rate ranges from 560 to 1000 Nm^3 per minute (20,000 to 35,000 scfm). A practical limit on the rate is often the volume of the furnace and the capacity of the gas collection and cleaning system to handle the gaseous reaction product and fume. A typical example of the oxygen batches is summarized in Table 9.2.

Table 9.2 Example of Oxygen Batches in a BOF

Batch No.	Lance Ht, in.	Oxygen Volume at Lance Change, Nm^3
1	150	850
2	120	1700
3 (main)	90	balance (to approx 14,200)

The first batch lance height is very high to avoid the possibility of lance tip contact with the scrap and to safely establish the oxidizing, heat generating reactions. If the lance would contact the pile of scrap in the furnace, a serious water leak could result causing a dangerous steam explosion.

The second batch lance height is usually approximately 20 to 30 inches lower than the first batch and approximately 20 to 30 inches higher than the main batch. The purpose here is to increase the reaction rate and control the early slag formation. This second or middle batch generates some early iron oxide to increase proper slag formation.

The main batch is where most of the action occurs—it is by far the longest batch. The lance height is an empirical compromise between achieving faster carbon removal rates and proper slag making. Some shops have more than three batches. Some change oxygen conditions (blow rate and lance height) nearly continuously. Other shops will raise the lance and change the blow rate near the end of the main batch to control the viscosity and chemical reactivity of the slag by raising its FeO content.

The position of the lance is very important for proper functioning of the process. If the lance is too high, the slag will be over stirred and over-oxidized with higher FeO percentages. This will cause higher than normal yield losses and lower tap alloy efficiencies due to oxidation losses. Further, the rate of carbon removal is reduced and becomes erratic. Slag volume increases and there is an increased chance of slopping, which is an uncontrolled slag drooling or spilling over the top of the furnace. When the lance is too low, carbon removal increases somewhat, slag formation, slag reactivity, and FeO are reduced and sulfur and phosphorus removal problems often occur. If the lance is very low, then spitting of metal droplets or sparking occurs which cause severe and dangerous metallic deposits, called skulls, on the lance and the lower waste gas hood.

Obviously, there is a correct lance height. It varies from shop to shop and depends on furnace configuration, lance configuration and oxygen supply pressure or flow rate. Each shop must find its own best lance height and comply with it. The problem is how to measure it. It can change quickly and significantly as a result of changes in furnace refractory shapes. Traditional measurement methods have been burn-off tests, where a pipe is wedged in an oxygen port. The lance is then lowered and the pipe is allowed to melt off at the slag-metal interface. This is done while waiting for a chemistry just before tap. This test has been erratic due to the temperature and fluidity of the slag, time of immersion, and initial protection of the pipe. This test has fallen out of favor because it is dangerous for the operator to attach the pipe to the lance.

More modern techniques include mathematically integrating the furnace volume from a refractory lining laser scan or determining the distance to the bath/slag using a radar unit mounted above the furnace. Generally, the radar method measures the height of the slag surface after it has collapsed at the end of a low carbon blow. There is uncertainty about the location of the slag steel interface but the measurement is considered better than none.

9.2.2.7 Flux Additions

Soon after the oxygen is turned on, flux additions are started and are usually completed at the end of the second batch of oxygen. The fluxes control the chemistry and sulfur and phosphorus capacity of the slag. The principle active ingredients from the fluxes are CaO (from burnt lime) and MgO (from dolomitic lime). The CaO component is used principally to control sulfur and phosphorous.

The dolomitic lime is used to saturate the slag with MgO. The principle ingredient of the furnace refractories is MgO. Steelmaking slags without it are very corrosive to the lining. The corrosion rate is reduced dramatically when MgO is added to saturate the slag. It is much cheaper to satisfy the slag's appetite for MgO from dolomitic lime than by dissolving it from the lining.

Another flux addition sometimes used in high carbon heats is fluorspar (CaF_2, or spar). This mineral is charged to dissolve the lime and to reduce the viscosity of the slag. It is used for making high carbon heats, (>0.30%C at the end of blow) because the iron oxide concentrations are low on these heats. Iron oxides help dissolve lime in lower carbon heats but these oxides are present in low concentrations in high carbon heats. To compensate for less FeO, many shops use spar to dissolve the lime. However, spar is used very sparingly because it is very corrosive on refractory linings. Unfortunately there is no corrosion inhibiting practice or ingredient to stop the corrosive effects of spar. In addition, spar forms hydroflouric acids in the gas cleaning system that seriously corrode any metal surfaces in the hood and cleaning systems. Finally, significant fluoride emissions are serious pollution and health hazards.

Coolants are other additions often made at about the same time as fluxes. There are several types of coolants. Iron ore, either lump or pellets, are the most common type. Varieties of limestone (calcium and/or magnesium carbonates) are often used but the cooling effects are less dependable than ores. Some shops use pre-reduced pellets which contain about 93% iron and thus behave similar to scrap. The coolant amounts are calculated by the computer. Ore (iron oxide) should be added as soon as possible to achieve early lime dissolution and to reduce the possibility of vigorous reactions and slopping at mid blow.

9.2.2.8 Final Oxygen Adjustments and Dynamic Sensors

The third or main batch is usually blown at 80 in. to 95 in. lance height above the bath depending on furnace design, practice and available oxygen pressure.

In many shops, the oxygen lance height is changed near the end of the blow to control the iron oxide (FeO) in the slag. Some shops activate their dynamic control systems to measure carbon and temperatures at this point. Oxygen is turned off based on either the static charge calculation or based on a modified result calculated from the dynamic sensor(s).

9.2.2.9 Turndown and Testing

After the blow is finished, the furnace is then rotated towards the charging side. Often the slag is very foamy and fills up the upper volume of the furnace. This foam will often take several minutes to collapse and settle down on its own. Thus, the operators will often toss pieces of wood or cardboard or scrapped rubber tires onto the bath to increase the collapse of the foam.

There is a heat shield on the charging floor on a track that is positioned in front of the furnace during the blow. The mouth of the furnace is rotated toward the charging side nearly 90° so the operator can look inside the furnace and sample the heat for chemical analysis and temperature measurement. Here, he also assesses the furnace condition to determine when and if any special maintenance is required.

9.2.2.10 Corrective Actions

Based on the chemical laboratory results, the melter decides if the heat is ready for tap or requires corrective action—a reblow and/or coolant. If a corrective action is required, the furnace is set upright. A reblow of additional oxygen may be required, with or without coolants or fluxes, to arrive at the desired (aim) chemistry and temperature. Usually, after a corrective action, another furnace turndown is required, adding five to eight minutes to the heat time. When the heat is ready, the furnace is rotated upward and over toward the tap side.

9.2.2.11 Quick Tap Procedures

Japan and some European shops reduce sampling and testing times to one to three minutes by using a quick tap procedure. Most of these shops use sub-lances to measure temperature and carbon by the liquidus thermal arrest technique. This testing is done without moving the furnace from the upright position. Success of quick-tap depends on consistently meeting the sulfur and phosphorus specification. This procedure can save three to six minutes of lab analysis time. These shops simply proceed immediately to tapping the furnace. Consequently, such shops often turn out 60 heats per day from two active furnaces.

Some North American shops have adopted a simplified variation of the quick-tap practice. A few have sub-lances. Others use the bomb drop-in thermocouples with or without oxygen sensors. Here a heavy cast iron bomb assembly with a specially wound and protected lead wire is dropped into the furnace. The wire lasts long enough to get a reading. The readings are more accurate if the oxygen is stopped but some shops get a usable reading during the blow. Again, tramp elements, S, P and other residuals are assumed to be acceptable and tapping proceeds immediately. Some two furnace shops use this technique to minimize production losses when one of the furnaces is being relined or repaired.

9.2.2.12 Tapping

For tapping, the furnace is rotated to the tap side and the steel flows through a taphole into a ladle sitting on a car below. The slag floats on top of the steel bath inside the furnace. Near the end of tapping (four to ten minutes) a vortex may develop near the draining taphole and entrain some of the slag into the ladle. There are various devices used to minimize or detect the onset of slag. Heavy uncontrolled slag entrainment into the ladle has a significant adverse effect on production costs and steel quality.

During tapping, alloys are added to adjust the composition to the final levels or to concentrations suitable for further ladle treatment processes. Typically, 2000 to 6000 lbs of alloys are added at tap. After tap, the ladle may be transported for further processing to a ladle arc furnace and/or a degasser. Some shops and some grades permit transport to ingot teeming or to the caster without any further treatments.

An increasing number of grades require limiting the amount of slag carryover to the ladle and close control of slag viscosity and chemical composition. Various devices have been developed to minimize slag draining from the furnace. There are two main techniques. One method consists of slowing down the pouring stream at the end of tap with a refractory plug. Usually a ball-shaped device, called a ball, or a cone shaped device, called a dart, is dropped into the taphole using a carefully positioned boom near the end of tap. These devices have a controlled density, between steel and slag, causing them to float at the slag-steel interface. Thus, it plugs the tap at about the time steel is drained out. These units can be very erratic depending on the geometry of the furnace, device shape and slag characteristics. But many shops have successfully minimized slag carryover into the ladle. Another approach is to detect slag carryover with a sensor coil installed around the taphole refractory. With suitable instrumentation, this system gives the operator an accurate and early warning of slag draining through the taphole at which time tapping is stopped by raising the furnace. The net result of these slag control/detection practices is to reduce furnace slag in the steel ladle, thereby improving chemical consistency and reducing the extent of post-tapping treatments and additions.

The condition and maintenance of the taphole and the furnace wall around it can influence alloy recovery consistency and metallic yield. Poor taphole maintenance and practice can lead to a burn-through in either the furnace shell or the taphole support frame. A very small taphole can significantly increase the tap time, reducing productivity, steel temperature, and nitrogen pickup in the ladle. A very large taphole will not allow enough time to add and mix the alloy additions in the ladle. Further, aged tapholes have ragged streams with higher surface areas that will entrain air, which in turn dissolves more oxygen and makes control of oxygen levels in the steel difficult.

A newly installed taphole yields a tap time of seven or eight minutes. Tapholes are generally replaced when the tap time falls below four minutes. A very important aspect of the melter/operator's job is to carefully monitor the condition and performance of the tap hole.

Steel is often lost to the slag pot, a yield loss, when a pocket or depression develops near or around the tap opening. Such a depression can prevent several tons of steel from being drained into the ladle. Again, the operator must carefully monitor yields and furnace condition and make repairs to prevent this problem.

9.2.2.13 Slagging Off and Furnace Maintenance

After tapping, the furnace is rotated back upright to prepare for furnace maintenance. The remaining slag is either immediately dumped into a slag pot toward the charging side or it is splashed on the walls of the furnace to coat the lining and thereby extend its life. This slag splashing (coating) maintenance is done by blowing nitrogen through the oxygen lance for two to three minutes, see Section 4.2. Often, the furnace is simply rocked back-and-forth to coat or build up the bottom, charge and tap pad areas. Dolomitic stone or dolomitic lime additions are made to stiffen the slag for splashing or to freeze the slag to the bottom. Frequently, repairs by spraying a cement-like refractory slurry are made as required prior to the next charge.

The furnace is then ready for the next heat.

9.2.3 Shop Manning

9.2.3.1 Introduction

Another way of looking at the steelmaking process is to look at how the operators and workers run the process. The shop crew is very much a coordinated team. Virtually all oxygen steelmaking shops operate around the clock—21 shifts per week. Some are able to schedule a down or repair turn once per week or every other week in conjunction with a similar scheduled blast furnace outage. Similarly, the down stream caster facility will also schedule a repair outage.

When manning is summarized, the manning levels are expressed on a per turn basis. To get the totals for 21 turns per week (three turns per day times seven days per week) multiply the turn manning by four and add the management, staff and maintenance forces (who work only five day turns).

The example here will cover a two furnace shop producing about 40 to 45 heats per day. Three furnace shop issues and exceptions will be noted in the discussion. The actual figures in a given shop may vary from the example presented here as different shops have different numbers of furnaces (two or three) and different demands (steel grades and production levels).

9.2.3.2 Hot Metal

Transporting the hot metal subs to the BOF requires one man if the engine has radio control. The pouring station has an operator who weighs and pours and an assistant who plugs in electric power to tilt the subs and moves the train locally and changes the subs at the pouring station. If there is desulfurizing, there will be one or two other operators to desulfurize, skim and maintain the equipment. A three furnace shop will double this crew to operate two or three pouring and treatment stations.

9.2.3.3 Charging Crane

A charging crane operator is required, often with a relief operator. Busy shops may operate two charging cranes with a relief operator serving the charge and scrap cranes. Often one relief crane operator serves both the charging and teeming or the casting crane.

9.2.3.4 Scrap

Scrap is often premixed by a contractor or a plant crew at a separate facility nearby. Gondolas bearing scrap are transported into the BOF by the plant railroad. There will be one or two scrap crane operators (depending on how many heats the shop makes) loading and weighing the boxes. The charging crane then picks up the scrap boxes and charges them into the furnace or puts them on a charging machine.

9.2.3.5 Furnace and Charging Floor

The overall responsibility for production of the shop during the turn belongs to the turn supervisor. There is a furnace crew consisting of a melter (who is the crew leader and is responsible for making the heat), pulpit operator who operates the computer, lance and flux systems, and two or three furnace operators. In small, lean shops, the melter is also the turn supervisor. The furnace operators actually position the furnace for charging, (operate a scrap charging machine if there is one) and test the heat by sampling the heats for chemistry and temperature. They also weigh up alloys, tap the heat and maintain the furnace refractory—all under the melter's supervision. In larger, high productivity shops, there may be one of these crews and a melter for each active furnace. In this case, the turn supervisor coordinates all shop activities.

9.2.3.6 Flux Handling

There is a track-hopper operator who controls equipment that unloads the fluxes and alloys from rail cars and trucks. Often there is an assistant at the unloading station at ground level, with the operator on the top floor where the conveyer dumper is located. Generally, this operation occurs on day turn in smaller shops or day and afternoon turns in larger shops.

9.2.3.7 Maintenance

There is an on-turn maintenance crew which responds to delay and breakdown calls and performs certain preventative maintenance procedures that are safe during furnace operation. Usually, the day time crew is larger to handle calls and do preventative maintenance. The maintenance crew also load and unload oxygen lances and maintain the oxygen system. Turn maintenance people consist of mechanical millwrights and electricians (three to ten millwrights and electricians). The day turn crew also includes instrument calibration, instrument repair and electronic skillspersons.

9.2.3.8 General Labor

Often, there is a small day turn labor crew (one to three persons) which operates a forklift trucks to transport various alloy additions for tapping and assists getting spray materials ready for furnace maintenance. The operators also run a mechanical sweeper and hand sweep around the shop to maintain good housekeeping.

9.2.3.9 Chemistry Lab

In older shops there is a vacuum conveyer systems where samples are transported to the chemistry lab. Laboratory manning will vary widely depending on how centralized the lab is to the whole plant. An increasing number of shops are installing robot labs for faster and more convenient service of chemical analysis. However, the robot lab requires an operator whose responsibility is supervising calibration, changing grinding media, and lubrication and maintenance of mechanical and optical equipment. The advantages of a well maintained robot lab are speed and consistency in sample preparation and analysis rather than labor savings.

9.2.3.10 Refractory Maintenance

Often furnace maintenance is done by the floor furnace crew. However, larger and busier shops will have an outside contractor or a separate crew to spray repair the furnaces. This crew may operate only during day turns and consists of two or three persons.

9.2.3.11 Relines and Major Repairs

Maintenance may schedule a crew to make hood and furnace shell repairs while relining is going on. Often, special code welding skills (usually hood or thick shell repairs) are required and are often done by an outside contractor. With the growth of slag splashing, which causes infrequent relines, hood repair work must be scheduled several times between relines.

9.2.3.12 Ladle Liners

Hot metal and steel ladles are a very important component of steelmaking. Hot metal ladle life is typically 1500 pours and a typical small shop will carry a fleet of four or five ladles of which two or three are active. Steel ladles last from 60 to 150 heats. Steel ladle fleets may vary from eight to 25 depending on the size of the shop. Relining of ladles is usually supervised by the Teeming Supervisor. Ladle lining crew sizes vary widely.

9.2.3.13 Management and Clerical

Finally, management and clerical staffs vary widely from shop to shop. The General Supervisor (sometimes called Area Manager) supervises the day-to-day operation and plans the course of the facility. Often there are separate area supervisors (five day turns per week only) covering hot metal pouring and treatment, furnace floor operations, scrap and flux loading, and maintenance. Usually there is a small clerical staff (one to five), a practice engineer, and a small, assigned crew of metallurgical or quality assurance (QA) people.

In summary, even in a small shop, the steelmaking team brings together many different skills working in concert. In addition to the production crew, there are numerous support groups, principally maintenance, that keep the shop running smoothly. Table 9.3 summarizes the manning of a small BOF shop.

Table 9.3 Example of Manning a Small Two Furnace BOF Shop

Position	Area	Turns	Number
General Supervisor	Supv	Day	1
Hot Metal Supervisor	Supv	Day	1
Furnace Supervisor	Supv	Day	1
Scrap & Flux Supervisor	Supv	Day	1
Maintenance Supervisor	Supv	Day	1
Melter	Supv	All	4
Maintenance Supervisor	Turn	All	4
Hot Metal Pourer	Turn	All	4
Hot Metal Assistant	Turn	All	4
Desulf and Skim	Turn	All	4
Charging Crane	Turn	All	8
Charging Crane Relief	Turn	All	4
Pulpit Operator	Turn	All	4
Furnace Operator	Turn	All	12
Scrap Crane	Turn	All	4
Track Hopper	Turn	Day	2
Track Hopper Assistant	Turn	Day	2
Labor	Turn	All	4
Maintenance	Turn	All	Varies
Ladle liners		Day	Varies
Metallurgist	Staff	Day	2
Clerical	Staff	Day	2

9.3 Raw Materials

9.3.1 Introduction

The basic raw materials required to make steel in the oxygen steelmaking process include: hot metal from the blast furnace, steel scrap and/or any other metallic iron source (such as DRI), ore (Fe_2O_3), and fluxes such as burnt lime (CaO), dolomitic lime (CaO–MgO), dolomitic stone ($MgCO_3$–$CaCO_3$) and fluorspar (CaF_2).

Scrap, charged from a scrap box, is the first material to be charged into the furnace. The hot metal is then poured into the vessel from a ladle, after which the oxygen blow is started. The fluxes, usually in lump form, are charged into the furnace through a bin system after the start of the oxygen blow. The fluxes can also be injected into the furnace in powder form through bottom tuyeres.

The composition and amounts of raw materials used in the steelmaking process vary from one shop to another, depending on their availability and the economics of the process. The basic raw materials used in the oxygen steelmaking process are described below.

9.3.2 Hot Metal

The hot metal, or liquid pig iron, is the primary source of iron units and energy in the oxygen steelmaking process. Hot metal is usually produced in blast furnaces, where it is cast into submarine shaped torpedo cars and transported either to a desulfurization station or directly to the steelmaking shop.

9.3.2.1 Composition

The chemical composition of hot metal can vary substantially, but typically it contains about 4.0–4.5% carbon, 0.3–1.5% silicon, 0.25–2.2% manganese, 0.04–0.20% phosphorus and

0.03–0.08% sulfur (before hot metal desulfurization).[1] The sulfur level in desulfurized hot metal can be as low as 0.001%. The composition of the hot metal depends on the practice and charge in the blast furnace. Generally, there is a decrease in the silicon content and an increase in the sulfur of the hot metal with colder blast furnace practices. The phosphorus contents of the hot metal increases if the BOF slag is recycled at the sinter plant.

Carbon and silicon are the chief contributors of energy. The hot metal silicon affects the amount of scrap charged in the heat. For example, if the hot metal silicon is high, there will be greater amounts of heat generated due to its oxidation, hence more scrap can be charged in the heat. Hot metal silicon also affects the slag volume, and therefore the lime consumption and resultant iron yield.

9.3.2.2 Determination of Carbon and Temperature

The hot metal is saturated with carbon, and its carbon concentration depends on the temperature and the concentration of other solute elements such as silicon and manganese. The carbon content of the hot metal increases with increasing temperature and manganese content, and decreases with increasing silicon content.

It is important to know the temperature and the carbon content of hot metal at the time it is poured into the BOF for steelmaking process control. The hot metal temperature is normally measured at the hot metal desulfurizer or at the time it is poured into the transfer ladle from the torpedo cars. If the hot metal temperature has not been measured close to the time of its charge into the BOF, then it can be estimated using the last hot metal temperature measurement, in conjunction with a knowledge of the rate of the hot metal ladle temperature loss with time, and the time elapsed between the last temperature measurement and the BOF charge. Typically, the temperature of the hot metal is in the range of 1315–1370°C (2400–2500°F). Once the temperature is measured or calculated, the carbon content of the hot metal at the time of charge can be estimated using a regression equation, which is mainly in terms of the temperature, hot metal silicon and manganese. Calculating the carbon in this manner has turned out to be as accurate as analyzing it chemically while saving considerable lab effort.

9.3.2.3 Hot Metal Treatment

Desulfurization is favored at high temperatures and low oxygen potentials. Also, the presence of other solute elements in the metal such as carbon and silicon increases the activity of sulfur, which in turn enhances desulfurization. Thus low oxygen potential and high carbon and silicon contents make conditions more favorable to remove sulfur from hot metal rather than from steel in the BOF.

Not all hot metal is desulfurized. Hot metal used for making steel grades with stringent sulfur specifications is desulfurized in the hot metal desulfurizer. The hot metal is poured into a transfer ladle from a torpedo car. It is then transported to the desulfurization station where the desulfurizer can reduce hot metal sulfur to as low as 0.001%, but more typically to 0.004 or 0.005%.

Typical desulfurizing reagents include lime-magnesium, and calcium carbide. Powdered reagents are generally injected using nitrogen gas. Apart from reducing sulfur to low levels, a hot metal desulfurizer can also allow the blast furnace operator to increase productivity by reducing the limestone burden and thereby producing higher sulfur hot metal.

It is important that the slag produced after hot metal desulfurization is removed effectively through slag skimming. This slag contains high amounts of sulfur, and any slag carried over into the BOF, where conditions are not good for desulfurization, will cause sulfur pickup in the steel.

9.3.2.4 Weighing

The weighing of the hot metal is done on a scale while it is being poured into the transfer ladle. It is very important that the weight of the hot metal is accurately known, as any error can cause problems in turndown chemistry, temperature and heat size in the BOF. This weight is an important input to the static charge model.

9.3.3 Scrap

Scrap is the second largest source of iron units in the steelmaking operation after hot metal. Scrap is basically recycled iron or steel, that is either generated within the mill (e.g. slab crops, pit scrap, cold iron or home scrap), or purchased from an outside source.

The scrap is weighed when loaded in the scrap box. The crane operator loads the box based on the weight and mix requirements of the upcoming heat. Then the box is transported to the BOF. It is important that the crane operator loads correct amounts and types of scrap (the scrap mix) as indicated by the computer or a fixed schedule. Otherwise the turndown performance of the heat will be adversely affected. Some typical types of scrap used in a BOF heat, and few of their properties are listed in Table 9.4.

Normally, the lighter scrap is loaded in the front, and the heavier scrap in the rear end of the box. This causes the lighter scrap to land first in the furnace as the scrap box is tilted. It is preferable that the lighter scrap fall on the refractory lining first, before the heavier scrap, to minimize refractory damage. Also, since heavy scrap is more difficult to melt than light scrap, it is preferable that it sits on top so that it is closest to the area of oxygen jet impingement and hence melt faster.

Scrap pieces that are too large to be charged into the furnace are cut into smaller pieces by means of shears or flame cutting. Thin, small pieces of scrap such as sheet shearings and punchings are compressed into block like bundles called bales using special hydraulic presses. Normally, larger, heavier pieces of scrap are more difficult to melt than lighter, smaller ones.

Unmelted scrap can cause significant problems in process control. It may result in high temperatures or missed chemistries at turndown. Bottom or mixed blowing, which can significantly enhance the mixing characteristics in the furnace, improves scrap melting of larger pieces.

Stable elements present in scrap, such as copper, molybdenum, tin and nickel cannot be oxidized and hence cannot be removed from metal. These elements can only be diluted. Detinned bundles, where tin is removed by shredding and treating with NaOH and then rebaled, are available but at considerably higher cost. Elements such as aluminum, silicon and zirconium can be fully oxidized from scrap and become incorporated in the slag. Elements which fall in the middle category in terms of their tendency to react, such as phosphorus, manganese and chromium distribute themselves between the metal and slag. Zinc and lead are mostly removed from scrap the bath as vapor.

Most steelmaking shops typically use about 20 to 35% of their total metallic charge as scrap, with the exact amount depending on the capacity of the steelmaking process. Much of this capacity depends on factors like the silicon, carbon and temperature of the hot metal, use of a post combustion lance, and external fuels charged, such as anthracite coal. The scrap ratio is also influenced by the relative cost of scrap and hot metal.

9.3.4 High Metallic Alternative Feeds

Direct reduced iron (DRI) is used in some steelmaking shops as a coolant as well as a source of iron units. DRI typically contains about 88–94% total iron (about 85–95% metallization), 0.5–3%C, 1–5% SiO_2, 3–8% FeO and small amounts of CaO, MgO and Al_2O_3.[2] DRI may contain phosphorus in the range of 0.005 to 0.09%, sulfur in the range 0.001 to 0.03% and low concentrations of nitrogen (usually less than 20 ppm).

DRI is normally fed into the BOF in briquetted form size at approximately 1 in. The DRI briquettes are passivated (by coating or binder) to eliminate any tendency to pyrophoricity (spontaneous burning) so that they can be handled conveniently in the steelmaking shop. DRI is usually fed into the steelmaking furnace through the bin system.

Certain elements such as nickel, copper and molybdenum can be added to the heat with the scrap charge. These elements do not oxidize to any significant level and they dissolve evenly in the metal during the oxygen blow. These additions can also be made after the oxygen blow, or in the ladle during tapping.

Table 9.4 Types of Scrap used in the BOF and Their Characteristics and Chemistry From Ref. 1, updated from Ref. 9.

Type of Scrap	Melting	Relative Cost	Bulk Density	Yield	Fe	C	Si	Mn	P	S	Cr	Cu	Ni	Mo	Sn
Plate and structural	easy	moderate	45	94.6	95.5	0.25	0.10	0.10	0.025	0.025	0.09	0.13	0.09	0.02	0.025
Punching and plate	easy	moderate	50	94.0	96.0	0.20	0.10	0.30	0.015	0.025	0.06	0.09	0.06	0.01	0.008
No.1 Heavy melting	easy	average	50	93.3	94.5	0.25	0.10	0.30	0.020	0.040	0.10	0.25	0.09	0.03	0.025
No. 2 Bundles	easy	inexpensive	50	88.0	90.0	0.25	0.10	0.30	0.030	0.090	0.18	0.50	0.10	0.03	0.100
No. 1 Busheling	easy	very expensive	60	95.7	98.0	0.15	0.01	0.30	0.010	0.020	0.04	0.07	0.03	0.01	.008
Ironmaking slag scrap	difficult	expensive	125	90.0	91.5	4.50	1.50	1.20	0.130	0.040	0.05	0.06	0.02	0.01	0.005
Briquetted iron borings	reactive at turndown	cheap	180	88.9	90.0	3.00	1.80	0.65	0.100	0.090	0.40	0.20	0.40	0.02	0.015
Home pit	fair	cheap	75	83.2	83.0	0.05	0.10	0.50	0.020	0.025	0.06	0.04	0.08	0.01	0.005
Home crops and skulls	fair to difficult	cheap	125	92.5	94.0	0.10	0.10	0.45	0.020	0.014	0.06	0.04	0.08	0.01	0.005
Hot metal	—	moderate	—	90.7	93.8	4.50	0.50	0.50	0.065	0.040	0.02	0.02	0.02	0.01	0.005

White areas indicate areas of potential trouble.

9.3.5 Oxide Additions

9.3.5.1 Iron Oxide Materials

Iron ore is usually charged into the BOF as a coolant and it is often used as a scrap substitute. Iron ores are available in the form of lumps or pellets, and their chemical compositions vary from different deposits as shown in Table 9.5.[3] Iron ores are useful scrap substitutes as they contain lower amounts of residual elements such as copper, zinc, nickel, and molybdenum. The cooling effect of iron ore is about three times higher than scrap. The reduction of the iron oxide in the ore is endothermic and higher amounts of hot metal and lower amounts of scrap are required when ore is used for cooling. Iron ores must be charged early in the blow when the carbon content in the bath is high to effectively reduce the iron oxide. The reduction of the iron oxides in the ore produces significant amounts of gas, and consequently increases slag foaming and the tendency to slop. Late ore additions have a detrimental affect on iron yield and end point slag chemistry. If only ore is used as a coolant just before tap, the slag becomes highly oxidized and fluid, enhancing slag carryover into the ladle. The delay in the cooling reaction from the unreduced ore causes a sudden decrease in temperature or a violent ladle reaction resulting in over-oxidation of the steel.

Table 9.5 Iron Ore Chemical Compositions *From Ref. 3.*

Ore	Country	Fe	SiO_2	Al_2O_3	CaO	MgO	P	S	Mn
Minnesota	USA	54.3	6.8	0.4	0.25	0.10	0.23	—	1.0
Carol Lake	Canada	64.7	3.9	0.35	—	—	.005	—	0.2
Cerro Bolivar	Venezuela	63.7	0.75	1.0	0.3	0.25	0.09	0.03	0.02
Goa	India	57.8	2.5	6.5	0.7	0.3	0.04	0.02	—
Itabira	Brazil	68.9	0.35	0.6	—	—	0.03	0.01	0.05
Tula	USSR	52.2	10.1	1.25	0.3	0.1	0.06	0.1	0.35

9.3.5.2 Waste Oxides

Economic and environmental issues have driven steel producers to recycle the waste iron oxides generated in the process. The increasing price of scrap, in addition to the increasing costs involved in the environmentally safe disposal of waste oxides, have encouraged steelmakers to recycle these materials back into the steelmaking process. Throughout the plant, various waste oxides and mill scales are collected and used in the sinter plant to produce some of the feed for the blast furnace. However, this does not consume all available oxides. In recent times, methods have been developed to substitute waste oxides in the BOF in place of ore. Mill scale has been used as a coolant in the BOF in amounts ranging from 5,000 to 25,000 lb.[4] Mill scale was found to be very effective in increasing the hot metal to scrap ratio; however, it causes heavy slopping during the process. Mill scale and other iron oxide additions are reduced during the main blow releasing iron and oxygen. This additional oxygen becomes available for carbon removal thus speeding up the overall reaction. Slopping is likely caused by the increased slag volume associated with using more hot metal (more pounds of Si and C generate more SiO_2 and CO, respectively) and by the increased reaction rate.

Waste oxide briquettes (WOB) containing steelmaking sludges, grit, and mill scale have also been charged into the furnace as a scrap substitute.[5] The waste oxides collected from the BOF fumes during the blow are high in iron content, typically more than 60 wt.%. These fumes, fines (sludge) and coarse (grit) waste oxide particles are blended, dried, mixed with lime and binders, and pressed into pillow-shaped briquettes. The briquettes are then cured for over 48 hours to remove their moisture. A typical composition of WOBs is 35 wt.% sludge, 20 wt.% grit, and 45 wt.% mill scale, see Table 9.6.[5]

Table 9.6 WOB Chemical Compositions *From Ref. 5.*

Composition	Wt. %
Total Fe	55–62
Metallic Fe	3–5
FeO	38–46
Fe_2O_3	29–32

Additions of WOBs are made early in the oxygen blow when the carbon content in the bath is high to ensure the reduction of ferrous, ferric, and manganese oxides to metallic iron and manganese. If the WOBs are added late in the blow, the oxides are likely to stay unreduced, resulting in yield loss, slopping, and a highly oxidized slag at turndown. WOBs are about two times better coolants than scrap, because their oxide reduction is endothermic, and therefore a higher hot metal ratio is required when WOBs are used for cooling—a situation similar to using ore. Various studies show that using WOBs causes no adverse effects on lining wear, molten iron yield, turndown performance and ladle slag FeO in the BOF.

9.3.6 Fluxes

9.3.6.1 Burnt Lime

In basic oxygen steelmaking, burnt lime consumption ranges from 40 to 100 lb. per net ton of steel produced. The amount consumed depends on the hot metal silicon, the proportion of hot metal to scrap, the initial (hot metal) and final (steel aim) sulfur and phosphorus contents. Burnt lime is produced by calcining limestone ($CaCO_3$) in rotary, shaft, or rotary hearth type kilns.[1] The calcining reaction is given below:

$$CaCO_3 + Heat \rightarrow CaO + CO_2 \qquad (9.3.1)$$

The calcination of high-calcium limestone will produce burnt lime containing about 96 wt.% CaO, 1 wt.% MgO, and 1 wt.% SiO_2. The sulfur content in burnt lime ranges from 0.03 to 0.1 wt.%. Most shops require less than 0.04 wt.% S in the lime to produce low sulfur steels.[1,6] Since an enormous amount of burnt lime is charged into the BOF within a short period of time, careful selection of the lime quality is important to improve its dissolution in the slag. In general, small lump sizes (1/2–1 in.) with high porosity have higher reactivity and promote rapid slag formation. The most common quality problems with either burnt or dolomitic lime are uncalcined inner cores, excess fines and too low a reactivity (calcined too hot or too long).

9.3.6.2 Dolomitic Lime

Dolomitic lime is charged with the burnt lime to saturate the slag with MgO, and reduce the dissolution of dolomite furnace refractories into the slag. Typically dolomitic lime contains about 36–42 wt.% MgO and 55–59 wt.% CaO.[1] Similarly, the dolomitic stone contains about 40% $MgCO_3$. The dolomitic lime charge into the BOF ranges from 30 to 80 lb. per net ton of steel produced, which represents about 25 to 50% of the total flux charge into the furnace (burnt plus dolomitic lime). The large variation in these additions strongly depends on experience and adjustments made by the steelmakers. These are based on observations of chemical attack of the slag on furnace refractories. Most of the dolomitic lime produced in the United States is obtained by calcining dolomitic stone in rotary kilns. The calcining reaction of the dolomitic stone is similar to that of limestone:

$$MgCO_3 + Heat \rightarrow MgO + CO_2 \qquad (9.3.2)$$

In some BOF operations dolomitic stone is added directly into the furnace as a coolant, and as a source of MgO to saturate the slag. It can also be added to stiffen the slag prior to slag splashing. It is important for the steelmaker to control the chemistry and size of the dolomitic lime.

9.3.6.3 Limestone

In most BOF shops limestone ($CaCO_3$) or dolomitic stone, ($CaCO_3 \cdot MgCO_3$) is frequently used as a coolant rather than as a flux. Limestone is commonly used to cool the bath if the turndown temperature is higher than the specified aim. When limestone is heated, the endothermic calcining reaction occurs producing CaO and CO_2, causing a temperature drop in the furnace. The extent of the temperature drop just before tap depends on the furnace size and slag conditions and is known for each shop. For example, in a 300 ton heat, 1000 lb. of limestone will drop the temperature of the bath by about 6°C (10°F).

9.3.6.4 Fluorspar

Calcium fluoride or fluorspar (CaF_2) is a slag fluidizer that reduces the viscosity of the slag. When added to the BOF it promotes rapid lime (CaO) dissolution in the slag by dissolving the dicalcium silicate ($2CaO \cdot SiO_2$) layer formed around the lime particles which retards the dissolution of the lime in the slag. In recent times, fluorspar has been used very sparingly because of its very corrosive attack of all types of refractories, including both furnace and ladle. Also, the fluorides form strong acids in the waste gas collection system which corrode structural parts and are undesirable emissions.

9.3.7 Oxygen

In modern oxygen steelmaking processes a water-cooled lance is used to inject oxygen at very high velocities onto a molten bath to produce steel. With the increasing demands to produce higher quality steels with lower impurity levels, oxygen of very high purity must be supplied. Therefore, the oxygen for steelmaking must be at least 99.5% pure, and ideally 99.7 to 99.8% pure. The remaining parts are 0.005 to 0.01% nitrogen and the rest is argon.[1]

In top-blown converters, the oxygen is jetted at supersonic velocities (Mach>1) with convergent-divergent nozzles at the tip of the water-cooled lance. A forceful gas jet penetrates the slag and impinges onto the metal surface to refine the steel. Today, most BOFs operate with lance tips containing four to five nozzles and oxygen flow rates, (in 230 to 300 ton converters), that range from 640 to 900 Nm^3/min (22,500 to 31,500 scfm).[7] Figure 9.6 shows an schematic of a typical five-nozzle lance tip. The tip is made of a high thermal conductivity cast copper alloy with precisely machined nozzles to achieve the desired jet parameters. The nozzles are angled about 12° to the centerline of the lance pipe and equally spaced around the tip. The tip is welded to a 12 inch seamless steel pipe (lance barrel) about 60 feet long. Cooling water is essential in these lances to keep them from burning up in the furnace. At the top of the lance, armored rubber hoses are connected to a pressure-regulated oxygen source and to a supply of recirculated cooling water. Details of the convergent-divergent nozzles are also shown in Fig. 9.6. As the oxygen passes the converging section it is accelerated and reaches sonic velocity (Mach =1) in the cylindrical throat section. Then it expands in the diverging sec-

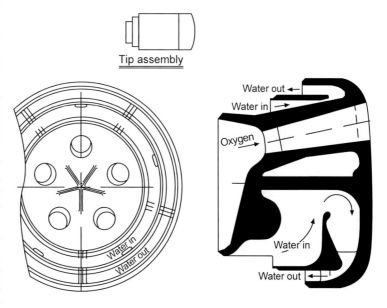

Fig. 9.6 Convergent-divergent nozzles. *From Ref. 1.*

tion and its temperature and pressure decreases while its velocity increases to supersonic levels (Mach > 1). The supersonic jets are at an angle of about 12° so that they do not interfere with each other.

In bottom-blown converters, the oxygen is injected through the bottom of the vessel using a series of tuyeres. About 14 to 22 tuyeres are used to blow about 4.0 to 4.5 Nm^3 of oxygen per minute per ton of steel.[8] Powdered lime, mixed with the oxygen, is usually injected through the liquid bath to improve lime dissolution and hence slag formation during the blow. The tuyeres consist of two concentric pipes, where oxygen flows through the center pipe and a hydrocarbon fluid, such as natural gas or propane, used as a coolant, flows through the annular space between both pipes.

9.4 Process Reactions and Energy Balance

9.4.1 Refining Reactions in BOF Steelmaking

In the oxygen steelmaking process, impurities such as carbon (C), silicon (Si), and manganese (Mn) dissolved in the hot metal are removed by oxidation to produce liquid steel. Hot metal and scrap are charged into the furnace and high-purity oxygen gas is injected at high flow rates, through a lance or tuyeres, to react with the metal bath. The oxygen injection process, known as the blow, lasts for about 16 to 25 minutes and the oxidation reactions result in the formation of CO, CO_2, SiO_2, MnO, and iron oxides. Most of these oxides are dissolved with the fluxes added to the furnace, primarily lime (CaO), to form a liquid slag that is able to remove sulfur (S) and phosphorus (P) from the metal. The gaseous oxides, composed of about 90% CO and 10% CO_2, exit the furnace carrying small amounts of iron oxide and lime dust. Typical oxygen flow rates during the blow range between 2–3.5 Nm^3 per minute per ton of steel (70–123 scfm per ton), and in general the rate of oxygen injection is limited either by the capacity of the hood and gas cleaning system or by the available oxygen pressure

The commercial success of oxygen steelmaking is mainly due to two important characteristics. First, the process is autogenous meaning that no external heat sources are required. The oxidation reactions during the blow provide the energy necessary to melt the fluxes and scrap, and achieve the desired temperature of the steel product. Second, the process is capable of refining steel at high production rates. The fast reaction rates are due to the extremely large surface area available for reactions. When oxygen is injected onto the metal bath a tremendous amount of gas is evolved forming an emulsion with the liquid slag and with metal droplets sheared from the bath surface by the impingement of the oxygen jet.[10] This gas-metal-slag emulsion, shown in Fig. 9.7, generates large surface areas that increase the rates of the refining reactions. The reaction mechanisms and the slag-metal reactions are discussed in detail in Chapter 2. Therefore, only a brief discussion of the sequence of these reactions during the blow are provided in this section.

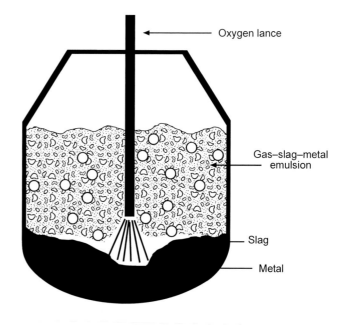

Fig. 9.7 Physical state of the BOF in the middle of the blow.

9.4.1.1 Carbon Oxidation

Decarburization is the most extensive and important reaction during oxygen steelmaking. About 4.5 wt% carbon in the hot metal is oxidized to CO and CO_2 during the oxygen blow, and steel with less than 0.1 wt% carbon is produced. The change in the carbon content during the blow is illustrated in Fig. 9.8,[11] which shows three distinct stages. The first stage, occurring during the first few minutes of the blow, shows a slow decarburization rate as nearly all the oxygen supplied reacts with the silicon in the metal. The second stage, occurring at high carbon contents in the metal, shows a constant higher rate of decarburization and its controlled by the rate of supplied oxygen. Finally, the third stage occurs at carbon contents below about 0.3 wt.%, where the decarburization rate drops as carbon becomes less available to react with all the oxygen supplied. At this stage, the rate is controlled by mass transfer of carbon, and the oxygen will mostly react with iron to form iron oxide. Also in this stage, the generation of CO drops and the flame over the mouth of the furnace becomes less luminous, and practically disappears below about 0.1 wt.% carbon.

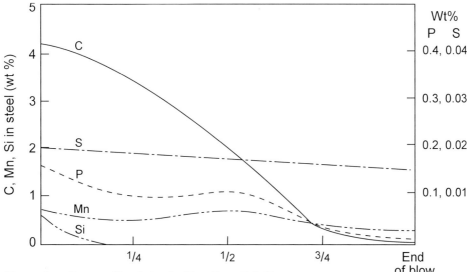

Fig. 9.8 Change in melt composition during the blow. *From Ref. 11.*

9.4.1.2 Silicon Oxidation

The strong affinity of oxygen for Silicon will result in the removal of almost all the Si early in the blow. The Si dissolved in the hot metal (0.25–1.3 wt.%) is oxidized to very low levels (<0.005 wt.%) in the first three to five minutes of the blow as shown in Fig. 9.8. The oxidation of Si to silica (SiO_2) is exothermic producing significant amounts of heat which raises the temperature of the bath. It also forms a silicate slag that reacts with the added lime (CaO) and dolomitic lime (MgO) to form the basic steelmaking slag. The amount of Si in the hot metal is very important since its oxidation is a major heat source to the process and it strongly affects the amount of scrap that can be melted. It also determines the slag volume and consequently affects the iron yield and dephosphorization of the metal. In general, more slag causes less yield but lower phosphorus.

9.4.1.3 Manganese Oxidation

The reaction involving the oxidation of Manganese in steelmaking is complex. In top-blown processes Mn is oxidized to MnO early in the blow and after most of the silicon has been oxidized, the Mn reverts into the metal.[12] Finally, as shown in Fig. 9.8, towards the end of the blow the Mn in the metal decreases as more oxygen is available for its oxidation. In bottom-blown processes, such as the OBM (Q-BOP), a similar pattern is found, but the residual Mn content of the steel is higher than for top-blown processes due to better stirring.[13,14]

9.4.1.4 Phosphorus Oxidation

Dephosphorization is favored by the oxidizing conditions in the furnace. The dephosphorization reaction between liquid iron and slag can be expressed by reaction 9.4.1.[15] Phosphorus removal is favored by low temperatures, high slag basicity (high CaO/SiO_2 ratio), high slag FeO, high slag fluidity, and good stirring.[12,15] The change in the phosphorus content of the metal during blow is shown in Fig. 9.8. The phosphorus in the metal decreases at the beginning of the blow, then it reverts into the metal when the FeO is reduced during the main decarburization period, and finally decreases at the end of the blow.[12] Stirring improves slag-metal mixing, which increases the rate of dephosphorization. Good stirring with additions of fluxing agents, such as fluorspar, also improves dephosphorization by increasing the dissolution of lime, resulting in a highly basic and fluid liquid slag.

$$\underline{P} + 2.5(FeO) = (PO_{2.5}) + 2.5Fe \qquad (9.4.1)$$

9.4.1.5 Sulfur Reaction

The BOF is not very effective for sulfur removal due to its highly oxidizing conditions. Sulfur distribution ratios in the BOF (% S slag /% S metal ~4–8) are much lower than the ratios in the steel ladle (% S slag /% S metal ~300–500) during secondary ladle practices. In the BOF, about 10 to 20% of sulfur in the metal reacts directly with oxygen to form gaseous SO_2.[15] The rest of the sulfur is removed by the following slag-metal reaction

$$\underline{S} + (CaO) + Fe = (CaS) + (FeO) \qquad (9.4.2)$$

Sulfur removal by the slag is favored by high slag basicities (high CaO/SiO_2 ratio), and low FeO contents. The final sulfur content of steel is also affected by the sulfur contained in the furnace charge materials, such as hot metal and scrap. The sulfur content in the hot metal supplied from the blast furnace generally ranges from 0.020 to 0.040 wt.%, and if the hot metal is desulfurized before steelmaking the sulfur content in the hot metal can be as low as 0.002 wt.%. Heavy pieces of scrap containing high sulfur contents must be avoided if low sulfur alloys with less than 60 ppm (0.006%) of sulfur are being produced. For example a slab crop of 2273 kg (5000 lb.) containing 0.25 wt.% S (12.5 lb. of sulfur) can increase the sulfur content of steel by about 15 ppm (0.0015%) in a 300 ton BOF.

9.4.2 Slag Formation in BOF Steelmaking

Fluxes are charged into the furnace early in the blow and they dissolve with the developing oxides to form a liquid slag. The rate of dissolution of these fluxes strongly affects the slag-metal reactions occurring during the blow. Therefore, it is important to understand the evolution of slag during the blow. Several investigators have studied slag formation in oxygen steelmaking,[16,17] and a detailed review of these investigations is given by Turkdogan,[11] and Deo and Boom.[18]

At the beginning of the blow, the tip of the oxygen lance is kept high above the bath surface, at about 3.5 m (12 ft), which results in the formation of an initial slag rich in SiO_2 and FeO. During this period large amounts of burnt lime and dolomitic lime are charged into the furnace. The lance is then lowered and the slag starts to foam at around one third of the blow due to the reduction of the FeO in the slag in conjunction with CO formation. The drop in the FeO content in the slag is shown in Fig. 9.9. Also, as the blow progresses, the CaO dissolves in the slag, and the active slag weight increases. Finally, after three quarters into the blow, the FeO content in the slag increases because of a decrease in the rate of decarburization. The resulting slag at turndown in top-blown converters (BOF or BOP) have typical ranges: 42–55 wt.% CaO, 2–8 wt.% MgO, 10–30 wt.% FeO_T, 3–8 wt.% MnO, 10–25 wt.% SiO_2, 1–5 wt.% P_2O_5, 1–2 wt.% Al_2O_3, 0.1–0.3 wt.% S. In the OBM (Q-BOP) or bottom-blown converter, the total FeO in the turndown slag is lower, ranging between 4–22 wt.%

During the blow, the temperature of the metal gradually increases from about 1350°C (2450°F) to 1650°C (3000°F) at turndown, and the slag temperature is about 50°C (120°F) higher than that of

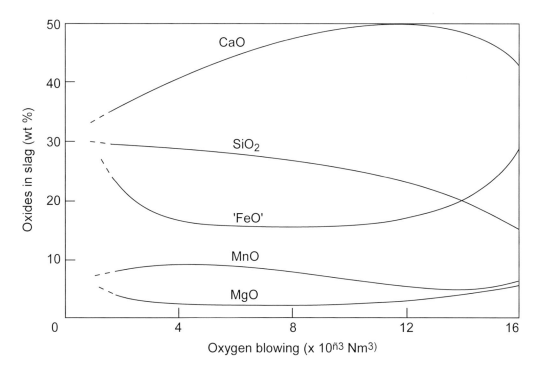

Fig. 9.9 Change in slag composition during the blow. *From Ref. 11.*

the metal.[19] The slag at turndown may contain regions of undissolved lime mixed with the liquid slag, since the dissolution of lime is limited by the presence of dicalcium silicate ($2CaO \cdot SiO_2$) coating, which is solid at steelmaking temperatures and prevents rapid dissolution.[11] The presence of MgO in the lime weakens the coating. Thus, charging MgO early speeds up slag forming due to quicker solution of lime.

9.4.3 Mass and Energy Balances

As shown in Fig. 9.10, hot metal, scrap, and iron ore are charged with the fluxes, such as burnt and dolomitic lime, into the furnace. Oxygen is injected at high flow rates and gases, such as CO and CO_2, and iron oxide fumes (Fe_2O_3) exit from the mouth of the furnace. At turndown, liquid steel and slag are the remaining products of the process. The oxidation reactions occurring during the blow produce more energy than required to simply raise the temperature of the hot metal, from about 1350°C (2450°F) to the desired turndown temperature, and to melt the fluxes. Most of the excess heat is used to increase the amount of steel produced by melting cold scrap and by reducing iron ore to metal. Some heat is also lost by conduction, convection, and radiation to the surroundings.

It is important to exactly determine the amount of each material to charge and the amount of oxygen to blow to produce steel of desired temperature and chemistry. The specific method for determining these amounts varies with each BOF shop; however, in general these computations are based on mass and energy balance calculations. A simple mass and energy balance is presented in this text for illustration purposes. A more detailed treatment can be found in the literature.[1]

Consider the production of 1000 kg of steel. Fluxes, such as burnt and dolomitic lime, are added to the furnace with the iron ore early in the blow. Before any calculation can be made it is required to specify the compositions and temperatures of the input materials, such as hot metal, scrap, iron ore, and fluxes, and also the temperature and chemistry specifications of the steel product. Table 9.7 shows typical compositions. The sequence of calculations required to determine the amounts of input materials necessary to produce 1000 kg of steel product is summarized as follows:

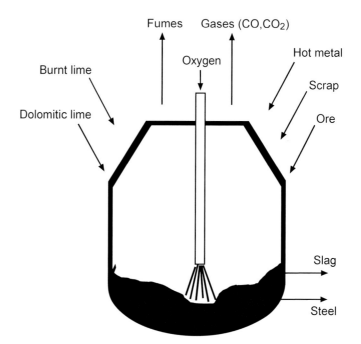

Fig. 9.10 Input and output materials in the BOF.

Table 9.7 Chemistries of Input and Output Materials in the BOF

Element	Hot Metal	Scrap	Steel	
% Fe	93.61	99.493	99.797	
% C	4.65	0.09	0.040	
% Si	0.60	0.020	0.005	
% Mn	0.45	0.360	0.138	
% P	0.06	0.012	0.010	
% S	0.01	0.025	0.010	
Other	0.61			
Compound	Burnt lime	Dolo-lime	Slag	Iron Ore
%CaO	96.0	58.0	47.86	—
%SiO$_2$	1.0	0.8	12.00	0.77
%MgO	1.0	40.5	6.30	—
%FeO$_T$	—	—	26.38	—
%Fe			—	65.8
%Al$_2$O$_3$	0.5	0.4	1.30	—
%MnO	—	—	5.00	0.20
%H$_2$O	1.3	1.4	—	—
%P$_2$O$_5$	—	—	1.16	—

9.4.3.1 Determination of the Flux Additions

The fluxes added to the process strongly depend on the hot metal silicon, the weight of hot metal, the lime to silica ratio (%CaO/%SiO$_2$), and the amount of MgO needed in the slag to avoid the wear of furnace refractories. The lime to silica ratio should range from two to four to achieve a basic slag during the blow. Also, approximately 6–12 wt.% of MgO is required, depending on slag temperature and chemistry, to saturate the slag and consequently retard dissolution of the furnace refractories.[7]

For a typical lime to silica ratio of four, each kg of SiO_2 in the slag requires 4 kg of CaO. For the example shown in Table 9.8, about 11.26 kg (24.78 lb.) of SiO_2 are produced from the oxidation of the hot metal silicon, and 47.92 kg (105.42 lb.) of CaO per metric ton of steel are required to neutralize the SiO_2 in the slag. The amounts of burnt lime and dolomitic lime needed are computed from the CaO and MgO requirements as shown in Table 9.8. In actual BOF operations, higher dolomitic lime additions are made than those predicted by the present example to ensure MgO saturation.

Table 9.8 Material and Energy Balance for the Production of 1000 kg (2200 lb.) of Steel

Input			Output		
Hot Metal	kg	lb.	**Steel**	kg	lb.
Fe	821.60	1807.5	Fe	997.97	2195.5
C	40.81	89.78	C	0.40	0.88
Si	5.26	11.58	Si	0.05	0.11
Mn	3.95	8.69	Mn	1.38	3.04
P	0.53	1.16	P	0.10	0.22
S	0.09	0.19	S	0.10	0.22
BF slag	5.40	11.88			
Total	877.64	1930.8	Total	1000.0	2200.0
Scrap	kg	lb.	**Slag**		
Fe	200.53	441.16	CaO	47.92	105.42
C	0.18	0.40	SiO_2	12.01	26.42
Si	0.04	0.09	MgO	6.35	13.97
Mn	0.72	1.60	FeO_T	26.42	58.12
P	0.02	0.05	MnO	5.00	11.01
S	0.05	0.11	P_2O_5	1.13	2.49
Total	201.55	443.41	Al_2O_3	1.3	2.86
			Total	100.13	220.29
Ore	kg	lb.	**Fumes**	kg	lb.
Fe	11.52	25.35	Fe	16.02	35.24
SiO_2	0.130	0.285	Total	27.52	60.61
MnO	0.034	0.075			
Other	5.134	11.294			
Total	16.818	37.0			
Fluxes	kg	lb.	**Gases:**	kg	lb.
CaO	47.92	105.42	CO (90%)	85.26	187.57
MgO	6.35	13.97	CO_2 (10%)	14.89	32.75
SiO_2	0.53	1.17			
Other	1.65	3.62			
Total Burnt time	40.78	89.70			
Total Dolo-lime	15.67	34.48			
Oxygen Gas	kg	lb.			
	52.8 Nm^3	1849.7 scf			
	@298K	@70°F			
Total O_2	75.34	165.75			
Total Inputs	1227.8	2701.2	**Total Outputs**	1227.8	2701.2

9.4.3.2 Determination of Oxygen Requirements

The volume of oxygen gas blown into the converter must be sufficient to oxidize the C, Si, Mn, and P during the blow, and it is computed from an oxygen balance as shown below. For the present example the oxygen required during the blow is about 52.8 Nm³ (1849.7 scf) per metric ton of steel produced.

$$\begin{bmatrix} \text{Oxygen} \\ \text{injected} \end{bmatrix} = \begin{bmatrix} \text{Oxygen for the} \\ \text{oxidation reactions} \end{bmatrix} - \begin{bmatrix} \text{Oxygen supplied} \\ \text{by iron ore} \end{bmatrix} - \begin{bmatrix} \text{Oxygen dissolved} \\ \text{in steel at turndown} \end{bmatrix} \quad (9.4.3)$$

9.4.3.3 Determination of the Weight of Iron-Bearing Materials

In general four distinct iron-bearing materials are involved in oxygen steelmaking: hot metal, scrap, iron ore, and the steel product. Slag and fume are usually considered heat and iron losses. The simultaneous solution of an iron mass balance and an energy balance permits the determination of the weights of two of the iron-bearing materials with a knowledge of the weights of the other two. For the example here, the product weight (1000 kg), and the weight of the iron ore (16.8 kg) are assumed to be known. Then the weights of the hot metal and scrap are computed to be 877.64 kg (1930.8 lb.) and 201.55 kg (443.41 lb.) respectively from the mass and energy balance shown below:

Mass balance for iron: (IRON INPUT = IRON OUTPUT)

$$\text{IRON INPUT} = \begin{bmatrix} \text{Weight of Fe} \\ \text{in Hot Metal} \end{bmatrix} + \begin{bmatrix} \text{Weight of} \\ \text{Fe in scrap} \end{bmatrix} + \begin{bmatrix} \text{Weight of Fe} \\ \text{in Iron Ore} \end{bmatrix} \quad (9.4.4)$$

$$\text{IRON OUTPUT} = \begin{bmatrix} \text{Weight of Fe} \\ \text{in Steel} \end{bmatrix} + \begin{bmatrix} \text{Weight of} \\ \text{Fe in Slag} \end{bmatrix} + \begin{bmatrix} \text{Weight of Fe} \\ \text{in fumes} \end{bmatrix}$$

Heat balance: (HEAT INPUT = HEAT OUTPUT)

$$\text{HEAT INPUT} = \begin{bmatrix} \text{Heat content in} \\ \text{the Hot Metal} \end{bmatrix} + \begin{bmatrix} \text{Heats of} \\ \text{reaction} \end{bmatrix} + \begin{bmatrix} \text{Heat of slag} \\ \text{formation} \end{bmatrix}$$

$$\text{HEAT OUTPUT} = \begin{bmatrix} \text{Sensible heat} \\ \text{of steel} \end{bmatrix} + \begin{bmatrix} \text{Sensible heat} \\ \text{of slag} \end{bmatrix} + \begin{bmatrix} \text{Sensible heat in} \\ \text{gas and fume} \end{bmatrix} + \begin{bmatrix} \text{Heat} \\ \text{losses} \end{bmatrix} \quad (9.4.5)$$

The heat added to the process comes from the heat content or enthalpy in the hot metal charged into the furnace at about 1343°C (2450°F), the heats of oxidation of elements, such as Fe, C, Si, Mn, P, and S, whose enthalpies are shown in Table 9.9, and the heats of formation of the different compounds in the slag.

Table 9.9. Enthalpies or Heats of Reactions *From Ref. 1.*

Oxidation Reactions	Heats of Reaction			
	Kilojoule	per mole of	BTU	per lb. of
$\underline{C} + \tfrac{1}{2}O_2 = CO$	4173	C	3952	C
$\underline{C} + O_2 = CO_2$	14884	C	14096	C
$CO + \tfrac{1}{2}O_2 = CO_2$	4593	CO	4350	CO
$\underline{Si} + O_2 = SiO_2$	13927	Si	13190	Si
$Fe + \tfrac{1}{2}O_2 = FeO$	2198	Fe	2082	Fe
$\underline{Mn} + \tfrac{1}{2}O_2 = MnO$	3326	Mn	3150	Mn

These sources will provide the heat necessary to raise temperature of the steel and slag to the aim turndown temperature, and also to heat up the gases and fumes leaving the furnace. Furthermore, there is enough energy generated to overcome the heat losses during the process, to heat and melt coolants such as scrap and iron ore, and to reduce the iron oxide in the ore.

9.4.3.4 Determination of the Gases and Fumes Produced

The amounts of CO and CO_2 produced from decarburization are determined from a mass balance for carbon. The carbon removed from the bath is converted to approximately 90% CO and 10% CO_2. With the gases about 1 to 1.5 wt.% of iron is lost in the form of iron oxide fumes that exit from the mouth of the furnace.

9.4.3.5 Determination of the FeO in the Slag

The FeO in the slag is generally determined from empirical correlations, developed by each shop, between the slag FeO and the aim carbon and the lime to silica ratio. Other parameters are generally of much lower significance. This empirical relationship is one of the larger error sources in a material and energy balance algorithm, arising from analytical errors of iron oxides and slag sample preparation problems.

9.4.4 Tapping Practices and Ladle Additions

When the blow is completed, the lance is removed from the furnace and the vessel is rotated to a horizontal position towards the charging side for sampling. A steel sample is withdrawn from the bath for chemical analysis, and an expendable immersion-type thermocouple is used to measure the temperature of the melt. The steel sample is analyzed with a mass spectrometer, and the concentrations of the elements present in the steel are determined in approximately three to five minutes. If the steel is too hot, meaning that the measured temperature is higher than the aim temperature, it can be cooled by rocking the vessel, or by adding coolants such as iron ore or limestone. If the steel is too cold, or if the measured concentrations of elements such as carbon, phosphorus, and sulfur are higher than the aim concentrations specified, additional oxygen is blown into the furnace (reblow) for approximately one to three minutes. Once the heat meets the temperature and chemistry requirements, the furnace is rotated towards the taphole side and the steel is tapped or poured into a ladle.

Tapping a 300 ton heat takes from four to seven minutes and the time strongly depends on the conditions or diameter of the taphole. A good tapping practice is necessary to maximize yield, or the amount of steel poured into the ladle. Slag carryover from the BOF into the ladle must be minimized. Furnace slag contains high FeO, which reduces desulfurization in the ladle, and enhances the formation of alumina inclusions. Also, the P_2O_5 present in BOF slags is a source of phosphorus carried into the ladle. Therefore, over the years, extensive work has been done to develop slag free tapping techniques, and the most commonly used are described here. Good tap hole maintenance, combined with the ability of the operator to rotate the furnace quick enough when all the steel has been tapped, will reduce the amount of slag carryover into the ladle. Slag free tapping devices are now commonly used to help the operators reduce slag carryover. Different types of taphole plugs, such as balls and darts, (discussed earlier in section 9.2.2.12) shown in Fig. 9.11, are dropped into the furnace at tap. These Devices float at the slag-metal interface, and plug the taphole when the steel has emptied but before the slag can exit the furnace. There has been much debate over the effectiveness of these devices. Electromagnetic slag detection sensors installed around the taphole will detect the presence of slag in the stream and send a signal to alarm the furnace operator. One of the problems with these devices is that they can give false alarms from slag entrainment within the vortex of the steel stream and they require maintaining a taper in the taphole to work well.

With current steelmaking alloying practices, most of the alloys are added to the ladle. However, large amounts of non-oxidizable alloys such as nickel, molybdenum and copper are usually charged with the scrap as they resist oxidation during the blow. This practice will prevent big temperature drops in the ladle. In aluminum killed steels aluminum (Al) is used to deoxidize the steel and reduce the dissolved oxygen from approximately 500–1000 ppm to less than 5 ppm, and is generally the first addition made into the ladle during tap. For example, a heat with a Celox reading of 750 ppm of dissolved oxygen requires approximately 365 kg (800 lb.) of aluminum at tap. In semi-killed steels, deoxidation is done with ferrosilicon, and the dissolved oxygen in steel is only lowered to about 50 ppm. Burnt lime is also added with the aluminum to satisfy a lime to alumina (CaO/Al_2O_3) ratio of 0.8 to 1.2. This produces a liquid slag over the molten metal that thermally insulates the melt to avoid excess temperature losses, protects the melt from reoxidation from air, desulfurizes the steel, and removes alumina inclusions from the melt.

Ferromanganese is added via chutes located over the ladle, in large quantities, after the steel has been deoxidized by aluminum or silicon. The general rule of thumb is that the aluminum is added when the melt

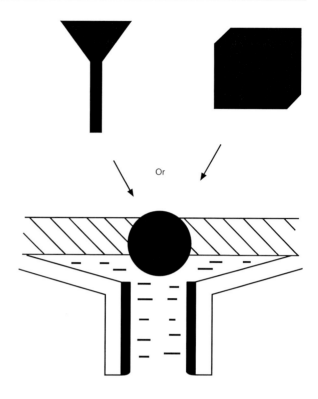

Fig. 9.11 Schematic of taphole plugs for BOF steelmaking. From Ref. 21.

reaches approximately 1/3 of the ladle's height, and all the alloys should be added by the time the melt reaches 2/3 of the full ladle height. Slag modifiers, containing about 50 wt.% aluminum, are added to the slag near the end of tap to reduce the FeO content in the ladle slag originating from furnace slag carryover. Chapter 11 contains a detailed discussion of the refining processes occurring in the ladle after steelmaking.

9.5 Process Variations

9.5.1 The Bottom-Blown Oxygen Steelmaking or OBM (Q-BOP) Process

The successful development and application of the shrouded oxygen tuyere in the late 1960s led to the development of the OBM (Q-BOP) process in the early 1970s. Oxygen in this process is injected into the bath through tuyeres inserted in the bottom of the furnace. Each tuyere is made from two concentric tubes forming an inner nozzle and an outer annulus. Oxygen and powdered lime are injected through the central portion of the tuyeres, while a hydrocarbon gas, typically natural gas or propane, is injected through the annular section between the two concentric pipes, as shown in Fig. 9.12. The endothermic decomposition of the hydrocarbon gas and the sensible heat required to bring the products of the decomposition up to steelmaking temperatures result in localized cooling at the tip of the tuyere. The localized cooling is enough to chill the liquid metal and form a porous mushroom on the tip of the tuyere and part of the surrounding refractory. This mushroom reduces the burn back rate of the tuyere, and the wear of the surrounding refractory. The injected lime provides additional cooling to the tuyere, and results in better slag refining characteristics.

Top lances in OBM (Q-BOP) furnaces have also been adopted, mainly for the purposes of increasing the post-combustion of the offgases within the furnace, and to control the buildup of slag and metal in the furnace cone area. Top lances used in OBM (Q-BOP) furnaces are normally stationary, since they are not used for refining purposes. Tuyeres, located in the upper cone area of furnaces with a heat size larger than approximately 150 tonnes have also been used, but typically

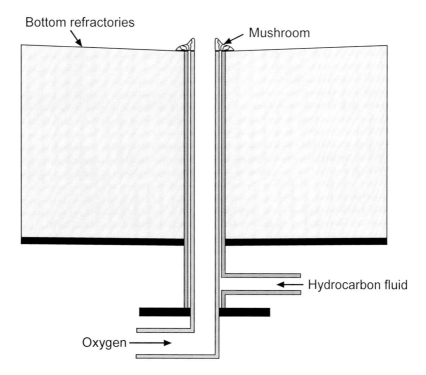

Fig. 9.12 Schematic drawing of an OBM (Q-BOP) tuyere.

result in higher refractory wear. For this reason, their application has been limited to shops which require increased scrap melting capabilities (resulting in shorter lining lives), and with a heat size smaller than 150 tonnes.

9.5.1.1 Plant Equipment

The injection of oxygen through the bottom in the OBM (Q-BOP) process, with a fraction of the total oxygen through a stationary top lance, results in the need for a low building, and consequently in lower greenfield construction costs.

Oxygen, a hydrocarbon fluid (natural gas or propane), nitrogen and argon are gases used in the OBM (Q-BOP) process. These gases are metered and controlled, and introduced through rotary joints located in the trunnion pins. The oxygen, aside from being used as the main process gas, is also used as the transport gas for the pulverized burnt lime. A high-pressure injector contains the burnt lime, which is transported by the oxygen through one of the trunnion pins into a lime and oxygen distributor, and then to the individual tuyeres in the furnace bottom. The hydrocarbon fluid is transported to the bottom of the furnace through the opposite trunnion, to avoid the possibility of leakage and of its mixing with oxygen in the transport line. The hydrocarbon fluid is then distributed to each tuyere. In some instances, the flow of the hydrocarbon fluid is controlled individually for each tuyere prior to entering the rotary joint. Nitrogen is used to protect the tuyeres from plugging during furnace rotation. It can also be used to increase the nitrogen content in the steel. Argon can be used to minimize the nitrogen pickup in the steel, and to produce lower carbon steels than in the BOF process, without excessive yield losses, and with low FeO contents in the slag.

Gas flow through the tuyeres has to be maintained above sonic flow to prevent penetration of steel into the tuyeres, and subsequent plugging. For this reason, the sonic flow is maintained during rotation of the furnace for turndown and for tapping of the steel. The ejections of metal and slag resulting from the injection of this high flow through the tuyeres during rotation, require the complete enclosure of the OBM (Q-BOP) furnaces. Movable doors are used in the charge side for this purpose. The complete enclosure also results in lower fugitive emissions during turndown and tapping than in the BOF furnaces, which usually are not enclosed.

Typically 12 to 18 tuyeres are used in the bottom, depending on the furnace capacity. The tuyeres are located in the refractory bottom on two rows that run from one trunnion to the other. The locations of these two rows are selected so that they are above the slag line during turndown and tapping to allow the reduction of the gas flow during these periods. The inside of the inner pipe of the tuyere is typically lined with a mullite sleeve to prevent excessive wear of the pipe by the burnt lime. The inside diameter of the ceramic liner is 1 to 1.5 inches. Stainless steel is normally used for the inner pipe, and carbon steel for the outer pipe, although copper has also been used.

The higher refractory wear observed in the vicinity of the tuyeres, due to the high temperature gradients experienced by the refractories during a heat cycle and the high temperature generated around them, results in bottom lives of 800 to 2500 heats, depending on tap temperature, turndown carbon content, etc. Since barrel lives approach 4000 to 6000 heats, the furnace is designed to include a replaceable bottom. When the bottom thickness is too thin for safe operation, it is removed and a bottom with new refractories and tuyeres is installed. The furnace can be back in operation with a new bottom in less than 24 hours.

9.5.1.2 Raw Materials

A distinct advantage of the OBM (Q-BOP) process is its capability to melt bigger and thicker pieces of scrap than the BOF process. Sections with thicknesses of up to two feet are melted routinely. This expands significantly the types of scrap that can be used, and lowers their preparation costs. There is no unmelted scrap at the end of the blow in the OBM (Q-BOP) process.

The burnt lime used for slag formation is pulverized and screened to less than 0.1 millimeters. It is sometimes treated to improve its flowability during pneumatic transport with oxygen. The dolomitic lime in the OBM (Q-BOP) process is essentially similar to that used in the BOF process, and is charged through top bins, if available. In shops where overhead bins are not available, a pulverized blend of burnt lime and dolomitic lime is injected through the bottom tuyeres to achieve the desired MgO content in the slag. The rest of the raw materials are the same as for the BOF process.

9.5.1.3 Sequence of Operations

After the steel is tapped, the furnace is rotated to the vertical position, and nitrogen is blown to splash the slag onto the furnace walls. This results in a coating that extends the life of the furnace barrel. The furnace is also rocked to coat the bottom with slag. This operation can be done with the slag as is, or with conditioned slag. The furnace is then ready to receive the scrap and hot metal. Nitrogen is injected at sonic flow to protect the tuyeres during the hot metal charge. The furnace is rotated to its vertical position, and the bottom and top oxygen blow are started. Burnt lime is injected with the oxygen through the bottom, and the dolomitic lime is added through the top at the beginning of the blow. Typically the lime is injected within the first half of the oxygen blow. When the calculated oxygen amount has been injected, the gas is switched to nitrogen or argon and the furnace is rotated for sampling. A sample for chemical analysis is taken, and the temperature and oxygen activity are measured. If the desired temperature and chemistry have been obtained, the heat can then be tapped. If necessary, small adjustments in temperature and chemistry can be made by injecting additional oxygen through the bottom, by injecting more lime, or by cooling the heat with ore or raw dolomite. Since the process is very reproducible, the heat is normally tapped after these adjustments, without making another temperature measurement or taking another sample for chemical analysis. Shops so equipped can take a sample for chemical analysis, temperature and carbon content measurement with sub lances a few minutes before the end of the oxygen blow. Any necessary adjustments can be made during the blow to obtain the aim steel temperature and chemistry. The turndown step can thus be avoided, decreasing the tap-to-tap time, and increasing the productivity of the shop.

9.5.1.4 Process Characteristics

The injection of the oxygen and hydrocarbons through bottom tuyeres results in distinct process characteristics. The oxygen reacts directly with the carbon and silicon in the liquid iron melt, resulting in lower oxidation levels in the metal and slag at the end of the blow. The bottom injection also results in very strong bath mixing. Steel decarburization is enhanced by the strong bath agitation, particularly during the last portion of the blow, when mass transfer of the carbon in the melt controls the rate of decarburization at carbon contents below 0.3%. This results in less iron being oxidized and lost to the slag, as shown in Fig. 9.13, and in less dissolved oxygen in the steel at turndown, as shown in Fig. 9.14. The manganese content at turndown is also higher than in the top-blown vessels, as shown in Fig. 9.15, due to the lower bath and slag oxidation. The variability in blow behavior introduced by the top lance in top-blown vessels is eliminated. By injecting the oxygen and lime through the bottom tuyeres in a controlled manner, a highly reproducible process control is obtained.

9.5.1.5 Product Characteristics

In general, the same steel grades that can be produced in top-blown vessels can be produced by the OBM (Q-BOP). Additionally, the better mixing attained in the bath allows the production of steels with carbon contents of 0.015–0.020% without excessive bath and slag oxidation and without vacuum decarburization. The slightly higher hydrogen contents obtained at turndown can be lowered by flushing the bath with nitrogen, argon, or a mixture of argon and oxygen at the end of the blow. For the production of low-nitrogen steels, argon can be used to minimize the nitrogen pickup during rotation, or by flushing the bath at the end of the blow. Nitrogen can also be alloyed into the steel by purging the melt with nitrogen at the end of the oxygen blow.

9.5.2 Mixed-Blowing Processes

Virtually every company has invented or modified a version of the oxygen steelmaking process to suit its own situation. Consequently, there are many designations for fairly similar processes. Table 9.10 attempts to translate and relate some of the more common acronyms. This table is divided into four broad categories: top blowing, mixed blowing with inert gases, mixed blowing with bottom oxygen, and bottom blowing.

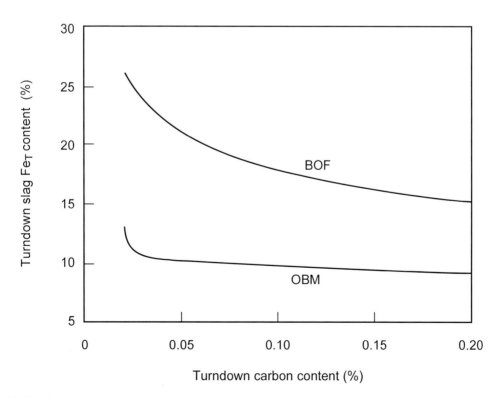

Fig. 9.13 Slag Fe_T content as a function of the carbon content at turndown in the U.S. Steel Gary Works BOF and OBM (Q-BOP) shops.

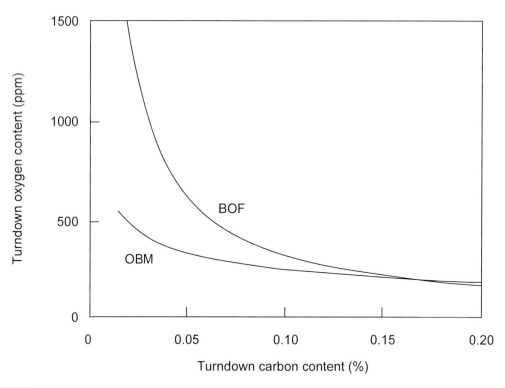

Fig. 9.14 Dissolved oxygen content as a function of carbon content at turndown in the U.S. Steel Gary Works BOF and OBM (Q-BOP) shops.

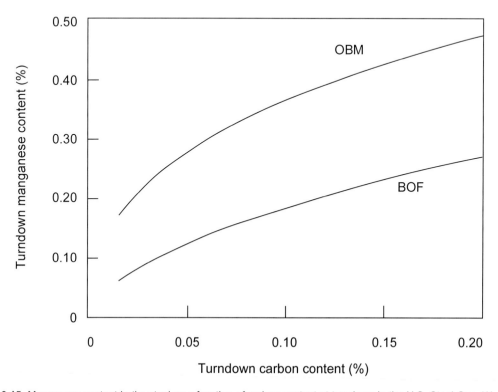

Fig. 9.15 Manganese content in the steel as a function of carbon content at turndown in the U.S. Steel Gary Works BOF and OBM (Q-Bop) Shops.

Table 9.10 Various Oxygen Steelmaking Process Designations
From Ref. 8, updated from Refs. 31–36.

Process Name	Origin	Description
Obsolete bottom-blown processes that preceded modern furnace configurations:		
Bessemer Converter		An early bottom-blown converter developed by Henry Bessemer in 19th century England, blowing air through simple tuyeres using acid refractories.
Thomas Converter		Similar to bessemer but using basic refractories in US.
All Top Blowing:		
LD	Voest, Austria	Linz-Donawitz. First top-blown process with water-cooled lance, lump lime.
BOF	Worldwide	Basic Oxygen Furnace. Common term for LD top lance-blown, lump lime.
BOP	USX and others	Basic Oxygen Process. Same as LD and BOF
LD-AC	ARBED/CRM Luxembourg; France	Similar to LD with powdered lime added through lance, for high P hot metal.
LD-CL	NKK, Japan	LD but lance rotates.
LD-PJ	Italsider	LD with pulsed jets, not in current use.
ALCI	ARBED, Luxembourg	LD basically with Ar /N_2 through lance. Post combustion ports and coal injection from top for higher scrap melting.
LD-GTL	Linde/National Steel, USA	LD with Ar or N_2 through top lance to limit over-oxidation, lump lime.
AOB	Inland, Union Carbide	Similar to LD-GTL.
Z-BOP	ZapSib Russia	Basically LD or BOF. Various process methods of adding additonal fuels to increase scrap melting. Some include preheat cycles. Can melt 100% scrap.
Mixed Blowing, Inert Stirring Gases:		
LBE	ARBED, Luxembourg IRSID, France	Lance Bubbling Equilibrium. LD with permeable plugs on bottom for inert gas. Lump lime.
LD-KG	Kawasaki, Japan	LD with small bottom tuyeres Ar and/or N_2.
LD-KGC	Kawasaki, Japan	LD with number of small nozzles using Ar, N_2CO for inert gas bottom stirring. Unique in that it uses recycled CO as a stirring gas. Lump lime.
LD-OTB	Kobe, Japan	Similar to LD-KG
LD-AB	Nippon Steel, Japan	LD with simple tuyeres to inject inert gas. Lump lime
NK-CB	NKK, Japan	Top-blown LD with simple bottom tuyere or porous plugs to introduce Ar/CO_2/N_2, lump lime

Table 9.10 Various Oxygen Steelmaking Process Designations *(continued)*

Mixed Blowing with Oxygen and/or Inert Bottom Gases:

OBM-S	Maxhutte, Germany Klockner, Germany	Mostly bottom OBM type with top oxygen through natural gas shrouded side tuyere, powdered lime through bottom.
K-BOP	Kawasaki Japan	Top and bottom blowing. Natural gas shrouded bottom tuyeres, powdered lime through tuyeres.
TBM	Thyssen, Germany	Top and bottom stirring with bottom nozzles and N_2/Ar.
LET	Solmer, France	Lance Equilibrium Tuyeres. Top blowing with 15 to 35% bottom-blown with fuel oil shrouded tuyeres.
LD-OB	Nippon Steel, Japan	OBM tuyeres (natural gas shrouded) on bottom with top lance, lump lime.
STB	Sumitomo, Japan	Mostly top-blown with lance, with special tuyere on bottom. Innerpipe O_2/CO_2, Outer pipe CO_2/N_2/Ar. Lump lime.
STB-P	Sumitomo, Japan	Similar to STB except powdered lime through top lance for phosphorus control.

All Bottom Blowing:

OBM	Maxhutte, Germany	Original 100% bottom-blown. Natural gas shrouded tuyeres, powdered lime through tuyeres
Q-BOP	USX, USA	OBM type 100% bottom-blown. Natural gas shrouded tuyeres, powdered lime through tuyeres.
KMS	Klockner, Germany	Similar to OBM. Early trials of oil shrouded tuyeres, now use natural gas. Can inject powdered coal from bottom for more scrap melting.
KS	Klockner, Germany	Similar to KMS only modified for 100% scrap melting.

(Note:**Bold** lettered processes are still in regular use today.)

The BOF is the overwhelmingly popular process selection for oxygen steelmaking. The usual reason to modify the BOF configuration is to lower the operating cost through better stirring action in the steel bath. It has been found by a number of companies and investigators that additional stirring of the bath from the furnace bottom reduces FeO in the slag. Lower FeO results in higher yield, and fewer oxidation losses of metallics. Bottom stirring increases slag forming, particularly if powdered limes are injected into the bath.

9.5.2.1 Bottom Stirring Practices

Inhomogenieties in chemical composition and temperature are created in the melt during the oxygen blow in the top-blown BOF[25] process due to lack of proper mixing in the metal bath. There is a relatively dead zone directly underneath the jet cavity in the BOF.

Bottom stirring practices using inert gases such as nitrogen and argon are being used extensively to improve the mixing conditions in the BOF. The inert gases are introduced at the bottom of the furnace by means of permeable elements (LBE process) or tuyeres. In a typical practice, nitrogen gas is introduced through tuyeres or permeable elements in the first 60 to 80% of the oxygen blow, and argon gas is switched on in the last 40 to 20% of the blow. The rapid evolution of CO in the first part of the oxygen blow prevents nitrogen pickup in the steel.

Some of the effects of bottom stirring and the resulting improved mixing include:

(a) *Decreased FeO content in slag.* Better mixing conditions in the vessel causes the FeO in the slag to be closer to equilibrium conditions, which results in lower concentrations of FeO in the slag. Plant studies have shown that for low carbon heats, bottom stirring can cause a reduction in the FeO level in slag by approximately 5%. This results in better metallic yield, lower FeO level in the ladle slag and reduced slag attack on the refractories. Improvements in iron yield by as much as 1.5% or more have been reported.[26] Lower levels of FeO in the steelmaking slag reduces the amount of heat generated during the oxygen blow, and hence reduces the maximum amount of scrap that can be charged in a heat.

(b) *Reduced dissolved oxygen in metal.* A study shows that bottom stirring can reduce the dissolved oxygen level in a low carbon heat by approximately 225 ppm. This lowering of dissolved oxygen leads to lower aluminum consumption in the ladle. Studies have shown aluminum savings of about 0.3 lb./ton due to bottom stirring.

(c) *Higher manganese content in the metal at turndown.* An increase of approximately 0.03% in the turndown manganese content of the metal has been shown. This leads to a reduction in the consumption of ferro-manganese.

(d) *Sulfur and Phosphorus removal.* Bottom stirring has been found to enhance desulfurization due to improved stirring. However, phosphorus removal has not been found to improve substantially in some studies. Although bottom stirring drives the dephosphorization reaction towards equilibrium, the reduced levels of FeO in steelmaking slags tend to decrease the equilibrium phosphorus partition ratio ((%P in slag)/[%P in metal]).

9.5.2.2 Bottom Stirring Maintenance Problems

The price for the bottom stirring benefits is in maintaining a more complicated process. This complexity varies depending on the type of bottom stirring devices installed. Generally, the simpler the device is to maintain, the lesser is the benefit obtained.

Starting at the simple end, porous plugs or permeable elements do not require maintaining gas pressure when the stirring action is not required. While gas under pressure can permeate these devices, steel does not penetrate, even when the gas is turned off. The disadvantage of the porous plugs is that they are not very effective stirrers. Two factors influence their poor performance. First, only a relatively small amount of gas can be introduced per plug. More gas flow requires more plugs thus adding complexity. Secondly, if the furnace bottom tends to build up, the plugs are covered over with a lime/slag mush and gas does not stir the steel bath. Rather, it escapes between the bricks and the protective mush. Another problem is at higher flow rates, the originally installed plugs do not last very long (1500 to 3000 heats) and replacements are shorter lived (200 to 2000 heats). The reasons for the shorter replacement plug life is not clear.

The use of the bottom stirring tuyeres (rather than plugs) can lead to stirring reliability and effectiveness problems. The tuyeres are known to get blocked early in the lining campaign when there is a buildup at the bottom of the furnace. The tuyere blockage can be prevented by maintaining proper gas flow rates through the tuyeres at different stages of the heat, by using high quality refractory bricks in the area surrounding the stirring elements or tuyeres, and by properly maintaining the bottom of the furnace by minimizing buildup.

9.5.2.3 Bottom Plug/Nozzle Configurations

There are basically three types of tuyeres used for mixed blowing. First, there is a refractory element that behaves much like porous plugs. This unit is made of compacted bricks with small slits. Like most tuyeres, it needs sufficient gas pressure to prevent steel penetration. This unit is more penetrating than porous plugs. Second, an uncooled tuyere is used to introduce large amounts of inert gases per nozzle. This results in local heavy stirring, which can more easily penetrate the buildup. Air or oxygen cannot be used because there is no coolant and the heat generated would make tuyere life too short to be practical. The third type is a fully cooled tuyere similar to the OBM (Q-BOP). Here, either inert gas or oxygen can be blown, causing very strong stirring and almost no problems penetrating bottom buildup.

In all cases the gas piping is routed through the furnace trunnions using rotary joints or seals to allow full rotation of the furnace.

9.5.3 Oxygen Steelmaking Practice Variations

9.5.3.1 Post-Combustion Lance Practices

Post combustion involves burning the CO gas produced by the reaction of carbon in the metal with jet oxygen, to CO_2 within the furnace. The heat generated as a result of this oxidation reaction can be used for any of the following purposes: (1) to increase the scrap-to-hot metal ratio, (2) to minimize the formation of skulls formed at the mouth of the vessel, (3) to minimize the formation of lance skulls and (4) to reduce the formation of "kidney" skulls in the upper cone region inside the BOF.

In a normal BOF operation, the percentage of CO in the off gas is approximately 80 to 90% during the middle portion of the blow, when the rate of decarburization is maximum, Fig. 9.16.[27] The oxidation of the CO gas to CO_2 generates 12 Mcal per Nm^3 CO (1260 BTU per scf CO)), which can melt more scrap and/or reduce skulling problems.

Post combustion lances have primary nozzles as well as secondary nozzles which typically are located at about nine feet from the tip of the lance, Fig. 9.17. The primary nozzles in PC lances are identical to those of the standard lances. They consist of four or five nozzles at the tip of the lance with inclination angles varying from 10 to 14 degrees. Usually, there are eight secondary nozzles above the primary ones, directed out, at an angle of about 35 degrees from the vertical. The oxygen flow rate through the secondary nozzles is considerably less than through the primary ones, usually 1.5 to 15% of the total oxygen flow.

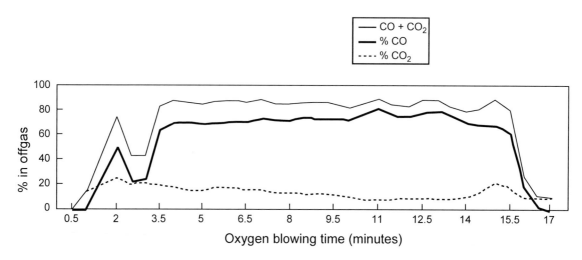

Fig. 9.16 The percentage of CO and CO_2 gases in the BOF offgas. *From Ref. 27.*

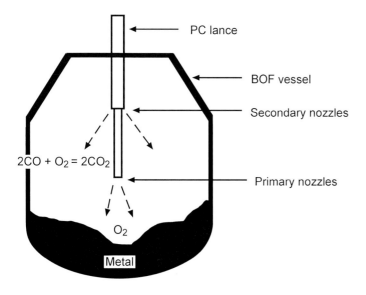

Fig. 9.17 Schematic of a post combustion lance in a top-blown BOF.

Studies show that post combustion has been successfully used to melt additional scrap in the BOF (achieving scrap-to-hot metal ratios of 25% or more), and to reduce lance and furnace mouth skulling problems. It has been found to increase the turndown temperature, if additional scrap or coolant is not added, to compensate for the extra heat generated. No effect on slag FeO[28] and turndown manganese in metal has been found. There has been no effect on refractories when post-combustion is used at low flow rates in the skull control mode. However, if post-combustion is used in high flow, scrap melting mode, some care in design and operating parameters must be observed to avoid damaging the cone.

9.5.3.2 High Scrap Melting Practices

The ability to vary the scrap-to-hot metal ratio in a BOF is important because of the variations in scrap prices and the possibility of having hot metal shortages from time to time due to problems in the blast furnace. Several shops add anthracite coal early in the oxygen blow as an additional fuel to melt more of scrap. Such processes can be used in conjunction with post combustion lances.

A recent example of the scrap stretch is a series of steelmaking processes called Z-BOP that were developed in Russia to melt higher scrap charge.[29,30] These processes include Z-BOP-30, Z-BOP-50, Z-BOP-75 and Z-BOP-100, where the numbers indicate the percentage of scrap charge in the BOP. The Z-BOP-100 process utilizes 100% scrap charge with no hot metal, using the conventional BOP with virtually no equipment modifications.

In the Z-BOP processes, the additional energy required to melt scrap comes from lump coal (anthracite and high volatile bituminous), which can be fed using a bin system. In the Z-BOP-100 process, the first batch of scrap is fed into the furnace using a scrap box, and coal is added in small batches while blowing oxygen. This process continues with additional scrap being charged. The number of scrap boxes charged can vary from three to five. The main blow commences when all the scrap is charged in the furnace, and small additions of coal are made throughout the main blow.

The typical tap-to-tap time for the Z-BOP-100 process is 64 to 72 minutes. No one has used these high scrap percent processes on a sustained basis because of damage to the furnace linings. Typical iron yield of the Z-BOP 100 process is about 80%, (91–92% is normal for the BOF) and the slag FeO content is very high (45 to 72%; normal for the BOF is approximately 25%). The sulfur content of tapped steel was not found to be higher than that of the scrap itself.

The major disadvantages of this process include decreased yield, increased slag FeO which can adversely affect the refractory lining, presence of tramp elements similar to EAF, and presence of fugitive emissions (SO_2 and NO_x). This process may be used during hot metal shortages, but is not proven for continuous use. Some shops have used a 70% or 80% scrap charge to accommodate a blast furnace reline and stretch a limited amount of hot metal from the remaining blast furnace(s).

9.5.3.3 Slag Splashing Refractory Maintenance

Any improvement in the furnace lining life is a boon for steelmakers due to the increased furnace availability and reduced refractory costs. Slag splashing is a technology which uses high pressure nitrogen through the oxygen lance after tapping the heat, and splashes the remaining slag onto the refractory lining, Fig. 9.18. The slag coating thus formed cools and solidifies on the existing refractory, and serves as the consumable refractory coating in the next heat.

Slag splashing is currently being used by several BOF shops, and has extended refractory lining life to record levels. A typical slag splash operation following the tapping of a heat is as follows:

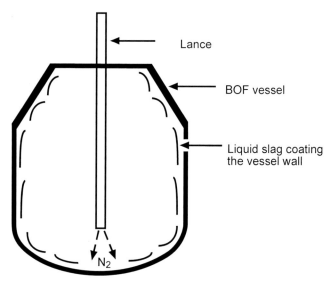

Fig. 9.18 Schematic of slag splashing in the BOF.

1. The melter observes the slag, and determines whether any conditioning material such as limestone, dolomite or coal must be added. High FeO levels in the slag (estimated by a simple correlation with turn-down carbon and manganese values) are undesirable, and these slags are either not splashed or are treated with conditioners to make them more viscous.

2. The vessel is rocked back and forth to coat the charge and tap pads. The furnace is then uprighted.

3. The nitrogen blow is started when the oxygen lance is lowered in the furnace to a predetermined height. The positioning of the lance and the duration of the nitrogen blow are determined by the melter, and depend on the condition of the slag and the area of the furnace that has to be coated.

4. At the end of the nitrogen blow, the furnace is tilted and excess slag is dumped into the slag pot.

Not only has slag splashing been successful in increasing the furnace lining life, but it has also significantly reduced the gunning costs. One major company has reported a reduction in gunning cost by 66%, and an increase in furnace availability from 78 to 97%. No detrimental effects on the turn-down performance (chemical and thermal accuracy) of the furnace have been found.

Slag splashing has reduced the frequency of furnace relines considerably. This creates a problem in maintaining hoods, which is normally done during long furnace reline periods. Now, planned outages are needed for hood maintenance without any work on the furnace lining.

9.5.3.4 Effects of Desulfurized vs. Non-Desulfurized Hot Metal

Generally, heats that are made for sulfur sensitive grades of steel (0.010% or lower) must be desulfurized. Grades that do not have stringent sulfur specifications (approximately 0.018%) may not require hot metal desulfurization. The high sulfur content of the hot metal (approximately 0.06%) has to be reduced to acceptable levels in the hot metal desulfurizer, the BOF and/or in the ladle. The hot metal desulfurizing station can reduce the sulfur levels in the hot metal to as low as 0.001% through injection of lime and magnesium.

The BOF is not a very good desulfurizer because of the oxidizing conditions in the furnace. Heats using non-desulfurized hot metal can be desulfurized to approximately 0.012% sulfur in the BOF. However, heats using desulfurized hot metal may pick up sulfur if the slag in the desulfurizer was not skimmed properly or if a high sulfur scrap was used.

Some shops desulfurize virtually 100% of the hot metal to enable the blast furnace to run with a lower flux charge. This practice also reduces the heat burden on the blast furnace and saves coke. The net effect is a reduction in hot metal costs that is larger than the expense of desulfurizing.

9.6 Process Control Strategies

9.6.1 Introduction

Process control is an important part of the oxygen steelmaking operation as the heat production times are affected by it. Several steelmaking process control strategies are available today, and steel plants use strategies depending on their facilities and needs. Process control schemes can be broadly divided into two categories: static and dynamic. Static models determine the amount of oxygen to be blown and the charge to the furnace, given the initial and final information about the heat, but yield no information about the process variables during the oxygen blow. Static models are basically like shooting for a hole-in-one on the golf course: there is no further control once the ball leaves the face of the club. Dynamic models, on the other hand, make adjustments during the oxygen blow based on certain in-blow measurements. The dynamic system is like a guided missile in a military situation; there is provision for in-flight correction for better accuracy to hit the target.

Lance height control is an important factor in any control scheme as it influences the path of the process reactions. Detailed descriptions of static and dynamic models currently in use by the steel industry, along with a brief discussion of lance height control, follow.

9.6.2 Static Models

A static charge model is normally a computer program in an on-site steelmaking process computer. Plants have static models that depend on the type of operation and product mix. The static charge model uses initial and final information about the heat (e.g. the amount of hot metal and scrap, the aim carbon and temperature) to calculate the amount of charge and the amount of oxygen required. The relevant information is collected in the process computer, and a static charge model calculation is performed at the beginning of the heat. The output from the model determines the amount of oxygen to be blown and the amount of fluxes to be added to attain the desired (aim) carbon and temperature for that heat.

Performance of the charge model depends on a number of factors: the accuracy of the model; the accuracy of the inputs (aims, weights, chemistry of charge ingredients) to the process computer; consistency in the quality of materials used and of practices; and conformance to the static charge model.

The objective of the static charge model is to have improved control of carbon, temperature, sulfur and phosphorus. An accurate control over these variables using the model will result in fewer reblows and steel bath cooling, which are time consuming and expensive. Improved control will also result in better iron yield, greater productivity, better refractory lining life and an overall lower cost of production. The OBM (Q-BOP) static charge model is essentially the same as the BOF model, except for slight modifications arising out of differences in slag chemistry and natural gas injection. Most of the static charge models have an OK-to-tap performance of approximately 50 to 80%.

There is one BOF shop in Europe that achieves over 90% OK-to-tap performance. It is a two-furnace LBE type shop with inert gas bottom stirring. They use only one furnace at a time while the other is being relined. Their success is attributable to discipline, attention to detail, pushing the one furnace very hard which results in consistent heat losses, almost zero furnace maintenance with resultant short lining lives, and well-developed computer models. Their resulting tap-to-tap times are very short and production rates are very high. There is no dynamic control system nor sensor sub-lance.

9.6.2.1 Fundamentals of the Static Charge Model

The charge model consists of a comprehensive heat and mass balance involving different chemical species that participate in the steelmaking reactions. The hot metal poured into the steelmaking vessel typically contains 4.5%C, 0.5%Si, 0.5%Mn and 0.06%P, apart from other impurities. These elements, along with Fe, get oxidized during the blow to form gases and slag, and these reactions generate the heat required to run the steelmaking process.

The static model also contains information on the heat of oxidation of different elements, such as carbon and iron, as a function of the oxygen blow. This information, along with a comprehensive mass balance involving different chemical species, determines the amount of heat generated from the steelmaking reactions as a function of total oxygen blown.

The primary coolants used in oxygen steelmaking processes are scrap, iron ore (sintered or lump), DRI and limestone (or dolomitic stone). The heat generated from the oxidizing reactions must be balanced with the use of these coolants to achieve the temperature aim of the heat. The model determines the amount of hot metal and scrap to be charged in the furnace, depending on the size and the type of heat to be made.

The rate of decarburization as a function of oxygen (available from the lance as well as from the ore) must be known to determine the duration of the blow, so that the heat turns down at the desired carbon level. The model also computes slag basicities (normally in the range of two to five) for effective dephosphorization and desulfurization. The amount of silica in the slag is calculated from hot metal silicon, assuming that it gets oxidized entirely to silica. The amount of lime (CaO) to be added is calculated from the desired basicity and the amount of silica formed from oxidation.

The charge model also determines the amount of dolomitic lime (CaO–MgO) to be charged into the furnace from an estimate of the overall slag chemistry. It is important to saturate the steelmaking slag with MgO to protect the furnace refractory lining.

9.6.2.2 Operation of the Static Charge Model

Many different kinds of charge models exist. In one typical model, charge calculations are made in three separate steps at the beginning of a heat. In the first step, called the hot metal calculation, the steelmaking operator makes a charge calculation of the amount of hot metal and scrap needed for the heat, given the product size, and the aim carbon and temperature of the heat. The hot metal analysis may or may not be available at this time. If not, an estimate based the on previous hot metal analysis is used.

Having fixed the hot metal and scrap weight, the operator runs the second part of the calculation, which is called the product calculation. For this, the hot metal analysis is required. The amount of fluxes to be added and the amount of oxygen to be blown are determined. This part of the calculation takes place before the start of the oxygen blow.

The third part, the oxygen trim calculation, takes place during the early stages of the oxygen blow and after the fluxes have been added to the furnace. In this step, the charge model utilizes the actual amounts of fluxes added to the furnace (which can be different from the calculated amounts), and provides the new trimmed value of total oxygen. This step is required to correct for any variations in the process due to weighing inaccuracies in the amount of flux charge.

9.6.3 Statistical and Neural Network Models

Over the last few years, a significant amount of attention has been given to statistical methods designed to improve the performance of the static charge model. The turn-down performance of the static model does not only depend on the accuracy of the mass and thermal balance of the steelmaking system, but also on the inherent variability of the process which arises out of a number of factors that cannot be accurately quantified. Steelmaking melters use the concept of bias adjustments to correct for unexplained inconsistencies in temperature or carbon performance. For example, if the melter finds a series of heats that were hot for unexplained reasons, he makes a temperature bias correction for the following heat, hoping that it will turn down close to the aim temperature. This bias correction, in effect, alters the aim temperature of the following heat for the purpose of charge calculation. The problem here is that different operators may react to a given situation differently and therefore the manual adjustments lack consistency. In addition, many operators are tempted to tamper (make adjustments too frequently), thereby introducing another variable.

Statistical models, in effect, use the same concept of bias to make adjustments in the charge calculation for a heat but do so with better consistency. The model tracks the turn-down performance of heats, and by statistical means, determines the amount of unexplained deviations in the performance of carbon, temperature etc. The model, then instructs the charge model to make appropriate modifications in the inputs for charge calculation for the following heat, so that these unexplained deviations may be countered. Studies indicate that the statistical models have been successful in fine tuning the static models to some extent.

Neural network algorithms are also being developed and used to improve the turndown performance. In this technique, the dependent variables (such as turndown carbon and temperature) are linked to the independent variables of a heat (such as the amount of hot metal, scrap and ore) through a network scheme. The network may consist of a number of layers of nodes, connected to each other in a linear fashion. As the input signals enter the network, they are multiplied by certain weights at the nodes and the products are summed as these signals are transferred from one layer to the next. The weights are determined by training the neural network model based on available historical data. The outputs from the model are calculated by adding the product of the weights and the signals in the last layer of the model.[37]

Researchers have obtained reasonable success in estimating process variables using neural network modeling.[38] However, the success of this technique depends on the quality and quantity of input signals that are fed into the model. In simple terms, the old computer proverb is still valid: "garbage in, garbage out."

9.6.4 Dynamic Control Schemes

Several steelmaking shops are currently using dynamic control schemes in combination with static models to improve their turndown performance. Dynamic control models use in-blow measurements of variables such as carbon and temperature, which can be used in conjunction with a static charge model, to fine tune the oxygen blow. A few dynamic control schemes are discussed in the following sections.

9.6.4.1 Gas Monitoring Schemes

This scheme is based on a continuous carbon balance of the entire BOF system. A gas monitoring system, which normally consists of a mass spectrometer, continuously analyzes dust free offgas samples for CO and CO_2 and determines the amount of carbon that has been oxidized at any given stage of the blow. Given the amount of carbon present in the system at the beginning of the heat (calculated from hot metal and scrap), one can calculate the level of carbon in the metal dynamically during the oxygen blow.

Although this method has a sound theoretical basis,[40] it has not been very successful at the commercial level. One problem is that it is not possible to accurately determine the initial amount of carbon present in the hot metal and scrap. Even a small amount of error in determining this initial carbon content could result in serious process control problems.[39] Another problem is the difficulty in continuous and accurate determination of the amount of CO and CO_2 exiting the system as offgas. Often the volume of filtering apparatus and tubing and the time to analyze a sample causes a delay of more than 60 seconds before the result is displayed or available for control.

9.6.4.2 Optical and Laser Based Sensors

A few shops in North America have started using light sensors to dynamically estimate carbon levels in the low carbon heats. The light meter system continuously measures the intensity of light emitted from the mouth of the steelmaking vessel during the oxygen blow.[43] The system then correlates characteristics of the light intensity curve with the carbon content towards the end of the oxygen blow when the estimated carbon is around 0.06%. This system has been been quite successful in dynamically estimating carbon levels in low carbon heats (0.06% or less), and adjusting the oxygen blow for improving the turndown carbon accuracy. However, this system cannot be used for heats with aim carbon greater than 0.06%.

A few oxygen steelmaking shops have tried using laser based sensors to dynamically estimate the bath temperature. Some laboratory researchers have attempted to devise a probe that would take in-blow metal samples and provide instant chemical analysis. However most of these techniques are still in the developmental stage and are not yet commercially available.

9.6.4.3 Sensor or Sub-Lances

Sensor or sub-lances are effective tools for controlling both the carbon and temperature of the metal.[41] In this technique a water-cooled lance containing expendable carbon and temperature sensors is lowered into the bath about two or three minutes before the end of the oxygen blow. The sensor determines the temperature of the metal bath at that time of blow and the carbon content of the steel by the liquidus arrest temperature method. The carbon and temperature readings thus taken are fed into the process computer which determines the additional amount of oxygen to be blown and the amount of coolants to be added.

The sensor lance has proven to be an extremely effective process control tool, with an OK-to-tap performance of 90% or more. The drawbacks of using sensor lances are substantial capital costs and engineering and maintenance problems. There are many oxygen steelmaking shops in Japan, and a few in North America, that are currently using the sensor lance technique to improve their turndown performance.

9.6.4.4 Drop-in Thermocouples for Quick-Tap

Drop-in thermocouples or bomb thermocouples are effective tools for measuring the temperature of the metal bath without turning down the steelmaking vessel. The thermocouple is contained in a heavy cylindrical casing, which is dropped into an upright vessel from the top of the furnace. The thermocouple probe has a sheathed wire attached to it which can convey a reliable emf reading before it burns up. The generated emf is conducted to a converter card and a computer, which converts the analog emf signal to a temperature value. The temperatures recorded by drop-in thermocouples have been found to be in good agreement with those measured using immersion thermocouples.

Drop-in thermocouples can be used to quick-tap heats, i.e. tap heats without turning the furnace down for chemical analysis. Certain grades of steel do not have stringent requirements for phosphorus or sulfur. In such a case, all that the steelmaking operator has to worry about at the end of the blow is temperature and carbon. If the operator feels confident that the carbon level in the bath is below the upper limit as specified by the grade (by either looking at the flame at the end of the oxygen blow, or by reading light meter carbon estimates, or by any other means), then he can use a drop-in thermocouple to determine the temperature of the metal bath without turning the vessel down. If the temperature recorded by the drop-in thermocouple is close enough to the aim temperature, then the heat can be quick-tapped.

Quick-tapping is becoming increasingly popular in North America with the advent of better devices for measuring in-blow carbon and bath temperatures. Quick-tapping has the obvious advantage of saving the production time that is otherwise consumed in turndowns. However, there is an obvious risk that a quick-tapped heat may be later found to have one or more of the elements present above their specification limits, in which case the heat may have to be regraded or scrapped. Another disadvantage is that the sensors cost between $0.05 to $0.10 per ton of steel.

9.6.4.5 Sonic Analysis

Several studies have been carried out in the past to correlate the audio emissions from the steelmaking furnace to the decarburization reaction and slag/foam formation. Investigators found an increase in the sound intensity at the beginning of the blow due to the establishment of the decarburization reaction.[42] The sound intensity reportedly decreased with the onset of slag and foam formation in the slag. The sonic device can be used to control carbon and slag formation, which in turn can be used to control dephosphorization. These units can be affected by extraneous noise. Control is done by altering lance height or blow rate.

9.6.5 Lance Height Control

Lance height is defined as the vertical distance between the slag-metal interface in the furnace and the tip of the oxygen lance. The consequences of poor lance height control were discussed earlier in Section 9.2.2.6 on Oxygen Blow. It is an extremely important parameter as the height from which oxygen is blown affects the overall fluid flow of metal and slag during the blow. It is important to be able to measure it and to keep it consistent to achieve good process control of the furnace.

9.6.5.1 Measurement of Lance Height

Measurement of lance height was also discussed earlier. However, the best practice in measuring lance height is to do it often. If the furnace bottom rapidly changes, due to either excess wear or to rapid bottom buildup, the lance height should be measured frequently (once per shift). If the furnace lining shape is stable, then less frequent determinations (once per day) will suffice. Lance height has a significant effect on slag formation, furnace and lance skull formation and decarburization rate. Generally, the safest, reasonably accurate method is calculating the bath height from refractory wear laser data.

9.7 Environmental Issues

9.7.1 Basic Concerns

Environmental or pollution control issues are becoming increasingly difficult and costly. State jurisdictions are setting increasingly stringent standards for new site permits or for significant modification to existing processes. Nearly every process has become a source of emissions. Usually, these must be controlled to a legal standard which is measured in pounds or tons per unit time. Sometimes, tradeoffs are permitted as long as a net major improvement or emissions reduction is achieved.

There are at least five major characteristic sources of environmental pollution. These include airborne emissions, water borne emissions, solid waste, shop work environment (usually burning and welding) and safety, and noise. Air emissions are the major issues in a BOF shop. Water borne pollutants, generated by the scrubber system, are clarified by settling and then the clean water is recirculated. Solid waste is generated by oxide bearing materials collected from the scrubber, electrostatic precipitators or baghouse. While much solid matter is recycled, the rest is put into long range storage for byproduct use (i.e., aggregate) and future recycle (for example, Waste Oxide Briquettes). Noise generally is not a major concern in a BOF (i.e., compared to an arc furnace). Shop environmental and safety is always a major concern for personnel protection and employee morale. This discussion will concentrate principally on issues of air pollution.

9.7.2 Sources of Air Pollution

There are two broad areas: undesirable gases such as CO, fluorides or zinc vapors; and particulates such as oxide dusts. A brief summary of the various sources is discussed below.

9.7.2.1 Hot Metal Reladling

Pouring of hot metal from the torpedo car into the transfer ladle results in plumes of fine iron oxides and carbon flakes. This mixture, called kish is the major source of dust and dirt inside the shop. The graphite particles are generated because the liquid hot metal is saturated with carbon and when the temperature drops during pouring into the ladle, the carbon precipitates out as tiny graphite flakes. The usual method of control is to pour inside an enclosure (hood) and to collect the fume in a baghouse.

9.7.2.2 Desulfurization and Skimming of Hot Metal

Pneumatic injection of the liquid iron with nitrogen and magnesium with a lance stirs the metal and generates a fume similar to that of reladling. Often, the reladling hood is designed to accommodate the desulfurizing lance operation. During slag skimming, the splash of slag and metal is collected in a vessel (usually a slag pot) which also generates fine oxide fume. This is frequently collected in a separate hood over the skimming collection pot.

9.7.2.3 Charging the BOF

There is some fume generated when the scrap hits the bottom of the furnace. However, the major emission is generated while pouring hot metal into the furnace. Here, very dense oxide clouds, kish, and heavy flames rise quickly. Some shops have suction hoods on the charging side above the furnace that collect fume and divert the heat away from the crane. However, many shops are not so equipped and must rely on slow-pouring to limit the fume emission from the roof monitors to comply with regulations. Pouring too fast results in heavy flame reactions that have been known to anneal the crane cables, causing spillage accidents.

9.7.2.4 Blowing (Melting and Refining)

By far the largest mass of fume is generated during the main blow. Approximately 90% by weight of total sources is generated at this point. The fume consists of hot gases at temperatures, over 1650°C (3000°F) and very heavy concentrations of iron oxide particles. The particulates can contain heavy metal oxides such as chromium, zinc, lead, cadmium, copper and others, depending on the scrap mix. The gas composition is approximately 80 to 95% CO and the rest is CO_2. In open combustion hoods, where air is induced just above the furnace into the cleaning system, the temperatures can be as high as 1925°C (3500°F) because of further combustion of CO. The fume is mostly fine iron oxides with some other oxides and dusts from flux additions.

The collection systems for this fume are of two types. Open hoods draft enough air to completely burn the CO before it hits the filtering device. Closed or suppressed combustion hoods either eliminate or reduce the induction of air to very low levels (<15%). This reduces the required fan horsepower and filtering capacity. However, consistent sealing between the furnace and hood is required to prevent the generating and igniting of explosive mixtures. Generally, the capital and operating costs for closed systems are less than for open systems. Thus, the most recently built shops are characterized by closed hood systems.

Another controlling characteristic is the filtering device. Two types have been successful in BOFs: electrostatic precipitators and venturi scrubbers.

The electrostatic precipitator draws the gas and fume into chambers. These are used only with open combustion hoods. These contain many parallel plates or alternating plates and wires spaced closely together within a few inches. Alternate elements (usually wires) are charged to a high potential. As the fume nears or contacts the highly charged element it becomes charged and is attracted to the other element (of opposite charge). Periodically the charge is dropped and the elements are vibrated, which releases the fume into a hopper and transport system below. Precipitators require gas temperatures less than 370°C (700°F) and some moisture (>6%) to be efficient. A common problem is plate warpage, due to excessive heating, which results in electrical short-circuits.

The venturi scrubber, used with either open or suppressed combustion hoods, induces the gas/fume through a violent spray which washes and separates the fume from the gas. The oxide laden water is then subjected to a series of separation processes, which are a combination of centrifugal, chemical and settling operations. The settled filter cake is dredged and dried for recycling. Most facilities now use venturi scrubbers.

9.7.2.5 Sampling and Testing

Turning the furnace down for testing and temperature sampling generates fine oxide fumes at a relatively low rate. Most of this fume is drafted into the hood above. However, many shops running on limited fan capacity close off suction in the hood on the testing furnace and start blowing the next heat on the other furnace. The sampling/testing fume rates for a bottom-blown process are much heavier than top-blown furnaces due to continued blowing through the tuyeres to keep them clear. Most bottom-blown facilities use a charge-side enclosure, or doghouse, to effectively collect these emissions into the main hood. In effect, the doghouse is a set of doors that extends the hood down to the charging floor.

9.7.2.6 Tapping

Some fume comes directly from the furnace but most comes from steel colliding with the bottom of the ladle or other liquid steel. The addition of alloys increases the smoke during tap. Some fume is collected in the main hood, but much escapes through the roof monitors. Some shops use a tap collection hood and route the fume to a baghouse.

9.7.2.7 Materials Handling

The handling of fluxes, alloys and treatment reagents can be a significant source of fume. Often the handling facilities are equipped with a collection system that leads to a baghouse. Materials used in small amounts per heat can be transported in bags or super sacks to eliminate fugitive dust during transfer. Sometimes materials quality control specifications are required to control the generation of dust during handling and transfers. Reducing burnt lime fines specification is an example.

9.7.2.8 Teeming

Ingot teeming generally does not require special collection systems unless a hazardous element, such as lead, is involved. Free machining bar stock and plates alloyed with lead are common examples. Here lead is added while teeming ingots. A mobile hood collects the lead laden fume while teeming each ingot. It is both a hazardous material for the workers (vapors and particulate) and a solid waste. Further, workers' lead contaminated protective clothing and equipment are collected in special containers for proper disposal.

9.7.2.9 Maintenance and Skull Burning

The BOF operations generate skulls on equipment such as the furnace mouth, ladles, oxygen lances, metallic spills etc. These must be processed or removed by oxy-propane burning or oxygen lancing. Generally, this is a small emission source.

9.7.3 Relative Amounts of Fumes Generated

Most of the fume in the overall process is generated during the main blow. Table 9.11 summarizes the relative amounts of fume generated from various BOF sources. Overall, about 31 lbs of fume are generated per ton of steel. Over 90% is from the main oxygen blow and reblows. The OBM (Q-BOP) and other major bottom blowing processes generate somewhat lesser amounts of fume with larger particle sizes which are easier to collect than the case of 100% top-blown. Desulfurizing and tapping each generate about one pound per ton.

Table 9.11 also shows how much is removed by a reasonably good filtering system. A well designed system will reduce the fugitive emissions in a shop to less than 0.2 lbs/ton, or by 99.4%

There is at least one consultant who has developed an effective modeling technique to test and optimize emission collection designs. A test is done in a water tank using small jets of salt water as pollution sources. The salt water is more dense than fresh water, and when put into a fresh water tank and modeled upside down, it generates a plume startlingly like the real thing. This technique has been very successful for both conformance to emission standards and for improving shop work

Table 9.11 Summary of Major Emission Sources of a BOF Shop. *From Ref. 44*

Emission Source	Approx. Amt. (lb/ton, steel)
BOF hot metal transfer, source	0.190
Building monitor	0.056
Hot metal desulfurization, uncontrolled	1.090
Controlled by baghouse	0.009
BOF charging, at source	0.600
Building monitor (roof top)	0.142
Controlled by baghouse	0.001
BOF refining, uncontrolled	28.050
Controlled with open hood, scrubber	0.130
Controlled with closed hood, scrubber	0.090
OBM (Q-BOP) refining, closed hood scrubber	0.056
BOF tapping, at source	0.920
Building monitor	0.290
Controlled by baghouse	0.003
Total at source or uncontrolled, BOF	30.850
Total controlled (scrubber, baghouse)	0.158
Total controlled (scrubber, monitors)	0.497

environment. Most jurisdictions look favorably on such modeling to check the viability of candidate collection system designs.

9.7.4 Other Pollution Sources

9.7.4.1 BOF Slag

BOF slag is generated at a rate of approximately 240 pounds per ton of steel. Currently about half of this quantity gets recycled within the plant to the sinter plant or is used directly in the blast furnace. Such in-plant slag recycling has been declining because of higher steel quality demands, e.g., lower phosphorus. Other uses such a aggregate and agricultural purposes are being explored.

9.7.4.2 BOF Dust and Sludge

BOF dust and sludge is generated at a rate of approximately 36 pounds per ton of steel. Small amounts are sold to other industries, such as the cement industry, while small amounts are recycled to the sinter plant for metal recovery. There is a growing interest in using this material in the BOF as an iron source and coolant, in the form of WOBs or waste oxide briquettes.

The major driving forces for recycling come from government mandated pressures to reuse byproducts and not to land fill them. In many cases there is an economic opportunity to use a low cost iron or coolant source. Problems to be overcome are buildup of phosphorus during recycling, zinc content, moisture, and the use of safe and suitable binders to improve handling characteristics.

9.7.5 Summary

Studying emission sources is a way to understand the process from another point of view. It is of increasing social and legal importance. Different jurisdictions do not always agree on acceptable levels, but they are all headed toward tougher environmental standards in the future. Accordingly, pollution control measures and evolving technologies will increasingly influence the design, capital, work environment, and operating costs of shops. Considerable thought by the steel industry is now being given to controlling emissions through engineering.

References

Raw Materials
1. R. D. Pehlke, ed., et. al., *BOF Steelmaking*, ISS-AIME 4 (1977): 86–95.
2. D. A. Goldstein, Ph.D. thesis, Carnegie Mellon University, 1996.
3. A. K. Biswas, *Principles of Blast Furnace Ironmaking* (SBA Publications, 1984), 190.
4. D. A. Dukelow, J. P. Werner, and N. H. Smith, in *Steelmaking Conference Proceedings*, ISS, Nashville 1995, 67–72.
5. R. Balajee, P. E. Callaway Jr., and L. M. Keilman, *Steelmaking Conference Proceedings*, ISS, Nashville 1995, 51–65.
6. R. Boom, R. Beisser, and W. van der Knoop, *Proceedings of the Second International Symposium on Metallurgical Slags and Fluxes*, ed. H. A. Fine and D. R. Gaskell, AIME, 1984, 1041–60.
7. Brahma Deo and Rob Boom, *Fundamentals of Steelmaking Metallurgy* (UK: Prentice Hall International, 1993), 161–88.
8. Chatterjee, C. Manrique, and P. Nilles, "Fundamentals of Steelmaking Metallurgy," *Ironmaking and Steelmaking*, vol. 11, no. 3 (1984): 117–31.
9. ICRI Report 517, "Ferrous Scrap Materials Manual," Iron Casting Research Institute, 1995.

Process Reactions
10. H. W. Meyer et al., *Journal of Metals* (July 1968), 35–42.
11. "Fundamentals of Steelmaking", E. T. Turkdogan, The Institute of Materials (UK), 1996, 209–43.
12. W. Lange, Fifth International Iron and Steel Congress, ISS, April 1986, 231–48.
13. J. Papinchak and A. W. Hutnik, "Basic Oxygen Steelmaking-a New Technology Emerges," *The Metals Society*: 46–52.
14. M. Kamana, "Pure Oxygen Bottom Blown Converter Steelmaking Process," *Tekko Kai,* (January 1978), Japan.
15. R. J. Fruehan, in *Advanced Physical Chemistry for Process Metallurgy,* ed. N. Sano, W.-K. Lu and P. Ribaud (London: Academic Press, 1997).
16. A. I. Van Hoorn, J. T. van Konijnenburgh, and P. J. Kreijger, "The Role of Slag in Basic Oxygen Steelmaking Processes," ed. W.-K. Lu (Ontario: McMaster University Press, 1976), 2.1–.22.
17. F. Bardenheurer, H. vom Ende, and K. G. Speith, in *Arch. Eisenhuttenwes* 39 (1968): 571.
18. R. Boom, R. Beisser, and W. van der Knoop, *op. cit.*, 161–88.
19. A. Masui, K. Yamada, and K. Takahashi, "The Role of Slag in Basic Oxygen Steelmaking Processes," ed. W.-K. Lu (Ontario: McMaster University Press, 1976), 3.1–.31.
20. R. D. Pehlke, ed., et. al., *op. cit.*, 101–70.
21. R. J. Fruehan, "Ladle Metallurgy Principles and Practices", *Iron and Steel Society*, 1985: 22–3.

Process Variations
22. H. N. Hubbard, Jr. and W. T. Lankford, Jr., "The Development and Operation of the OBM (Q-BOP) Process in the United States Steel Corporation," *Proceedings of the 57th National Open Hearth and Basic Oxygen Steel Conference,* (Atlantic City, April 28–May 1), AIME, vol. 57, 1974, 258–74.
23. Y. Sahai, C. Xu, and R. I. L. Guthrie, "The Formation and Growth of Thermal Accretions in Bottom Blown/Combined Blown Steelmaking Operations," *Proceedings of the International Symposium on the Physical Chemistry of Iron and Steelmaking,* (Toronto, August 29-September 2, 1982), CIM, 1982: VI38–VI45.
24. M. J. Papinchak and A. W. Hutnik, "OBM (Q-BOP) Practice and Comparisons," *Proceedings of Basic Oxygen Steelmaking—a New Technology Emerges*, (London, May 4–5, 1978), The Metals Society, 1979, book 197, 46–52.

25. J. Jimenez, C. T. Simison, and H. J. Rossi, in *Steelmaking Conference Proceedings* (Warrendale, PA: Iron and Steel Society, 1991), 447–51.
26. T. H. Bieniosek, in *Steelmaking Conference Proceedings* (Warrendale, PA: Iron and Steel Society, 1988), 281–7.
27. B. Sarma, R. C. Novak, and C. L. Bermel, in *Steelmaking Conference Proceedings* (Warrendale, PA: Iron and Steel Society, 1996), 115–22.
28. J. J. Repasch et al., in *Steelmaking Conference Proceedings* (Warrendale, PA: Iron and Steel Society, 1995, 225–35.
29. G. S. Galperin et al., in *Steelmaking Conference Proceedings* (Warrendale, PA: Iron and Steel Society, 1993, 47–54.
30. G. S. Galperin et al., in *Steelmaking Conference Proceedings* (Warrendale, PA: Iron and Steel Society, 1994, 73–91.
31. R. B. Mullen and F. J. Goetz, "Experience with Lance Bubbling Equilibrium and Post-Combustion at DOFASCO, Inc.," *Proceedings, 68th Steelmaking Conference,* (Detroit, April 1985), AIME, vol. 68, 293–7.
32. M. Ohji et al., "Operation Technology for Long and Steady Refractory Life of LD Converter with Top and Bottom Blowing Type," *Proceedings, International Oxygen Steelmaking Congress*, AIME (Linz, May, 1987), 399–412.
33. H. Iso et.al., "Development of Bottom-Blowing Nozzle for Combined Blowing Converter," *Transaction of the Iron and Steel Institute of Japan* 28:1 (1988): 49–58.
34. Emoto et al., "Combined Blowing Processes of Kawasaki Steel," *Proceedings, 70th Steelmaking Conference,* AIME (Pittsburgh, 1987), vol. 70, 347–53.
35. T. H. Bieniosek, "LD-KGC Bottom Stirring at LTV Cleveland #2 BOF," *Proceedings, 71st Steelmaking Conference,* AIME (Toronto, 1988), vol. 71, 281–7.
36. H. Yamana et al., "CO Gas Bottom Blowing in the Top and Bottom Blowing Converter," *Proceedings, 70th Steelmaking Conference,* AIME, (Pittsburgh, 1987), vol. 70, 339–46.

Process Control Strategies

37. Philip D. Wasserman, *Neural Computing: Theory and Practice* (New York: Van Nostrand Reinhold, 1989).
38. S. Y. Yun, K. S. Chang, and S. M. Byun, *Iron and Steelmaker* (August 1996): 37–42.
39. R. D. Pehlke, ed., et. al., *op. cit.*, 265–6.
40. P. Hahlin, *Steel Times Supplement* (June 1993): 4–8.
41. D. W. Kern and P. D. Stelts, 76th General Meeting of AISI, New York, 1968.
42. A. E. Parsons and D. Shewring, *Journal of Iron and Steel Institute*, (May 1964): 401–5.
43. A. Sharan, "Light Sensors for BOF Carbon Control in Low Carbon Heat," *Proceedings 81st Steelmaking Conference,* Toronto, 1998, Vol 81, Iron and Steel Society, pp. 337–45.

Environmental Issues

44. *Air Pollution Engineering Manual,* 1990, Air & Waste Management Association.
45. "Steel Industry Technology Roadmap," 1997, American Iron & Steel Institute.

Chapter 10

Electric Furnace Steelmaking

J.A.T. Jones, Consultant
B. Bowman, Senior Corporate Fellow, UCAR Carbon Co.
P.A. Lefrank, Director, Applications Technology, SGL Carbon Corp.

Over the past 20 years the use of the electric arc furnace (EAF) for the production of steel has grown considerably. There have been many reasons for this but primarily they all relate back to product cost and advances in technology. The capital cost per ton of annual installed capacity generally runs in the range of $140–200/ton for an EAF based operation. For a similar blast furnace–BOF based operation the cost is approximately $1000 per annual ton of installed capacity. As a result EAF based operations have gradually moved into production areas that were traditionally made through the integrated route. The first of these areas was long products—reinforcing bar and merchant bar. This was followed by advances into heavy structural and plate products and most recently into the flat products area with the advancement of thin slab casting. At the current time, approximately 40% of the steel in North America is made via the EAF route. As the EAF producers attempt to further displace the integrated mills, several issues come into play such as residual levels in the steel (essentially elements contained in the steel that are not removed during melting or refining) and dissolved gases in the steel (nitrogen, hydrogen, oxygen).

Both of these have a great effect on the quality of the steel and must be controlled carefully if EAF steelmakers are to successfully enter into the production of higher quality steels.

There have been many advances in EAF technology that have allowed the EAF to compete more successfully with the integrated mills. Most of these have dealt with increases in productivity leading to lower cost steel production. These are described in the detailed process sections.

10.1 Furnace Design

The design of electric arc furnaces has changed considerably in the past decade. Emphasis has been placed on making furnaces larger, increasing power input rates to the furnace and increasing the speed of furnace movements in order to minimize power-off time in furnace operations.

10.1.1 EAF Mechanical Design

Many of the advances made in EAF productivity over the last 20 years are related to increased electrical power input and alternative forms of energy input (oxygen lancing, oxy-fuel burners) into the furnace. These increased energy input rates have only been made possible through improvements in the mechanical design of the EAF. In addition, improvements to components which allow for faster furnace movement have reduced the amount of time which the furnace stands idle. Thus the objective has been to maximize the furnace power-on time, resulting in maximum productivity.

10.1.1.1 Furnace Structural Support

In the 1960s and 1970s it was common to install electric arc furnaces at grade level. These furnaces would have pits dug out at the front for tapping and at the back for pouring slag off into slag pots. This configuration lead to many interferences and delays and is no longer recommended for large scale commercial operations. Modern furnace shops usually employ a mezzanine furnace installation. Thus the furnace sits on an upper level above the shop floor. The furnace is supported on a platform which can take on several different configurations. In the half platform configuration, the electrode column support and roof lifting gantry is hinged to the tiltable platform during operation and tapping. When charging the furnace, the complete assembly is lifted and swiveled. This design allows for the shortest electrode arm configuration. In the full platform design, the electrode column support and roof lifting assembly is completely supported on the platform. These configurations are shown in Fig. 10.1.

10.1.1.2 General Furnace Features

The EAF is composed of several components as shown in Fig. 10.2. These components fall into the functional groups of furnace structures for containment of the scrap and molten steel, components

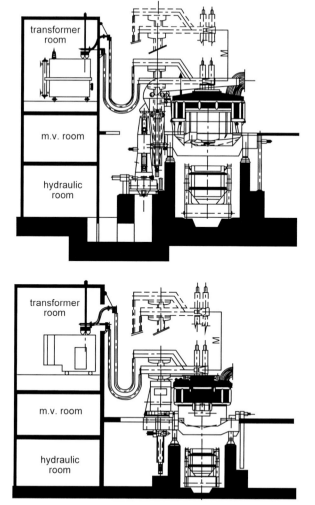

Fig. 10.1 Furnace platform configurations. *(Courtesy of Danieli.)*

which allow for movement of the furnace and its main structural pieces, components that support supply of electrical power to the EAF, and auxiliary process equipment which may reside on the furnace or around its periphery.

The EAF is cylindrical in shape. The furnace bottom consists of a spherically shaped bottom dish. The shell sitting on top of this is cylindrical and the furnace roof is a flattened sphere. Most modern furnaces are of the split shell variety. This means that the upper portion of the furnace shell can be quickly decoupled and removed from the bottom. This greatly minimizes down time due to changeout of the top shell. Once the top shell is removed, the furnace bottom can also be changed out fairly quickly. Some shops now follow a practice where the shell is changed out on a regular basis every few weeks during an eight hour downshift.

The furnace sidewall above the slag line usually consists of water-cooled panels. These panels are hung on a water-cooled cage which supports them. The furnace roof also consists of water-cooled panels. The center section of the roof which surrounds the electrode ports is called the roof delta and is a cast section of refractory which may be water cooled. The furnace bottom consists of a steel shell with several layers of refractory. This is discussed in more detail in Section 10.1.2. Fig. 10.3 shows a modern EBT EAF.

Electric Furnace Steelmaking

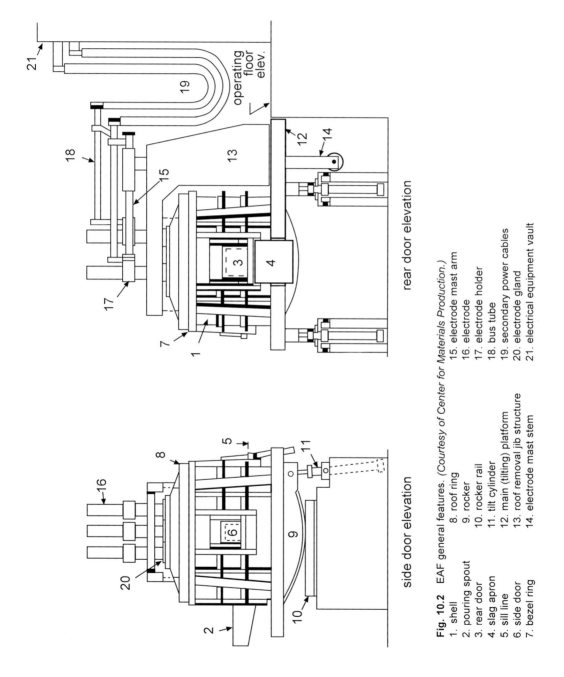

Fig. 10.2 EAF general features. *(Courtesy of Center for Materials Production.)*

1. shell
2. pouring spout
3. rear door
4. slag apron
5. sill line
6. side door
7. bezel ring
8. roof ring
9. rocker
10. rocker rail
11. tilt cylinder
12. main (tilting) platform
13. roof removal jib structure
14. electrode mast stem
15. electrode mast arm
16. electrode
17. electrode holder
18. bus tube
19. secondary power cables
20. electrode gland
21. electrical equipment vault

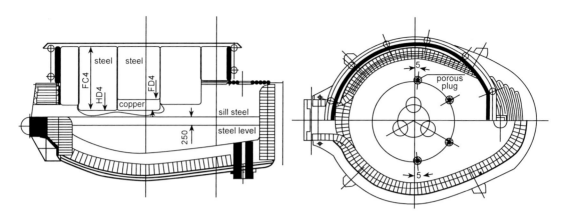

Fig. 10.3 Plan and section views of a modern EBT EAF. *(Courtesy of Danieli.)*

10.1.1.3 Water-Cooled Side Panels

One of the most important innovations in EAF design was water cooling, Fig. 10.4. Although this was used to a limited extent in older furnace designs for cooling of the roof ring and door jambs, modern EAFs are largely made up of water-cooled panels which are supported on a water-cooled cage. This allows for individual replacement of panels with a minimum of downtime. By water cooling the cage structure, it can be ensured that thermal expansion of the cage does not occur. Thus warping of the cage due to thermal stresses is avoided as are the resulting large gaps between the panels. Water-cooled panels allow very large heat inputs to the furnace without damaging the furnace structure. In the older EAF designs, these high power input rates would have resulted in increased refractory erosion rates and damage to the furnace shell.

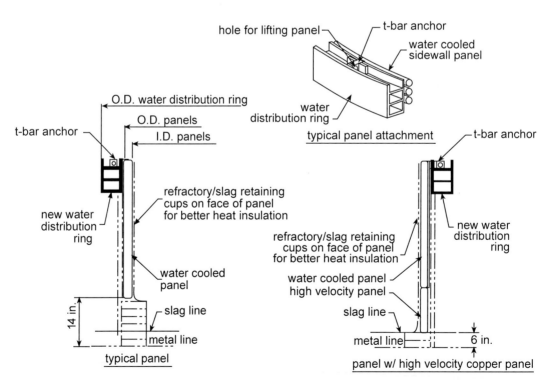

Fig. 10.4 Water-cooled panel designs. *(Courtesy of Fuchs.)*

Parameters which have a strong influence on panel life include water quantity and quality, water flow rate and velocity, inlet water pressure and pressure drop across the panel, pipe/panel construction material, and pipe diameter.

Water-cooled panels must withstand both high thermal loads as well as high mechanical loads. The greatest mechanical load occurs during furnace charging. Scrap can strike the panels causing denting of pipes or even splitting and rupture. Thus the pipe selection must allow for a wall thickness which can withstand these forces. At the same time, a minimum wall thickness is desired in order to maximize heat transfer to the cooling water. This tradeoff must be evaluated in order to arrive at an optimum pipe thickness. Generally, the minimum pipe thickness used is 8 mm. Maximum pipe thickness will depend on the thermal load to be removed and the amount of thermal cycling. In practical application the wall thickness is usually 8–10 mm.

In most applications, boiler tube grade A steel is used for water-cooled panels. These steel grades are generally reasonably priced, are easy to work with and provide a suitable thermal conductivity for heat transfer (approximately 50 $Wm^{-1}K^{-1}$) up to thermal loads of 7 million $kJ\ m^{-2}hr^{-1}$ (10,000 $BTU\ ft^{-2}min^{-1}$). However, this material is also susceptible to fatigue caused by thermal cycling.

In areas which are exposed to extremely high heat loads, copper panels might be used. Copper has a thermal conductivity (383 $Wm^{-1}K^{-1}$) approximately seven times that of boiler tube enabling it to handle very high thermal loads up to 21 million $kJ\ m^{-2}hr^{-1}$ (approximately 31,000 $BTU\ ft^{-2}min^{-1}$). Copper panels will result in much greater heat transfer to the cooling water and thus to optimize energy consumption in the EAF, copper panels should only be used in areas where excessive heats loads are encountered (e.g. close to the bath level, oxy-fuel burner ports etc.). This will of course be dependent on the amount of slag buildup over these panels. Generally, slag buildup will be thicker on copper panels due to the greater heat transfer capacity. Alternatively, the distance between the start of water-cooled panels and the sill level can be increased if steel panels are to be used in the lower portion of the water-cooled shell.

For most boiler tube grade water-cooled panels, a pipe diameter of 70–90 mm is usually selected. For copper panels, the choice of tube diameter will usually be a cost based decision but frequently falls into the same range. In some high temperature zones (e.g. lower slag zone), a smaller diameter copper pipe is used with higher water velocity to prevent steam formation.

Cooling panel life is primarily dependent on the amount of thermal cycling that the panel is exposed to. During flat bath conditions, the panels can be exposed to very high radiant heat fluxes. Following charging, the panels are in contact with cold scrap. Thus the amount of thermal cycling can be considerable. The optimum solution to this problem involves providing maximum cooling while minimizing the heat flux directly to the panel. In practice it has been found that the best method for achieving these two goals is to promote the buildup of a slag coating on the panel surface exposed to the interior of the furnace. With a thermal conductivity of 0.12–0.13 $Wm^{-1}K^{-1}$, slag is an excellent insulator. Cups or bolts are welded to the surface of the panels in order to promote adhesion of the slag to the panels.

It is important that the temperature difference across the pipe wall in the panels does not get too high. If a large temperature differential occurs, mechanical stresses (both due to tension and compression) will build up in the pipe wall. Most critically, it is important that the yield stress of the pipe material not be exceeded. If the yield stress is exceeded, the pipe will deform and will not return to its original shape when it cools. Repeated cycling of this type will lead to transverse cracks on the pipe surface and failure of the panel.

Panel layout is related to several factors including thermal load, desired temperature rise for the cooling water, desired pressure drop across the panel, and desired pipe diameter.

Generally, designers aim for a water velocity which will result in turbulent flow within the pipe. Thus the minimum required water velocity will vary depending on the inside diameter of the pipe which is used. A minimum flow velocity of approximately 1.2 $m\ s^{-1}$ is quoted by several vendors.[1,2,3] The maximum velocity will depend on the desired pressure drop but will usually be less than 3 m s^{-1}. The range quoted by several vendors is 1.2–2.5 m s^{-1}. This will help to minimize the possibility of steam formation in the pipe. Some systems are designed so that a water velocity of at least 2.5 m s^{-1} is provided to ensure that the steam bubbles are flushed from the panel. Failure to remove smaller steam bubbles can result in a larger bubble forming. Steam generation could lead to greatly reduced heat transfer (by a factor of ten) and as a result possible panel overheating and failure in the area where the steam is trapped. In extremely high temperature regions such as the lower slag line, water velocities in excess of 5 m s^{-1} are used to ensure that steam bubbles do not accumulate in the panel.

A vertical configuration is preferred in most modern designs because it minimizes scrap holdup and also can help to minimize pressure drop across the panel. Generally water cooling systems on EAFs are designed for an average water temperature rise (i.e. outlet − inlet) of 8–17°C with a maximum temperature rise of 28°C. Most furnace control systems have interlocks which will interrupt power to the furnace if the cooling water temperature rise exceeds the maximum limit.

In high powered furnaces Fuchs Systems Inc. (FSI) recommends that a water flow rate of 150 l min^{-1} m^{-2} (3.65 gpm ft^{-2}) for a side wall panel or 170 l min^{-1}m^{-2} (4.14 gpm ft^{-2}) for a roof panel should be available. For DC furnaces an additional 10 l min^{-1}m^{-2} should be made available.

Pressure drop across the panel is a parameter frequently ignored by most furnace operators. In order to get uniform flow to each panel, it is imperative that the pressure drop across panels be as even as possible. FSI recommends that the water pressure exiting the panels be at least 20 psi. Frequently, panels of different design are mixed within the same supply system. This occurs because replacement panels are frequently sourced from local fabricators as opposed to the original furnace vendor. Slight changes in materials, dimensions and configuration can lead to big changes in the panel pressure drop. Mismatched panels within the same supply circuit will lead to some panels receiving insufficient flow and ultimately panel failure.

Most water-cooled panels are designed to be airtight. However, replacement panels are sometimes made without complete welds between pipes. This may be a cost saving when considering the fabrication of the panel but will lead to increased operating costs and lost production. The EAF is operated under negative pressure and as a result, air will be pulled into the furnace through any openings. Thus openings should be minimized, especially those in and around the water-cooled panels.

10.1.1.4 Spray-Cooled Equipment

Spray cooling has also shown itself to be a viable means for cooling EAF equipment. Spray-cooled components are typically used for roof and sidewall cooling. The main components of a spray-cooled system are the inner shell, spray nozzles and water supply system. Water is sprayed against the hot surface (inner shell) of the cooling element. The outer shell acts purely for containment of the water and steam. The inner shell is fabricated from rolled plate. A uniform spray pattern is essential for proper heat removal from the hot face. Maintenance of a slag coating on the exposed side of the hot face is also integral to regulating heat removal and protection of the cooling panel. In general, at least half an inch of slag coating should be maintained.

The water supply system for spray-cooled panels is simpler than that used for conventional water-cooled panels because the operation is carried out at atmospheric pressure. Strainers are provided to ensure that suspended solid material is not carried through the circuit where nozzle clogging could become a concern.

10.1.1.5 Furnace Bottom

The furnace bottom consists of a spherical plate section. This section is refractory lined and this lining will generally consist of a safety lining with a rammed working lining on top. The plate thickness is usually between 25 and 50 mm thick depending on the furnace diameter. Refractory lining thickness is generally a function of operating practices and change out schedule. For AC fur-

Electric Furnace Steelmaking

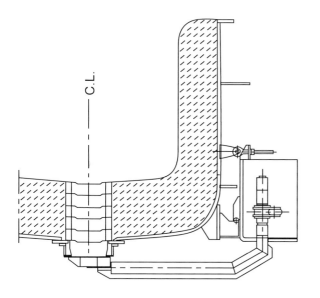

Fig. 10.5 Eccentric bottom tapping configuration. *(Courtesy of Danieli.)*

nace operations, the typical refractory thickness may range between 500 and 700 mm. DC furnaces typically have thicker working linings and the resulting refractory thickness can range from 750–950 mm. Furnace refractories are described in more detail in Section 10.1.2.

In some cases gas stirring elements are installed in the bottom of the furnace. If this is the case, special pocket blocks will be installed during installation of the brick safety lining. Alternatively, stirring elements are lowered into place and refractory is rammed around them.

The furnace bottom section will also contain the tapping mechanism. Most modern furnaces employ bottom tapping. The reasons for this are several, including slag free tapping, decreased steel temperature losses during tapping, hot heel practices, and mechanical considerations.

Bottom tapping requires that the ladle be moved into position under the furnace on a ladle car or using a ladle turret arm. This is beneficial as it frees up the crane for other duties.

For bottom tapping furnaces, the taphole is located in the upper section of the oval shell as shown in Fig. 10.5. The taphole is filled with sand following tapping, prior to tilting the furnace upright for charging. The sand is held in place with a slide gate or a flapper type valve which retracts in the horizontal plane.

Many furnaces with tapping spouts are in operation in North America and in the case of pit furnace operations, offer the only option. If a submerged siphon spout is used, slag free tapping can be achieved though maintenance can be an issue. A leveled spout allows for tapping of steel and slag together and is common in the stainless steel industry. Some furnace spouts have now been retrofitted with slide gates on the spout so that quick shutoff of the tap stream can be achieved. This also allows for slag free tapping. Fig. 10.6 shows several different spout configurations.

Typically, during furnace operation, the bath depth will not exceed 1 m in depth. Bath depth will be a function of furnace diameter and typically ranges from 0.7–1.0 m. During bottom tapping, it is important to maintain a minimum height of steel above the taphole to prevent vortexing which will cause slag entrainment. The typical recommendation is that the height of liquid metal be 2.5 times the taphole diameter.

  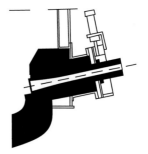

Fig. 10.6 Tapping spout configurations. *(Courtesy of Danieli.)*

10.1.1.6 Furnace Movements

In the course of electric arc furnace operations it is necessary for several of the furnace components to move. Typical requirements for movement include roof raise/rotation to allow for scrap charging, electrode raise/lower and swing to allow for scrap charging, electrode raise/lower for arc regulation, furnace tilt forwards for tapping, slag door up/down for deslagging operations, furnace tilt backwards for slag removal, and electrode clamp/unclamp to adjust the working length of the electrode.

Most furnace movements are made using a central hydraulic system which is usually sized for the maximum flow requirement of the individual task requiring the greatest flow rate of hydraulic fluid. Thus the size of the system can be kept to a minimum while still meeting the requirements for furnace operations.

10.1.1.7 Furnace Tilting

The EAF is tilted for both tapping and slag removal. In the case of furnace tapping, the maximum forward tilting angle will be dependent on the type of furnace bottom. For conventional spout tapping, it may be necessary to tilt to an angle of 45° to fully tap the furnace. For bottom tapping furnaces the maximum tilt angle is usually 15–20°. An important requirement of slag free tapping is that the furnace can be tilted back quickly as soon as slag begins to carry over into the ladle. The typical maximum forward tilting speed is 1° per second. Most modern furnaces have a tilt back speed of 3–4° per second. Any attempt to tilt back faster than this will place undue stress on the furnace structure. Many furnace designs now arrange the center of the cradle radius so that the furnace is balanced slightly back tilted towards the deslagging side. Many furnaces are also designed so that if hydraulic power is lost, the furnace center of gravity will cause the furnace to tilt back up into its resting position.

The furnace movement can be controlled by using either one or two double acting cylinders. The furnace bottom sits on a cradle arm which has a curved segment with geared teeth. This segment sits on a rail which also has geared teeth. As the tilt cylinder is extended, the furnace rocks forward to tap the furnace.

For removing slag from the furnace, the furnace must be tilted backwards. The tilt cylinder is contracted fully causing the furnace to tilt backwards. The slag pours off the top of the steel bath and out through the slag door. The slag may be collected in pots or may be poured onto the ground where it will be removed using a bulldozer. The typical tilt angle for deslagging is 15–20°.

One newer design of furnace bottom has the furnace cradle sitting on a bogie wheel. A cylinder is mounted on an angle as shown in Fig. 10.7. In this configuration the taphole moves towards the deslagging side of the furnace during tapping. The required tilting force is greatly reduced for this configuration.

10.1.1.8 Furnace Roof

The furnace roof can be dome shaped or as is more common with water-cooled roofs used in modern practice, the roof resembles a shallow cone section. The roof consists of a water-cooled roof ring which forms the outer perimeter of the roof cage. This cage acts as part of the lifting structure for the roof. Water-cooled panels insert into this cage which has a cylindrical opening at the center. The refractory delta section is inserted to fill this opening. This delta section is designed to allow the minimum opening around the electrodes without risk of arcing between the electrodes and the water-cooled panels. The whole furnace roof is cantilevered off the roof lift column. Typically, roof and electrode support can be swiveled together or independently. The electrode stroke allows the electrodes to be swiveled with the roof resting on the furnace shell which allows for removal and replacement of the delta section without removing the roof. Typically, for a full platform design, a swiveling support with pivot bearing, bogie wheel and gantry arm is employed. For large furnaces, a roof lifting gantry is required. The swiveling can be achieved using either a

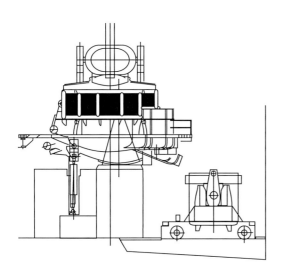
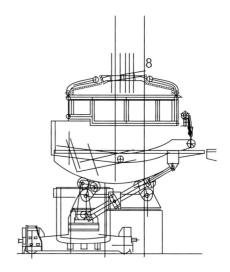

Fig. 10.7 Furnace tilting configurations. *(Courtesy of Danieli.)*

bogie wheel support or a thrust bearing. Twin shell furnace arrangements where the roof must pivot between furnace shells sometimes employ a fork shaped gantry. Typically most modern furnaces can raise and swing the roof in about 20–30 seconds. Roof swing speed is usually 4–5° per second. This minimizes the amount of dead time during charging operations and increases furnace melting availability. Typical roof configurations are shown in Fig. 10.8.

10.1.1.9 Electrode Arms and Lifting Column

Electrode control performance is limited by the lowest natural frequency in the positioning system. It is therefore very important to ensure sufficient stiffness in the columns with respect to torsion and bending. Several possible configurations for the guiding of the columns is shown in Fig. 10.9. The choice of configuration is usually specific to a particular operation. However, the main objective is to avoid friction in the roller system while arranging the roller design to be compact yet rigid. It is important to note that each arm must be capable of individual movement to allow for electrode regulation. For lateral movement of the electrodes a common swing mechanism is used. The conventional design uses a hydraulic cylinder to move the swing column. Some newer designs

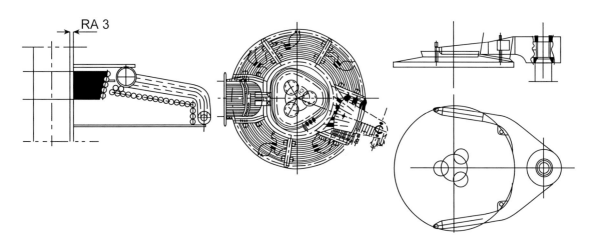

Fig. 10.8 EAF roof and roof lifting configurations. *(Courtesy of Danieli.)*

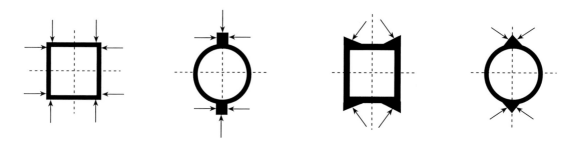

Fig. 10.9 Electrode arm column guiding configurations. *(Courtesy of Danieli.)*

now use a hydraulic motor inside the roller bearing. This cleans up the space around the cylinder but makes the mechanism less accessible for maintenance.

Several options are available for electrode arms in modern EAF operations. In conventional EAF design, the current is carried to the electrodes via bus tubes as shown in Fig. 10.10. These bus tubes usually contribute approximately 35% of the total reactance of the secondary electrical system.[4] Care must be taken to avoid induced current loops which can cause local overheating of the arm components. Thus the electrode clamping mechanism requires several insulation details.

Current conducting arms combine the mechanical and electrical functions into one unit as shown in Fig. 10.11. These arms carry the secondary current through the arm structure instead of through bus tubes. This results in a reduction of the resistance and reactance in the secondary circuit which allows an increase in power input rate without modification of the furnace transformer. Productivity is also increased. Current conducting arms are constructed of copper clad steel or from aluminium. Most new furnace installations now use current conducting arms.

Typical maximum electrode speed is approximately 30–35 cm per second when operating in automatic arc regulation mode. When operating in manual raise/lower mode the maximum speed is usually 15–20 cm per second.

10.1.1.10 EAF Secondary Electrical Circuit Components

The electric arc furnace secondary electrical circuit consists of several components including the delta closure, power cables, bus tubes, electrode holder(s), and the electrode(s).

Power exits the furnace transformer via low voltage bushings which are connected to the delta closure. The delta closure usually consists of either a series of copper plates, copper tubes or in some cases both. These are arranged so that the individual secondary windings of the transformer are joined to form a closed circuit. Usually the delta closure is contained in the transformer vault and protrudes through the vault wall to connect with the power cables. The delta closure is anchored using insulated material to minimize distortion which may be caused by vibration and/or electromagnetic forces. A delta closure is shown in Fig. 10.12.

The power cables are the only flexible part of the secondary electrical system. These allow for raising, lowering and swinging of the electrodes. Furnaces usually operate with between two and four

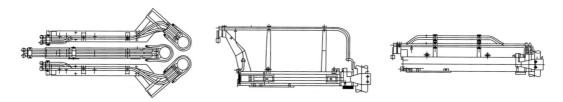

Fig. 10.10 Conventional bus tube electrode arms. *(Courtesy of Danieli.)*

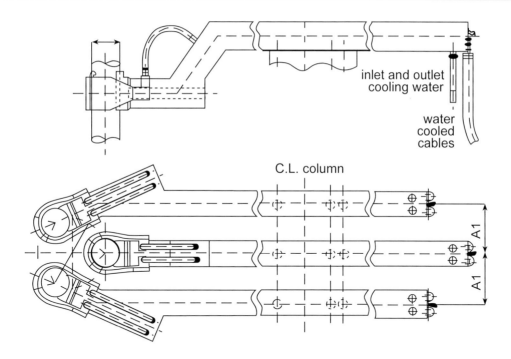

Fig. 10.11 Current conducting electrode arms. *(Courtesy of Danieli.)*

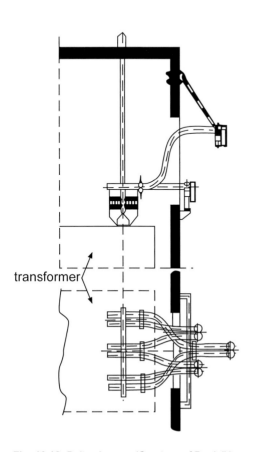

Fig. 10.12 Delta closure. *(Courtesy of Danieli.)*

cables per phase. The power cable generally consists of multiple copper wire strands which are soldered to copper terminals at either end of the cable. The copper cable is covered with a rubber compound which is secured at the terminals with nonmagnetic stainless steel bands. Water cooling is critical to the life of the cable. Usually, a dedicated closed loop cooling water circuit supplies cooling to all of the critical electrical components including the transformer, the delta, the power cables, bus bars and electrode clamp. Cooling water requirements range from 11–95 l min^{-1} depending on the cable size.[5]

The power cables are attached to bus tubes which are mounted on the electrode arm. Bus tubes consist of rigid round copper pipe and provide the electrical connection between the power cables and the electrode holder. These bus tubes are also water cooled.

The electrode clamping mechanism provides the link between the electrode arm and the electrode. It is very important that the electrode be held firmly in place during furnace operations. For current conducting arms, it is not necessary to insulate the clamping mechanism from the arm because the arm is part of the electrical circuit. In the case of conventional electrode arms with bus tubes, the arm does not form part of the electrical circuit and must be electrically insulated from the clamp. Thus the electrode clamp is bolted onto the arm with a layer of insulation separating the two. The clamp casing which has contact shoes ensures that the electrode is making electrical contact. The copper bus bars which are supported on the arm are connected to the contact shoe. Clamping systems can be mounted external to the mast arm or can be mounted inside the arm. Most are now mounted inside the arm as this provides protection for the mechanism. A spring mechanism ensures that the shoes contact the electrode and a mobile yoke holds the electrode in place. The spring assembly is retracted using a hydraulic cylinder. This allows for release of the electrode for slipping operations or to remove an electrode column.

The traditional electrode holder is annular or horseshoe-shaped and consists of a single piece copper construction. Some more recent designs have incorporated copper contact pads attached to a steel or nonmagnetic steel body. Electrode heads and contact pads are the most common failure point in the secondary power system. Electrode contact surface failures are usually related to dirt buildup between the contact face and the electrode. Insufficient clamping pressure can also contribute to failures. Contact pads/electrode holders should be changed out and cleaned during scheduled maintenance downtimes. Fig. 10.13 shows an electrode holder.

Fig. 10.13 Electrode holders. *(Courtesy of Fuchs.)*

Water cooling is integral to service life for electrode holders. Several designs are common for electrode holders, one of the most common having a cast-in pipe which allows circulation of the water. Contact pads are usually castings or forgings with drilled holes to provide the cooling water channels.

Clamping force is a function of the electrode diameter. A general rule of thumb is that the clamping force in Newtons equals 850 times the square of the electrode diameter, in inches. Typical values quoted by vendors are 235 kN for 16 in. diameter electrodes and 460 kN for electrode diameters of 24 in.[6]

10.1.1.11 Electrode Spray Cooling

It has been recognized for some time that sidewall oxidation of the graphite electrodes is related to heating of the electrode surface. Because the current flow is down through the electrode into the bath, only the portion of the electrode below the electrode holder will heat up. Some early methods for reducing the amount of heating of the electrode involved a water-cooled graphite segment located at the clamping level. This method did reduce the amount of sidewall oxidation but required changing out the whole column once the bottom graphite segment was consumed. Other methods for reducing sidewall consumption involved special coatings which would oxidize at a higher temperature.

The most practical solution has now been adopted by most high powered furnace operations. This involves installation of a spray ring which is installed at the bottom of the electrode holder. The concept is that water will run down the surface of the electrode and will evaporate once it enters the furnace. Ideally, the water should be evaporated within the roof port where it will also provide some protection to the delta refractory. The required flow rate will vary depending on furnace operating conditions but typical flow rate for a 24 inch electrode is 7–15 l min^{-1} per electrode. Typical savings in electrode consumption amount to 5–15%. This is however, highly dependent on the type of furnace operation (i.e. DC, high impedance AC, etc.).

Electric Furnace Steelmaking

10.1.1.12 Shell Openings

Several openings are usually provided for furnace operations. The most obvious are the three electrode ports which allow the electrodes to penetrate into the furnace through the roof. In addition, a fourth hole is provided in the furnace roof to allow for extraction of the furnace fume. A fifth hole may be provided for several reasons such as continuous DRI/HBI feed, coal injection or lime injection. These holes all occur high up in the furnace and as a result do not affect air infiltration into the furnace as much as lower openings.

The lower openings in the furnace include the taphole which is filled with sand and the slag door. The slag door was originally provided to allow decanting of the slag from the furnace. In most modern operations, it is also used to provide access to the furnace for oxy-fuel burners and oxygen lances. Several ports may also be provided around the circumference of the furnace shell for burners. Occasionally, an opening may be provided high up on the furnace sidewall to allow a water-cooled decarburization lance access to the furnace. Other openings may be provided low in the furnace sidewall or actually in the furnace hearth to allow for injection of inert gases, oxygen, lime or carbon.

10.1.1.13 Auxiliary Equipment

Auxiliary equipment is usually associated with injection of solids and gases into the furnace. Oxy-fuel or air-fuel burners are used to provide additional heat to the cold spots within the furnace as shown in Fig. 10.14. Typically, the cold spots occur between the electrodes and also in the nose of

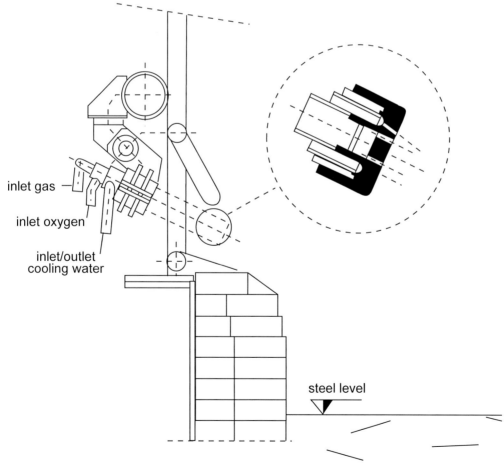

Fig. 10.14 Sidewall oxy-fuel burner. *(Courtesy of Danieli.)*

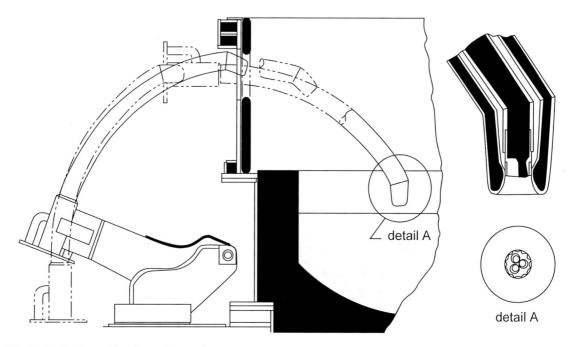

Fig. 10.15 Unilance. *(Courtesy of Empco.)*

the EBT furnace because it is further away from the electrodes. Burner efficiency is highest when the scrap is cold and decreases as the scrap heats up. This is described in more detail in the section on EAF operations. Burners are usually constructed of high alloy steel or stainless steel and are water cooled. The burner tips are frequently constructed out of copper. Burners must be capable of withstanding high heat fluxes especially early in the meltdown cycle when it is likely that the flame will flash back if its passage is obstructed with scrap. Most sidewall burners are stationary. Slag door burners tend to be retractable. The so-called banana lance supplied by Empco actually follows the scrap down into the furnace as it melts as shown in Fig. 10.15. This unit can operate in either oxy-fuel burner or oxygen lance mode. Burners are operated at different firing rates depending on the furnace operating phase. A minimum low firing rate is maintained at all times to ensure that the burner is not plugged by splashed slag.

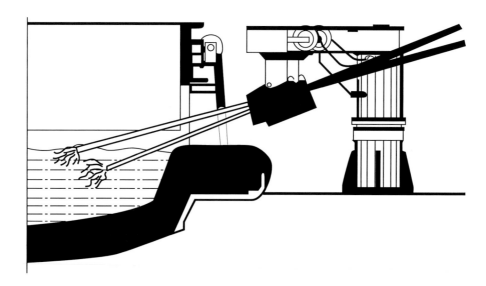

Fig. 10.16 Operation of consumable oxygen lance. *(Courtesy of Badische Stahl Technology.)*

Fig. 10.17 Consumable oxygen lance manipulator. *(Courtesy of Badische Stahl Technology.)*

Oxygen lances come in many configurations but generally fall into two categories; consumable pipe lances and non-consumable water-cooled lances. Many oxygen lances are now multi-port lances and are also capable of injecting lime and carbon. Consumable oxygen lances are capable of actually penetrating the slag/bath interface as shown in Fig. 10.16. The consumable pipe burns back during operation and periodically, a new section of pipe must be inserted. A consumable lance manipulator is shown in Fig. 10.17. Water-cooled lances do not actually penetrate the bath. Instead they inject oxygen at an angle through the slag at high pressure so that the oxygen penetrates the bath. A water-cooled oxygen lance manipulator is shown in Fig. 10.18.

Another important auxiliary function is sampling the bath and obtaining a bath temperature. This information is critical for the furnace operator so that adequate bulk additions can be made at tap.

Fig. 10.18 Water-cooled oxygen lance. *(Courtesy of Fuchs.)*

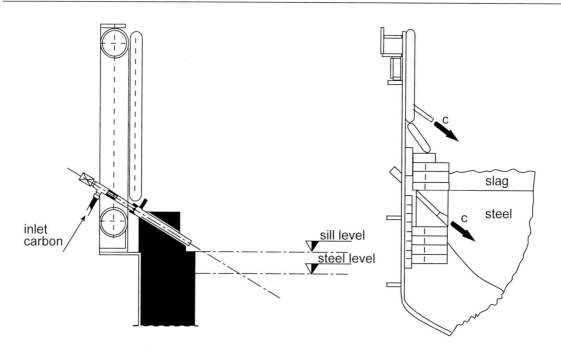

Fig. 10.19 Carbon injector configurations, submerged and above the bath. *(Courtesy of Danieli.)*

Typically, bath temperature measurements are taken manually using a lance with a disposable cardboard thermocouple tip. The lance is connected to an electronic display which converts a voltage signal into a temperature. Combined function units are also capable of measuring bath oxygen levels. The bath sample is usually taken manually using a lollipop type sampler. The sample is cooled and sent to the lab for chemical analysis. Some recent installations have automated bath temperature measurement and sampling using an automated lance. This reduces the time which operators have to spend out on the shop floor and also maintains a consistent sample location.

Many operations are now providing for carbon and lime injection at multiple points around the furnace shell. It was recognized that with injection solely through the slag door, slag foaming tended to be localized. By providing multiple injectors, it is possible to maintain more uniform slag foaming across the entire bath. Several possible configurations are shown in Fig. 10.19.

10.1.1.14 Bottom Stirring

For conventional AC EAF scrap melting there is little natural convection within the bath. As a result high temperature and concentration gradients can exist within the bath. These can lead to increased energy consumption, reduced reaction rates and over or under reaction of some portions of the bath. Bath stirring can help to reduce these gradients and can be accomplished either electromagnetically or by using inert gas injection to stir the bath. Most modern operations opt for the latter.

Typically, the elements used to introduce the gas into the bath can be tuyeres, porous plugs or combinations of porous plugs with permeable refractory rammed into place over top of the plug. Non-contact elements (i.e. those which do not actually contact the bath directly) have the advantage that they can be operated at higher turndown ratios without the fear of plugging. The choice of gas used for stirring seems to be primarily argon or nitrogen though some trials with natural gas and with carbon dioxide have also been attempted. Stirring elements are mounted within the refractory lining of the EAF. The number of elements used can vary greatly based on the design and gas flow rate but is usually in the range of one to six. The size of the furnace and its design will also create different demands for the number of stirring elements to be used.

Electric Furnace Steelmaking

10.1.1.15 Special DC Furnace Design Considerations

DC furnaces have probably been the biggest innovation in EAF technology in the past ten years. The concept of operating an EAF on DC current is not new but only recently did the costs for rectification units drop to the point where DC furnaces became economical. DC furnaces have several unique requirements over AC furnaces in addition to the obvious differences in electrical power supply which are discussed in the section on furnace electrics. A typical layout for a DC furnace is given in Fig. 10.20. DC furnaces have only one electrode mast arm and a single graphite electrode. This electrode acts as the cathode. Thus the top of the furnace is much less cluttered for the DC case and in general has fewer components to maintain as compared to AC designs. The DC furnace however, requires a return electrode, the anode, to complete the electrical circuit. This anode is commonly referred to as the bottom electrode because it is located in the bottom of the furnace shell. Several different designs exist for the bottom return electrode including metal pin return electrodes with non-conductive refractories, billet electrode, metal fin electrodes, and conductive bottom refractory. These are shown in Fig. 10.21.

In the case of current conducting refractory contact, the refractory lining at the center of the furnace bottom acts as the anode. The bottom has a circular flange which rests inside a circular channel which is welded to the furnace shell. Inside the channel, the flange is supported by fiber reinforced ceramic blocks. The space between the channel, support blocks and flange is filled with a refractory ramming compound. This isolates the bottom electrically from the rest of the furnace shell as shown in Fig. 10.22.

The spherical furnace bottom is constructed of high temperature steel. A circular copper plate is bolted directly to the furnace bottom. Four copper terminals extend down through the furnace bottom from the copper plate and connect to flexible cables which in turn are connected to the bus tubes. The conductive refractory bricks are installed over top of the copper plate. Heat flow from the bottom of the furnace (usually about 15 kWm^{-2}) is removed by forced air cooling. Due to the large surface area of the bottom electrode, the current density tends to be quite low, typically about 5 kAm^{-2}. However, it has been necessary in some installations to use nonconductive patching material in the center of the furnace in order to force the current to distribute more evenly

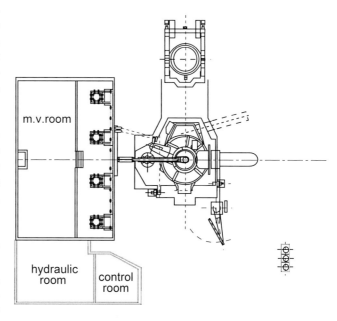

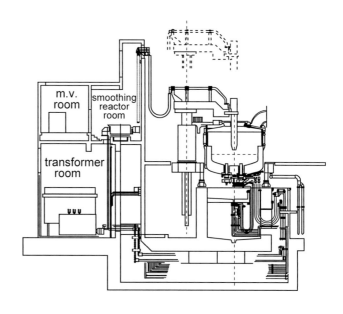

Fig. 10.20 Typical DC furnace configurations. *(Courtesy of Danieli.)*

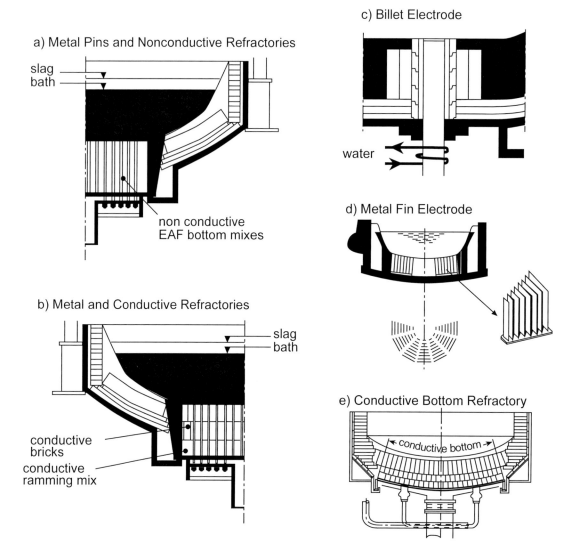

Fig. 10.21 DC furnace bottom anode configurations. *(Courtesy of ABB.)*

over the whole bottom. Failure to achieve proper distribution of the current resulted in hot spots in the center of the furnace.

The billet return electrode configuration employs from one to four large steel billets (on the order of 10–15 cm diameter but can be as large as 25 cm diameter) depending on the size of the furnace. Usually, the design aims for a current of 40–45 kA per bottom electrode. The billets are in contact with the bath at the top surface and will melt back. The degree to which the billet melts back is controlled by water cooling. The billet is inserted into a copper housing through which cooling water is circulated. By providing sufficient cooling, it can be ensured that the billet will not melt back completely. Thermocouples monitor the bottom billet temperature and the cooling water temperature. An insulating sheath isolates the copper housing from the billet. The billet is connected to a copper base. The copper base provides the connection to a power cable. Fig. 10.23 shows the Kvaerner-Clecim billet electrode.

The pin type of return electrode uses multiple metal pins of 2.5–5.0 cm in diameter to provide the return path for the electrical flow. These pins are configured vertically and actually penetrate the refractory. The pins extend down to the bottom of the furnace where they are fixed in position by two metal plates. The bottom ends of the pins are anchored to the lower power conductor plate. The

Electric Furnace Steelmaking

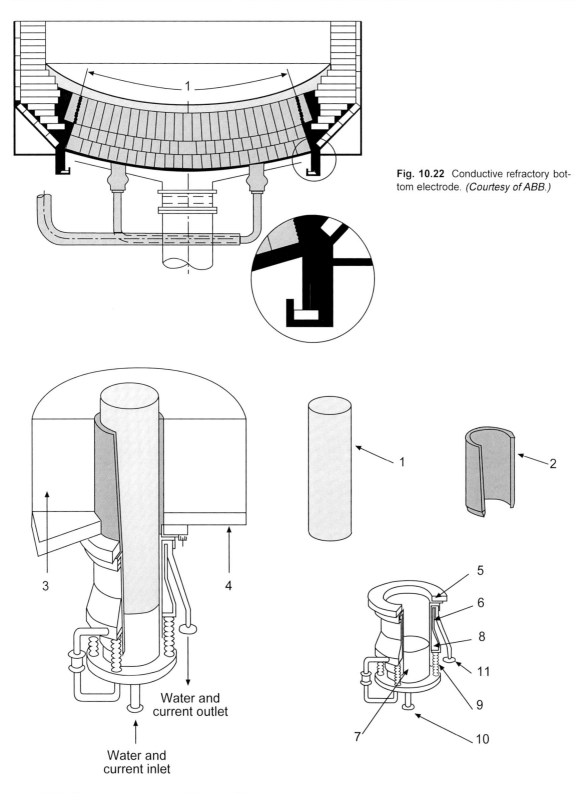

Fig. 10.22 Conductive refractory bottom electrode. *(Courtesy of ABB.)*

Fig. 10.23 Billet anode. *(Courtesy of Kvaerner Metals.)*
1. steel billet
2. special refractory
3. hearth
4. bottom shell plate
5. insulation material
6. copper sleeve
7. copper cap
8. water jacket
9. compensator
10. water and current inlet
11. water and current outlet

bottom contact plate is air-cooled and is located in the center of the furnace bottom. The top portion of the pins are flush with the working lining in the furnace. The pins are in direct contact with the bath, and melt back as the working lining wears away. A return power cable is attached to the bottom conductor plate.

An extensive temperature monitoring system is provided to track lining wear and bottom electrode life. This enables scheduled change out of the bottom electrode. The integral cartridge design which has evolved allows for quick change out of the bottom electrode over a scheduled eight hour maintenance outage. The sequence of change-out steps is as follows:

1. Following the last heat prior to change-out, the slag is drained from the furnace and refractory surfaces are sprayed with water to accelerate cooling.
2. Electrical connections to the bottom electrode are decoupled and thermocouples are disconnected.
3. The integral cartridge is pushed up by six hydraulic cylinders which are located around the perimeter of the bottom contact plate.
4. Once the cartridge is broken free of the furnace bottom, it can be removed by crane.
5. A new cartridge is then lowered into place and electrical connections are recoupled.
6. A small amount of scrap is charged and an arc is struck to test the system. Following a successful test, the furnace is charged with scrap and melting is commenced at reduced electrical current. Once a liquid heel is established, regular operation is resumed.

A typical anode removal device is shown in Fig. 10.24. Fig. 10.25 shows a typical refractory lining for a UNARC furnace.

The steel fin return electrode uses steel fins arranged in a ring in the furnace bottom to form several sectors. Each sector consists of a horizontal ground plate and several welded on steel fins

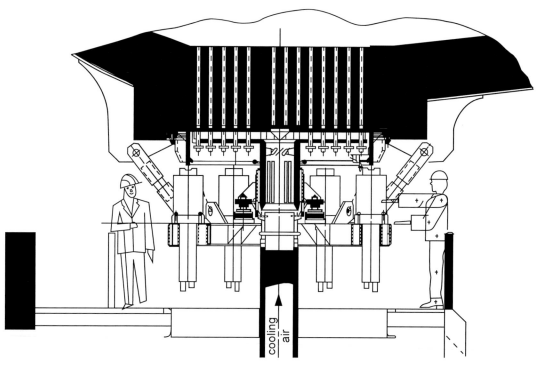

Fig. 10.24 Anode removal device. *(Courtesy of SMS GHH.)*

Electric Furnace Steelmaking

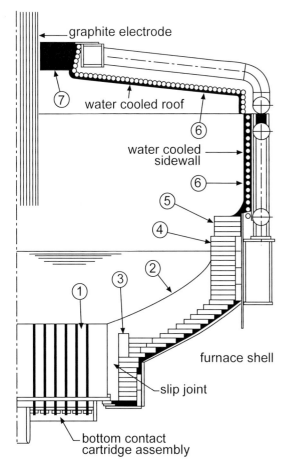

Fig. 10.25 Refractory details of UNARC furnace. *(Courtesy of SMS GHH.)*

which protrude upwards through the refractory. The fins are approximately 0.16 cm thick and are about 9 cm apart. The sectors are bolted onto an air-cooled bottom shell which is electrically insulated from ground and is connected to four copper conductors.

Most DC furnaces are operated with long arcs, typically two to three times those encountered in conventional UHP furnace operations. As a result, many furnace manufacturers specify higher water flow rates for water-cooled panels. Some typical values proposed by Fuchs Systems Inc. are $160 \; l \; min^{-1} \; m^{-2}$ (3.90 gpm ft^{-2}) for a sidewall panel and $180 \; l \; min^{-1} \; m^{-2}$ (4.38 gpm ft^{-2}) for a roof panel.

10.1.2 EAF Refractories

Refractories are materials which withstand high temperature without a significant change in chemical or physical properties. Refractory materials are very important to EAF operation as they allow for containment of the liquid steel in the furnace hearth without damaging the furnace structure. Refractories in general are discussed in Chapters 3 and for EAF use are discussed in Section 4.3.

10.1.2.1 Steelmaking Refractories

The electric furnace requires a variety of refractory products. Most important are the refractories in direct contact with the steel. Today's electric furnaces and EBTs generally use magnesite or magnesite carbon products in the steel contact zones. Specialized refractories with good thermal shock resistance are generally used in the roof/delta, the taphole, and the spout or runner of the furnace. The following descriptions give general refractory recommendations.

10.1.2.1.1 Sub-bottom Magnesite or tar-impregnated magnesite brick are generally used in the sub-bottom due to good resistance to steel and slag in the event the working hearth is penetrated. These refractories are expected to last six months, one year, or even several years if the bottom hearth material is well maintained.

10.1.2.1.2 Hearth Most electric furnace operators prefer a monolithic magnesite hearth which is easy to install and maintain. Monolithic hearth materials are generally dry, vibratable high MgO materials which develop strength by sintering in place during burn-in of the bottom. These hearth materials generally have impurity oxides, like iron oxide, to facilitate sintering. Other operators prefer all brick hearths, where the initial construction is totally brick, and monolithic materials are only used to repair holes in the brick hearth after some period of operation. Modern EAF bottom designs often include gas stirring elements which require porous refractory materials or tuyeres through dense refractories to deliver the stirring gas to the molten steel. These stirring gases can cause erosion of the hearth refractories and require regular maintenance.

10.1.2.1.3 Lower Sidewall The lower sidewall generally uses the same brick quality as the sub-hearth, high MgO, plain or tar-impregnated. These brick form the base for the slagline and upper sidewall and are protected by the banks of the hearth material. They are replaced along with the monolithic hearth.

10.1.2.1.4 Slagline The slagline area of most furnaces uses magnesite carbon brick or tar impregnated magnesite brick. Refractory selection in the slagline must be carefully coordinated with expected slag chemistries. Carbon steel shops generate lime rich, $FeO–SiO_2$ slags with a 2:1+ lime:silica ratio, which demand basic refractories. In stainless operations, a more neutral fused grain magnesite chrome brick may be utilized.

10.1.2.1.5 Sidewall/Hot Spots Above the slagline, the refractory lining sees high temperature arc flare, slag splashing, scrap impingement, and flame impingement from burners or lances. Refractoriness, strength, and thermal shock resistance are the critical parameters for electric furnace sidewall refractories. In furnaces with water-cooled panels, magnesite carbon brick is the most common choice, with fused grain magnesite carbon brick in the hot spots.

10.1.2.1.6 Roof and Delta If the roof is water-cooled, a precast delta of high alumina or alumina chrome materials is a likely choice. Greater economy can sometimes be obtained by bricking the electrode ports and ramming a high alumina plastic in the balance of the delta section. Roof deltas are subjected to extreme thermal shock during roof swings for charging. Basic materials would have better resistance to the furnace atmosphere, but cannot generally withstand the thermal cycling. For complete refractory roofs, 70–80% alumina brick annular rings composing about two-thirds of the roof are most common; with high alumina plastic or alumina chrome plastic rammed in the delta section, again with high purity alumina or alumina chrome brick electrode rings.

10.1.2.1.7 Taphole For conventional tapping electric furnaces, a taphole module or taphole block with a predrilled hole is the most common construction. Other operators will ram high alumina or basic ramming mixes around a steel pipe to form their taphole, while others will gun basic gunning mixes around a steel pipe for the initial taphole. Some taphole designs will last the life of the furnace campaign, while other operators regularly maintain their taphole by gunning around an inserted metal pipe. For EBTs, dense magnesite or magnesite carbon shapes form the vertical taphole. These are often encased within larger rectangular shapes to form a permanent taphole well in the nose of the furnace. Again, some EBT tapholes last the life of the furnace while other operators patch the furnace with resin bonded patching mixes or perform regular maintenance by installing new taphole shapes as needed. EBT tapholes require a well-fill sand between heats to prevent steel freezing in the taphole; this sand is usually olivine, magnesite or chromite based.

10.1.2.1.8 Spout or Runner The spout or runner of conventional electric furnaces may be brick shapes, rammed plastic, or a large, preformed spout shape. Due to the extreme thermal shock on this runner or spout, high alumina materials are preferred. Occasionally, a combined basic/high alumina system will be utilized to maximize performance.

10.1.2.1.9 Refractory Considerations for DC Furnaces DC electric furnaces have special refractory requirements due to the fact that the return electrode is usually installed in the bottom of the furnace (some DC furnaces use an alternative arrangement with two graphite electrodes).

In the case of a current conducting bottom, the refractory lining at the center of the furnace bottom acts as the anode. A copper plate is usually connected below the conductive refractory and the return copper bus bar is connected to the plate. In this case special requirements for the refractory are low electrical resistance (preferably <0.5 milliOhms per meter), low thermal conductivity, and high wear resistance.

A typical configuration uses a six inch thick working lining consisting of carbon bonded magnesia mixes containing 5–10 wt% carbon. These materials can be installed either hot or cold. Below the working lining a three layer magnesia carbon brick is installed. The residual carbon content of the bricks ranges from 10–14 wt%. With regular maintenance, this bottom electrode configuration has achieved a bottom life of up to 4000 heats.

The billet return electrode configuration employs from one to four large steel billets (on the order of 10 inches in diameter) depending on the size of the furnace. The billets are embedded in the bottom refractory. The billets are surrounded with a basic refractory brick. The remainder of the hearth is rammed with a special magnesia ramming mix. Magnesia ramming mix is used to maintain the brick area around the electrode. This return electrode configuration has achieved in excess of 1500 heats on the furnace bottom.

The pin type of return electrode uses multiple metal pins of 25–50 cm in diameter to provide the return path for the electrical flow. These pins actually penetrate the refractory down to the bottom of the furnace where they are attached to a metal plate. Dry magnesia ramming mix is used for the entire hearth lining. This mix is rammed between the metallic pins. Alternatively magnesia carbon brick can be used in the area around the anode. This helps to improve the furnace bottom life but is more costly. Typical bottom life ranges from 2000 to 4000 heats depending on the refractory materials used.

The steel fin return electrode uses steel fins arranged in a ring in the furnace bottom to form several sectors. Each sector consists of a horizontal ground plate and several welded steel fins which protrude upwards through the refractory. Dry magnesia ramming mix is used between the fins. The hearth is also lined with this material.

The important points to consider during installation are refractory zoning pattern, hearth contour, slagline location, furnace steel capacity, taphole location, taphole size and angle, roof/delta orientation, expansion allowances, burner port location, slag door construction, bottom stirring elements, and DC furnace bottom electrode.

10.1.2.2 Electric Furnace Lining Installation

Typical procedures for a complete new electric furnace or EBT lining follow.[7]

10.1.2.2.1 Sub-bottom and Monolithic Hearth The furnace must be level and the shell cleaned from all debris prior to starting. It helps to locate the exact center of the furnace and to punch a mark in the bottom center of the shell. The preferred construction is rectangular magnesite brick laid on flat using a basic granular material as fill to provide a flat surface against the rounded steel shell. After locating the EBT taphole seating blocks, any bottom stirring elements and allowing for the DC furnace bottom electrode, the first course of sub-hearth brick is laid dry, tight and LEVEL, and a dry magnesite mortar is swept into the brick joints. Additional fill material is placed around the perimeter of this first course and leveled out. The second course is laid at a 45° angle to the first course and dry magnesite mortar again is swept into the joints. This process is repeated for three or four flat courses. (There is an alternate, less preferable, sub-hearth design which lays two to four flat courses which follow the curved contour of the steel shell. This may be used in furnaces where the shell has a small spherical radius and gives more uniform thickness in the monolithic hearth material.) At the proper elevation, key shaped brick are used to begin the first stadium course. It is preferable to start with the largest ring one inch from the shell and work towards the center. To close the ring, a key brick is cut on a brick saw to the exact dimensions required to close the ring. If the cut shape will be less than a half brick, two cut shapes should be utilized. The void at the end of the course up to the shell is filled with granular magnesite material. The next stadium ring is installed in similar fashion. The contour of the stadium hearth shown on the drawing must be carefully followed to leave sufficient room to add the monolithic hearth material at the appropriate thickness.

For EBT furnaces, it is even more critical to follow the refractory bottom drawing exactly. There are partial rings of brick with varying radii extending out into the nose section of the furnace which must be kept level. One way to facilitate this is to drill a hole in the top flat course of bottom brick in the exact center of the furnace and then utilize a broomstick with a nail on it extending up from this center brick as a mandrel to draw circles and arcs for the stadium rings and partial rings extending into the nose.

10.1.2.2.2 All Brick Hearths If a monolithic hearth material is not utilized, the final course or final two courses in the all-brick hearth are laid in rowlock (on edge) or soldier (on end) construction. Rowlock or soldier construction gives much greater brick-to-brick contact and minimizes heaving of the hearth in service. Again, all courses in the hearth and stadium are laid dry and swept with magnesite mortar to fill the joints.

10.1.2.2.3 Slagline and Sidewalls Once the stadium rings are completed, the slagline brick are installed course by course using the same keying up concept utilized for the stadium rings. The slagline should also be installed in excess of one inch away from the steel shell to permit thermal expansion without spalling or heaving the brick. Brick rings or partial rings should be continued up into the sidewall and hot spots until the water-cooled panels or the top of the furnace is reached.

10.1.2.2.4 Door Jambs The door jambs are a critical design area for the refractory lining. Several operators simply utilize regular key shapes in interlocked courses as their door jamb, and they have a successful practice. Other operators utilize special door jamb shapes which have greater surface area for better interlocking between courses and a sharper angle which opens up the door opening and eliminates or reduces refractory damage when slagging off. Still other door jam designs involve brick or precast shape assemblies which are welded or bolted to the steel shell. These are generally installed first and the slagline and sidewall brick laid directly against these assemblies, with sidewall rings keyed up halfway between the door and the taphole.

10.1.2.2.5 Taphole Conventional tilting electric furnaces generally use taphole module shapes set with a crane at the proper elevation prior to bricking the slagline and sidewalls. One alternative is to leave an opening in the sidewall rings and then ram or gun around a steel pipe forming the taphole. This pipe is then melted out on the first heat. Refractory taphole shapes can also be utilized in this same manner with monolithic material holding them in place at the proper elevation and angle.

10.1.2.2.6 Roofs and Deltas The electric furnace roof or the delta section in a water-cooled roof are generally installed in a refractory reline area, and a finished roof is waiting for the furnace to be rebuilt. With water-cooled roofs and precast deltas, a castable or plastic refractory is often placed around the perimeter of the precast shape to lock it in place against the water-cooled roof. For brick refractory roofs, a roof form is required (generally concrete) which creates the appropriate dome shape for the inside contour of the roof. The mandrel(s) are set for the electrodes. Triple tapered electric furnace roof shapes are laid in concentric rings against the roof ring for the outer two-thirds of the roof. Often partial rings of brick are laid in a wedge pattern between electrodes. Electrode ring brick surround each mandrel and are held in place with steel bands. A castable or plastic is cast or rammed into place in the cavity between the outer rings and electrode ring brick.

10.1.2.2.7 Monolithic Hearth Most electric furnace operators use a monolithic hearth material. This is a high magnesite, self sintering product which is granular in nature. After installing any EBT taphole seating blocks, bottom stirring elements, and DC bottom electrode forms; monolithic hearth construction is started. The hearth material comes ready to use in large bulk bags; a crane holds the bulk bag in position over the brick subhearth while the bag is split and the material is shoveled into place. After two or three bulk bags are in the furnace, several workers using shovels or pitch forks repeatedly jab the granular material in order to remove air and densify the refractory hearth. As the material densifies, the workers further compact it with mechanical vibrators or by simply walking on the hearth to achieve the proper contour. Additional bags of hearth material are added and de-aired and densified until the final contour is reached (usually measured with chains or a form). The new hearth sinters in place during the initial heat.

10.1.2.2.8 Heatup Schedule On a new electric furnace lining with a completely new monolithic hearth, steel plates or light scrap are generally lowered by magnet into the bottom of the furnace to provide protection for the unsintered hearth material. After this cushioning scrap is in place the first bucket is charged and the arc is struck, utilizing a long arc to avoid boring down into the new monolithic bottom. The bottom is usually sintered after the first heat, although it is important to inspect the bottom and banks for any holes or erosion due to unsintered material leaving a void in the lining. Since the new electric furnace lining has very little moisture in it, no special precautions are required during the initial heatup, other than using a long arc to avoid eroding the bottom prior to sintering.

Electric furnace refractory linings are maintained by gunning, fettling, and patching with brick.

10.1.2.2.9 Gunning Maintenance Gunning maintenance consists of mixing water with a magnesite based gunning mix and spraying this mixture onto the refractory lining. Gunning is used to maintain hot spots, slagline erosion, tapholes, the door breast area, or any other portion of the lining which experiences selective refractory wear. Gunning material is usually a temporary measure and will have to be regunned in the same place within the next several heats. Gunning maintenance, while temporary, does offer balanced life by evening out the highly selective wear pattern in the electric furnace lining. That is, refractory wear in AC furnaces is usually greater in the sidewall closest to the mast electrode; and gunning this area maximizes overall lining performance. Most refractory gunned maintenance is done with a pressurized chamber batch gun of roughly 900–1800 kg capacity. This gun delivers dry material pneumatically to a water mixing nozzle, and the air pressure sprays the wet gunning mix onto the surface of the lining. The nozzle operator's training and skill is a factor in the quality of the gunned patch, and there is a tendency to empty the gun, which can increase refractory costs. Gunning can be automated by using a mechanical center-throw gunning device which shoots gun material in a circular pattern while suspended from a crane. This mechanical gunning is faster and easier on people, but often wastes material by placing it where it may not be needed.

Basic gunning mixes range from 40–95% MgO in quality. High temperature operations and high power furnaces generally use higher MgO content gun mixes; while moderate temperature operations utilize lower MgO content gun mixes. Carbon steel producers use between 1.4–7.5 kg of gun mix per ton of steel.

10.1.2.2.10 Fettling Maintenance Fettling maintenance is the technique used to patch holes in the monolithic bottom. A rapid sintering version of the granular hearth material, or the original product, are used for fettling. The dry material is shoveled or dropped by crane wherever there is a hole or divot in the monolithic bottom; or a mechanical chute suspended by crane delivers material onto the sloped banks of the hearth. Occasionally the magnet is used to level this patch material, which then sinters in place during the next heat. With this fettling technique, monolithic bottoms often last anywhere from three months to one year or longer.

10.1.2.2.11 Brick Patching After several weeks of operation, gunning maintenance becomes less efficient in maintaining the refractory lining. Most operators will then take the furnace cold and dig out anywhere from 30–80% of the sidewall and hot spots. The rubble is then removed from the furnace and workers install new brick in all areas of the hot spots and sidewall which were removed. Often the same refractory quality and thickness are reinstalled as was used in the initial lining. Alternatively, lesser quality or thinner linings are installed during this patch and slightly less lining life is anticipated from the patch as would be received from a new complete lining. The brick patch is completed by gunning an MgO gunning mix into all the voids and cracks in the patched brickwork.

Usually during a brick patch, the taphole is completely replaced or repaired with ramming mix or gunning material, if not replaced with brick work. Most electric furnace operators will develop a regular brick patch schedule for their electric furnace. This could be one or two intermediate patches for every complete sidewall job. Or, some operators have utilized the continuous patching concept where some brick are replaced every two weeks or even every week, and the complete lining is totally replaced only once or twice per year. These maintenance and patching decisions are usually dictated by the severity of the operating conditions as well as the company's economic maintenance philosophy.

10.1.2.2.12 Miscellaneous Refractory Maintenance Each EAF has unique features or conditions requiring specific refractory maintenance. Roof delta sections must be replaced at failure or on a regular schedule. Tapholes must be replaced or repaired when the tap time gets too short or slag carryover starts. Conventional EAF tapholes are usually replaced with a high MgO gunning mix shot around a steel pipe, while EBT tapholes are knocked out and a new assembly or one-piece tube inserted by crane from above. Bottom stirring elements and bottom electrodes each require specialized maintenance procedures which vary with design.

10.1.2.3 Operating Philosophy

The operational philosophy often has a significant impact on refractory performance. If maximum production is the key operating philosophy, decisions will be aimed at maximum production and less than optimum refractory maintenance can result. Maximizing production usually requires a high quality initial refractory lining that is expected to last a long time without much refractory maintenance. There is a crossover point where a lack of maintenance actually reduces productivity, so the maximum production philosophy often requires a delicate balance on refractory lining design and maintenance. The minimum cost philosophy is also common. Here the initial refractory lining is inexpensive, generic brands which are anticipated to last a long time through patching or gunned maintenance. Minimum cost linings generally have little zoning and often utilize the same brand and lining thickness throughout the furnace. This philosophy can be effective for low intensity operations.

10.1.2.3.1 Predictable Campaign This philosophy generally requires a set period between refractory relines or patches. In this scenario the electric furnace operator schedules full production for a two week or three week period, knowing the furnace lining will make it that long. Then a patch or reline is scheduled and another predictable campaign is begun. This philosophy requires close coordination and optimization of the refractory design with shop operations.

10.1.2.3.2 Minimum Downtime Philosophy This scenario is similar to maximum production, but without the same zeal for total production. Here the refractory lining must deliver predictable operations with very little maintenance, so a high quality initial lining is usually required. There is not much time for downtime or maintenance until the scheduled down turn or down day. Predictable maintenance is acceptable, but must be on a regularly scheduled basis.

10.1.2.4 Future Considerations

Electric furnace refractory requirements have changed significantly in the past several years and are faced with even more change as electric melting technology and processes progress. The increasing use of scrap substitutes, direct reduced iron (DRI) and hot briquetted iron (HBI) requires modified furnace linings. Post-combustion of carbon monoxide and growing use of oxy-fuel burners both place demands on furnace refractory performance. Increased water cooling in areas of the door jambs and burner openings will be required. There will be pressure to reduce refractory consumption of chromium containing products and to ensure that refractories from all steelmaking processes are recycled.

10.2 Furnace Electric System and Power Generation
10.2.1 Electrical Power Supply

An arc furnace requires a very high power. Each ton of steel melted requires on the order of 400 kWh (0.4 MWh), thus to melt 100 tons 40 MWh of electrical energy is required. For a meltshop producing at a rate of one million tons per year it is necessary to produce at the rate of 100 tons per hour or more. To reach this level, 40 MWh needs to be put into the furnace in about half an hour of power-on time to achieve the desired productivity. This means that the average power level in MW is on the order of 80 MW.

EAF meltshops typically require power levels in the 20–200 MW range, (enough for a small town). Such large power levels can only be supplied by the utility from their high voltage network, where voltages in the range 100–500 kV are present. This is a three-phase system (the three-phase choice, rather than some other number of phases, was made early in the 20th century) and for this reason the traditional AC furnace has three electrodes.

Inside the furnace the physics of electric arcs dictate arc voltages in the range 100–600 volts (AC furnaces); above these levels the arcs become too long to be manageable. In order therefore to obtain arc powers of the required MW it is necessary to operate with currents in the kiloampere range; typically 20–80 kA (AC furnaces).

The power flows from the utility's generators through their network and arrives at the steel plant at very high voltage and must therefore be converted to low voltage suitable for arcs. This task is performed by transformers.

10.2.1.1 Transformer System

For a variety of reasons the task of transforming the power from the kV level at the incoming utility line to the voltage level needed by the arcs is usually done in two stages. A first transformer (occasionally two transformers in parallel) steps the voltage down from the high voltage line to a medium voltage level which is generally standardized for each country, Fig. 10.26. In the U.S. this medium voltage is usually 34.5 kV, while in Europe, Japan and other areas the voltages are not very different, often 30–33 kV.

As the steel plant requires electrical power for other sectors, for example a caster or rolling mill, there will be several transformers connected at the 34.5 kV level and thus it is common to have a small 34.5 kV substation within the meltshop. This arrangement is one of the reasons for making the transformation from utility to arc furnace in two steps.

From the 34.5 kV busbar the arc furnace is powered by a special, heavy duty furnace transformer. The secondary voltage of this furnace transformer is designed to allow operation of the arcs in the desired range of arc voltages and currents. However, because there are varying requirements of arc voltage/current combinations through a heat—for boredown, melting and refining—it is necessary to have a choice of secondary voltages. The furnace transformer is equipped with a tap changer for this purpose.

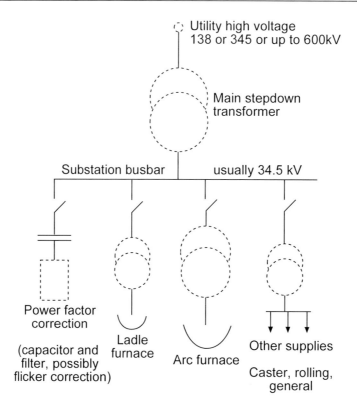

Fig. 10.26 Power transformation from high voltage line to the arc furnace. *(Courtesy of UCAR.)*

10.2.1.2 Transformer Basics

The basic principle of operation of a transformer is illustrated in the sketch of Fig. 10.27. A primary winding is wrapped around one leg of a closed, silicon steel iron core. Around another leg there is the secondary coil of a lower number of turns. The energy flows through the core transported by the magnetic field which is common to both windings. Because of the common field the product of amps × turns is the same for each coil. Thus if there are, for example, 100 turns on the primary coil and ten on the secondary the secondary current will be ten times the primary current.

At the same time the voltage induced in each coil is determined by the number of turns and the magnetic field. Again the magnetic field is common so the voltage ratio of the primary coil to the secondary coil is also determined by the ratio of primary turns to secondary turns.

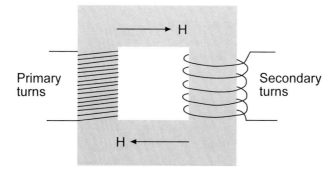

Fig. 10.27 Relationships for a basic transformer. *(Courtesy of UCAR.)*

Mathematically we have:

$$\frac{\text{Primary voltage}}{\text{Secondary voltage}} = \frac{\text{Primary turns}}{\text{Secondary turns}} = \frac{\text{Secondary Amps}}{\text{Primary Amps}} \qquad (10.2.1)$$

The arrangement of a three-phase transformer is sketched in Fig. 10.28. The closed iron core consists of three vertical legs, one for each phase. Since it is not necessary to have primary and secondary windings on different legs they are usually wrapped over the same leg. Because the secondary carries a high current it is practical to place it outside of the primary.

10.2.1.3 Core or Shell Designs

As sketched the windings are shown as cylinders enclosing the iron core—the core-type transformer design. An alternative used in some transformers is to wind the coils as spiral pancakes which are arranged like a stack of disks over each leg. This design is known as a shell-type transformer and was chosen by some furnace manufacturers/steelmakers for its lower reactance (see Section 10.2.4 for discussion of reactance). However in latter years the search for low reactance has been replaced by the high reactance design so the use of shell-type transformers is diminishing.

10.2.1.4 Tap Changer

The purpose of a tap changer is to allow a choice of different combinations of volts and amps for different stages of a heat. This is achieved by changing the number of turns of primary coil. (The primary takes lower current so it is simpler to change the number of turns on this coil rather than the high current secondary coil). Basically it takes the form of a motorized box of contacts which switch the primary current to different parts of the coil around the iron core. A schematic diagram of the connection to the primary coil is given in Fig. 10.29.

Most tap changers are designed to operate on-load. This means switching primary current, usually in the 2 kA range, at 34.5 kV. A make-before-break contact movement is used to avoid current interruption. Accordingly these contacts are subject to heavy erosion due to arcing and therefore require preventative maintenance.

Some steelmakers choose an off-load tap changer in order to avoid the heavy duty of on-load switching. However, such a tap changer requires that the steelmaker break the arc by lifting

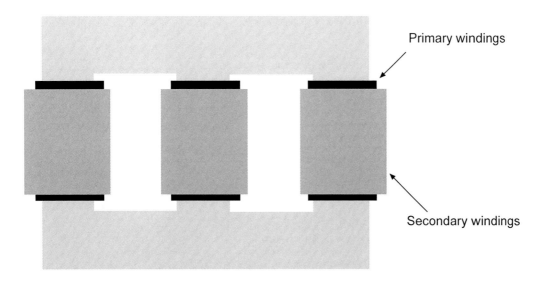

Fig. 10.28 Arrangement of a three-phase transformer. *(Courtesy of UCAR.)*

electrodes and this procedure may take as long as one minute. Today such a delay each time a tap is changed is intolerable and as such this design is becoming rare.

10.2.1.5 Transformer Generalities

The maximum magnetic field strength within the core is determined by the properties of the silicon steel laminations from which the core is built. This field maximum therefore dictates the size of the core which typically weighs several tons. Several more tons are contributed by the copper windings of which the secondary may be capable of 60 kA operation. Cooling of these coils, heated by Ohmic losses, is achieved by forced oil circulation, the oil itself being cooled by water in a heat exchanger usually outside the transformer tank. Together with its oil, a typical furnace transformer may weigh in the 50–100 ton range.

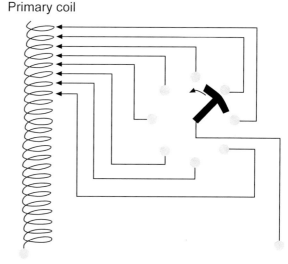

Fig. 10.29 Schematic arrangement of a tap changer. *(Courtesy of UCAR.)*

Transformers are rated in terms of MVA which is a measure of the Ohmic heating of the coils and therefore the temperatures reached by the windings. Persistent overloading of the transformer by the steelmaker will lead to higher than design coil temperatures and a corresponding decrease in lifetime of the insulating material separating coils from each other and the core.

The duty of a furnace transformer is arduous. Arcs are unsteady loads and are often short-circuited and sometimes open-circuited by scrap movements, leading to large current fluctuations. The secondary coils are subject to severe magnetic force shocks which must be absorbed by the insulated clamping system holding the coils in place. Any loosening of these clamps could result in failure of the insulation.

The primary coil sees a variety of voltage spikes due to switching the transformer on or off, which may happen up to 100 times per day. Although the coils are rated at approximately 34.5 kV, between them high frequency voltage surges can reach several times this level.

Because of the critical role of the furnace transformer in steel production there are a variety of alarms and indicators intended to allow monitoring and preventative maintenance.

Tap changers tend to be attached outside of the transformer tank to allow for easy access for contact maintenance.

10.2.2 Furnace Secondary System

The secondary coil terminations from the furnace transformer exit the tank as a set of copper busbars. To step up to the high currents required in the furnace these exit busbars are then connected into a delta closure usually affected outside the transformer, see Fig. 10.30. (Some suppliers have put this delta closure inside the tank to improve cooling.)

As there is always a slight risk that the transformer oil may be involved with a fire hazard it is standard

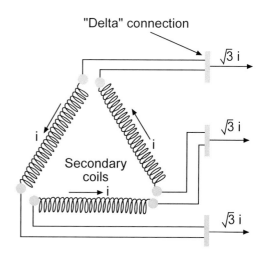

Fig. 10.30 Secondary coil terminations are connected in delta, usually outside the tank. *(Courtesy of UCAR.)*

practice to place the transformer and some other ancillary equipment in its own building, the transformer house, which simplifies fire control measures associated with the transformer. Thus the high current busbars powering the furnace pass through a wall of this house and it is at this junction where the flexible cables are connected.

10.2.2.1 Cables

Cables are necessary to allow the electrodes to follow the scrap level lowering during meltdown and to allow for furnace movements such as roof swing and tapping. Usually several cables are required on each phase to handle the current.

10.2.2.2 Furnace Arms and Busbars

The mobile ends of the cables are attached to the horizontal arm/busbar system. More and more the manufacturers of furnaces pass the current through an electrically conductive arm to the electrodes rather than through separate busbars. Such conductive arms are either copper clad, steel box constructions or aluminum alloy fabrications. In both cases the whole arm runs at voltage and must therefore be insulated from its supports. In the older separate busbar arrangement both the busbars and the arms were individually isolated electrically from one another.

Current passes to the electrode via the holder. The most common design here is to have one or two fixed contact surfaces, which also act to align the electrode, and a separate pressure pad. The latter is pressed against the electrode in fail-safe fashion with a spring. This spring is compressed hydraulically to open the holder.

10.2.3 Regulation

10.2.3.1 Hydraulic Components

The electrode/arm/mast/cable assembly weighs typically in the range of 20 tons. This is moved vertically for control purposes by a hydraulic cylinder incorporated in the mast. (In some furnaces the movement is effected by an electric motor/cable winch arrangement). As the arc length is dependent, amongst other things, on the level of scrap or liquid under the electrode, and this level changes through the heat, it is necessary to have an automatic control over electrode position—the regulation system.

The flow of the inflammable fluid to the mast cylinder is under the control of a hydraulic spool valve. Flow control is achieved by covering and uncovering ports by displacement of the spool over a stroke in the range of 10 mm. It is pushed by an hydraulic amplifier, a device which generates the necessary power from a separate valve operating with lubricating oil under pressure. An electrical signal enters this amplifying valve at the level of milliamps or a few volts. Thus the system consists of a low power electrical signal, amplified by a hydraulic valve causing displacement of the main spool valve. Fig. 10.31 shows a section of a typical combined amplifier/spool valve.

10.2.3.2 Electronic Control of the Hydraulic Valve

Modern valves incorporate electronic control of several parameters. The most basic parameter is the proportionality between signal level and electrode speed—the gain. Spool valves usually exhibit a dead band over which a signal change produces no flow and to some extent this can be compensated in the electronic system. It is also common practice to control maximum speed electronically rather than through separate flow impedances in the hydraulic lines.

Further signals which can be used by the electronic part of the regulation can be provided from hydraulic pressure and flow rate, allowing speed feedback to be employed.

10.2.3.3 Electrical Control Philosophies

There are significant variations in the way the furnace electrical signals are manipulated in order to achieve the control of arc furnaces. These variations reflect the different philosophies and opinions

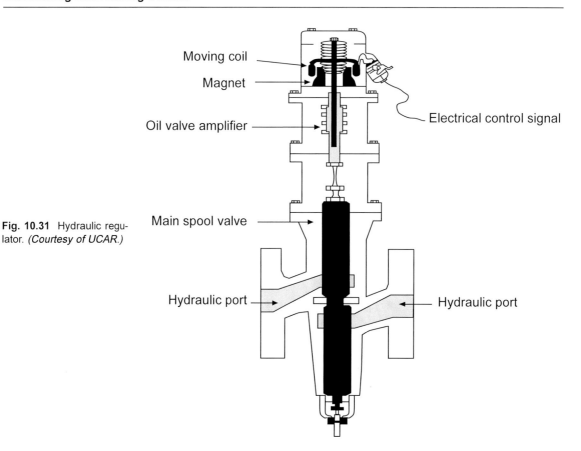

Fig. 10.31 Hydraulic regulator. *(Courtesy of UCAR.)*

amongst suppliers and steelmakers on how best to achieve control. It is the regulation system which influences many important aspects of furnace performance, such as MW input, mean current, arc stability, scrap melting pattern, energy losses to water-cooled panels, energy, and electrode and refractory consumptions. As all these parameters are interrelated in a complex and not well understood manner, it is not surprising that there are many differences of opinion on optimum control strategies.

There is a limit to the acceleration to which the cantilevered mast/arm/electrode can be moved; too high can lead to excessive forces on the system. If, however, the acceleration is too low the time taken to move the electrode becomes excessive. As a general rule the response time (time to reach 90% of full speed) falls between 0.2 and 0.5 seconds. In contrast, arc voltage and current changes are very fast; times are less than one cycle (1/60 or 1/50th second). Because of this inherent mechanical limitation it is not possible to move the system fast enough to correct the fast arc changes; only the lower speed electrical variations can be controlled by the electrode regulation system.

The accepted standard handling of the electrical signals is to form an impedance control. A voltage signal taken from the phase to ground and a current signal are each separately rectified and their DC values are compared back-to-back, as shown in Fig. 10.32. If the voltage and current are each at a desired level—the set point chosen by the

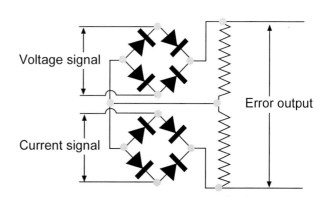

Fig. 10.32 Method of forming a control signal. *(Courtesy of UCAR.)*

steelmaker—the output from this comparison of signals is arranged to be zero. If however the current exceeds this level its signal increases and simultaneously the voltage decreases. Then the two back-to-back voltages do not balance and an output voltage is generated. This signal goes to the regulating valve in such a way to command the electrode to raise, aiming to reduce current. This method of control attempts to hold the ratio of voltage to current constant, hence its description as impedance control.

Alternatives to control of impedance are control over current, arc resistance, arc power or arc voltage. In each case the system attempts to keep the relevant parameter constant. In the cases of arc resistance, arc power and arc voltage controls it is necessary for the supplier to obtain an arc voltage signal. This is not straightforward but the techniques for doing so have been known for some time.[8] The extraction of an arc voltage signal requires that inductive voltages proportional to the rate of change of current be subtracted from the easily accessible secondary phase-to-ground voltages. Rogowski coils are used for this purpose. Indeed such coils, with integration of the output, are commonly used for current measurement.

During the course of a heat the electrical characteristics of the furnace vary. As a consequence the mean electrical parameters will drift with time through the heat if the set points are constant. And the different control philosophies—impedance, current, resistance, etc.—drift through the heat in different ways. Another strategy open to the steelmaker therefore is to vary the set points through the heat in such a way as to achieve real-time optimization.

10.2.4 Electrical Considerations for AC Furnaces

10.2.4.1 Importance of Reactance

In high current systems like arc furnaces reactance plays an important role in the electrical characteristics.

Fig. 10.33 illustrates a high current loop consisting of two bus tubes 15 m (50 ft) long, separated by about 1 m (3 ft). These dimensions are chosen to represent two phases of a three-phase furnace secondary from transformer to arc. If the current is DC the voltage necessary to drive a current of, say, 60 kA around the loop is merely that required to overcome Ohmic resistance. For typical furnace busbars this voltage would only be about 2.5 volts. But if we wish to vary the current we must also overcome the induced voltage generated by the change in magnetic field linking the loop. High currents mean high fields so the induced voltage is also high and is always opposed to the source of current variation. In this example if we wish to drive 60 kA AC current around this loop we would need a much higher voltage—about 300 volts! This inductive impedance to alternating current flow is called reactance.

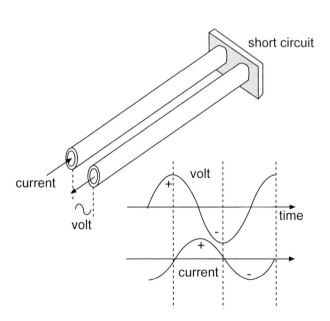

Fig. 10.33 Representation of two phases of an arc furnace secondary current circuit. *(Courtesy of UCAR.)*

For a sinusoidal current waveform the voltage waveform is also sinusoidal but one quarter cycle ahead in time (90 electrical degrees), as shown in Fig. 10.33. This means that the voltage maximum occurs as the current passes through zero. This feature of high voltage around current zero is critical to arc stability when an arc is present.

10.2.4.2 Arc Characteristics

The power flow from the utility's generator, through the net and set of transformers, through cables, arms and electrodes finally reaches the arcs where the useful work of melting and heating occurs. Approximately 95% of the power taken from the utility is released in the arc so its characteristics dominate the functioning of the whole system.

As an electrical element the arc is complicated; it is not a simple fixed Ohmic resistor. Current is transported by free electrons and positive ions in a high temperature plasma in which temperatures may be of the order of 10,000–15,000°C in the hottest regions, dropping to local ambient of, say, 2000°C at the boundary. The diameter of this conducting column depends on current and is of the order of 10–20 cm (4–8 in.), much less than that of the graphite electrode. The voltage depends principally on the length of the column and is of the order of 10 V/cm (25 V/in.); a 300 V arc may therefore typically be 30 cm (12 in.) long.

Because of its high temperature, density is low and for this reason the arc is very easily displaced at high speed by magnetic fields, especially local fields produced by the current flow in pieces of scrap. These effects can result in arc movement over the end of the graphite electrode in milliseconds (ms). In doing so the length of the arc generally changes for geometrical reasons and this in turn results in rapid voltage variations.

Another consequence of the low mass of an arc is that its thermal capacity is also very low. If the current is switched off the ionized plasma column disappears with a characteristic time of about one millisecond. Thus when the current is alternating, as in an AC furnace, the arc has a tendency to cool off each time the current drops to zero. Re-establishment of current flow in the reverse direction may take several hundred volts, depending on ambient conditions, mainly how hot are the graphite and steel arc terminations. This re-establishment of the arc as the current passes through zero is assisted greatly by the presence of reactance in the circuit. The more reactance present the greater is the available voltage at current zero and the smoother is the current polarity transition.

Fig. 10.34 shows arc voltage waveforms under various conditions during a heat. At the beginning when the arc is running between the comparatively cool graphite electrode and cold pieces of scrap, instability of arc voltage is at a maximum. During the course of melting the arc voltage waveform becomes smoother as temperatures rise. A further improvement occurs when the steel termination of the arc is liquid. At the extreme, when the arcs are submerged in foaming slag, the voltage waveforms are almost sinusoidal.

This large excursion in the stability and harmonic content of the arc voltage is reflected in the electrical characteristics of an arc furnace. In the early periods of a heat, effective MW input is lower for a given regulated set point than at later periods. Fig. 10.35 shows generalized electrical characteristics both as MW versus kA and as MW versus MVAR. As stability improves all the curves move to a higher level of MW, tending towards the limit case where the arc voltages would be perfectly sinusoidal and steady. In this hypothetical case of perfectly Ohmic arcs the MW–MVAR characteristic becomes the familiar semicircle, the magnitude of which depends on the transformer voltage and the circuit reactance. Mathematically the effect of real arcs is similar to introducing more reactance into the circuit than physically exists. One can model empirically the changing characteristics by

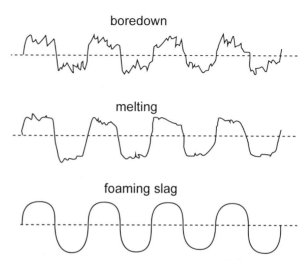

Fig. 10.34 Arc voltage waveforms at three stages of a heat. *(Courtesy of UCAR.)*

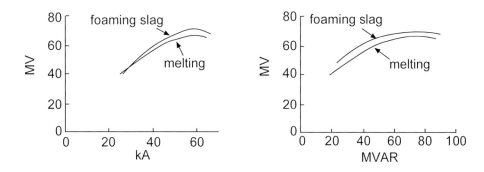

Fig. 10.35 Example of electrical characteristics for an AC furnace. *(Courtesy of UCAR.)*

considering the total circuit to have an operating reactance which is higher than the actual and which decreases with time through the heat.

10.2.4.3 Supplementary Reactance; the High Reactance AC Furnace

The geometrical constraints of the layout of arc furnaces in the 80–160 ton range result in systems which tend to have the total reactance—transformer plus secondary system—falling in a limited range of 2.5–4 milliOhms (per phase). In order to operate with arc voltages in the 400–500 V range this natural reactance level is too low to guarantee smooth transition through current zero during scrap melting. It is now therefore common practice to add supplementary reactance to the circuit. This is achieved conveniently by adding a reactor on the line supplying the primary of the furnace transformer where currents are lower.

Such reactors come in two varieties; one is the air-cored coil design, generally placed outdoors of the meltshop. This design is simple and inexpensive. The alternative is to use an iron-cored reactor, in which the magnetic field is restrained to the restricted volume of the core. Consequently this design takes up less space and is suitable for installation next to the furnace transformer, or, in some cases, within the transformer tank.

The usual magnitude of the additional reactance is such as to push the total reactance per phase into the 5–10 mOhm range (as viewed from the secondary voltage of the transformer). Since the total system impedance is higher in this design of furnace it is necessary to have a corresponding higher transformer voltage to obtain the desired characteristics.

The effect of operating with higher reactance and higher secondary voltage is illustrated in Fig. 10.36. With the increased reactance the secondary voltage waveform is much higher during the

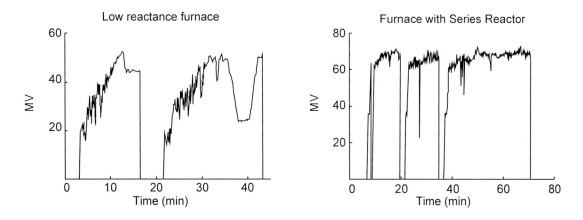

Fig. 10.36 Improvement of stability with additional reactance.

passage through current zero. As a consequence the risk of current interruption is greatly reduced; current waveforms are smoother and MW are therefore increased.

10.2.5 Electrical Considerations for DC Furnaces

10.2.5.1 DC Supply

The required high power again is supplied from a high voltage three-phase AC network. This is converted to DC by rectification of the output of the furnace transformer.

Rectification is achieved by bridge-connected thyristors. A single, three-phase bridge connection is illustrated in Fig. 10.37. The voltage waveforms for different thyristors firing angles are shown in Fig. 10.38 (simplified). This output is described as six-pulse. Normally 12, 18 or 24-pulse supplies are used in arc furnaces, obtained by multiple, parallel transformers electrically displaced one from another so that their individual pulses overlap uniformly. This electrical displacement, of 15, 10 or 7.5°, corresponding to the 12, 18 or 24-pulse systems, is made by various coil connections within the transformer. For this reason the transformers used for DC furnaces are quite different to those for AC and are generally unsuitable for AC furnace operation.

The volt/amp characteristic of a DC supply consists of a weakly declining drop of DC voltage as the DC current increases. The slope of this line is on the order of 1 volt per kA and is determined by the commutating reactance of the transformer/rectifier combination, not by the arc furnace. In order therefore to limit wide current excursions due to widely different arc voltages thyristors are used in preference to diodes. The conducting instant after current zero (firing angle delay) is under the control of the gate terminal. Each thyristor can, in principle, be turned off within half a cycle.

Even so, within the several millisecond delay between an arc voltage change (e.g. a short circuit) and the control of the thyristors, currents could increase significantly. To reduce the rate of rise of the current it is normal to add a reactor within the DC current loop, the natural reactance of the high current DC loop being inadequate.

These reactors are sized to have an inductance in the 100–400 microH range. Since they take the full DC current, Ohmic losses are significant and can only be maintained within acceptable bounds by employing an adequate section of the copper or aluminum making up the coils.

Thyristors are each capable of handling a few kA and a few kV of reverse polarity. An arrangement of series and parallel connected thyristors makes up each leg. Fuses and voltage balancing resistors are used as protective measures. Cooling is affected by de-ionized water.

10.2.5.2 Electrical Characteristics of DC Furnaces

The thyristor control is normally chosen to hold current constant. Thus the AC current before the rectifier is also constant, as is the primary current. Considering powers on the AC primary we see that constant current means MVA is constant. The characteristic of MW as a function of MVAR is therefore a quadrant of a circle for which

$$MW^2 + MVAR^2 = MVA^2 = constant \qquad (10.2.2)$$

The corresponding DC voltage/DC current relationship is illustrated in Fig. 10.39 where an example of a 1000 V maximum is shown. Generally the slope of the volt/amp line is linear and drops typically 100 V in 100 kA.

Thus at 100 kA, for example, the thyristor control can hold constant current over an arc voltage range from about 900 V down to short circuit by varying the firing angle.

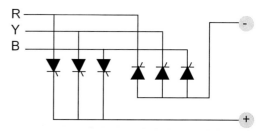

Fig. 10.37 Six-pulse bridge rectifier using thyristors. *(Courtesy of UCAR.)*

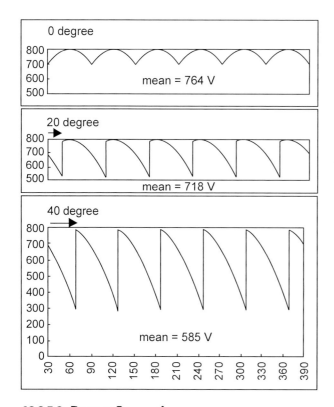

Fig. 10.38 Three-phase bridge waveforms (simplified); effect of firing angle on waveform and mean voltage. *(Courtesy of UCAR.)*

10.2.5.3 Bottom Connections

In order to operate with a single DC arc it is necessary to make an electrical connection—the positive anode—to the steel charge. Various furnace manufacturers have developed several solutions to this problem. The ideas are illustrated in Fig. 10.40.

In one type the anode current is shared amongst many steel rods embedded in a rammed refractory block. The rods, with a diameter of about one inch, could be one meter long and are linked by a copper plate below the furnace shell. The whole anode block may measure 1–2 m in diameter.

A variation on the pin type is to use thin steel sheets, again embedded in refractory.

Another variation is to employ a steel billet of diameter 400 mm (8 in.) passing through an insulated sleeve, leading to a cooled copper connection below the furnace shell.

In all three of these designs (pin, sheet or billet) the top of the steel conductor melts through the course of the heat. It resolidifies during power-off and after scrap charging.

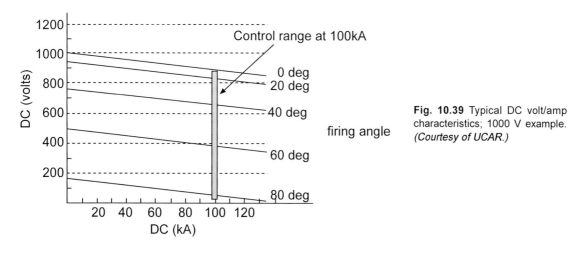

Fig. 10.39 Typical DC volt/amp characteristics; 1000 V example. *(Courtesy of UCAR.)*

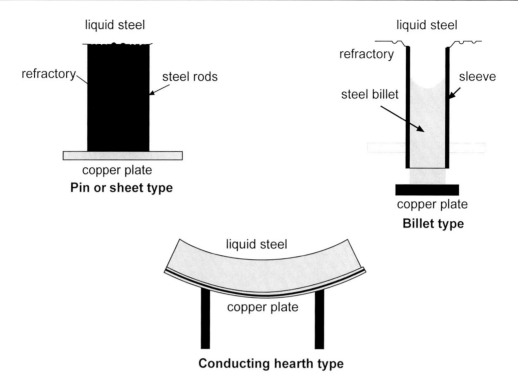

Fig. 10.40 Sketches of types of bottom connection for DC furnaces. *(Courtesy of UCAR.)*

An alternative to the steel-to-steel current designs is one where the current is taken through conductive refractories to a large diameter, copper bottom plate.

In all bottom connection types there must be insulation between the anode connection and the furnace shell. This is to reduce the likelihood of current passing through the shell directly to the anode busbars.

10.3 Graphite Electrodes

Graphite electrodes play an important part in electric arc furnace operation, allowing for the transfer of electrical energy from the power supply to the furnace bath. Electrodes must be capable of withstanding large temperature swings during furnace operation while at the same time providing for continuous and uniform power supply to the process.

As electric furnaces have evolved over the past 20 years, advances in graphite electrode technology have played an important part. As furnace designs were improved to allow higher power input rates, graphite electrode properties were improved to allow greater supply of electrical power to the furnace. Thus as EAF designs and operating practices have evolved, graphite electrode technology has kept pace, allowing for continued advances in EAF steelmaking capability. It is now common for new arc furnace installations to have transformers rated at in excess of 100 MVA. DC furnace operations have also placed additional requirements on electrode characteristics.

10.3.1 Electrode Manufacture

Graphite electrodes are typically made using premium petroleum needle coke, coal tar pitch, and selected proprietary additives. Electrodes are formed in the shape of cylindrical sections that are fit together using screw-in connecting pin sections. Electrodes are available in a variety of diameters, the requirements being dependent on the required current carrying capacity. Typically, diameters greater than 24 inches carry a premium price due to the requirement for different process technologies. Higher quality raw materials, increased energy amounts, extended processing schedules, and the installation of larger, heavy duty equipment are necessary to produce these electrodes.

Processing and manufacture of graphite electrodes consists of the following steps as shown in Fig.10.41:

1. Milling and mixing of petroleum needle coke with coal tar pitch and selected additives.

2. The mixture is then extruded and cut to cylindrical, green electrode sections.

3. The green electrodes are placed in saggers (stainless steel cans) in which the electrodes are covered with a protective packing medium like sand. The filled saggers are loaded onto railcar flatbeds which are moved into large gas fired car bottom kilns where the green electrodes are baked to approximately 800°C. The bituminous, green electrode material is transformed into amorphous, brittle carbon which is abrasive and difficult to machine. This process requires careful control to ensure

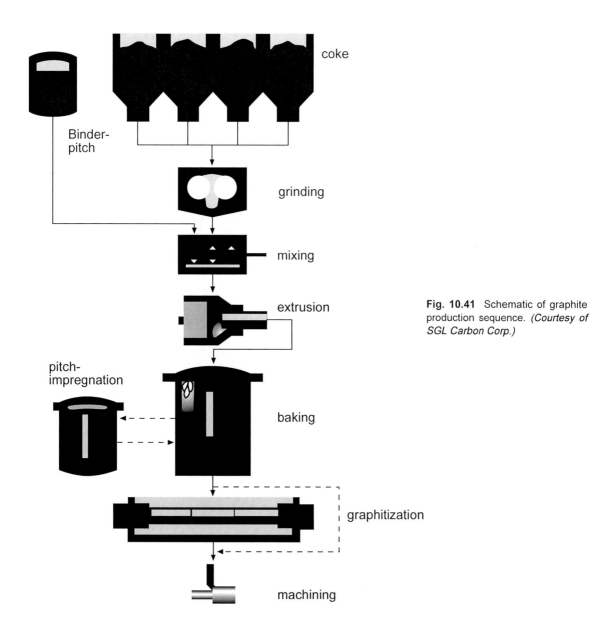

Fig. 10.41 Schematic of graphite production sequence. *(Courtesy of SGL Carbon Corp.)*

that thermal gradients remain small and rapid gas buildup does not occur. Thermally induced stresses or rapid gas buildup can result in cracking, distortion, or excessive porosity which can not be tolerated in the finished product. For this reason, bake cycles are long and take between three to four weeks.

4. The baked carbon sections are impregnated with petroleum pitch in order to increase strength and density. This also improves the end product electrical conductivity.

5. The impregnated carbon sections are again loaded into car bottom kilns and re-baked so that the petroleum pitch is converted to carbon.

6. The re-baked carbon sections are assembled to columns of eight to ten electrodes and loaded into large, electrically powered graphitizing furnaces. Direct current of more than 100kA is passed through the electrode columns heating them to approximately 3000°C. The intense heating causes the crystalline structure to change from the random amorphous form to the ordered layer structure of graphite. This modification increases machinability of the material as well as greatly improving electrical, thermal and mechanical properties. The graphitizing process is very energy intensive and requires more than 3000 kWhr per ton of graphite.

7. Finally, the graphitized sections are machined to the required diameter and length on large lathes. Tapered sockets are machined into each end to accommodate screw-in connecting pins which are used to attach the electrode sections end-to-end. The total production process from extrusion to shipping is quite time consuming and takes approximately three months.

10.3.2 Electrode Properties

Electrode physical properties vary somewhat from manufacturer to manufacturer. However, for most ultra high power (UHP) furnace operations these properties will be similar regardless of manufacturer. Following in Table 10.1 are typical electrode physical properties:

Table 10.1 Typical Electrode Physical Properties.

Property	Units	Electrode Diameter (in.)		
		18–20	22–24	26–28
Bulk Density	g/cm^3	1.69–1.76	1.68–1.76	1.68–1.76
Real Density	g/cm^3	2.22–2.25	2.22–2.25	2.22–2.25
Porosity	%	21–25	21–25	21–25
Specific Resistivity	10^{-5} Ohm in.	17.0–21.5	17.0–21.5	17.0–21.5
Bending Strength min.	psi	1200	1150	1000
Bending Strength max.	psi	1500	1450	1400
Young's Modulus	psi x 10^6	1.1–1.5	1.0–1.4	1.0–1.3
Coefficient of Thermal Expansion	10^{-6}/°C	1.25–1.65	1.25–1.65	1.25–1.65
Ash Content	%	0.1–0.6	0.1–0.6	0.1–0.6

10.3.3 Electrode Wear Mechanisms

Considerable research has been carried out over the years to evaluate the factors affecting electrode consumption. Generally speaking, electrode consumption occurs as both continuous consumption and discontinuous consumption. Continuous consumption usually accounts for 90% or greater of the total consumption and results from tip consumption (sublimation) and sidewall oxidation. Discontinuous consumption mechanisms include electrode breakage, butt losses, and electrode tip spalling.

Prior to implementation of electrode water spray cooling, the two continuous consumption mechanisms were believed to be approximately equal. Water spray cooling results in reduced electrode sidewall temperatures above the furnace roof which helps to reduce the amount of sidewall oxidation. Thus a shift to 50% of consumption due to tip consumption and 40% due to sidewall oxidation has been observed in operations using electrode spray cooling.

10.3.3.1 Original Bowman Correlations

Correlations have been developed empirically to estimate electrode consumption for various operating practices. It was found that electrode tip consumption is proportional to the square of the current and to the time over which the tip is consumed, which is the power-on time. The correlation is as follows:

$$C_{TIP} = R_{SUB} * \left(\frac{I^2 * t_{PO}}{P} \right) \quad (10.3.1)$$

or emphasizing the importance of productivity in tons per hour:

$$C_{TIP} = R_{SUB} * \left(\frac{I^2 * TU}{p} \right) \quad (10.3.2)$$

where

C_{TIP} = graphite tip consumption (lbs/ton)
R_{SUB} = sublimation rate (lbs/kA^2 per hr)
t_{PO} = power-on time (hrs)
I = current per phase (kA)
P = furnace productivity (tons/heat)
TU = time utilization = t_{PO}/t_{TAP}
p = productivity (tons/hr)

For electrode sidewall consumption, it was found that it is proportional to the oxidizing electrode surface area and to the relevant time period which is the tap-to-tap time:

$$C_{SIDE} = R_{OX} * \left(\frac{A_{OX} * t_{TAP}}{P} \right) \quad (10.3.3)$$

where

C_{SIDE} = graphite sidewall consumption (lbs/ton)
R_{OX} = oxidation rate (lbs/ft^2 per hr)
A_{OX} = oxidizing electrode surface area (square feet)
t_{TAP} = tap-to-tap time (hrs)
P = furnace productivity (tons/heat)

Below the limit of the water cooling it is sufficiently accurate to approximate the shape of each oxidizing column to a cone, the diameter at the top being the full electrode diameter, D, tapering to the tip diameter, D_t, at the bottom and the mean of these two diameters, D_{AV}. If the length of this cone is L_{OX} then the oxidizing area per column is

$$A_{OX} = \pi D_{AV} L_{OX} \quad (10.3.4)$$

With optimum water cooling the oxidizing length, L_{OX}, is close to the length of the column inside the furnace at flat bath. This increases with furnace size and is typically in the 2–4 m (7–12 ft) range.

Typical average values for oxidation and sublimation rates are:

R_{OX} = average oxidation rate = 8 kg/m² per hr (1.64 lbs/ft² per hr) for the oxidizing cone of water spray cooled electrodes for operations using around 25 Nm³ (900 scf) lance oxygen and no post-combustion.

$R_{sub\ AC}$ = average sublimation rate per electrode for AC operation = 0.0135 kg/kA² per hr (0.0298 lbs/kA² per hr).

$R_{sub\ DC}$ = average sublimation rate per electrode for DC operation = 0.0124 kg/kA² per hr (0.0273 lbs/kA² per hr).

10.3.3.2 Updated Bowman Correlations

In 1995 Bowman published an update to the above correlations in order to fit modern arc furnace practice.[9] The loss of accuracy of the old model seems to have occurred with the move to longer arcs allowed by the application of water-cooled panels. This suggests that arc length, i.e. arc voltage, is an important parameter which was not included in the old model. Other significant changes in arc furnace practice which influence electrode consumption, particularly oxygen consumption and its method of application needed to be considered in order to establish a new, empirical model for oxidation consumption.

10.3.3.2.1 Tip Consumption The old model for tip consumption, contained current squared and a function for the tip diameter. The link with tip diameter was necessary in order to conserve a relationship with the square of the current, which seemed to be correct on both theoretical and experimental grounds. The relation suggested that with larger electrodes, consumption rates decreased. However, larger electrodes are also associated generally with higher voltages. Thus a link to voltage is used in the new model and the link with electrode diameter has been dropped.

The evidence for a dependence on the square of the current is very strong; the data analysis in the original report showed a very good fit over a very large range of values, from 1–60 kA and almost four orders of magnitude of consumption rate. In order to identify the form of the dependence on arc voltage, the strong effect of the current can be eliminated by dividing measured tip consumption rates by kA² and plotting the resulting data as a function of arc voltage. The data for the old model were grouped mostly below the 200 arc voltage limit. Newer data covers higher arc voltages approaching 400 volts. Below an arc voltage of about 250 V there is an increase in the specific consumption rate; above 250 V the rate is more or less constant.

Long periods of arcing onto a slag covered bath are a common feature of operations with more than 70% continuously charged DRI or continuously charged processes such as Consteel. Measurements show that such operations consistently yield figures in the 0.010 kg/kA² (0.0221 lbs/kA²) per hour range. Some furnaces running with 100% scrap also achieve these low figures; others, however, reach up to 0.016 kg/kA² (0.0353 lbs/kA²) per hour. The difference between these two groups of furnaces seems to show up in the angle of the electrode tip. Operations with the higher levels of specific consumption rate are characterized by steep tip angles, up to 45° on meltdown; the lower levels seem to go along with tip angles in the 25–30° range. Bowman offers the following hypotheses to answer the questions why the tip consumption rate depends on tip angle and why the tip angles vary between furnaces.

10.3.3.2.1.1 Dependence of Tip Consumption Rate on Tip Angle For ladle furnaces, increasing tip consumption rate with shorter arcs can be explained by the dissolution of graphite due to splashing by the liquid steel. In the case of melting furnaces, the geometry of the arcing volume around the electrode tip illustrates that for a given average arc voltage the lower part of the tip comes closer to the liquid as the angle increases. Thus, at a tip angle of 40° the gap between graphite and steel is only about two inches at an average arc of 200 volt. In contrast, at 25° it is over three inches at 200 V. At 300 V the corresponding gaps are about five inches and seven inches respectively. The probability of graphite contact with steel is therefore increased with greater tip angles and lower arc voltages.

10.3.3.2.1.2 Variation of Tip Angles between Furnaces Arc blowout, the cause of tip angling, is a magnetic phenomenon. It depends on the proximity of the electrodes, i.e. pitch circle diameter (PCD), the distribution of the ferromagnetic scrap and the current distribution within the charge, between the arcs. It has also been suggested that deep, foaming slag can offer magnetic field protection. For a given current, the parameters which can generate low or high magnetic fields in the arc regions can be summarized in Table 10.2.

Table 10.2 Magnetic Field Generation

Influencing Parameters	High Field Strength	Low Field Strength
PCD	small	large
Bath	shallow	deep
Slag	shallow	deep

Thus a liquid heel operation with maximum time on deep, foaming slag promotes the lower tip angles. Those operations without a hot heel and with shallow slag depths such as stainless steel or acid iron production, are likely to generate steeper tip angles during meltdown.

In long-arc-only operations tip angle tends to stay constant. Tip consumption rates on DC arc furnaces are associated with long arcs and tip angles are usually below 10°. These parameters, as discussed above for AC, should not therefore play a role in consumption.

Based on DC operating data covering the current range 10–130 kA and arc voltages from about 250–550 volts, Bowman found that the specific tip consumption rate is essentially constant at 0.0124 kg/kA^2 (0.0273 lbs/kA^2) per hour, taking into account the experimental errors. No dependence on arc voltage was apparent, as expected from the AC analysis above.

Arc voltage is normally not available on most arc furnaces. Bowman indicates that an accurate value can, however, be calculated from readily available electrical data, in this case MW and kA as follows

$$\text{average arc voltage} = \frac{MW * 1000}{3 * kA} - r * kA \qquad (10.3.5)$$

where the r represents the phase resistance, in milliOhm. Typical values, if not known from a short circuit test, vary with furnace size as shown in Table 10.3.

Table 10.3 Typical Phase Resistance Values

Furnace Electrode Diameter, in.	Phase Resistance, milliOhm
16	0.55
20	0.45
24	0.40

Summarizing, Table 10.4 presents the following specific sublimation rates taking the varying voltage ranges and tip angles into account:

Table 10.4 Sublimation Rates by Furnace Condition

Furnace Condition	R_{SUB}, kg/kA² (lbs/kA²) per hr
AC, arc voltage>250 V	0.0130 (0.0287) × 3 electrodes
AC, tip angle<30°	0.0100 (0.0221) × 3 electrodes
AC, tip angle>45°	0.0160 (0.0353) × 3 electrodes
DC	0.0124 (0.0273)

10.3.3.2.2 Sidewall Oxidation A number of studies have been published which show that there are important differences in the degree to which furnace practice exposes the graphite electrodes to oxidation. These differences are reflected in varying specific rates of oxidation, expressed as kg/m² per hr (lbs/ft² per hr). Low rates occur with a low oxygen consumption practice and low ingress of air. A continuously charged DRI operation of relatively long duration with a closed furnace and no oxygen lancing is a good example. At the other extreme an operation using large volumes of oxygen, perhaps with burners and post-combustion injection, tends to produce higher specific rates of oxidation. Surface area exposed to oxidation is a major parameter which influences oxidation consumption. Direct water cooling of the electrodes has helped to reduce this area in AC furnaces. In DC furnaces the use of only one electrode further reduces the area significantly.

With such varying oxygen practices in modern operations, it is necessary to assume a varying specific rate of oxidation, kg/m² per hr (lbs/ft² per hr), based on furnace practice. Oxidation rates are dependent on several features of the use of oxygen (and burners if used) which are often not recorded or not under control: lance angle, lance direction relative to the electrodes, oxygen flow rate and lance diameter, position of the lance tip relative to steel and slag levels, burner excess oxygen levels and duration. In many furnaces the lower parts of the graphite columns are heavily splashed by FeO-rich slag or incompletely reacted oxygen. Post-combustion of CO by injection of oxygen above the slag level can also increase electrode oxidation rates because CO_2 also reacts with graphite.

To account for such varying practices Table 10.5 presents the following range of specific oxidation rates that have been calculated from measurements on various furnaces:

Table 10.5 Specific Oxidation Rates by Practice

Practice	Specific oxidation rate, kg/m² (lbs/ft²) per hr
Ladle furnaces	3–4 (0.62–0.82)
Closed furnace, <5 Nm³ (<200 scf) per ton of O_2	3–4 (0.62–0.82)
Closed furnace, 5–15 Nm³ (~200 to ~550 scf) per ton of O_2	5–6 (1.03–1.23)
Closed furnace, 25–45 Nm³ (~900 to ~1600 scf) per ton of O_2	5–8 (1.03–1.64)
Closed furnace, 25–45 Nm³ (~900 to ~1600 scf) per ton of O_2 with post-combustion	8–10 (1.64–2.05)

Water cooling of the electrodes above the roof is now standard practice and has helped to reduce wasteful oxidation of electrodes outside the furnace. However, excessive flows of water will cause an increase in energy consumption, the cost of which normally exceeds any gain in graphite savings. Ideally the water should convert to steam within the roof port, providing some protection to the refractory delta insert. The flow rate to achieve this condition depends on escaping flame intensity and current, but typically may require between 380–680 litres (100–180 gal) per hour per one 24 inch electrode. If the flow rate is in excess of the optimum then water is effectively pouring into the furnace, exiting via the extraction system as steam at high temperature.

10.3.4 Current Carrying Capacity

Limits for current carrying capacity have been developed with the objective of minimizing electrode consumption. Generally, exceeding these limits will result in increased consumption via the discontinuous consumption mechanisms, electrode breakage, butt losses, and spalling. Thus the decision to exceed these limits must weigh increased graphite consumption against increased EAF productivity. Current carrying capacities are given in Fig. 10.42.

The current carrying capacity is greater for DC than for AC electrodes. Under AC conditions, the current is forced towards the peripheral region of the electrode. This is known as skin effect. Under DC conditions, the full electrode cross section is available for the current flow. That is why the electrode resistance under DC conditions, R_{DC}, is apparently lower than the resistance, R_{AC}, under AC conditions. Consequently, the current can be increased under DC over that for AC conditions without increasing the resulting Joule's heating in the electrode. Some typical examples are given in Table 10.6.

Besides electrical influences, the current carrying capacity is mainly affected by thermal and mechanical boundary conditions and by the material characteristics of graphite. Again, exceeding these boundary conditions will lead to an increase of the discontinuous consumption processes.

10.3.5 Discontinuous Consumption Processes

Discontinuous electrode consumption has been studied extensively by Lefrank et al. with the objective to improve electrode performance specifically for DC operations.[10] The discontinuous

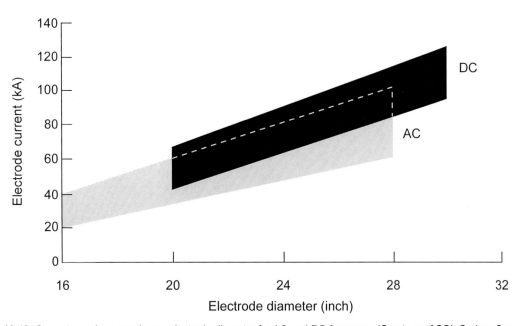

Fig. 10.42 Current carrying capacity vs. electrode diameter for AC and DC furnaces. *(Courtesy of SGL Carbon Corp.)*

Table 10.6 R_{DC} and R_{AC} of 110 in. Electrodes

	Electrode Diameter, (in.)			
	20	24	28	32
R_{DC}, microOhm	70	49	36	27
R_{AC}, microOhm, at 60 Hz	83	65	53	44
I_{DC}/I_{AC}	1.09	1.15	1.21	1.28

consumption processes are results of mechanically or thermally induced stresses which exceed the electrode strength limits. Stress levels of this magnitude occur as electrode tip stresses at the arc foot point or as hoop stresses around the electrode joints. Tip stress leads to tip spalling. Hoop stresses lead to column breakage and butt losses.

10.3.5.1 Butt Losses

Current, temperature, and stress distributions for a 28 inch electrode, operated at 120 kA under AC and DC conditions, are shown schematically in Fig. 10.43.

Under AC conditions, most of the current is flowing through the peripheral region of the electrode. Under DC conditions, more current will flow through the electrode center resulting in an almost evenly distributed, and lower maximum current density than under AC conditions. As a result the DC electrode generates more energy inside the electrode and therefore the temperature gradient is much higher in DC electrodes. Consequently, the hoop stress gradients are much greater for DC conditions. As the electrode diameter and current loads are increased this situation will be amplified further. This phenomenon is a result of the finite graphite thermal conductivity and the reduced capacity of the larger electrode for radiant heat transfer resulting from a reduced surface to volume ratio. Electrode joints are affected even more by the central flow of current in DC operations and potentially, the nipple can overheat resulting in joint opening and socket splitting.

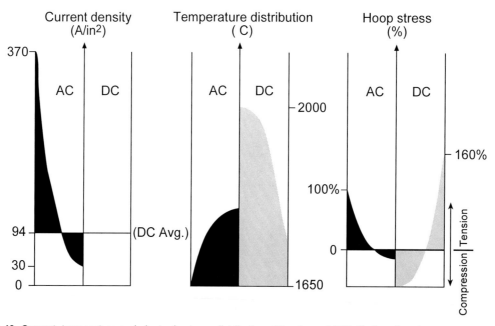

Fig. 10.43 Current, temperature and electrode stress distribution. *(Courtesy of SGL Carbon Corp.)*

Electric Furnace Steelmaking

10.3.5.2 Tip Spalling

Very high current densities develop at the arc spot on an electrode due to the self-magnetic pinch effect. For DC operations with only one electrode, the tip stresses are higher than for AC operations. The arc spot will move around the tip of the electrode in a random manner but sometimes in DC operation, the arc spot will tend to fix preferentially on one location. This will result in the generation of longitudinal splits and severe butt losses. Improvements in arc deflection control have improved this situation reducing butt losses significantly.

10.3.5.3 Breakage

State of the art AC operations experience few electrode breaks, typically less than two to three per month. However, frequent top joint opening and breakage some time after an electrode addition have been reported by some DC shops. It was hypothesized that a rotational arc movement in some DC operations might generate a clockwise tangential force to the electrode axis leading to the unwinding of the top column joint. For this reason, Lefrank et al. conducted a comprehensive high speed video motion analysis of the arc behavior at major DC furnace operations of different furnace design in the U.S.

The videos show that the DC arc spot is moving with high speed randomly over the electrode tip. The arc spot movement seems to be influenced by the combination of totally random events with controlled, directional forces.

The directional forces are the result of the powerful magnetic fields generated by the DC current loop of the furnace. They can restrict the arc spot to a limited area of the electrode tip and may direct the arc jet preferentially to one furnace side which is shown in Fig. 10.44. Depending on the furnace design, they can also force the arc to rotate either clockwise or counter clockwise for a few hundred milliseconds, but then the arc spot will again move in a totally random and unpredictable way over the electrode tip for prolonged time periods.

The results of this study clearly indicate that a stable DC arc direction or continuous arc rotation did not exist in any of the investigated DC furnaces. Therefore, the phenomenon of top joint opening of DC electrode columns must be generated mainly by the absence of the self tightening,

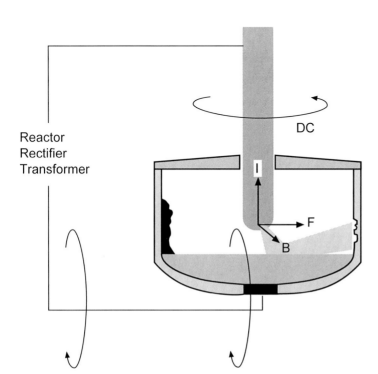

Fig. 10.44 Influence of external magnetic field on DC arc furnace without electromagnetic compensation. *(Courtesy of SGL Carbon Corp.)*

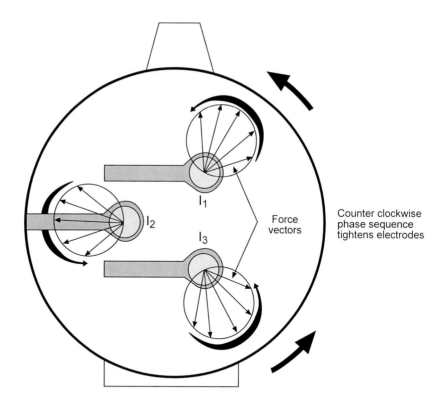

Fig. 10.45 Electromagnetic forces in an AC electric arc furnace. *(Courtesy of SGL Carbon Corp.)*

counter clockwise, electromagnetic force which helps AC operation to maintain tight joints as shown in Fig. 10.45.

Lower powered DC operations suffer specifically from the heat sink effect which is caused by every electrode addition as demonstrated in Fig. 10.46. At low power levels, the necessary thermal nipple expansion supporting the tightness of the top joint does not happen fast enough. The absence of the self-tightening AC effect will be aggravated under these conditions through furnace vibrations, excessive water spray cooling, and, if pitch plugs are used, through joint lubrication by the liquid pitch between 150–250°C.

For higher powered DC operations, the failure mechanism seems to be different. In this case, the contact resistance between the electrode end faces may become too high for the extreme current levels. This phenomenon can be the combined result of the missing self-tightening AC effect, questionable joint design, inadequate tightening torque, or poor electrode addition practice. As a consequence, the current will flow preferentially through the nipple. This can lead to overheating of the nipple to its creep temperature which will result in joint breakage.

A number of counter measures can be taken to prevent these failures. The design of electrodes and nipples should be adjusted by the electrode producer to the demanding DC joint conditions (contact resistance as a function of surface finish, expansion coefficients, and dimensions). DC electrode joints should be locked by nailing after each electrode addition. DC operations should not use the electrode water spray system for at least twenty minutes after each electrode addition. DC shops, with an abnormal number of top joint breaks should apply automated, off- furnace electrode additions or use the automatic on-furnace addition devices available in the marketplace. DC shops should apply the recommended, high tightening torque moments. Table 10.7 presents published torque levels.

10.3.6 Comparison of AC and DC Electrode Consumption

There has been much speculation over the past few years as to the true electrode savings for DC operation compared with AC operation. An analysis by Bowman indicated the following. For a

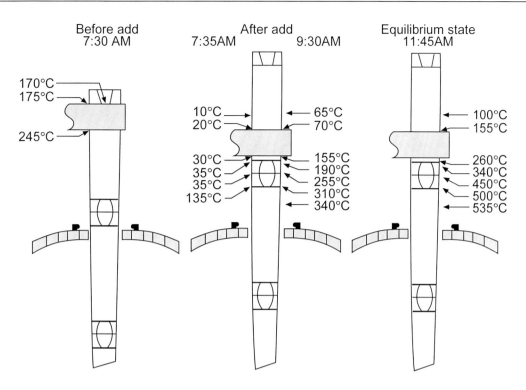

Fig. 10.46 Influence of electrode additions on temperature distribution. *(Courtesy of SGL Carbon Corp.)*

Table 10.7 Recommended Tightening Torque Moments

Electrode Diameter, in.	Recommended Tightening Torque, ft-lbs
24	3000
26	3700
28	4400
30	5500

given size of operation, the DC electrode tip consumption is larger than the combined tip consumption of all three AC electrodes. This is due to the much higher operating current for the DC electrode. DC electrode consumption savings result mainly from reduced sidewall oxidation. Overall savings can be expected to be approximately 25%.

Lefrank et al. have calculated continuous electrode consumption for the DC furnace shops operational in the US in 1995.[10] Then, they compared the calculated consumption figures with the actually observed consumption as shown in Table 10.8. They assumed that the difference between calculated and observed consumption would be equivalent to the amount of discontinuous consumption: breakage, butt losses, and tip spalling. The resulting dependence of discontinuous consumption on average phase current is shown in Fig. 10.47. It is important to note that the data points for the current levels above 100 kA are averages for 28 in. and 30 in., with the 28 in. performing worst and the 30 in. performing best.

With this data, DC operations in the United States can be divided into three major electrode consumption groups:

1. <u>DC furnaces operating below 110 kA</u>: Predominantly continuous, low electrode consumption is achieved with electrodes of 28 in. and smaller diameter. Overall

Fig. 10.47 Discontinuous consumption versus kA for DC EAF meltshops. *(Courtesy of SGL Carbon Corp.)*

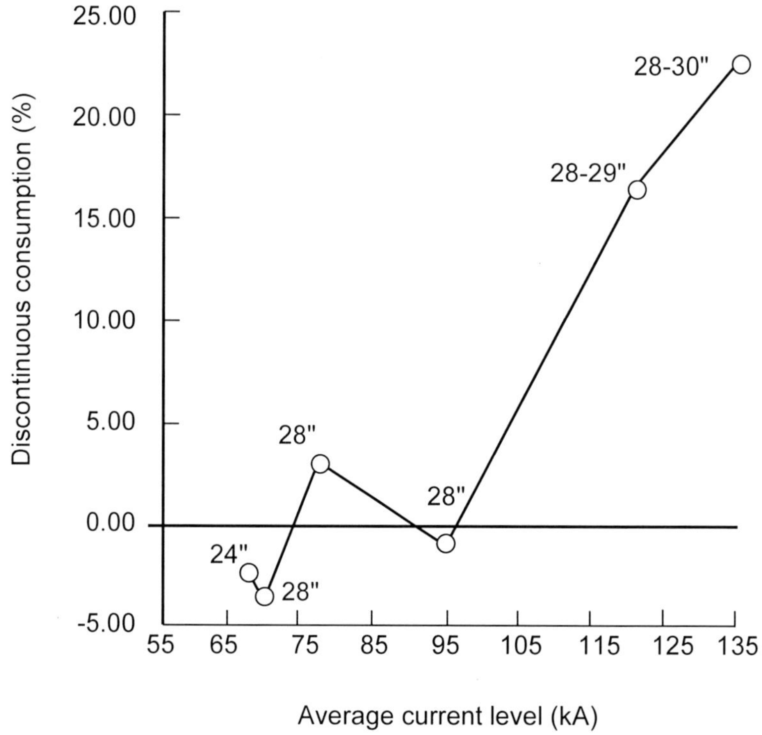

Table 10.8 Calculated vs. Observed Electrode Consumption for DC EAFs.

EAF Meltshop	Ave. Current, kA	Calculated Side Cons., lbs/ton	Calculated Tip Cons., lbs/ton	Calculated Total Cons., lbs/ton	Measured Total Cons., lbs/ton	Discont. Cons., (%)
A	68	1.62	2.60	4.21	4.10	−2.67
B	70	1.26	1.24	2.50	2.40	−3.91
C	77	1.26	1.56	2.83	2.90	2.62
D	95	0.96	1.96	2.92	2.90	−0.63
E	120	0.82	2.49	3.31	3.85	16.42
F	134	0.81	2.38	3.18	3.90	22.55

electrode savings of 25%, as estimated in Bowman's AC versus DC comparison, are realistic for these operations.

2. <u>DC furnaces operating between 110 and 130 kA</u>: Discontinuous consumption processes increase dramatically with phase currents above 110 kA. Electrodes of 30 in. diameter are necessary to control breaks, butt losses, and spalling. Electrode consumption savings using 28 in. electrodes are marginal at best when compared to high impedance AC shops of similar capacity.

3. <u>DC furnaces operating above 130 kA</u>: 130 kA is currently the maximum recommended current carrying capacity for 30 in. electrodes. DC shops operating above this level are realizing electrode consumption figures higher than high impedance AC shops of similar capacity.

10.3.7 Development of Special DC Electrode Grades

In response to demanding DC furnace conditions, graphite producers are developing special DC electrode grades. SGL Carbon Corp. has published information on a finite element analysis (FEA) model for the evaluation of the influence of graphite property changes on thermal stress generation in electrode columns under DC conditions.[11] Using the information provided by the model along with actual data from DC operations, new electrode nipple systems are being produced for DC applications.

The modifications evaluated have concentrated on:

1. Reduction of joint resistance in combination with high strength nipples with lower thermal expansion coefficients to reduce thermal joint stress and failure
2. Reduction of specific electrode resistivity to lower the generation of thermal stresses resulting from Joule's heating of the electrode center.
3. Enhancement of transverse thermal conductivity to transport the heat from the electrode center faster to the surface.
4. Reduction of coefficients of thermal electrode expansion to lower thermal stress from electrode expansion.
5. Improvement of homogeneity and coarseness of electrodes to increase resistance to crack initiation and propagation.

Shown in Fig 10.48 is an example of the resulting stress levels calculated by the FEA program from the current flow pattern inside the electrode column and the resulting temperature distribution. The material properties of Table 10.9 were used to generate this FEA plot. All columns in this example are loaded with 86 kA.

The FEA results confirm that the highest tangential stress levels are being developed around the bottom joint. Going from the 24 in. electrode with AC graphite properties to the 28 in. electrode with DC graphite properties reduces the joint stress from 2465 to 1160 psi.

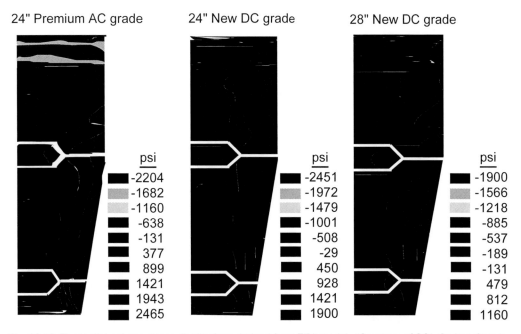

Fig. 10.48 Tangential column stress distributions derived from FEA model. *(Courtesy of SGL Carbon Corp.)*

Table 10.9 Properties of AC and DC Electrodes.

Average Property	Electrodes		Nipples	
	AC	New DC	AC	New DC
Apparent density (g/cm³)	1.71	1.73	1.76	1.78
Special electrical (Ω in. × 10^{-5}) ∥	20.5	18.5	13.4	12.6
resistance (Ω in. × 10^{-5}) ⊥	33.1	30.3	27.5	26.8
Young's (psi × 10^6) ∥	1.35	1.51	2.65	2.54
modulus (psi × 10^6) ⊥	0.7	0.75	0.96	0.94
Coefficient of (ppm/°C) ∥	0.6	0.5	0.32	0.2
thermal expansion (ppm/°C) ⊥	1.4	1.6	1.84	1.7
Thermal (WK^{-1}m^{-1}) ∥	230	250	315	320
conductivity (WK^{-1}m^{-1}) ⊥	150	175	175	200

The maximum tangential joint stress from the FEA calculations as a function of phase current is shown in Fig. 10.49. In addition, the strength range for electrode graphite at 2000°C is shown as a bar. Below this level the DC electrodes will mainly be consumed in a continuous way. Above this limit, discontinuous consumption will gradually commence. The FEA data coincide well with electrode performance observations in the field.

It is interesting to note that the FEA model predicts start of discontinuous consumption around 110 kA even for a 30 in. electrode with improved properties. Some DC operations are designed to use operating currents up to 160 kA. Thus, further substantial improvements in electrode material properties will be necessary to control breakage, butt losses, and splitting at such operating currents.

Increasing DC electrode diameters beyond 30 in. would improve the current carrying capacity, but with decreasing efficiency because the thermomechanical stresses in these electrodes will increase exponentially with increasing diameter. Alternatively, the current load for high powered DC operations can be distributed over two electrodes with the objective to reduce electrode stress levels. Two DC arc furnaces with double electrode systems have recently started operation. Reliable performance comparisons are not yet known.

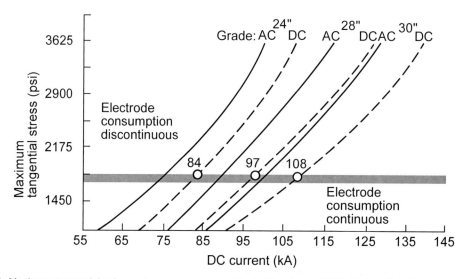

Fig. 10.49 Maximum tangential column stress versus current load. *(Courtesy of SGL Carbon Corp.)*

10.4 Gas Collection and Cleaning

Over the past 30 years electric arc furnace fume systems have evolved considerably from very simple systems aimed at improving the ambient work environment around the furnace to sophisticated systems aimed at controlling not only the emission of particulate but also the control of toxic gases. Modern fume systems are now designed to minimize the formation of toxic gases and to ensure that others are destroyed before exiting the system. Thus gas cleaning in the modern day fume system entails considerably more than trapping and collection of particulate.

Initially, electric arc furnaces operated without any emissions control. In the earlier days, operators merely sought to improve workplace conditions in order to increase productivity. The advent of environmental regulations saw the progress of primary fume control on electric arc furnaces from side draft hoods (SDH), through furnace roof hoods, to rudimentary fourth hole extraction systems, culminating in today's sophisticated direct evacuation systems (DES).

10.4.1 Early Fume Control Methods

As mentioned above, early electric arc furnaces operated without any emissions control and operators merely sought to improve workplace conditions. Most of the EAFs were relatively small in size and in some cases the roof was stationary. As a result many of the early fume control systems involved a roof mounted type of furnace hood. These allowed for effective fume control without requiring elaborate pressure control in the furnace. In addition fume collection did not adversely affect the steel metallurgy. For larger furnaces (>5 metre diameter), furnace hoods were cumbersome and not generally recommended. Typical examples of furnace hoods included the side draft hood, the full roof type hood and the close fitting (donut) hood. Another method that was sometimes used was a mobile canopy hood that is local to the EAF. These are discussed in more detail in the following sections.

10.4.1.1 Side Draft Hood (SDH)

A side draft hood is depicted in Fig. 10.50(a). This type of furnace hood picked up the furnace fume as it exited the electrode ports. Additional branches were sometimes used to collect emissions from the slag door and from the tapping spout. Gas cooling was accomplished by bleed-in air. Typical advantages of this type of hood include: minimal air volume required to give effective control, no interference with electrode stroke, no effect on steel metallurgy, hood does not need to be removed when changing the roof, and complete accessibility to furnace roof components.[12] Disadvantages include: cumbersome for large furnaces, fume escapes at roof ring if there is not a good seal, and difficult to get good capture at B phase due to electrode configuration.

10.4.1.2 Full Roof Type Hood

The full roof type hood configuration is shown in Fig.10.50(b). The full roof type hood encloses the roof entirely allowing for capture of any fume exiting the upper portion of the EAF. This hood design is often less expensive than the side draft design but has several disadvantages: electrode stroke is decreased by 15–24 in. depending on the furnace size, parts of the hood must be removed when the furnace roof is changed, hood covers entire top of EAF making maintenance difficult, openings around the periphery of the hood make fume control difficult, hood has difficulty handling very large offgas volumes, and requires refractory lined or water-cooled sections above the electrode openings in the EAF.[13]

The Pangborn modified full roof hood is shown in Fig. 10.50(c). This has the following advantages over the standard full hood configuration: better utilization of ventilation air (lower volumes than full hood), some fume pickup during furnace charging, less weight than a full hood, and less time required for roof change.[14]

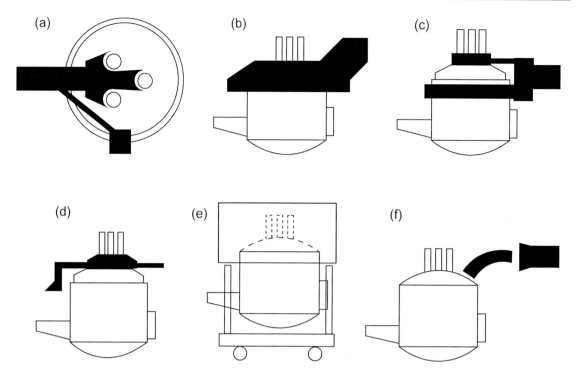

Fig. 10.50 Hood configurations: (a) side draft, (b) full roof, (c) Pangborn modified full roof, (d) close fitting, (e) mobile canopy, (f) snorkel system.

10.4.1.3 Close Fitting Hood

The close fitting hood is shown in Fig. 10.50(d). This type of hood is only really applicable for small furnaces. This hood can only be used if the roof is not raised during oxygen lancing. The hood is generally constructed of stainless steel, heavy reinforced carbon plate or may be water-cooled. The main section has a built-in frame that is mounted to the furnace roof ring. The section over the slag door is permanently attached to the bezel ring. This hood arrangement is relatively tight and gives excellent fume control. Disadvantages include reduction of the electrode stroke and difficulty in accessing the roof. This type of design has been incorporated into ladle furnace roofs for fume control.

10.4.1.4 Mobile Canopy Hood

The mobile canopy hood is as its name suggests, a mobile hood that is moved into position above the furnace to capture fume as it rises from the furnace. This type of hood is only suitable for very small furnace operations (less than 10 tons). A mobile canopy hood is shown in Fig. 10.50(e).

10.4.1.5 Snorkel System

The snorkel system originated in Germany and is used on larger furnaces melting open hearth grade steels.[12] The snorkel system configuration is shown in Fig. 10.50(f). The system employs a fourth hole in the furnace roof followed by a water-cooled elbow. The snorkel is a funnel shaped section of duct that can be retracted with a motor or hydraulic system. The snorkel captures fume as it exits the furnace elbow. The gap between the elbow and the snorkel, 5–120 cm (2–48 in.), allows air to enter the system. This air provides oxygen for combustion and also provides dilution cooling for the system. The fourth hole serves only as a natural pressure release for the furnace. During periods of heavy fume emission, the snorkel is moved closer to the elbow to provide greater suction. The furnace atmosphere remains undisturbed and theoretically operates at a positive pressure at all times.

10.4.2 Modern EAF Fume Control

Essentially all phases of normal EAF operation result in either primary or secondary emissions. Primary emissions are those that are produced during EAF melting and refining operations. Secondary emissions are those that result from charging, tapping and also from escape of fume from the EAF (usually from the electrode ports and/or roof ring). Primary emissions are generally controlled using a direct evacuation system. Secondary emissions are captured using canopy hoods and in some cases auxiliary tapping hoods.

10.4.2.1 Direct Evacuation Systems

The Direct Evacuation System (DES) or fourth hole system as it is commonly known, is probably the most prevalent fume control system in meltshops today. An electric arc furnace (EAF) offgas system is comprised of containment, cooling and collecting elements.[15] Total offgas handling usually involves a direct evacuation system (DES) and a secondary fugitive capture system consisting of a canopy hood in the shop roof.

In a typical DES system, containment components include the fourth hole, system ducting and the fan. Cooling is achieved by elements such as water-cooled ducting, dilution air cooling, evaporative cooling and forced or natural draft heat exchangers.[15] The latter three provide gas cooling without increasing the collector size.

Particulate collection is typically achieved with a baghouse, though scrubbers and electrostatic precipitators are used in some cases.

A good offgas system is one that combines the containment, cooling and collecting elements economically to provide effective fume control within regulatory standards, for both the meltshop and external environments.

A modern DES addresses not only particulate capture but also control of CO, NO_x and VOC emissions.

The main advantages of a DES system are: it requires the lowest offtake volume for furnace fume control, it provides the most effective method for fume capture in UHP steelmaking, it allows low interference with meltshop operations, it requires minimal space on the furnace roof, and it addresses the issues of CO and VOC combustion. The main disadvantages of DES systems are: excessive in-draft leads to increased power requirements and undesired metallurgical effects, and it is difficult to control the furnace freeboard pressure when offgas surges occur. A typical system flowsheet is given in Fig. 10.51.

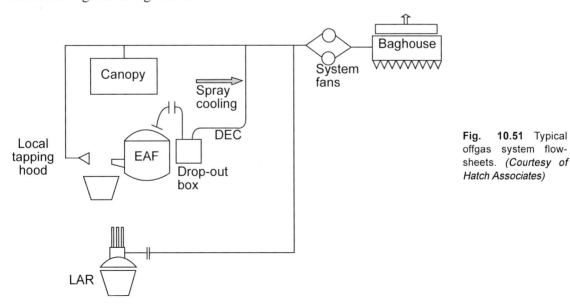

Fig. 10.51 Typical offgas system flowsheets. *(Courtesy of Hatch Associates)*

10.4.2.2 DES Design Methodology

Design of an offgas system requires consideration of several steps including containment of furnace fume, the reaction processes going on in the furnace, and processes through the offgas system itself. In the past there has not been enough collaboration of the furnace operator with the offgas system designer. As a result it has been difficult to provide systems which matched operations within the meltshop. Without a clear understanding of the furnace operations, it is not possible to design an appropriate offgas system.

Essentially, the design of the offgas system involves a detailed mass and energy balance around the EAF to provide the basis for gas quantity and heat content leaving the EAF. There are a variety of furnace process factors that influence the peak offgas flow rate and the peak heat load contained in the offgas. These factors include the charge weight, the scrap mix (its oil, combustible material and moisture contents), furnace power and power-on time, tap-to-tap time, furnace openings and associated air infiltration (slag door, electrode ports, roof ring, auxiliary ports), oxy-fuel burner and oxygen lance rates and duration of use, carbon injection rates, and use of alternative iron sources and feed rate.

Both the gas quantity and its heat content affect the sizing of the DES system. It should be noted that the system fan is a constant volume device. The fan is capable of moving a greater mass of offgas if the offgas is cooler (the cooler the offgas, the lower its volume). At the same time, the fan motor has a limited power rating. As a result, to provide maximum evacuation for a given DES system configuration, it is desired to cool the offgas as much as possible but at the same time ensure that we are staying within the amp current rating of the fan. Thus, it is common to modulate the fan damper position based on the fan motor current and an acceptable setpoint.

In order to keep the gas cleaning system cost economical, it is usual to cool the offgas prior to the baghouse. This allows for use of lower cost bag materials and also keeps the volume of gas to be cleaned to a minimum.

It is important to note that not all of the heat content in the offgas is necessarily sensible heat. If the offgas contains non-combusted species such as CO, a significant amount of energy can be contained in the offgas. This not only represents increased heat removal duty for the offgas system, but also energy lost to the melting process and hence an inefficiency.

The furnace heat cycle includes phases such as flash-off (the first few minutes of melting when the moisture and volatile components of the scrap mix are driven off by the applied heat), melting, which can have several operational phases depending upon the use of oxy-fuel burners, oxygen lancing or DRI additions, and refining, which can be broken down into periods of oxygen lancing, carbon injection and foamy slag practices. It is important to analyze all of these phases to determine the peak offgas flow rate and peak heat load.

Frequently the peak load in terms of gas volume occurs during the first few minutes of melting, when the volatile materials burn and flash-off. However, refining is often seen to result in the highest offgas temperatures during oxygen lancing for carbon removal. Operations where DRI is added continuously can lead to high evolution rates of CO gas. CO evolution rates can be even greater during iron carbide addition. If this CO does not burn in the furnace, high heat loads to the offgas system will result.

Typical peak load offgas characteristics at the fourth hole for effective EAF emission control are an offgas flow rate of 4–6 Nm^3 (150–200 scf) per ton per hour tapped and an offgas temperature of 1370–1925°C (2500–3500°F).

Fourth hole sizes are selected on the basis of a design offgas velocity in the range of 2400–3000 m/min (8000–10,000 ft/min). The gas velocity at the fourth hole is important because it will affect the system pressure drop at the fourth hole and also will affect the amount of material which will carry over from the furnace to the offgas system. In some cases, slag and scrap will be carried over into the offgas system.

A good understanding of the furnace processes and how they affect the offgas characteristics is necessary for the process engineer to design an efficient and cost effective DES. Once the flow rate and heat content for emission control at the fourth hole has been determined for the EAF, the required performance of the DES has been defined. If some of the components of an existing system are to be retained (e.g. system fan, baghouse) then certain limitations on the total allowable system pressure drop will also be fixed. Definition of the heat content of the offgases also fixes the amount of cooling required prior to routing the offgas to a given baghouse filter operation. The fourth hole serves as the union between the furnace and the DES.

In order to control fume emissions, it is necessary to maintain a slight negative pressure within the furnace. This pressure is typically in the range of –0.09 to –0.18 mm Hg (–0.05 to –0.1 in. W.G.). It is important to limit the resultant in-draft to the minimum required to contain fume generated by the melting process. Higher in-draft results in heat losses from the furnace and higher offgas flow rates. Test results presented by Bender indicate that a negative pressure of –0.09 mm Hg (–0.05 in. W.G.) will usually prevent electrode emissions when scrap cave-ins occur.[16] Bender shows, based on measurements at Von Moos Stahl (Swiss Steel AG), that operating with too high a negative pressure can increase the electrical power requirement by approximately 45 kWh/ton. Previous trials on several furnaces in Japan indicated that running the furnace at positive pressure gave power savings of 10–20 kWh per ton.

Furnace openings of significance are usually the electrode ports, the roof ring gap, the slag door and auxiliary ports in the furnace sidewall (decarburization oxygen lance, solids injection etc.).

The minimizing of these openings improves the efficiency of both the furnace and the emission control system. Openings that occur higher up in the furnace are less critical than those which occur low in the furnace due to the pressure profile across the height of the furnace. The pressure tap which is used to measure the pressure used for furnace draft control is located in the furnace roof. If the system controls to the pressure setpoint at the roof level, the negative pressure at the low furnace openings such as the slag door will be much greater. The flowrate of air into the furnace through openings is proportional to the square root of the differential between the furnace pressure and the ambient pressure. As a result the infiltration rate will be much higher at the openings lower in the furnace.

Early in the meltdown phase, furnace openings will not result in much air infiltration because the scrap will act as a resistance to infiltration flow. However, later in the cycle when oxygen lances are inserted through the slag door, considerable air infiltration can result.

The fourth hole serves as the union between furnace operations and the DES. Gases exiting the fourth hole enter the water-cooled elbow mounted on the furnace roof. Generally, there is a gap between the furnace roof and the water-cooled elbow which admits air due to the negative pressure. The air mixing with the offgases provides opportunity for further combustion of offgas and/or dilution cooling. In general the mass of air drawn into this first gap matches the mass of gas flowing out of the fourth hole.

A second gap that allows the furnace to tilt and the furnace roof to swing is normally situated between the water-cooled elbow exit and the entrance to the water-cooled duct. This gap admits additional infiltration air which may provide some further combustion as well as dilution cooling. In some cases a duct section similar to a snorkel is located following the gap. This section is adjustable typically to give a gap of 2.5–12 cm (1–5 in.). This allows better control of the negative pressure in the furnace freeboard and is often necessary in providing sufficient retraction to enable the furnace roof to swing without interference. Again, in general, the mass of air entering the second gap is similar to the total mass exiting the elbow.

Thus the DES typically exhausts a mass flow that is about four times that which actually exits the furnace at the fourth hole. Generally the designer prefers that these final exhaust gases contain 100% of oxygen in excess of the combustion needs for all combustibles generated in the furnace process.

There have been many innovations to the water-cooled elbow design in recent years. While it is recognized that sufficient combustion air needs to be introduced to ensure combustion of all VOCs and CO in the water-cooled duct, it is also necessary to ensure that excessive air is not introduced which could lead to over cooling of the offgas and non-combusted material downstream in the system. Flanges (also known as elephant ears), are sometimes welded around the water-cooled duct and the elbow so that the amount of air infiltration can be better controlled. The water-cooled duct is sometimes sized larger than the elbow so that efficient capture can still be achieved even when the furnace is tilted ±5°. One of the more recent innovations is to use a very steep elbow configuration which will help to reduce slag buildup. Bender recommends that this type of elbow be coupled with a high velocity water-cooled duct entrance followed by a sharp bend downward.[16] This configuration provides for good turbulence for mixing and combustion of the CO and slag can fall out into a dropout box located below. Good mixing of the gas is very important to ensure complete combustion.

Water-cooled ducting is used to cool the offgas and also to contain the offgas combustion reaction heat which would otherwise damage carbon steel ductwork. In some older systems, refractory lined ductwork was used but with improved design and lower costs; water-cooled duct is now the preferred choice. Water-cooled duct is also generally less maintenance intensive. It is important however, to ensure that water-cooled duct is accessible so that routine maintenance can be performed. In many of the newer meltshop installations, the water-cooled duct has a sharp downward bend following the combustion gap. A dropout box is located at the bottom of this bend and then a horizontal run of water-cooled duct is located at grade. This makes it easier to use forklift trucks to remove sections of the duct for maintenance. This also makes the duct accessible for periodic cleanout of any dust buildup which might occur. Some planning is required to ensure that the layout of the ducting does not interfere with other operations.

The gas temperature at the end of the water-cooled ducting is typically 650–760°C (1200–1400°F). Below this range, water-cooled ducting is no longer an efficient method of cooling due to low heat transfer coefficients. Typically offgas flowrates at the water-cooled duct exit for effective EAF emission control are 14–23 Nm^3 (500–800 scf) per ton per hour tapped.

10.4.2.3 Gas Cooling

Prior to entering the carbon steel dry ducting, dilution, evaporative or non-contact cooling is generally used to cool the gas so that the duct wall temperature is well below 425°C (800°F). This generally equates to a gas temperature of about 650–760°C (1200–1400°F).

Evaporative cooling consists of spraying atomized water into the offgas stream. The water will evaporate, absorbing a large quantity of energy. Evaporative cooling is usually carried out following the water-cooled duct. This can be carried out in a spray chamber or in the ductwork. If carried out in ductwork, care must be taken that the water stream evaporates fully before it strikes the duct wall. The evaporation rate is dependent on the water flowrate, the spray angle, quantity of atomizing air, the water droplet size and the presence of contaminant films on the droplet. Evaporative cooling will result in a decrease in offgas volume and is usually the most cost effective method for achieving this because it involves the minimum amount of equipment. The only potential downside to this technology is the possibility of condensation occurring if the offgas becomes saturated. It is important to evaluate the total moisture content of the combined DES and secondary offgases entering the baghouse. If a dewpoint problem exists, condensation will occur which can affect bag performance and life. If this is the case, an alternative to evaporative cooling may have to be selected.

Dilution cooling consists of adding air to the offgas stream to cool it. This will always result in a larger gas volume and increases the required capacity of the fume extraction system. This is the lowest cost method for cooling of the offgas but may adversely affect the cost of system components downstream.

Non-contact cooling is carried out using forced draft coolers or hair pin coolers. The gas is cooled by transferring heat to the wall of the cooler which in turn transfers heat to the outer surface of the tube or duct wall. The net result is to cool the offgas without an increase in offgas volume. Though this is an efficient method for gas cooling which does not result in increased gas volume, it is the most expensive method and requires the greatest amount of equipment.

Further cooling is provided by dilution air drawn in through the canopy hood and mixed with the DES offgases to reduce the combined gas temperature to around 120°C (250°F) before entering the baghouse. The reason for this is to allow polyester filter media to be used. Polyester is an economic filter media commonly used in EAF air pollution control (APC) systems with a design temperature limit of 120°C (250°F) for continuous operation.

Gas velocities in the DES system ductwork are important for many reasons. If the gas velocity is too high, large system pressure drops will result and there is the potential for increased duct wear. If the gas velocity is too low, the particulate in the offgas will settle out in the ductwork. This will also lead to higher system pressure drop and poses a maintenance problem. It has been found in practice that a design velocity of 4000–5000 fpm at standard conditions is optimum for meeting both dust entrainment and economic criteria.

For any configuration, a detailed mass and energy balance must be performed on each DES system component to yield gas flows, temperatures, velocities and pressure losses throughout the entire system. Analysis of various system configurations can be a tedious and time consuming process. As a result, several system designers have incorporated all of the necessary calculations into a computer model. Thus, the effect of changes in process and system design can be evaluated quickly allowing for greater optimization of system design. A typical DES is shown in Fig. 10.52.

10.4.3 Secondary Emissions Control

Control of secondary fume emissions from furnace charging, tapping and slagging operations became strictly regulated subsequent to regulations for primary control. Secondary emissions also result from offgas escaping the furnace around the electrodes and at the roof ring. Initial treatment of secondary emissions were provided by generally inefficient canopy hoods at roof truss level.

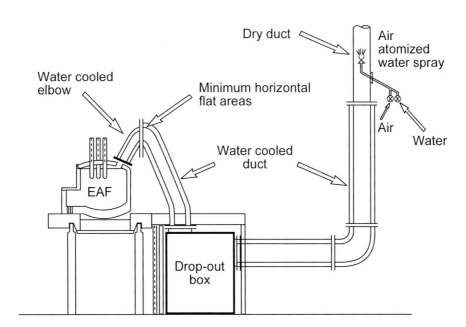

Fig. 10.52 Typical DES configuration. *(Courtesy of Hatch Associates.)*

Modern systems tend to provide much greater evacuation capacity. The sizing of the secondary fume control system is the controlling factor of the overall sizing of the gas cleaning system because secondary extraction tends to be four to six times that required for the DES.

10.4.3.1 Canopy Hoods

Canopy hoods were installed initially to capture primary furnace emissions. As furnace operations became larger, a need arose for control of secondary fume emissions during charging and tapping operations. The canopy hood was chosen to fulfill this need because it could capture these emissions without interfering with operations. Canopy hoods are located in the building trusses above the crane runway. In North America, canopy hoods now form an integral part of modern meltshop fume control systems. The canopy system is throttled back during furnace operation and provides air for dilution cooling of the DES offgases prior to entering the baghouse. During charging and tapping operations the DES does not draw any fume and full extraction capacity is diverted to the canopy hood.

The ability of a canopy hood to capture fume depends on the face area of the canopy, the canopy shape, the canopy volume, the amount of interference presented by the charging/tapping crane, the air flowrate at the canopy face, heat sources within the meltshop (which may cause errant drafts), the height of the canopy hood above the furnace, and the absence of cross drafts within the meltshop.

Many canopy systems are compartmentalized so that one portion of the canopy is used for charging emissions control while another section is used for tapping emissions. With the move towards the use of bottom tapping furnaces, it is impractical to attempt fume control during tapping with a canopy and it is now common to use a side hood for capture of tapping fumes. Hood capture can be greatly affected by cross drafts in the shop and by thermal gradients within the shop. Thus it is important to try to keep doorways closed and to eliminate heats sources as much as possible. One simple way of reducing thermal gradients is to duct ladle preheater gases outside the building. Removal of full slag pots from the building can also provide beneficial effects.

10.4.3.2 Deep Storage Canopy Hoods

Many canopy hood designs are incapable of handling the large plumes generated, particularly during charging. The solution is to provide for temporary storage of this fume until it can be extracted by the fume system. The result is a deep storage type canopy hood which has a storage volume of five to ten times that of a conventional canopy hood.[17] It provides a large storage capacity that is capable of temporarily storing the large surge in fume emissions experienced during furnace charging and allows the fume to be extracted by the gas cleaning system over an extended period of time. This allows the total extraction capacity of the fume system to be lower and yet still meet surge requirements that occur during furnace charging. The design has been verified using physical fluid dynamic modeling and shows greatly improved capture characteristics as compared to conventional technology. There are several successful installations in North America. The Irish Steel installation was the first of its kind in Europe.[17] The concept is illustrated in Fig. 10.53.

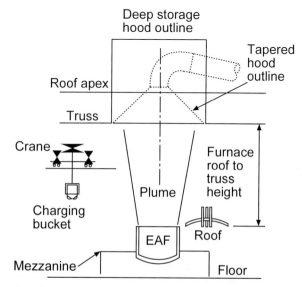

Fig. 10.53 Deep storage canopy hood concept. *(Courtesy of Hatch Associates.)*

10.4.3.3 Furnace Enclosures

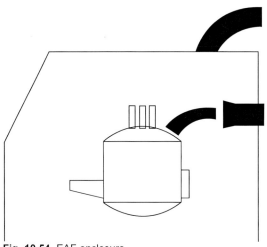

Fig. 10.54 EAF enclosure.

Furnace enclosures (also known as doghouses, snuff boxes or clean houses) are located on the operating floor of the melt shop and totally enclose the furnace. The enclosure generally provides sufficient room for the roof swing and working space around the furnace. A typical configuration is shown in Fig. 10.54. The charge bucket is brought in above the furnace through access doors in one side of the enclosure. The enclosure is low enough that the crane can pass over it. In some cases there is an air curtain sealed slot in the roof of the enclosure that allows the crane cables access when charging to the furnace. Fume is extracted at the roof level adjacent to one of the enclosure walls. Makeup air enters through openings in the operating floor. Tapping ladles are usually brought in on a transfer car. Most modern enclosure systems also incorporate a fourth hole system.

Enclosures provide the benefit that all fume is captured and the extraction volume is considerably less than that required for an equivalent DES/Canopy Hood system. In addition enclosures have no detrimental effects on metallurgy in the EAF and noise levels are reduced substantially (10–25 dB A).[18] It has been reported that energy savings of up to 20 kWh/ton may be achieved due to operation of the furnace at positive pressure. Savings due to lower evacuation volumes have been reported in the range of 10–15 kWh/ton.[4]

Enclosures offer several advantages including cooler, cleaner, quieter and safer meltshop working environments, reduced crane maintenance, lower total emissions, and lower fume system capacity requirements. Some disadvantages include restricted furnace accessibility for furnace maintenance, retrofit applications may be impractical due to the constraints of the existing meltshop layout, and furnace charging operations are sometimes slower than for a conventional meltshop.

In some cases some of the benefits of enclosures can be achieved without fully enclosing the EAF. These are called partial enclosures and consist of free standing walls which do not restrict crane access to the furnace. These baffles help to dampen out noise and minimize the effect of cross drafts within the meltshop.

10.4.3.4 Enclosed Meltshop

An alternative to the furnace enclosure is to close the meltshop and provide total meltshop evacuation. Obviously, the required evacuation capacity for a conventional meltshop layout using this method is much greater than would be required for a canopy/DES configuration. In addition, inadequate extraction could lead to high ambient temperatures in the meltshop and unacceptable working conditions if dust were to drop out within the shop. The solution to these concerns was to provide a compact meltshop layout which provides only enough space to accommodate the EAF and the charging crane. This is shown conceptually in Fig. 10.55. This configuration isolates the furnace melting operations from all other operations within the steelmaking facility. Scrap buckets are transported into the enclosed area using transfer cars. Likewise, the tapping ladle is moved under the furnace on a transfer car. Slag is collected in slag pots which are removed using pot haulers through a dedicated doorway. This emission control configuration has been implemented in several meltshops in Europe and in North America and has been shown to be a cost effective method of providing improved fume control for EAF operation.

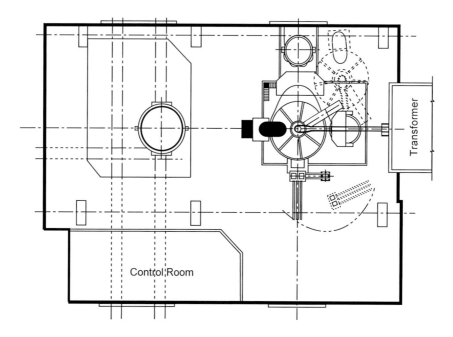

Fig. 10.55 Enclosed EAF meltshop. *(Courtesy of Bender Corp.)*

10.4.4 Gas Cleaning

Following evaluation of primary and secondary gas extraction requirements, the gas cleaning system including system fans, the baghouse and the material handling system can be specified. Some important aspects of equipment selection and design related to specific parts of the gas cleaning system are covered in the following sections.

10.4.4.1 System Control

Many operations within the meltshop have fairly constant exhaust requirements which are independent of the furnace operation. Such sources (ladle furnace, secondary dedusting) can be effectively controlled using booster fans to maintain constant exhaust flows when required, independent of the influence of furnace operations, which may switch large exhaust volumes from one source to another throughout the normal course of furnace operation.

Fume collection systems typically consume 60–100 kWh per ton of good billets produced. As a result, energy conservation has become an important parameter for melt shop profitability. More effort is now placed on identifying potential energy savings in emission control systems. Canopy hood exhaust during charging is by far the largest single exhaust requirement of typical emission control systems. Outside of the charging period, most systems have capacity in excess of that required for DES and other secondary requirements. This is usually applied to the canopies as a simple method of utilizing fan capacity. Some systems achieve significant energy savings by using variable speed fans. In such a configuration fans are turned down during non-charging periods.

Care must be taken, however, because exhausting excess non-charging capacity through the canopies is a way of providing general building ventilation, removing excess heat and fugitive emissions not captured at source. Elimination of this excess capacity may result in higher ambient temperatures and a dirtier shop. Adjustable speed fans are better suited to operations using furnace enclosures, which inherently contain and shield the meltshop from heat and fugitive emissions related to the furnace operations.

10.4.4.2 Fans

Proper specification of fan design parameters is essential to achieve desired fan performance.[19] A typical centrifugal fan performance curve is illustrated in Fig. 10.56. The shape of the system curve, initiating at the origin of the graph, demonstrates the squared relationship between fan static pressure (FSP) and volume flow (i.e. FSP varies directly with the square of Q). The system curve reflects how the system losses change with variations in flow (note this applies to a single damper arrangement in the system because changing damper positions can change the system curve). The fan curve, which intersects the system curve, shows the design FSP produced under different volume flows. The point of intersection of the two curves represents the operating point for the system. If the operating point occurs near the top of the fan curve (point of maximum FSP), unstable surging can result. To avoid this, it is recommended to select an FSP that does not exceed 80% of the maximum FSP. Fans selected near the peak on the pressure curve will be smaller and therefore cheaper but will be at lower efficiency and have no margin.

Due to the system curve relationship between FSP and volume flow, the recommended margins for specifying performance are an additional 10% on design flow and 21% on design FSP. The margins are recommended because fans frequently perform at slightly less than the published fan curve. Some margin should also be allowed for baghouse pressure drop creep above the design performance level.

It is important to remember that the definition of fan static pressure generally used on fan performance curves is not simply the difference in static pressure measured across the fan. One way to define FSP is:

$$FSP = FTP - VP_{OUT} \qquad (10.4.1)$$

In equation 10.4.1, FTP is fan total pressure or system total pressure loss and VP_{OUT} is velocity pressure at the fan outlet.

Poor selection and maintenance of baghouses, which can lead to high baghouse pressure drop, can restrict the flow achievable by the fan. A fan selected near the peak pressure output has no chance of meeting the needs of such a system.

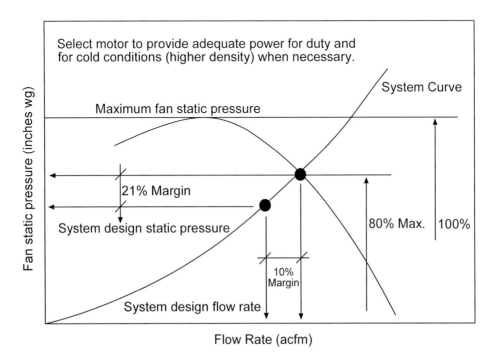

Fig. 10.56 Fan curve and selection margins. *(Courtesy of Hatch Associates.)*

10.4.4.3 Baghouses

A variety of baghouse types and arrangements are available depending on the specific requirements of the particular emission control system. Net refers to the operating condition where one compartment is off-line (such as during online cleaning or maintenance of a compartment). Key baghouse design parameters by baghouse type are given here.

Reverse air baghouses are economical for gas volumes greater than 5600 Nm^3/min. (200,000 scfm).[19] Operating at positive pressure, this design offers cost savings over negative pressure designs. The maximum recommended air to cloth ratio is 2.0–2.5, net. Reverse air baghouses are generally very large in size.

Shaker baghouses are generally similar in application as the reverse air design, but tend to require more maintenance than reverse air designs due to more mechanical movement. It is difficult to achieve the same level of gas cleaning as achieved by reverse air designs. The maximum recommended air to cloth ratio is 1.5–2.0, net.[19]

Pulse jet baghouses are economical for gas volumes less than 5600 Nm^3/min. (200,000 scfm), are of modular construction and operate under negative pressure. The maximum recommended air to cloth ratio is 4.0, net.[19]

These design parameters are general and are meant to be guidelines for baghouse selection. Individual requirements must be determined on a site by site basis.

Reverse air baghouses can be both positive or negative pressure designs. Positive pressure baghouses are cheaper but use dirty-side fans. Dirty-side fan applications have been successfully applied in many operations, but attention is required in impeller selection to minimize wear.

Negative pressure reverse air baghouses are more expensive than positive pressure baghouses because they need to have air-tight construction to avoid leakage. This is generally not an issue with positive pressure baghouses where the baghouse shell encloses the clean-side of the bags and exhausts to atmosphere (only hoppers need to be gas-tight because they see the dirty gas). Also, leakages in positive pressure systems are noticeable, while negative pressure system leakages are not evident and frequently remain undetected. Resulting differences between inlet and outlet flows as high as 15% have been measured.

Large pulse jet baghouses, due to the sheer number of valves and associated equipment, require more maintenance. Their modular construction tends to make them economical for smaller volume requirements (less than 5600 Nm^3/min. or 200,000 scfm).

Insufficient baghouse capacity translates into insufficient cloth area in the baghouse. The type of baghouse and cloth material used determines the maximum flow that can realistically be drawn through the bag filter.

A common design consideration in all baghouses is that the hoppers should never be used for dust storage. Hoppers should be continuously emptied. Allowing dust to build up in the hopper provides an opportunity for dust re-entrainment, increasing the load on the bags, the propensity for fires and potential for spark damage.

Bag damage can occur for a number of reasons including wear, direct impaction of particulate, spark burn, hot flame burn, and high temperature operation. Wear can be a result of poor gas distribution causing bags to swing (for pulse jet baghouses with cage-mounted bags below the tube sheet) or improper tensioning of bags allowing them to rub against each other or compartment walls (for reverse air or shaker baghouses with spring-mounted bags above the tube sheet).

Direct impaction is related to dirty gas distribution, where particulate can follow a direct path from the gas inlet to the bags. The way in which the dirty gas stream is introduced to the bags is a critical aspect of hopper design. Direct impaction of dirty gas on the bags can result in bag wear and premature bag failure. One method of avoiding this problem is to design a high profile hopper with a series of vertical baffle plates as illustrated in Fig. 10.57. This forces an extreme change in direc-

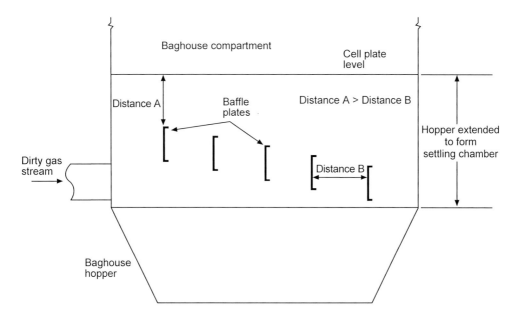

Fig. 10.57 Recommended baghouse hopper baffle plate arrangement. *(Courtesy of Hatch Associates.)*

tion for the particulate causing it to lose its momentum before reaching the bags. This method of dirty gas distribution can be achieved with minimal pressure drop.

Spark burns are normally associated with particulate still combusting as it reaches the baghouse. Hopper baffle plates can help reduce this problem as it is normally associated with larger particulate. An adequately sized dropout box can also remove these large particulate from the gas stream prior to the baghouse inlet.

Hot flame burn can occur when the duct length between the baghouse and canopy is relatively short. Without major modifications to the system configuration, protection from fireballs reaching the baghouse can be achieved using thermocouples tied in to control of emergency dilution dampers.

High temperature operation for extended periods can gradually weaken bag fabric leading to premature failures.

10.4.4.4 Material Handling

Material handling in emission control systems is achieved with a variety of equipment types.

Pneumatic conveying systems can be either dense or dilute phase design. For EAF dust transport, it is recommended to use dilute phase. The pneumatic lines should be a minimum 15 cm (6 in.) diameter.[19] Well designed systems can provide dependable service with minimal maintenance requirements.

Air slide/air lift systems operate by fluidizing the dust to be conveyed. In the case of slides, the equipment includes an enclosed chute mounted diagonally (minimum 7° slope) to allow the dust to flow down the length of the slide as it is fluidized.[19] The chute configuration increases the overall height of the baghouse. Air lifts incorporate low velocity vertical conveying of the collected particulate in a pipe over a fluidizing ejector. Such systems have low wear and low maintenance.

Mechanical conveyors can include any such types as screw conveyors, drag chain conveyors and bucket elevators. It is important that these systems be adequately sized (both in terms of quantity throughput and motor power) for surge conditions. Conveyors discharging from baghouse hoppers should be capable of starting under full hopper conditions.

Mechanical conveying systems will often require dedusting exhaust capacity. In order to avoid fugitive emissions (or in the case of negative pressure systems, in-leakage which robs capacity and re-entrains dust) it is important for the equipment to be well sealed. Routine maintenance is essential for maintaining such systems in good working condition.

For convenient operation, storage silos should be sized for a minimum three day storage capacity. This is so that material discharge can be carried out during day time shifts. It also allows the EAF operation to continue over a long weekend without bringing in staff for dust disposal.

Typically dust control is required for clean operation of silos. This can be accomplished by either an independent bin vent filter mounted on the silo, or a direct connection from the silo to the inlet side of the baghouse (in the case of positive pressure baghouses this connection would be on the inlet side of the fans). In either case, the silo is maintained under a negative pressure so that any openings in the silo will see an infiltration of air rather than a puffing out of dust.

Silo dischargers are also candidates for dust control. One effective design is a telescopic discharge which incorporates a concentric flexible discharge with exhaust applied to the outer section. The source of exhaust can be either of the methods described for the silo itself.

10.4.5 Mechanisms of EAF Dust Formation

Currently about 600,000 tons of EAF dust are generated each year in the U.S. The amount of dust generated is usually in the range of 9–18 kg (20–40 lbs) per ton of scrap melted. The reason for the wide range is due to differences in scrap quality, operating practices, dust capture efficiency and methods of introducing raw materials into the EAF. Currently the Center for Materials Production, (a group sponsored by the Electrical Power Research Institute) is conducting a major study on the mechanisms related to fume generation and hopes to be able to identify methods by which fume generation may be reduced. Much of the previous work done in this field deals with dust and fume generation in the BOF. However, as oxygen use in the EAF continues to rise, analogous mechanisms can be expected. Some of the key fume generation mechanisms are expanded upon.

10.4.5.1 Entrainment of Particles

Entrainment of particles occurs as CaO, rust and dirt in the scrap are pulled out of the EAF in the offgas system. It has been shown that the amount of CaO carryover is highly related to the method by which it is fed to the EAF.

10.4.5.2 Volatilization Of Volatile Metals

Certain metals volatilize as the scrap heats up. These include Zn, Pb, Cd, Na, Mn and Fe. These metals tend to form oxides following volatilization and show up as such in the EAF dust. Obviously, one way of reducing the total amount of dust generated would be to ensure that these materials were not included in the scrap charge. However, the recycle of coated scrap is on the increase and as such this portion of the dust generation is likely to increase. The degree of volatilization is likely related to hot spots in the furnace such as those under the arc and those which occur in oxygen lancing and oxy-fuel burner operations.

10.4.5.3 CO Bubble Bursting

CO bubbles are generated during oxygen lancing and slag foaming operations. Some liquid steel is usually associated with these bubbles and when they burst, this steel is ejected into the furnace atmosphere where it forms very fine particles of iron oxide.

10.4.6 Future Environmental Concerns

With the increased focus on environmental concerns, the focus on meltshop fume control has shifted to include not only particulate capture but also to control the emission of toxic gases from the fume system. Several gases have been scrutinized over the past five years and in 1991, the EPA revised its list of hazardous emissions to address control of emissions of these gases. Some of those gases that are associated with electric furnace operations are CO/CO_2, NO_x, VOCs and dioxins. In addition, emission standards for particulate matter continue to become more stringent. These issues are discussed in detail in the following sections.

10.4.6.1 Comparison of Environmental Regulations

Over the past few years environmental regulations have been made much more stringent as the public has become more aware of the hazards posed by various industrial emissions. In Europe the regulations are even tougher than those proposed in North America. This is probably due to a greater population density that results in industrial facilities frequently being located in heavily populated areas. The European Community has recently adopted regulations that closely align to the German Federal Air Emission Directive (17th BImSch V) of 1990. Table 10.10 shows the German standards.

Table 10.10 German Federal Air Emission Directive, 17th BImSch V

Pollutant	Standard
Dust	10 mg/m^3
CO	
NO_x	200 mg/m^3
SO_x	50 mg/m^3
VOCs	
Dioxins	0.1 ng/m^3 total equivalent

It is likely that North American standards will closely match these in the future.

10.4.6.2 Dust Emissions

Typically, electric furnace operations generate approximately 9–18 kg (20–40 lbs) of dust for every ton of steel produced. The preferred method of capture appears to be via a fourth hole system feeding to a baghouse for capture of the dust. Most modern systems claim discharge levels of 5–50 mg/m^3 in the discharge gas. Of a greater concern in most meltshops is the level of fugitive dust emissions to the shop environment. This can be addressed by several methods such as increasing fourth hole extraction (though this can have a negative effect on productivity), increasing canopy hood extraction (this is the most expensive option), and providing a furnace enclosure.

Of the three options, the latter appears to be gaining popularity because it results in a low volume gas cleaning system without side effects on the process.

10.4.6.3 CO

CO and CO_2 result from several operations in the EAF. When oxygen is lanced into the furnace it reacts with carbon to produce CO. If substantial amounts of CO are not captured by the DES system, ambient levels in the work environment may not be acceptable. Typically up to 10% of the CO unburned in the furnace reports to the secondary fume capture system during meltdown.

CO and CO_2 are of concern due to the fact that they are greenhouse gases and contribute to global warming. As oxygen use has increased in the EAF, so have the quantities of CO and CO_2 emitted from furnace operations. CO generation has also increased considerably with the adoption of foamy slag practices. Much of the CO generated within the EAF does not burn to CO_2 until it is within the fume system. This makes the fume system requirements greater due to the increased heat load. In addition, the benefit of CO post-combustion as an energy source for steelmaking is not utilized.

If however, the CO is burned in the freeboard it is possible to recover heat within the furnace thus reducing the heat load that the offgas system must handle. Results are discussed in Section 10.7.5.1.

If greater energy efficiencies can be achieved in the furnace via post-combustion, the net effect is to decrease the generation of CO_2 considerably. The combustion of carbon to CO produces only one-third of the energy of the combustion of CO. Thus even if only 50% of the energy available from post-combustion of CO is transferred back to the bath, a reduction of oxygen lancing into the bath of 50% could be made. By utilizing the energy from the post-combustion of CO, it will not be necessary to lance as much carbon out of the bath. This will lead to reduced CO/CO_2 emissions as well as considerable cost savings.

Other factors which can reduce CO emissions have been identified and include:

1. Use a low velocity water-cooled elbow exit coupled with a high velocity water-cooled duct inlet.
2. Ensure proper combustion air gap setting.
3. Provide accurate and reliable furnace draft control.
4. Provide sufficient DES capacity.
5. Ensure that carbon for slag foaming is injected at the bath/slag interface.
6. Melt as fast as possible.
7. Apply post-combustion to burn the CO in the furnace.
8. Begin oxygen lancing as early as possible—this spreads out CO generation throughout the melting cycle and reduces the maximum heat load on the DES.

10.4.6.4 NO_x

NO_x is formed in furnace operations when nitrogen passes through the arc between electrodes. Some thermal NO_x is also generated from burner use in EAFs. Typical levels of NO_x reported are in the range of 36–90 g NO_x per ton of steel.[18] Thermal NO_x can be addressed using any of the conventional methods. Typically, thermal NO_x is reduced by improving the burner design and providing good mixing of the pre-combustion gases. NO_x resulting from nitrogen passing through the arc can be reduced by reducing the amount of nitrogen available in the furnace. This can be achieved by closing up furnace gaps and by closing the slag door whenever possible. Many operations lance oxygen through the door and as a result large a large volume of air enters the furnace. This can be reduced by providing a shield close to the door or by hanging chains close to the opening. Foamy slag practice can be beneficial since the slag foams up partially blocking the opening. Also, since foamy slag helps to bury the arc, it is more difficult for nitrogen to pass through the arc and be ionized.

Thermal NO_x is also formed in the water-cooled duct following the combustion air gap as any combustible materials burn with the oxygen which has entered the ductwork in the combustion air. If all of the combustible material burns quickly, the gas temperature will reach a level where thermal NO_x is formed. One option which is now being investigated is to close up the combustion gap somewhat and to supply combustion air at various points in the water-cooled duct downstream of the combustion gap through injectors. Thus the combustion will be staged along the first two-thirds of the water-cooled duct and will avoid the temperatures associated with thermal NO_x generation.

10.4.6.5 VOCs

Most scrap mixes contain organic compounds to some extent. When scrap is charged to the furnace, some of these organic compounds burn off contributing to the charging plume. In most cases organic materials continue to burn off for five to ten minutes following charging. If there is insufficient oxygen available for combustion, these hydrocarbons will report to the offgas system where they may or may not be combusted. The total emissions of hydrocarbons appears to be related to the amount of chlorine in the EAF.[20] Scrap preheating tends to produce greater emissions of hydrocarbons because these do not burn due to the gas temperature range and as a result report to the offgas system.

The best methods for reducing VOCs from EAF operations is to either destroy them or to limit the amount entering the operation. If burners are operated with excess oxygen during the flash off period it is possible to burn most of the organic compounds within the furnace. This is beneficial because some of the heat is transferred back to the scrap thus aiding melting operations. It is also beneficial in that the resulting heat load to the offgas is lower.

Some EAF operations try to limit the amount of oily scrap that is charged to the furnace. In some cases the scrap is washed to remove some of the oil and grease. This is particularly true in the case of turnings and borings. If the amount of organic material in the scrap is reduced the level of VOC emissions is also reduced.

Some processes employ scrap preheating followed by an afterburner to destroy any VOCs. This requires additional equipment and increases operating costs but ensures that emissions of VOCs are kept to a minimum.

10.4.6.6 Dioxins

Dioxins and furans have become a major concern over the past few years in the incineration process. Dioxins and furans are combustion byproducts and the prevention of these emissions depends strongly on control of the combustion process. The term dioxins generally refers to 17 specific dioxin and furan congeners that are extremely toxic, the most toxic of which is tetrachloro–dibenzo–p–dioxin, see Fig. 10.58. These compounds are collectively referred to as dioxins and regulations are specified as a tetrachloro–dibenzo–p–dioxin (TCDD) toxic equivalent (TEQ). It is only recently that dioxin emissions in steelmaking operations have been monitored in Europe.

Dioxins are cyclic hydrocarbons that are usually formed in combustion processes. Dioxins do not have any known technical use and are not produced intentionally. They are formed as trace amounts in some chemical processes and during combustion of fuels, especially materials containing chlorinated materials. In incineration operations, dioxins are believed to form at a temperature of approximately 500°C (932°F). Dioxins tend to be formed downstream of incineration operations as the flue gases are cooled. This usually occurs in the temperature range of 250–400°C (482°F–752°F) by reactions between oxygen, water, hydrochloric gas and products of incomplete combustion.

In studies related to the formation of dioxins it was found that several factors had an affect. High carbon and high chlorine concentrations favor dioxin formation. Carbon acts as a sorbent for precursor compounds for dioxin formation. Water vapor helps to inhibit chlorine and thus reduces dioxin formation. Copper acts as a strong catalyst for dioxin formation. Al_2O_3 increases the effectiveness of copper to catalyze the reaction. Fe, Zn, Mn and Cu along with their oxides help to catalyze dioxin formation. Injection of ammonia into the offgas stream helps inhibit dioxin formation. Activated carbon can be used to strip dioxins from the gas stream.

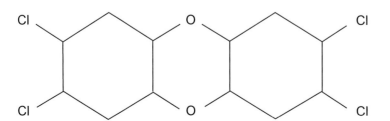

Fig. 10.58 Tetrachloro–dibenzo–p–dioxin structure.

From the above information it is apparent that the components of the offgas stream in the EAF are ideal for the formation of dioxins. In material published by MEFOS it was shown that the biggest contributor to dioxins in scrap steel melting was PVC coated scrap followed by oily scrap.[20] It was also found that the baghouse emissions of dioxins ranged from 20–60% of the offgas stream concentration. This indicates that dioxins were being captured in the baghouse. This is supported by research that indicates that 20–30% of dioxins are bound to particulate.[21] The Swedish paper concludes that the following factors can reduce dioxin emissions: scrap sorting, scrap pretreatment, post-combustion of organics in the furnace, and absorbent injection in the offgas.[20]

Tests conducted at several European steelmakers have indicated that dioxin formation can be controlled by rapid quenching of the offgas through the temperature zone in which dioxin formation usually takes place. In these operations offgas is cooled rapidly from approximately 800°C (1470°F) to less than 300°C (570°F) using spray water. This has been shown to limit the dioxin emissions to 0.1–0.2 ng/Nm^3 TEQ. For most DES systems this would involve spray cooling of the offgas following the water-cooled duct portion of the system. This should not pose problems for system operation as long as the amount of spray cooling does not create dewpoint problems in the baghouse.

10.4.7 Conclusions

Electric arc furnace fume systems have evolved considerably from very simple systems aimed at improving the ambient work environment around the furnace to sophisticated systems aimed at controlling not only the emission of particulate but also the control of toxic gases. Modern fume systems are now designed to minimize the formation of toxic gases and to ensure that others are destroyed before exiting the system. Thus gas cleaning in the modern day fume system entails much more than trapping and collection of particulate matter. With the tighter environmental restrictions expected in the future, it is expected that electric furnace operations will have to look at environmental concerns in conjunction with furnace operations. A variety of factors will affect the costs and processes necessary to maintain production and meet environmental restrictions. Without doubt the cheapest method of meeting emissions restrictions is to prevent these compounds from being present in the first place. Capture and treatment of pollutants is an expensive way of dealing with the problem. Of course, tighter restrictions of the feed materials to the furnace will also have an affect on productivity and operating costs. It will become necessary to weigh environmental considerations with operations concerns in order to find an optimum that allows for low cost steel production while meeting emissions requirements. This means that we must integrate environmental thinking into operations in order to arrive at a low cost, environmentally friendly steel production.

10.5 Raw Materials

The main raw material for EAF steelmaking is steel scrap. Scrap is an energy intensive and valuable commodity and comes primarily from three main sources: reclaimed scrap (also known as obsolete scrap) which is obtained from old cars, demolished buildings, discarded machinery and domestic objects; industrial scrap (also known as prompt scrap) which is generated by industries using steel within their manufacturing processes; and revert scrap (also known as home scrap) which is generated within the steelmaking and forming processes (e.g. crop ends from rolling operations, metallic losses in slag etc.).

The latter two forms of scrap tend to be clean, i.e. they are close in chemical composition to the desired molten steel composition and thus are ideal for recycle. Reclaimed/obsolete scrap frequently has a quite variable composition and quite often contains contaminants that are undesirable for steelmaking. Levels of residual elements such as Cu, Sn, Ni, Cr, and Mo are high in obsolete scrap and can affect casting operations and product quality if they are not diluted. Thus a facility which has a need for very low residual levels in the steel will be forced to use higher quality prompt scrap but at a much higher cost. The alternative is to use a combination of the contaminated obsolete scrap along with what are generally referred to as clean iron units or virgin iron units. These are materials which contain little or no residual elements. Clean iron units are typically in the form of direct reduced iron (DRI), hot briquetted iron (HBI), iron carbide, pig iron, and molten pig iron (hot metal).

It is possible to use lower grade scrap which contains residual elements, if this scrap is blended with clean iron units so that the resulting residual levels in the steel following melting meet the requirements for flat rolled products.

Obsolete scrap is much more readily available than prompt scrap and thus the use of clean iron units is expected to increase as shortages of prompt scrap continue to grow.

In addition to classification of scrap into the above three groups, scrap is also classified based on its physical size, its source and the way in which it is prepared. For example the following categories are those commonly used in North America: No. 1 bundles, No. 1 factory bundles, No. 1 shredded, No. 1 heavy melt, No. 2 heavy melt, No. 2 bundles, No. 2 shredded, busheling, turnings, shredded auto, structural/plate 3 ft., structural/plate 5 ft., rail crops, and rail wheels.

In addition to the residual elements contained in the scrap, there are also several other undesirable components including, oil, grease, paint coatings, zinc coatings, water, oxidized material and dirt. The lower the grade of scrap, the more likely it is to contain greater quantities of these materials. As a result this scrap may sell at a discount but the yield of liquid steel may be considerably lower than that obtained when using a higher grade scrap. In addition, these undesirable components may result in higher energy requirements and environmental problems. Thus the decision for scrap mix to be used within a particular operation will frequently depend on several factors including availability, scrap cost, melting cost, yield, and the effect on operations (based on scrap density, oil and grease content, etc.).

In practice, most operations buy several different types of scrap and blend them to yield the most desirable effects for EAF operations.

10.6 Fluxes and Additives

Carbon is essential to the manufacture of steel. Carbon is one of the key elements which give various steel grades their properties. Carbon is also important in steelmaking refining operations and can contribute a sizable quantity of the energy required in steelmaking operations. In BOF steelmaking, carbon is present in the hot metal that is produced in the blast furnace. In electric furnace steelmaking, some carbon will be contained in the scrap feed, in DRI, HBI or other alternative iron furnace feeds. The amount of carbon contained in these EAF feeds will generally be considerably lower than that contained in hot metal and typically, some additional carbon is charged to the EAF. In the past carbon was charged to the furnace to ensure that the melt-in carbon level was above that desired in the final product. As higher oxygen utilization has developed as standard EAF practice, more carbon is required in EAF operations. The reaction of carbon with oxygen within the bath to produce carbon monoxide results in a significant energy input to the process and has lead to substantial reductions in electrical power consumption in EAF operations. The generation of CO within the bath is also key to achieving low concentrations of dissolved gases (nitrogen and hydrogen) in the steel as these are flushed out with the carbon monoxide. In addition, oxide inclusions are flushed from the steel into the slag.

In oxygen injection operations, some iron is oxidized and reports to the slag. Oxy-fuel burner operations will also result in some scrap oxidation and this too will report to the slag once the scrap melts in. Dissolved carbon in the steel will react with FeO at the slag/bath interface to produce CO and recover iron units to the bath.

The amount of charge carbon used will be dependent on several factors including carbon content of scrap feed, projected oxygen consumption, desired tap carbon, and the economics of iron yield versus carbon cost. In general, the amount used will correspond to a carbon/oxygen balance as the steelmaker will try to maximize the iron yield. Typical charge carbon rates for medium carbon steel production lie in the range of 2–12 kg per ton of liquid steel.

Generally, the three types of carbonaceous material used as charge carbon in EAF operations are anthracite coal, metallurgical coke and green petroleum coke. Most anthracite coal used in North American steelmaking operations is mined in eastern Pennsylvania. This material has a general composition of 3–8% moisture content, 11–18% ash content and 0.4–0.7% sulfur.

The high variation in ash content translates into wide variations in fixed carbon content and in general, EAF operations strive to keep ash content to a minimum. The ash consists primarily of silica. Thus increased ash input to the EAF will require additional lime addition in order to maintain the desired V-ratio in the slag. The best grades of anthracite coal have fixed carbon contents of 87–89%. Low grade anthracite coals may have fixed carbon levels as low as 50%.

Anthracite coal is available in a wide variety of sizes ranging from 4×8 in. down to 3/64 in. $\times$ 100 mesh. The most popular sizes for use as charge carbon are nut (1 5/8 $\times$ 13/16 in.), pea (13/16 $\times$ 9/16 in.), and buckwheat (9/16 $\times$ 5/16 in.).

Metallurgical coke is produced primarily in integrated steel operations and is used in the blast furnace. However, some coke is used as EAF charge carbon. Generally, this material has a composition of 1–2% moisture content, 1–3.5% volatile material, 86–88% fixed carbon, 9–12% ash content and 0.88–1.2% sulfur.

Usually coke breeze with a size of $-1/2 \times 0$ is used as charge carbon. Coarser material can be used but is more expensive.

Green petroleum coke is a byproduct of crude oil processing. Its properties and composition vary considerably and are dependent on the crude oil feedstock from which it is derived. Several coking processes are used in commercial operation and these will produce considerably different types of coke.

Sponge coke results from delayed coker operations and is porous in nature. It may be used as a fuel or may be processed into electrodes or anodes depending on sulfur content and impurity levels. This material is sometimes available as charge carbon.

Needle coke is produced using a special application of the delayed coker process. It is made from high grade feedstocks and is the prime ingredient for the production of carbon and graphite electrodes. This material is generally too expensive to be used as charge carbon.

Shot coke is a hard, pebble-like material resulting from delayed coker operation under conditions which minimize coke byproduct generation. It is generally used as a fuel and is cost competitive as charge carbon.

Fluid coke is produced in a fluid coker by spraying the residue onto hot coke particles. It is usually high in sulfur content and is used in anode baking furnaces. It can also be used as a recarburizer if it is calcined.

Over the past decade, many operations have adopted foamy slag practices. At the start of meltdown the radiation from the arc to the sidewalls is negligible while the electrodes are surrounded by the scrap. As melting proceeds the efficiency of heat transfer to the scrap and bath drops off and more heat is radiated from the arc to the sidewalls. By covering the arc in a layer of slag, the arc is shielded and the energy is transferred to the bath. Oxygen is injected with coal to foam the slag by producing CO gas in the slag. In some cases only carbon is injected and the carbon reacts with FeO in the slag to produce CO gas. When foamed, the slag cover increases from 10 to 30 cm thick.[22] In some cases the slag is foamed to such an extent that it comes out of the electrode ports. Claims for the increase in energy transfer efficiency range from an efficiency of 60–90% with slag foaming compared to 40% without.[23,24] It has been reported that at least 0.3% carbon should be removed from the bath using oxygen in order to achieve a good foamy slag practice.[25]

The effectiveness of slag foaming is dependent on several process parameters as described in the section on furnace technologies. Typical carbon injection rates for slag foaming are 2–5 kg per ton of liquid steel for low to medium powered furnaces. Higher powered furnaces and DC furnaces will tend to use 5–10 kg of carbon per ton liquid steel. This is due to the fact that the arc length is much greater than low powered AC operations and therefore greater slag cover is required to bury the arc.

Lime is the most common flux used in modern EAF operations. Most operations now use basic refractories and as a result, the steelmaker must maintain a basic slag in the furnace in order to minimize refractory consumption. Slag basicity has also been shown to have a major effect on slag foaming capabilities. Thus lime tends to be added both in the charge and also via injection directly into the furnace. Lime addition practices can vary greatly due to variances in scrap composition. As elements in the bath are oxidized (e.g. P, Al, Si, Mn) they contribute acidic components to the slag. Thus basic slag components must be added to offset these acidic contributions. If silica levels in the slag are allowed to get too high, significant refractory erosion will result. In addition, FeO levels in the slag will increase because FeO has greater solubility in higher silica slags. This can lead to higher yield losses in the EAF.

The generation of slag also allows these materials which have been stripped from the bath to be removed from the steel by pouring slag out of the furnace through the slag door which is located at the back of the EAF. This is known as slagging off. If the slag is not removed but is instead allowed to carry over to the ladle it is possible for slag reversion to take place. This occurs when metallic oxides are reduced out of the slag by a more reactive metallic present in the steel. When steel is tapped it is frequently killed by adding either silicon of aluminum during tapping. The purpose of these additions is to lower the oxygen content in the steel. If however P_2O_5 is carried over into the ladle, it is possible that it will react with the alloy additions producing silica or alumina and phosphorus which will go back into solution in the steel.

Sometimes magnesium lime is added to the furnace either purely as MgO or as a mixture of MgO and CaO. Basic refractories are predominantly MgO, thus by adding a small amount of MgO to the furnace, the slag can quickly become saturated with MgO and thus less refractory erosion is likely to take place.

10.7 Electric Furnace Technology

The electric arc furnace has evolved considerably over the past 20 years. Gone are the days when electric power was the only source of energy for scrap melting. Previously tap-to-tap times in the range of 3–8 hours were common.[26] With advances in technology it is now possible to make heats in under one hour with electrical energy consumptions in the range of 380–400 kWh/ton.[23] The electric furnace has evolved into a fast and low cost melter of scrap where the major criterion is higher productivity in order to reduce fixed costs. Innovations which helped to achieve the higher production rates include oxy-fuel burners, oxygen lancing, carbon/lime injection, foamy slag practices, post-combustion in the EAF freeboard, EAF bath stirring, modified electrical supply (series reactors etc.), current conducting electrode arms, DC furnace technology and other innovative process technologies (scrap preheat, continuous charging etc.).

10.7.1 Oxygen Use in the EAF

Much of the productivity gain achieved over the past 10–15 years was related to oxygen use in the furnace. Exothermic reactions were used to replace a substantial portion of the energy input in the EAF. Whereas oxygen utilization of 9 Nm^3/ton (300 scf/ton) was considered ordinary just 10 years ago, some operations now use as much as 40 Nm^3/ton (1300 scf/ton) for lancing operations.[27,28] With post-combustion, rates as high as 70 Nm^3/ton (2500 scf/ton) have been implemented. It is now common for 30–40% of the total energy input to the EAF to come from oxy-fuel burners and oxygen lancing.[23,28] By the early 1980s, more than 80% of the EAFs in Japan employed oxy-fuel burners.[23] In North America it was estimated that in 1990, only 24% of EAF operations were using such burners.[29] Since that time, a large percentage of North American operations have looked to the use of increased oxygen levels in their furnaces in order to increase productivity and decrease electrical energy consumption. High levels of oxygen input are standard on most new EAF installations.

The IISI 1990 electric arc furnace report indicates that most advanced EAF operations utilize at least 22 Nm^3/ton (770 scf/ton) of oxygen.[18] In addition oxygen supplies 20–32% of the total power input in conventional furnace operations. This has grown with the use of alternative iron sources in the EAF, many of which contain elevated carbon contents (1–3%). In some cases, electrical energy now accounts for less than 50% of the total power input for steelmaking.

One of the best examples of the progressive increase in oxygen use within the EAF is the meltshop operation at Badische Stahlwerke (BSW). Between 1978 and 1990, oxygen use was increased from 9 Nm3/ton to almost 27 Nm^3/ton. Productivity increased from 32 ton per hour to 85 ton per hour while power consumption decreased from 494 kWh/ton to 357 kWh/ton.[30] During this period, BSW developed their own manipulator for the automatic injection of oxygen and carbon. In 1993, BSW installed the ALARC post-combustion system developed by Air Liquide and increased oxygen consumption in the furnace to 41.5 Nm^3/ton. The corresponding power consumption is 315 kWh/ton with a tap-to-tap time of 48 minutes. This operation has truly been one of the pioneers for increased chemical energy use in the EAF.

10.7.2 Oxy-Fuel Burner Application in the EAF

Oxy-fuel burners are now almost standard equipment on electric arc furnaces in many parts of the world. The first use of burners was for melting the scrap at the slag door where arc heating was fairly inefficient. As furnace power was increased, burners were installed to help melt at the cold spots common to UHP operation. This resulted in more uniform melting and decreased the melting time necessary to reach a flat bath. It was quickly realized that productivity increases could be achieved by installing more burner power. Typical productivity increases reported in the literature have been in the range of 5–20%.[22,31–33] In recent years oxy-fuel burners have been of greater interest due to the increase in the cost of electrodes and electricity. Thus natural gas potentially provides a cheaper source of energy for melting. Fig. 10.59 shows oxy-fuel burners in operation in an EAF.

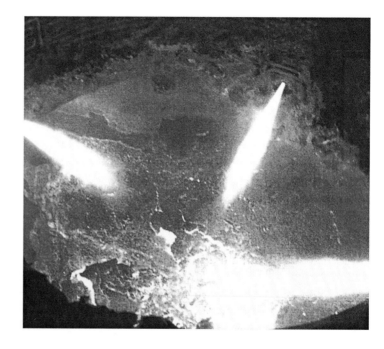

Fig. 10.59 Oxy-fuel burners in EAF operations.

Typically burners are located in either the slag door, sidewall or roof. Slag door burners are generally used for small to medium sized furnaces where a single burner can reach all of the cold spots.[22,33] Door burners have the advantage that they can be removed when not in use. For larger furnaces, three or four sidewall mounted burners are more effective for cold spot penetration. However these are vulnerable to attack from slag, especially if employing a foamy slag practice. In such cases the burners are sometimes mounted in the roof and are fired tangentially through the furnace cold spots.

Oxy-fuel burners aid in scrap melting by transferring heat to the scrap. This heat transfer takes place via three modes; either forced convection from the combustion products, radiation from the combustion products or conduction from carbon or metal oxidation and from scrap to other scrap.[33]

Primarily heat transfer is via the first two modes except when the burners are operated with excess oxygen. Heat transfer by these two modes is highly dependent on the temperature difference between the scrap and the flame and on the surface area of the scrap exposed for heat transfer. As a result oxy-fuel burners are most efficient at the start of a melt-in period when the scrap is cold. As melting proceeds the efficiency will drop off as the scrap surface in contact with the flame decreases and due to the fact that the scrap temperature also increases. It is generally recommended that burners be discontinued after 50% of the meltdown period is completed so that reasonable efficiencies are achieved.[23] An added complication is that once the scrap heats up it is possible for iron to react with the water formed by combustion to produce iron oxide and hydrogen. This results in yield loss and the hydrogen must be combusted downstream in the offgas system. Usually the point at which burner use should be discontinued is marked by a rise in offgas temperature(indicating that more heat is being retained in the offgas).[33] In some operations the temperature of the furnace side panels adjacent to the burner is used to track burner efficiency. Once the efficiency drops below a set point the burners are shut off.[29]

Heat transfer by conduction occurs when excess oxygen reacts with material in the charge. This will result in lower yield if the material burned is iron or alloys and as a result is not generally recommended. However in the period immediately following charging when volatile and combustible materials in the scrap flash off, additional oxygen in the furnace is beneficial as it allows this material to be burned inside the furnace and thus results in heat transfer back to the scrap. This is also beneficial for the operation of the offgas system as it not required to remove this heat downstream.

Fig. 10.60 indicates that 0.133 MW of burner rating should be supplied per ton of furnace capacity.[34] Other references recommend a minimum of 30 kWh/ton of burner power to eliminate cold spots in a UHP furnace and 55–90 kWh/ton of burner power for low powered furnaces.[23]

A burner/lance has been developed by Empco, Fig. 10.15, that has the capability of following the scrap level as it melts. The lance is similar to a BOF design (water-cooled) and is capable of operation at various oxygen to natural gas ratios or with oxygen alone for decarburization. The lance is banana shaped and follows the scrap as its melts back. Thus a high heat transfer efficiency can be maintained for a longer period of time. Typical efficiencies reported for the Unilance are in the range of 65–70%.[35]

Heat transfer efficiencies reported in the literature vary greatly in the range of 50–75%.[24] Data published by L'Air Liquide shows efficiencies of 64–80% based on energy savings. Krupp indicates an energy efficiency based on energy savings alone of 78%,[36] and also gives an indication of why such high efficiencies are reported when looking at energy replacement alone. According to Krupp, an increase of one minute in meltdown time corresponds to an increase in power consumption of 2 kWh/ton. Thus the decrease in tap-to-tap time must be taken into account when calculating burner efficiency. For typically reported energy efficiencies, once this is taken into account burner efficiencies lie in the range of 45–65% in all cases. This is supported somewhat by theoretical calculations by Danieli that indicated that only 20–30% of the energy from the burners went to the scrap;[37] as much as 40–60% of the heat went to the offgas and the remainder was lost to the water-cooled furnace panels.

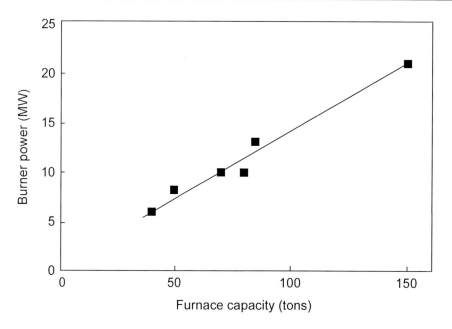

Fig. 10.60 Oxy-fuel burner rating versus furnace capacity.

Fig. 10.61 shows burner efficiency as a function of operating time based on actual furnace offgas measurements.[34] This shows that burner efficiency drops off rapidly after 40–50% of the melting time. By 60% into the melting time burner efficiency has dropped off to below 30%. It is apparent that a cumulative efficiency of 50–60% is achieved over the first half of the meltdown period and drops off rapidly afterwards. As a result, typical operating practice for a three bucket charge is to run the burners for two-thirds of the first meltdown, one-half of the second meltdown and one-third of the third meltdown. For operations with only one backcharge, burners are typically run for 50% of each meltdown phase.[32]

The amount of power input to the furnace has a small effect on the increase in heatload and offgas volume. The major factor is the rate at which the power is put into the furnace. Thus for a low pow-

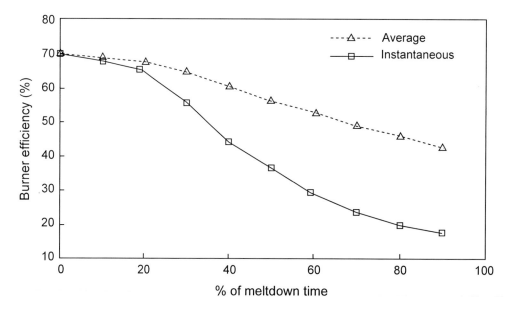

Fig. 10.61 Oxy-fuel burner efficiency versus time into meltdown.

ered furnace(high tap-to-tap time), the total burner input to the EAF may be in the range of 80–100 kWh/ton but the net effect on the offgas evacuation requirements may not change much as the burners are run for a longer period of time. Likewise for a high powered furnace, due to the short tap-to-tap time, the burner input rate may be quite high even though the burners supply only 30 kWh/Ton to the furnace.

Bender estimates that offgas flowrates can increase by a factor of 1.5 and offgas heatload by as high as a factor of 2.5 for operation with burners and high oxygen lance rates.[37]

10.7.3 Application of Oxygen Lancing in the EAF

Over the past ten years oxygen lancing has become an integral part of EAF melting operations. It has been recognized in the past that productivity improvements in the open hearth furnace and in the BOF were possible through the use of oxygen to supply fuel for exothermic reactions. Whereas previously oxygen was used primarily only for decarburization in the EAF at levels of 3–8 Nm^3 per ton (96–250 scf/ton),[23,32] in modern operations anywhere from 10–30% of the total energy input is supplied via exothermic bath reactions.[23,24,28,38,29] For a typical UHP furnace oxygen consumption is in the range of 18–27 Nm^3/ton (640–960 scf/ton).[28] Oxygen utilization in the EAF is much higher in Japan and in Europe where electricity costs are higher. Oxygen injection can provide a substantial power input—a lance rate of 30 Nm^3/min. (1050 scfm)is equivalent to a power input of 11 MW based on the theoretical reaction heat from combustion of C to CO.[23]

Oxygen lances can be of two forms. Water-cooled lances are generally used for decarburization though in some cases they are now use for scrap cutting as well. The conventional water-cooled lance was mounted on a platform and penetrated into the side of the furnace through a panel. Water-cooled lances do not actually penetrate the bath though they sometimes penetrate into the slag layer. Consumable lances are designed to penetrate into the bath or the slag layer. They consist of consumable pipe which is adjusted as it burns away to give sufficient working length. The first consumable lances were operated manually through the slag door. Badische Stahl Engineering developed a robotic manipulator to automate the process. This manipulator is used to control two lances automatically.[22] Various other manipulators have been developed recently and now have the capability to inject carbon and lime for slag foaming simultaneously with oxygen lancing. One major disadvantage of lancing through the slag door is that it can increase air infiltration into the furnace by 100–200%. This not only has a negative impact on furnace productivity but also increases offgas system evacuation requirements substantially. As a result, not all of the fume is captured and a significant amount escapes from the furnace to the shop. This can be a significant problem if substantial quantities of CO escape to the shop due to its rapid cooling and subsequent incomplete combustion to CO_2. Thus background levels of CO in the work environment may become an issue. To reduce the amount of air infiltration to the EAF, some operations insert the lance through the furnace sidewall as shown in Fig. 10.62.

Energy savings due to oxygen lancing arise from both exothermic reactions (oxidation of carbon and iron) and due to stirring of the bath with leads to temperature and composition homogeneity of the bath. The product of scrap cutting is liquid iron and iron oxide. Thus most of the heat is retained in the bath. The theoretical energy input for oxygen reactions in the bath is as follows[23]:

$$Fe + 1/2\ O_2 \rightarrow FeO, \text{ heat input} = 6.0 \text{ kWh/Nm}^3\ O_2 \qquad (10.7.1)$$

$$C + 1/2\ O_2 \rightarrow CO, \text{ heat input} = 2.8 \text{ kWh/Nm}^3\ O_2 \qquad (10.7.2)$$

Thus it is apparent that much more energy is available if iron is burned to produce FeO. Naturally though, this will impact negatively on productivity. Studies have shown that the optimum use of oxygen for conventional lancing operations is in the range of 30–40 Nm^3/ton (1000–1250 scf/ton).[27] Above this level yield losses are excessive and it is no longer economical to add oxygen. Typical operating results have given energy replacement values for oxygen in the range of 2–4 kWh/Nm^3 O_2 (0.056–0.125 kWh/scf O_2), with an average of 3.5 kWh/Nm^3 O_2 (0.1 kWh/scf O_2).[24,27,28,31,32,32,40,41] These values show that it is likely that both carbon and iron are reacting. In

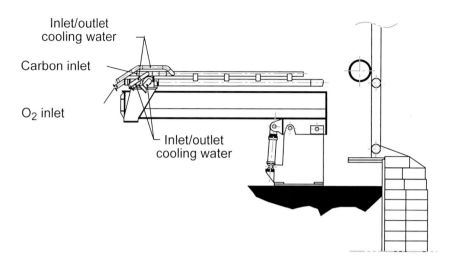

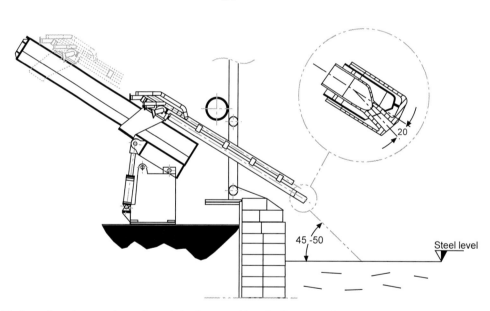

Fig. 10.62 Operation of oxygen lance through the furnace sidewall. *(Courtesy of Danieli.)*

addition, some studies have shown that the oxygen yield (i.e. the amount reacting with carbon) is in the range of 70–80%.[34,41] This would support the theory that both carbon and iron are reacting. During scrap cutting operations, the oxygen reacts primarily with the iron. Later when a molten pool has formed the FeO is reduced out of the slag by carbon in the bath. Thus the net effect is to produce CO gas from the oxygen that is lanced.

Based on the information cited in the preceding section, it can be expected that for every Nm^3 of oxygen lanced, 0.75 Nm^3 will react with carbon to produce 1.5 Nm^3 of CO (based on the average energy replacement value of 3.5 kWh/ Nm³ O_2). If in addition the stirring effect of the lancing brings bath carbon or injected carbon into contact with FeO in the slag, an even greater quantity of CO may result. That this occurs is supported by data that indicates a decarburization efficiency of greater than 100%. Thus during the decarburization period up to 2.5 Nm³ of CO may result for every Nm³ of oxygen injected. Typical oxygen rates are in the range of 30–100 Nm³/min. (1000–3500 scfm) and are usually limited by the ability of the fourth hole system to evacuate the furnace fume. Recommended lance rates for various furnace sizes are shown in Fig. 10.63 and indicate a rate of approximately 0.78–0.85 Nm³/ton (25–30 scfm/ton) of furnace capacity. In some newer processes where feed materials are very high in carbon content, oxygen lance rates equivalent to 0.1% decarburization per minute are required. In such cases, the lance rates may be as high as 280 Nm³/min (10,000 scfm) which is similar to BOF lance rates.

The major drawback to high oxygen lance rates is the effect on fume system control and the production of NO_x. Offgas volumes are greatly increased and the amount of CO generated is much greater. This must be taken into account when contemplating increased oxygen use.

The use of oxygen lancing throughout the heat can be achieved in operations using a hot heel in the furnace. Oxygen is lanced at a lower rate throughout the heat to foam the slag. This gives better shielding of the arc leading to better electrical efficiency. It also gives lower peak flowrates of CO to the offgas system, thus it reduces the extraction requirement of the offgas system.

High generation rates of CO may necessitate a post-combustion chamber in the DES system. If substantial amounts of CO are not captured by the DES system, ambient levels in the work environment may not be acceptable. Typically up to 10% of the CO unburned in the furnace reports to the secondary fume capture system during meltdown.

Operating with the slag door open increases the overall offgas evacuation requirements substantially. If possible oxygen lances should penetrate the furnace higher up in the shell. Another factor to consider is that the increased amount of nitrogen in the furnace will likely lead to increased NO_x.

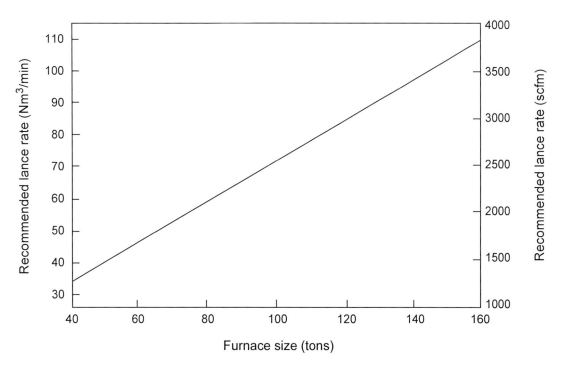

Fig. 10.63 Recommended lance rate versus furnace size.

10.7.4 Foamy Slag Practice

In recent years more EAF operations have begun to use a foamy slag practice. Foamy slag was initially associated with DRI melting operations where FeO and carbon from the DRI would react in the bath to produce CO which would foam the slag. At the start of meltdown the radiation from the arc to the sidewalls is negligible because the electrodes are surrounded by the scrap. As melting proceeds the efficiency of heat transfer to the scrap and bath drops off and more heat is radiated from the arc to the sidewalls. By covering the arc in a layer of slag, the arc is shielded and the energy is transferred to the bath as shown in Fig. 10.64.

Oxygen is injected with coal to foam up the slag by producing CO gas in the slag. In some cases only carbon is injected and the carbon reacts with FeO in the slag to produce CO gas. When foamed, the slag cover increases from four inches thick to twelve inches.[22] In some cases the slag is foamed to such an extent that it comes out of the electrode ports.[42] Claims for the increase in efficiency range from an efficiency of 60–90% with slag foaming compared to 40% without.[23,24] This is shown in Fig. 10.65 and Fig. 10.66. It has been reported that at least 0.3% carbon should be removed using oxygen in order to achieve a good foamy slag practice.[25] If a deep foamy slag is achieved it is possible to increase the arc voltage considerably. This allows a greater rate of power input. Slag foaming is usually carried out once a flat bath is achieved. However, with hot heel operations it is possible to start slag foaming much sooner.

Some of the benefits attributed to foamy slag are decreased heat losses to the sidewalls, improved heat transfer from the arcs to the steel allows for higher rate of power input, reduced power and voltage fluctuations, reduced electrical and audible noise, increased arc length (up to 100%) without increasing heat loss and reduced electrode and refractory consumption.[4,18,24]

Several factors have been identified that promote slag foaming. These are oxygen and carbon availability, increased slag viscosity, decreased surface tension, slag basicity > 2.5 and FeO in slag at 15–20% to sustain the reaction.[4,25,43] Fig. 10.67 and Fig. 10.68 show several of these effects graphically.

The only negative side of foamy slag practice is that a large quantity of CO is produced in the EAF. Bender estimates that offgas flowrates can increase by a factor of 1.5 and offgas heatload by as much as a factor of 2.5 for high slag foaming rates.[37] In many operations, a large amount of carbon is removed from the bath in order to generate chemical energy input for the operation. If this CO is to be generated anyway, the operator is well advised to ensure that slag conditions are such that foaming will result in order to benefit from the CO generation.

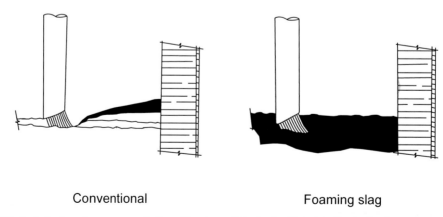

Conventional   Foaming slag

Fig. 10.64 Effect of slag foaming on arc radiation. *(Courtesy of Center for Materials Production.)*

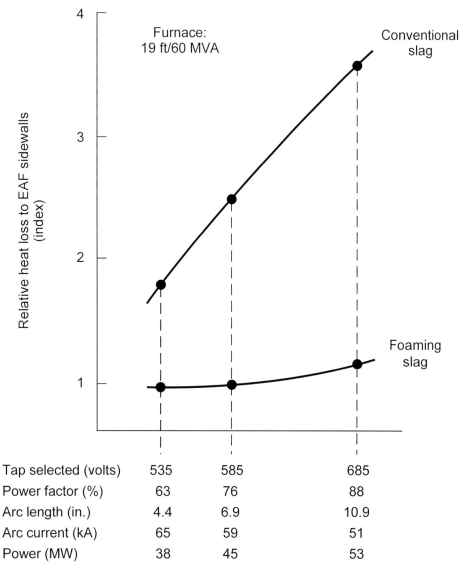

Fig. 10.65 Effect of slag foaming on heat loss to the furnace sidewalls. *(Courtesy of Center for Materials Production.)*

10.7.5 CO Post-Combustion

The 1990s have seen steelmakers advance further in lowering production costs of liquid steel. Higher electrical input rates and increased oxygen and natural gas consumption has led to short tap-to-tap times and high throughputs. Thus, energy losses are minimized and up to 60% of the total power input ends up in the steel. This has not come without cost, as water-cooled panels and roofs are required to operate at higher heat fluxes. Typically 8–10% of the power input is lost to the cooling water and offgas temperatures are extremely high, with losses of approximately 20% of the power input to the offgas. As EAF steelmakers attempt to lower their energy inputs further they have begun to consider the heat contained in the offgas. One way in which this can be recaptured is to use the offgas to preheat scrap. This results in recovery of the sensible heat but does not address the calorific heat which can represent as much as 50–60% of the energy in the offgas.

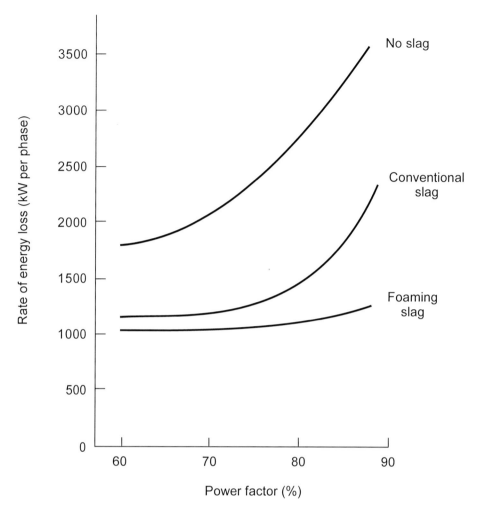

Fig. 10.66 Effect of slag cover on heat loss in the EAF. *(Courtesy of Center for Materials Production.)*

10.7.5.1 Introduction

Generically, post-combustion refers to the burning of any partially combusted compounds. In EAF operations both CO and H_2 are present. CO gas is produced in large quantities in the EAF both from oxygen lancing and slag foaming activities and from the use of pig iron or DRI in the charge. Large amounts of CO and H_2 are generated at the start of meltdown as oil, grease and other combustible materials evolve from the surface of the scrap. If there is sufficient oxygen present, these compounds will burn to completion. In most cases there is insufficient oxygen for complete combustion and high levels of CO result. Tests conducted at Vallourec by Air Liquide showed that the offgas from the furnace could contain considerable amounts of non-combusted CO when there was insufficient oxygen present.[44]

The heat of combustion of CO to CO_2 is three times greater than that of C to CO (for dissolved carbon in the bath). This represents a very large potential energy source for the EAF. Studies at Irsid (Usinor SA) have shown that the potential energy saving is significant and could be a much as 72 kWh/ton.[45] If the CO is burned in the freeboard it is possible to recover heat within the furnace. Some of the expected benefits and concerns regarding post-combustion are given in Table 10.11. As more oxygen is used to reduce electrical consumption, there will be greater need for improved oxygen utilization. In addition, environmental regulations may limit CO_2 emissions. As a result it will be necessary to obtain the maximum benefit of oxygen in the furnace. This can only be achieved if most CO is burned in the furnace.

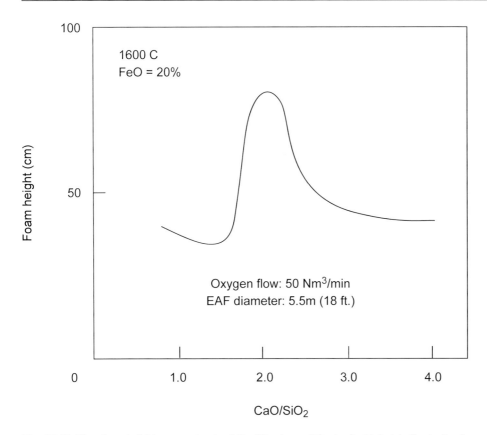

Fig. 10.67 Slag foam height versus slag basicity. *(Courtesy of Center for Materials Production.)*

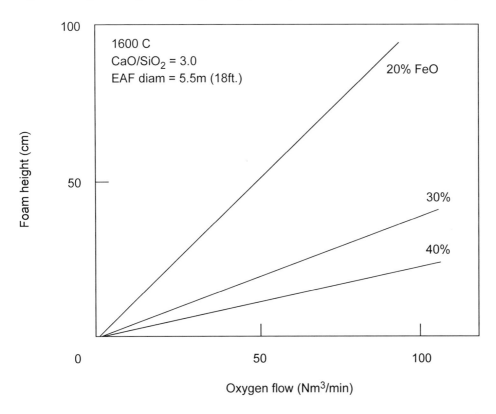

Fig. 10.68 Slag foam height versus oxygen injection rate for various FeO levels in the slag. *(Courtesy of Center for Materials Production.)*

Table 10.11 Benefits and Concerns for Post-Combustion

Benefits	Concerns
Decreased heat load to the offgas system	Increased electrode consumption
Decreased CO emissions to the meltshop and baghouse	Increased heat load to the water-cooled panels and roof
Higher heat transfer due to higher radiation from combustion products	Decreased iron yield
Decreased water-cooled duct requirement	Economics of additional oxygen
Increased utilization of energy from oxygen and carbon	
Reduced electrical power consumption	
Decreased NO_x emissions from the EAF	
Increased productivity without increased offgas system requirements	

In order to maintain consistency of the results presented the following definitions are made for EAF trials:

$$\text{Post combustion ratio (PCR)} = \frac{CO_2}{CO + CO_2} \quad (10.7.3)$$

$$\text{Heat transfer efficiency (HTE)} = \frac{\text{reduction in kWh to steel}}{\text{theoretical energy of PC for CO}} \quad (10.7.4)$$

10.7.5.2 Post-Combustion in Electric Arc Furnaces

Several trials have been run using post-combustion in the EAF. In some of the current processes, oxygen is injected into the furnace above the slag to post-combust CO. Some processes involve injection of oxygen into the slag to post-combust the CO before it enters the furnace freeboard. Most of these trials were inspired by an offgas analysis which showed large quantities of CO leaving the EAF. Fig. 10.69 shows a typical post-combustion system where oxygen is added in the freeboard of the furnace. Thus combustion products directly contact the cold scrap. Most of the heat transfer in this case is radiative. Fig. 10.70 shows the approach where post-combustion is carried out low in the furnace or in the slag itself. Heat transfer is accomplished via the circulation of slag and metal droplets within the slag. Post-combustion oxygen is introduced at very low velocities into the slag. Heat transfer is predominantly convective for this mode of post-combustion. Some other systems have incorporated bottom blown oxygen (via tuyeres in the furnace hearth) along with injection of oxygen low in the furnace. In fact the first two processes to employ extensive post-combustion as part of the operation (K-ES, EOF) both took this approach. These processes are discussed in detail in Section 10.9 on future developments.

Some of the results that have been obtained are summarized in Table 10.12 along with their theoretical limits. Note that in many cases the energy savings has been stated in kWh/Nm³ of post-combustion oxygen added. The theoretical limit for post-combustion of CO at bath temperatures (1600°C) is 5.8 kWh/Nm³ of oxygen. PCRs have been calculated based on the offgas analysis presented in these references. HTE is rarely measured in EAF post-combustion trials and is generally backed out based on electrical energy replacement due to post-combustion. This does not necessarily represent the true post-combustion HTE because the oxygen can react with compounds other than CO.

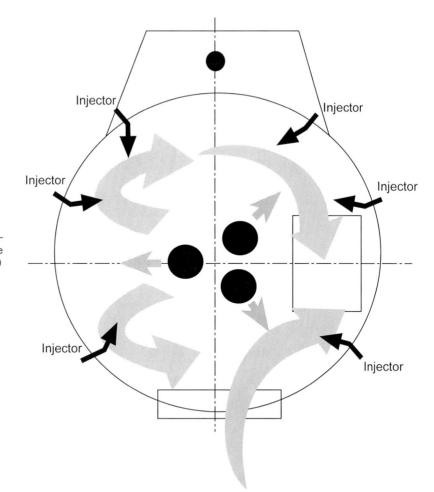

Fig. 10.69 Injection of post-combustion oxygen above the slag. *(Courtesy of Air Liquide.)*

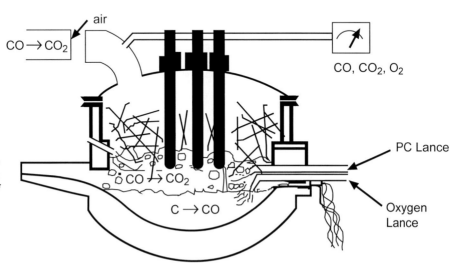

Fig. 10.70 Post-combustion in the slag layer. *(Courtesy of Praxair.)*

- Maximimum degree of post combustion and heat transfer
- PC in scrap melting and flat bath periods
- Unit power savings > 4 kWh/Nm³ O_2
- 8–10% power savings, 6–8% productivity increase

Table 10.12 Typical Post-Combustion Results for Scrap Based Operations

Location	Method	Net Efficiency	HTE %	PCR %	kWh/Nm³ O₂	Reference
Theoretical		*100%*			*5.8*	
Acciaierie Venete SpA	K-ES	40–50%		80		46
Commonwealth Steel Co Ltd	Independent	30%				42
Ferriere Nord SpA	K-ES/DANARC	40%	50+	72–75	3–4	47,48,49
Vallourec	Air Liquide	60%		66–71		50,51
Ovako Steel AB	AGA	61%		77–87	4.3	52
Nucor	Praxair	80%(lance)	80	42–75	5.04	53,54,55
		60%(burners)			3.78	53,54,55
UES Aldwerke	Air Products/ Independent	64%			3.7	56,57
Von Roll (Swiss Steel AG)	Air Liquide				2.53	58
Ferriera Valsabbia SpA	Air Liquide	Burners			1.2–1.6	58
		ALARC PC			3.0–3.5	58
Atlantic Steel	Praxair			60–70	3.7–5.13	60
Cascade Steel	Air Liquide				2.5–3.0	61
Badische Stahlwerke GmbH	Air Liquide	64%		83	3.69	58,59
Cia Siderurgica Pains SA	Korf EOF	60%		80–100		18,62,63

10.7.5.3 Theoretical Analysis of Post-Combustion

Due to the scatter of reported results to date for EAF post-combustion it is necessary to discuss the theoretical analysis to put the results into perspective. In addition, the trials in converters and smelters can help to draw conclusions.

10.7.5.3.1 Post-Combustion and Heat Transfer Mechanisms Heat transfer to a mass of solid scrap is relatively efficient due to its large surface area and the inherent large difference in temperature. However, heat transfer to the bath is more difficult and two methods are proposed. Interaction of metal and slag droplets occurs in the Hismelt process while heat transfer is through the slag phase in deep slag (DIOS and AISI) processes. In bath smelting, post-combustion energy is consumed and is not transferred to the bath. Exact mechanisms are not known at this time.

10.7.5.3.1.1 Post-Combustion in the Furnace Freeboard All of the post-combustion results indicate that high efficiency can only be achieved by effectively transferring the heat from the post-combustion products to some other material. This has been verified in several BOF post-combustion trials where scrap additions during post-combustion resulted in higher post-combustion levels.[64,65] If the combustion products do not transfer their heat, they will tend to dissociate. The best opportunity during scrap melting is to transfer the heat to cold scrap. The adiabatic flame temperature for the combustion of CO with oxygen levels off at 2800°C.[66] The adiabatic flame temperature does not increase proportionally to the amount of CO burned. Thus the amount of heat transferred from the flame does not increase proportionally.[66] The adiabatic flame temperature for oxygen/natural gas combustion is in the range of 2600–2700°C.[67,68] Natural gas combustion produces a greater volume per unit oxygen and therefore more contact with the scrap should occur. Trials with oxy-fuel and oxy-coal burners indicate that efficiencies of 75–80% can be obtained at the start of meltdown, but that efficiency drops off quickly as the scrap heats and an average efficiency of 65% can be expected.[34,69] For post-combustion, the efficiency can be expected to be similar or lower, depending on how long the process is carried out. Burner efficiency over a flat bath is 20 to 30%, setting a lower limit and as a result, heat flux to the furnace panels will increase. Air Liquide has reported PCR = 70% and HTE = 50% for flat bath operations with post-combustion. This equates to a net efficiency of 35%.

Increased electrode consumption is a possibility. High bath stir rates may help to recover more of this energy. If the heat is not transferred from the post-combustion gases then a high PCR can not be expected since some of the CO_2 will dissociate.[70,71,72] During scrap melting oil and grease burn off the scrap. CO and H_2 concentration spikes occur when scrap falls into the bath. The hydrogen from these organics tend to form hydoxyl ions which enhance the rate at which CO can be post-combusted.[67,73,74] The reaction between CO and O_2 is believed to be a branched chain reaction whose rate is greatly enhanced if a small quantity of H_2 is present. Thus higher PCRs in the EAF should be achieved during scrap meltdown. This benefit will not occur for post-combustion carried out in the slag though oxygen injected into the slag will continue to react in the furnace freeboard once it leaves the slag.

Nippon Steel found in their trials that for high offgas temperatures, radiation was the dominant heat transfer mechanism but the net amount of heat that was transferred was low.[75] For temperatures <1765°C, heat transfer by radiation and convection accounted for 30% of the heat transfer. The remainder was due to circulation of materials in the slag.

10.7.5.3.1.2 Post-Combustion in the Slag Layer Praxair has taken the approach of using subsonic or soft blowing of post-combustion oxygen through a multi-nozzle lance. The post-combustion lance is positioned close to the primary oxygen injection lance. The impact of the supersonic primary oxygen jet produces an emulsion of metal droplets in the slag. For post-combustion operations in an EAF, Praxair estimates that 0.25% of the steel bath needs to be emulsified in order to transfer all of the post-combustion heat. Praxair has found no evidence that post-combustion oxygen reacts with the metal droplets or with carbon in the slag. Post-combustion is carried out both during scrap melting and during a flat bath. During the start of scrap melting, post-combustion is difficult to achieve because oxygen lances (especially water-cooled lances) cannot be introduced into the furnace until sufficient scrap has meted back from the slag door. Information supplied by Praxair indicates that they do not attempt to burn all of the CO which is generated. Rather, they attempt to burn only a portion of the CO ranging from 10–50% of that which is generated from primary oxygen lancing.

The injection of oxygen must be localized and well controlled to avoid oxygen reaction with the electrodes. Previous trials for post-combustion in EAF slags have used a piggyback lance over top of the decarburizing oxygen lance. Thus an attempt is made to burn the CO in the slag as it leaves the bath.

The slag layer must be thick in order to avoid the reaction of CO_2 with the bath to produce CO which would make a high PCR difficult to achieve. In converter and bath smelting operations, the slag layer is considerably thicker than in the EAF. Sumitomo proposed that a deep slag layer be used with post-combustion in the upper layer of the slag where the oxygen would not react with metal droplets.[76] The temperature gradient in the slag layer was minimized with side blown oxygen into the slag. From Nippon Steel indications were that moderate bath stirring is necessary to accelerate reduction of iron ore and to enhance heat transfer from post-combustion.[75] A thick slag layer was used in order to separate the metal bath from the oxygen and the post-combustion products. A thick slag layer also helped to increase post-combustion and decrease dust formation. It might prove economical to couple slag post-combustion in the EAF with addition of DRI/HBI following scrap melting operations although DRI addition typically results in much slag foaming. Post-combustion could take place in the upper layer of the slag. Intense stirring would transfer heat back to the slag/bath interface where the DRI/HBI would be melted. This could lead to optimum energy recovery for post-combustion in the slag. Bath smelting results indicate that a net efficiency of about 50% can be achieved when CO is post-combusted in the slag.[75,77,78,79]

During scrap melting operations, post-combustion above the slag low in the furnace freeboard should give the best results. During flat bath operations, it may be advantageous to burn the CO in or just above the slag. However, post-combustion in the slag will be a much more complicated process and will require a greater degree of control over the post-combustion oxygen dispersion pattern in the slag in order to prevent undesirable reactions.

10.7.5.3.2 Electrode Consumption Previous studies have attributed electrode consumption to two mechanisms: sidewall consumption and tip erosion. Tip erosion is dependent on the electrical current being carried by the electrode. Sidewall consumption is dependent on the surface temperature of the electrode and can be controlled by providing water cooling of the electrode (sprays). During the initial stages of meltdown, the scrap helps to shield the electrodes and the hot surface will not react with oxygen present in the furnace. Once the scrap melts back from the electrodes, the hot carbon surface will react with any oxygen that is present. The source of the oxygen can be injected oxygen or oxygen that is present in air that is pulled into the EAF through the slag door.

If the electrodes are not shielded from the combustion products of post-combustion, carbon can react with CO_2 to form CO.[53] The electrode will also react with any available oxygen. Both mechanisms will increase electrode consumption. At high offgas temperatures, the amount of electrode consumption can increase drastically.

Readers are refered to Section 10.3.3 for electrode consumption factors developed by Bowman. These indicate an increase in electrode consumption for operations using large amounts of oxygen and those using post-combustion in the furnace.

10.7.5.3.3 Heatload to the EAF Shell The usefulness of the heat generated by post-combustion will be highly dependent on the effective heat transfer to the steel scrap and the bath. The theoretical adiabatic flame temperature for CO combustion with oxygen (2800°C) is similar to that for combustion of natural gas with oxygen. It is well known that when oxy-fuel burners are fired at very high rates on dense scrap, the flame can blow back against the water-cooled panels and overheat them. If the heat from post-combustion is not rapidly transferred to the scrap, overheating of the panels will also occur. If the combustion products do not penetrate the scrap, then most of the heat transfer will be radiative and the controlling rate mechanism will be conductive heat transfer within the scrap. This will be very slow compared to the radiative heat transfer to the scrap surface and will result in local overheating.

10.7.5.3.4 Iron Yield Oxy-fuel burner use can lead to yield losses and increased electrode consumption as some combustion products react with iron to form FeO. Trials run by Leary and Philbrook on the preheating of scrap with oxy-fuel burners showed that above scrap temperatures of 760°C, 2–3% yield loss occurred.[80] This is supported by work that shows that the thermodynamic equilibrium between iron and CO_2 at temperatures greater that 1377°C is 24% CO_2. Gas temperatures exceeding 1800°C are possible in the EAF and the corresponding equilibrium at this temperature is 8.6% CO_2. Though the gas residence time in the EAF probably will not allow for equilibrium to be reached, some oxidation will occur. If additional carbon is not supplied a yield loss will occur. This is likely to be the case for post-combustion. An iron yield loss of 1% equates to a power input of 12 kWh/ton. This can have a significant effect on the overall post-combustion heat balance resulting in a fictitiously high HTE for post-combustion.

As Fe is oxidized to FeO, a protective layer can form on the scrap. Once the FeO layer is formed, oxygen must diffuse through the layer in order to react with the iron underneath. This will help to protect the scrap from further oxidation if this layer does not peel off exposing the iron. At temperatures above 1300°C the FeO will tend to melt and this protective layer will no longer exist.

10.7.5.3.5 Limits on Potential Gains from Post-Combustion Bender indicates that CO emissions typically range between 0.3 and 2.6 kg/ton.[81] Air infiltration into the furnace will result in natural post-combustion. This is indicated in several EAF post-combustion studies where CO2 levels prior to the trials are in the range of 15–40% and PCR ranges from 30–60%.[44,50,53,57,59] With a post-combustion system, CO levels typically dropped to below 10%. The starting level of CO has a big effect on the efficiency obtained once post-combustion is installed. These results are presented in Table 10.13.

Table 10.13 PCR With and Without Injection of Post-Combustion Oxygen

Location	PCR prior to post-combustion of O_2	PCR after post-combustion of O_2
Vallourec	33–50%	66–71%
BSW	58–61%	83%
Nucor Plymouth	30–50%	42–75%
UES Aldewerke	40–62.5%	
Ferrierre Nord		71–75%
Venete		80%
Co-Steel Sheerness	22–62.5%	
AGA	44–66%	77–87%

CO_2 has a higher heat capacity than CO. As a result for the same offgas temperature, the CO_2 will remove more heat per Nm^3. Thus a portion of the post-combustion energy will be removed from the furnace in the CO_2. For an offgas temperature range of 1200–1400°C the maximum HTE is limited to 93.6–92%. If the slag is heated by 110°C the heat contained is equivalent to 0.9–1.2 kWh/ton (assumes 4–5% slag). These factors should be taken into account.

Some benefits are generally unaccounted for when post-combustion is implemented. A decrease of one minute in tap-to-tap time can save 2–3 kWh/ton.[82] A more recent study indicates a savings of 1 kWh/ton for highly efficient furnaces.[83] This is dependent on the efficiency of the current operation. Decreased cycle times should be considered when backing out the true heat transfer efficiency.

10.7.5.3.6 Environmental Benefits Increased volumes of injected gas will tend to decrease the amount of furnace infiltrated air. A positive pressure operation can save from 9–18 kWh/ton and some savings can be expected to result from a reduction in air infiltration depending on the operation of the offgas system. However, a reduction in NO_x due to reduced air infiltration has yet to be shown in EAF post-combustion operations. If we consider the amount of air infiltration at the start of meltdown we can see that the scrap will help to retard airflow into the furnace. Thus at the start of meltdown, perhaps the effective opening area is only 25% of the slag door area. By the end of meltdown this area is almost entirely open so now the effective opening might be 90% of the slag door area. Thus the amount of air infiltration into the furnace increases drastically over the course of the meltdown period. This will have a big affect on the effectiveness of auxilliary post-combustion oxygen added to the furnace.

Post-combustion helps the fume system capacity because the amount of heat that must be removed by the water-cooled duct is decreased. Several operations report decreased offgas temperatures when using post-combustion.[49,58] CO emissions at the baghouse are a function of offgas system design (sizing of the combustion gap, evacuation rate). Burning of some of the CO in the EAF will help to ensure that less CO enters the canopy system which will help to reduce CO emissions. Lower CO emissions have been reported in several EAF post-combustion operations.[55,58] If CO levels are reduced below 5% in the furnace, it may be difficult to complete CO combustion in the offgas system following addition of dilution air at the combustion gap. Thus controlling CO levels leaving the furnace to approximately 10% is a good practice. Alternatively a post-combustion chamber may be used within the offgas system.

It has been shown that CO tends to react with NO at high temperatures to form nitrogen and carbon dioxide as follows:[74]

$$2CO + 2NO \rightarrow 2CO_2 + N_2 \qquad (10.7.5)$$

Thus it is beneficial to have a certain amount of CO present in the furnace in order to reduce the NO emissions. This is another good reason why total post-combustion of the CO is not desireable.

Trials carried out at Dofasco on their K-OBM showed that when excessive slag foaming occurred, particulate emissions from the furnace increased by as much as 59%.[84] Nippon Steel found that a thick slag layer helped to increase post-combustion and decrease dust formation.[75] For post-combustion in the slag in the EAF, high dust generation rates should be expected unless a thick slag layer is used. Excessive stirring of the bath will also lead to increased dust generation if metal droplets react with the post-combustion oxygen.

10.7.5.3.7 Need for a Post-Combustion Chamber If an attempt is made to burn close to all of the CO in the furnace, it is likely that any CO exiting the furnace will not burn at the combustion gap because the concentration will be below the lower flammability limit. If good heat transfer is not achieved in the furnace, some of the CO_2 will dissociate to CO and O_2 and as a result the CO concentration in the offgas will be higher than anticipated. If the combustion gap is not designed accordingly, there will be insufficient combustion air and some of the CO will report downstream to the gas cleaning equipment. If chlorine and metallic oxides (catalyst) are present in the offgas stream, there is a possibility of dioxin formation. The best way to ensure that this does not happen is to install a combustion chamber followed by a water spray quench to cool the gases below the temperature at which they will dissociate and react to form dioxins. This will also help to ensure low levels of CO downstream in the offgas system.

10.7.5.4 Conclusions

The following conclusions are drawn.

1. High levels of PCR in excess of 80% have been demonstrated for EAF operations.

2. An upper limit of 65% can be expected for HTE from post-combustion when there is cold scrap present. For post-combustion above the slag this will drop to 20–30% when there is no scrap present. For post-combustion in the slag an HTE of 80% has been reported. A net efficiency of 50–60% (PCR × HTE) may be achieved if DRI/HBI is added to absorb the energy released by post-combustion or if good heat transfer is achieved via circulation of metal droplets in the slag. During scrap melting operations, post-combustion above the slag should give the best results. During flat bath operations, it appears to be advantageous to burn the CO in or just above the slag.

3. Environmental benefits due to post-combustion have been demonstrated, but additional work needs to be carried out to better understand and optimize these benefits.

4. Potential gains due to post-combustion are highly dependent on the individual EAF operation efficiencies. Most applications of post-combustion in the EAF achieve an energy savings of 20–40 kWh/ton.

5. The economic lower limit for CO level leaving the EAF needs to be established but will likely be in the range of 5–10% based on environmental considerations.

6. If it is attempted to post-combust all of the CO in the EAF, yield losses and increased electrode consumption will occur.

7. Attempts to post-combust all of the CO in the furnace will likely have a negative effect on NO_x levels.

8. Oxygen injection into the bath should start early so that post-combustion can be carried out while the scrap is still relatively cold and is capable of absorbing the heat generated. In order to be most effective decarburization oxygen needs to be distributed throughout the bath. This will help to reduce local iron oxidation in the bath (and therefore the amount of EAF dust generated) and will also distribute the CO that is generated throughout the furnace which will help to maximize energy recovery once it is post-combusted.

9. Staged post-combustion similar to that carried out in the EOF preheat chambers has the greatest potential for capturing the energy generated through post-combustion. This is because the energy from the post-combustion reaction can be transferred to the scrap thus cooling the offgas and avoiding dissociation. A similar effect can probably be achieved in the shaft furnace. In the conventional EAF it will not be possible to recover as much of this energy, and some gas dissociation will likely take place.

10. Post-combustion in the slag can result in yield losses and an increase in the amount of EAF dust generated unless a thick slag layer is used. Alternatively partial post-combustion of the CO in the slag may prove to be more effective than complete post-combustion. The additional fluxes and energy consumed to provide a thick slag layer might offset any savings from post-combustion. In some operations where scrap preheating takes place, the dust is captured on the scrap and a conventional slag cover may be acceptable for post-combustion in the slag.

11. The optimum post-combustion strategy will vary from one operation to the next. Careful deliberation must be undertaken to determine the most cost effective means of applying post-combustion to EAF operations. Frequently, the optimum approach will involve selecting specific portions of the operating cycle in which post-combustion can be applied to give the highest returns. Other issues such as operability and maintenance requirements will also help to determine the complexity of the strategy employed.

Complete post-combustion of CO will be difficult to achieve and is likely uneconomical in light of possible detrimental effects (yield loss, refractory wear, electrode consumption, damage to furnace panels). Thus a complete analysis should be carried out on each installation to determine the best post-combustion practice for that location. Such a program should include CO, CO_2, H_2, H_2O, N_2, NO_x, SO_x gas analysis at the elbow, offgas flowrate and temperature exiting the EAF and at the baghouse, offgas analysis at the baghouse, analysis of baghouse dust before and after post-combustion for increases in iron content, monitoring of iron yield, electrode consumption and other consumables, monitoring EAF cooling water temperatures, slag analysis (check iron levels), and monitoring alloy consumption (check alloy yield).

Post-combustion can be an effective tool for the EAF operator but an economical operating practice must be established based on individual site criteria. There is no universal recipe which can apply for every facility. The way in which post-combustion can be applied is highly dependent on raw materials and operating practices and these must be thoroughly evaluated when implementing a post-combustion practice.

10.7.6 EAF Bottom Stirring

For conventional AC melting of scrap there is little natural convection within the bath. Temperature gradients have been reported in the range of 40–70°C.[43] If there is limited bath movement, large pieces of scrap can take considerably longer to melt unless they are cut up as discussed previously under oxygen lancing operations. Concentration gradients within the bath can also lead to reduced reaction rates and over or under reaction of some portions of the bath.

The concept of stirring the bath is not a new one and records indicate that electromagnetic coils were used for stirring trials as early as 1933.[85] Japanese studies indicated that flow velocities are much lower for stainless steels as compared to carbon steels for electromagnetic stirring.[86] Studies indicate that electromagnetic stirring is capable of supplying sufficient stirring in some cases. However, the cost for electromagnetic stirring is high and it is difficult to retrofit into an existing operation.

Most EAF stirring operations presently in use employ gas as the stirring medium.[87,88] These operations use contact or non-contact porous plugs to introduce the gas into the furnace. In some cases tuyeres are still used. The choice of gas used for stirring seems to be primarily argon or nitrogen though some trials with natural gas and with carbon dioxide have also been attempted.[89] In a conventional EAF three plugs are located midway between the electrodes.[87,88] For smaller furnaces a single plug centrally located appears to be sufficient. In EBT and other bottom tapping operations, the furnace tends to be elliptical and the nose of the furnace tends to be a cold spot. A stirring element is commonly located in this part of the furnace to promote mixing and aid in meltdown. Some operations have also found it beneficial to inject inert gas during tapping to help push the slag back and prevent slag entrainment in the tap stream. For common steel grades the gas flowrate for a contact system is typically 0.03–0.17 Nm^3/min. (1–6 scfm) with a total consumption of 0.1–0.6 Nm^3/ton (3–20 scf/ton).[4] Non-contact systems appear to use higher gas flow rates. Service life for contact systems is in the range of 300–500 heats. Some non-contact systems have demonstrated lives of more than 4000 heats.[90] Fig. 10.71 shows several commercially available stirring elements.

Some of the main advantages attributed to bottom stirring include: reduction in carbon boils and cold bottoms, yield increases of 0.5–1%, time savings of 1–16 minutes (typical is 5 min.) per heat,

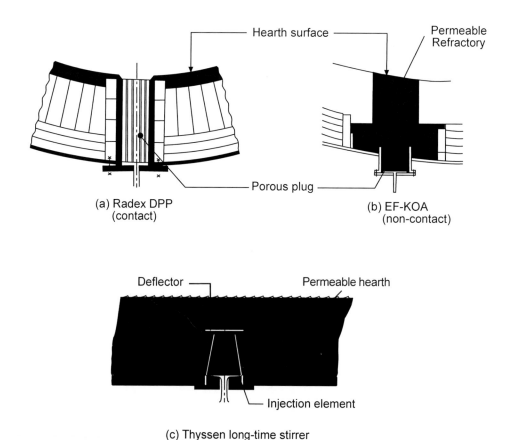

Fig. 10.71 Stirring element configurations. *(Courtesy of Center for Materials Production.)*

energy savings of up to 43 kWh/ton (typical is 10–20 kWh/ton), improved alloy recovery, increased sulfur and phosphorus removal, and reduced electrode consumption.[43,87–91]

In various operations surveyed the cost savings have ranged from $0.90 to $2.30 per ton.[89,90]

10.7.7 Furnace Electrics

In addition to increased oxygen use in the EAF, considerable effort has been devoted towards maximizing electrical efficiency. This has been partly due to the fact that there are practical limitations as to the amount of oxygen used in any one operation (due to environmental concerns). In addition, it has been realized that by using longer arc, lower current operations, it is possible to achieve much more efficient power input to the furnace. Several innovations have contributed to increased efficiencies in this area and are discussed in the following sections.

10.7.7.1 Electrode Regulation

In the days when furnaces melted small sized heats (5–20 tons), electrode regulation was not a major concern. As furnaces became bigger and operating voltages increased however, it became necessary to control the electrodes more closely in order to maximize the efficiency of power input to the furnace. Over the past five years, there have been some substantial advances made in electrode regulation.[92,93] These coupled with advances in furnace hydraulics (thus allowing for faster electrode response) have lead to considerable improvements in EAF operation. For one installation the following benefits were obtained following an upgrade of the electrode regulation system: power consumption decreased by 5%, flicker decreased by 10%, broken electrodes decreased by 90%, electrode consumption decreased by 8.5%, tap-to-tap time decreased by 18.5%, and average power input increased by 8.5 %.[92]

10.7.7.2 Current Conducting Arms

In conventional EAF design, the current is carried to the electrodes via bus tubes. These bus tubes usually contribute approximately 35% of the total reactance of the secondary electrical system.[4] Current conducting arms combine the mechanical and electrical functions into one unit. These arms carry the secondary current through the arm structure instead of through bus tubes. This results in a reduction of the resistance and reactance in the secondary circuit which allows an increase in power input rate without modification of the furnace transformer. Productivity is also increased. Current conducting arms are constructed of copper clad steel or from aluminum.[94–97] Some of the benefits attributed to current conducting arms include increased productivity, increased power input rate (5–10%), reduced maintenance and increased reliability, lower electrode consumption, and reduced electrode pitch circle diameter with a subsequent reduction in radiation to sidewalls.[94–97]

The aluminum current conducting arms, Fig. 10.72, are up to 50% lighter than conventional or copper clad steel arms. Several additional benefits are claimed including higher electrode speeds resulting in improved electrode regulation, reduced strain on electrode column components, and less mechanical wear of components.[94,97]

In comparison to the overall weight of arms and column, the weight of the arms should not be a significant factor.

10.7.8 High Voltage AC Operations

In recent years a number of EAF operations have retrofitted new electric power supplies in order to supply higher operating voltages.[34,98–101] Energy losses in the secondary circuit are dependent on the secondary circuit reactance and to a greater extent on the secondary circuit current. If power can be supplied at a higher voltage, the current will be lower for the same power input rate. Operation with a lower secondary circuit current will also give lower electrode consumption.[34] Thus it is advantageous to operate at as high a secondary voltage as is practical. Of course this is

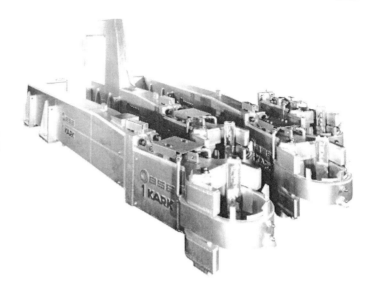

Fig. 10.72 Aluminum current conducting electrode arms. *(Courtesy of Badische Stahl Technology.)*

limited by arc flare to the sidewall and the existing furnace electrics. A good foamy slag practice can allow voltage increases of up to 100% without adversely affecting flare to the furnace sidewalls.[4] Energy losses can be minimized when reactance is associated with the primary circuit.

Supplementary reactance is not a new technology. In the past, supplementary reactors were used to increase arc stability in small furnaces where there was insufficient secondary reactance. However, in the past few years this method has been used to increase the operating voltages on the EAF secondary circuit. This is achieved by connecting a reactor in series with the primary windings of the EAF transformer.[98–100] This allows operation at a power factor of approximately $\sqrt{2}/2$ which is the theoretical optimum for maximum circuit power. This is made possible because the arc sees a large storage device in front of it in the circuit, which in effect acts as an electrical flywheel during operation. The insertion of the series reactor drops the secondary voltage to limit the amount of power transferred to the arc. In order to compensate for this, the furnace transformer secondary voltage is increased into the 900–1200V range allowing operation at higher arc voltages and lower electrode currents. Some of the benefits attributed to this type of operation are a more stable arc than for standard operations, electrode consumption reduced by 10%, secondary voltage increased by 60–80%, power savings of 10–20 kWh/ton, system power factor of approximately 0.72, furnace power factor of approximately 0.90, lower electrical losses due to lower operating current, and voltage flicker is reduced up to 40%.[4,98–101]

10.7.9 DC EAF Operations

The progress in high power semiconductor switching technology brought into existence low cost efficient DC power supplies. Due to these advances, the high power DC furnace operation became possible. North American interest in DC furnace technology is growing with several existing installations and others that are currently under installation.[102–105] The DC arc furnace is characterized by rectification of three phase furnace transformer voltages by thyristor controlled rectifiers. These devices are capable of continuously modulating and controlling the magnitude of the DC arc current in order to achieve steady operation. DC furnaces use only one graphite electrode with the return electrode integrated into the furnace bottom. There are several types of bottom electrodes: conductive hearth bottom, conductive pin bottom, single or multiple billet, and conductive fins in a monolithic magnesite hearth.[102]

All of these bottom return electrode designs have been proven. The ones that appear to be used most often are the conductive pin bottom where a number of pins are attached to a plate and form the return path and the bottom billet design. The bottom electrode is air cooled in the case of the pin type and water-cooled in the case of the billet design. The area between pins is filled with ramming mix and the tip of the pins is at the same level as the inner furnace lining. As the refractory wears, the pins also melt back. DC furnaces operate with a hot heel in order to ensure an electrical path to the return electrode. During startup from cold conditions, a mixture of scrap and slag is used to provide an initial electrical path. Once this is melted in, the furnace can be charged with scrap.

Some of the early benefits achieved with DC operation included reduced electrode consumption (20% lower than high voltage AC, 50% lower than conventional AC), reduced voltage flicker (50–60% of conventional AC operation) and reduced power consumption (5–10% lower than for AC).[102,106–110]

These results were mainly achieved on smaller furnaces which were retrofitted from AC to DC operation. However, some larger DC furnace installations did not immediately achieve the claimed benefits. Notably, two areas of concern emerged: electrode consumption and refractory consumption.

Several DC furnace operations found that the decrease in electrode consumption expected under DC operation did not occur. Much analysis by the electrode manufacturers indicated that physical conditions within the electrodes was different for AC and DC operations. As a result, for large DC electrodes carrying very large current, an increased amount of cracking and spalling was observed as compared to AC operations. Therefore, it was necessary to develop electrodes with physical properties better suited to DC operation. This is discussed in greater detail in Section 10.3 on electrodes. The economical maximum size for DC furnaces tends to be a function of limitations due to electrode size and current carrying capacity. At the present time the maximum economical size for a single graphite electrode DC furnace appears to be about 165 tons. Larger furnace sizes can be accommodated by using more than one graphite electrode. Fig. 10.73 shows furnace dimensions for a Kvaerner Clecim DC EAF.

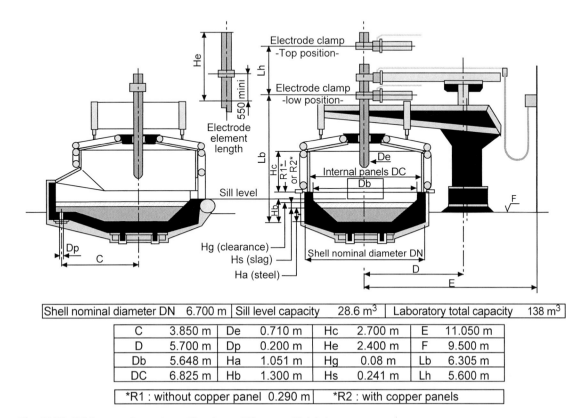

Fig. 10.73 DC furnace dimensions. *(Courtesy of Kvaerner Metals.)*

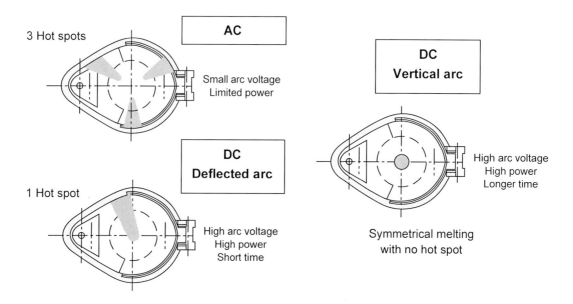

Fig. 10.74 Electric arc behavior. *(Courtesy of Kvaerner Metals.)*

Several of the early DC operations experienced problems with refractory wear and bottom electrode life. These problems were directly related to arc flare within the furnace. The anode design has the greatest influence on the arc flare. In all DC furnaces, the electric arc is deflected in the direction opposite to the power supply due to assymetries in magnetic fields which are generated by the DC circuit. Thus the arc tends to concentrate on one area within the furnace creating a hot spot and resulting in excessive refractory wear. This is shown in Fig. 10.74. Several solutions have been developed to control or eliminate arc flare. All commercial bottom electrode designs are now configured to force the arc to the center of the furnace. In the case of bottom conductive refractory and the pin type bottom, it is necessary to provide split feed lines to the bottom anode or a bottom coil which helps to modify the net magnetic field generated. In the billet bottom design, the amount of current to each billet is controlled along with the direction of anode supply in order to control the arc. The bottom fin design utilizes the fact that electrical feed occurs at several points in order control arc deflection. Quadrants located further from the rectifier are supplied with higher current than those located closer to the rectifier.

Some feel that the possibility for increased automation of EAF activities is greater for the DC furnace.[109,110] This is because with only one electrode, there is increased space both on top and within the furnace. DC furnace installations can be expected to cost from 10–35% more than a comparable AC installation.[106,107] However, calculations on payback indicate that this additional cost can be recovered in one to two years due to lower operating costs.[106,107]

Bowman conducted an analysis comparing AC and DC furnace operations and found that the electrical losses amount to approximately 4% in AC operations and 5.5% in DC operations; the difference in absolute terms is relatively insignificant.[9] The difference in total energy consumption between AC and DC furnaces is likely less than 9 kWh/ton in favor of the DC furnace, however many other variables influence the power consumption and it is difficult to develop accurate figures. DC furnaces experience roughly 25% less electrode consumption than AC furnaces, this correlating to typically 0.4 kg/ton. This difference appears to be greater for smaller AC furnaces. Flicker is approximately 60% lower for DC operations, however, advances in AC power system configurations (additional reactance) may reduce this difference to 40%.

Some typical results which have been presented for large DC EAF operations are electrode consumption of 1–2 kg/ton liquid steel, power consumption at 350–500 kWh/ton liquid steel, tap-to-tap times of 45–120 minutes, and bottom life of 1500–4000 heats. It is important to remember however, that power consumption is highly dependent on operating practices, tap temperature, use of auxiliary fuels, scrap type etc.

10.7.10 Use of Alternative Iron Sources in the EAF

Hot metal production is a standard part of operations in integrated steelmaking. Hot metal is produced in the blast furnace from iron ore pellets. This hot metal is then refined in basic oxygen furnaces to produce steel. However, several operations which were previously integrated operations are now charging hot metal to the EAF. One such installation is Cockerill Sambre in Belgium where up to 40% of the total charge weight is hot metal. This installation gets its hot metal from a blast furnace. In several other operations, hot metal is provided via Corex units, mini blast furnaces, or cupolas. In the case of the Saldahna Steel facility currently under construction in South Africa, the EAF feed will consist of 45% hot metal and 55% DRI.

10.7.10.1 Effect of Feedstocks on EAF Operation

There is a wide range of tabulated effects for various iron alternatives in the EAF. This is primarily due to the fact that within any given product such as DRI or HBI, a number of process parameters may vary quite considerably. These include % metallization, % carbon, % gangue, etc. All of these parameters will have an affect on the energy requirement to melt the material. If there is sufficient carbon to balance the amount of FeO in the DRI, the total iron content can be recovered. Approximately 1% carbon is required to balance out 6% FeO. If insufficient carbon is present, yield loss will result unless another source of carbon is added to the bath.. If excess carbon is present, it can be used as an energy source in conjunction with oxygen injection in order to reduce electrical power requirements. Generally speaking, DRI requires 100–200 additional kWh per ton to melt as compared to scrap melting. If up to 25% DRI is to be used in the charge makeup, it can be added in the bucket. If a larger percentage is to be used it can be fed continuously through the roof. One advantage of DRI is that it can be fed continuously with power on and therefore no thermal losses are incurred by opening the roof. HBI is more dense than DRI and as a result can be charged in the scrap bucket without increasing the number of charges required. DRI tends to float at the slag bath interface while HBI, which has a much higher density, tends to sink into the bath and melt in a manner similar to pig iron. The amount of silica present in the DRI will have a large effect on the economics of steel production from DRI. Silica will attack refractories unless sufficient lime is present to neutralize its effect. In general, a V-ratio of 2.5–3.0 is desired for good slag foaming. Thus the lime requirement increases greatly if silica levels in the DRI are high. Melting power requirements increase accordingly.

In the case of cold pig iron, a power savings should result from using 10–15% pig iron in the charge. This is due to the silicon and carbon contained in the pig iron. These act as a source of chemical heat in the bath when oxygen is injected. Pig iron typically contains up to 0.65% silicon which reacts with oxygen to produce silica which reports to the slag. This requires some additional lime addition in order to maintain the slag basicity. Usually, a maximum of 20% cold pig iron is used in the EAF because it takes longer to melt in than scrap, especially if it is supplied in large pieces. Small sized pieces are preferable. The pig iron can contain up to 4% carbon which results in a very high bath carbon level. Removal of this carbon with oxygen generates much heat but also requires increased blowing times because practical limits exist on the rate at which oxygen can be blown into the steel.

Iron carbide can be charged into the furnace in sacks or it can be injected. Injection is the preferred method of introducing the material into the bath as recovery is maximized. This however creates some practical limitations as to the quantity of iron carbide used since limitations on the injection rate exist. At Nucor Crawfordsville, the maximum rate achieved so far is 2500 kg per minute. Iron carbide dissolves into the steel bath and as the carbon goes into solution, it reacts with FeO or dissolved oxygen in the bath producing a very fine dispersion of carbon monoxide bubbles. These bubbles are very beneficial because they help to strip nitrogen from the bath. Depending on the degree of metallization in the iron carbide, the energy requirements for dissolution of the iron carbide also vary.

The charging of hot metal to the EAF sounds like a simple proposition though it is in fact quite complex. Care must be taken that the hot metal which is charged does not react with the highly oxidized slag which is still in the EAF. Some operations charge hot metal to the EAF by swinging the roof and pouring it into the furnace. This causes very rapid mixing of the hot heel and the highly oxidized slag in the EAF with the hot metal and sometimes explosions do occur. Thus for this mode of operation it is recommended that a slag deoxidizer be added prior to hot metal addition. Typical deoxidizers are silicon fines, aluminum fines and calcium carbide. An alternative method of charging the hot metal to the EAF is to pour it down a launder which is inserted into the side of the EAF. This method requires more time for charging of the hot metal but results in a much safer operation. Paul Wurth has recently developed a side charging system whereby the hot metal can be charged while power is on and thus the charging time is not an issue.

10.7.11 Conclusions

It is apparent that there are many technologies available for improving EAF operating efficiency. The general results for these have been listed in an attempt to provide a starting point for those in the process of upgrading operations. Of course results will vary from one installation to the next. However if a conservative approach is taken and the median of the reported results is used for calculation purposes, the results can be expected to be achievable. It is important to evaluate the effects of certain processes both on furnace operations and on other systems such as fume control. The use of substitute fuels (oxygen and natural gas) may be limited by the capacity of the fume system. If upgrades are to be made, one must also evaluate the need for upgrades to auxiliary systems. The data presented in this section provides a starting point for the person evaluating process changes to improve EAF efficiency. The technologies that have been reviewed are well proven. The interaction between these processes has not been evaluated, though in some cases the blending of these operations can prove to be most beneficial (eg. oxygen lancing, slag foaming and CO post-combustion). When evaluating upgrade requirements for a particular operation, it is necessary to clearly list the objectives and then match these with suitable technologies. It is important to maintain a global perspective regarding overall costs and operations in order to arrive at true optimal operating efficiency in the EAF.

10.8 Furnace Operations

10.8.1 EAF Operating Cycle

The electric arc furnace operates as a batch process. Each batch of steel that is produced is known as a heat. The electric arc furnace operating cycle is known as the tap-to-tap cycle. The tap-to-tap cycle is made up of the following operations: furnace charging, melting, refining, de-slagging, tapping and furnace turnaround. Modern operations aim for a tap-to-tap cycle of less than 60 minutes. With the advance of EAF steelmaking into the flat products arena, tap-to-tap times of 35–40 minutes are now being sought with twin shell furnace operations.

A typical 60 minute tap-to-tap cycle is :

first charge	3 minutes
first meltdown	20 minutes
second charge	3 minutes
second meltdown	14 minutes
refining	10 minutes
tapping	3 minutes
turnaround	<u>7 minutes</u>
Total	**60 minutes**

10.8.2 Furnace Charging

The first step in the production of any heat is to select the grade of steel to be made. Usually a heat schedule is developed prior to each production shift. Thus the melter will know in advance the schedule for the shift. The scrap yard operators will batch buckets of scrap according to the needs of the melter. Preparation of the charge bucket is an important operation, not only to ensure proper melt-in chemistry but also to ensure good melting conditions. The scrap must be layered in the bucket according to size and density in order to ensure rapid formation of a liquid pool in the hearth while also providing protection of the sidewalls and roof from arc radiation. Other considerations include minimization of scrap cave-ins which can break electrodes and ensuring that large heavy pieces of scrap do not lie directly in front of burner ports which would result in blow-back of the flame onto the water-cooled panels. The charge can include lime and carbon or these can be injected into the furnace during the heat. Many operations add some lime and carbon in the scrap bucket and supplement this with injection.

The first step in any tap-to-tap cycle is charging of the scrap. The roof and electrodes are raised and are swung out to the side of the furnace to allow the scrap charging crane to move a full bucket of scrap into place over the furnace. The bucket bottom is usually a clam shell design—i.e. the bucket opens up by retracting two segments on the bottom of the bucket, see Fig. 10.75. Another common configuration is the "orange peel" design. The scrap falls into the furnace and the scrap crane removes the scrap bucket. The roof and electrodes swing back into place over the furnace. The roof is lowered and then the electrodes are lowered to strike an arc on the scrap. This commences the melting portion of the cycle. The number of charge buckets of scrap required to produce a heat of steel is dependent primarily on the volume of the furnace and the scrap density. Most modern furnaces are designed to operate with a minimum of backcharges. This is advantageous because charging is dead time, whereby the furnace does not have power on and therefore is not melting. Minimizing these dead times helps to maximize the productivity of the furnace. In addition, energy is lost each time the furnace roof is opened. This can amount to 10–20 kWh/ton for each occurrence. Most operations aim for 2–3 buckets of scrap per heat and will attempt to blend their scrap to meet this requirement. Some operations achieve a single bucket charge. Continuous charging operations such as Consteel and the Fuchs shaft furnace eliminate the charging cycle.

Fig. 10.75 Clam shell bucket charging scrap to the EAF. *(Courtesy of SMS GHH.)*

10.8.3 Melting

The melting period is the heart of EAF operations. The EAF has evolved into a highly efficient melting apparatus and modern designs are focused on maximizing its melting capacity. Melting is accomplished by supplying energy to the furnace interior. This energy can be electrical or chemical. Electrical energy is supplied via the graphite electrodes and is usually the largest contributor in melting operations. Initially, an intermediate voltage tap is selected until the electrodes can bore into the scrap. Usually, light scrap is placed on top of the charge to accelerate bore-in. After a few minutes, the electrodes will have penetrated the scrap sufficiently that a long arc (high voltage) tap can be used without fear of radiation damage to the roof. The long arc maximizes the transfer of power to the scrap and a liquid pool of metal will form in the furnace hearth. Approximately 15% of the scrap is melted during the initial bore-in period. At the start of melting the arc is erratic and unstable. Wide swings in current are observed accompanied by rapid movement of the electrodes. As the furnace atmosphere heats up the arcing tends to stabilize and once the molten pool is formed, the arc becomes quite stable and the average power input increases.

Chemical energy can be supplied via several sources such as oxy-fuel burners and oxygen lancing. Oxy-fuel burners burn natural gas using oxygen or a blend of oxygen and air. Heat is transferred to the scrap by radiation and convection. Heat is transferred within the scrap by conduction. In some operations, oxygen is used to cut scrap. Large pieces of scrap take longer to melt into the bath than smaller pieces. A consumable pipe lance can be used to cut the scrap. The oxygen reacts with the hot scrap and burns iron to produce intense heat for cutting the scrap. Once a molten pool of steel is generated in the furnace, oxygen can be lanced directly into the bath. This oxygen will react with several components in the bath including, aluminum, silicon, manganese, phosphorus, carbon and iron. All of these reactions are exothermic (i.e. they generate heat) and will supply energy to aid in the melting of the scrap. The metallic oxides which are formed will eventually reside in the slag. The reaction of oxygen with carbon in the bath will produce carbon monoxide which may burn in the furnace if there is oxygen available. Otherwise the carbon monoxide will carry over to the direct evacuation system. Auxiliary fuel operations are discussed in more detail in Section 10.7.

Once enough scrap has been melted to accommodate the second charge, the charging process is repeated. After the final scrap charge is melted, the furnace sidewalls can be exposed to high radiation from the arc. As a result, the voltage must be reduced. Alternatively, creation of a foamy slag will allow the arc to be buried and will protect the furnace shell. In addition, a greater amount of energy will be retained in the slag and is transferred to the bath resulting in greater energy efficiency. When the final scrap charge is fully melted, flat bath conditions are reached. At this point, a bath temperature and a chemical analysis sample will be taken. The analysis of the bath chemistry will allow the melter to determine the amount of oxygen to be blown during refining. The melter can also start to arrange for the bulk tap alloy additions to be made. These quantities are confirmed following refining.

10.8.4 Refining

Refining operations in the electric arc furnace have traditionally involved the removal of phosphorus, sulfur, aluminum, silicon, manganese and carbon. In recent times, dissolved gases in the bath have also become a concern, especially nitrogen and hydrogen levels. Traditionally, refining operations were carried out following meltdown , i.e. once a flat bath was achieved. These refining reactions are all dependent upon oxygen being available. Oxygen was lanced at the end of meltdown to lower the bath carbon content to the desired level for tapping. Most of the compounds which are to be removed during refining have a higher affinity for oxygen than carbon. Thus the oxygen will preferentially react with these elements to form oxides which will report to the slag.

In modern EAF operations, especially those operating with a hot heel, oxygen may be blown into the bath throughout the heat cycle. As a result, some of the refining operations occur concurrent with melting.

Most impurities such as phosphorous, sulfur, silicon, aluminum and chromium are partially removed by transfer to the slag. These refining reactions are discussed in detail in Chapter 2. In particular the equilibrium partition ratio between metal and slag are given as functions of slag chemistry and temperature.

The slag in an EAF operation will, in general, have a lower basicity than that for oxygen steelmaking. In addition, the quantity of slag per ton of steel will also be lower in the EAF. Therefore the removal of impurities in the EAF is limited. A typical slag composition is presented in Table 10.14.

Table 10.14 Typical Slag Constituents

Component	Source	Composition Range
CaO	Charged	40–60%
SiO_2	Oxidation product	5–15
FeO	Oxidation product	10–30%
MgO	Charged as dolomite	3–8%
CaF_2	Charged slag fluidizer	
MnO	Oxidation product	2–5%
S	Absorbed from steel	
P	Oxidation product	

Once these materials enter into the slag phase they will not necessarily stay there. Phosphorus retention in the slag is a function of the bath temperature, the slag basicity and FeO levels in the slag. At higher temperature or low FeO levels, the phosphorus will revert from the slag back into the bath. Phosphorus removal is usually carried out as early as possible in the heat. Hot heel practice is very beneficial for phosphorus removal because oxygen can be lanced into the bath while the bath temperature is quite low. Early in the heat the slag will contain high FeO levels carried over from the previous heat. This will also aid in phosphorus removal. High slag basicity (i.e. high lime content) is also beneficial for phosphorus removal but care must be taken not to saturate the slag with lime. This will lead to an increase in slag viscosity which will make the slag less effective for phosphorus removal. Sometimes fluorspar is added to help fluidize the slag. Gas stirring is also beneficial because it will renew the slag/metal interface which will improve the reaction kinetics.

In general, if low phosphorus levels are a requirement for a particular steel grade, the scrap is selected to give a low level at melt-in. The partition ratio of phosphorus in the slag to phosphorus in the bath ranges from 5.0–15.0. Usually the phosphorus is reduced by 20–50% in the EAF. However, the phosphorous in the scrap is low compared to hot metal and therefore this level of removal is acceptable. For oxygen steelmaking higher slag basicity and FeO levels give a partition ratio of 100 and with greater slag weight up to 90% of the phosphorous is removed.

Sulfur is removed mainly as a sulfide dissolved in the slag. The sulfur partition between the slag and metal is dependent on the chemical analysis and temperature of the slag (high basicity is better, low FeO content is better), slag fluidity (high fluidity is better), the oxidation level of the steel (which should be as low as possible), and the bath composition. Generally the partition ratio is 3.0–5.0 for EAF operations.

It can be seen that removal of sulfur in the EAF will be difficult especially given modern practices where the oxidation level of the bath is quite high. If high lime content is to be achieved in the slag, it may be necessary to add fluxing agents to keep the slag fluid. Usually the meltdown slag must be removed and a second slag built. Most operations have found it to be more effective to carry out desulfurization during the reducing phase of steelmaking. This means that desulfurization is commonly carried out during tapping (where a calcium aluminate slag is built) and during ladle furnace operations. For reducing conditions where the bath has a much lower oxygen activity, distribution ratios for sulfur of 20–100 can be achieved.

Control of the metallic constituents in the bath is important as it determines the properties of the final product. Usually, the melter will aim for lower levels in the bath than are specified for the final product. Oxygen reacts with aluminum, silicon and manganese to form metallic oxides which are slag components. These metallics tend to react before the carbon in the bath begins to react with the oxygen, see Table 2.1 in Chapter 2. These metallics will also react with FeO resulting in a recovery of iron units to the bath. For example:

$$Mn + FeO = MnO + Fe \qquad (10.8.1)$$

Manganese will typically be lowered to about 0.06% in the bath.

The reaction of carbon with oxygen in the bath to produce CO is important as it supplies energy to the bath and also carries out several important refining reactions at the same time. In modern EAF operations, the combination of oxygen with carbon can supply between 30 and 40% of the net heat input to the furnace. Evolution of carbon monoxide is very important for slag foaming. Coupled with a basic slag, CO bubbles will help to inflate the slag which will help to submerge the arc. This gives greatly improved thermal efficiency and allows the furnace to operate at high arc voltages even after a flat bath is reached. Submerging the arc helps to prevent nitrogen from being exposed to the arc where it will dissociate and become dissolved in the steel.

If the CO is evolved within the bath, it will also remove nitrogen and hydrogen from the steel. The capacity for nitrogen removal is dependent on the amount of CO generated in the metal. Nitrogen levels as low as 50 ppm can be achieved in the furnace prior to tap. Bottom tapping is beneficial for maintaining low nitrogen levels as tapping is fast and a tight tap stream is maintained. A high oxygen content in the steel is beneficial for reducing nitrogen pickup at tap as compared to deoxidation of the steel at tap.

At 1600°C, the maximum solubility of nitrogen in pure iron is 450 ppm. Typically, the nitrogen levels in the steel following tapping are 80–100 ppm. An equation describing the solubility of nitrogen in the steel is given in Chapter 2.

Decarburization is also beneficial for the removal of hydrogen. It has been demonstrated that decarburizing at a rate of 1% per hour can lower hydrogen levels in the steel from 8 ppm down to 2 ppm in ten minutes.

At the end of refining, a bath temperature measurement and a bath sample are taken. If the temperature is too low, power may be applied to the bath. This is not a big concern in modern meltshops where temperature adjustment is carried out in the ladle furnace.

10.8.5 Deslagging

Deslagging operations are carried out to remove impurities from the furnace. During melting and refining operations, some of the undesirable materials within the bath are oxidized and enter the slag phase.

Thus it is advantageous to remove as much phosphorus into the slag as early in the heat as possible (i.e. while the bath temperature is still low). Then the slag is poured out of the furnace through the slag door. Removal of the slag eliminates the possibility of phosphorus reversion.

During slag foaming operations, carbon may be injected into the slag where it will reduce FeO to metallic iron and will generate carbon monoxide which helps to inflate the slag. If the high phosphorus slag has not been removed prior to this operation, phosphorus reversion will occur.

Fig. 10.76 EBT furnace during tapping. *(Courtesy of Fuchs.)*

10.8.6 Tapping

Once the desired bath composition and temperature are achieved in the furnace, the taphole is opened and the furnace is tilted so that the steel can be poured into a ladle for transfer to the next batch operation (usually a ladle furnace or ladle station). During the tapping process bulk alloy additions are made based on the bath analysis and the desired steel grade. Deoxidizers may be added to the steel to lower the oxygen content prior to further processing. This is commonly referred to as blocking the heat or killing the steel. Common deoxidizers are aluminum or silicon in the form of ferrosilicon or silicomanganese. Most carbon steel operations aim for minimal slag carryover. A new slag cover is built during tapping. For ladle furnace operations, a calcium aluminate slag is a good choice for sulfur control. Slag forming compounds are added in the ladle at tap so that a slag cover is formed prior to transfer to the ladle furnace. Additional slag materials may be added at the ladle furnace if the slag cover is insufficient. Fig. 10.76 shows an EBT furnace tapping into a ladle.

10.8.7 Furnace Turnaround

Furnace turnaround is the period following completion of tapping until the first scrap charge is dropped in the furnace for the next heat. During this period, the electrodes and roof are raised and the furnace lining is inspected for refractory damage. If necessary, repairs are made to the hearth, slagline, taphole and spout. In the case of a bottom tapping furnace, the taphole is filled with sand. Repairs to the furnace are made using gunned refractories or mud slingers. In most modern furnaces, the increased use of water-cooled panels has reduced the amount of patching or fettling required between heats. Many operations now switch out the furnace bottom on a regular basis (every 2–6 weeks) and perform the maintenance off-line. This reduces the power-off time for the

EAF and maximizes furnace productivity. Furnace turnaround time is generally the largest dead time in the tap-to-tap cycle. With advances in furnace practices this has been reduced from 20 minutes to less than five minutes in some newer operations.

10.8.8 Furnace Heat Balance

To melt steel scrap, it takes a theoretical minimum of 300 kWh/ton. To provide superheat above the melting point of 1520°C (2768°F) requires additional energy and for typical tap temperature requirements, the total theoretical energy required usually lies in the range of 350–370 kWh/ton. However, EAF steelmaking is only 55–65% efficient and as a result the total equivalent energy input is usually in the range of 560–680 kWh/ton for most modern operations. This energy can be supplied from a number of sources including electricity, oxy-fuel burners and chemical bath reactions. The typical distribution is 60–65%, 5–10% and 30–40% respectively. The distribution selection will be highly dependent on local material and consumable costs and tends to be unique to the specific meltshop operation. A typical balance for both older and more modern EAFs is given in the Table 10.15.

Table 10.15 Typical Energy Balance for EAFs

		UHP Furnace	Low to Medium Power Furnace
INPUTS	Electrical Energy	50–60%	75–85%
	Burners	5–10%	
	Chemical Reactions	30–40%	15–25%
	TOTAL INPUTS	**100%**	**100%**
OUTPUTS	Steel	55–60%	50–55%
	Slag	8–10%	8–12%
	Cooling Water	8–10%	5–6%
	Miscellaneous	1–3%	17–30%
	Offgas	17–28%	7–10%
	TOTAL OUPUTS	**100%**	**100%**

Several factors are immediately apparent from these balances. Much more chemical energy is being employed in the EAF and correspondingly, electrical power consumption has been reduced. Furnace efficiency has improved with UHP operation as indicated by the greater percentage of energy being retained in the steel. Losses to cooling water are higher in UHP operation due to the greater use of water-cooled panels. Miscellaneous losses such as electrical inefficiencies were much greater for older, low powered operations. Energy loss to the furnace offgas is much greater in UHP furnace operation due to greater rates of power input and shorter tap-to-tap times.

Of course the figures in Table 10.15 are highly dependent on the individual operation and can vary considerably from one facility to another. Factors such as raw material composition, power input rates and operating practices (e.g. post-combustion, scrap preheating) can greatly alter the energy balance. In operations utilizing a large amount of charge carbon or high carbon feed materials, up to 60% of the energy contained in the offgas may be calorific due to large quantities of uncombusted carbon monoxide. Recovery of this energy in the EAF could increase energy input by 8–10%. Thus it is important to consider such factors when evaluating the energy balance for a given furnace operation.

The International Iron and Steel Institute (IISI) classifies EAFs based on the power supplied per ton of furnace capacity. For most modern operations, the design would allow for at least 500 kVA per ton of capacity. The IISI report on electric furnaces indicates that most new installations allow for 900–1000 kVA per ton of furnace capacity.[18] Most furnaces operate at a maximum power factor of 0.85. Thus the above transformer ratings would correspond to a maximum power input of 0.75–0.85 MW per ton of furnace capacity.

10.9 New Scrap Melting Processes

Over the past ten years the steelmaking world has seen many changes in operating practices and the utilization of new process concepts in an attempt to lower operating costs and to improve product quality. In addition many new alternatives have been presented as lower cost alternatives to conventional AC EAF melting. Some of the specific objectives of these processes include lowering specific capital costs, increasing productivity, and improving process flexibility.

All of these processes share one or more of the following features in common. Energy from the waste offgas is used to preheat the scrap. Carbon is added to the bath and is later removed by oxygen injection in order to supply energy to the process. An attempt is made to combust the CO generated in the process to maximize energy recovery. An attempt is made to maximize power-on time and minimize turnaround time.

The EOF was one of the first scrap melting processes to use hot metal, post-combustion and scrap preheating and is described in Chapter 13.

Many of these process have their roots associated with scrap preheating. It is only fitting that this be discussed first in order to provide the groundwork for the development of these processes.

10.9.1 Scrap Preheating

Scrap preheating has been used for over 30 years to offset electrical steel melting requirements primarily in regions with high electricity costs such as Japan and Europe. Scrap preheating involves the use of hot gas to heat scrap in the bucket prior to charging. The source of the hot gases can be either offgases from the EAF or gases from a burner. The primary energy requirement for the EAF is for heating of the scrap to its melting point. Thus energy can be saved if scrap is charged to the furnace hot. Preheating of scrap also eliminates the possibility of charging wet scrap which eliminates the possibility of furnace explosions. Scrap preheating can reduce electrical consumption and increase EAF productivity.

Some suppliers have noted that there is a maximum preheat temperature beyond which further efforts to heat the scrap lead to diminished returns. This temperature lies in the range of 540–650°C. It is estimated that by preheating the scrap to a temperature of 425–540°C, a total of 63–72 kWh/ton of electrical energy can be saved. Early scrap preheaters used independent heat sources. The scrap was usually heated in the scrap bucket. Energy savings reported from this type of preheating were as high as 40 kWh/ton with associated reductions in electrode and refractory consumption due to reduced tap-to-tap times.[4,18]

As fourth hole offgas systems were developed, attempts were made to use the EAF offgas for scrap preheating. A side benefit that was reported was that the amount of baghouse dust decreased because the dust was sticking to the scrap during preheating. Scrap preheating with furnace offgas is difficult to control due to the variation in offgas temperature throughout the melting cycle. In addition a temperature gradient forms within the scrap. Temperatures must be controlled to prevent damage to the scrap bucket and in order to prevent burning or sticking of fine scrap within the bucket. Scrap temperatures can reach 315–450°C (600–850°F) though this will only occur at the hot end where the offgas first enters the preheater. Savings are typically only in the neighborhood of 18–23 kWh/ton.[4] In addition, as operations become more efficient and tap-to-tap times are decreased, scrap preheating operations become more and more difficult to maintain. Scrap handling operations can actually lead to reduced productivity and increased maintenance costs. At Badische Stalwerke the energy savings due to scrap preheating were decreased by 50% when tap-to-tap time was reduced by one third.

Some of the benefits attributed to scrap preheating are increased productivity by 10–20%, reduced electrical consumption, removal of moisture from the scrap, and reduced electrode and refractory consumption per unit production. Some drawbacks to scrap preheating are that volatiles are removed from the scrap, creating odors and necessitating a post-combustion chamber downstream. In addition spray quenching following post-combustion is required to prevent recombination of dioxins and furans. Depending on the preheat temperature, buckets may have to be refractory lined.

10.9.2 Preheating With Offgas

Preheating with offgas from the EAF requires that the offgas be rerouted to preheat chambers which contain loaded scrap buckets. The hot gases are passed through the buckets thus preheating the scrap. For tap-to-tap times less than 70 minutes the logistics of scrap preheating lead to minimal energy savings that will not justify the capital expense of a preheating system. Typical savings are in the range of 15–20 kWh/ton. Some examples of preheating systems are given in Fig. 10.77 and Fig. 10.78.

10.9.3 Natural Gas Scrap Preheating

Natural gas scrap preheating originated in the 1960s and usually involves a burner mounted in a refractory lined roof which sits over the top of the scrap bucket. Scrap is typically preheated to 540–650°C. Above 650°C, scrap oxidation becomes a problem and yield loss becomes a factor. Advantages of this form of scrap preheating are that the preheating process is decoupled from EAF operations and as a result the process is unaffected by tap-to-tap time. However, heat is provided by natural gas as opposed to offgas and as a result an additional cost is incurred.

One of the primary concerns with scrap preheating is that oil and other organic materials associated with the scrap tend to evaporate off during preheating. This can lead to discharge of hydro-

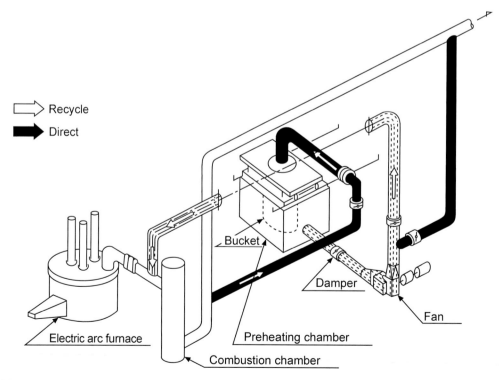

Fig. 10.77 NKK scrap preheater. *(Courtesy of NKK.)*

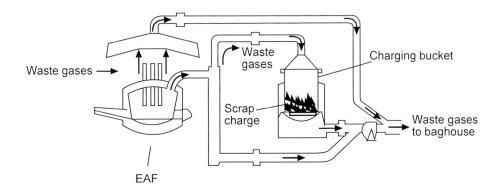

Fig. 10.78 Scrap preheating in the scrap bucket. *(Courtesy of Center for Materials Production.)*

carbons to the atmosphere and foul odors in the shop environment. In some Japanese operations, this has been remedied by installing a post-combustion chamber following scrap preheating operations. In one such operation scrap preheating is conducted in conjunction with a furnace enclosure. It is reported that power savings are 36–40 kWh/ton and electrode consumption is reduced by 0.4–0.6 kg/ton. It is not specified what the additional capital costs and maintenance costs were for this system. In North America, most operations find that the savings offered by conventional scrap preheating do not compensate for the additional handling operations and additional maintenance requirements.

10.9.4 K-ES

The K-ES process is a technology developed jointly by Klockner Technology Group and Tokyo Steel Manufacturing group. Subsequently the rights to the process were acquired by VAI. The process uses pulverized or lumpy coal in the bath as a source of primary energy. Oxygen is injected into the bath to combust the coal to CO gas. The CO gas is post-combusted in the furnace freeboard with additional oxygen to produce CO_2. Thus a large portion of the calorific heat in the process is recovered and is transferred to the bath. In addition the stirring action caused by bottom gas injection results in better bath mixing and as a result accelerated melting of the scrap. The process is shown schematically in Fig. 10.79. Fig. 10.80 shows the carbon and oxygen injectors used in K-ES. Fig. 10.81 presents oxygen consumption versus electrical power consumption.

10.9.4.1 Historical Development

The first K-ES installation was at Tokyo steel in 1986, where a 30 ton EAF was converted to run trials on the process. A productivity increase of 20% and an electrical savings of approximately 110 kWh/ton resulted. Thus the process concept was proven on an industrial scale.[18]

In 1988, Ferriere Nord in Italy decided to install the K-ES process on its 88 ton EAF in Osoppo. At that time Ferriere Nord was producing approximately 550,000 tons per year of steel on a 1975 vintage EAF that was originally designed to produce 220,000 tons per year. This had been accomplished through a combination of long arc/foamy slag practice, high oxygen utilization (32 Nm^3/ton), water-cooled furnace roof and walls and oxy-fuel burners. The first K-ES heat took place early in 1989 following the installation of post-combustion lances, coal injection equipment and bottom injection tuyeres. This process continues to operate at Ferriere Nord.

In December 1989 another K-ES facility was installed on a 82 ton oval tapping furnace at Acciaierie Venete in Padua, Italy. This operation differs in that the A phase electrode is hollow and is used for carbon injection. This installation has seven post-combustion lances that are mounted into the furnace wall, through the water-cooled panels. Five bottom tuyeres are located to form a pitch circle positioned between the furnace wall and the electrode pitch circle. The operation at

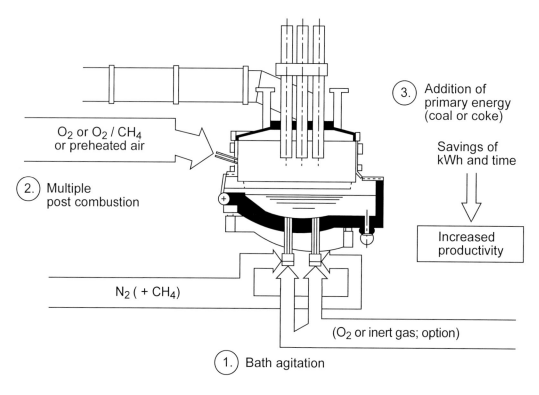

Fig. 10.79 K-ES process. *(Courtesy of Voest Alpine Ind.)*

Venete has achieved a productivity of 2.2 tons/MW. Tap-to-tap times average 54 minutes even though the transformer is only capable of supply a maximum of 45 MW.[41,46]

10.9.4.2 Results

At Tokyo Steel the average power consumption for a set of trials was 255 kWh/ton with an average coal injection rate of 24 kg/ton. This gave an electrical power replacement of 4–5.5 kWh/kg of coal injected. In addition the melting time was reduced from 60 minutes to 48–50 minutes. Ferrosilicon and ferromanganese consumption was also decreased by 1.1 kg/ton. Results are presented in Table 10.16.

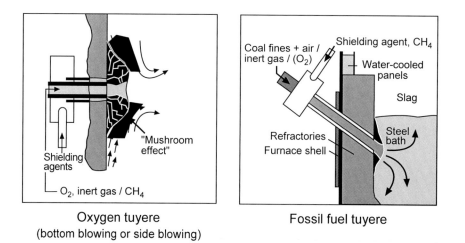

Fig. 10.80 K-ES carbon and oxygen injectors. *(Courtesy of Voest Alpine Ind.)*

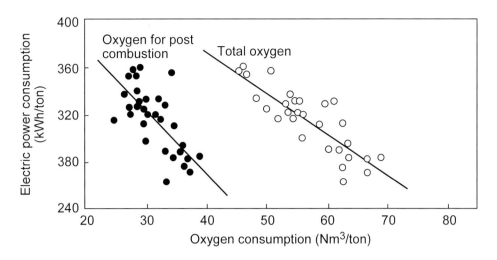

Fig. 10.81 Total oxygen consumption versus electrical power consumption for K-ES operations.

Table 10.16 K-ES Analysis Based on Tokyo Steel Results

Material	Consumption/ton liquid steel	Difference
Power on time	58 minutes	−26 minutes
Tap-to-tap time	79 minutes	−26 minutes
Production	476,000 tons/year	97,000 tons/year
Coal	32 kg/ton	+27 kg/ton
Oxygen	63 Nm³/ton	+53 Nm³/ton
Electricity	330 kWh/ton	−150 kWh/ton
Lime		+5.0 kg/ton
Electrodes		−0.3 kg/ton
FeMn		−1.0 kg/ton
Inert Gas	6.0 Nm³/ton	+6.0 Nm³/ton
Production Increase		+33%

Following installation and testing of the K-ES system at Tokyo Steel, an 88 ton furnace at Ferriere Nord was converted to K-ES operation. Lumpy coal is added in the charge and additional powdered coal is injected into the bath. Oxygen consumption is 50 Nm^3/ton. The distribution of this oxygen is as follows: 13 Nm^3/ton is injected to the bath via tuyeres, 13 Nm^3/ton is injected to the bath via an oxygen lance and 18 Nm^3/ton is injected into the freeboard through post-combustion lances. Typically carbon consumption is on the order of 20 kg/ton unless the charge is 30% hot pig metal, in which case the carbon consumption is 13 kg/ton. Electrical energy consumption decreased by 60 kWh/ton with the K-ES operation. An energy balance showed that the electrical savings were highly dependent on efficient post-combustion of the process gases in the freeboard.

Best results at Ferriere Nord were achieved at conditions of 180 kWh/ton, 53 Nm^3 O$_2$/ton (1850 scf O$_2$/ton), 30% pig iron and tap-to-tap time of 35 minutes and at conditions of 255 kWh/ton, 40 Nm^3 O$_2$/ton (1400 scf O$_2$/ton), 100% scrap and tap-to-tap time of 53 minutes.

Similar results have been obtained at Acciaerie Venete using a Fuchs OBT furnace with K-ES.

10.9.5 Danarc Process

The Danieli Danarc process combines high impedance technology with high chemical energy input to the furnace in order to achieve high productivity and energy efficiency. The first installation of this technology was at Ferriere Nord in Italy. Features for chemical energy input are very similar to K-ES. Bottom tuyeres are used to inject oxygen and carbon. Sidewall lances are also used. Post-combustion oxygen is supplied via burners. Fig. 10.82 shows a top view of the Danarc furnace at Ferriere Nord.

The purpose of tuyeres installed in the furnace bottom is to distribute the oxygen throughout the furnace in order to maximize the decarburization rate. With just oxygen lancing, the area around the lance becomes depleted of carbon and some of the oxygen reacts with iron. CO is generated mostly in one part of the furnace around the injection point. With multiple tuyeres, CO generation is spread out within the furnace and the potential for heat recovery via post-combustion is greater. Injection at several points also gives good bath mixing, an added benefit. Multiple sidewall carbon injectors allow for good control of slag foaming across the whole surface of the bath.

Natural gas, nitrogen and carbon dioxide are used as shroud gas for the oxygen tuyeres. Wear rates are approximately 0.5 mm per heat.[47,49]

Traditionally, a series reactor has been installed on the primary side of the furnace transformer to allow for high impedance operation. This allows the furnace to operate at long arc (i.e. high voltage) and low electrode current which gives high arc stability, improved power input to the bath and reduced electrode consumption.

Alternatively, the saturable reactor provides a method to reduce both current and reactive power fluctuation. This reduces the electrodynamic stresses on the furnace transformer secondary circuit and reduces the flicker level in the supply network. The main purpose of the saturable reactor is to control the reactance so as to minimize current fluctuations. The saturable reactor acts as a variable reactance controlled by the excitation current. The current control performance becomes similar to a DC furnace with thyristor controlled current. Several advantages resulting from the use of the saturable reactor are that current fluctuations and electrodynamic stresses are limited, reactive power

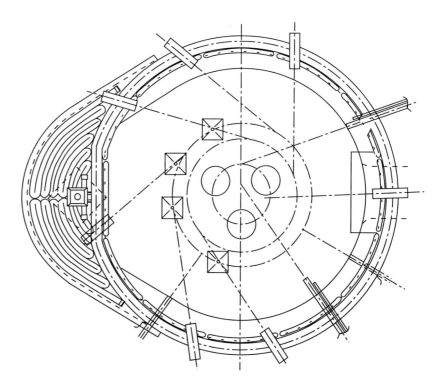

Fig. 10.82 Top view of a Danarc furnace. *(Courtesy of Danieli.)*

fluctuations are reduced resulting in less flicker, energy transfer to the melt is improved, reduced electrode consumption, and tap changing is not necessarily required.

The following results are reported for Ferriere Nord for a charge makeup of 82% scrap and 18% cold pig iron. Tap-to-tap time of 50 minutes was achieved, with an aggregate power-on time of 40 minutes. Power consumption was 270 kWh/ton liquid. Electrode consumption was 1.6 kg/ton liquid. Total oxygen consumption was 50 Nm^3/ton liquid while total natural gas consumption was 10 Nm^3/ton liquid. Total carbon introduced was 10 kg/ton liquid. The resultant tap temperature was 1640°C. For 100% scrap operation, the power consumption increases by 23 kWh/ton liquid.[47,49]

Trials have also been run using hot metal, with 70% scrap and 30% hot pig iron as part of the charge. The results were a tap-to-tap time of 45 minutes and a power-on time of 38 minutes. Power consumption was 160 kWh/ton liquid. Electrode consumption was 1.0 kg/ton liquid. Total oxygen consumption was 40 Nm^3/ton liquid and total natural gas consumption was 2.2 Nm^3/ton liquid. Air supplied to burners was 9 Nm^3/ton liquid and total carbon injected was 14.9 kg/ton liquid. Resultant tap temperature was 1680°C.[47,49]

10.9.6 Fuchs Shaft Furnace

The need to reduce the amount of power input into EAF operations lead to a prototype study of a shaft furnace at Danish Steel Works Ltd (DDS).[18,112] The concept was to load scrap into a shaft where it would be preheated by offgases exiting the EAF. The scrap sat in a column at one end of the furnace and was constantly fed into the furnace as the scrap at the bottom of the column melted away.

In January 1990, a production shaft preheater was retrofitted to one of DDS's two 125 ton EAFs. DDS stopped using the shaft after less than two years of operation. This was partly due to the problems of maintaining a suitable scrap flow in the shaft. In addition, Danish tariff regulations on electricity made a three shift per day operation uneconomical. As a result DDS could not move to a single furnace operation which would make full use of the preheating shaft.

10.9.6.1 Single Shaft Furnace

The Fuchs shaft furnace concept was installed at Co-Steel Sheerness[113,114] in England in 1992. This installation is the outcome of work done in Denmark at DET Danske Stalvalsevaerk on first and second versions of the process. The process has a shaft with a conventional oval EAF bottom with three phase AC arcs. Fig. 10.83 illustrates the shaft. The relatively short shaft is scrap bucket fed. Unlike the EOF, there are no movable fingers and the scrap descends continuously into the bath where the arcs and oxygen produce the unrefined metal. The Fuchs shaft preheater at Sheerness consists of a reverse taper (larger at bottom) shaft which sits on the furnace roof, offset from the furnace centerline.

The furnace shell is mounted on a frame and is tilted by means of four hydraulic cylinders mounted in each corner of the frame, enabling the furnace shell to be lowered by one foot to the first bucket charging position or to be tilted into the tapping or deslagging positions. The furnace runs on four tracks, two in the center and two on the outside. The furnace water-cooled roof, the shaft and the electrode gantry are fixed and do not tilt.

Energy input to the furnace is from an 80 MVA transformer, 6 MW oxy-fuel burners and a water-cooled oxygen lance.

A typical operating cycle is as follows. The heat cycle is started by charging the first basket of scrap. The furnace is lowered one foot onto a bumper and the furnace car is moved west to the charging position where the first bucket is held on a bale arm support structure. The bucket is opened remotely via claw-type actuators. The first charge is approximately 44 tons of scrap.

Once the furnace has been charged it is moved back under the roof and is raised back to the upper position. The second bucket is then charged to the shaft. The second charge is approximately 35 tons. The complete charge cycle for the first two buckets of scrap is typically less than two minutes. Meltdown commences and after approximately four minutes the electrodes are raised and the

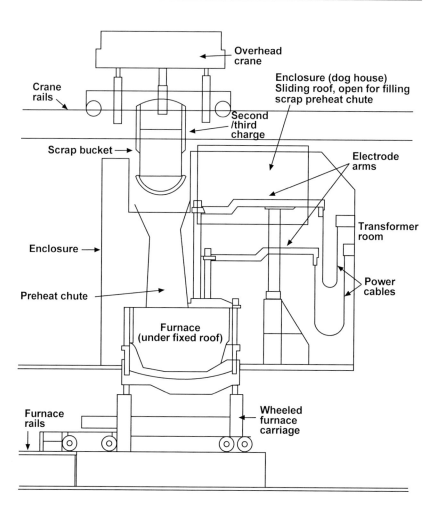

Fig. 10.83 Co-Steel Sheerness shaft furnace.

third and final scrap bucket is charged to the shaft. Meltdown then proceeds uninterrupted and the furnace taps 99 tons at 1640°C with an average power-on time of 34 minutes.

The Fuchs shaft has claimed benefits ranging from reduced EAF dust (due to dust sticking to the scrap in the shaft) to reduced exhaust fan requirements due to lower gas volumes. As all of the scrap is melting in, the zinc quickly volatizes, resulting in furnace dust that is high in zinc content. For a 120 ton furnace with an 80 MVA transformer, Fuchs predicts a tap-to-tap time of 56 minutes with an electrical power consumption of approximately 310 kWh/ton. The projections anticipate that the equivalent of 72 kWh/ton of energy will be recovered from the offgas.

The results reported by Co-Steel Sheerness PLC have indicated the following:[115]

1. Liquid steel yield, due to high FeO reversion from the slag, resulted in an average of 93.5%.

2. The volume of flue dust produced by the shaft furnace was 20% lower than for the conventional furnace (14 kg/ton billet compared to 18 kg/ton billet).

3. The flue dust chemistry was changed with shaft furnace operation. The zinc oxide content rose from 22% to 30%. In addition the lime content was decreased from 13% to 5%.

4. Fan power requirements for the fume extraction fans were decreased from 19.3 kWh/ton billet to 10 kWh/ton billet due to lower gas volumes.

Table 10.17 presents results for various trials conducted at Co-Steel Sheerness. As can be seen, the amount of burner power was increased considerably in the most recent set of trials. The burner efficiency that was achieved was much higher than that in a conventional furnace as the flame contacted cold scrap for a longer duration. In addition the CO content in the offgas was monitored in order to adjust the oxygen flow to the burners to promote CO post-combustion at the base of the shaft.[116]

Table 10.17 Results of Shaft Furnace Operating Trials at Co-Steel Sheerness

	Old Furnace	Standard Shaft Charge	Trial Shaft Charge
Electrical energy (kWh/ton billet)	467	327	250
Burner oxygen (Nm^3/ton billet)	3.0	20.0	3.2
Lance oxygen (Nm^3/ton billet)	10.9	10.0	5.0
Burner natural gas (Nm^3/ton billet)	1.5	10.0	16.0

Fuchs has other single shaft furnaces, of similar design, operating in Turkey, China, Malaysia and at North Star Steel in Arizona. The latter, Fig. 10.84, is powered with a DC transformer and is equipped with an ABB DC bottom.

Fig. 10.84 Shaft furnace in operation at North Star Steel, Kingman, Arizona.

10.9.6.2 The Double Shaft Furnace (DSF)

In order to increase the production capacity of a furnace with one transformer to over 1,000,000 tons per year, the double shaft furnace was developed. There are currently two in operation in Europe, SAM in France and ARBED in Luxembourg, Fig. 10.85. Both are 95 ton AC high impedance furnaces which, when fully utilized, deliver a liquid steel capacity of over 1,000,000 tons per year. The double shaft furnace utilizes one transformer and one set of electrodes for both furnace shells.

North Star BHP Steel Ltd. started up the first double shaft furnace in the United States. This furnace, with a tap weight of 180 tons, has a rated capacity of 1,700,000 tons per year. The AC transformer is rated at 140 MVA and will operate with a secondary voltage of 1200–1300 V. All (100%) of the scrap charged can be preheated. A savings in electric energy consumption of between 100–120 kWh/ton is projected in comparison with a single shell EAF based on 100% scrap charge. North Star BHP Steel plans to utilize DRI/HBI and/or pig iron as part of the charge. This type furnace would also be well suited for the use of iron carbide.

The trend toward the use of more fossil energy by electric arc furnace steelmakers will result in more latent energy in the offgas, primarily in the form of CO. Measurements indicate a value of energy in the offgas of 170–190 kWh/ton in a standard single shell EAF operation. In addition, instead of using energy to circulate water in the duct system to cool the hot offgases, this energy could instead be utilized to heat and melt the scrap, as is done in shaft furnaces.

The double shaft furnace at SAM, Table 10.18, currently operates with a typical scrap charge that is 25% heavy scrap, 30% shredded, 5% municipal recycled, 15% galvanized sheet, 10% turnings

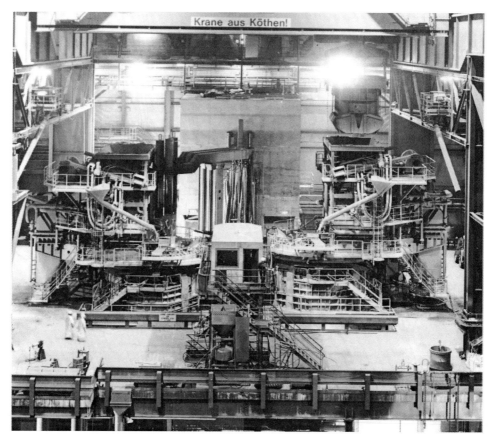

Fig. 10.85 Double shaft furnace in operation at ARBED. *(Courtesy of Fuchs.)*

and 15% light scrap. Three charge buckets are used. A round bucket holding 50 % of the charge is used to charge the furnace. Two rectangular buckets each containing 25 % of the charge are used to charge the shaft. While one shell (A) is tapping, the electrodes are moved to the other shell (B) to begin meltdown. Bore-in proceeds at a voltage of 750V. Meltdown then proceeds at 900 V. Six burners located at the bottom of the shaft aid meltdown. Vessel A, once finished tapping, is charged with 75% of the total charge. The burners in this vessel are placed on high fire to help preheat the scrap. Vessel B completes meltdown and its offgases are directed to the vessel A to aid in scrap preheating. The remaining 25% of the scrap is now charged to the shaft. Once vessel B is ready to tap, the electrodes are moved back over to vessel A to start another meltdown cycle.[117]

Table 10.18 Furnace Data of the AC Double Shaft Furnace at SAM

Furnace Capacity	115 tons
Furnace Diameter	6.3 m
Tapping Weight	99 tons
Liquid Heel	11 tons
Transformer Capacity	95 MVA
Active Power	60 MW
Electrode Diameter	600 mm
Oxy-Gas Burners	7 × 3 MW (ea)
Oxy-Carbon Lances	2 × 35 Nm^3/min
Bottom Stirring	5 elements — nitrogen

Consumption figures reported by SAM and ARBED are presented in Table 10.19 and Table 10.20, respectively.[117]

Table 10.19 Reported Consumption Figures at SAM

	Best Day	Best 4 Days
Charge Mix	Scrap 100%	Scrap 100%
Electrical Energy	320 kWh/ton	338 kWh/ton
Electrode Consumption	1.3 kg/ton	1.45 kg/ton
Oxygen Lance	10.3 Nm^3/ton	11.8 Nm^3/ton
Oxygen Burners	11.8 m^3/ton	12.7 Nm^3/ton
Gas	6.5 Nm^3/ton	6.2 Nm^3/ton
Carbon Charge	7.5 kg/ton	7.6 kg/ton
Carbon Foamy Slag	6.7 kg/ton	5.4 kg/ton
Lime	35.2 kg/ton	39.5 kg/ton
Power-On Time	40 min.	45.5 min. Liquid
Yield	91.5%	

Table 10.20 Reported Consumption Figures at ARES Schifflange

Electrical Energy	298 kWh/ton
Electrode Consumption	1.23 kg/ton
Oxygen	22.0 Nm^3/ton
Gas	5.9 Nm^3/ton
Power-On Time	39.2 min.

10.9.6.3 The Finger Shaft Furnace

Various studies have shown that the greatest proportion of the energy leaving the furnace in the offgas occurs during flat bath operations. This is due to the absence of scrap to absorb energy from the offgas as it exits the furnace. A large potential for energy recovery exists if the offgas can be used to preheat scrap. For operations with continuous feed of high carbon iron units (e.g. DRI or iron carbide), the amount of energy escaping in the offgas will be higher yet due to the higher levels of carbon monoxide in the offgas. Finger shaft furnace installations with tap weights ranging from 77–165 tons are located at Hylsa, Cockerill Sambre, Swiss Steel AG and Birmingham Steel.

To use the energy of the offgas efficiently and to reduce the electrical power requirements needed to obtain short tap-to-tap times, Fuchs developed the finger shaft furnace at Hylsa in Monterrey, Mexico. This furnace, tapping 150 tons, maintains a 44 ton liquid heel and is powered with a 156 MVA DC transformer utilizing the Nippon Steel water-cooled DC bottom electrode (3 billet bottom electrode). Reported consumption figures are presented in Table 10.21. The furnace charge is 45% scrap and 55% DRI charged during the 50 minute power-on time of the furnace. Half of the scrap is charged onto a water-cooled finger system in the shaft during the refining period of the previous heat. The other half of the scrap is charged onto the fingers after the first charge has been dropped into the furnace and power is applied.

Table 10.21 Reported Consumption Figures at Hylsa. *From Ref 117.*

Charge Mix	50% Scrap, 50% DRI
Electrical Energy	394 kWh/ton
Electrode Consumption	0.9 kg/ton
Oxygen Lance	20 Nm^3/ton
Oxygen Burners	8.1 Nm^3/ton
Natural Gas	3.6 Nm^3/ton
Carbon Charge	9 kg/ton
Carbon Foamy Slag	4.5 kg/ton
Power-On Time	48 min.
Tap Temperature	1620°C

The furnace roof and shaft sit on a trolley which moves on rails mounted on the slag door and tapping sides of the furnace. To initiate a heat, the electrode is raised and rotated away from the furnace. The shaft is moved into position over the furnace center and the fingers are retracted so that the preheated scrap is charged onto the hot heel. The roof is then moved back and the electrode is returned to its operating position. A second scrap charge is added to the shaft and DRI is fed continuously throughout meltdown. Once the first scrap charge is melted, the second charge is added to the center of the furnace. Initial bore-in utilizes a voltage of 400V and is increased to 550V once the electrode has penetrated to about one metre. During meltdown a voltage of 635V is used. During meltdown of the second scrap charge and refining, the first charge for the next heat is preheated in the shaft. The DRI, containing 2.3% carbon, is cold and continuously charged. Oxy-fuel burners are used during scrap meltdown. Future plans call for running the burners with excess oxygen for post-combustion of CO.

A finger shaft furnace is now operating at Swiss Steel AG (Von Roll). This furnace was converted in only five weeks. It is powered with a high impedance AC transformer. They have seen a 55% improvement in productivity and are tapping 79 tons of steel every 37 minutes. Reported consumption figures are presented in Table 10.22.

Table 10.22 Reported Consumption Figures at Swiss Steel AG. *From Ref 117.*

Charge Mix	100% Scrap
Electrical Energy	260 kWh/ton
Electrode Consumption	1.3 kg/ton
Oxygen Lance	14.4 Nm^3/ton
Oxygen Burners	11.8 Nm^3/ton
Natural Gas	5.1 Nm^3/ton
Carbon Charge	5.5 kg/ton
Carbon Foamy Slag	2.7 kg/ton
Power-On Time	28 min.
Tap Temperature	1620°C

This operating data indicates that 30 minute tap-to-tap times are likely to be achieved in the future.

Paul Wurth S.A. has installed a 154 ton DC finger shaft furnace at Cockerill Sambre in Belgium under license from Fuchs. The unique feature of this finger shaft furnace is that 20–50% of the charge can be liquid iron with a carbon content of 4%. The operation utilizes the Paul Wurth proprietary hot metal charging system.[118]

This operation utilizes four water-cooled billets as the bottom electrode. The hot metal charging system is a technology developed by Paul Wurth based on trials conducted at various ISCOR facilities in South Africa. Hot metal is charged through a side launder in the furnace as shown in Fig. 10.86. This system allows for continuous charging of 44–60 tons of hot metal to the furnace over a period of 15–20 minutes. The furnace is located within an enclosure to minimize fume emissions to the shop. Fig. 10.87 shows the hot metal ladle charging system operating sequence

Reported consumption figures at Cockerill Sambre are reported in Table 10.23.

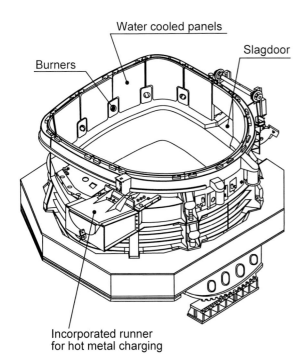

Fig. 10.86 Shaft furnace bottom with runner for hot metal charging. *(Courtesy of Paul Wurth, S.A.)*

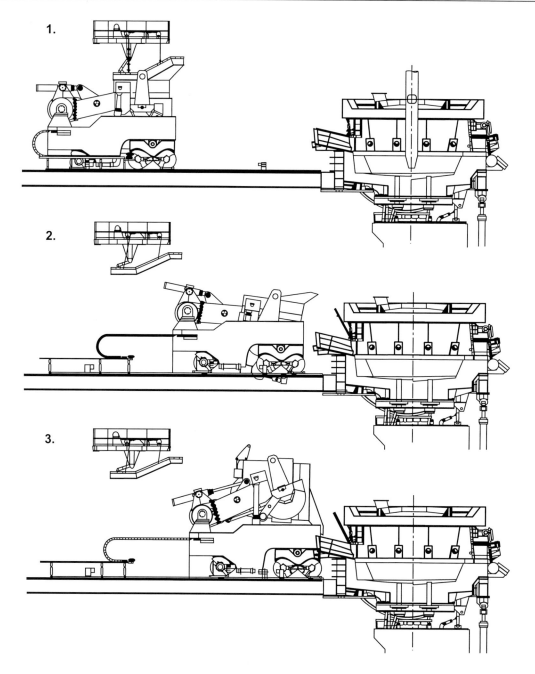

Fig. 10.87 Operating sequence to charge hot metal. 1. Ladle with cover in place. 2. Cover removed and ladle to furnace. 3. Charging hot metal. *(Courtesy of Paul Wurth, S.A.)*

A replacement equation for the equivalent value of hot metal in the furnace has been developed by Paul Wurth based on various plant trials and is presented below.[118]

1 ton hot metal + 22.7 kg lime = 0.92 ton scrap + 45 kg coal + 300 kWh electricity

10.9.7 Consteel Process

The Consteel process was developed by Intersteel Technology Inc. which is located in Charlotte, North Carolina. This is another process that is based on recovering heat from the offgas. In this

Table 10.23 Reported Consumption Figures at Cockerill Sambre. *From Ref 118.*

	34% Hot Metal Charge	100% Scrap Charge
Electrical Energy	187 kWh/ton	290–310 kWh/ton
Power-On To Tap	<40 min.	
Oxygen		23–27 Nm^3/ton
Natural Gas		5.5–7.3 Nm^3/ton
Carbon Charge		9–13.5 kg/ton
Power-On Time		50–55 min.
Tap Temperature		1620°C

case, the scrap enters a long preheater tunnel and is preheated by the offgas as it travels to the furnace. The scrap is moved through the tunnel on a conveyor and is fed continuously to the EAF. The offgas flows counter-current to the scrap. The EAF maintains a liquid heel following tapping. Fig. 10.88 illustrates the key system components.

10.9.7.1 Historical Development

The Nucor Steel Corporation was the first company to commit to an industrial application of the Consteel technology. In 1985 Consteel was retrofitted into an existing Nucor plant located in Darlington, South Carolina. Due to space restrictions the installation was designed with a 90° bend in the scrap conveyor and a relatively short preheater section. This was reported to result in scrap feeding and preheating problems. The unit was operated for eighteen months and was then shut down.

Several key concepts were confirmed by this prototype which was put into operation in 1987. A consistent heel of hot metal acted as a thermal flywheel, increasing efficiencies of scrap melting. It was demonstrated that keeping the bath temperature within a proper range ensured a constant equilibrium between metal and slag and a continuous carbon boil, which resulted in a bath which was homogeneous in temperature and composition. The foaming of the slag could be continuously and precisely controlled and was very important for successful operation.

The first greenfield demonstration of the Consteel process was at AmeriSteel in Charlotte, North Carolina. This layout included a new meltshop that was built parallel to an existing one. In this case the continuous scrap feed system and the scrap preheater were all in line. The scrap was taken from

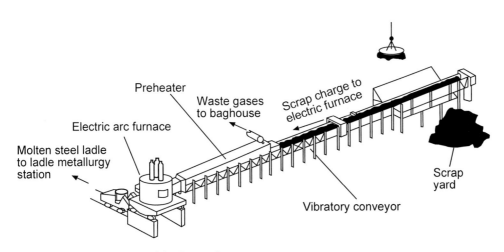

Fig. 10.88 Key components of the Consteel process.

railcars to a loading station through a preheater to a connecting car and then into the furnace. The preheater design included an afterburner for control of carbon monoxide emissions though this feature was never put into operation. The scrap preheater was designed to heat scrap up to 700°C. Particular emphasis was placed on providing a tight seal on the scrap preheater. This was accomplished by providing a water seal in the preheater. The key characteristics of this facility have been described.[119]

The Contifeeding system consists of three conveyors in cascade, each 1.5 metres (5 ft.) × 0.3 metres (1 ft.) deep × 60 metres (200 ft.) long. A refractory lined tunnel covers the conveyor and a water seal prevents outside air from leaking between the cover and the conveyor pans. Shredder scrap, #1 scrap, turnings and light structural scrap are charged to the furnace on the conveyor. A leveller bar at the scrap charge end maintains a maximum scrap height of 0.45 metres (18 in.) on the conveyor.

The scrap preheater is 24 metres (80 ft.) in length with 60 natural gas burners of 7.0 Nm^3/min. (250 scfm) capacity mounted in the preheater roof. Preheat temperatures up to 700°C are achieved using the furnace offgas and the burners. The burners were later removed as sufficient preheating was achieved using only furnace offgas.

The furnace proper is a 75 ton EBT EAF designed to tap 40 tons and retain a 30–35 ton heel. The scrap feed rate is approximately 680 kg/min. (1500 lbs/min.). For cold startup the furnace is top charged and this charge is melted in to provide the liquid heel; a slag door burner is used to accelerate meltdown for these heats. Electrical energy is supplied by a 30 MVA transformer through 20 inch electrodes, with resultant tap-to-tap times of approximately 45 minutes. Other facilities include carbon and oxygen bath injection for producing a foamy slag and a pneumatic lime injection system.

In addition to the operation at AmeriSteel, several other plants have installed Consteel at their facilities including Kyoei Steel (Nagoya), Nucor Steel Darlington and New Jersey Steel.

10.9.7.2 Key Operating Considerations

The key to obtaining good operating results with Consteel is based on controlling several operating parameters simultaneously. These are bath temperature, scrap feed rate and scrap composition, oxygen injection rate, bath carbon levels, and slag composition.

It can be seen that any type of imbalance in any of these operating parameters will have a ripple effect through the whole process. Generally, the hot heel size is 1.4 tons times the power input in MW. Some operations use process models to track the process inputs. Slag composition is very important because without good slag foaming it will be difficult to bury the arc. This will result in large heat losses and potentially furnace damage. Continuous evolution of CO from the bath is crucial to maintaining scrap preheat temperatures. Approximately 70–75% of the CO generated in the furnace is available as fuel in the preheater. A reducing atmosphere is maintained in the first 30% of the preheater in order to control scrap oxidation. Complete combustion of CO and VOCs evolving from the scrap is achieved prior to the fluegas exiting the preheater. The FeO content of the slag is maintained around 15% which is suitable for good slag foaming.

A variety of scrap types can be fed to the Consteel operation. The primary stipulation is that the size is less than the conveyor width in order to minimize scrap bridging in the feed system. Bundles can be used but will not benefit from preheating. Best results are obtained when using loose, shredded scrap which is light and has a large surface area for heat transfer from the hot offgases. High carbon scrap such as pig iron can be used but must be distributed within the charge in order to maintain a fairly uniform carbon level in the bath. HBI can be charged to the preheater along with scrap. DRI should only be charged into the section of the preheater with a reducing atmosphere. The key to good operation in all cases is to maintain homogeneity of the scrap feed so that bath composition remains within preset limits. Scrap density must also be optimized so that the desired residence time in the preheater shaft can be attained thus ensuring sufficient preheating of the scrap. Charge permeability also affects the amount of preheat achieved.

The following summary, Table 10.24, from Intersteel lists the latest data on existing and future installations.

Table 10.24 Consteel Operating Results. *From Ref 120*

Facility	AmeriSteel Charlotte	Kyoei Nagoya	Nucor Darlington	New Jersey Steel	ORI Martin Italy	N.S.M. Thailand
Start-up Date	Dec. 1989	Oct. 1992	Sept. 1993	May 1994	On Hold	End 1997
Productivity	54 ton/hr	140 ton/hr	110 ton/hr	95 ton/hr	87 ton/hr	229 ton/hr
EAF Type	AC	DC	DC	AC	AC	AC
New/Retrofit	New	New	New	Retrofit	Retrofit	New
Transformer Power	24 MW	47 MW	42 MW	40 MW	31 MW	95 MW
Power	370 kWh/ton	300 kWh/ton	325 kWh/ton	390 kWh/ton		
Oxygen	22 Nm^3/ton	39 Nm^3/ton	33 Nm^3/ton	23 Nm^3/ton		
Electrodes	1.75 kg/ton	1.14 kg/ton	1.0 kg/ton	1.85 kg/ton		
Yield	93.3	94	93	90		

Note that some of the above figures represent unit consumption records.

10.9.8 Twin Shell Electric Arc Furnace

10.9.8.1 Historical Development

One of the new technologies that seems to be generating much interest is the twin shell furnace. Essentially this technology is similar to conventional scrap preheating with the exception that scrap preheating actually takes place in a furnace shell as opposed to a scrap bucket.

The original work done on in-furnace scrap preheating was in Sweden, where SKF operated an installation which had a single power supply and two furnace shells. In the early 1980s Nippon Steel developed a twin shell process for the production of stainless steel, Fig. 10.89. At the current time, several furnace manufacturers are producing twin shell furnaces. The key goals of a twin shell operation are equal to those of other developing technologies, but in addition the cycle times are similar to those for BOF operations due to the minimization of power-off times.

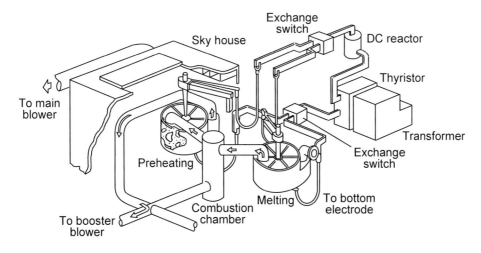

Fig. 10.89 Schematic of the DC twin shell EAF at Nippon Steel.

10.9.8.2 Process Description

This type of operation consists of two furnace shells and one set of electrode arms that are used alternately on one shell and then the other. The scrap is charged to the shell that is not melting and the scrap is preheated by the offgas from the melting furnace. Supplemental burner heat can also be used. The result is that the scrap is preheated in the furnace prior to melting. The more the scrap is preheated, the greater the energy savings.

Some early designs proposed a gantry that moved along a rail from one shell to the other. Most designs now use a furnace roof and electrode arms that can swing between two positions for the shells. Generally a twin shell installation will consist of two identical vessels with a lower shell, an upper shell and a roof and one set of electrode arms and lifting supports with one conventional power supply.

It is interesting to note that several of the recent orders for twin shell installations have opted for DC operation. This obviously has some inherent benefits as there is only one electrode arm to rotate between the two operating positions. Another variation introduced is that some operations have an electrode mast and arm for each furnace and have swithgear allowing the sharing of a common transformer.

There have been several operating modes suggested for the twin shell operation. The Nippon Steel operating cycle consists generally of two phases; scrap preheating and melting. The furnace is only charged one time and thus the heat size is smaller than the actual capacity of the furnace. In the case of the Nippon Steel installation at POSCO, the operation is a two charge operation where 60% of the total charge is preheating during the melting phase on the second furnace. Thus the second charge (40% of the total) is not preheated in the furnace.

The operating cycle proposed by Clecim is similar to that at POSCO. The Mannesmann Demag operating cycle is quite different in that the operation uses two charges but the power is alternated between furnaces between charges. Thus one furnace melts its first charge while the other preheats its first charge. Once the first charge is melted, the second charge is dropped and the power is diverted to the first preheated charge. Thus the second charge is preheated as well as the first during the preheating cycle for the second charge; oxygen lancing or oxy-fuel burners can also be used to accelerate preheating and melting. This operating method maximizes the recovery of heat from the offgas but requires very precise process control to keep from freezing the molten heel in the furnace during the preheat phase.

10.9.8.3 Operating Results

10.9.8.3.1 Nippon Steel In this operation, there is only one charge per heat. While one charge is being melted, a second charge is being heated in the second shell. Combustion gas from a burner on the second shell is mixed with offgas from the first shell to give a constant temperature of 1650°F for preheating in the second vessel. The remaining offgas from both shells is used to preheat scrap in a scrap bucket. The net electrical power input requirement is reported at 260 kWh/ton which is 29% lower than for the two conventional furnaces that were replaced by the twin shell operation.

10.9.8.3.2 Davy-Clecim The Clecim installation at Unimetal Gandrange tapped its first heat in July 1994. This twin shell operation is DC with the Clecim water-cooled billet bottom electrode. The shells have a nominal capacity of 165 tons with a nominal diameter of 7.3 metres. Each shell has a working volume of approximately 200 m^3 which enables the furnace to operate with a single charge each heat. The rated installed transformer capacity is 150 MVA. The maximum secondary voltage is 850V. Maximum power input is 110 MW. Each shell has four bottom billet electrodes. The graphite electrode swings between the two furnaces as required. Each shell has a manipulator with consumable oxygen and carbon injection lances. Initially, integrated steelmaking operations continued to operate which led to large furnace delays due to the logistics of movements within the

Fig. 10.90 Twin shell furnace operation at ProfilArbed.

shop. This operation experienced problems with arc flare due to the configuration of the anode bus tubes. The configuration was altered and this problem was eliminated.

The ProfilArbed installation started up at the end of 1994 and is shown in Fig. 10.90.

Several recent North American twin shell installations include Steel Dynamics, Gallatin Steel, Tuscaloosa Steel and Nucor-Berkeley. Fig. 10.91 shows the operation at Tuscaloosa Steel.

Fig. 10.91 Twin shell furnace operation at Tuscaloosa Steel.

10.9.9 Processes Under Development

Based on the success of several of the scrap preheat technologies (Consteel, Fuchs Shaft Furnace, EOF), several new shaft type technologies are currently in the development stage. These are outlined in this section.

10.9.9.1 Ishikawajima-Harima Heavy Industries (IHI)

IHI is currently developing a shaft type preheat furnace based on twin electrode DC technology. The first commercial installation has started up at the Utsunomiya plant of Tokyo Steel. The DC furnace itself is oval in shape with two graphite electrodes and two bottom electrodes consisting of conductive hearth brick (as per ABB's DC furnace design). There are two DC power supplies which are individually controlled. The power feeding bus is arranged so that the two arcs will deflect towards the center of the furnace, thus the energy of the arcs will be concentrated at the center of the furnace and the thermal load to the furnace walls will be low compared to a conventional furnace. As a result refractory walls are used instead of water-cooled panels thus reducing heat loss. Scrap is charged to the furnace between the electrodes. The furnace will maintain a large hot heel (110 ton heel, 140 ton tap weight) so that uniform operating conditions can be maintained (similar to the Consteel concept). Steel is tapped out periodically via a bottom taphole in the furnace. A diagram of the furnace is given in Fig. 10.92.

The scrap charging system consists of two main components, the preheat chamber and the charging equipment. The scrap is fed into the upper part of the chamber from a receiving hopper. The exhaust gas from the furnace flows up through the chamber, preheating the scrap. In the pilot plant, scrap preheat temperatures as high as 800°C were achieved. Gas exit temperatures from the chamber were as low as 200°C. At the base of the preheat chamber are two pushers. These operate in two stages, allowing scrap to feed into the furnace at a constant rate. Offgas leaves the top of the preheat chamber and flows to a bag filter. Some gas can be recycled to the furnace to regulate the inlet gas temperature to the preheater.

10.9.9.1.1 Operation Scrap is fed continuously to the furnace until the desired bath weight is achieved. This is followed by a short refining or heating period leading to tapping of the heat. Power input is expected to be almost uniform throughout the heat. Most furnace operations will be fully automated. Charging of scrap into the preheater will be fully automated based on the scrap height

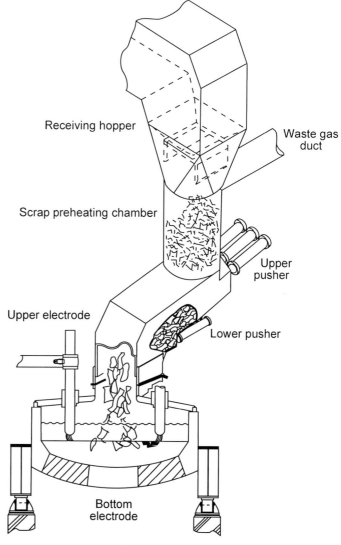

Fig. 10.92 IHI shaft furnace.

in the chamber. Carbon and oxygen injection will be controlled based on the depth of foamy slag.

The results presented in Table 10.25 have been reported for the IHI shaft furnace.[121]

Table 10.25 Reported Results at Tokyo Steel.

	Utonomiya Plant	Takamatsu Plant
Tap-to-tap time	60 min.	45 min.
Power-on time	55 min.	40 min.
Power consumption	236 kWh/ton liquid	236 kWh/ton liquid
Electrode consumption	1.0 kg/ton liquid	1.0 kg/ton liquid
Oxygen consumption	28 Nm^3/ton liquid	25 Nm^3/ton liquid
Carbon consumption	27 kg/ton liquid	25 kg/ton liquid

10.9.9.2 VAI Comelt

VAI has been working on a new furnace design they call Comelt. This furnace features a shaft and four individually moveable graphite electrodes which protrude into the furnace at an angle. A bottom anode is installed in the hearth center. The power supply is DC. The lower part of the shaft contains openings for lime and coke additions via a bin system. The upper part of the shaft contains a lateral door for scrap charging and an opening through which the furnace offgases flow. Scrap is charged to the shaft via a conveyor belt.

The furnace consists of a tiltable furnace vessel and a fixed shaft which are connected with a movable shaft ring. The furnace employs eccentric bottom tapping. The whole vessel structure rests on a tiltable support frame. The upper part of the vessel, the shaft and shaft ring are all lined with water-cooled panels. The shaft is enclosed with a steel structure which sits on a carriage. The shaft can be moved off the furnace shell via this carriage. The shaft ring provides a tight connection between the furnace vessel and the shaft. Key furnace components are shown in Fig. 10.93.

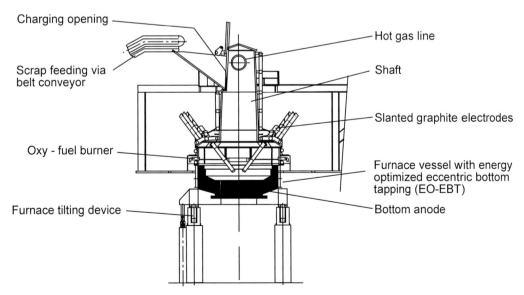

Fig. 10.93 Key components of the VAI Comelt furnace.

10.9.9.2.1 Process Description Following tapping, the furnace is charged through the shaft with up to 80% of the total charge weight, lime and carbon. After charging, the charging door in the top of the shaft is closed and the electrodes are moved into working position. Oxy-fuel door burners are used to clear an area for insertion of oxygen lances which immediately begin to inject oxygen into the heel. Additional lances higher up in the furnace provide post-combustion oxygen. As the scrap melts in and the shaft begins to empty, additional scrap is fed to the shaft. The electrode arrangement allows the flow of gases to penetrate down into the scrap and then rise up through the scrap column thus maximizing heat recovery.

VAI has projected that the capital cost for Comelt will be greater than that for an equivalent DC operation for small to medium heat sizes. This is due to the large investment required for the electrical systems for the electrodes. For large heat sizes however, the projections indicate that Comelt should be less expensive than conventional DC technology.[122]

Though this process is still in the pilot plant stage, projections indicate that this furnace will have lower power (304 kWh/ton liquid) and electrode (0.8 kg/ton) consumption than either conventional AC or DC furnaces.[122]

10.9.9.3 Mannesmann Demag Huettentechnik Conarc

The MDH Conarc is based on an operation which combines converter and EAF technologies, namely a <u>con</u>verter <u>arc</u> furnace. This technology is based on the growing use of hot metal in the EAF and is aimed at optimizing energy recovery and maximizing productivity in such an operation. The use of hot metal in the EAF is limited by the maximum oxygen blowing rates which are dependent in turn on furnace size. The basic concept of Conarc is to carry out decarburization in one vessel and electric melting in another vessel. The Conarc system consists of two furnace shells, one slewable electrode structure serving both shells, one electric power supply for both shells, and one slewable top oxygen lance serving both shells

Thus one shell operates in the converter mode using the top lance while the other shell operates in the arc furnace mode. The first order for a Conarc was placed by Nippon Dendro Ispat, Dolvi, India. This operation will be based on the use of hot metal, scrap, DRI and pig iron. Projections for this operation are an energy consumption below 181 kWh/ton when operating primarily with hot metal and DRI. Once complete, this facility will contain two independent twin shell furnaces each with a tap weight of 164 tons. Fig. 10.94 shows the key system components for a Conarc facility.

A Conarc has also been ordered for the Saldahna Steel facility which is being built in Saldahna Bay, South Africa. This operation will be based on 45% Corex hot metal and 55% DRI feed to the furnaces. Hot metal will be charged to one shell and the top lance will be used for decarburization. Simultaneously, DRI will be added to recover the heat generated. Once the aim carbon level is achieved, the electrodes will be moved to this shell and additional DRI will be fed to make up the balance of the heat.

Large quantities of heat are generated during the oxygen blowing cycle. As a result it is important that charge materials be added during this period so that some of this energy can be recovered. This will also help to protect the furnace shell from overheating.

10.9.9.4 Mannesmann Demag Huettentechnik Contiarc

The Contiarc EAF is a technology that has been proposed by MDH. It is based on a stationary ring shaft furnace with a centrally located DC arc heating system. Scrap is fed continuously by conveyor to the ring shaft. There the scrap is picked up by a series of magnets and is distributed evenly throughout the ring shaft area. The magnet train runs on a circular track below the furnace roof. Fig. 10.95 shows the main components of the Contiarc.

The scrap settles in the ring shaft and is preheated as furnace offgas rises up through the furnace. The descending scrap column is monitored using a level measuring system. If the charge sinks unevenly, additional scrap is fed to the lower points to restore an even scrap height profile. Scrap

Electric Furnace Steelmaking

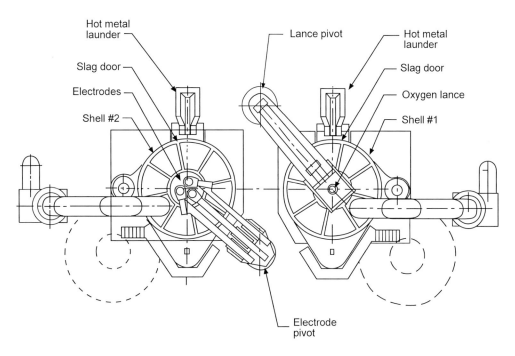

Fig. 10.94 Schematic of the Conarc process components.

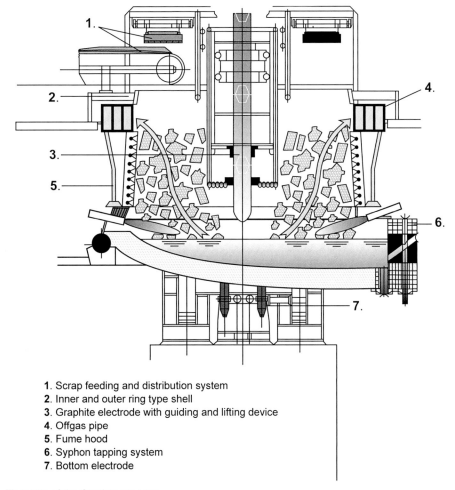

1. Scrap feeding and distribution system
2. Inner and outer ring type shell
3. Graphite electrode with guiding and lifting device
4. Offgas pipe
5. Fume hood
6. Syphon tapping system
7. Bottom electrode

Fig. 10.95 Elements of the Contiarc process.

is always present to protect the furnace walls so the arc can run continuously at maximum power without fear of damage to the furnace sidewalls.

The graphite electrode is located in a protective inner vessel to protect it from falling scrap. The inner vessel has a wear guard and a cooling water system in the lower hot zone. An electrically insulated ceramic electrode bushing is provided at the point where the electrode penetrates out into the furnace. Electrode regulation is performed hydraulically.

Chemical energy input, such as burners, is highly efficient because gas residence times in the Contiarc are greater than for conventional furnaces. Burners are located near the taphole to super-heat the steel prior to tapping. Burners are also located near the slag door to facilitate slagging off. Slag free tapping is performed intermittently using a syphon tapping system.

Energy efficiency is extremely high using Contiarc as heat losses to the shell are minimized due to the continuous cover provided by the scrap. The elimination of top charging also reduces heat losses. The furnace is essentially airtight so that all offgas rises up through the scrap and is collected at the ring header at the top of the shaft. Dust losses are reduced 40% over conventional EAF operations as the scrap acts as a filter, trapping the dust as the offgas rises through the scrap. MDH projections indicate that the total energy input requirement for Contiarc will be only 62% of that required by an equivalent conventional EAF operation.[123]

10.9.9.5 Future Melting Furnace Design

Without doubt, current trends in EAF design indicate that high levels of both electrical and chemical energy are likely to be employed in future furnace designs. As so many have pointed out, we are headed towards the oxy-electric furnace. The degree to which one form of energy is used over another will be dependent on the cost and availability of the various energy forms in a particular location. Raw materials and their cost will also affect the choice of energy source. The use of alternative iron sources containing high levels of carbon will necessitate the use of high oxygen blowing rates which equate to high levels of chemical energy use. A word of caution is in order though for such operations, as some thought must enter into how to maximize energy recovery from the offgases generated. In addition, high levels of materials such as cold pig iron can sometimes lead to extended tap-to-tap times which will reduce furnace productivity. If hot metal is used, some method must be provided to recover energy from the offgases. Materials which are also high in silicon will provide additional energy to the bath but at the cost of additional flux requirements and greater slag quantities generated.

A closed furnace design appears to be imperative for complete control of reactions in the furnace freeboard. Minimization of the offgas volume exiting the furnace will help to minimize offgas system requirements. If air infiltration is eliminated, offgas temperatures will be quite high which will give more efficient heat transfer in scrap preheating operations. If high levels of hydrogen and CO are present in the cooled gas, it could be used as a low grade fuel. Alternatively, if the CO is post-combusted in several preheater stages, maximum energy recovery to the scrap could be achieved.

The main disadvantage to operation of a closed furnace is that conventional oxygen lancing and coal injection operations would be difficult to carry out. These can however be carried out in a closed furnace if a series of submerged and side injectors are installed in the furnace shell and sidewalls similar to those used in conventional BOFs. The required injection rates would likely require that the furnace bath depth be increased in order to minimize blow through. A deeper bath with intense mixing will be beneficial for slag bath mixing which should improve steel quality and maximize recovery of iron units. Generation of CO in the bath will help to flush nitrogen and hydrogen out of the bath.

Some form of scrap preheating should be utilized to recover heat from the offgas. The best method for scrap preheating is still under debate but staged preheating coupled with post-combustion in each stage could help to minimize the need for a secondary post-combustion system downstream in the offgas system.

If a slanted electrode configuration similar to that used in Comelt is employed, it would be conceivable to retract the electrodes if necessary during part of the tap-to-tap cycle. This would allow preheated scrap to be charged periodically if a finger type preheat shaft were used. Alternatively the scrap could be continuously charged into a shaft by conveyor or a preheat tunnel conveyor similar to that used in Consteel, could be installed. A DC arc could be directed at the point where scrap feed falls into the furnace similar to the IHI furnace concept. This would help to maximize heat transfer from the arc to the scrap. If a scrap buildup on the walls of the furnace can be maintained, heat losses to the furnace shell can also be minimized.

10.9.9.6 Conclusions

There are many new processes for steelmaking which are now being commercialized. In almost all cases the goal is to minimize the electrical energy input and to maximize the energy efficiency in the process. Thus several technologies have attempted to maximize the use of chemical energy into the process (EOF, K-ES, LSF, etc.). These processes are highly dependent on achieving a pseudo-equilibrium where oxygen has completely reacted with fuel components (carbon, CO, natural gas, etc.) to give the maximum achievable energy input to the process. Other processes have attempted to maximize the use of the energy that is input to the furnace by recovering energy in the offgases (Fuchs shaft furnace, Consteel, EOF, IHI Shaft). These processes are highly dependent on good heat transfer from the offgas to the scrap. This requires that the scrap and the offgas contact each other in an optimal way.

All of these processes have been able to demonstrate some benefits. The key is to develop a process that will show process and environmental benefits without having a high degree of complexity and without affecting productivity. There is no perfect solution that will meet the needs of all steelmaking operations. Rather, steelmakers must prioritize their objectives and then match these to the attributes of various furnace designs. It is important to maintain focus on the following criteria:

1. To provide process flexibility.
2. To increase productivity while improving energy efficiency.
3. To improve the quality of the finished product.
4. To meet environmental requirements at a minimum cost.

With these factors in mind, the following conclusions are drawn :

1. The correct furnace selection will be one that meets the specific requirements of the individual facility. Factors entering into the decision will likely include availability of raw materials, availability and cost of energy sources, desired product mix, level of post furnace treatment/refining available, capital cost and availability of a trained workforce.

2. Various forms of energy input should be balanced in order to give the operation the maximum amount of flexibility. This will help to minimize energy costs in the long run, i.e. the capability of running with high electrical input and low oxygen or the converse.

3. Energy input into the furnace needs to be well distributed in order to minimize total energy requirements. Good mixing of the bath will help to achieve this goal.

4. Oxygen injection should be distributed evenly throughout the tap-to-tap cycle in order to minimize fluctuations in offgas temperature and composition. Thus post-combustion operations can be optimized and the size of the offgas system can be minimized. In addition, fume generation will be minimized and slag/bath approach to equilibrium will be greater.

5. Injection of solids into the bath and into the slag layer should be distributed across the bath surface in order to maximize the efficiency of slag foaming operations. This will also enable the slag and bath to move closer to equilibrium. This in turn will help to minimize flux requirements and will improve the quality of the steel.

6. Submerged injection of both gases and solids should be maximized so that the beneficial effects of bath stirring can be realized. Slag fluxes could be injected along with oxygen. The high temperatures achieved in the fireball coupled with the high oxygen potential will help the lime flux into the slag faster, and as a result, dephosphorization and desulfurization operations can proceed more optimally.

7. If injection of solids and gases is increased, it will likely be beneficial to increase the bath depth. This will also be beneficial for steel quality. For operations using high levels of solids injection (those feeding high levels of iron carbide), a deeper bath will help to reduce blow through (solids exiting the bath) and could also allow higher injection rates.

8. The melting vessel should be closed up as much as possible in order to minimize the amount of air infiltration. This will minimize the volume of offgas exiting the furnace leading to smaller fume system requirements. In the case of post-combustion operations coupled with scrap preheating, the gas volume will be minimized while maximizing offgas temperature for efficient heat transfer to the scrap. If secondary post-combustion is required it can be achieved most effectively at minimum cost by minimizing the volume of offgas to be treated.

9. If high carbon alternative iron sources are used as feed in scrap melting operations, some form of post-combustion is imperative in order to recover energy from the high levels of CO contained in the offgas. For these operations, energy recovery will likely be maximized by coupling post-combustion with some form of scrap preheating.

10. Scrap preheating provides the most likely option for heat recovery from the offgas. For processes using a high degree of chemical energy in the furnace, this becomes even more important, as more energy is contained in the offgas for these operations. In order to maximize recovery of chemical energy contained in the offgas, it will be necessary to perform post-combustion. Achieving high post-combustion efficiencies throughout the heat will be difficult. Staged post-combustion in scrap preheat operations could optimize heat recovery further.

11. Alternatively, operations which generate high levels of hydrogen and carbon monoxide may find it cost effective to try to recover the calorific energy contained in the offgas in the same manner as some BOF operations which cool and clean the gas prior to using it as a low grade fuel. For this type of operation, it will be necessary to operate with a closed furnace. To minimize fuel gas storage requirements, generation of CO in the furnace should be balanced throughout the cycle. This can be achieved by using side shell or bottom oxygen injectors and by injecting oxygen at constant levels throughout the heat. If this form of operation is coupled with scrap preheating, any VOCs resulting from the scrap will form part of the offgas stream. A greater scrap column height can be utilized because secondary post-combustion downstream in the offgas system is not necessary. The high scrap column will also help strip the fume and dust from the offgas stream thus minimizing cleaning requirements. The resulting fume system requirement will be considerably smaller than that for a conventional furnace with similar melting capacity.

12. In the future, it will likely be necessary to recycle furnace dust and mill scale back into the steelmaking vessel. Preliminary trials indicate that this material is a good slag foaming agent. Key drawbacks are handling the dust prior to recycle and its concentration buildup of zinc and lead over time.

13. Twin shell furnace configurations will continue to be used in order to reduce cycle times below 45 minutes (comparable to BOF operations) and to optimize power-on time. In the case of operations using a high percentage of hot metal, operations similar to Conarc will be used, thus employing oxygen blowing into a hot heel with

scrap addition. Thought must be given to maximizing recovery of energy from the hot offgases generated during periods of high oxygen injection rates.

14. The use of hot metal in the EAF will increase as more high tonnage EAF operations attempt to achieve cycle times of 35–40 minutes. Hot metal should be added to the heat gradually so that large fluctuations in the bath chemistry do not occur which could result in explosions. In addition, high decarburization efficiency can be achieved if the bath carbon level is kept at approximately 0.3% until final refining takes place (blowing down the desired tap carbon level at the end of the heat). Decarburizing from high carbon levels has been shown to generate high iron losses to fuming. At very low bath carbon levels, large amounts of FeO are generated and report to the slag.

15. DC furnaces will continue to be of interest for high efficiency melting operations. The additional space available on the furnace roof due to a single electrode operation will allow for much greater automation of operating functions such as bath sampling, temperature sampling, solids/gas injection, etc. The concept of using side slanted electrodes as in the case of Comelt affords several advantages to furnace operation and could allow for electrodes to be retracted in some phases of the of the tap-to-tap period. This would allow scrap to be charged from a preheating shaft without fear of electrode damage. The use of DC furnaces might also allow for smaller furnace diameters coupled with greater bath depth.

16. Operations which desire maximum flexibility at minimum cost will result in more hybrid furnace designs. These designs will take into account flexibility in feed materials and will continue to aim for high energy efficiency coupled with high productivity. For example operations with high solids injection, iron carbide or DRI fines, may choose designs which would increase the flat bath period in order to spread out the solids injection cycle. Alternatively, a deeper bath may be used so that higher injection rates can be used without risk of blow through.

17. Operating practices will continue to evolve and will not only seek to optimize energy efficiency in the EAF but will seek to discover the overall optimum for the whole steelmaking facility. Universally, the most important factor is to optimize operating costs for the entire facility and not necessarily one operation in the overall process chain.

Along with added process flexibility comes greater process complexity. This in turn will require greater process understanding so that the process may be better controlled. Much more thought consequently must enter into the selection of electric furnace designs and it can be expected that many new designs will result in the years ahead. As long as there is electric furnace steelmaking, the optimal design will always be strived for.

References

1. Fuchs Systems, Inc., personal communication.
2. Kvaerner Metals, personal communication.
3. Danieli of America, Inc., personal communication.
4. E. H. McIntyre and E. R. Landry, *Electric Arc Furnace Efficiency* (Pittsburgh: Center for Materials Production Report 92–10), 9:11–14.
5. D. A. Surgeon and J. R. Tober, *Maintenance of Secondary Electrical Power and Structural Systems for Electric Furnaces*, Electric Furnace Conference Proceedings 53 (Warrendale: Iron and Steel Society, 1995): 445–450.
6. P. F. Hammers, *Maintaining and Explaining the Secondary Circuit of the Arc Furnace*, Electric Furnace Conference Proceedings 50 (Warrendale: Iron and Steel Society, 1992): 541–545.

7. H. L. Vernon, AISE Electric Furnace Training Seminar, Ypsilanti, MI, 1996.
8. K. Bretthauer and K. Timm, "On the measurement of electrical parameters on the high current side of AC furnaces"(in German), *Elektrowaerme Int.* 29 (1971): 381–7.
9. B. Bowman, *An Update of the Electrode Consumption Model*, UCAR SNC, 4 Place des Etats Unis, F 94518, Rungis, CEDEX, France (1995).
10. A. Lefrank, W. J. Jones, and R. G. Wetter, *DC Steelmaking Conditions and Electrode Performance,* Electric Furnace Conference Proceedings 53 (Warrendale, PA: Iron and Steel Society, 1995), 337–346.
11. D. Klein and K. Wimmer, "DC Electrodes—A Key Factor for Progress in EAF Production," *Metallurgical Plant and Technology International* 18:4 (1995): 54–63.
12. L. L. Teoh, "Improving Environmental Performance in Mini-Mills: Part 1," *Steel Times International*, 15:1 (1991): 29–31.
13. A. D Bennett, "Methods of Controlling Fume Emissions From Electric Arc Furnaces," *Steel Technology International* (1993): 32–8.
14. Promotional material on EAF fume control, Carborundum.
15. P. G. A. Brand et al., "Application of Computer Analysis to Electric Arc Furnace Fume Control," *Iron and Steel Engineer* 69:1 (1992): 56–9.
16. M. Bender, Bender Corp., personal communication.
17. "Environmental Roundup," *33 Metalproducing* 31:7 (1993).
18. *The Electric Arc Furnace–1990*, (Brussels, Belgium: International Iron and Steel Institute, 1990).
19. P. Brand, Hatch Associates, Ltd., personal communication.
20. B. Lindblad, E Burstrom., *A Scandinavian View On Coated Scrap and the Environment* (Lulea, Sweden: MEFOS)
21. P. Acharya, S. G. DeCicco, and R. G. Novak, "Factors that Can Influence and Control the Emissions of Dioxins and Furans From Hazardous Waste Incinerators," *J. Air Waste Management Assoc.* 41:12 (1991): 1605–15.
22. L. L. Teoh, "Electric Arc Furnace Technology: Recent Developments and Future Trends," *Ironmaking and Steelmaking* 16 (1989): 303–13.
23. H. Gripenberg, M. Brunner, and M. Petersson, "Optimal Distribution of Oxygen in High Efficiency Arc Furnaces," *Iron and Steel Engineer* 67:7 (1990): 33–7.
24. D. L. Schroeder, *Use of Energies in Electric Steelmaking Shops,* Electric Furnace Conference Proceedings 49 (Warrendale: Iron and Steel Society, 1991): 417–28.
25. R. J. Fruehan, "Scrap in Iron and Steelmaking: Technologies to Improve the Use of Scrap," *Iron & Steelmaker* 12:7 (1985): 31–36.
26. D. H. Zollner, H. J. Kuhn, and W. Redlich, "Present Trends for Graphite Operation in Electric Furnaces," *Steel & Metals Magazine* 27:8 (1989): 577–9.
27. E. Inagaki, I. Kikuma, and M. Ichikawa, "Integrated Oxygen Enrichment Control to Attain Maximum Overall Economy in Steelmaking Arc Furnaces," (Malaga, Spain: 11th International Electric Melting Conference, Oct. 3–7, 1988).
28. K. Klein, G. Paul, "Reflections on the Possibilities and Limitations of Cost Saving in Steel Production in Electric Arc Furnaces," *Metallurgical Plant and Technology* 1 (1989): 32–42.
29. J. W. Clayton and S. Severs, "Steelmaking for the 1990s—The New Scrap Handling System at Sheerness Steel," *Ironmaking and Steelmaking* 16 (1989): 277–88.
30. Badische Stahlwerke, personal communication.
31. Oxygen in Electric Arc Furnaces, Liquid Air Corp.
32. H. Adolph et al., "A New Concept for Using Oxy-Fuel Burners and Oxygen Lances to Optimize Electric Arc Furnace Operation," *Iron & Steelmaker* 16:2 (1989): 29–33.
33. M. B. Wells, F. A. Vonesh, "Oxy-Fuel Burner Technology for Electric Arc Furnaces," *Iron & Steelmaker* 13:11 (1986): 13–22.
34. K. Bergman and R. Gottardi, "Design Criteria for the Modern UHP Electric Arc Furnace With Auxiliaries," *Ironmaking and Steelmaking* 17:4 (1990), 282–7.
35. "Novel Sidewall Burner on Electric Arc Furnace," Best Practice Programme, U.K. Department of Energy, Final Profile 46, June 1991.

36. G. Schoenfelder, J. Pearce, and G. Kunze, *Application of State-of-the-art Technology for Modernization of Electric Arc Furnace Plants,* 929–38.
37. M. Bender, "Meltshop Emission Control in North America," *Steel Technology International,* (1988): 167–70.
38. H. Berger et al., "Methods, Aims and Objectives of Improved Electric Arc Furnace Steelmaking," *Steel & Metals Magazine* 27:8 (1989): 581–5.
39. D. Ameling et al., "Metallurgy and Process Technology of Electric Steelmaking—Development and Current Significance," *Metallurgical Plant and Technology International* 11:4 (1986): 49–65.
40. J. Goodwill, "Demand Side Management for Electric Arc Furnaces," Electric Furnace Conference Proceedings 49 (Warrendale: Iron and Steel Society, 1991): 81–5.
41. W. Ballandino et al, "The Oval Shaped Bottom Tapping Furnace at Acciaierie Venete," *Metallurgical Plant and Technology International* 13:5 (1989): 62–8.
42. M. Motlagh, "The Influence of Coke Injection on Power Consumption in an UHP Electric Furnace," *Iron & Steelmaker* 18:10 (1991): 73–8.
43. E. H. McIntyre and E. R. Landry, "EAF Steelmaking—Process and Practice Update," *Iron & Steelmaker* 17:5 (1993): 61–6.
44. N. Perrin et al., "Continuous Fume Analysis at Vallourec Saint-Saulve," *Electric Furnace Conference Proceedings* 49 (Warrendale, PA: Iron and Steel Society, 1991), 233–41.
45. A. J. Berthet and J-C. Grosjean, "The 90s Electric Arc Furnace Steelmaking Route: The Leap Forward," *Proceedings of the Sixth International Iron and Steel Congress*, (Nagoya, Japan: 1990), 180–9.
46. W. Ballandino and F. G. Hauck, "Steelmaking at Acciaierie Venete—Improved Performance by the KES Process," *Metallurgical Plant and Technology International* 16:5 (1992): 42–51.
47. I. Manzocco and L. Rizzani, "EAF Performance Improvement Program Through Multiple Balanced Injections of Technical Gases and Coal at Ferriere Nord," (Madrid, Spain: 4th European Electric Steel Congress, 1992): 215–24.
48. H. Berger et al., "Enhancement of EAF Performance by Injection Technology," *Proceedings, 22nd McMaster Symposium on Ironmaking and Steelmaking* (Hamilton, Ontario: McMaster University Press, 1994).
49. I. Manzocco et al., "Ferriere Nord EAF Process: Present Steelmaking Practice and Future Developments," *Danielli METEC Book* (1994), 18–24.
50. N. Perrin et al., "Application of the ALARC Post-combustion Oxygen Injection Practice In the Case of High Carbon Alternative Iron Sources," *Proceedings of the Alternative Iron Sources for the EAF Conference,* (Warrendale, PA: Iron and Steel Society, 1993).
51. P Boussard et al., "Industrial Practice of a New Injection Process of L'air Liquide For Post-combustion at Vallourec Saint-Saulve," *Electric Furnace Conference Proceedings* 50 (Warrendale, PA: Iron and Steel Society, 1992).
52. M. Aderup et al., "New Tools For Improved Operation Of High Efficiency Electric Arc Furnaces," *Electric Furnace Conference Proceedings* 50 (Warrendale, PA: Iron and Steel Society, 1992).
53. P. Mathur and G. Daughtridge, "Oxygen Injection For Effective Post-combustion in the EAF," *Electric Furnace Conference Proceedings* 51 (Warrendale, PA: Iron and Steel Society, 1993).
54. P. Mathur and G. Daughtridge, "High Efficiency Post-combustion in the Electric Arc Furnace," *Proceedings, 22nd McMaster Symposium on Ironmaking and Steelmaking* (Hamilton, Ontario: McMaster University Press, 1994).
55. P. Mathur and G. Daughtridge, "Recent Development in Post-combustion Technology at Nucor Plymouth," *Electric Furnace Conference Proceedings* 52 (Warrendale, PA: Iron and Steel Society, 1994).
56. M. Broxholme et al., "Practical Developments in Submerged Oxygen Tuyere Injection in EAF Steelmaking," *4th European Electric Steel Congress* (Madrid, Spain: 4th European Electric Steel Congress, 1992): 263–72.

57. M. Broxholme et al., "Practical Injection of Oxygen into the EAF Via Submerged Lances," *Proceedings, 22nd McMaster Symposium on Ironmaking and Steelmaking* (Hamilton, Ontario: McMaster University Press, 1994).
58. *ALARC Newsletter* (Houston, TX: Air Liquide, November 1994).
59. K-H. Klein et al., "The Rapid Melting Technology at BSW—A Combined Process of Oxygen, Gas and Coal Injection With ALARC-PC Post-combustion," *Proceedings, 22nd McMaster Symposium on Ironmaking and Steelmaking* (Hamilton, Ontario: McMaster University Press, 1994).
60. J. E. Rupert et al., "Creating a Competitive Edge Through New Technology," *Iron & Steelmaker* 23:10 (1996): 75–7.
61. D. S Gregory et al., "Results of ALARC-PC Post-combustion at Cascade Steel Rolling Mills Inc.," *Electric Furnace Conference Proceedings* 53 (Warrendale, PA: Iron and Steel Society, 1995), 211–7.
62. R. Weber, "Alternate Steelmaking Technologies," *6th Aachen Steel Colloquium* (1990).
63. R. Weber, *Steel Production With Optimized Raw Material and Energy Input*, (Metals Park, OH: ASM, 1991).
64. D. Huin et al., "Study of Post-combustion Mechanisms in a 6t Pilot Oxygen Converter," *Steelmaking Conference Proceedings* 71 (Warrendale, PA: Iron and Steel Society, 1988), 311–5.
65. Y. Kato et al., "Heat and Mass Transfer in a Combined Blowing Converter," 113th ISIJ Meeting (1987), lecture S213.
66. S. Sugiyama, "Heat Transfer Analysis of Post-combustion in LD Converter," 112th ISIJ Meeting (1987), lecture S1029.
67. L. J. Fischer, ed., *Combustion Engineers' Handbook*, (London: G. Newnes, 1961), 2–72.
68. *North American Combustion Handbook*, (Cleveland, OH: North American Manufacturing Company, 1952), 12.
69. N. Demukai, "Development of Pulverized Coal Burner for Scrap Heating," 114th ISIJ Meeting (1987), lecture S915.
70. S. Hornby-Anderson et al., *Cost and Quality Effectiveness of Carbon Dioxide in Steel Mills*, (Houston, TX: Air Liquide).
71. L. Zhang and F. Oeters, "A Model Of Post-combustion in Iron Bath Reactors, Part 1: Theoretical Basis," *Steel Research* 62:3 (1991): 95–106.
72. L. Zhang and F. Oeters, "A Model Of Post-combustion in Iron Bath Reactors, Part 1: Results for Combustion with Oxygen," *Steel Research* 62:3 (1991): 107–16.
73. M. Hirai et al., "Mechanism of Post-combustion in the Converter," *ISIJ International* 27 (1987): 805–13.
74. G. Gitman, American Combustion Inc., private communication.
75. H. Katayama et al., "Mechanism of Iron Oxide Reduction and Heat Transfer in the Smelting Reduction Process With a Thick Layer of Slag," *ISIJ International* 32:1 (1992): 95–101.
76. T. Hirata et al., "Stirring Effect in Bath Smelting Furnace With Combined Blowing of Top and Side Blown Oxygen and Bottom Blown Nitrogen," *ISIJ International* 32 (1992): 182–9.
77. K. Brotzman, "Post-combustion in Smelt Reduction," *Steel Times International* 16:1.
78. K. Takahashi et al., "Post-combustion Behavior in In-Bath Type Smelting Reduction Furnace," *ISIJ International* 32 (1992): 102–10.
79. B. A. Pollock, "BF or Alternative Iron?—A Review of the Ironmaking 2000 Symposium," *Iron & Steelmaker* 22:1 (1995): 33–5.
80. R. J. Leary and W. O. Philbrook, "An Efficient Preheater for Electric Furnace Charges," *Electric Furnace Conference Proceedings* 14 (Warrendale, PA: Iron and Steel Society, 1956), 300–13.
81. M. Bender, "Control of Carbon Monoxide Emissions From Electric Arc Furnaces," *PTD Conference Proceedings* 12 (Warrendale, PA: Iron and Steel Society, 1993), 335–42.
82. J. Clayton, Co-Steel Sheerness, private communication.
83. S. Kohle, "Variables Influencing Electric Energy and Electrode Consumption In Electric Arc Furnaces," *Metallurgical Plant and Technology International* 16:6 (1992): 48–53.

84. B. L. Farrand et al., "Post-combustion Trials at Dofasco's KOBM Furnace," *Steelmaking Conference Proceedings* 73 (Warrendale, PA: Iron and Steel Society, 1990), 331–6.
85. B. Hanas, "Improving Metallurgical Processes by Inductive Stirring," UIE, 9th International Congress (1980).
86. K. Ayata, T. Fujimoto, and T. Onoe, "Fluid Flow of Carbon and Stainless Steels by Electromagnetic Stirring," 111th ISIJ Meeting (1986), lecture S231.
87. S. Hornby-Anderson et al., "Worldwide Experience of Electric Arc Furnace Bottom Stirring," *Electric Furnace Conference Proceedings* 48 (Warrendale, PA: Iron and Steel Society, 1990), 167–81.
88. V. Pawliska et al., "Survey of Bottom Stirring in the Electric Arc Furnace," *Metallurgical Plant and Technology International* 15:5 (1992): 22–9.
89. C. T. Schade and R. Schmitt, "Evaluation of Bottom Stirring in Lukens' Electric Arc Furnace," *Electric Furnace Conference Proceedings* 48 (Warrendale, PA: Iron and Steel Society, 1990), 157–65.
90. Y. Kawazu et al., "New Long Life Bottom Stirring EF-KOA," *Electric Furnace Conference Proceedings* 49 (Warrendale, PA: Iron and Steel Society, 1991), 377–82.
91. M. F. Riley and S. K. Sharma, "An Evaluation of the Technical and Economic Benefits of Submerged Inert Gas Stirring in an Electric Arc Furnace," *Iron & Steelmaker* 14:6 (1987).
92. Data presented on SMI Seguin results with AMI regulator at the Iron and Steel Society 1993 Spring Globetrotters Meeting in San Antonio, Texas.
93. A. Sgro and A. Cozzi, "Application of Digital Technology in Electrode Control," *Metallurgical Plant and Technology International* 15:6 (1992): 56–8.
94. K. Bileman, Badische Stahl Engineering GmBH, private communication.
95. E. A. Elsner and H. Dung, "Advantages of Electric Furnace Operation with Low System Reactance," *Metallurgical Plant and Technology International* 14:1 (1991): 44–51.
96. J. Ehle et al., "Design and Operational Results of Current Conducting Electrode Arms for Electric Arc Furnaces," *Metallurgical Plant and Technology International* 8:6 (1985): 28–35.
97. K-H. Klein et al., "Experiences with a New Type Current Conducting Electrode Arm Made of Aluminum at Badische Stahlwerke AG, Kehlî, *International Steel and Metals,* Vol. 10 (1989).
98. K. Bergman, "System Design Characteristics and Experience from Operating EAFs with High Arc Voltages," *Electric Furnace Conference Proceedings* 46 (Warrendale, PA: Iron and Steel Society, 1988), 283–8.
99. K. Bergman, R. Gottardi, "Design Criteria for the Modern UHP Electric Arc Furnace With Auxiliaries," *Proceedings of the 3rd European Electric Steel Congress* (1989), 169–78.
100. K. Bergman, "Danieli High Impedance Electric Arc Furnace," *Iron and Steel Engineer* 69:7 (1992): 37–42.
101. G. Gensini and V. Garzitto, "New Developments in Electric Arc Furnace Technology," *Metallurgical Plant and Technology International* 14:1 (1991): 52–5.
102. E. G. Mueller and M. Schubert, "UNARC DC Steelmaking," (Charleston, SC: 13th Advanced Technology Symposium, Plasma 2000, April 1992)
103. "Direct Current EAF—A Review," *Steel Times* 7 (1990).
104. "Developments in EAF and Ladle Furnaces," *Steel Times* 7 (1990).
105. Presented at Seminar on DC Furnaces, Plant Operations Division, Steel Manufacturers Association, 1991 Spring Meeting.
106. T. Maki, "Updated Description of NKK-type DC Arc Furnace," *Steel World* (1992): 58–62.
107. K. Ishihara et al., "Benefits of Direct Current Arc Furnace Operation," *Steel Technology International,* (1993): 109–1.
108. "Japan's Largest DC Arc," *Steel Times International* 5 (1989): 17.
109. X. Xingnan, "DC Furnaces in Ascendancy," *Steel Times International* 11 (1991): 42–4.
110. T. Maki, "DC Furnaces Make the Running in Japan," *Steel Times International* 7 (1992): 11–3.
111. *Scrap Preheating and Melting in Steelmaking,* (Warrendale, PA: Iron and Steel Society, 1986), 169–72.
112. "Giving EAFs the Shaft to Recoup Energy," *33 Metal Producing* 28:11 (1990): 52.

113. M. Haissig, "New High Performance Electric Arc Furnace Concept," *Iron and Steel Engineer,* 69:7 (1990): 43–7.
114. J. Clayton, "The Sheerness Shaft Furnace," manuscript supplied from author.
115. J. Clayton et al., "The Sheerness Shaft Electric Furnace," *Steel Technology International,* (1993): 97–101.
116. Co-Steel Sheerness, private communication.
117. Fuchs Systems, Inc., private communication.
118. Robert Heard, Paul Wurth, Inc., private communication.
119. J. Bosley and D. Klesser, "The Consteel Scrap Preheating Process," CMP Report 91-9, (Pittsburgh: EPRI Center for Materials Production, 1991).
120. Intersteel Technology, private communication.
121. Results reported by IHI at the 25th ISS Advanced Technology Symposium in St. Petersburg, Florida.
122. H. Berger, Voest-Alpine Ind., private communication.
123. A. Kurzinski, Mannesmann Demag, private communication.

Chapter 11

Ladle Refining and Vacuum Degassing

G. J. W. Kor, Scientist, The Timken Co. (Retired)
P. C. Glaws, Senior Research Specialist, The Timken Co.

The treatment of steel in the ladle started approximately 45 years ago when the first ladle-to-ladle and ladle-to-ingot mold vacuum degassing processes for hydrogen removal appeared on the scene. In the late 1950s more efficient vacuum degassers such as the Dortmund Hoerder (DH) and Ruhrstahl-Heraeus (RH) processes became popular. In the middle 1960s degassing processes such as vacuum arc degassing (VAD), the ASEA-SKF process, and the vacuum oxygen decarburization (VOD) process for treating high-chromium steels were successfully implemented. Converter processes such as the argon oxygen decarburization (AOD) process were introduced in the early 1970s. The AOD process is now the preferred route in many specialty steel and stainless steel shops.

Granulated flux injection into the liquid steel, combined with argon stirring, started in the early 1970s. This was soon followed by the application of cored-wire feeding of alloying elements for better control of composition and inclusion morphology. A good overview of the various developments was given by Nijhawan[1], while an extensive review of the thermodynamic and kinetic principles underlying the various secondary steelmaking processes was prepared by Lange.[32]

All the aforementioned innovations have had a pronounced effect on the steelmaking process, particularly with respect to the vessel or the furnace in which the steel is produced. For example, the implementation of ladle metallurgy and its related aspects enabled electric furnace steelmakers to use their furnaces as fast melters without the need to perform any refining in the furnace. In addition, ladle refining and degassing make it possible for the steelmaker to exert much tighter control over the properties of the final product through improved accuracy in the composition of the final product as well as its cleanliness and by being able to control inclusion morphology.

The contents of this chapter are arranged according to the sequence of operations in a steelmaking shop, i.e. starting with tapping the furnace, followed by reheating, refining, inclusion modification and degassing. Where appropriate, the underlying metallurgical principles of each operation will be discussed in terms of reaction equilibria and kinetics as well as fluid dynamics. In preparing this chapter frequent reference has been made to a recently published book by Turkdogan.[2]

11.1 Tapping the Steel

11.1.1 Reactions Occurring During Tapping

During tapping of the steel, air bubbles are entrained into the steel where the tap stream enters the bath in the tap ladle. The quantity of air entrained into the steel increases with the increasing free fall height of the tap stream as was demonstrated with the aid of water model studies.[2] The entrainment of a gas such as air into a falling stream of liquid steel has been the subject of a number of studies.[3-6] However, a reliable prediction of the quantity of air entrained into a stream of liquid steel during tapping is difficult because of the assumptions that have to be made.

The nitrogen contained in the air entrained by the steel will be absorbed by the liquid steel depending on the extent to which the reaction

$$N_2 (g) \rightarrow 2 [N] \qquad (11.1.1)$$

will proceed to the right; the symbol within the square brackets refers to nitrogen dissolved in the steel. It is well-known that surface active solutes such as oxygen and sulfur impede the kinetics of nitrogen absorption by the steel. The higher the concentration of dissolved oxygen and/or sulfur, the lower is the extent of nitrogen absorption. This is illustrated in Fig. 11.1 where the effect of deoxidation practice on the nitrogen pickup during tapping of an electric furnace is shown.[7] For deoxidized steels the average nitrogen pickup during tap is significantly higher than for non-deoxidized steels. The same effect is shown in Fig. 11.2 where the nitrogen pickup during tapping of 220 tonne oxygen converter heats is depicted as a function of the dissolved oxygen content for steels containing approximately 0.01% sulfur.[2] The data in Fig. 11.1 and Fig. 11.2 are in complete accord.

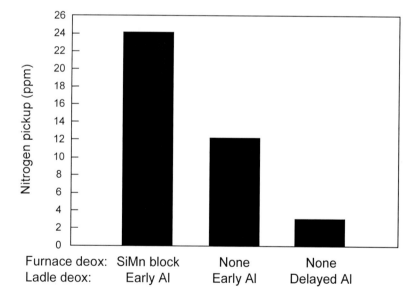

Fig. 11.1 Effect of deoxidation practice on nitrogen pickup during tap. *From Ref. 7*

Other sources contributing to the nitrogen pickup during or shortly after tap are: petroleum coke, when used for recarburization and various ferroalloys, particularly ferrotitanium, ferrovanadium and low and medium-carbon ferrochromium.

Ladle additions often contain moisture which reacts with the liquid steel according to the following reaction:

$$H_2O \rightarrow 2[H] + [O] \qquad (11.1.2)$$

Ladle Refining and Vacuum Degassing

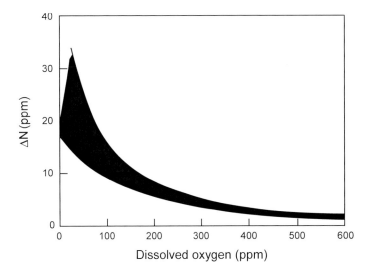

Fig. 11.2 Nitrogen pickup during tapping of 220 tonne heats as affected by the extent of deoxidation in the tap ladle. *From Ref. 2.*

It is seen from this equation that the extent of hydrogen pickup will be more pronounced for a fully deoxidized steel in which the dissolved oxygen content is low.

Of the various ferroalloys, ferromanganese is probably the major contributor of hydrogen.

11.1.2 Furnace Slag Carryover

It is generally unavoidable that a quantity of furnace slag is carried over into the tap ladle during tapping. The furnace slag generally contains a high concentration of FeO and MnO and therefore (in an untreated form) is not suitable for use as a refining slag. Accordingly, methods to minimize the amount of furnace slag carryover have been developed and implemented; Szekely et al. discuss various methods for BOFs and EAFs.[8] These include locating the taphole in the barrel of the converter sometimes in conjunction with the use of a ceramic sphere to block off the taphole towards the end of tap. In electric arc furnaces the first improvement was the use of submerged tapholes, later followed by the widespread implementation of eccentric bottom tapholes, which are now common in modern arc furnaces.

In older shops, inadequately equipped to control furnace slag carryover, slag raking is often practiced to remove the furnace slag. A good slag-free surface is attainable by careful raking. According to Hoeffken et al. raking is usually accompanied by a temperature loss of approximately 2.5°C (~5°F) per minute of treatment time and a metal loss of approximately 0.2%,[9] raking times are typically of the order of ten minutes. For best results it is recommended that the steel be tapped open, then raked and covered with the synthetic ladle slag and finally deoxidized.

As a result of furnace slag carryover into the tap ladle, oxidation of aluminum and silicon present in the ladle additions occurs through reactions with less stable oxides (e.g. iron oxide and manganese oxide) present in the furnace slag. Another consequence of furnace slag carryover is phosphorus reversion from the slag to the steel, particularly when the steel is fully deoxidized.

To be able to predict the aluminum and silicon losses as well as the anticipated degree of phosphorus reversion, it is necessary to know the quantity of furnace slag carried over into the tap ladle. Kracich et al. describe a sensor to measure the depth of the slag layer in a ladle.[10] These data provide accurate feedback to the melter and assist him in controlling the amount of furnace slag carryover.

11.1.2.1 Aluminum and Silicon Losses

The reactions of aluminum and silicon dissolved in the steel with the iron and manganese oxide in the slag and with the fallen converter skull may be represented by the following general reaction:

$$Fe(Mn)O_x + Al(Si) = Fe(Mn) + Al(Si)O_x \qquad (11.1.3)$$

Using average molecular masses and assuming 80% Fe_3O_4 for the composition of the skull, Turkdogan[11] derived the following approximate emperical relation for the percentages of aluminum and silicon lost to the ladle slag for a 200 tonne steel bath in the ladle:

$$[\%Al + \%Si]_s \approx 1.1 \times 10^{-6} \Delta(\%FeO + \%MnO) W_{fs} + 1.1 \times 10^{-4} W_{sk} \qquad (11.1.4)$$

where W_{fs} and W_{sk} are the weights (kg) of the carried over furnace slag and the fallen converter skull, respectively and Δ(FeO + %MnO) is the decrease in oxide contents of the furnace slag during tapping.

11.1.2.2 Phosphorus Reversion

Furnace slag carryover generally results in phosphorus reversion from the slag to the steel, particularly when the steel is fully deoxidized. The general relationship for the increase in steel phosphorus content, $\Delta[\%P]$, as a result of reversion from the slag is:

$$\Delta[\%P] = (\%P)(W_{fs} / W_b) \quad (11.1.5)$$

where (%P) is the phosphorus content of the furnace slag and W_{fs} and W_b are the weights of the carried-over furnace slag and the steel bath in the ladle, respectively.

For OBM (Q-BOP) heats and low-carbon heats made in EAFs with oxygen injection, typical values for (%P) and $\Delta[\%P]$ are approximately 0.3 and 0.003, respectively.[11,12] Substitution of these values into equation 11.1.5 gives $W_{fs}/W_b \approx 0.01$. In other words, when proper measures are taken to prevent excessive furnace slag carryover, the average quantity of carried-over converter or furnace slag is approximately 1% of the steel tapped. This quantity of slag carried over during tapping corresponds to a slag thickness in a 200 tonne ladle of 5.5 ± 3 cm, in general agreement with plant observations.[11]

Hoeffken et al.[9] observed that phosphorus reversion is more likely to occur when both the basicity, %CaO/%SiO$_2$, of the carryover slag is approximately 2 or lower and its iron oxide content is approximately 17% or lower. For iron oxide contents of approximately 25% or higher the phosphorus reversion is noticeably less, provided the slag basicity exceeds 2–2.5.

For heats that are tapped open, to which only ferromanganese and a small amount of aluminum are added, the steel is not sufficiently deoxidized to cause phosphorus reversion. In fact, in some cases of tapping open heats to which 0.3–0.6% manganese is added, the phosphorus content of the steel decreases by approximately 0.001% due to mixing of the carried-over furnace slag with the steel during tapping.[11]

11.1.3 Chilling Effect of Ladle Additions

Ferroalloys and fluxes added to the steel in the tap ladle affect the temperature of the steel in the ladle, usually resulting in a decrease in temperature. The effect of various alloying additions, including coke, on the change in temperature of the steel for an average bath temperature of 1650°C (3002°F) is summarized in Table 11.1. These data were calculated from the heat capacities and heats of solution of the various solutes.

Table 11.1 Effect of Alloying Additions on the Change in Temperature of the Steel in the Tap Ladle for an Average Bath Temperature of 1650°C (3002°F).

Addition to give 1% of alloying element at 100% recovery	Change in steel temperature ΔT, °C (°F)
Coke	–65 (–117)
FeCr (50%), high-C	–41 (–74)
FeCr (70%), low-C	–28 (–50)
FeMn, high-C	–30 (–54)
FeSi (50%)	~0 (~0)
FeSi (75%)	+14 (+25)

It can be seen from Table 11.1 that ferrosilicon is the only ferroalloy that, upon addition, does not result in a decrease in steel bath temperature; in fact, the use of FeSi (75%) results in an increase in temperature. This is a consequence of the fact that the dissolution of silicon into liquid iron is exothermic, i.e. heat is liberated. Although FeSi (75%) usually costs more per unit weight of contained silicon than FeSi (50%), the use of the former material can be justified under certain conditions, particularly when relatively large quantities must be added and when the shop has no or limited reheating facilities such as ladle furnaces.

For aluminum-killed steels, the exothermic heat of the deoxidation reaction must be taken into account when calculating the effect of an aluminum addition to the tap ladle on the change in steel temperature. For example, when a steel containing 600 ppm of dissolved oxygen is deoxidized with aluminum, the heat generated by the deoxidation reaction results in a change in steel temperature of +19°C (+34°F). In other words, when the steel in the tap ladle is deoxidized with aluminum, the decrease in steel temperature as a result of tapping will be less by 19°C (34°F).

Flux and slag conditioner additions decrease the temperature of the steel in the ladle. The effect of these additions on the change in steel temperature as determined from the heat capacity data is summarized in Table 11.2.

Table 11.2 Change in Steel Temperature in the Tap Ladle as a Result of Various Flux Additions at a Rate of 1 kg/tonne Steel (2.0 lbs/ton).

Flux added (1 kg/tonne)	Change in steel temperature ΔT, °C (°F)
SiO_2	−2.5 (−4.5)
CaO	−2.0 (−3.6)
MgO	−2.7 (−4.9)
CaO•MgO (dolomite)	−2.3 (−4.1)
CaO•Al_2O_3 (Ca-aluminate)	−2.4 (−4.3)
CaF_2	−3.2 (−5.8)

When tapping aluminum-killed steels to which typical ladle additions of 10 kg lime and pre-fused calcium aluminate per tonne of steel (20 lbs/ton) are made, the decrease in steel temperature during tapping to a pre-heated ladle is 55–75°C (99–135°F). The heat loss due to the flux addition is approximately balanced by the heat generated by the deoxidation reaction with aluminum. Thus, in this particular practice the heat losses are almost entirely from radiation and conduction into the ladle lining.

11.2 The Tap Ladle

11.2.1 Ladle Preheating

In most oxygen and many EAF steelmaking shops the ladle lining consists of high-alumina bricks (70–80%Al_2O_3) while the slag line consists of magnesia bricks, usually containing approximately 10% carbon and small amounts of metallic additions such as aluminum, magnesium or chromium to minimize the oxidation of carbon. For many EAF shops the lining is dolomite. The refractory materials used in tap ladles are discussed in detail in Chapter 4 and will not be discussed here further,

Table 11.3 Heat Storage Capacity per Unit Volume of Various Refractory Materials at 1200 and 1500°C Relative to Magnesia (=100%), after Tomazin et al.[14]

Material	1200°C (2192°F)	1500°C (2732°F)
70% alumina	83%	85%
50% alumina	79%	82%
fireclay	73%	74%

except to make reference to a series of articles by Engel et al.[13] which represents a comprehensive overview of refractory materials and configurations employed in secondary steelmaking.

The thermal properties of various refractories used in ladles are summarized in Table 11.3 and Fig. 11.3.

It is seen from Fig. 11.3 that magnesia has a significantly higher thermal conductivity than the other refractory materials. Magnesia also has a higher heat storage capacity than the other materials as shown in Table 11.3. The properties of dolomite are similar to those of magnesia.

Because of the relatively high thermal conductivity of the refractories used in ladles, preheating of the ladle prior to its use is required to avoid excessive heat losses during tapping and during subsequent refining operations. Tomazin et al.[14] studied the effect of ladle refractories and practices on steel temperature control and developed a mathematical model to simulate a ladle which is used to supply liquid steel to a continuous caster. The model was used to evaluate the

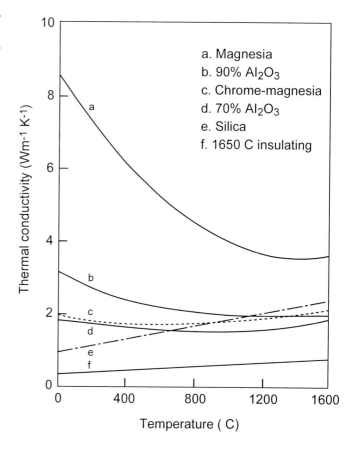

Fig.11.3 Thermal conductivity of refractory bricks used in ladle linings. *From Ref. 2.*

temperature increase of the hot and cold faces of a ladle during preheating. This is shown in Fig. 11.4 from which it is seen that the temperature of the hot face increases rapidly with time. However, the cold face temperature does not exceed 100°C (212°F) after preheating for approximately 5½ hours when the ladle is cold and dry initially.

The rate of rise of temperature of the hot face depends on the distance from the ladle top to the burner wall as well as on the thermal input from the preheater. A rapid heating rate should be avoided for the following reasons:

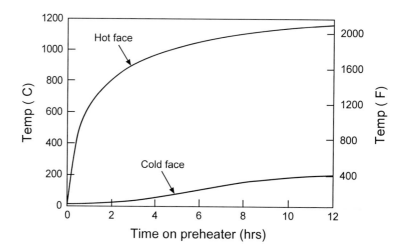

Fig. 11.4 Lining temperatures during preheating of a cold dry ladle. *From Ref. 14.*

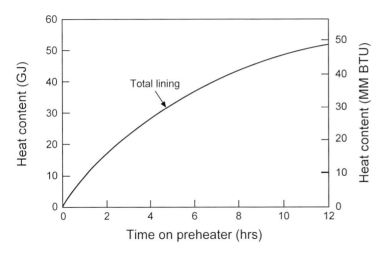

Fig. 11.5 Total heat content of a lining as a function of preheating time. *From Ref. 14.*

(i) Rapid heating results in a non-equilibrium temperature profile, i.e. a steep temperature gradient adjacent to the hot face.

(ii) Rapid heating causes extreme shell stresses.

(iii) The thermal shock resistance of the brick may not be high enough to withstand a rapid heating rate.

Another measure of the thermal condition of the ladle is the total heat content of the ladle brick. As shown in Fig. 11.5, the brick continues to absorb heat at a significant rate for up to twelve hours. At this time the ladle brick contains in excess of 90% of the maximum heat content that can be attained once steady state has been reached after 17–18 hours of preheating.

After preheating the ladle is moved to the converter or the furnace. This causes a decrease in the temperature of the hot face. The temperature decrease must be taken into account when adjusting the tap temperature. The required adjustment in tap temperature is shown as a function of the time elapsed between the end of preheating and the start of tapping for two hot face temperatures, Fig. 11.6.

11.2.2 Ladle Free Open Performance

Upon arrival of the ladle at the continuous caster tundish, it has to be opened to allow the steel to flow into the tundish. When the ladle slide gate is stroked open and steel starts to flow without operator assistance, the procedure is classified as 'free open'. If poking or oxygen lancing is nec-

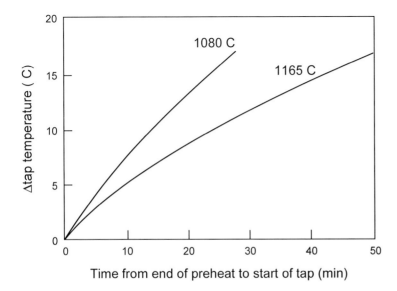

Fig. 11.6 The effect of cooling time for two different preheat temperatures on the tap temperature adjustment; ladle not covered between end of preheat and tap. *From Ref. 14.*

essary to open the ladle, the process is classified as 'assisted open'. When all attempts to open the ladle are unsuccessful, it is classified as 'non-free open'. These classifications were taken from Vitlip[15] who undertook a comprehensive study of ladle free open performance.

According to Vitlip[15] there are several factors that determine the opening performance of a ladle. The most dominant factor is the residence time of the steel in the ladle followed by the ladle preheat practice, the elapsed time between the end of stirring and the opening of the ladle at the caster, the cycle time for an empty ladle and the argon stirring practice. The relative effect of these factors on the percentage of free open ladles is summarized in Table 11.4.

Table 11.4 Percentage of Free Open Ladles as Affected by Various Factors, after Vitlip.[15]

Factor	Free Open	
Steel residence time in the ladle	< 5 hr, 85%	> 5 hr, 80.8%
Ladle preheat practice	normal cycle, 98.2%	repair, 90.9%
Time between end of stirring	< 20 min, 98.4%	> 20 min, 94.1%
Empty ladle cycle time	< 2 hr, 98.7%	> 2 hr, 97.0%
Argon stirring practice	bottom, 98.5%	top, 97.9%

11.2.2.1 Steel Residence Time in the Ladle

The residence time in the ladle is defined as the elapsed time between tapping and the opening of the ladle at the tundish. Under the operating conditions prevailing at the Wheeling-Pittsburgh Steel Corporation this time is 80 min.[15] However, longer residence times are not uncommon whenever problems are encountered at the caster. It can be seen from Table 11.4 that a residence time in the ladle longer than five hours results in a significant decrease in free open performance. Long residence times in the ladle cause partial sintering of the nozzle fill material due to extended exposure to liquid steel temperatures.

11.2.2.2 Ladle Preheat Practice

The ladle preheat practices at Wheeling-Pittsburgh were examined for ladles returned to service following repair.[15] The data in Table 11.4 show that the free open performance is worse for repaired ladles compared with those that follow the normal cycle, indicating an inadequate preheat practice for repaired ladles.

11.2.2.3 Elapsed Time Between End of Stirring and Opening

The data in Table 11.4 show that the time elapsed between the end of stirring at the trim station and the opening of the ladle at the tundish is important. The longer the time without stirring the worse is the free opening performance.

11.2.2.4 Empty Ladle Cycle Time

The empty ladle cycle time is the elapsed time between closing the ladle at the caster at the end of a cast and the time the ladle is filled again at tap. The practice at Wheeling-Pittsburgh was not to preheat the ladles when they were being rotated between the caster and the converter. However, the rate of rotation could vary significantly because of delays. The data in Table 11.4 show that an empty ladle cycle time longer than two hours leads to a decrease in free open performance.

11.2.2.5 Argon Stirring Practice

Today most ladles are equipped with bottom plugs for argon bubbling. Whenever the plug is out of commission, it may be necessary to provide the required stirring via a top lance. The data in Table 11.4 show that top stirring with argon results in a slightly decreased free open performance.

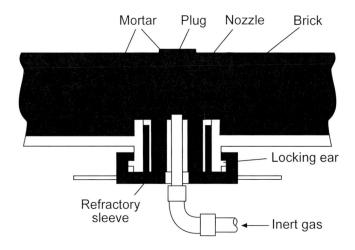

Fig. 11.7 Porous plug assembly in the bottom of a ladle. *From Ref. 2.*

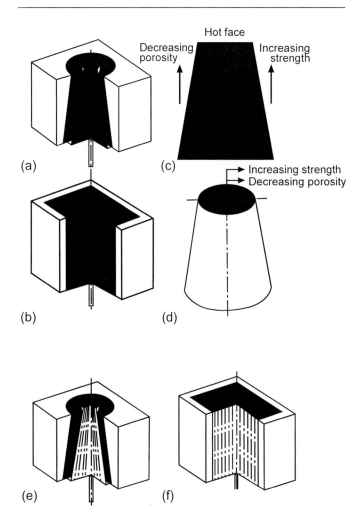

Fig. 11.8 Standard shapes of isotropic plugs: (a) and (b). Component plugs: sliced (e), concentric (d). Capillary plugs: conical (e) rectangular (f). *From Refs. 16 and 17*

Vitlip[15] studied the effect of a number of other factors on free open performance, the details of which may be found in the cited reference.

11.2.3 Stirring in Ladles

To achieve a homogeneous bath temperature and composition, the steel in the ladle is most often stirred by means of argon gas bubbling. For moderate gas bubbling rates, e.g. less than 0.6 Nm^3/min (~20scfm) porous refractory plugs are used, usually mounted in the bottom of the ladle. A schematic illustration of a porous plug assembly in the ladle bottom is shown in Fig. 11.7.

Anagbo and Brimacombe[16] discussed some typical examples of various plugs, shown in Fig. 11.8. As can be seen, porous plugs have either a conical or a rectangular shape. The conically shaped plug is easier to change should the plug wear out before the lining. Rectangular plugs are geometrically compatible with the surrounding bricks and can be used to advantage in cases where the plug life is comparable with that of the lining. The performance and life of isotropic plugs can be improved by producing the element in two or three components stacked together with metal inserts.[17] The primary advantage of the so-called directional-porosity or capillary plug, shown in Fig. 11.8(e) and (f), is that the plug can be made of the same dense refractory as the lining brick, or even denser. This results in increased hot compression strength, greater resistance to erosion and a longer service life. Disadvantages of capillary plugs are that they are more prone to infiltration by liquid steel upon loss of argon gas pressure. More details regarding the configurations of plugs and the modeling of porous plug operations can be found in the cited references.[16,17]

Some melt shops utilize electromagnetic induction stirring in the ladles.

Some reported features of induction stirring in the ladle include better stirring homogeneity (especially near the ladle bottom), the ability to reverse the direction of the stirring forces (useful for alloy additions) and stirring without breaking slag cover and exposing steel to the ambient oxidizing atmosphere. These benefits are offset by the high capital cost, including ladles equipped with stainless steel panels comprising at least 1/3 of the ladle shell and the need for auxiliary gas stirring for adequate hydrogen removal.

11.2.3.1 Stirring Power and Mixing Times

Homogenization of bath temperature and composition by gas bubbling is primarily caused by the dissipation of the buoyant energy of the injected gas. The thermodynamic relationship describing the effective stirring power of a gas was derived by Pluschkell.[18] The following equation for the stirring power is derived from Pluschkell's relationship:

$$\dot{\varepsilon} = 14.23 \left(\frac{\dot{V}T}{M} \right) \log \left(\frac{1+H}{1.48 P_o} \right) \quad (11.2.1)$$

where:

- $\dot{\varepsilon}$ = stirring power, W/tonne
- $\dot{V}$ = gas flowrate, Nm³/min
- T = bath temperature, K
- M = bath weight, tonne
- H = depth of gas injection, m
- P_o = gas pressure at the bath surface, atm

The stirring time to achieve 95% homogenization is defined as the mixing time τ. There have been numerous experimental and theoretical studies dealing with mixing phenomena in gas-stirred systems. Mazumdar and Guthrie[19] published a comprehensive review on the subject. The following relationship expressing the mixing time, τ, in terms of the stirring power, $\dot{\varepsilon}$ (W/tonne), ladle diameter, D(m), and depth of injection, H(m), was obtained from the work of Mazumdar and Guthrie.[20]

$$\tau(s) = 116 \, (\dot{\varepsilon})^{-1/3} (D^{5/3} H^{-1}) \quad (11.2.2)$$

Mixing times calculated from equations 11.2.1 and 11.2.2 are shown in Fig. 11.9 for the simplified case of D = H. The mixing times shown in Fig. 11.9 are in good agreement with those calculated from other correlations.[21,22] It can be seen from Fig. 11.9 that a 200 tonne heat will be homogenized in 2–2½ minutes after bubbling with argon at a flow rate of 0.2 Nm³/min (~7 scfm).

The effect of the location of the bottom stirring plug on mixing times was studied by several authors.[23,24] The general finding was that the mixing time is decreased by placing the bottom plug off-center, e.g. at mid-radius.

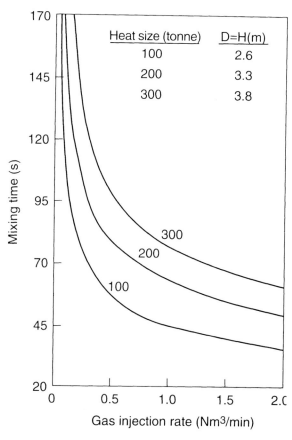

Fig. 11.9 Calculated mixing times for 100, 200 and 300 tonne heat sizes.

According to Mietz and Oeters[23] a stirring plug placed in the center of the ladle bottom generates a toroidal loop of metal flow in the upper part of the bath while a dead zone is created in the lower part, resulting in longer mixing times. Eccentrically located bottom plugs give rise to extensive circulation of metal throughout the entire bath, avoiding dead zones and leading to shorter mixing times.

11.2.3.2 Slag–Metal Reaction Rates in Gas-Stirred Melts

Numerous experimental studies to investigate slag–metal gas reactions in gas-stirred ladle systems under a variety of experimental conditions have been performed. Reviews on this subject were prepared by Mazumdar and Guthrie[19], Emi[25] and Asai et al.[26] For most slag–metal reactions the rates are controlled primarily by mass transfer of the reactants and products across the slag–metal interface. In stirred systems such as a steel bath in a ladle stirred by argon, the slag–metal interfacial area is affected by the degree of agitation in the bath which, in turn, is determined by the stirring power.

11.2.3.2.1 Desulfurization During desulfurization of the steel in the ladle, the mixing of slag and metal is achieved by argon bubbling and the rate of desulfurization is described by equation 2.10.19 in Chapter 2.

The overall rate constant is related to the average mass transfer coefficient, m_S, the slag–metal interfacial area, A, and the steel both volume, V, by the following expression:

$$k_S = m_S \left(\frac{A}{V}\right) \quad (11.2.3)$$

It has been shown that:

$$k_s \propto \dot{\varepsilon}^n \quad (11.2.4)$$

where the exponent n may vary between 0.25 and 0.30, depending on the specific system under consideration.[26]

From pilot plant tests with 2.5 tonne heats[26] to study desulfurization, it was found that at moderate gas bubbling rates, corresponding to $\dot{\varepsilon} < 60$ W/tonne, there was little or no slag–metal mixing, hence the rate of desulfurization was slow. For higher stirring rates corresponding to $\dot{\varepsilon} > 60$ W/tonne, better mixing of slag and metal was achieved and the rate constant for desulfurization increased accordingly. The results of these experiments are summarized in Fig. 11.10 from which the following approximate relationships between the overall rate constant and the stirring power are derived:

$$k_S \,(\text{min}^{-1}) \approx 0.013\, (\dot{\varepsilon})^{0.25} \text{ for } \dot{\varepsilon} < 60 \text{ W/tonne}$$

$$k_S \,(\text{min}^{-1}) \approx 8.10^{-6}\, (\dot{\varepsilon})^{2.1} \text{ for } \dot{\varepsilon} > 60 \text{ W/tonne} \quad (11.2.5)$$

It should be stressed that these are empirical correlations. The value of k_s depends on the energy dissipation per unit area at the slag–metal interface, the properties of the slag and the amount of the slag.

The abrupt change in overall rate constant for values of the stirring power at approximately 60 W/tonne is explained[8] by the fact that an increase in the energy input rate results in increased emulsification of slag and metal, leading to an increase in interfacial area, A, which, in turn, increases k_S.

11.2.3.2.2 Dephosphorization The removal of phosphorus from the steel by the ladle slag is governed by the same rate equation as that for sulfur removal, equation 2.10.19. Thus, the overall rate constant for dephosphorization is expected to have the same form as equation 11.2.4. That this is, in fact, the case was shown by Kikuchi et al.[27] who studied dephosphorization in the ladle with

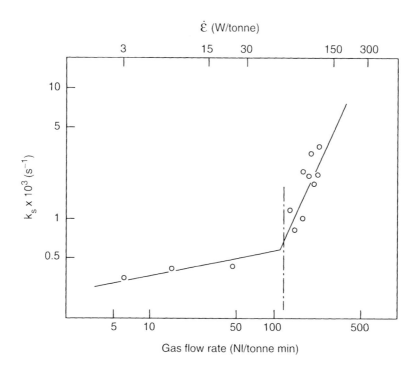

Fig. 11.10 Effect of gas flowrate and stirring power on the desulfurization rate constant. *From Ref. 26.*

CaO–CaF$_2$–FeO slags in a 50 tonne VAD/VOD as well as in a 250 tonne ladle furnace facility. The overall rate constant can be represented by the following approximate relationship:

$$k_P \, (\text{min}^{-1}) \approx 0.019 \, (\dot{\varepsilon})^{0.28} \tag{11.2.6}$$

This expression is similar to that for k_S, valid for $\dot{\varepsilon} < 60$ W/tonne, equation 11.2.5. It is not clear why no abrupt change in k_P was observed for $\dot{\varepsilon} > 60$ W/tonne, as in the case of desulfurization, Fig. 11.10.

It is interesting to note that Kim and Fruehan[28] as well as Mietz et al.[29] observed that mass transfer between metal and slag is impeded when the stirring plug in the ladle bottom is located off-center. A stirring plug located in the center results in increased slag–metal emulsification with increasing gas flowrate. Eccentrically located stirring plugs create a slag-free zone, the so-called eye, close to the ladle wall. This affects the detachment of slag particles from the main slag phase and results in decreased emulsification.[29] The ultimate choice of location of the stirring plug in the bottom of the ladle would, therefore, appear to be determined by which aspect of stirring is more important for a given operation: good mixing characteristics, or the ability to achieve rapid desulfurization and/or dephosphorization. In most cases a compromise will have to be struck.

11.2.4 Effect of Stirring on Inclusion Removal

One of the objectives of stirring the steel in the ladle is the removal of non-metallic inclusions. Nakanishi and Szekely[30] studied the deoxidation kinetics of aluminum-deoxidized steels in 20 kg melts as well as in a 50 tonne ASEA-SKF furnace. The authors developed a model for inclusion removal based on the postulate that the decrease in total oxygen content is determined by the coalescence of oxide particles as the rate controlling step. The model is in essence a combination of a coalescence theory and an algorithm for turbulent recirculatory flows.

Engh and Lindskog[31] also presented a fluid mechanical model for inclusion removal from liquid steel. According to their model the total oxygen content after stirring time, t, is given by

$$\frac{C_t - C_f}{C_i - C_f} = e^{-at} \qquad (11.2.7)$$

where

- C_t = total oxygen content after the stirring time, t,
- C_f = final total oxygen content after long stirring times (steady state),
- C_i = initial total oxygen content,
- α = the time constant for inclusion removal.

It must be stressed that equation 11.2.7 is an extremely simplified expression. The rate of inclusion removal depends on many factors including the inclusion type, refractory type, exact stirring conditions, etc.

From experiments with an inductively stirred 140 tonne melt[31] employing a range of values for the specific stirring power, $\dot{\varepsilon}$, the following approximate relationship can be obtained:

$$\alpha \,(\text{min}^{-1}) \approx \frac{\dot{\varepsilon}}{27} \qquad (11.2.8)$$

The above expression is an approximation and is valid only for moderate induction stirring. If it is assumed that the final steady state total oxygen content, C_f, is small compared to C_i and C_t, combination of equations 11.2.7 and 11.2.8 gives:

$$\frac{C_t}{C_i} \approx e^{-\dot{\varepsilon}\tau/27} \qquad (11.2.9)$$

11.3 Reheating of the Bath

The ever increasing pressure on steelmakers to lower operating costs and increase efficiency has made it necessary to make effective use of furnaces, BOF or EAF, and implement sequential continuous casting. These factors have prompted the installation of facilities for steel reheating, needed for the additional time required for steel refining and the adjustment of the temperature of the steel for uninterrupted sequential casting. The two methods for reheating steel in the ladle, arc reheating and injecting oxygen and aluminum or silicon, will be discussed separately.

11.3.1 Arc Reheating

Over time, several types of furnaces for arc reheating have been developed and commercialized. Examples are: ASEA-SKF, Daido-NKK, Finkl-Mohr, Fuchs, Lectromelt, MAN-GHH, Stein Heurty-S.A.F.E., etc. Some of these designs also have the capability for degassing of the steel. An important issue in arc reheating of a steel bath is whether the thermal energy which is supplied at, or near, the surface of the melt can be dispersed rapidly enough such that no significant temperature gradients are created within the steel in the ladle. Szekely[8] estimated that, in the absence of agitation, the Biot number is of the order of 300, indicative of significant temperature gradients in the steel bath. (The Biot number is defined as $N_{Bi} = hL/k_{eff}$, where h is the heat transfer coefficient between the arc and the bath, k_{eff} is the thermal conductivity of the steel and L is the bath depth.) In systems agitated either by induction stirring or gas bubbling, the Biot number is estimated to be of the order of 5.0, indicative of small temperature gradients even in agitated systems. Once the heat supply is discontinued, the temperature in gently agitated baths is expected to become uniform quite rapidly.[8]

Ruddlestone et al.[33] have compared the operating costs of an ASEA-SKF ladle furnace and a Fuchs ladle furnace and found that in both cases the primary cost factor is electric power, followed by electrode and refractory costs. This agrees with observations at The Timken Company.[34] These three cost categories will be discussed in more detail.

11.3.1.1 Electric Power

The heating efficiency, η, of arc heating is defined as:

$$\eta = \frac{\Delta T_{act}}{\Delta T_{th}} = 0.22\left(\frac{\Delta T_{act}}{E}\right) \tag{11.3.1}$$

where

ΔT_{act} = the actual temperature increase of the bath, °C,

ΔT_{th} = the theoretical temperature increase of the bath for 100% thermal efficiency, °C,

E = the energy consumption, kWh/tonne.

The heat capacity of liquid steel is 0.22 kWh/tonne°C; i.e. for 1 tonne of liquid steel $\Delta T_{th} = E/0.22$.

The heating efficiency increases with increasing bath weight, as shown in Fig. 11.11 obtained from data reported by Cotchen.[35] These data represent overall averages comprising a range of heating times. To minimize refractory consumption, heating times in ladle furnaces are kept as short as possible, typically around 15 min.[35] Further measures to shorten the reheating time and, thus, minimize refractory erosion are:[2] the use of a large capacity transformer, e.g. 35–40 MW for a 200 to 250 tonne heat, submerged arcing in the slag layer, argon stirring through a bottom porous plug at a flowrate of approximately 0.5 Nm³/min (~18 scfm), a slag layer thickness of approximately 1.3 times the length of the arc (values of the arc length as a function of the secondary current are given by Turkdogan[2]).

11.3.1.2 Electrode Consumption

In general, electrode consumption in ladle furnaces increases with increasing cross sectional current density and heating time. The trends are shown in Fig. 11.12 which is based on data from a worldwide survey cited by Cotchen.[35] For the Stein Heurtey-S.A.F.E. facility in operation at the Faircrest steel plant of The Timken Company[34] the average electrode consumption is 0.2 kg/tonne for typical total reheat times of 20 minutes and average current densities of approximately 35 A/cm².

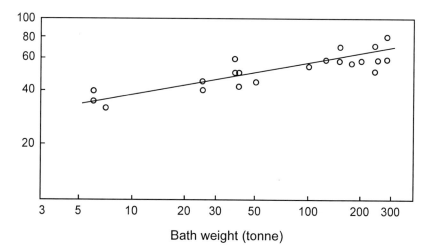

Fig. 11.11 Heating efficiency for various bath weights. *From Ref. 35.*

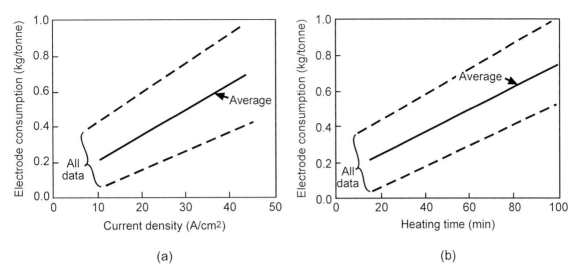

Fig. 11.12 Electrode consumption as affected by current density (a) and heating time (b). *From Ref. 35.*

11.3.1.3 Refractory Consumption

The refractory materials used in ladle furnace linings are similar for most steelmaking shops and applied in the same configurations, i.e. slagline, bottom and barrel. Thus, a comparison of refractory consumption data can be made for a variety of shops. The data in Fig. 11.13 show that the ladle life, expressed as the number of heats processed in a given ladle, increases with increasing bath weight.[35] At the Faircrest steel plant of The Timken Company, where 158 tonne (175 ton) heats are processed, the average ladle life is 45 heats.[34] It should be noted that the ladle refractory consumption in any given shop is strongly affected by the specific operating practice, as reflected by the considerable scatter in the data in Fig. 11.13. As mentioned before, submerged arc heating in the slag layer should result in lower refractory consumption.

11.3.2 Reheating by Oxygen Injection

Liquid steel can be reheated by oxidizing aluminum and/or silicon by means of oxygen injection through a lance. The heats generated for the reactions

$$2Al\ (R.T.) + \tfrac{3}{2}O_2\ (R.T.) \rightarrow Al_2O_3\ (1630°C) \tag{11.3.2}$$

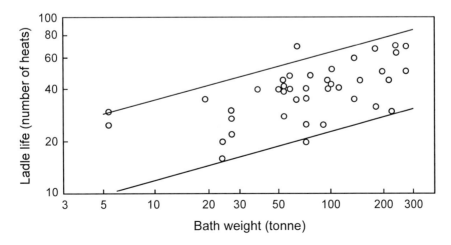

Fig. 11.13 Ladle life for various bath weights. *From Ref. 35.*

$$\text{Si (R.T.)} + \text{O}_2 \text{ (R.T.)} \rightarrow \text{SiO}_2 \text{ (1630°C)} \tag{11.3.3}$$

are:[2] 27,000 kJ/kg Al for reaction 11.3.2 and 28,500 kJ/kg Si in Fe–75% Si for reaction 11.3.3.

The enthalpies are calculated from the thermodynamic data, taking into account that the reagents, aluminum and oxygen, must be heated from room temperature (R.T.) to the temperature of the bath (1630°C = 2966°F). On the basis of 100% thermal efficiency the bath temperature can be raised by 50°C (90°F) when 1 Nm³O$_2$/tonne of steel is injected together with 1.46 kg Al/tonne or by injecting 1.2 Nm³O$_2$/tonne together with 1.85 kg Fe–75% Si/tonne[2].

Reheating of steel in the ladle with submerged oxygen injection is being practiced at the steel plants of the Bethlehem Steel Corporation. Barbus et al.[36] have published data on reheating by submerged injection of oxygen into 270 tonne heats. From their data the temperature increase as a function of the specific quantity of oxygen (Nm³/tonne of steel) injected can be obtained. This is shown in Fig. 11.14, line b. A comparison of the presented data with the maximum attainable temperature increase for 100% thermal efficiency (line a) indicates that reheating by means of submerged oxygen injection is approximately 70% efficient.

Miyashita and Kikuchi[37] presented data on the temperature increase in a 160 tonne RH-OB vessel. Their data are indicated by line c in Fig. 11.14 and indicate an average thermal efficiency of 20–30%. (Fruehan[68] quotes a reheating efficiency of approximately 80% for the RH-OB operation at the Oita works of the Nippon Steel Corporation.) Data for a 245 tonne RH-KTB vessel[38], in which the oxygen is supplied via a top lance instead of through submerged tuyeres as in the RH-OB, are indicated by line d in Fig. 11.14. The thermal efficiency for the RH-KTB process appears to be similar to that for submerged oxygen injection into the ladle.

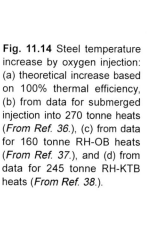

Fig. 11.14 Steel temperature increase by oxygen injection: (a) theoretical increase based on 100% thermal efficiency, (b) from data for submerged injection into 270 tonne heats (*From Ref. 36.*), (c) from data for 160 tonne RH-OB heats (*From Ref. 37.*), and (d) from data for 245 tonne RH-KTB heats (*From Ref. 38.*).

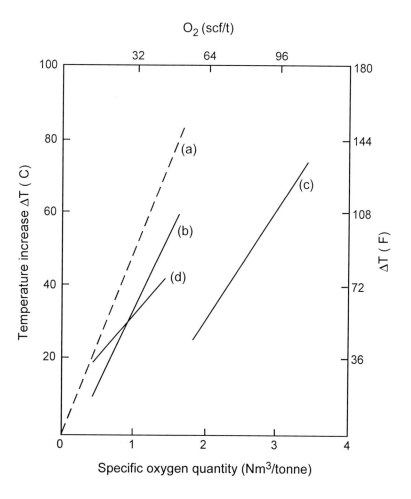

A comparison of total oxygen contents measured in slabs cast from oxygen-reheated heats and heats that were not reheated showed no significant differences between the two sets of values.[36] In addition, Griffing et al.[39] compared the inclusion ratings in rail-grade steels produced from oxygen-reheated steels and heats that were not reheated and found no significant differences between the two. These authors recommend the addition of a synthetic ladle slag after reheating, followed by a thorough argon rinse to float out the alumina inclusions so they may become dissolved in the slag.

Jung et al.[52] studied the effect of reheating with aluminum and oxygen in the RH-OB process on the cleanliness of the final product. An increase in the total oxygen content during and shortly after oxygen blowing was observed. However, the total oxygen content in the final product was found to be similar for steels treated with oxygen and aluminum compared with those treated without oxygen blowing in the RH, provided the total bath circulation time in the RH-OB was sufficiently long.

11.4 Refining in the Ladle

The refining of steel in the ladle is broadly defined here as comprising the following operations: deoxidation, desulfurization, dephosphorization, controlled additions of alloying elements and inclusion modification. Each of these operations will be discussed in detail in the following sections.

11.4.1 Deoxidation

The first step in the refining sequence in the ladle is usually the deoxidation of the steel with ferromanganese, ferrosilicon, silicomanganese and aluminum. There are three categories of steel deoxidation.

(a) Steel deoxidized with ferromanganese to yield 100–200 ppm dissolved oxygen; these are usually resulfurized steel grades.

(b) Semi-killed steels deoxidized with:

  (i) Si/Mn to yield 50–70 ppm dissolved oxygen,

  (ii) Si/Mn/Al to yield 25–40 ppm dissolved oxygen,

  (iii) Si/Mn/Ca to yield 15–20 ppm dissolved oxygen.

(c) Killed steels deoxidized with aluminum to yield 2–4 ppm dissolved oxygen.

The reaction equilibrium data for steel deoxidation are discussed in depth in Chapter 2, Section 2.10. In this section the focus will be on the practical aspects of deoxidation.

11.4.1.1 Deoxidation in the Presence of Synthetic Slags

The practice of refining steel in the ladle has made it possible to deoxidize the steel partially with Fe/Mn and/or Fe/Si later followed by a final deoxidation with aluminum. Such a practice has several advantages, including minimization of nitrogen pickup during tapping as discussed in Section 11.1.1, minimization of phosphorus reversion from the carried-over furnace slag, and minimization of aluminum losses due to reaction with carried-over furnace slag.

Today the use of synthetic slags in the ladle is an integral part of ladle metallurgy because of the requirements necessary to produce ultraclean steels, frequently combined with a demand for extra low sulfur contents. The concept of using synthetic slags in ladles dates back to the 1930s when the Perrin process was developed for the enhanced deoxidation of open hearth or Bessemer steel with ferromanganese or ferrosilicon by tapping the steel on a molten calcium aluminosilicate slag placed on the bottom of the tap ladle. The dissolution of the deoxidation products such as Mn(Fe)O or manganese silicates in the calcium aluminosilicate slag lowers their thermodynamic activity, thus increasing the extent of deoxidation.

11.4.1.1.1 Partial Deoxidation with Ferromanganese Turkdogan[11] described the results obtained for deoxidation with ferromanganese in several plant trials. When deoxidizing with ferromanganese the deoxidation product is Mn(Fe)O, the activity of which is lowered in the presence of a calcium

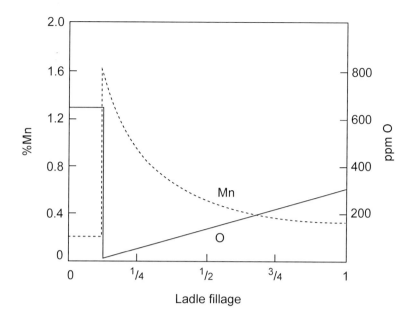

Fig. 11.15 Change in dissolved manganese and oxygen contents during tapping of a 200 tonne heat in the presence of 1800 kg ladle slag consisting of CaO-saturated calcium aluminate charged at 1/8 ladle fillage. *From Ref. 11.*

aluminate slag. The change in dissolved oxygen and manganese contents during tapping of a 200 tonne heat of steel to which 1800 kg of lime-saturated calcium aluminate and ferromanganese was added when the tap ladle was 1/8 full is shown schematically in Fig. 11.15. Upon addition of the ferromanganese, the small amount of steel present in the ladle is almost completely deoxidized, resulting in approximately 1.6% manganese in the steel. As the ladle is filled, the dissolved manganese is consumed by the deoxidation reaction and decreases to approximately 0.32% when the ladle is full and the residual dissolved oxygen content is reduced to approximately 300 ppm from the original 650 ppm at the beginning of tap.

The results obtained using this deoxidation practice in EAF and OBM (Q-BOP) shops[11] are depicted in Fig. 11.16 for steels containing less than 0.003 wt.% of aluminum and silicon each.

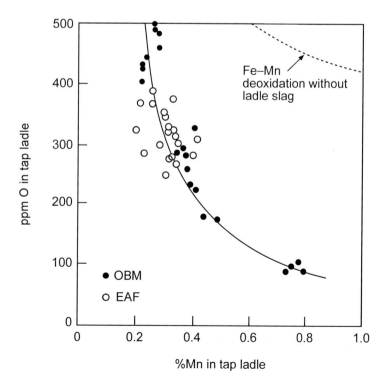

Fig. 11.16 Partial deoxidation of steel with ferromanganese and calcium aluminate slag during furnace tapping; steel containing less than 0.003% aluminum or silicon each. *From Ref. 11.*

Without calcium aluminate slag addition to the tap ladle, i.e. deoxidation with manganese and iron only and pure Mn(Fe)O as the deoxidation product, the concentration of dissolved oxygen in the steel would follow the broken line in Fig. 11.16. In the EAF trial heats there was no argon stirring in the ladle during furnace tapping. Yet, it was found that the slag-aided partial deoxidation of the steel attained during tap was close to the levels determined by the slag–metal equilibrium. This observation led Turkdogan[11] to conclude that there was sufficient mixing of slag and metal to promote relatively rapid deoxidation in the ladle during tap.

11.4.1.1.2 Deoxidation with Silicomanganese It is well known that the deoxidation of steel with manganese and silicon together leads to lower dissolved oxygen contents than the deoxidation with either of these elements alone. This is because the activities of the oxides in the deoxidation reaction

$$[Si] + 2(MnO) = 2[Mn] + (SiO_2) \qquad (11.4.1)$$

are less than unity. The symbols within the square brackets refer to species dissolved in the steel, those within parentheses to species in the manganese silicate phase. By making use of the oxide activity data in the $MnO-SiO_2$ system together with the thermodynamic data for reaction 11.4.1, Turkdogan[40] computed the equilibrium state pertaining to the deoxidation with silicomanganese, as shown in Fig. 2.127.

When the deoxidation with silicomanganese takes place in the presence of a small amount of aluminum dissolved in the steel the deoxidation product is molten manganese aluminosilicate and the resulting dissolved oxygen content is approximately 50 ppm for a steel containing roughly 0.8% manganese and 0.2% silicon.[11] This is approximately half the value in a steel deoxidized with silicomanganese and not containing aluminum. This is because the activities of MnO and SiO_2 are lowered further in the presence of the aluminosilicate phase. For example, it is possible to decrease the dissolved oxygen content to approximately 20 ppm by means of deoxidation with silicomanganese together with the addition of 1000 kg of prefused calcium aluminate to a 200 tonne heat of steel.[11]

For ladle slags containing a high percentage of alumina there will be some reduction of the alumina by the silicon in the steel. The data for steel containing aluminum and silicon in equilibrium with calcium aluminate slags containing approximately 5% silica are shown in Fig. 11.17.[2] As may be seen from this diagram, appreciable pickup of aluminum from the slag can be expected if the steel is initially low in aluminum, e.g. <0.01% Al, and contains approximately 0.2% silicon. The final stage of deoxidation of the steel in the ladle is determined by the amount of aluminum recovered from the slag.

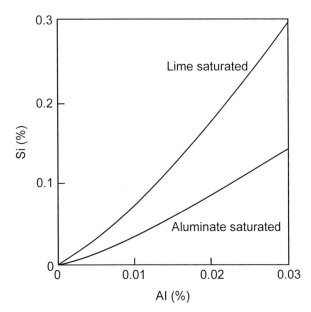

Fig. 11.17 Aluminum and silicon contents in steel in equilibrium at 1600°C with molten calcium aluminate slags containing approximately 5% SiO_2. *From Ref. 2.*

11.4.1.1.3 Deoxidation with Calcium/Silicon Semi-killed steel deoxidized with silicomanganese can be deoxidized further with calcium/silicon, e.g. by injection of Cal-Sil in the form of cored wire. Iyengar and Duderstadt[41] studied the deoxidation of 50 kg steel melts at 1620°C with silicomanganese and varying amounts of CaSi. Some of their results are shown in Fig. 11.18, from which it is seen that the dissolved oxygen content in a low carbon steel deoxidized with silicomanganese can be lowered from approximately 85 to approximately 55 ppm by adding 2.5 kg CaSi/tonne. Similar observations were made in a series of plant trials[42] in which Cal-Sil cored wire was injected into 60 tonne heats of steel. The lower dissolved oxygen content obtained after the addition of Cal-Sil is a result of the formation of calcium manganese silicate as the deoxidation product, further decreasing the activities of MnO and SiO_2.

11.4.1.1.4 Deoxidation with Aluminum

Numerous experimental data, obtained by the emf technique, exist on the solubility product of Al_2O_3 in pure liquid iron. The reported values for 1600°C are in the range[43]

$$[\%Al]^2 \cdot [\%O]^3 = 9.77 \times 10^{-15} \text{ to } 1.2 \times 10^{-13}$$
(11.4.2)

The higher values are from older work when the interference with the emf readings caused by partial electronic conduction in the electrolyte of the emf cell was not well recognized. In the most recent work by Dimitrov et al.[43] the emf readings were corrected for electronic conduction, leading to the value of 9.77×10^{-15} for the Al_2O_3 solubility product at 1600°C (2912°F).

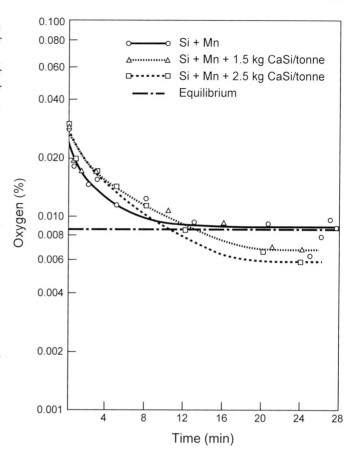

Fig. 11.18 Dissolved oxygen content in 0.05%C steel after deoxidation with SiMn and SiMn + CaSi at 1620°C; steel contained ~0.65% Mn and ~0.20% Si. *From Ref. 41.*

In addition, Dimitrov et al.[43] did a number of emf measurements in inductively stirred iron melts in contact with $CaO–Al_2O_3$ slag mixtures in which the activity of alumina was less than unity. They observed that the aluminum-oxygen relationship in these melts was indistinguishable from that in iron melts in equilibrium with pure alumina. Similar observations were reported by Schuermann et al.[44]

These experimental findings indicate that even in the presence of a calcium aluminate slag the dissolved oxygen content in steel is controlled by the alumina inclusions always present in the interior of the bath. In ladle metallurgy operations where the steel is frequently covered with a calcium aluminate slag containing minor amounts of magnesia and silica, it is therefore to be expected that the final dissolved oxygen content is controlled by the alumina inclusions dispersed throughout the bath. A measurable decrease in dissolved oxygen content as a result of treatment with calcium aluminate slag is not to be expected. A decrease in dissolved oxygen content can only be expected if substantially all the alumina inclusions have been modified to calcium aluminates, e.g. by calcium treatment; this will be discussed in more detail in Section 11.4.5.2

11.4.2 Desulfurization

In certain steel grades, such as those used in line pipe applications, a very low sulfur content is required, e.g. 20 ppm or less. These low sulfur contents can only be achieved by steel desulfurization in the ladle in the presence of a calcium aluminate slag when the steel is fully killed. Desulfurization is also discussed in depth in Chapter 2. The governing reaction is

$$\tfrac{2}{3}[Al] + [S] + (CaO) = (CaS) + \tfrac{1}{3}(Al_2O_3)$$
(11.4.3)

where the symbols within square brackets denote species dissolved in the steel and those within parentheses refer to species dissolved in the slag phase.

The change in free energy for reaction 11.4.3 based on the most recent data on the solubility product of alumina in liquid iron[43] is given by

$$\Delta G° = -319{,}343 + 111.3T, \text{ J/mole} \tag{11.4.4}$$

This gives for the equilibrium constant of reaction 11.4.3

$$\left. \begin{array}{l} \log K = \dfrac{16{,}680}{T} - 5.813 \\[2ex] \text{where} \\[2ex] K = \dfrac{a_{CaS}\ a_{Al_2O_3}^{1/3}}{a_{CaO}\ a_S\ a_{Al}^{2/3}} \end{array} \right\} \tag{11.4.5}$$

The oxide and sulfide activities are relative to the respective pure solid phases and the activities of aluminum and sulfur dissolved in liquid iron are defined such that for dilute solutions, below approximately 0.5 wt.%, the activities may be replaced by their respective concentrations in weight percent. For a given slag composition the activities of the oxides are fixed and may be incorporated in the equilibrium constant while the sulfur content of the slag, (%S), is proportional to the sulfide activity, a_{CaS}. Thus, the equilibrium constant may be replaced by a pseudo-equilibrium constant as follows

$$K_S = \frac{(\%S)}{[\%S]}[\%Al]^{-2/3} \tag{11.4.6}$$

For a given aluminum content of the steel the ratio $L_S = (\%S) / [\%S]$ is a function of the ladle slag composition at a given temperature. This is shown in Fig. 11.19 where lines of equal L_S-values are projected on the Al_2O_3–CaO–SiO_2 phase diagram for slags in equilibrium with steels containing 0.03% aluminum.[45] For different aluminum contents the L_S-values in Fig. 11.19 should be multiplied by the factor $(\%Al/0.03)^{2/3}$.

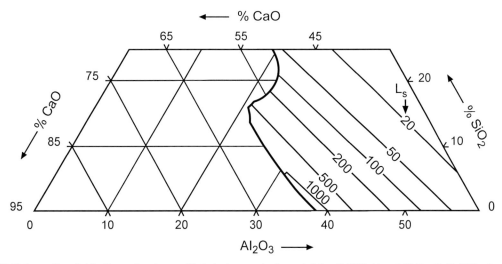

Fig. 11.19 Iso-sulfur distribution ratios for equilibria between steels containing 0.03% Al and Al_2O_3–CaO–SiO_2 slags containing 5% MgO at 1600°C. *From Ref. 45.*

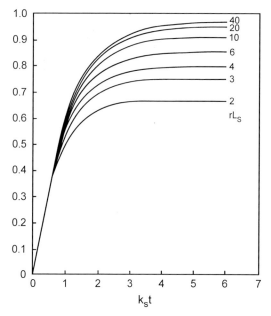

Fig. 11.20 Relative degree of desulfurization, (R=W_s/W_m) as affected by stirring power and time, ($k_s t$) and the product of specific quantity of ladle slag and sulfur partition ratio, (rL_S) as indicated by the numbers near the curves. *From Ref. 45.*

Depending on the specific operating conditions, the range of compositions of ladle slags commonly used is: 20–40% Al_2O_3, 35–55% CaO, 8–15% MgO and 10–15% SiO_2 together with minor amounts of FeO and MnO. In some cases CaF_2 is added to the ladle slag.

11.4.2.1 Desulfurization Rate

For reaction 11.4.3 to proceed rapidly such that the required degree of desulfurization can take place within a practical time span, good mixing of steel and slag is essential. The rate at which sulfur can be removed is, therefore, strongly affected by the gas flowrate or the stirring power density. As the capacity of typical ladle slags to absorb sulfur is high, the rate of desulfurization is controlled by mass transfer in the liquid steel and the rate of desulfurization is described by equation 2.10.19.

Riboud and Vasse[45] calculated the relative sulfur removal, R, for various conditions; the results are shown in Fig. 11.20. This diagram may be used to estimate the specific quantity of ladle slag, r, required to give the desired degree of sulfur removal, as illustrated by the following example.

Consider a shop where hard stirring, equivalent to a stirring power density of 100 W/tonne, is practiced for desulfurization with a ladle slag with a composition equivalent to L_S=500. From Fig. 11.10 the value $k_S \sim 0.13$ min^{-1} is obtained for $\dot{\varepsilon}$ = 100 W/tonne. Assume that the total time to be reserved for desulfurization is 15 min to achieve 80% desulfurization, e.g. from 0.01 to 0.002% S. From Fig. 11.20 it is found that R = 0.80 can be achieved in the given time, provided rL_S=10 or r=0.02, equivalent with a specific quantity of ladle slag of 20 kg/tonne of steel (40 lb/ton). For a stirring power density of approximately 50 W/tonne the value of k_S is estimated to be approximately 0.03 min^{-1} (Fig. 11.10) and a treatment time of 65 min would be required to achieve 80% desulfurization. This example illustrates the importance of hard stirring for effective desulfurization to low sulfur levels.

The rate of desulfurization depends on stirring rate, slag chemistry which affects L_S and Al content, which also effects L_S. Application of equation 2.10.19 indicates desulfurization equilibrium in a well-stirred ladle using 0.85–1.10 Nm³ (30–40 scfm)Ar occurs in approximately 10–15 min.

To achieve very low sulfur contents the injection of fluxes into the ladle is often practiced. Hara et al.[46] describe results obtained by injecting 70% CaO–30% CaF_2 power mixtures into 150 tonne heats of low-carbon Al-Si killed steels. Their results are shown in Fig. 11.21. It is noted that powder injection results in a desulfurization rate that is, on average, approximately 15% faster than desulfurization with a top slag only, combined with gas stirring. This implies that the contribution of the so-called transitory reaction with the powder mixture as it ascends the bath is minor compared with the reaction with the top slag, in general agreement with a mathematical model developed by Sawada et al.[47]

Desulfurization of steel in the ladle is accompanied by a decrease in the temperature of the steel bath. Today most steelmaking shops are equipped with facilities for reheating the steel, either by electric arc or by injection of oxygen and aluminum. If no facilities for reheating are available an where the symbols within square brackets denote species dissolved in the steel and those within parentheses refer to species dissolved in the slag phase.

The change in free energy for reaction 11.4.3 based on the most recent data on the solubility product of alumina in liquid iron[43] is given by

$$\Delta G° = -319,343 + 111.3T, \text{ J/mole} \tag{11.4.4}$$

This gives for the equilibrium constant of reaction 11.4.3

$$\left. \begin{array}{c} \log K = \dfrac{16,680}{T} - 5.813 \\ \\ \text{where} \\ \\ K = \dfrac{a_{CaS} \; a_{Al_2O_3}^{1/3}}{a_{CaO} \; a_S \; a_{Al}^{2/3}} \end{array} \right\} \tag{11.4.5}$$

The oxide and sulfide activities are relative to the respective pure solid phases and the activities of aluminum and sulfur dissolved in liquid iron are defined such that for dilute solutions, below approximately 0.5 wt.%, the activities may be replaced by their respective concentrations in weight percent. For a given slag composition the activities of the oxides are fixed and may be incorporated in the equilibrium constant while the sulfur content of the slag, (%S), is proportional to the sulfide activity, a_{CaS}. Thus, the equilibrium constant may be replaced by a pseudo-equilibrium constant as follows

$$K_S = \dfrac{(\%S)}{[\%S]} [\%Al]^{-2/3} \tag{11.4.6}$$

For a given aluminum content of the steel the ratio $L_S = (\%S) / [\%S]$ is a function of the ladle slag composition at a given temperature. This is shown in Fig. 11.19 where lines of equal L_S-values are projected on the Al_2O_3–CaO–SiO_2 phase diagram for slags in equilibrium with steels containing 0.03% aluminum.[45] For different aluminum contents the L_S-values in Fig. 11.19 should be multiplied by the factor $(\%Al/0.03)^{2/3}$.

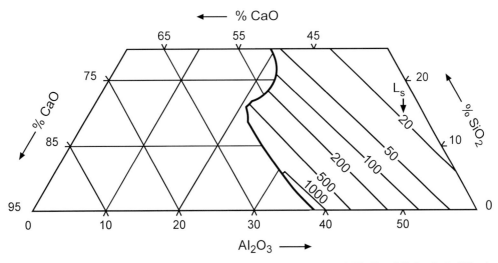

Fig. 11.19 Iso-sulfur distribution ratios for equilibria between steels containing 0.03% Al and Al_2O_3–CaO–SiO_2 slags containing 5% MgO at 1600°C. From Ref. 45.

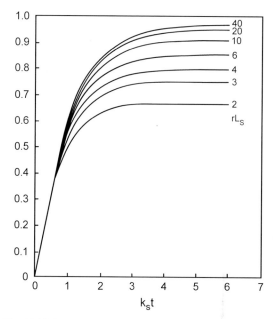

Fig. 11.20 Relative degree of desulfurization, (R=W_s/W_m) as affected by stirring power and time, ($k_s t$) and the product of specific quantity of ladle slag and sulfur partition ratio, (rL_S) as indicated by the numbers near the curves. *From Ref. 45.*

Depending on the specific operating conditions, the range of compositions of ladle slags commonly used is: 20–40% Al_2O_3, 35–55% CaO, 8–15% MgO and 10–15% SiO_2 together with minor amounts of FeO and MnO. In some cases CaF_2 is added to the ladle slag.

11.4.2.1 Desulfurization Rate

For reaction 11.4.3 to proceed rapidly such that the required degree of desulfurization can take place within a practical time span, good mixing of steel and slag is essential. The rate at which sulfur can be removed is, therefore, strongly affected by the gas flowrate or the stirring power density. As the capacity of typical ladle slags to absorb sulfur is high, the rate of desulfurization is controlled by mass transfer in the liquid steel and the rate of desulfurization is described by equation 2.10.19.

Riboud and Vasse[45] calculated the relative sulfur removal, R, for various conditions; the results are shown in Fig. 11.20. This diagram may be used to estimate the specific quantity of ladle slag, r, required to give the desired degree of sulfur removal, as illustrated by the following example.

Consider a shop where hard stirring, equivalent to a stirring power density of 100 W/tonne, is practiced for desulfurization with a ladle slag with a composition equivalent to L_S=500. From Fig. 11.10 the value $k_S \sim 0.13$ min^{-1} is obtained for $\dot{\varepsilon} = 100$ W/tonne. Assume that the total time to be reserved for desulfurization is 15 min to achieve 80% desulfurization, e.g. from 0.01 to 0.002% S. From Fig. 11.20 it is found that R = 0.80 can be achieved in the given time, provided rL_S=10 or r=0.02, equivalent with a specific quantity of ladle slag of 20 kg/tonne of steel (40 lb/ton). For a stirring power density of approximately 50 W/tonne the value of k_S is estimated to be approximately 0.03 min^{-1} (Fig. 11.10) and a treatment time of 65 min would be required to achieve 80% desulfurization. This example illustrates the importance of hard stirring for effective desulfurization to low sulfur levels.

The rate of desulfurization depends on stirring rate, slag chemistry which affects L_S and Al content, which also effects L_S. Application of equation 2.10.19 indicates desulfurization equilibrium in a well-stirred ladle using 0.85–1.10 Nm^3 (30–40 scfm)Ar occurs in approximately 10–15 min.

To achieve very low sulfur contents the injection of fluxes into the ladle is often practiced. Hara et al.[46] describe results obtained by injecting 70% CaO–30% CaF_2 power mixtures into 150 tonne heats of low-carbon Al-Si killed steels. Their results are shown in Fig. 11.21. It is noted that powder injection results in a desulfurization rate that is, on average, approximately 15% faster than desulfurization with a top slag only, combined with gas stirring. This implies that the contribution of the so-called transitory reaction with the powder mixture as it ascends the bath is minor compared with the reaction with the top slag, in general agreement with a mathematical model developed by Sawada et al.[47]

Desulfurization of steel in the ladle is accompanied by a decrease in the temperature of the steel bath. Today most steelmaking shops are equipped with facilities for reheating the steel, either by electric arc or by injection of oxygen and aluminum. If no facilities for reheating are available an

Ladle Refining and Vacuum Degassing

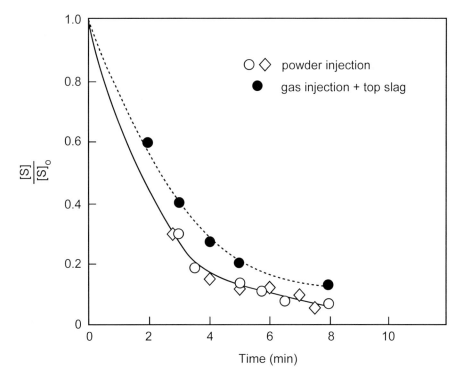

Fig. 11.21 Desulfurization of 150 tonne heats of Al-Si killed steel with mixtures of 70% CaO–30% CaF$_2$ (open symbols) and by gas stirring + top slag –FeO equilibrium. *From Ref. 46.*

exothermic mixture consisting of 58% burnt lime, 30% hematite and 12% aluminum powder can be used for desulfurization.[11] Further details about the use of such mixtures and the results obtained in extensive plant trials were presented by Turkdogan.[2]

11.4.3 Dephosphorization

In general it is preferred to remove phosphorus from steel under the oxidizing conditions prevalent in the BOF or in EAFs with oxygen injection. In older EAF shops equipped with furnaces with inadequate or no oxygen injection capability the need for steel dephosphorization in the ladle may

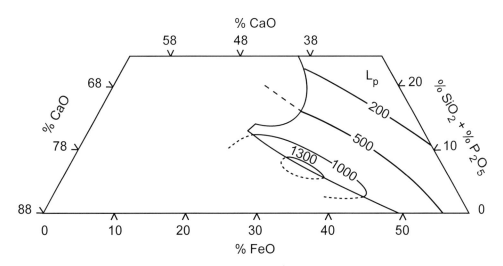

Fig. 11.22 Iso-phosphorus distribution ratios, L_P = (%P) / [%P], for equilibrium at 1600°C between steel and CaO–FeO–(SiO$_2$ + P$_2$O$_5$) slags containing 2% MgO, 6% MnO and 4% Na$_2$O; the dissolved oxygen is controlled by the Fe(filled symbols). *From Ref. 49.*

arise. Also, ladle dephosphorization may be necessary in BOF shops in which hot metal with a high phosphorus content is charged and where there is no capability of dephosphorizing the hot metal prior to charging to the BOF.

Removal of phosphorus from the steel in the ladle is achieved by treating the steel with lime-based oxidizing slags containing iron oxide. The fundamentals of the phosphorus reaction are discussed in Chapter 2.

For a steel with a given dissolved oxygen content the ratio $L_p = (\%P) / [\%P]$ is a function only of the slag composition and the temperature. This is shown in Fig. 11.22 where lines of equal L_p-values are projected on the $CaO-FeO-(SiO_2 + P_2O_5)$ phase diagram[49], the dissolved oxygen contents being controlled by the Fe–FeO equilibrium.

Dephosphorization in the ladle during tapping of the BOF converter was studied by Becker et al.[50] who used varying quantities of a mixture of 50% CaO, 30% iron oxide (FeO_x) and 20% CaF_2. Approximately 30–40% of the mixture was placed on the bottom of the tap ladle while the remainder was added during tap. The phosphorus content of the steel tapped from the converter varied between approximately 0.01 and 0.035%. The results of these plant trials are summarized in Fig. 11.23, reproduced from the data by Becker et al.[50] It is seen from this diagram that approximately 75% of the phosphorus was removed when 12 kg/tonne of the aforementioned mixture was used. In the practice described by Becker et al.[50] the high-phosphorus slag was removed by reladling followed by reheating in a ladle furnace.

Another example of steel dephosphorization in the ladle at tap is given by Bannenberg and Lachmund.[51] In this practice the steel was tapped open while, depending on the anticipated quantity of converter slag carryover, varying amounts of lime, ore and sometimes fluorspar were added during tap to produce a lime-saturated dephosphorizing slag high in iron oxide. The results of these trials indicate 60–70% phosphorous removal.

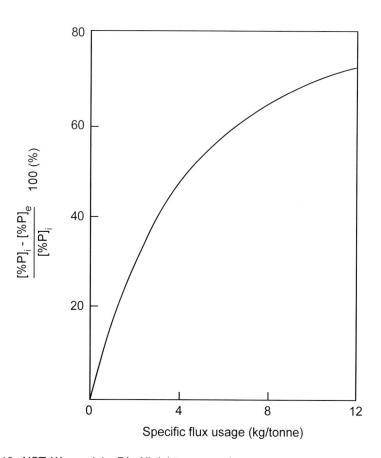

Fig. 11.23 Degree of dephosphorization during tap as affected by the specific quantity of 50% CaO–30% FeO_x–20%CaF_2 used, reproduced from data by Becker et al.[50]

Because of the oxidizing conditions prevailing during dephosphorization, manganese and chromium losses are to be expected, as discussed by Bannenberg and Lachmund.[51] The authors derived the following relationship between the loss in chromium or manganese, η_X, and the degree of dephosphorization, η_P

$$\eta_X = \frac{100\ \eta_P}{100\ K_X - \eta_P(K_X - 1)} \tag{11.4.7}$$

where

X	=	Cr or Mn and
η_X	=	$\{[\%X]_i - [\%X]_f\} / [\%X]_i$
η_P	=	$\{[\%P]_i - [\%P]_f\} / [\%P]_i$

From the plant data Bannenberg and Lachmund[51] derived the following values for K_X: K_{Cr} = 6.6 and K_{Mn} = 2.2. For example, for 50% phosphorus removal (η_P = 50) the chromium loss would be 13% (η_{Cr} = 13) while the manganese loss would be 31% (η_{Mn} = 31), as found from equation 11.4.7.

11.4.4 Alloy Additions

Metals and alloys can be added to liquid steel at various stages in the steelmaking process, e.g. inside the furnace, during furnace tapping, in the ladle furnace, during vacuum treatment, etc. The timing of the additions depends on the process route, the shop logistics and on certain characteristics of the addition in question such as its melting point, volatility and its susceptibility to oxidation. For example, nickel can be added to the EAF at any time as nickel oxide, which is easily reduced. In the oxygen steelmaking process route alloying additions such as ferrosilicon and ferromanganese are made during furnace tapping while the other alloys are added in subsequent stages of secondary steelmaking.

Argyropoulos and Guthrie[53] were perhaps the first to undertake a systematic study on the dissolution kinetics of ferroalloys. They defined two broad categories of ferroalloys: class I ferroalloys with melting points below the temperature of liquid steel, and class II ferroalloys with melting points higher than the liquid steel temperature. The thermophysical properties of the class I and class II ferroalloys are summarized in Tables 11.5 and 11.6, respectively.

When a ferroalloy is added to liquid steel a solidified shell of steel forms around the alloy particle as a result of the local chilling effect. As time progresses, the shell melts while the ferroalloy inside the shell is heated to its melting point. The complete dissolution is governed by convective heat transfer processes in the bath as well as the size of the ferroalloy added.

The class II ferroalloys, listed in Table 11.6 all have melting points higher than the temperature of the liquid steel. These alloys dissolve at a slower rate than the class I alloys, their dissolution rate being controlled by mass transfer in the liquid steel, even in agitated baths. It is, therefore, important to ensure that their size is within 3–10 mm in order to obtain good mixing, fast dissolution and high recovery rates. Compacted powder mixtures of ferroalloys such as ferrovanadium, ferrotungsten and ferromolybdenum dissolve faster than solid pieces of similar size. Autoexothermic alloys, which generate heat upon melting, can also be used for faster melting and dissolution as well as improved recovery. Argyropoulos and Guthrie[53] present a number of predicted dissolution times for an assortment of ferroalloys of varying sizes under a number of different conditions such as bath temperature and convection within the bath.

In a later investigation Lee et al.[54] made an extensive study of the dissolution kinetics in liquid steel as well as in slags of the most widely used ferroalloys such as ferrosilicon (75 FeSi), silicomanganese (SiMn), high-carbon ferromanganese (HCFeMn) and high-carbon ferrochrome (HCFeCr). Their findings corroborated the earlier results of Argyropoulos and Guthrie.[53]

Table 11.5 Physical and Thermal Properties Relevant to Class I Ferroalloys. From Ref. 53.

Material	Density kg m^{-3}	Heat Capacity J kg^{-1} K^{-1}	Thermal Conductivity W m^{-1} K^{-1}	Latent Heat kJ kg K^{-1}	$T_{solidus}$ K	$T_{liquidus}$ K
Ferromanganese Mn = 79.5% C = 6.4% Si = 0.27% Fe: balance	7200	700.0	7.53	534,654	1344	1539
Silicomanganese Mn = 65.96% Si = 17.07% C = 1.96% Fe: balance	5600	628.0	6.28	578,783	1361	1489
50% Ferrosilicon Si = 49.03% Al = 1.20% max Fe: balance	4460	586.0	9.62	908,200	1483	1500
Ferrochrome Cr = 50–58% C = 0.25% max Si = 1.5% max Mn = 0.50% max Al = 1.50% max	6860	670.0	6.50	324,518	1677	1755

Table 11.6 Physical and Thermal Properties Relevant to Class II Ferroalloys. From Ref. 53.

Material, A.	Density kg m^{-3} (1873K)	Heat Capacity J kg^{-1} K^{-1}	Thermal Conductivity W m^{-1}K^{-1}	Diffusivity $D_{A/Fe} \times 10^9$ m$^{2-1}$
Molybdenum	10000	310.0	100.0	3.2
Vanadium	5700	400.0	50.0	4.1
Niobium	8600	290.0	64.0	4.6
Tungsten	19300	140.0	115.0	5.9

Several methods of alloy addition are practiced. Examples are: throwing of filled bags, adding with a shovel or via mechanized chutes, wire feeding, powder injection, bullet shooting, etc. A special process for making alloy additions is the so-called CAS process (Composition Adjustment by Sealed argon bubbling). In this process a refractory-lined snorkel is partially immersed in the steel bath in such a manner that it envelopes the ascending gas plume created by the injection of argon through the porous plug in the ladle bottom. Alloy additions are made onto the liquid steel surface within the area covered by the snorkel. The plume eye within the snorkel is filled with argon, thus has a low oxygen partial pressure preventing oxidation of the alloy addition. Melting and distribution rates are high as a result of the agitation brought about by the ascending gas bubbles. Mazumdar and Guthrie[55] have made water model studies to investigate the subsurface motion of both buoyant and sinking additions in the CAS process. The study showed that buoyant additions such as aluminum and ferrosilicon dissolve more readily into the steel bath rather than react partially with the slag as in conventional addition methods, thus giving improved recovery. The authors further recommend that ferromanganese and ferroniobium be crushed to an average size of approximately 5 mm to obtain better control.

Wirefeeding of alloys by means of the cored-wire techniques, developed primarily for the addition of calcium to steel, is practiced for adding elements that are less dense than steel or have a limited solubility, high vapor pressure and high affinity for oxygen. Wirefeeding is also used in cases where the element to be added is toxic or when very small additions are required. The cored-wire technique permits the quantity of alloy or elements being fed into the steel to be adjusted with high precision and to trim the composition of the steel within narrow limits. For example, ferroboron or tellurium additions can be made in precise and minute quantities by wirefeeding. Excessive additions of these elements may cause hot-shortness.

It is also possible to wirefeed aluminum with the same wirefeeding equipment used for cored-wire. Advantages of aluminum wire additions include: higher recovery, better control of aluminum content, and improved cleanliness.[56] Herbert et al.[57] give examples of improved control of the steel aluminum content by wirefeeding at the Lackenby plant of the British Steel Corporation.

Schade et al.[58] studied the dissolution characteristics of cored-wire additions of ferromolybdenum, ferroniobium and ferrochromium modified with minor quantities of silicon—so-called microexothermic alloys—to achieve improved dissolution into liquid steels. The exothermicity exhibited by these modified alloys is based on the formation of an intermetallic compound (a silicide) which is accompanied by the release of heat. The enthalpy released is sufficient to melt the compound, thus allowing rapid dissolution of the ferroalloy into the liquid steel. The enhanced dissolution rate of these modified ferroalloys makes them well-suited for tundish additions where the relatively short residence time requires a rapid dissolution of additions.

11.4.5 Calcium Treatment and Inclusion Modification

The addition of calcium to steel goes back a long time, Watts[59] being the first to add CaSi to a steel melt. The widespread practice of calcium additions to steel melts did not start until the 1960s with the development of improved addition methods and composite calcium-bearing alloys. Today calcium

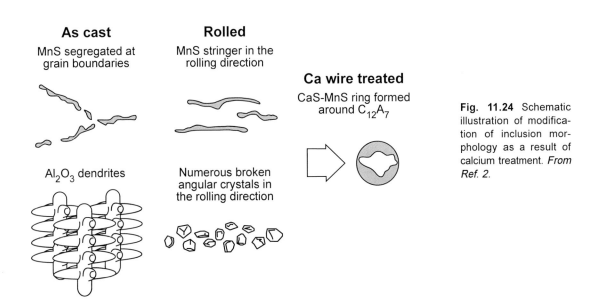

Fig. 11.24 Schematic illustration of modification of inclusion morphology as a result of calcium treatment. *From Ref. 2.*

treatment of steel is a common practice, with particular emphasis on the modification of alumina inclusions in aluminum-killed steels to prevent nozzle clogging during continuous casting operations.

As a result of the treatment with calcium, the alumina and silica inclusions are converted to liquid calcium aluminates or calcium silicates. These liquid inclusions are globular in shape because of surface tension effects. This change in inclusion composition and shape is commonly known as inclusion morphology control or modification. The effect of calcium treatment on inclusion morphology is illustrated schematically in Fig. 11.24.

It is seen from Fig. 11.24 that few or no sulfide stringers are expected to be present after rolling steel that was successfully treated with calcium. This phenomenon is known as sulfide shape control by calcium treatment, the underlying fundamental principles of which will be discussed later.

Examples of other metallurgical advantages brought about by the modification of oxide and sulfide inclusions by calcium treatment of steel are:[2] improvement of castability in continuous casting operations through minimization or prevention of nozzle clogging; decreasing inclusion-related surface defects in billet, bloom and slab castings; improving the machinability of the final product at high cutting speeds; and minimization of the susceptibility of high-strength low-alloy (HSLA) line pipe steels to hydrogen-induced cracking (HIC) in sour gas or oil environments.

11.4.5.1 Addition of Calcium to Steel Melts

The boiling point of calcium is 1491°C (2716°F), accordingly calcium is a vapor at steelmaking temperatures. Thus, when adding calcium to liquid steel special measures must be taken to ensure its proper recovery in the steel bath. Recently developed processes for adding calcium to a liquid steel bath are all based on the principle of introducing the calcium or calcium alloy into the bath at the greatest possible depth so as to make use of the increased pressure from the ferrostatic head to prevent the calcium from evaporating.

Ototani[60] gives details regarding the TN (Thyssen Niederrhein) process for injecting calcium with argon as a carrier gas as well as the SCAT process, also known as the bullet shooting method. Today the majority of the steel producers add calcium by wirefeeding. The principle is similar to wirefeeding of ferroalloys and aluminum, discussed in Section 11.4.4.

When wirefeeding calcium in the conventional manner, there is a possibility that the wire does not travel in a straight downward line after entering the bath, thus causing the calcium to be released at a shallow bath depth and decreasing the calcium recovery in the steel. To prevent this, the so-called wire lance (WL) method for adding calcium was developed. A schematic illustration is depicted in Fig. 11.25.

The wire lance method ensures that the calcium wire travels in a straight downward line after entering the bath while it is claimed that the dispersion of calcium throughout the bath is improved by the argon which is injected simultaneously. The calcium recovery or yield observed for additions with the wire lance process is compared with that observed for conventional wirefeeding in Fig. 11.26. The better recovery of calcium obtained with the wire lance method is especially pronounced for calcium addition rates less than approximately 0.2 kg/tonne (0.4 lb/ton).

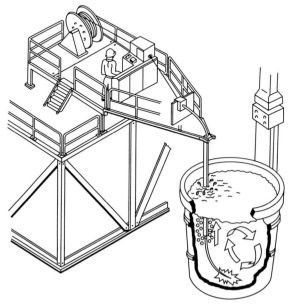

Fig. 11.25 Wire lance method for adding calcium in which calcium wire is fed through a refractory lance immersed in the bath with argon flowing through the lance during wirefeeding. From Ref. 60.

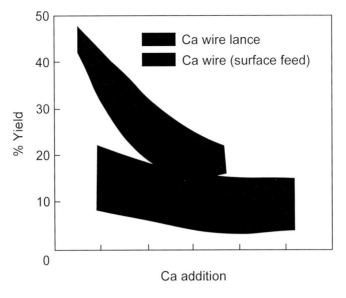

Fig. 11.26 Comparison of calcium recoveries (yields) obtained with the wire lance addition method and those obtained with conventional wirefeeding. *From Ref. 60.*

11.4.5.1.1 Calcium Usage Efficiency

The material balance for calcium consumption is expressed as follows

$$W_i = W_b + W_o + W_o' + W_s + W_v \qquad (11.4.8)$$

where

W_i = amount of calcium injected,

W_b = amount of calcium dissolved in the bath,

W_o = amount of calcium present in aluminates and sulfides,

W_o' = amount of calcium reacted with alumina and subsequently floated out,

W_s = amount of calcium reacted with the slag,

W_v = amount of calcium escaped via the vapor phase and subsequently burnt at the bath surface.

It is generally accepted[2] that $W_b \ll W_o$, thus giving for the efficiency of calcium usage

$$\eta(Ca)_u = \frac{W_o + W_o'}{W_i} * 100\% \qquad (11.4.9)$$

while the efficiency of calcium retention in the steel is given by

$$\eta(Ca)_r = \frac{W_o}{W_i} * 100\% \qquad (11.4.10)$$

Experience obtained in numerous plant trials has shown that the calcium retention efficiency decreases with increasing quantity of calcium injected. The amount of calcium to be injected has to be adjusted in accordance with the degree of cleanliness of the steel or its total oxygen content. Obviously, injecting more calcium than can react with the available inclusions leads to a low calcium retention efficiency. Furthermore, it is to be expected that the calcium retention efficiency in the continuous casting mold or in the teeming ingot will be lower than the retention efficiency in the ladle because of flotation of calcium-containing inclusions out of the bath in the time interval prior to casting or teeming. Turkdogan[2] quotes the following calcium retention efficiencies in the ladle and the tundish for aluminum-killed steels initially containing 50 to 80 ppm oxygen as alumina inclusions

Ca injected (kg/tonne)	Ladle $\eta\,(Ca)_r$	Tundish $\eta\,(Ca)_r$
0.16	24–30%	12–15%
0.36	12–18%	6–9%

11.4.5.2 Reactions of Calcium in Steel and Inclusion Modification

Regardless of the form in which calcium is added to liquid steel, e.g. as Cal-Sil, Ca-Al or as pure calcium admixed with nickel or iron powder, the subsequent reactions taking place in the bath are the same. The following series of reactions is expected to occur to varying extents in aluminum-killed steels containing alumina inclusions and sulfur

$$Ca\,(l) \rightarrow Ca\,(g) \qquad (11.4.11)$$

$$Ca\,(g) \rightarrow [Ca] \qquad (11.4.12)$$

$$[Ca] + [O] \rightarrow CaO \qquad (11.4.13)$$

$$[Ca] + [S] \rightarrow CaS \qquad (11.4.14)$$

$$[Ca] + (x + 1/3)Al_2O_3 \rightarrow CaO \cdot xAl_2O_3 + 2/3[Al] \qquad (11.4.15)$$

$$(CaO) + 2/3[Al] + [S] \rightarrow (CaS) + 1/3(Al_2O_3) \qquad (11.4.16)$$

The symbols within square brackets refer to species dissolved in the steel, those within parentheses are dissolved in the aluminate phase.

Observations by a number of investigators have indicated that the extent to which reaction 11.4.13 occurs is negligible. For steels with sufficiently low sulfur contents reaction 11.4.15 will take place first, followed by reaction 11.4.16. The critical question is for which sulfur content reaction 11.4.14 predominates such that, for a given quantity of calcium added, there is insufficient calcium available to modify the alumina inclusions according to reaction 11.4.15. Adequate modification of the solid alumina inclusions into liquid calcium aluminates is essential in order to prevent nozzle clogging during continuous casting operations.

By combining the data for the activities of the oxides in aluminate melts[61,62] with the equilibrium constant for reaction 11.4.16 as given by equation 11.4.5, the critical sulfur content for the formation of liquid calcium aluminates can be evaluated.[63] The results are shown in Fig. 11.27 and Fig. 11.28 where the critical sulfur and aluminum contents for the formation of liquid alumina-saturated calcium aluminate and liquid $12CaO \cdot 7Al_2O_3$ are shown.

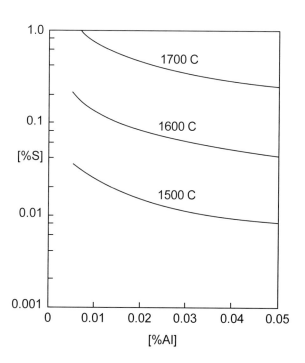

Fig. 11.27 Critical steel sulfur and aluminum contents below which liquid alumina-saturated calcium aluminate is formed at the indicated temperatures; $a_{CaS} = 1$. *From Ref. 63.*

Whenever the aluminum and sulfur contents in the steel fall below a curve for a given temperature, the formation of a liquid calcium aluminate is favored.

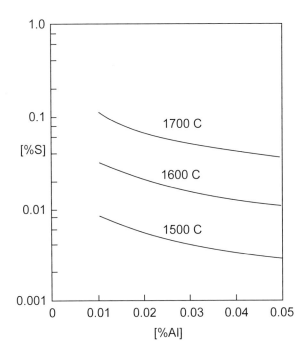

Fig. 11.28 Critical steel sulfur and aluminum contents below which liquid $12Ca \cdot 7Al_2O_3$ is formed at the indicated temperatures; $a_{CaS} = 1$. From Ref. 63.

The underlying assumption made in the calculations of the diagrams in Fig. 11.27 and Fig. 11.28 was that the activity of calcium sulfide equals unity. However, the presence of manganese in most steels causes the sulfur to precipitate as calcium manganese sulfides, Ca(Mn)S, in which the activity of calcium sulfide in less than unity, thus decreasing the critical sulfur content for the formation of liquid calcium aluminate. Kor[63] has shown that the presence of up to 2% manganese in the steel has only a small effect on the critical sulfur content. The relationship between aluminum and sulfur for the formation of liquid calcium aluminates for $a_{CaS} = 0.75$ is shown in Fig. 2.139.

Larsen and Fruehan[64] studied the modification of oxide inclusions by calcium in a number of samples obtained from laboratory melts as well as from steelmaking operations. The results from this study are summarized in Fig. 11.29 from which it may be seen that the agreement with the theoretical predictions, indicated by the curves labeled CA and $C_{12}A_7$, is generally good.

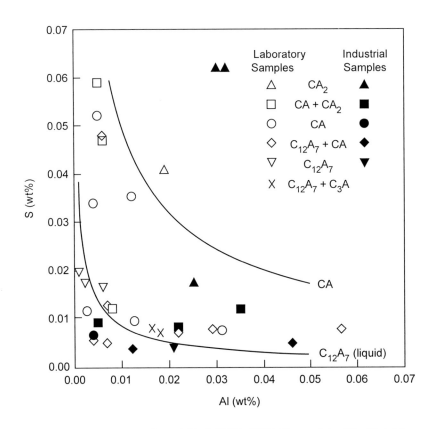

Fig. 11.29 Composition of inclusions found in laboratory and plant samples compared with theoretical predictions. From Ref. 64.

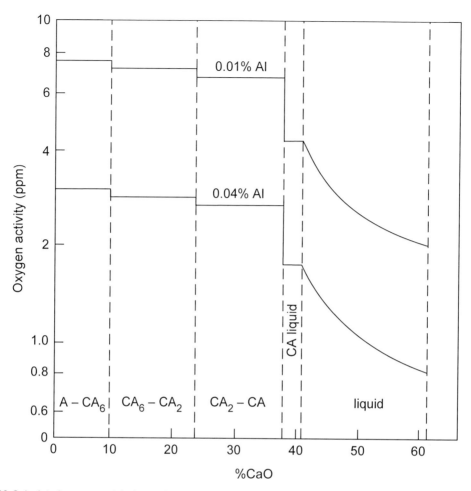

Fig. 11.30 Calculated oxygen activity in steels containing 0.01 and 0.04% Al in equilibrium with the indicated calcium aluminate at 1600°C. *From Ref. 64.*

In general, it is difficult to assess whether the injection of calcium into the steel has resulted in the desired degree of inclusion modification. Larsen and Fruehan[64] have pointed out that, in theory, the degree of inclusion modification can be determined by measuring the activity of oxygen in the steel by means of an oxygen sensor. In general, the activity of alumina in calcium aluminate inclusions is less than unity. Thus, the activity of oxygen in a steel in equilibrium with a calcium aluminate is less than that in a steel in equilibrium with alumina. As the inclusions are modified from alumina to calcium-rich calcium aluminate, the activity of oxygen in the steel decreases, provided the aluminum content of the steel is essentially constant. This is shown schematically in Fig. 11.30 from which it can be seen that the decrease in oxygen activity (in ppm) is significant whenever complete modification to liquid calcium aluminate has occurred. Thus, an oxygen sensor measurement before and after calcium treatment should, in principle, indicate how effective the treatment was in terms of inclusion modification.

In many samples obtained from steelmaking operations the oxide inclusions contain varying amounts of magnesia. Kor[63] has estimated the effect of the presence of magnesia in the oxide inclusions on the critical sulfur content for the formation of liquid calcium magnesium aluminate inclusions for a given aluminum content of the steel. It was found that for inclusions containing less than 10% magnesia, the critical sulfur content is somewhat higher than that for inclusions not containing magnesia.

11.4.5.3 Sulfide Shape Control

In steels not treated with calcium, the sulfur precipitates as finely dispersed manganese sulfide particles in the interdendritic liquid that freezes last. The manganese sulfides delineate the prior austenitic grain boundaries in the as-cast structure. During hot rolling the manganese sulfide particles are deformed, resulting in stringers in the rolled product. These stringers make the final product susceptible to, for example, hydrogen-induced cracking in sour gas or oil environments.

In calcium-treated low-sulfur steels the grain boundary precipitation of MnS during solidification is suppressed as a result of the precipitation of sulfur as a Ca(Mn)S complex on the calcium aluminate inclusions as indicated by the following reaction

$$(CaO) + 2[S] + [Mn] + 2/3[Al] \rightarrow (CaS \cdot MnS) + 1/3(Al_2O_3) \qquad (11.4.17)$$

The extent of sulfide shape control that can be achieved during solidification of calcium-treated steel depends on the total oxygen, sulfur and calcium contents of the steel. This is described by a model based on the reactions occurring in the impurity enriched interdendritic liquid during solidification.[65] On the basis of this model, the following criteria can be derived for the tundish compositions of aluminum-killed steels to give adequate sulfide shape control in the final product.[2]

Table 11.7 Tundish Composition Ranges for Al-Killed Steels to Achieve Acceptable Sulfide Shape Control. *From Ref. 2.*

O (ppm) as aluminate inclusions	Ca (ppm)	Mn (%)	S (ppm)
25	20–30	0.4–0.6	<20
25	20–30	1.3–1.5	<30
12	15–20	0.4–0.6	<10
12	15–20	1.3–1.5	<15

In steels with a total oxygen content of 10 ppm or less and relatively high sulfur contents, e.g. > 100 ppm, sulfide shape control by means of calcium treatment is obviously not feasible. To minimize the occurrence of sulfide stringers in such steels, the addition of tellurium or sometimes selenium has been found to be beneficial. Due to the strong effect of both these elements on the interfacial tension between sulfides and steel, the tendency of sulfide stringer formation during rolling is decreased. The result is that after rolling the sulfides are ellipsoidal in shape with a length-to-width ratio that depends on the Te/S-ratio in the steel.[63] Tellurium is usually added to liquid steel either by powder injection or by wirefeeding.

11.5 Vacuum Degassing

Vacuum degassing of steel has an even longer history than the treatment of steel with calcium, Aitken[66] possibly being the first to have proposed an arrangement for the ladle degassing of a heat of steel. An overview of the various processes in use until 1965 was given by Flux.[67] Since the 1950s and 1960s many new developments have taken place in regard to equipment for the vacuum treatment of steel as well as the technology of steel refining in vacuum degassing facilities. A more recent overview dealing with vacuum degassing was prepared by Fruehan.[68]

Initially, vacuum degassing was used primarily for hydrogen removal. However, during the last twenty years or so there has been an increased use of vacuum degassing for the production of ultra-low-carbon (ULC) steels with carbon contents of 30 ppm or less. Furthermore, a relatively new family of steel grades, the so-called interstitial-free (IF) steels with carbon and nitrogen contents of 30 ppm or less, has appeared on the scene. To achieve these low carbon and nitrogen contents, a treatment under vacuum is mandatory. Presently, almost every high-quality steel producer has installed a vacuum treatment facility.

11.5.1 General Process Descriptions

More detailed schematic illustrations of the most popular degassing systems are presented in Section 11.6. For the purpose of the present discussion a brief description of the salient features is given here. There are two principal types of degassers: recirculating systems such as RH, RH-OB, RH-KTB and DH; and non-recirculating systems such as ladle or tank degassers, including VAD (vacuum arc degassing) and VOD (vacuum oxygen decarburization), and stream degassers.

In both recirculating and non-recirculating systems argon is used as the lifting or stirring gas. In recirculating systems the argon is used as the so-called lifting gas to lower the apparent density of the liquid steel to be lifted up from the ladle into the vacuum vessel. In non-recirculating systems argon is used as the stirring gas to promote the removal of hydrogen and/or nitrogen and to homogenize the bath.

The decision which degassing system, recirculating or non-recirculating, to install in a given shop is largely determined by the product mix to be produced. If a relatively large number of heats has to be decarburized to very low levels to produce ULC or IF steels, a recirculating system such as the RH or one of its modifications is usually preferred. For example, a carbon content of 25 ppm can easily be attained in an RH or RH-OB (KTB) degasser whereas in a tank degasser, such as a VOD, such low carbon contents cannot be attained within a practical time span.

As will be discussed later, there is not much difference between recirculating and non-recirculating systems in terms of the effectiveness with which hydrogen or nitrogen can be removed. Thus, if the primary function of the degasser is to remove hydrogen and sometimes nitrogen, the choice of system will be determined primarily by the desired match between the steel melting vessel (BOF or EAF) and the caster as well as by considerations in regard to capital and operating costs.

11.5.2 Vacuum Carbon Deoxidation

One of the purposes to treat steel in an RH or RH-OB (KTB) degasser is to lower the dissolved oxygen content of the steel by means of carbon deoxidation before adding aluminum to kill the steel completely. With such a carbon deoxidation practice there are considerable cost savings as a result of the decreased usage of aluminum.

Vacuum carbon deoxidation is described by the following reaction

$$[C] + [O] \rightarrow CO (g) \qquad (11.5.1)$$

where the carbon and oxygen are dissolved in the steel bath. The carbon-oxygen relationship during the vacuum decarburization treatment is schematically illustrated in Fig. 2.138. In the RH process, decarburization proceeds nearly to the stoichiometrically related decrease in carbon and oxygen contents: $\Delta ([O] = (16/12) \Delta[C]$. This is also called the self-decarburization process.

Because in the RH-OB (KTB) process oxygen is supplied from an outside source, decarburization initially takes place without a simultaneous decrease in the steel oxygen content, the so-called forced decarburization. In the later stages decarburization follows the path of self-decarburization. One of the advantages of an RH-OB (KTB) over a conventional RH is that the steel can be tapped at a higher carbon content, thus decreasing converter processing time and increasing the iron yield (lower slag FeO).

Plant data obtained for the carbon and oxygen contents of the steel before and after RH treatment are shown in Fig. 11.31, reproduced from data quoted by Turkdogan.[2] Similar results are obtained with the tank degasser.[69] Although the pressure in the vacuum vessel was approximately 0.001 atm, the final carbon and oxygen contents correspond to CO pressures varying from 0.06 to 0.08 atm, Fig. 11.31.

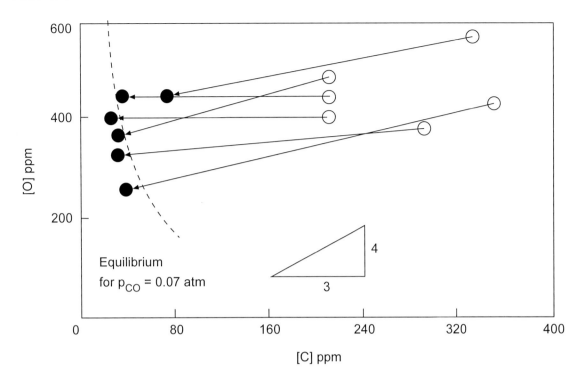

Fig 11.31 Carbon and oxygen contents of steel before (open symbols) and after (filled symbols) RH treatment, from data reported by Turkdogan.[2]

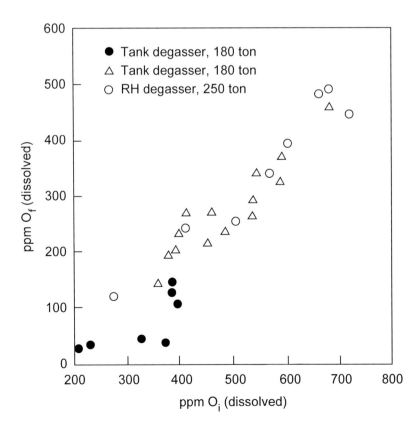

Fig. 11.32 Oxygen contents before (O_i) and after (O_f) vacuum decarburization. *From Ref. 2.*

After approximately 20 min treatment time, the final oxygen content of the steel is always high when the initial content is high (low initial carbon content) in both the tank degasser and the RH, Fig. 11.32.[2]

It can be seen from Fig. 11.32 that the decrease in oxygen content of the steel as a result of the vacuum decarburization treatment is less than that expected from the stoichiometry of reaction 11.5.1. This is because there is oxygen transfer from the ladle slag to the steel during vacuum decarburization. Another source of oxygen is the iron oxide-rich skull which builds up on the inside of the vacuum vessel as a result of previous operations. Thus, some decarburization via the following reaction takes place also

$$(FeO) + [C] \rightarrow Fe + CO\,(g) \tag{11.5.2}$$

From the stoichiometry of the reactions, the material balance gives the following relation for the quantity of oxygen transferred to the steel from the ladle slag and the oxidized skull inside the vacuum vessel[2]

$$\Delta O\,(slag) = (16/12)\,\Delta([C] - \Delta[O] \tag{11.5.3}$$

where

$$\Delta[C] = [\%C]_i - [\%C]_f$$

$$\Delta[O] = [ppm\,O]_i - [ppm\,O]_f$$

The values of ΔO (slag) derived from the plant data using equation 11.5.3 are shown in Fig. 11.33.[2] It is seen that during decarburization the amount of oxygen transferred from slag to steel is higher the higher the initial carbon content. For initial carbon contents of 200 ppm or less there is no more oxygen pickup from the slag during vacuum decarburization and the carbon content decreases according to the stoichiometry of reaction 11.5.1.

11.5.2.1 Rate of Decarburization

The rate of decarburization is expressed by the following relationship

$$\ln\left\{\frac{[\%C]_f}{[\%C]_i}\right\} = -k_C t \tag{11.5.4}$$

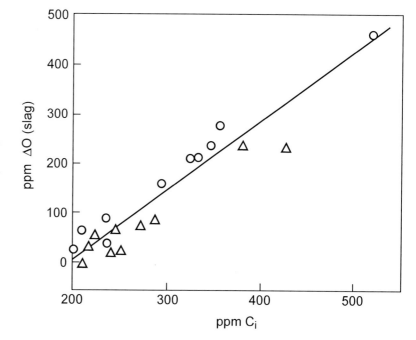

Fig. 11.33 Transfer of oxygen from ladle slag to steel during decarburization in tank degassers (triangles) and in RH degassers (circles). *From Ref. 2.*

where $[\%C]_i$ and $[\%C]_f$ are the carbon contents before and after decarburization, respectively and k_C is the rate constant for decarburization. For RH degassers the rate constant is given by the following relationship

$$k_C = \frac{Q}{V_b \rho} \frac{q}{\frac{Q}{\rho} + q}, \text{ min}^{-1} \quad (11.5.5)$$

where

- Q = circulation rate of liquid steel, kg/min
- V_b = volume of the steel bath in the ladle, m^3
- ρ = density of liquid steel, kg/m^3
- q = volumetric mass transfer coefficient of decarburization, m^3/min

According to Kuwabara et al.[70] the circulation rate of the liquid steel in the RH vessel is given by

$$Q \text{ (tonne/min)} = 11.4 (\dot{V})^{1/3} (D)^{4/3} \left[\ln\left(\frac{P_1}{P_0}\right) \right]^{1/3} \quad (11.5.6)$$

where

- $\dot{V}$ = flowrate of argon injected into the up-leg snorkel, Nl/min
- D = inside diameter of the up-leg snorkel, m
- P_1, P_0 = pressure at the argon injection point and at the bath surface, respectively, Pa.

The volumetric mass transfer coefficient of decarburization, q, is proportional to the cross sectional area, A_v, of the vessel which is equivalent to the surface area of the bath. From plant observations on 240 to 300 tonne RH vessels Kato et al.[71] found the following approximate empirical relationship for q:

$$q = 0.26 \, Q^{0.64} \, A_v \, [\%C] \quad (11.5.7)$$

valid for $0.0025 " [\%C] " 0.01$.

It should be noted that the actual rate is very complex. The reaction occurs at various sites including the argon bubble surface, refractory surfaces, metal free surfaces, and homogeneously in the melt. Therefore, equation 11.5.7 should only be used for similar conditions for which it was developed.

According to the above equations, the rate of decarburization will increase with snorkel diameter and vessel diameter, which was confirmed by actual plant data.[71]

In Fig. 11.34 the decarburization rate constant, k_C, is shown as a function of the specific flowrate (Nm^3/min tonne) of the stirring gas for recirculating systems such as RH and DH and for non-recirculating systems such as VAD.[69] Because of the lower specific flowrates for the stirring gas used in non-recirculating systems, the time required to remove 50% of the carbon is approximately 7 min, whereas in the RH this time can be as short as 3 to 4 min, Fig. 11.34.

Several methods to enhance the decarburization rate in the RH have been reported.[25] Kuwabara et al.[70] were able to increase the decarburization rate by injecting argon through nozzles installed in the hearth of the RH vessel. By injecting argon at a rate of 400–500 Nl/min the carbon content in a 100 tonne heat of steel was decreased from 200 to approximately 10 ppm in 20 min, corresponding to $k_C \approx 0.15$ min^{-1} as found from equation 11.5.4. This is approximately 50% higher than k_C for a conventional RH, Fig. 11.34.

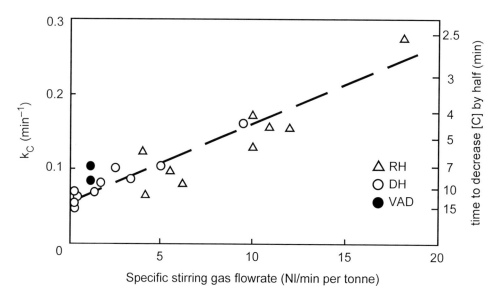

Fig. 11.34 Decarburization rate constant as affected by the specific flowrate of the stirring gas for recirculating systems (RH, DH) and non-recirculating systems (VOD). *From Ref. 69.*

Yamaguchi et al.[72] co-injected hydrogen and argon into the up-leg of the RH to attain 6 ppm hydrogen in the steel. The vessel was then evacuated while the injection of hydrogen continued. The evolution of hydrogen gas bubbles within the bath resulted in a final carbon content of 5–10 ppm.

Okada et al.[73] reported that top blowing of powders onto the surface of the bath in the RH—the so-called RH-PB (powder blowing) process—is effective for attaining ultra low carbon, nitrogen and sulfur contents. For example, by blowing 20–60 kg/min of iron ore powder (–100 mesh) through a top lance positioned 2–3m above the bath surface the final carbon content attained was less than 5 ppm.

Whenever the initial carbon content of the steel is relatively high, the decarburization rate may be limited by the supply of oxygen. To remedy this, the RH-OB (oxygen blowing) and RH-KTB (Kawasaki Top Blowing) processes were developed. In the RH-OB process the oxygen is supplied via tuyeres installed in the sidewalls in the lower part of the RH vessel.[74] In the RH-KTB process the oxygen is supplied via a lance situated in the RH vacuum vessel.[25] In these configurations of the RH process skull formation inside the vessel has been minimized by post combustion of the CO by the injected oxygen.

Yamaguchi et al.[75] developed a reaction model for decarburization in the RH and RH-KTB processes which is based on a mixed control mechanism involving the mass transfer of carbon and oxygen in the liquid steel present in the vacuum vessel as well as on the transport of carbon and oxygen by the recirculating steel. The model satisfactorily takes into account the effect of the concentration of oxygen on the decarburization rate in the conventional RH as well as in the RH-KTB process.

11.5.3 Hydrogen Removal

The rate of hydrogen removal during degassing is controlled by mass transfer in the liquid steel for which the rate equation is given by

$$\ln\left\{\frac{[H]_f - [H]_e}{[H]_i - [H]_e}\right\} = -k_H t \qquad (11.5.8)$$

where

[H]$_f$ = the hydrogen content after degassing

[H]$_i$ = the initial hydrogen content

[H]$_e$ = the equilibrium hydrogen content as determined by the pressure in the system

k$_H$ = the overall rate constant for hydrogen removal.

In the majority of modern degassers the attainable pressure is below 0.01 atm (~10 torr) and, consequently, [H]$_e$ can be neglected with respect to [H]$_i$ and [H]$_f$. This gives the following simplified rate equation

$$\ln\left\{\frac{[H]_f}{[H]_i}\right\} = -k_H t \tag{11.5.9}$$

Bannenberg et al.[76] developed a mathematical model for hydrogen removal in a 185 tonne tank degasser. The agreement between the hydrogen content measured after degassing and that calculated from the model is excellent, as shown in Fig. 11.35. The model is based on fundamental principles. The most critical parameter is the bubble size, which is extremely difficult to predict. Therefore, the model should only be used to make comparisons for similar operating conditions.

Using the model, Bannenberg et al.[76] calculated hydrogen removal rates for initial hydrogen contents varying between 3 and 7 ppm and argon flowrates of 0.9 and 1.8 Nm³/min. The results are summarized in Fig. 11.36 from which it can be seen that it takes 2 to 3 min longer to achieve a given degree of hydrogen removal for steels initially containing 3 ppm hydrogen than for steels initially containing 7 ppm hydrogen. Moreover, it is noted that doubling the flowrate of the argon stirring

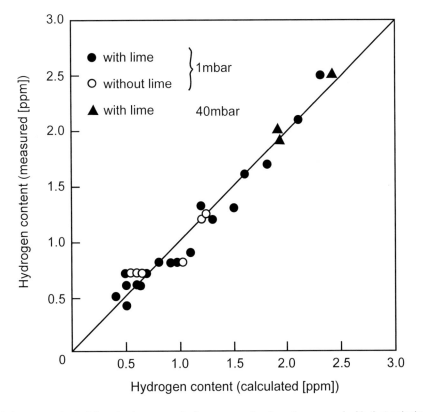

Fig. 11.35 Hydrogen content of the steel measured after vacuum treatment compared with that calculated from a model developed for tank degassing. *From Ref. 76.*

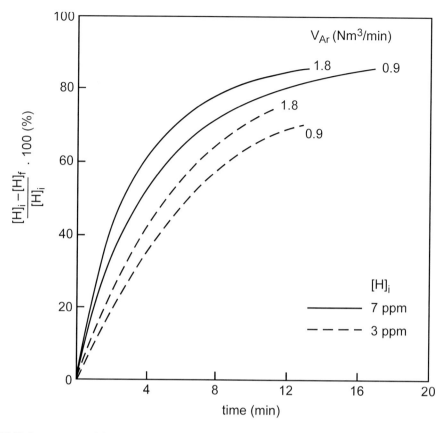

Fig. 11.36 Hydrogen removal for two values of the initial hydrogen content and two argon flowrates in a 185 tonne tank degasser, based on a model developed by Bannenberg et al.[76]

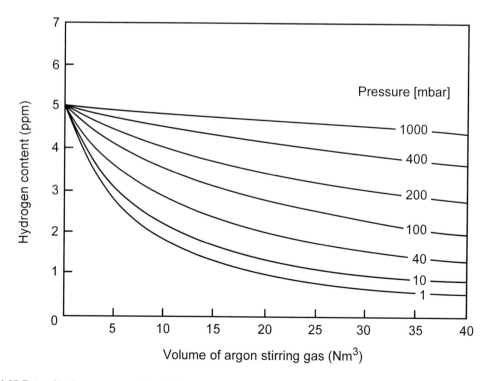

Fig. 11.37 Rate of hydrogen removal in a 185 tonne tank degasser as affected by the tank pressure. *From Ref. 76.*

gas results in a marginally shorter treatment time, e.g. approximately 2 min shorter to achieve 70% hydrogen removal.

Hydrogen removal rates for tank pressures ranging from 1 to 1000 mbar (~10^{-1} atm) are shown as a function of the total volume of argon stirring gas flowing at 1.8 Nm^3/min in Fig. 11.37. It is seen that the final hydrogen content of the steel is essentially unaffected by the tank pressure for pressures up to 10 mbar (~10^{-2} atm).

From the data depicted in Fig. 11.36 it is found that the overall rate constant for hydrogen removal, k_H, increases from approximately 0.09 to 0.16 min^{-1}, when the argon flowrate increases from 0.9 to 1.8 Nm^3/min. The higher k_H value is comparable to the value of 0.13 min^{-1} observed for a RH vessel[2] with a 600 mm diameter snorkel and a steel circulation rate of approximately 140 tonne/min. These values indicate that the efficiencies of recirculating and non-recirculating systems for the removal of hydrogen are similar.

11.5.3.1 Hydris Probe

In shops where hydrogen-sensitive steel grades such as, for example, large bars, are produced, it is important to know the hydrogen content of the steel bath before it is delivered to the caster or the teeming platform. Under such circumstances an in-site determination of the hydrogen content of the bath in the ladle may be desirable.

Around ten years ago, Plessers et al.[79] described an immersion system, the Hydris probe, for the rapid in-situ determination of hydrogen in a bath of liquid steel. The principle of the measurement is based on equilibrating a known volume of argon, being passed through the liquid steel, with the hydrogen dissolved in the steel. Thus, the argon–hydrogen gas mixture leaving the steel after equilibration has a partial pressure of hydrogen that, via Sieverts' law (equation 2.4.8), can be related to the hydrogen content of the steel. The Hydris probe was tested extensively by Frigm et al.[80] who found that it gave reliable readings of the hydrogen content of the liquid steel with an uncertainty of approximately ±5%.

The Hydris probe has been in regular use at the Faircrest steel plant of The Timken Company for a number of years and has been found to be rugged and reliable. The cost associated with the use of the probe has to be weighed against cost savings made possible by its use such as, for example, a significant decrease in degassing time to attain the required hydrogen content in the steel.

11.5.4 Nitrogen Removal

Some nitrogen removal from liquid steel during vacuum degassing is possible, provided the steel is fully killed and has a low sulfur content. Bannenberg et al.[76] developed a rate equation for nitrogen removal in a 185 tonne tank degasser which was based on a mixed-control model, i.e. liquid-phase mass transfer of nitrogen to the argon bubbles coupled with chemical reaction control at the liquid-gas bubble interface. As shown in Fig. 11.38, the nitrogen contents after degassing calculated from the model and indicated by the solid lines, are in good agreement with the measured nitrogen contents indicated by the different symbols. Equally good agreement between calculated and measured values was found for steels containing between 20 and 200 ppm sulfur.[77]

The rate of nitrogen removal for various initial nitrogen contents as calculated from Bannenberg's model is shown in Fig. 11.39 for a killed steel containing 2 ppm dissolved oxygen and 10 ppm sulfur and a tank pressure of 1 mbar (~10^{-3} atm). It can be seen that under these conditions approximately 50% of the nitrogen can be removed in roughly 15 min, provided the initial nitrogen content is 50 ppm or higher.

The effect of the steel sulfur content on the rate of nitrogen removal as calculated from Bannenberg's model is shown in Fig. 11.40. As with the hydrogen case, the model must be used carefully due to the uncertainty in bubble size.

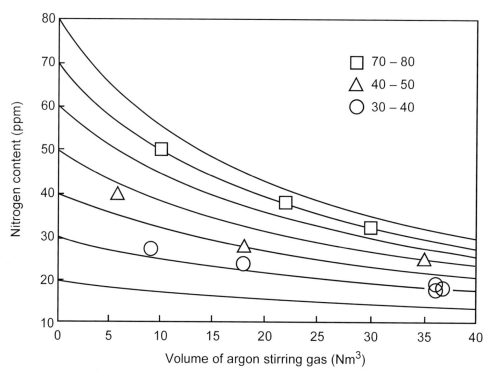

Fig. 11.38 Calculated rates of nitrogen removal in a 185 tonne tank degasser compared with measured data (symbols) for steels containing less than 20 ppm sulfur. *From Ref. 77.*

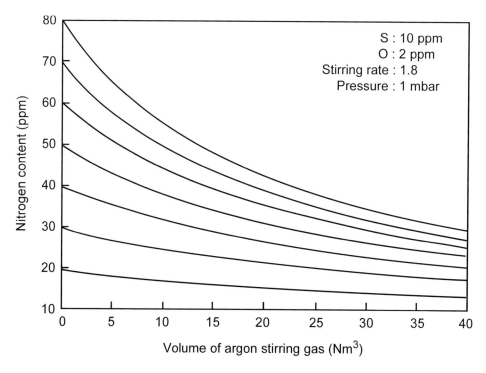

Fig. 11.39 Rates of nitrogen removal in a 185 tonne tank degasser calculated from a model. *From Ref. 76.*

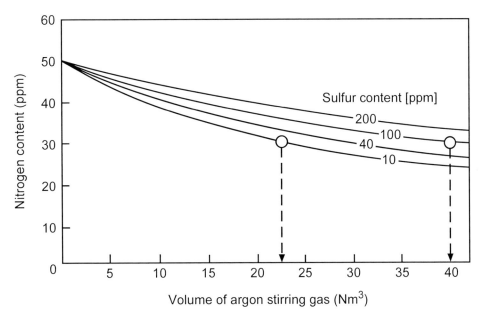

Fig. 11.40 Rates of nitrogen removal in a 185 tonne tank degasser as affected by the steel sulfur content. *From Ref. 78.*

A simplified form of the rate equation for nitrogen removal, equation 2.2.58 is:

$$\frac{1}{\text{ppm N}} - \frac{1}{\text{ppm N}_o} = k_N (1-\Theta) t \qquad (11.5.10)$$

where k_N is the apparent rate constant for denitrogenization for the limiting case in which both the oxygen and sulfur contents tend to zero. From the nitrogen removal rates shown in Fig. 11.40, together with equation 2.2.56 for $(1 - \Theta)$, the apparent rate constant is estimated to be $k_N \approx$ 0.0013 (ppm N min)$^{-1}$ for 185 tonne heats in a tank degasser at 1 mbar ($\sim 10^{-3}$ atm) pressure.

The overall rate constant for denitrogenization in a RH degasser[81] is shown as a function of the sulfur content of the steel in Fig. 11.41. It is noted that the value of k_N for sulfur contents approaching zero is very similar to the aforementioned value for a tank degasser. Thus, from these values it is concluded that recirculating and non-recirculating degassers are equally effective in removing nitrogen as well as hydrogen (Section 11.5.3).

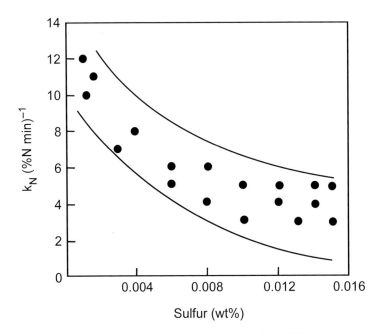

Fig. 11.41 Overall rate constant for nitrogen removal in a RH degasser as affected by the steel sulfur content. *From Ref. 81.*

Recently, an on-line method for the determination of the nitrogen content of liquid steel has been developed.[89] The system, called Nitris, is based on the

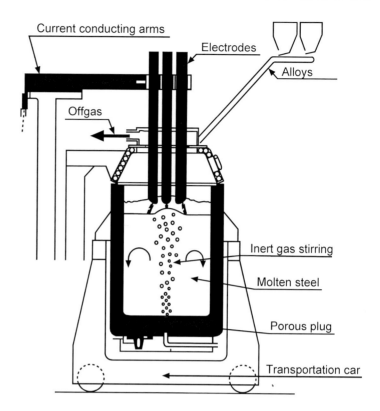

Fig. 11.42 Schematic illustration of a Daido ladle furnace as modified by Fuchs System-technik. *From Ref. 84.*

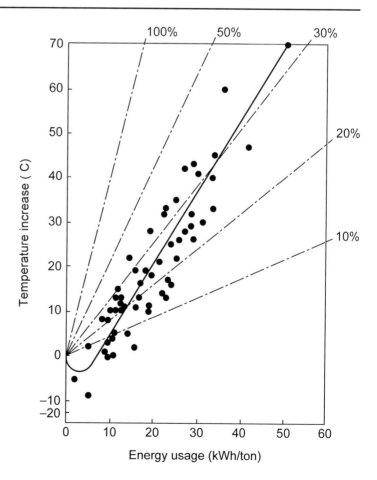

Fig. 11.43 Temperature increase in a 45 tonne ladle furnace as affected by the specific energy input. *From Ref. 84.*

same measuring principle as the Hydris probe, discussed in Section 11.5.3. Jungreithmeier et al.[89] presented details on plant data obtained with the Nitris system as well as a comparison with nitrogen contents obtained by means of the traditional combustion method.

11.6 Description of Selected Processes

In this section a more detailed description of the most popular processes used in secondary steelmaking operations will be presented. Other, less popular, processes are described in several review papers.[1, 68, 82] This section is concluded with some general remarks concerning process selection.

11.6.1 Ladle Furnace

Ladle furnaces are among the most widely used pieces of equipment in secondary steelmaking operations and range from relatively simple retrofitted installations[83] to elaborately equipped facilities. An example of the latter category is the ladle furnace as originally developed by Daido Steel Co., shown schematically in Fig. 11.42.[84]

The ladle furnace illustrated in Fig. 11.42 is lined with a basic lining and covered with a water-cooled roof. The bath in the ladle is heated with the aid of three electrodes which are supported by current conducting arms. The process is usually operated with a slag cover on the bath, thus avoiding excessive wear of the ladle lining due to arc radiation. Another advantage of this mode of operation is that a relatively long arc can be employed, resulting in increased energy efficiency and lower specific electrode consumption. During reheating the bath is continually stirred by means of inert gas supplied via a porous plug in the ladle bottom.

The efficiency of reheating a 45 tonne heat of steel in a ladle furnace is shown in Fig. 11.43.[84] The reheating efficiency is between 20 and 30%, depending on the specific energy input. For larger heat sizes the efficiency can be expected to be higher.

An example of the use of a ladle furnace in conjunction with a degasser is given by Bieniosek.[85] The installation and operation of a low budget ladle metallurgy facility at USS/Kobe Steel is discussed by Mobberly and Diederich.[86]

11.6.2 Tank Degasser

Steel can be treated in a tank degasser without arc reheating. This is shown schematically in Fig. 11.44 for two different stirring systems. An inductively stirred bath is shown in Fig. 11.44(a) while in Fig. 11.44(b) the bath is stirred by bubbling argon through a porous plug located in the ladle bottom.

11.6.3 Vacuum Arc Degasser

A vacuum arc degasser (VAD) is a tank degasser with electrodes added for the purpose of reheating the steel. A schematic illustration of a VAD unit is shown in Fig. 11.45.

Whittaker[87] described a VAD process used at Atlas Specialty Steels in which two ladle covers are used sequentially, shown schematically in Fig. 11.46. During reheating the ladle is placed underneath a water cooled steel roof fitted with three electrodes. The cover is not designed to operate under vacuum. For degassing the ladle is placed underneath a roof fitted with a sight port, vacuum offtake and a water-cooled O-ring seal. This seal makes the system gastight such that the final pressure inside the ladle can be less than 1 torr (~0.0013 atm). For processing stainless steels the oxygen lance is used and the unit is operated as a VOD.

Another VAD unit that operates with a gastight ladle cover is the Stein-Heurtey–S.A.F.E. System.[88] Two such units are in operation at The Timken Company's steel plants where heat sizes between approximately 120 and 160 tonnes are processed.

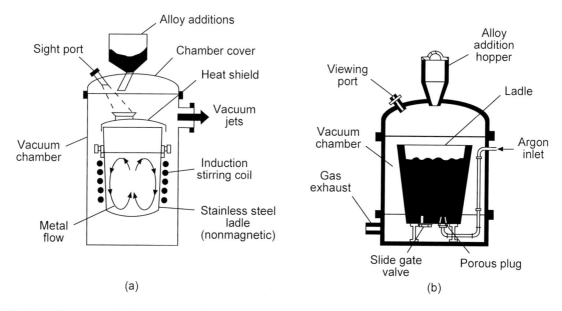

Fig. 11.44 Two types of tank degasser: (a) induction coil stirring, (b) porous plug for argon bubbling. *From Ref. 2.*

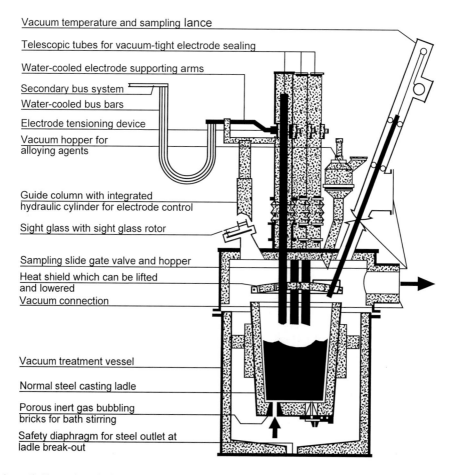

Fig. 11.45 Schematic illustration of a VAD unit. *From Ref. 68.*

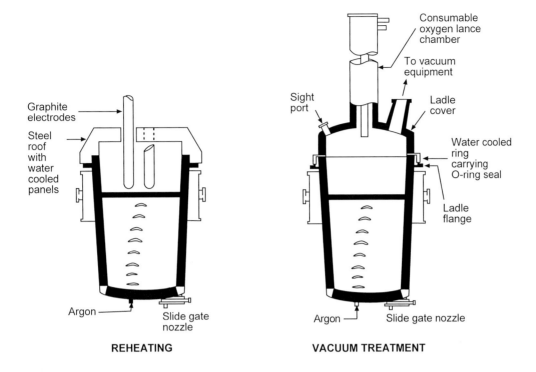

Fig. 11.46 Schematic illustration of the use of separate ladle covers for reheating and degassing of 68 tonne heats at Atlas Specialty Steels. *From Ref. 87.*

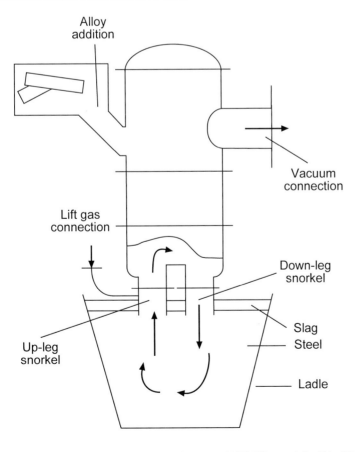

Fig. 11.47 Schematic illustration of the principle of the RH process. *From Ref. 92.*

11.6.4 RH Degasser

A schematic illustration of the principle of the RH process is depicted in Fig. 11.47[92] while a sketch of a RH unit with ancillary equipment is shown in Fig. 11.48.[93] The 145 tonne RH unit shown in Fig. 11.48 was designed for a monthly steel production of 40,000 tonne, approximately 30% of which represents ULC-IF steel grades.

To increase the availability of the RH installation, many steel shops operate the facility with two vessels, one of which will be in the operating position while the other is being repaired or relined. The operation of such a two-vessel facility at Inland Steel Co. was discussed by Schlichting and Dominik.[94]

Over the years the RH process was developed further with the aim to enhance the capabilities of the process. One of the important developments was the addition of oxygen to the RH vessel, either by injection through a tuyere mounted in the sidewall of the vessel (RH-OB) or via a top lance inside the vessel (RH-KTB).

Because of the violent splashing occurring inside the RH vessel, skull formation on its inside walls results. In the RH-KTB process the skull formation is minimized as a result of the heat generated by the post combustion of CO to CO_2, as schematically illustrated in Fig. 11.49.

To enhance the capabilities of the RH process even further, the RH-PB (powder blowing) process was developed. The addition of powder and/or fluxes makes it possible to desulfurize or dephosphorize the steel during the RH operation. Kuwabara et al[96] have reported on the injection of fluxes into the lower part of the ladle during the operation of a RH unit.

Fujii et al.[95] studied the effect of fluid flow on the decarburization rate in a RH degasser and related this to design parameters such as the snorkel and vessel diameters as well as the circulation flowrate. A control model for the RH degasser was discussed by Sewald[92] with emphasis on the

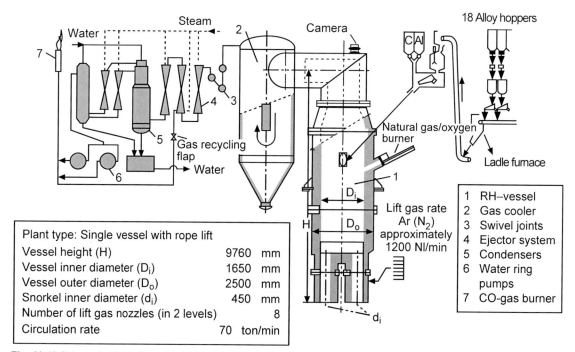

Fig. 11.48 Schematic illustration of a 145 tonne RH unit with ancillary equipment as installed in the LD steel plant of Voest-Alpine Stahl, Linz, Austria. *From Ref. 93.*

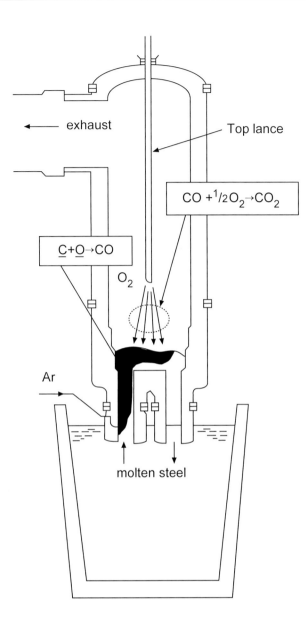

Fig. 11.49 Schematic illustration of the the RH-KTB process. *From Ref. 95.*

supervisory computer control system as well as the design and operational characteristics of the process model.

11.6.5 CAS-OB Process

The CAS-OB process (Composition Adjustment by Sealed argon bubbling with Oxygen Blowing) was developed by Nippon Steel Co. and was recently installed at the Steubenville plant of the Wheeling-Pittsburgh Steel Co.[97] A schematic illustration of a CAS-OB unit is shown in Fig. 11.50.

The main feature of the process is the refractory snorkel or bell underneath which alloy additions to the bath are made. The ladle is positioned such that the snorkel is situated right above the porous stirring plug, Fig. 11.50. This ensures that the agitated surface of the steel bath is confined to the area underneath the bell. Additional argon stirring, if necessary, is obtained via a specially shaped submerged lance, Fig. 11.50. Reheating of the steel is accomplished by injecting oxygen through the top lance in conjunction with aluminum additions. For low-carbon steel the rate of reheating in the

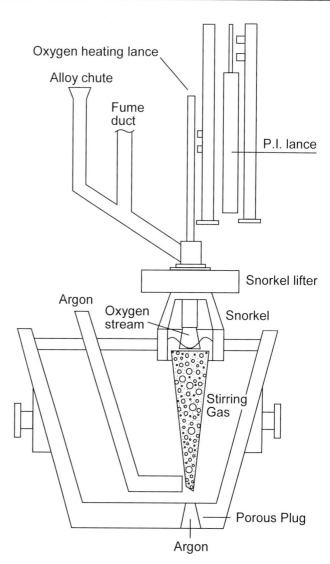

Fig. 11.50 Schematic illustration of a CAS-OB process. *From Ref. 97.*

CAS-OB was found to be approximately 10°C/min[97] (~20°F/min). This is in broad agreement with data reported by Palchetti et al.[98] who observed a similar reheating rate for an oxygen blowing rate of approximately 70 Nm³/min (~2470 scfm) for 300 tonne heats.

11.6.6 Process Selection and Comparison

Fruehan[68] has addressed the various issues that are involved in the selection of a process, particularly a degassing unit, for a given steel plant. Some of the most important considerations are: future and current product mix; requirements with respect to carbon, hydrogen, nitrogen and sulfur (phosphorus) content of the final product; reheating capability (e.g. does the shop have a ladle furnace); effect on steelmaking facility (BOF or EAF); degassing time compatible with other operations, e.g. with sequential continuous casting; and capital and operating costs.

An important issue is to match the time requirement of the degassing unit to the processing times in the steelmaking furnace (BOF or EAF) and in the continuous caster. For example, to attain a carbon content of 0.005% or less requires 10–15 min in an RH-OB (KTB) compared with 15–20 min in a tank degasser such as a VOD. When such low carbon contents are a frequent requirement, the installation of an RH-OB (KTB) unit is usually warranted.

Another important issue is the demand on the steelmaking furnace. Ultra-low-carbon steels are easier to produce in an RH or tank degasser when the initial carbon content (the tap carbon content) is less than 0.025%. At these low carbon levels the steel contains enough dissolved oxygen to remove the required amount of carbon in a practical time span. However, having to tap the steel at such low carbon contents puts an extra load on the BOF with detrimental ramifications in regard to the vessel lining because of the higher temperature and FeO content of the slag. With an RH-OB (KTB) or VOD it is possible to tap the steel at higher carbon contents because supplemental oxygen is available for enhanced decarburization, as discussed in Section 11.5.2.1.

Table 11.8 may serve as a first approximation to a semi-quantitative comparison of the various vacuum degassing systems[68].

Table 11.8 Comparison of Various Degassing Systems. *From Ref. 68.*

	Type of Degasser				
	RH-OB	RH	VOD	Tank	Ladle
Decarburization level (ppm)	20	20	20–30	30–40	30–40
Decarburization rate	Highest	Satisfactory for low carbon	High	Approx. 2–30% slower than RH, RH-OB, VOD	
Decarburization time to 50 ppm min)	10–15	12–15	15–18	15–20	15–20
Hydrogen removal	All systems are reasonably satisfactory				
Inclusion removal	All systems can improve cleanliness, VOD, tank & ladle require a rinse cycle				
Desulfurization	Only possible with RH injection or RH-PB. (Rarely practiced)			Excellent desulfurization possible but must be separate from decarburization	
Aluminum heating	yes	no	yes	no	no
Relative Capital Cost (RH-OB = 1.0)	1.0	0.7–0.8	0.4–0.6	0.4–0.5	0.3–0.4
Maintenance Cost	Decreasing	⟶		⟶	

References

1. B. R. Nijhawan, *Proc. Emerging Technologies for New Mat. and the Steel Industry* (Cincinnati: October 1991), 215.
2. E. T. Turkdogan, *Fundamentals of Steelmaking*, (London: The Institute of Materials, 1996), Chapter 9.
3. J. Szekely, *Trans. Met. Soc. of AIME* 245 (1969): 314.
4. K. Schwerdtfeger and W. Wepner, *Met. Trans. B*, 8B (1977): 287.
5. P. Massard and K. W. Lange, *Arch. Eisenhuettenw.* 48 (1977): 521.
6. T. Choh, K. Iwata and M. Inouye, *Trans. ISIJ* 23 (1983): 598.
7. P. C. Glaws, G. J. W. Kor, and R. V. Fryan, *Proc. Electric Furnace Conference,* 47 (Warrendale, PA: Iron and Steel Society, 1989), 383.

8. J. Szekely, G. Carlsson, and L. Helle, in *Ladle Metallurgy*, Materials Research and Engineering Series, B. Ilschner and N. J. Grant, eds. (New York: Springer-Verlag, 1988).
9. E. Hoeffken, H. D. Pflipsen, and W. Florin, *Proc. Int. Conf. On Secondary Metallurgy* (Aachen, 1987): 124.
10. R. E. Kracich and K. Goodson, *Iron and Steelmaker* 23, 7 (1996): 41.
11. E. T. Turkdogan, *Ironmaking and Steelmaking* 15 (1988): 311.
12. P. C. Glaws, G. J. W. Kor, The Timken Company, internal reports.
13. R. Engel, R. Marr, and E. Pretorius, *Keeping Current* series, *Iron and Steelmaker*, 23–24 (1996–97).
14. C. E. Tomazin, E. A. Upton, and R. A. Wallis, *Steelmaking Conf. Proc.*, 69 (Warrendale, PA: Iron and Steel Society, 1986), 223.
15. C. Vitlip, *Iron and Steelmaker* 23, 7 (1996): 55.
16. P. E. Anagbo and J. K. Brimacombe, *Iron and Steelmaker* 15, 10 (1988): 38.
17. P. E. Anagbo and J. K. Brimacombe, *Iron and Steelmaker* 15, 11 (1988): 41.
18. W. Pluschkell, *Stahl und Eisen* 101 (1981): 97.
19. D. Mazumdar and R. I. L. Guthrie, *ISIJ Int.*, 35 (1995): 1.
20. D. Mazumdar and R. I. L. Guthrie, *Met Trans. B*, 17B (1986): 725.
21. M. Neifer, S. Roedl, and D. Sucker, *Steel Research*, 64 (1993): 54.
22. M. Sano and K. Mori, *Trans. ISIJ*, 23 (1983): 43.
23. J. Mietz and F. Oeters, *Steel Research*, 60 (1989): 387.
24. S. Joo and R. I. L. Guthrie, *Met. Trans. B*, 23B (1992): 765
25. T. Emi, *Proceedings,* Scaninject VII, part I (1995): 225.
26. S. Asai, M. Kawachi, and I. Muchi, *Proceedings,* Scaninject III, part I (1983): 12:1.
27. Y. Kikuchi et al., *Proceedings,* Scaninject III, part I (1983): 13:1.
28. S. Kim and R. J. Fruehan, *Met. Trans. B*, Vol. 18B (1987): 381.
29. J. Mietz, S. Schneider, and F. Oeters, *Steel Research* 62 (1991): 10.
30. K. Nakanishi and J. Szekely, *Trans. ISIJ* 15 (1975): 522.
31. T. A. Engh and N. Lindskog, Scand. *J. Met.* 4 (1975): 49.
32. K. W. Lange, *Int. Materials Reviews* 33 (1988): 53.
33. R. Ruddelston et al., *Ironmaking and Steelmaking* 18 (1991): 41.
34. J. A. Eckel, The Timken Company, private communication.
35. J. K. Cotchen, *Int. Symp. On Ladle Steelmaking and Furnaces,* (Montreal: The Met. Soc. of CIM, 1988): 111.
36. J. A. Barbus et al., *Steelmaking Conf. Proc.*, 72 (Warrendale, PA: Iron and Steel Society, 1989), 23.
37. Y. Miyashita and Y. Kikuchi, *Proc. Int. Conf. On Secondary Metallurgy* (Aachen, 1987), 195.
38. C. J. Bingel, LTV Steel Co., private communication, 1997.
39. N. R. Griffing et al., *Proc. Electric Furnace Conf.*, Vol. 46 (Warrendale, PA: Iron and Steel Society, 1988), 301.
40. E. T. Turkdogan, *JISI* 210 (1972): 21.
41. R. K. Iyengar and G. C. Duderstadt, *Trans. Met. Soc. AIME* 245 (1969): 807.
42. A. Thomas, F. Villette and F. J. Piton, *Iron and Steel Engineer* 63, 2 (1986): 45.
43. S. Dimitrov, A. Weyl and D. Janke, *Steel Research* 66 (1995): 3.
44. E. Schuermann et al., *Arch. Eisenhuettenw* 50 (1979): 139.
45. P. Riboud and R. Vasse, *Rev. de Métallurgie–C.I.T.,* 82 (1985): 801.
46. Y. Hara et al., *Proceedings,* Scaninject IV, part I (1986), 18:1.
47. I. Sawada, T. Kitamura, and T. Ohashi, *Proceedings,* Scaninject IV, part I (1986), 12:1.
48. C. A. Abel, R. J. Fruehan, and A. Vassilicos, *Trans ISS, Iron & Steelmaker* 22, 8, (1995): 49.
49. P. V. Riboud and C. Gatelier, *Ironmaking and Steelmaking*, 12 (1985): 79.
50. G. Becker et al., *Rev. de Métallurgie–C.I.T.*, 81 (1984): 857.
51. N. Bannenberg and H. Lachmund, *Rev. de Métallurgie–C.I.T.*, 91 (1994):1043.
52. G. Jung et al., *Process Technology Proc.*, 14 (Warrendale, PA: Iron and Steel Society, 1995), 109.
53. S. A. Argyropoulos and R. I. L. Guthrie, *Steelmaking Conf. Proc.*, 65 (Warrendale, PA: Iron and Steel Society, 1982), 156.

54. Y. E. Lee, H. Berg, and B. Jensen, *Ironmaking and Steelmaking* 22 (1995): 486.
55. D. Mazumdar and R. I. L. Guthrie, *Met. Trans. B* 24B (1993): 649.
56. A. L. Gueussier et al., *Iron and Steel Engineer* 60, 10 (1983): 35.
57. A. Herbert et al., *Ironmaking and Steelmaking* 14 (1987): 10.
58. J. Schade, S. A. Argyropoulos, and A. McLean, *ISS Trans.* 12 (1991): 19.
59. O. P. Watts, *J. Am. Chem. Soc.* 28 (1906): 1152.
60. T. Ototani, in *Calcium Clean Steel*, Materials Research and Engineering Series (B. Ilschner and N. J. Grant, eds.),(New York: Springer Verlag, 1986).
61. R. H. Rein and J. Chipman, *Trans. Met. Soc. AIME* 233 (1965): 415.
62. B. Ozturk and E. T. Turkdogan, *Met. Sci.* 18 (1984): 299.
63. G. J. W. Kor, Elliott Symp. Proc. (P. J. Koros and G. R. St. Pierre, eds.), (Warrendale, PA: Iron and Steel Society, 1990), 479.
64. K. Larsen and R. J. Fruehan, *ISS Trans.* 12 (1991): 125.
65. E. T. Turkdogan, *Process Technology Proc.* 6 (Warrendale, PA: Iron and Steel Society, 1986), 767.
66. R. Aitken, *JISI*, 1886, I, 109.
67. J. H. Flux, *Vacuum Degassing of Steel*, Special Report No. 92, (London: Iron and Steel Institute, 1965).
68. R. J. Fruehan, *Vacuum Degassing of Steel*, (Warrendale, PA: Iron and Steel Society, 1990).
69. N. Bannenberg, P. Chapellier, and M. Nadif, *Stahl und Eisen*, Vol. 113, No. 9 (1993), 75.
70. T. Kuwabara, et al., *Trans. ISIJ* 28 (1988): 305.
71. Y. Kato et al., *Tetsu-to-Hagané* 79 (1993): 1248.
72. K. Yamaguchi, T. Sakuraya, and K. Hamagami, *Kawasaki Steel Giho* 25 (1993), 283.
73. Y. Okada et al, *Proc. 8th Japan-Germany Seminar* (Sendai, October 1993): Tokyo, *ISIJ,* 81
74. B. A. Otterman, R. J. Hale, and R. Merk, *Steelmaking Conf. Proc.* 73 (Warrendale, PA: Iron and Steel Society, 1990), 69.
75. K. Yamaguchi et al., *ISIJ Int.* 32 (1992): 126.
76. N. Bannenberg, B. Bergmann, and H. Gaye, *Steel Research* 63 (1992): 431.
77. B. Bergmann, N. Bannenberg, and H. Gaye, *Rev. de Métallurgie–C.I.T.,* 86 (1989): 907.
78. N. Bannenberg et al., *Proc. 6th Int. Iron and Steel Congr.* 3 (Nagoya, 1990), *ISIJ,* 603.
79. J. Plessers, R. Maes and E. Vangelooven, *Stahl und Eisen* 108 (1988): 451.
80. G. Frigm et al., *Proc. Electric Furnace Conf.* 48 (Warrendale, PA: Iron and Steel Society, 1990), 83.
81. T. Kishida et al. (Daido Steel Co.), quoted in *Stahl und Eisen* 107 (1987): 894
82. B. Grabner and H. Hoeffgen, *Radex Rundschau* 3 (1985): 581.
83. G. Mischenko and S. V. McMaster, *Iron and Steel Engineer* 71, No. 5 (1994): 49.
84. H. Knapp, *Radex Rundschau* 2 (1987): 356.
85. T. H. Bieniosek, *Iron and Steel Engineer* 72, 7 (1995): 17.
86. M. E. Mobberly and D. J. Diederich, *Iron and Steel Engineer*, Vol. 67, No. 5 (1990): 48.
87. D. A. Whittaker, *Int. Symp. on Ladle Steelmaking and Furnaces,* (Montreal: CIM, 1988), 235.
88. R. J. Fruehan, *Ladle Metallurgy Principles and Practices,* (Warrendale, PA: Iron and Steel Society, 1985).
89. Jungreithmeier, K. Jandl, A. Viertauer and F. Sikula, *Steel Technology Int.*, (1996–97):69.
90. D. L. Brown, J. P. Dixon, and J. N. Stacey, *Iron and Steel Engineer* 67, 5 (1990): 30.
91. R. G. Rada and C. R. Clarkson, *Iron and Steel Engineer* 67, 7 (1990): 38.
92. K. E. Sewald, *Iron and Steel Engineer* 68, 7 (1991): 25.
93. H. Presslinger et al., *Radex Rundschau* 4 (1992): 171.
94. M. Schlichting and M. Dominik, *Iron and Steel Engineer* 66, 3 (1989): 47.
95. T. Fujii, Y. Kato, and T. Sakuraya, *Proceedings,* Scaninject VI (1992), 317.
96. T. Kuwabara et al., *Steelmaking Conf. Proc.* 69 (1986), 293.
97. T. J. Cauchie and T. S. Landon, *Iron and Steel Engineer* 73, 5 (1996): 20.
98. M. Palchetti, P. Buglione, and D. Tegas, *Proc. 6th Int. Iron and Steel Congr.* 3 (Nagoya), *ISIJ,* 159.

Chapter 12

Refining of Stainless Steels

B.V. Patil, Manager, Process Research & Development, Allegheny Ludlum Corp.
A.H. Chan, Manager of AOD Process Technology, Praxair Inc.
R.J. Choulet, Consultant to Praxair, Inc.

12.1 Introduction

Stainless steels contain from 10 to 30% chromium. Varying amounts of nickel, molybdenum, copper, sulfur, titanium, niobium, etc. may be added to obtain desired mechanical properties and service life. Stainless steels are primarily classified as austenitic, ferritic, martensitic, duplex or precipitation hardening grades. The first stainless steel grades that were commercially produced in the United States were similar to 302, 410, 420, 430, and 446. These five alloys were produced and marketed prior to 1930. In subsequent years, the grades similar to 303, 304, 316, 321, 347, 416, and 440 were brought to market.

In the early stages of their production, stainless steels were melted using an electric arc furnace. Stainless steels were made to contain no more than 0.12% carbon. A large fraction of production of stainless steels was melted to a maximum carbon content of 0.07 or 0.08%.

In those early days, carbon steel scrap, iron ore, and burnt lime were charged into an electric arc furnace. After the scrap was molten, carbon was removed by adding ore until the carbon content reached 0.02%. The electrodes were then raised and the slag removed as completely as possible. Desired amounts of ferrosilicon, burnt lime, and fluorspar were added and the temperature of the bath was raised so that a large amount of low carbon ferrochromium could be added to achieve the aim chromium level. The desired amount of the low carbon ferrochromium was added in two or three separate batches. The bath had to be mixed thoroughly by rabbling or readling and slag had to be kept fluid by continuous additions of ferrosilicon, lime and fluorspar. After all the desired chromium was in the alloy, a sample was taken for preliminary analysis. Final additions were made and the heat was brought to the desired tapping temperature and tapped.[1]

Until about 1970, the majority of stainless steel was produced in the arc furnace. With the advent of tonnage oxygen production, the electric furnace stainless steel melting practice changed from the above. Gaseous oxygen could be used to improve the rate of decarburization. This could be achieved by injecting oxygen gas into the liquid steel using a water-cooled lance. The faster oxidation of carbon with high oxygen potential was accompanied by the adverse reaction of extensive oxidation of chromium to the slag. This necessitated a well-defined reduction period in which ferrosilicon was used to reduce the oxidized chromium from the furnace slags.

In the late 1960s, a number of laboratory studies were performed to understand the thermochemistry of the stainless steels.[2] One of these studies was by Krivsky, who studied the carbon-chromium-temperature relationships. His experiments involved blowing oxygen onto the surface of the baths of molten chromium alloys. He was trying to perform the experiments under isothermal conditions but found it difficult because of the exothermic nature of the oxidation reactions. Krivsky added argon to oxygen in order to control the temperature. He found that with argon dilution he could decarburize the melt to even lower levels of carbon without excessive oxidation of chromium.[3]

Krivsky's observations led to initial experiments where argon-oxygen mixtures were injected through a lance into the bath in the arc furnace. It was found that argon injection in the wide and shallow bath of an arc furnace did not influence the decarburization reaction completely as predicted. Hence, after many experiments, the developers decided that a separate refining vessel was necessary to develop a commercial process. At Joslyn Steel (now Slater Steels), a 15-ton converter with three tuyeres was built. The first successful heat was made in October 1967. These successful trials led to patents for the argon-oxygen decarburization (AOD) process for the refining of stainless steels and other specialty alloys[4] by the industrial gases division (now Praxair, Inc.) of the Union Carbide Corporation.

The AOD process revolutionized stainless steelmaking. It lowered the cost of production of stainless steels significantly. It allowed operators to use electric arc furnaces for melting down of

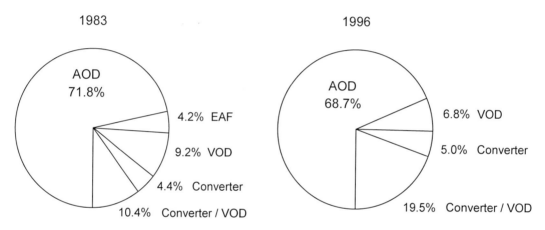

Fig. 12.1 Stainless steel production by process for two selected years.

stainless and carbon steel scraps with desired amounts of low cost high carbon ferrochromium. The decarburization operation was moved out of the electric furnace and into the newly designed converter. The oxidation-reduction operation could be conducted at very high productivity rates. Additionally, the quality of the alloys produced was improved. The process was adopted by major stainless producers at a very rapid rate. As shown in Fig. 12.1, the AOD process is the predominant method for making stainless steel in the world.[5] A schematic of a modern AOD vessel is shown in Fig. 12.2.

Duplex processes are used for making stainless steels. There is an electric arc furnace or similar melting unit that melts down scrap, ferroalloys and other raw materials to produce the hot metal. The hot metal, which contains most

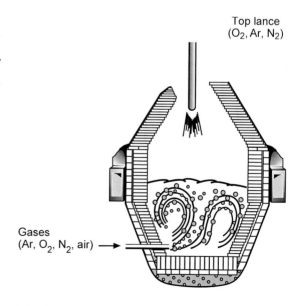

Fig. 12.2 AOD vessel.

Refining of Stainless Steels

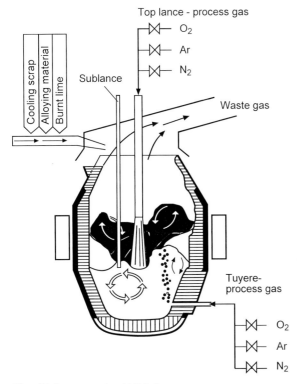

Fig. 12.3 Schematic of KCB-S process.

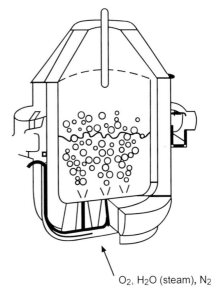

Fig. 12.4 CLU vessel.

of the chromium and nickel as well as some other alloying elements, is the charge to the converters. The converters are used to achieve low carbon stainless steels which may be tapped into a teeming ladle. The EAF-AOD process is one such duplex route. The versatility of the AOD process led steelmakers to re-examine the use of different converters for melting of stainless steels. This led to the development of several other converters for duplex processes. These include: KCB-S[6] process developed by Krupp Stahl, shown in Fig. 12.3; K-BOP[7] process used by Kawasaki Steel Corporation; K-OBM-S[8] promoted by Voest Alpine; metal refining process (MRP)[9] developed by Mannesmann Demag; Creusot-Loire-Uddeholm (CLU)[10] process, shown in Fig. 12.4; Sumitomo top and bottom (STB)[11] blowing process by Sumitomo Metals; top mixed bottom inert (TMBI)[12] process used by Allegheny Ludlum Corporation; VODC[13] process tried by Thyssen where vacuum is applied to the converter; and AOD/VCR[14] process developed by Daido, shown in Fig. 12.5.

The development work to make stainless steels using conventional BOF converters had begun in the late 1950s and early 1960s. By the mid-1960s, some steelmakers were using existing BOF converters for a partial decarburization followed by decarburization in a ladle under vacuum to make the low carbon stainless steels. These processes are known as triplex processes because three process units, such as electric arc furnace, a converter for pre-blowing, and a vacuum decarburization unit for final refining, are involved. The steels undergo

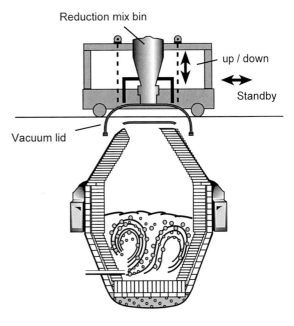

Fig. 12.5 AOD/VCR vessel.

treatment for final decarburization, final trimming, homogenization and flotation of inclusions before the ladle is taken to the teeming operation. In almost all of these triplex processes, vacuum processing steels in the teeming ladle is the final step before casting. The vacuum oxygen decarburization (VOD) process was developed by Thyssen in West Germany.[15] The other processes using vacuum include the use of RH-OB for making stainless steels at Nippon Steel Corporation[16] and the use of an SS-VOD process[17] by Kawasaki Steel Corporation. These processes are shown in Fig. 12.6.

With all these developments over several decades, now there are many different processes to make stainless steels. The available processes can be divided into three groups: the converter processes, converter with vacuum processes, and vacuum processes. Table 12.1 shows the variety of processes.

Table 12.1 Different Process Routes for Making Stainless Steels

Process	Tuyere Location	Bottom Gases	Top Gases
AOD	Side	O_2, N_2, Ar, Air, CO_2	O_2, N_2, Ar
KCB-S	Side	O_2, N_2, Ar	O_2, N_2, Ar
K-BOP/K-OBM-S	Bottom or Side	O_2, N_2, Ar, Hydrocarbons	O_2, N_2, Ar
MRP, ASM	Bottom	O_2, N_2, Ar	
CLU	Bottom	O_2, Steam, N_2, Ar	O_2, N_2, Ar
AOD-VCR (vacuum)	Side	O_2, N, Ar_2, N_2, Ar	
VOD (vacuum)	Bottom (bubbler)	Ar, O_2	

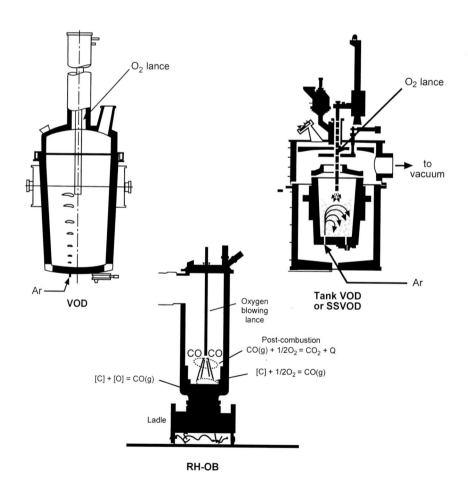

Fig. 12.6 Vacuum processes for refining stainless steel.

A worldwide survey of the steel companies using one of these processes to make stainless steels is given in Table 12.2. Only the companies with process vessels larger than 20 tons are considered. A number in parentheses indicates the number of similar converter vessels currently operated by the company. A slash indicates the company also uses another process. Companies that have known vacuum processing capability are noted. This list is by no means comprehensive.

Table 12.2 Survey of Steel Companies and Processes Used

Company	Converter	Vacuum
Acenor Inoxydales Olarra	AOD	
Aceria de Alava	AOD	
Acerinox S.A.	(2) AOD	
Acesita	AOD/MRP-L	VOD
Aichi	(2) AOD	
Allegheny Ludlum Corp.	AOD	
ALZ	MRP	VOD
Armco Advanced Materials	(2) AOD	
Ashoka	AOD	
Atlas Specialty Steels	AOD	(2) VOD
Aubert et Duval	AOD	
Avesta-Sheffield	(3) AOD/CLU	
Boschgotthardshutte	AOD	
Carpenter Technology Corp.	(3) AOD	
Cheliabinsk	KCB-S	
China Steel	LD	VOD
Cogne	AOD	
Columbus	AOD/CLU	
Crucible Materials Corp.	AOD	
Daido	(2) AOD/AOD-VCR	
Electralloy Corp.	(2) AOD	
Erasteel Commentry	AOD	
Ferro Alloys Corp. (Facor)	ASM	
Ferrometals	CLU	
FirstMiss Steel Inc.	AOD	VOD
Foroni	AOD	
Inco Alloys Inter., Inc.	AOD	
Isibars	ASM	
ISPL Industries Ltd.	ASM	
J&L Specialty Products	AOD	
Jindal Strips	(2) ASM	VOD
Jorgensen Forge	AOD	
Kawasaki	K-BOP	VOD
Keshari Steels	ASM	
Krupp Thyssen Nirosta Terni	(5) AOD	
Kusum Ingots & Alloys	ASM	
Latrobe Steel Co. (Timken)	AOD	
Lukens	AOD	VOD
Maharashtra Elektro	CLU	
Mardia Ispat Ltd.	AOD	
Mardia Steel II Ltd.	ASM	
Metarom	AOD	VOD
Microsteel	K-OBM-S	VOD
Mukand Ltd.	ASM	VOD

continued

Table 12.2 Survey of Steel Companies and Processes Used (continued)

Company	Converter	Vacuum
Nippon Metal Industries	(2) AOD	(2) VOD
Nippon Steel Corp.	AOD	
Nippon Yakin Kogyo	AOD	VOD
Nisshin Steel	LD	VOD
Nova Udyog Ltd.	OTB	
Nucor Steel Corp.	AOD	
Outokumpu Oy	(2) AOD	
Pacific Metal (PAMCO)	(3) AOD	
Paliwal Ministeel	AOD	
Panchmahal	ASM	
Pohang Iron and Steel (POSCO)	AOD/K-OBM-S	VOD
Prakash Industries	ASM	
Rathi Alloys and Steel Ltd.	AOD	
Rathi Ispat	AOD	
Republic Engineered Steels	AOD	
Sandvik, A.B.	AOD	
Shanghai #3		VOD
Shri Ishar	MRP	
Slater Steel Corp. (Joslyn)	AOD	
Sumitomo Metals Industry	(2) AOD	VOD
Tang Eng Iron Works (TESSP)	AOD	
Ugine	(3) AOD	VOD
Universal Stainless	AOD	
Valbruna	AOD/K-OBM-S	
Vardhmaw Special	ASM	
Walsin-Cartech	MRP-L	VOD
Yieh United	MRP-L	VOD
Yodogawa Steel Works	AOD	

In the following sections, the various processes for making stainless steels are discussed. First, there is a brief discussion of the general principles of refining stainless steels, followed by an examination of process routes. Then the melting of scrap and raw materials is described. Melting methods are common to all the duplex and triplex routes for making stainless steels. Most steel plants use electric arc furnaces for melting scrap and raw materials. Alternative technologies such as a direct current furnace, a plasma furnace or a similar melting unit can serve the same purpose. The balance of the chapter is devoted to a discussion of refining methods.

12.2 Special Considerations in Refining Stainless Steels

As discussed in Chapter 2, stainless steel decarburization must account for or minimize the oxidation of chromium. It is generally accepted, that when oxygen is injected into stainless steel, a mixture of chromium and iron is oxidized.[18,19] Decarburization occurs when dissolved carbon reduces the chromium and iron oxides that form. The decarburization sequence is thus:

$$3/2 O_2(g) + 2\underline{Cr} \rightarrow Cr_2O_3 \qquad (12.2.1)$$

$$Cr_2O_3 + 3\underline{C} \rightarrow 2\underline{Cr} + 3CO(g) \qquad (12.2.2)$$

Decarburization occurs on the surface of rising bubbles that form from either the inert gas that is injected or on the suface of chromium oxide particles that are being reduced and generating CO.

The decarburization of stainless steel involves techniques to minimize chromium oxidation. There are three basic techniques: temperature, dilution, and vacuum.

The temperature technique was used by electric arc furnace stainless steelmaking before the development of duplex methods. As shown in Chapter 2, Section 2.9.1, as the temperature increases, the equilibrium carbon content at a particular chromium content decreases. As discussed above, however, this leads to operational difficulties and high costs.

The dilution technique is that used by the AOD and all converter processes. The injection of inert gas (argon or nitrogen) lowers the partial pressure of CO in the bath, thus allowing higher chromium contents to be in equilibrium with lower carbon contents, as shown in Chapter 2, Section 2.9.1.

Applying a vacuum to the metal bath also removes CO, allowing high chromium contents to be in equilibrium with low carbon contents. It is especially effective when the carbon content is low.

Careful manipulation of the slag, as it participates in the reaction, is important. Any chromium oxide not reduced by carbon ends up in the slag, which can form a complex spinel. Subsequent processing (called reduction) is required to recover oxidized elements such as chromium, iron, manganese, etc. The effectiveness of the reduction step is dependent on many factors including slag basicity and composition, temperature, mixing conditions in the vessel and solid addition dissolution kinetics.[20] Chapter 2, Section 2.9.3 describes these reactions in more detail.

12.3 Selection of a Process Route

As outlined earlier, there are many different process routes available for stainless steel melting and refining. The steelmaker has to choose from these routes based on many factors which determine capital costs as well as operating costs. The choice of process route is influenced by raw material availability, desired product, downstream processing, existing shop logistics, and capital economics. In general, some degree of flexibility in process route is desirable, as these factors can change rapidly. In general, stainless steelmaking process flow can be classified as duplex or triplex. The comparison of these two processes is shown in Fig. 12.7. Duplex refining, where electric arc furnace

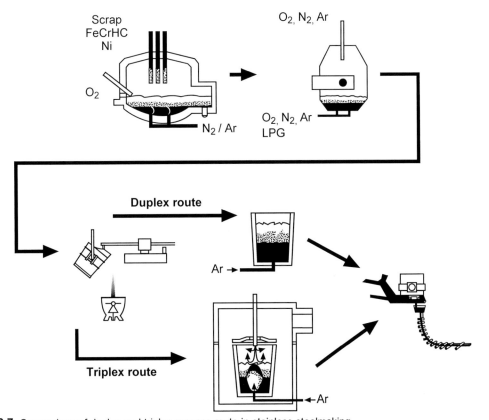

Fig. 12.7 Comparison of duplex and triplex process route in stainless steelmaking.

melting is followed by refining in a converter, tends to be flexible with respect to raw material selection. Triplex refining, where electric arc furnace melting and converter refining is followed by refining with a vacuum system, is often desirable when the final product has very low carbon and nitrogen specifications.[21] Triplex refining tends to have overall cycle times longer than duplex refining because there is an extra transfer from refining converter to vacuum unit. It also tends to have slightly higher refractory costs because there are two vessels performing decarburization.

Raw material availability affects the choice of process route. For example, high argon costs shift the economics of stainless production away from dilution type processes in favor of vacuum processes. A lack of scrap forces the use of more high carbon ferrochrome, increasing the carbon load in the charge. In such a case, process routes that have high decarburization rates would be favored. Lack of scrap, which can be used for cooling the bath, also can favor processes that are not thermally efficient, as there is no economical addition for cooling the bath. Where scrap and ferroalloys are expensive or unavailable, process routes have been developed that use chromium ore and blast furnace or similar hot metal.[22, 23] These processes typically incorporate a smelting unit, often a converter, to reduce the ore, followed by decarburization in another converter, possibly followed by vacuum processing.

The desired product mix affects the choice of process. High production levels of ultra low carbon and nitrogen stainless steels tend to favor vacuum processes to avoid high argon consumption. For typical levels of carbon and nitrogen in stainless steels, the processing cost are usually lower for the converter processes than vacuum or triplex processes.

Downstream processing considerations also will affect the choice of process. Sequence continuous casting requires a stainless steel refining route that will provide feed metal at the proper temperature, composition and time to sequence cast.

Availability of ladle furnaces and ladle stations can shift some of the non-decarburization time from the refining vessel to the ladle furnace or station. Some high productivity shops tap the heat out of the converter just after the reduction stage. The final additions are made to the ladle. This frees up the refining vessel for the next heat.

If existing melting capacity is insufficient for desired production levels, the refining process may need to melt large amounts of cold material. This would favor processes that are thermally efficient or ones that can generate heat for melting. If melting is primarily by induction furnaces, raw material choices may be limited. Blast furnace or hot metal availability would also influence the type of refining process. Available vacuum or baghouse capacity may influence the choice between vacuum or dilution methods. Crane capacity may limit vessel size. Crane traffic and shop layout will affect process flow. Existing ladle cars can help or hinder material movement. An existing continuous caster also would affect process choice.

Cost of capital always affects process choices. Systems for making alloy additions can be very expensive. Baghouse and vacuum capacity is also expensive. There is often a balance between capital and operational cost. The process with the lower operating cost sometimes has the highest capital cost.

In some recent installations, capital spending considerations have eclipsed those of all other factors, including flexibility and operating costs, in the selection of process route. For major turnkey installations, which may include steel melting, refining, casting, hot and cold rolling, anneal and pickle lines, as well as the associated technology, the choice of equipment suppliers is limited to three or four large firms. Each of these suppliers aggressively promotes its own secondary steelmaking process as superior relative to operating costs, quality, productivity, etc. However, several of these suppliers will manufacture and install a version of any other type of secondary steelmaking, if the customer insists. Depending on geographical location, order backlogs, previous equipment sales, manpower availability, etc., any one of these suppliers may be in a better position to manufacture and install equipment at a substantially lower price than the others.

For maximum flexibility, a shop should have a process flow that can incorporate a duplex or triplex route. This offers the most flexibility in raw materials, production capability, and process flow. In this case, only products that require vacuum refining for economic and/or quality reasons will use a triplex route.

12.4 Raw Materials

Raw materials used in all stainless steel making operations can be divided into two main groups; nonmetallic raw materials and metallic raw materials. The nonmetallic materials typically needed are burnt lime, dolomitic lime and flurospar. The metallic raw materials needed are stainless steel scrap, carbon steel scrap and a variety of alloying materials and deoxidizers. The major alloying materials used are chromium, nickel, manganese and silicon additives.

Ferrochromium is the prime source of chromium and it comes in a range of addition agents. The common chromium alloying materials charged to the electric arc furnace and AOD are given in Table 12.3.

Table 12.3 Raw Materials Used to Supply Chromium Units

Chromium Additive	Carbon	Chromium	Silicon	Nitrogen	Manganese
Low Carbon Ferrochromium	0.10	69	0.9		
Medium Carbon Ferrochromium	0.13	70.7	0.9		0.2
Low C, Low N_2 Ferrochromium	0.045	67.8	0.63	0.01	0.45
Simplex Ferrochromium	0.01	69	0.28		
Intermediate Carbon Ferrochromium	2	55	0.1		0.15
High Carbon Ferrochromium	7.4	68.7	0.85		0.22
High Carbon Ferrochromium	7.7	69.2	0.17		0.16
Ferrochromium-Silicon	0.055	37	40		0.27

Nickel is present in significant amounts in the AISI 200 and 300 series stainless steels. In addition to austenitic stainless steel scrap, nickel is available as electrolytic nickel, nickel powder, nickel oxide, briquetted nickel and ferronickel. Typical analyses of nickel containing raw materials are given in Table 12.4.

Table 12.4 Raw Materials Used to Supply Nickel Units

Nickel Additives	Carbon	Nickel	Silicon	Cobalt	Sulfur	Chromium	Copper
Nickel Briquettes		99.9					
Low Carbon Nickel Shot	0.01	50	0.02	0.78	0.007	0.007	
High Carbon Nickel Shot	1.64	28	1.4	0.6	0.06	0.7	
Nickel Pig	0.07	39	0.4	0.7	0.015	0.15	
Nickel Copper Shot	0.3	51	0.1			1.2	26.5
Nickel Oxide Sinter		75					

Manganese is present in significant amounts in the AISI 200 series of alloys. Sources of manganese are given in Table 12.5.

Table 12.5 Raw Materials Used to Supply Manganese Units

Manganese Additives	Carbon	Manganese	Silicon
Electrolytic Manganese		99.99	
Ferromanganese	7.1	79	0.05
Manganese-Silicon	0.07	61	31

Silicon is used in all the stainless steel making processes to chemically reduce the chromium which is present in the slag at the end of the decarburization period. For this purpose, silicon is added in the reduction mix as ferrosilicon or ferrochromiumsilicon. Additional silicon would be needed to meet any silicon specification on many of these stainless steels. The most common raw materials used to supply the silicon are its ferroalloys which contain nominally either 50% silicon or 75% silicon.

Molybdenum is added to some of the stainless steel grades. The materials which can be used to add molybdenum include molybdenum oxide in powder or briquette form and ferromolybdenum. Aluminum is a common deoxidant and high purity aluminum pig or bar is added for this purpose. Titanium is a very frequent addition to many stainless steel grades for the purpose of stabilization (see Chapter 11). The most commonly used alloys of titanium are 70%Ti–30%Fe and 90%Ti–4%V–5%%Al alloy. The other elements which are generally added in the ladle include carbon, sulfur, nitrogen, niobium (columbium), tungsten and copper. Selenium and tantalum are added to a very limited number of grades.

In stainless steelmaking, most of the cost of the final product is due to raw materials. Over more than the past two decades, computer programs, most of which use linear programming techniques, have been developed and used to utilize the raw materials in such a way that the total raw material cost of production is minimized.

12.5 Melting

For the electric arc furnace, typical charges are scrap and ferroalloys. As there is further refining, least cost charges incorporate high carbon ferroalloys for chromium and manganese units. Usually, the least cost source of material is used; for scrap this may consist of oily turnings and furnace skulls.

Sometimes, hot metal from the blast furnace or other iron source is available. In this case, the electric arc furnace can be used to melt only ferroalloys with the resulting melt then mixed with the molten iron in a converter process for refining.[22] The electric arc furnace can also be bypassed altogether.

When hot metal is the source and stainless scrap or ferroalloys are unavailable or too expensive, then chromium, manganese and nickel ores can be used as sources of material.[23,24]

12.5.1 Electric Arc Furnace Melting

The delivery of molten material for further processing requires melting facilities of some kind. The electric arc furnace is by far the most popular choice due to its flexibility of raw material sources and widespread use. Induction melting is popular in smaller shops and foundries; however, it is less capable of handling some scrap sources, such as turnings and oily scrap.

The job of the melter, be it electric arc furnace or other method, is to melt down the basic raw materials as quickly and economically as possible. The potential of chromium oxidation places some limits on the use of oxygen injection into the arc furnace. Carbon injection and foamy slag practices, while common in the carbon steel industry, are rarely employed in stainless steels. The slags that form in the melting of stainless are high in chromium oxide and are not as fluid as carbon steel slags. The stainless steel melt slags do not foam as easily.[25] CO generation for formation of the foamy slags also does not occur if the chromium content is high.

Stainless steel melt charges in the United States are primarily scrap based. Scrap stainless steel is often a cheaper source of chromium and nickel units than virgin material such as ferrochromium and nickel.

Melting of stainless steel is a choice of the least cost sources of metal units. There is usually some consideration of impurities such as copper and phosphorus, which are not removed during subsequent processing.

For a 300 series stainless steel, a typical melt-in mixture might consist of: 50% 300 series scrap (18%Cr, 8%Ni, 1%Mn); 30% carbon steel scrap;14% high carbon ferrochrome (7%C, 65%Cr); 4% nickel (commercial purity); and 1% high carbon ferromanganese (7%C, 65%Mn).

At melt-in, this mixture would yield about 18%Cr, 8%Ni, 1%Mn and 1%C.

In the early stages of melting, some shops will use burners and oxygen lancing. The added energy input increases the melt-in rate in the early stages of melting. After the second bucket however, burner and oxygen use is limited to avoid chromium oxidation. At tap, ferrosilicon is often added to recover oxidized chromium from the slag.

12.5.2 Converter Melting

Converter melting requires oxygen and carbon injection coupled with post-combustion to generate heat. Converter melting is rarely economical but high electrical power costs or lack of electric melting capacity, combined with hot metal availability, may make this option viable.

In converter melting, molten hot metal is charged into a smelting converter and the ores added. Carbon is also injected or charged as the reductant and as fuel. Oxygen is injected by a lance for burning carbon and post-combustion. This generates the heat necessary for the smelting reaction. Ores typically contain some amount of gangue, so appropriate slag conditioners are also added. The slag is decanted and the metal, now close to the desired composition, is transferred to another converter for refining.

12.6 Dilution Refining Processes

The melting processes supply the liquid stainless steel, which now contains desired amounts of chromium, nickel and other alloying elements, to the next process step. The latter is directed at decarburizing the stainless steel alloy to the desired carbon content. It is accompanied by oxidation of chromium, iron, silicon, aluminum, titanium and manganese which are present in the charge. In a converter, decarburization is carried out using the dilution principle. There are different converters based on the gases used. Another difference is whether the vessel is side or bottom-blown. Proponents of side blowing argue that side blowing results in higher carbon removal efficiencies (amount of oxygen reacting with carbon divided by the total amount of gases blown) in the range of 0.1 to 0.005% carbon, due to longer inert gas bubble residence time, and improved desulfurization in the range of 0.005% to less than 0.001% sulfur due to improved mixing.

The most popular converters for making stainless steels are AOD, KCB-S, K-BOP/K-OBM-S, MRP and CLU. Each of these are discussed in the following sections.

12.6.1 Argon-Oxygen Decarburization (AOD) Converter Process

The AOD process, shown in Fig. 12.2, is used for the production of stainless steel in the second step of a duplex process. The molten steel, which contains most of the chromium and nickel needed to meet the final heat composition, is tapped from the electric arc furnace into a transfer ladle. The transfer ladle is lifted with a crane and the liquid steel is poured into the AOD vessel. The vessel can be rotated downwards so that the side mounted tuyeres are above the bath level during charging of the liquid steel. In the early days of the AOD process, the vessel was tilted for making raw material additions as well as for taking samples and temperature measurements using immersion thermocouples. The desire to increase the productivity has led to continuous charging of raw materials during the blow periods as well as reduction periods. Modern instrumentation has been developed which can take melt samples as well as steel temperatures using a specially designed sub-lance with the vessel in the upright position.

Typical consumptions[26] of an 80-ton AOD making AISI 304 are shown in the Table 12.6.

Table 12.6 Typical Consumptions

	Units	Typical	Best
Argon	Nm³/ton	12	9
Nitrogen	Nm³/ton	9–11	9
Oxygen	Nm³/ton	25–32	NA
Lime	kg/ton	50–60	42
Spar	kg/ton	3	2
Aluminum	kg/ton	2	1
Silicon (reduction)	kg/ton	8–9	6
Brick	kg/ton	5–9	2
Decarburization metallics	kg/ton	135	NA
Charge to tap time	min.	50–80	40
Total Cr yield	%EAF/AOD	96–97	99.5
Total Mn yield	%EAF/AOD	88	95
Total Metallic yield	%EAF/AOD	95	97

*1.8% start carbon, 0.05% Nitrogen, 0.005% Sulfur

A major modification of the original AOD process involves the use of a top-blowing lance in addition to the side blowing tuyeres. The lance can be used to inject oxygen at desired flow rates to increase decarburization and/or for post-combustion. The top lance can also be designed for blowing mixed gases such as inert-oxygen gas mixtures. The installation of a lance and introduction of oxygen in the early stages of decarburization can reduce the time for a heat. The technology can be used to increase productivity, in tons per hour, of a melt shop.

Another modification of the AOD involves applying vacuum on the converter to reduce the argon and silicon consumption as well as process time when making low carbon grades. This modification is identified as AOD-VCR and is used by Daido Steel, Japan.

12.6.2 K-BOP and K-OBM-S

Kawasaki Steel Corporation's K-BOP process began as a conventional top oxygen blown BOF. It was modified to have seven bottom tuyeres of the OBM (Q-BOP) type. These tuyeres could blow oxygen with propane for tuyere cooling. Powdered lime could also be injected through these tuyeres.

In its initial development stage, an 85 ton ultra-high-powered electric arc furnace (UHP-EAF) was used to supply the hot metal to the reactor. The top and bottom oxygen blowing was used to achieve high decarburization rates.[27]

Kawasaki Steel Corporation continued the developmental work where chromium ore and coke were added to the reactor. The reactor was thus used for reduction of chromium ore and after that switched over to decarburization operation. Additional developmental work at Kawasaki Steel led to a smelt reduction process for stainless steelmaking using separate reactors. Final decarburization (below 0.10%C) was carried out in a top-blown RH degassing unit (the KTB process). Eventually a process using separate converters and a VOD unit was developed. The later development is discussed later in Section 12.8 on direct stainless steelmaking.

The K-OBM-S process developed by Voest Alpine Industrieanlagenbau (VAI) and evolved from Kawasaki's K-BOP process. The K-OBM-S process initiated with tuyeres installed in the converter bottom. However, two recent installations, POSCO in Korea[28] and Bolzano in Italy, are side-blown reactors.[29] Thus, a K-OBM-S converter is top-blown with a lance and bottom or side-blown with tuyeres. It is very similar to a modern AOD. However, in the K-OBM-S process, hydrocarbons, such as natural gas or propane, are used for tuyere protection and this can be helpful in increasing the refractory life. Typical consumptions and conditions[21] for making AISI 304 are given in Table 12.7.

Table 12.7 Consumptions and Conditions for AISI 304

Condition/Addition	Units	Typical Values
Aim carbon	%	0.033
Aim nitrogen	ppm	300
Melt-in carbon	%	1.77
Melt-in silicon	%	0.13
Oxygen	Nm³/ton	29.5
Nitrogen	Nm³/ton	13.2
Argon	Nm³/ton	16.5
Silicon (reduction)	kg/ton	11.1
Lime	kg/ton	51
Dolomite	kg/ton	20
Fluorspar	kg/ton	7.9

12.6.3 Metal Refining Process (MRP) Converter

The metal refining process was developed by Mannesmann Demag Huttentechnik. It is also a duplex process where scrap and raw materials are melted in an electric arc furnace or similar unit. Molten metal, which contains chromium and nickel, is charged to the MRP converter. Decarburization is carried out using oxygen and inert gases. In early stages of development, the gases were alternately blown through the tuyeres in the bottom of the reactor.[30, 31] The oxygen is blown into the melt without dilution with any inert gas. The desired oxygen blow is followed by blowing with inert gas only. The cycle of oxygen blow followed by inert blow is called cyclic refining or pulsing and the developers claim that the flushing with pure inert gases can lead to achieving low CO partial pressure and faster decarburization and thus lower chromium oxidation and consumption of silicon for reduction. The original version of the converter has now evolved into the MRP-L process in which all oxygen is top-blown and inert gas is injected through the porous elements in the bottom.

The developers claim that the process can use higher blowing rates than those used in the AOD process, which has sidewall tuyeres. The bottom tuyeres can be replaced easily through the use of an exchangeable bottom. With bottom tuyeres, there is less likely to be erosion on the sidewalls of the vessel.

In recent years, the MRP-L units have been coupled with a vacuum unit as part of the triplex process for making stainless steels, especially those requiring lower carbon and nitrogen levels. In these plants, the heats are tapped at an intermediate carbon level appropriate for subsequent vacuum decarburization.

12.6.4 Creusot-Loire-Uddeholm (CLU) Converter

The CLU Process is similar to the AOD process for making stainless steels. It also uses liquid steel from an electric arc furnace or similar melter. The major impetus for its development was the idea to substitute steam as the diluting gas rather than argon. The process was developed by Uddeholm of Belgium and Creusot-Loire of France. The converter is bottom-blown thus differentiating it from the side-blown AOD converter. The first commercial plant using the CLU process was built in 1973 by Uddeholm, now a part of Avesta-Sheffield Stainless Co. The process is also used by Samancor in South Africa for the production of medium carbon ferrochrome. Columbus Stainless in Middleburg, South Africa chose this process, supplied by VAI, for their newest stainless plant, which came on stream in 1996.[32]

The decarburization period consists of injecting an oxygen-steam mixture. The process is energy inefficient as the reaction of steam with the molten steel bath is endothermic. Chromium oxidation is higher than in the AOD process when decarburization is continued below about 0.18% carbon. Although the original goal of reducing argon consumption can be met, the increased silicon requirement for the reduction step does not necessarily lead to overall cost savings. Further, the use of steam throughout the entire period was found to lead to undesirable hydrogen contents in the refined steel. Therefore, practices have evolved which use various amounts of steam, argon, and nitrogen in the process. Consumptions and conditions[32] for two grades are given in Table 12.8.

Table 12.8 Consumptions and Conditions for AISI 304 and AISI 409

Condition/Addition	Units	AISI 304	AISI 409
Aim carbon	%	0.03	0.01
Aim nitrogen	ppm	350	100
Melt-in carbon	%	1.65	0.96
Melt-in silicon	%	0.2	0.13
Oxygen	Nm^3/ton	27.7	22.4
Nitrogen	Nm^3/ton	13.5	1.7
Steam	Nm^3/ton	10.4	6.0
Hydrogen	ppm	5.9	3.8
Argon	Nm^3/ton	7.0	17.1
Silicon (reduction)	kg/ton	15.5	15.9

12.6.5 Krupp Combined Blowing-Stainless (KCB-S) Process

The production of stainless steels in the BOF converter using the top lance was being practiced prior to the advent of the AOD Process.[33, 34] After introduction of the AOD process, Krupp Stahl AG modified the converter at its Bochum Works so that combined blowing through the lance and tuyeres could be practiced for refining stainless steels. The process was named Krupp combined blowing-stainless or KCB-S. The simultaneous introduction of process gases helped them increase the decarburization rate. The blowing through a top lance and through the tuyeres below the bath surface helped achieve very high decarburization rates. The increased decarburization rate led to a reduction of up to 30% in the refining times compared to a conventional AOD alone.[28] The schematic diagram of the KCB-S process is shown in Fig. 12.3.

Liquid steel from an electric arc furnace is charged to the converter. At the start of the blow, pure oxygen is injected simultaneously through the lance and sidewall tuyeres. After a desired process temperature is reached, various additions are made during the blow. The additions consist of lime, ferroalloys, and scrap. After a critical carbon level is reached, the oxygen content of the process gas is reduced by using inert gases such as nitrogen or argon. Oxygen to inert gas ratios of 4:1, 2:1, 1:1, 1:2 and 1:4 are used as decarburization to lower levels is pursued. When the carbon content of 0.15% is reached, the use of the lance is discontinued and the process gases are introduced only through the tuyeres. When the desired aim carbon level is reached, the oxygen blow is discontinued and silicon is added as ferrosilicon to reduce the chromium oxide in the slag and to achieve the required silicon specification. The addition of lime and other fluxing agents with the ferrosilicon leads to lowering of the dissolved oxygen content and enhances the desulfurization.

12.6.6 Argon Secondary Melting (ASM) Converter

This process was developed by MAN GHH in Germany.[35] It is similar to the AOD process, except that the tuyeres are in the bottom of the vessel. When using top-blown oxygen, it is identified as the ASM-L process.

12.6.7 Sumitomo Top and Bottom Blowing Process (STB) Converter

Sumitomo Metal Industries developed the Sumitomo top and bottom blowing process (STB). It was developed to overcome the disadvantages of a pure top or pure bottom blowing process by combining the two concepts into one process.[36, 37] It also tried to overcome two disadvantages of the AOD process at the time: tuyere erosion and limited oxygen flow rate. The additional supply of oxygen rich gases from the top lance led to shorter decarburization times compared to the AOD process as practiced in late 1970s. The process was developed in a 250-ton converter in the No.2 steelmaking shop at Kashima Works. The process was renamed Sumitomo metal refining (SMR) and used for a short time in the 160 ton converter at Wakayama Works.

12.6.8 Top Mixed Bottom Inert (TMBI) Converter

Allegheny Ludlum Corporation used a BOF converter to make ferritic stainless steels whenever incremental stainless steel capacity was necessary.[38] In the early stage of development, oxygen, and oxygen-argon mixtures were introduced through a supersonic top lance. However, the process was less efficient than the AOD process. Consequently, one BOF converter was equipped with bottom tuyeres to inject only inert gases such as argon or nitrogen. The majority of the process gas was introduced through the top lance. The top lance could be used to introduce the desired mixture of gases. The process was called top mixed bottom inert (TMBI). The process is similar to the other processes that use combined gas blowing in a converter. The particular plant operated by Allegheny Ludlum has coreless induction furnaces that melt carbon steel scrap and supply chromium-free hot metal to the BOF converters.

12.6.9 Combined Converter and Vacuum Units

The converter processes discussed above have one disadvantage in that stainless steels with very low carbon and nitrogen residuals become difficult to produce. The decarburization period becomes longer while chromium oxidation and argon consumption increase as the desired carbon and nitrogen levels decrease. Some steelmakers have tried to overcome this disadvantage by applying vacuum to the converter at the very late stages of the decarburization process. The concept was tried in an evacuated BOF by Thyssen in the early 1970s. The concept was also promoted by Leybould-Heraus as an alternative to AOD or VOD. The concept of applying vacuum to a converter is being pursued by the installation of AOD/VCR by Daido at Shibukawa and Chita[39] plants and by Nippon Steel at its Hikari works.[40] A schematic of the converter was shown earlier in Fig. 12.5.

The AOD-VCR operates as a conventional AOD down to 0.08–0.10%C. The process is stopped for sampling and a vacuum lid is put into place. The lid is sealed to a flange located about half way up the conical section of the converter. A vacuum is pulled and used for the remainder of decarburization and reduction. Desulfurization is carried out in the transfer ladle prior to AOD charge. The major advantages of this process relative to converter processes are decreased argon and silicon consumption. Disadvantages include higher refractory consumption, decreased ability to melt scrap and added maintenance and costs associated with steam production.

When compared to separate converter and VOD units, the AOD-VCR has higher operating costs (silicon, refractory, and argon), lower productivity and higher nitrogen contents. Capital costs may be somewhat lower than having two separate units.

12.7 Vacuum Refining Processes

The use of vacuum for decarburization of steels was developed by different steelmakers in Germany. The early processes included RH degassing, DH degassing, and the Allegheny vacuum refining (AVR) as a second step in the duplex process. These processes involve lowering the pressure above the steel bath to promote evolution of carbon monoxide gas. The liquid stainless steels going into the vacuum process generally contain carbon of about 0.5% or lower. Most vacuum processes are performed in a chamber with a ladle full of metal as opposed to a separate refining vessel used in the dilution/converter processes.

In mid 1960s, the vacuum decarburization concept was used by Allegheny Ludlum Steel Corporation to develop the Allegheny vacuum refining process. It was used to make the regular grades of stainless steels and lowered the consumption of low carbon ferrochrome associated with the oxidation-reduction practice used in the electric arc furnace. This process became non-competitive with the introduction of the AOD process. In late 1970s, Allegheny Ludlum built a plant with an AOD reactor and discontinued the use of the AVR process.

Early duplex processes where vacuum processing was used as the second step were too slow and had very limited flexibility with respect to raw materials that could be used. The vacuum processes could not keep up with the improving productivity of electric arc melting furnaces and the operating costs were very high. Hence, later developments focused on the use of converters to decarburize the molten metal from electric arc furnaces as a second stage followed by vacuum degassing for the finishing stage. Such processes are known as triplex processes for making stainless steels as they use three processes to achieve the desired finished chemistry.

Nippon Steel Corporation introduced the RH-OB process for making of stainless steels in the Muroran Works.[41] Hot metal from a blast furnace was fed to a BOF converter where the metal was alloyed with chromium and blown down to a 0.5–0.6% carbon level. The final decarburization was conducted using the RH-OB process. Nippon Steel converted an existing RH degasser, which was used for carbon steels, so that oxygen could be injected under vacuum.

Edelstahlwerk Witten (now Thyssen) in Germany developed the vacuum oxygen decarburization (VOD) process in the mid-1960s. In the early stages of its development, VOD was used to decarburize molten alloy from the electric arc furnace. Later, preliminary decarburization was done in a BOF and the EAF-BOF-VOD triplex process became more productive. In early 1970s, Kawasaki Steel Corporation modified the VOD process using multiple porous plug bubblers in the ladle. They called this the SS-VOD (strong stirring VOD) process.[42] In 1988, ALZ in Belgium modified the facilities to make stainless steels by a triplex process consisting of electric arc furnace melting, metal refining process-lance (MRP-L) converter, and vacuum oxygen decarburization (VOD) process.[43]

The major advantages of the vacuum processes include low consumption of argon and low oxidation of chromium during the final decarburization to low carbon levels. The latter leads to less consumption of reduction elements for recovering chromium from slag. The teeming of steel from the ladle used in the vacuum processes eliminates the pickup of nitrogen and oxygen from air that is associated with tapping of the converters. The SS-VOD process, because of the strong stirring achieved using multiple bubblers in the ladle, further enhanced the ability to produce even lower levels of carbon, nitrogen and hydrogen at higher chromium levels.

A major disadvantage of VOD processing is that it is less flexible than an AOD or other converter process with respect to raw materials usage. Typical additions to the VOD are around 4–8% of tap weight. Typical vacuum treatment times are 50 to 70 minutes with start carbon contents of 0.3%, compared with 40 to 60 minutes with a converter starting at 1.5 to 2.5%C. This added time makes it difficult to sequence continuous casting of heats.

Many steelmakers have also realized that vacuum processes often have high operating costs and cannot compete with the ease of operating a converter process at atmospheric pressure. However, the vacuum processes, especially the SS-VOD process, have the unique ability of achieving lower carbon and nitrogen levels in stainless steels which cannot be easily achieved by the AOD process or other converter processes.

12.8 Direct Stainless Steelmaking

In recent years, there have been efforts to use chromium and nickel ores for stainless steelmaking in lieu of the ferroalloys. In Japan, a number of companies have developed and are using such

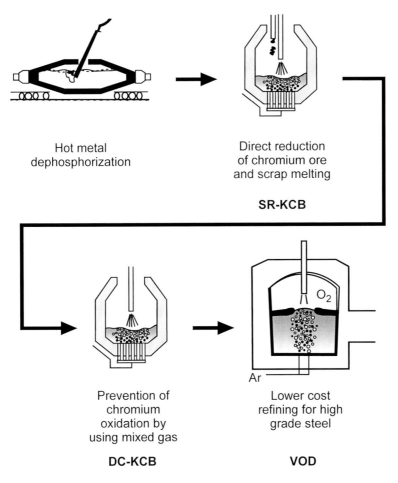

Fig.12.8 The KCS process for direct stainless steelmaking.

processes commercially. In particular, Kawasaki Steel Corporation (KSC) has developed a process which smelts chromite ore ($FeCr_2O_4$ plus other oxides). Stainless steel is produced with two 160 ton converters, without any EAF operation, but using dephosphorized hot metal, chromite ore and ferroalloys as the major charge materials. An outline of the KSC process is shown in Fig. 12.8. The hot metal is desulfurized and then fed into a K-BOP/K-OBM vessel (SR-KCB) which is described in Chapter 9. Chromite pellets, which are partially prereduced up to 60% chromium content by rotary kiln, are charged to the first reactor. Coke is also charged and serves as a heat source in the reactor. The charging rate of ore and coke and the oxygen blow rate are controlled to achieve suitable temperature for the melt.[44, 45]

The steel melt, which now contains about 11–16% Cr and 5–6% C is charged to the second K-BOP/K-OBM vessel. Primary decarburization (DC-KCB) is carried out using oxygen, argon and nitrogen gases in this vessel. Top and bottom oxygen blowing is used to lower the carbon content of the melt to about the 1% level. This is followed by mixed gas blowing only from the bottom. This is called the DC-KCB process and is similar to a modern AOD operation. The process reduces the carbon to about 0.15%. Final decarburization and reduction is carried out in a VOD. The process uses less amount of scraps than the EAF-AOD process route. However, it requires liquid hot metal from a blast furnace and more process steps and more capital investment than the EAF-AOD route.

NKK Corp. has developed and commercialized a similar process which also smelts both chromite and nickel ore.[46] After hot metal dephosphorization, the metal is transferred to a smelting furnace and nickel ore is smelted. As a nickel ore typically contains 1–3% nickel, a considerable amount of ore is added and the slag must be removed several times to achieve the desired nickel level. After reaching the required nickel content, chromite or chromium ore is smelted. The smelting takes about six hours; four for nickel and two for chromium. This is followed by the decarburization period. Similar to the KSC process, the NKK process uses less scrap and only a small amount of nickel metal and ferrochrome. However, the process requires hot metal and a very long process time.

These direct stainless steelmaking processes require less scrap and ferroalloys and may be considered when the stainless scrap and ferroalloys are not readily available in sufficient quantities. However, chromium and nickel ores are needed and can be used in an integrated steel plant where the hot metal is available. The processes are more capital intensive than the other processes. Therefore, their implementation in other plants and locations is likely to be limited.

12.9 Equipment for EAF-AOD Process

The previous three sections outlined the concepts of different processes to make stainless steels. The EAF-AOD, EAF-CLU and similar duplex process routes are the most popular for making stainless steels. This section describes the equipment used in the EAF-AOD process route, which is by far the most popular process in the world for making stainless steels.

For converter refining processes, the vessel consists of a refractory-lined steel shell. Several types and qualities of basic refractory have been used in converter vessels. Refer to Chapter 4 for more details of stainless steelmaking refractories.

Process gases are injected through submerged tuyeres that are installed in the sidewall or bottom of the vessel. Sidewall injection imparts maximum stirring energy to the bath for greatest efficiency of mixing. Bottom injection usually improves wear characteristics in the barrel section of the vessel. The number and relative positioning of tuyeres is determined in part by vessel size, range of heat sizes, process gas flow rates and the types of alloys refined. Process gases are oxygen, nitrogen, carbon dioxide, air, hydrocarbons and argon. Most recent converter installations include the use of a top lance for blowing oxygen.

The gas control system supplies the process gases at nominal rates of 1.0–3.0 Nm^3/min/ton (2000–6000 scfh/ton). The system accurately controls the flowrates and monitors the amount of gas injected into the bath to enable the operator to control the process and measure the total oxygen injected. In contrast, vacuum processes typically use much lower total input rates, e.g., 0.02–0.6 Nm^3/min/ton (typical top oxygen blow rate 0.3–0.4 Nm^3/min/ton) and seldom use gases other than oxygen, nitrogen and argon.

A high production shop will typically have three interchangeable vessels for 100% availability of the process. At any given time, one of the vessels is in the tiltable trunnion ring refining steel, a second newly lined vessel is at a preheating station, and the third vessel is at a reline station. The vessel in the trunnion ring typically can be replaced with a preheated vessel in less than an hour.

Process gas control, vessel activities and ancillary equipment can range from manual to fully automated. Most installations are equipped with a computer for calculating oxygen requirements as well as alloy additions. Some installations have computer controls capable of setting flow rates to the gas control system.

12.9.1 Vessel Size and Shape

The converter vessel consists of a refractory-lined steel shell.[47] With a removable, conical cover in place, the vessel outline is sometimes described as pear-shaped. Modified BOF vessels have also been used. As molten stainless steels do not generate foam, and most stainless steel refining processes are side or bottom-blown, the dimensions of a stainless refining converter are smaller than a comparable tonnage BOF or OBM (Q-BOP). Typical internal volumes of stainless steel refining converter vessels are in the range 0.4–0.8 m^3/metric ton bath weight.

For vessels that tap into a ladle held by a crane, a sliced cone top section is often used. The slice portion allows the crane to come close to the vessel mouth. Fig. 12.2 shows an AOD vessel with a sliced cone top section. Vessels that tap into a ladle car usually have a BOF-type concentric cone top section.

12.9.2 Refractories

High temperatures at the tuyere tip and high bath agitation place great demands on the vessel refractory.[48] While typical BOF refractory campaigns are months or years long, stainless converter campaigns are several days or weeks long. Refractory costs are a significant fraction of total operating costs.

There are two basic choices of refractory type, magnesite-chromite, and dolomite. The choice of refractory is dependent on the vessel operation pattern, final product specifications, and economics.

Magnesite-chromite refractories have high wear resistance but have a higher unit cost than dolomitic refractories. Chromium pickup from the brick is possible. Magnesite-chromite bricks are simultaneously acidic and basic and strict slag compositions must be maintained to prevent rapid wear.[49,50,51]

Dolomitic refractories are usually less costly than magnesite-chromite refractories and chromium pickup is not a factor. Desulfurization to very low levels is generally easier in dolomitic refractories because very basic slags can be used without detrimental effects to the brick.[52] Most major stainless producers in the United States and Europe use dolomitic refractories in the converter.

Vessels are typically zoned by thickness and brick quality to maximize lining life and minimize costs. High wear areas of the vessel, usually the tuyere wall, slag line, and transfer pad (where the metal stream strikes the vessel during transfer) are zoned thicker and with higher quality refractory than other parts of the vessel.

For more analysis of stainless steelmaking refractories, see Section 4.4.

12.9.3 Tuyeres and Plugs

The AOD process and most other converter processes have tuyeres mounted in the sidewall or in the bottom. These tuyeres typically consist of a copper tube with a stainless steel outer tube. An annulus is formed between the copper and stainless tubes. Cooling gases blown through the outer annulus (shroud) form a metal or oxide accretion (called a knurdle or a mushroom) at the tuyere tip. This accretion protects the tuyere and surrounding refractory. Process gases of oxygen/inert mixtures blow through the inner annulus. Special designs exist for normalizing the flow in the annular gap.[53] Tuyere size and number depend on specific process parameters. There are usually between two and nine tuyeres in an AOD vessel.

Sidewall mounted tuyeres are submerged while processing. When the vessel is rotated, the tuyeres are above the bath. At this point, the process gases can be shut off and a small cooling flow protects the tuyeres.

Bottom-blown vessels have a variety of tuyere configurations depending on flow rates required. There are usually two to four tuyeres in the bottom. Most have some variation of OBM (Q-BOP) tuyeres. Processes with low flow rates that use only inert gases, such as VOD, for the bottom-blown gases can use a plug instead of a tuyere.

12.9.4 Top Lances

A lance, while not required, greatly improves converter processes. In the high carbon region, a lance can increase the oxygen input rate, increasing the decarburization rate. A lance can be used for post-combustion of carbon monoxide, increasing the scrap melting rate or temperature of the bath. For small vessels, post-combustion lowers fueling requirements and increases process efficiency.

The amount of oxygen that reacts with the bath can range from 30 to 100% of the oxygen blown. This amount depends on lance height, nozzle design and gas pressure used.

There are two basic lance types, sonic/subsonic and supersonic. Sonic/subsonic lances consist of straight pipes, while supersonic lances use converging/diverging (Laval) nozzles. Supersonic lances are always water cooled. Sonic/subsonic lances might be water cooled depending on lance operating position. Smaller sonic/subsonic lances often are not water cooled. Many sonic/subsonic lances are simple pipes with nickel or calorized tips to increase heat resistance.

A water-cooled lance is typically positioned one to four metres from the bath surface. Non-water-cooled lances are positioned close to the vessel mouth.

Supersonic lances are used for high oxygen input rates where post-combustion is not required. A great majority of the oxygen from a supersonic lance will react with the bath.[54] As in BOF operation, the lance can have post-combustion ports.

With sonic lances, some of the oxygen will not react with the bath; some post-combustion is inevitable.[55] Heat balances indicate that 75 to 90% of available energy from post-combustion is transferred to the molten bath. The simplicity of a non-water-cooled sonic lance makes it attractive to smaller shops.

Inert gases are sometimes injected with the oxygen from the top lance to extend the period the top-blown gases are used. Argon top blowing can reduce air infiltration into the vessel and improve carbon removal efficiency in the late stages of the decarburization blow. This is especially true for vessels where high rate bottom argon blowing causes excessive splashing, vibration, or refractory wear.

A recent development in top lance technology is the development of top heating. An oxy-fuel burner in the top of the vessel provides heat to the vessel during operation.[56] This results in benefits similar to post-combustion.

12.9.5 Gases

Pure argon, oxygen, and nitrogen are used by virtually all shops. In rare cases, nitrogen is not used because the product mix or operation pattern does not warrant the extra cost of supplying nitrogen. Some other gases used are crude argon, high-pressure air, non-cryogenic nitrogen and oxygen, hydrocarbons, and carbon dioxide. These other gas choices depend on operational pattern, relative cost of supply and product mix.

Crude argon contains 95–97% argon, 1–3% oxygen, and 0.5–2% nitrogen. It can be used in the middle to late stages of decarburization and reduction, depending on final product specifications. The choice of using crude argon is dependent on availability, relative cost to pure argon, and product specifications.

High-pressure air can be used to partially replace nitrogen, but the capital and maintenance cost of compressors and associated equipment limit this option to larger melt shops. Non-cryogenic nitrogen and oxygen sources also require some compression for use in refining processes, and the choice of this supply method again depends on pure versus non-cryogenic prices, duty cycle, power costs and production rates compared to pure gases.[57]

Carbon dioxide and hydrocarbons can be used as process gases down to certain carbon levels. In stainless production, increased chromium oxidation below 0.3%C by carbon dioxide[58] or hydrogen and carbon pickup from hydrocarbons limits the use of these gases. The replacement of nitrogen by these gases depends on the relative cost of the gases. Low nitrogen specification grades of stainless steel, which typically are made using only argon as an inert gas, can use carbon dioxide as a replacement for argon in the early stages of decarburization.

12.9.6 Vessel Drive System

The drive system is similar to a BOF converter except that in most cases the vessel is removable from the trunnion. Torque requirements and safety interlocks are an important design criteria.

Some vessels are designed to rotate 360°. Side-blown vessels generally have replaceable shells. With removable vessels, crane access and capacity are considerations in the placement and size of the operating unit.

The trunnion bearings and drive must be designed to handle the vibration and rocking motion of the converter during processing.

12.9.7 Emissions Collection

Emissions collection strategies fall into two categories. Close capture hoods have a small gap, typically 0.2 to 0.4 metres, between the hood and vessel mouth. They are usually water cooled because the hot gases cannot entrain enough dilution air to cool the gas. Alloy addition chutes and lance positioning for alloy additions must be integrated with the hood. Other types of capture hoods are further from the vessel mouth. These are usually refractory lined and not water cooled.

Canopy systems have a substantial distance from the vessel mouth to the collection duct that is located in the roof of the shop. An accelerator stack is sometimes used as a chimney to direct the smoke plume to the collection duct.

Baghouse fan capacity, crane location, emissions regulations and cost dictate the choice of emission collection system. Several systems use wet or dry precipitators in place of bags for dust collection. Top lance placement can hamper optimal placement of the collection hood. The lance will have to penetrate the collection hood at some point. Close capture systems require less baghouse capacity and are likely to emit less smoke, but water cooling duct work can be expensive.

Canopy systems will not operate effectively if the crane must pass through the plume periodically. Baghouse requirements are usually higher because the baghouse must also take in dilution air that is entrained into the rising smoke plume. On the plus side, water cooling of the canopy is usually not necessary. However, placement of a lance with an accelerator stack system can be problematical.

The gas evolved from the converter is primarily CO along with argon or nitrogen, unless substantial post-combustion occurs. A major portion of the combustion and cooling takes place at the mouth exit and/or at the hood intake. Further cooling takes place in the duct. By the time the gas reaches the baghouse, the gas temperature is typically 120°C. The gas is composed primarily of CO_2 and inert gas. Solids emissions from the converter are primarily particulates of metallic (e.g., iron, chromium and manganese) and non-metallic oxides (e.g., lime and silica). Total solids emissions average between 6 to 10 kg/ton of metal. Dust loading is about 50 grams/cubic metre. In the United States, baghouse dust is normally handled by commercial processors.

For a new installation, the total cost of the fume system is typically about 40% of the total converter installation cost. In vacuum processes, the fume collection system is an integral part of the equipment and typically is used in conjunction with a water treatment plant.

12.10 Vessel Operation

No matter what converter-based process is used for refining stainless steel, there are several common steps during the refining process. Decarburization is the carbon removal step. As decarburization is not completely efficient, some metallic oxides are formed. Reduction recovers these oxides from the slag phase. Refining of the metal occurs throughout processing. Each of these steps will be described in more detail.

12.10.1 Decarburization

After the molten charge is in the vessel, an oxygen-inert gas (argon, nitrogen, or carbon dioxide) mixture is injected into the vessel. The initial oxygen to inert ratio generally varies in the range of 5:1 to 3:1 and is lowered with the progress of decarburization. The major benefit associated with the dilution process comes into play when the oxygen to inert gas ratio is 1:1 and then further reduced to 1:3 or lower. Oxidation of carbon continues, but oxidation of chromium is limited. The latter is due to the very low oxygen potential of the gas mixture, which minimizes chromium oxidation.

For a 300 series stainless, a typical stepped blow would be 1.5 Nm^3/min/ton oxygen mixed with 0.5 Nm^3/min/ton inert gas from the start carbon content to 0.35%C, then 0.5 Nm^3/min/ton oxygen mixed with 0.5 Nm^3/min/ton inert gas to 0.15%C, followed by 0.25 Nm^3/min/ton oxygen mixed with 0.75 Nm^3/min/ton inert gas to 0.05%C; and then 100% argon to final specification.

More recent blow programs consist of using a starting oxygen to inert gas ratio of higher than 3:1 and continuous adjustment of the oxygen to inert gas ratio during the blowing period. Depending on the aim nitrogen content, nitrogen can be used as the inert gas for a period of the oxygen blow.

Beginning carbon levels in converter processing may range from 0.7–4.5%C, with a typical range of 1.0–2.5%. Typical silicon levels range 0.2–0.4% before decarburization.

If a top lance is available, it is used during the high oxygen to inert gas periods. Some vessels use a top lance to add heat. The top heating lance is a modified oxy-fuel burner.

During decarburization, additions are made for obtaining the proper final chemical composition. These additions usually consist of desired amounts of high carbon ferrochromium, stainless steel scrap, carbon steel scrap, nickel, iron, high carbon ferromanganese, and molybdenum oxide. These additions serve to reduce the bath temperature as carbon and chromium oxidation are exothermic. In general, the bath temperature is controlled to less than 1720°C. Total alloy addition weights are in the range of 5–30% of tap weight.

Lime or dolomitic limes are usually added just before the oxygen blow to flux the transfer slag and silicon in the metal. During the oxygen blow, silicon is oxidized before carbon. Lime and dolomitic lime are sometimes added before the end of the blow to cool the bath and to reduce the volume of reduction additions. Slag fluxing additions, such as lime, dolomitic lime and spar, are typically in the range of 3–7% total bath weight.

It follows that the length of the blow period is determined by the starting carbon and silicon levels of the hot metal charged to the AOD. Decarburization times range from 20 to 35 minutes in modern converters (start carbon 1.5 to 2.5% and aim carbon 0.04%). Usually, the vessel is turned down to a horizontal position and a sample of the melt is taken for analyses at a carbon level of about 0.1%. Some high productivity shops can sample without rotating the vessel by using a sub-lance. The steel sample is analyzed for carbon. The temperature of the steel bath is also measured.

The next step is the reduction step, in which the reduction additions are charged and stirred with an inert gas for a desired time. The reduction mix consists of silicon alloys, such as ferrosilicon or chromium-silicon, and/or aluminum, which are added for the reduction of metallic oxides from the slag and fluxing agents such as limestone, lime, dolomitic lime, and fluorspar. The bath is then stirred with inert gas, typically for only for five to eight minutes. An example of a reduction mix is given in Chapter 2, Section 2.9.3.

The formation of a high basicity slag and the reduction of oxygen potential in the metal bath are good conditions for sulfur removal. For example, with a start sulfur of 0.03%, a reduction treatment of 2–3 kg aluminum/ton, 2–3 kg spar/ton, final slag basicity of about 1.7, and temperature of 1700°C, finish sulfur contents of 0.003–0.005% can be obtained.

If the grade to be produced requires an extra low sulfur level, the bath is deslagged after the reduction step and another basic slag is added. The liquid steel and the fluxes are then mixed to complete the desulfurization reaction. In modern practices a sulfur level of 0.001% or less is easily achieved with this double slag practice.

Ideally, at this stage of the process, the chemistry of the liquid steel should meet final specifications so that the heat can be tapped. If necessary, additional raw materials may need to be charged for small chemistry adjustments. Depending on the grade, additional deoxidation or alloying additions maybe needed. Such final additions are stirred in the AOD just before tap or are added in the tap ladle.

Following tap, the ladle is often stirred for composition homogenization and temperature uniformity along with flotation of inclusions. This is done in a ladle equipped with stirring facilities with or without the use of a ladle furnace. After the ladle treatment, the steel is ready to be cast.

12.10.2 Refining

The amount of stirring energy from the gas blown through the subsurface tuyeres and the formation of carbon monoxide deep within the metal bath results in the converter processes being among the most intensely stirred metallurgical reactors. The intimate gas–metal contact and excellent slag–metal mixing facilitate refining reactions.

Sulfur removal is a slag–metal reaction that occurs during the reduction phase of the process. Phosphorus, which requires oxidizing conditions, cannot be removed in the converter processing.

Nitrogen control is a gas–metal reaction. Depending on final nitrogen specifications for the stainless grade, the inert gas during the initial stages of decarburization can be nitrogen. After a certain carbon level is achieved, the nitrogen gas is replaced by argon. Such an approach is practiced by all steelmakers to reduce argon usage and costs and still achieve a desired nitrogen specification. After the change from nitrogen to argon, nitrogen is removed from the bath by both evolved carbon monoxide and argon.[59,60] Volatile elements with high vapor pressures, such as lead, zinc, and bismuth, are removed during the decarburization period.[61]

12.10.3 Process Control

The nature and number of reactions makes mathematical modeling and process control of stainless steel refining more complicated than standard pneumatic steelmaking.

Static calculation (i.e. no feedback) methods were used in the early days of stainless processing. These calculations were standardized on worksheets that allowed operators to estimate temperature changes, oxygen to inert gas ratio changes and nitrogen to argon changes. Reduction materials calculations were also performed.

With the introduction of powerful and inexpensive computers, dynamic calculations with automated or manual correction can be employed. These allow the incorporation of sophisticated models for temperature, blow program, nitrogen, and alloy addition control. Coupled with automated additions systems, process variability has been much reduced and process efficiency has steadily improved.

Process control computer models (static, dynamic or hybrid) may utilize artificial intelligence to control endpoint carbon and temperature levels and may eventually eliminate the need for sublances and/or offgas analyses. At a minimum, these programs advise the operator of blowing procedures, alloy and flux addition requirements and nitrogen switch-points, as well as provide data logging. Many operators integrate the process control computer model with the gas control system (valve rack) and some with other vessel functions (vessel rotation, additions, and fume system interfaces, sampling, etc.). These programs can be integrated with automated sampling equipment and/or equipment for offgas analyses. The degree of control and process algorithm sophistication varies.

12.10.4 Post-Vessel Treatments

Post-converter vessel treatments after converter processing are similar to treatments for other steel grades. Post-treatment facilities have several benefits: reduced converter tap-to-tap time, improved quality, and an inventory buffer of metal between the converter and the continuous caster. The use of ladle metallurgy furnaces for post-vessel treatments is increasing, especially in shops with continuous casting facilities.

Ladle treatment stations normally include an automated additions system, an injection lance or porous plug for stirring with inert gas, and wire feeding equipment. Some producers use lances to inject lime, fluorspar, and calcium alloys for quality improvements such as reduced inclusion content, inclusion shape control, improved weldability, and decreased dissolved oxygen levels. Ladle additions of deoxidizing compounds are common. Wire feeding is used to add alloys of titanium, aluminum, calcium, and sulfur to improve alloy recoveries and quality.

Some operators use ladle furnaces to provide additional heat to perform the above operations. Carbon pickup from the electrode arcs can be minimized by close control of slag thickness, arc length and stirring rates, even when treating ultra-low carbon grades.

Shops having vacuum refining units would normally use these for the last stage of decarburization on ultra-low carbon (ULC) and ultra-low nitrogen (ULN) grades or as part of the triplex (EAF-Converter-VOD) process.

12.11 Summary

Stainless steels were developed at the beginning of the 20th century. Their unique properties have led to their use in many different applications. Fundamental research in the thermodynamics and kinetics led to understanding of the phenomenon associated with chromium oxidation and how to minimize it. The development of the argon oxygen decarburization (AOD) process revolutionized stainless steelmaking. This process, based on the dilution principle, allowed steelmakers to use lower cost raw materials and to shorten the steelmaking times compared to the other processes in use at that time. This led to lowering the cost of producing stainless steels. This, in turn, led to dramatic increases in the applications for stainless steels. Over the years, other processes have been developed for making stainless steels. Most of these use vessels or reactors in which gases are introduced to carry out the necessary decarburization reactions. Some of these processes for making stainless steels utilize decarburization under vacuum.

Of the many specific processes available for stainless steel refining, argon oxygen decarburization and its converter-based variants are by far the most commercially popular. For maximum flexibility to produce a very wide range of stainless steel grades, a modern melt shop may have a source of molten metal, a converter that can produce most grades economically (duplex processing) and a vacuum unit for further processing of extra low carbon and nitrogen grades (triplex processing).

References

1. D. C. Hilty, H. P. Rassbach, and Walter Crafts, "Observations of Stainless Steel Melting Practice," *Journal of the Iron and Steel Institute*, (June 1955): 116–28.
2. W .E. Dennis and F. D. Richardson, "The Equilibrium Controlling the Decarburization of Iron-Chromium-Carbon Melts," *Journal of the Iron and Steel Institute*, (November 1953): 264–6.
3. W. A. Krivsky, "The Linde Argon-Oxygen Process for Stainless Steel; A Case Study of Major Innovation in a Basic Industry," The 1973 Extractive Metallurgy Lecture, 102nd TMS-AIME Annual Meeting, Metallurgical Transactions, vol. 4, (June 1973): 1439–47.
4. R. J. Choulet, F. S. Death, and R. N. Dokken, "Argon-Oxygen Refining of Stainless Steel," *Canadian Metallurgical Quarterly* 10:2: 129–36.
5. R. J. Choulet and S. K. Mehlman, "Status of Stainless Refining," *Metal Bulletin, 1st International Stainless Steel Conference*, November 12, 1984.
6. R. Heinke et al., "Developments In Refining Stainless Steel Melts By the Krupp Combined Blowing Process," presented at the International Oxygen Steelmaking Congress, Linz, 1987.
7. K. Taoka et al., "Progress in Stainless Steel Production by Top and Bottom-Blown Converter," *Kawasaki Technical Report No. 14*, March 1986, 12–22.
8. E. Fritz and J. Steins, "Outline of Stainless Steel Production with an Emphasis on the K-OBM-S Process," *BHM*, 140 Jg. (1995), Heft 5: 227–34.
9. E. Wagener and K. M. Sinha, "Construction and Operation of Metal Refining Reactor for Foundries and Mini Steel Mills," *CPT—Casting Plant and Technology* 4 (1985): 24–36.

10. P. J. Mullins, "Steelmaking Costs Drop with CLU Process," *Iron Age,* (Feb. 4, 1974): 48–9.
11. T. Aoki et al., "Development of Top and Bottom Blowing Process in BOF," presented at 101st ISIJ Meeting, April 1981. Abstract in *Transactions of the Iron and Steel Institute of Japan* 21:9 (1981): B-388.
12. J. W. Tommaney, "Method For Refining Molten Metal Bath To Control Nitrogen," U.S. Patent 4,615,730, October 7, 1986.
13. K. F. Behrens, H. Bauer, and M. Walter, "The Vacuum Decarburization of High Chromium Heats in an Evacuated BOF," *Proceedings of the Third International Iron and Steel Congress,* (Chicago, April 1978): 609–14.
14. T. Kishida et al., "Development of New Stainless Steelmaking Processes," *Scandinavian Journal of Metallurgy,* 22 (1993): 173–80.
15. H.-D. Scholer and H. Maas, "Operation Technique and Plant Design of the VAD/VOD Process," *Metallurgical Plant and Technology* 6 (1985): 36–45.
16. Y. Suzuki and T. Kuwabara, "Secondary Steelmaking: Review of Present Situation in Japan," *Ironmaking and Steelmaking* 2 (1978): 80–8.
17. S. Iwaoka et al., "Development of Modified VOD Process for Making Ultra-Low Interstitial Ferritic Steels," *Stainless Steel* 77 (London, England): 139–56.
18. R. J. Fruehan, "Reaction Model for the AOD Process," *Ironmaking and Steelmaking* 3 (1976): 153–58.
19. J. Reichel and J. Szekely, "Mathematical Models and Experimental Verification in the Decarburization of Industrial Scale Stainless Steel Melts," *Iron and Steelmaker* 22:5 (1995): 41–8.
20. Y. E. Lee, O. S. Klevan, and R. I. L. Guthrie, "A Model Study to Examine the Reactions During Reduction Period of AOD Process," *Proceedings of the 11th Process Technology Division Conference,* Iron and Steel Society (Atlanta, 1992): 131–8.
21. J. Steins, E. Fritz, and L. Gould, "Outline and Application of VAI Stainless Steelmaking Technology," *Iron & Steel Review* 40:10 (March 1997).
22. K. Yamada et al., "Direct Stainless Steelmaking from Molten Ferronickel and Ferrochromium by the AOD Process with Top Blowing," *Iron and Steelmaker* 11:10 (October 1984): 35–40.
23. K. Taoka et al., "Production of Stainless Steel with Smelting Reduction of Chromium Ore by Two Combined Blowing Converters," *Proceedings of the Sixth International Iron and Steel Congress,* Iron and Steel Institute of Japan (1990, Nagoya Japan): 110–7.
24. M. Nishikori et al., "Development of Stainless Steel Manufacturing Techniques Using Cr Ore in Actual Production Units. Improvements in the Stainless Steel Refining Process Using a Large Amount of Cr Pellets", *CAMP-ISIJ* 4 (1991): 281.
25. K. Ogino, S. Hara, and H. Kawai, "Foaming of Molten Slags in Pyrometallurgical Processes", *Process Technology Proceedings,* Iron and Steel Society, (1988): 23–9.
26. R. J. Choulet, "Stainless Steel Refining." Paper presented at AISE Modern Steelmaking Seminar, Detroit, MI, June 5, 1997.
27. K. Taoka et al., "Progress in Stainless Steel Production by Top and Bottom-Blown Converter," *Kawasaki Technical Report No. 14*, March 1986, 12–22.
28. Y. Lee et al., "Start-up of Stainless Plant No. 2 at POSCO," *Steel Times International* (May 1997): 31–6.
29. P. Dinelli et al., "Supply of the K-OBM_S Stainless Steel Converter to Acciaierie di Bolzano," *Iron & Steel Review* 40:10 (March 1997).
30. E. Wagener and K. M. Sinha "Construction and Operation of Metal Refining Reactor for Foundries and Mini Steel Mills," *CPT—Casting Plant and Technology* 4 (1985): 24–36.
31. H. Kappes, "Metallurgical Possibilities of MRP and Chief Distinctive Features as compared with other Pneumatic Converter Processes," *Mannesmann Demag Huttentechnik Publication,* 1980.
32. M. Fisher et al., "The New Stainless Steelmaking Plant of Columbus Joint Venture," *Iron and Steelmaker* 23:7 (July 1996): 47–52.
33. M. Schmidt et al., "Production of High Alloy Special Steels in Basic Oxygen Converter," *Stahl u. Eisen* 88 (1968): 153–68.

34. "Stainless By Combined Blowing," *Steel Times International* (September 1989): 42–4.
35. U. Glasmeyer, "The Production of Stainless Steel," *Metallurgical Plant and Technology* 3 (1989): 32–7.
36. Y. Umeda et al., "Investigation of Special Converter Process—STB Process," Sumitomo Metals Industries Publication.
37. S. Masuda et al., "Development of Oxygen Top Blowing and Argon Bottom Blowing Method in Refining of Stainless Steel," *Tetsu-to-Hagane* 72 (1986): 71–8.
38. G. J. McManus, "Allegheny Ludlum and Kawasaki go back to basics in making stainless steel," *Iron Age* (December 5, 1983): 19–21.
39. M. Shinkai et al., "Development of AOD-VCR for a New Stainless Steel Refining Process", *Proceedings of the 14th Process Technology Division Conference,* Iron and Steel Society (Orlando, 1995):37–43.
40. Nippon Steel Hikari Works AOD-VCR information.
41. N. Kamii, S. Okubo and B. Ero, "Development of a New Stainless Steel Melting and Refining Process (LD-RHOB Process)," *Transactions ISIJ* 19 (1979): 58-63.
42. "On the Production of extra low interstitial 30Cr-2Mo and other stainless steels by SS-VOD Process," Kawasaki Steel Corporation publication, 1982.
43. W. D. Huskonen, "New Melt Shop in Belgium Features 100-Tonne Converter," *33 Metal Producing* (February 1989): 31–2.
44. Kawasaki Steel Corporation, "Chiba's New Converter Shop Concentrates on Stainless," *Steel Times International* (July 1995): 43.
45. H. Suzuki et al., "Production of Stainless Steel By Combined Decarburization Process," *Proceedings of the 75th Steelmaking Conference*, Iron and Steel Society, (Toronto *1992):* 199–204.
46. H. Nakamura et al., "Operational Results and Reaction Analysis for a New Stainless Steel Refining Furnace," *CAMP-ISIJ* 7 (1994): 1056–9.
47. Joseph M. Saccomano and John D. Ellis, "Molten Metal Reactor Vessel," U.S. Patent 3,724,830, April 3, 1973.
48. Stewart K. Mehlman, ed., *Pneumatic Steelmaking Volume Two: The AOD Process* (Iron and Steel Society, 1991): 64–8.
49. D. A. Whitworth, "Refractory Considerations for the A.O.D. Process," *Industrial Heating*, (April 1975): 16–22.
50. D. A. Whitworth and E. L. Bedell, "AOD Refractory Practices, Wear Patterns and Methods to Enhance Vessel Life," *Proceedings, 37th Electric Furnace Conference*, Iron and Steel Society (Detroit, 1979): 86–98.
51. J. W. Kaufman and C. E. Aguirre, "Refractory Wear by Transfer, Decarburization, and Reduction Slags in the AOD Process," *Proceedings, 35th Electric Furnace Conference,* Iron and Steel Society (Chicago, 1977): 74–9.
52. E. B. Pretorius and R. Marr, "The Effect of Slag Modeling to Improve Steelmaking Processes," *Proceedings, 53rd Electric Furnace Conference,* Iron and Steel Society (Orlando, 1995): 407–15.
53. Ronald E. Fuhrhop and Allan R. Wandelt, "Metallurgical Tuyere and Method for Calibrating Same," U.S. Patent 4,795,138, January 3, 1989.
54. H. Takano et al., "Improvement of Refining Technologies by AOD Combined Blowing Process," *Proceedings of the 11th Process Technology Division Conference,* Iron and Steel Society (Atlanta, 1992): 37–48.
55. I. F. Masterson, "Method for Controlling Secondary Top-Blown Oxygen in Subsurface Pneumatic Steel Refining," U.S. Patent 4,599,107, July 8, 1986.
56. T. Saito et al, "Application of a Heat Supply Technique to a Small AOD Furnace," *CAMP-JISI* 3 (1990): 223.
57. R. J. Selines and C. F. Gottzmann, "Industrial Gases for Process Metallurgy," *Proceedings of the International Symposium on Injection in Process Metallurgy, TMS-AIME* (1991): 43–63.
58. B. H. Heise and R. N. Dokken, "Use of CO_2 in Argon-Oxygen Refining of Molten Metal," U.S. Patent 3,861,888, January 21, 1975.

59. J. M. Saccomano et al., "Nitrogen Control in Argon-Oxygen Refining of Molten Metal," U.S. Patent 3,754,894, August 28, 1973.
60. R. J. Fruehan, "Nitrogen Control in Stainless Steels," *Proceedings of the 11th Process Technology Division Conference,* Iron and Steel Society (Atlanta, 1992): 125–30.
61. Stewart K. Mehlman, ed., *Pneumatic Steelmaking, Volume Two: The AOD Process* (Iron and Steel Society, 1991): 26.

Chapter 13

Alternative Oxygen Steelmaking Processes

R. J. Fruehan, Professor, Carnegie Mellon University
C. L. Nassaralla, Assistant Professor, Michigan Technological University

13.1 Introduction

Virtually all of the steel in the United States and the world is produced either in an oxygen steelmaking converter such as a BOF, LD, OBM (Q-BOP) etc. or an electric arc furnace (EAF). The only exception are the few remaining open hearth shops and a negligible amount in other processes such as the Energy Optimizing Furnace (EOF). This is understandable since much of the steelmaking capacity was built or rebuilt from 1955 to 1975; the BOF and EAF were the best technologies available at the time and were compatible with large blast furnace production and relatively inexpensive scrap. However, these conditions are changing. It may be possible to produce iron economically on a small scale, 0.5 million tons per year, in new processes such as COREX and bath smelting. Steelmakers desire more flexibility to use scrap and hot metal and other forms of iron such as direct reduced iron (DRI) and iron carbide. Therefore, it is of value to re-examine steelmaking options.

There is currently a large amount of development work related to the electric arc furnace including scrap preheating, continuous melting and the use of fossil fuels and hot metal. These are discussed in detail in Chapter 10. In particular electric arc furnaces will use more oxygen to oxidize carbon and for post-combusting CO to CO_2. The EAF will act more like a BOF using fossil fuel such as carbon in iron or in direct reduced (DRI) products. They will also use more virgin (non scrap) iron such as DRI and liquid hot metal. The alternative steelmaking processes examined in this chapter are those using no electrical energy.

Many alternative oxygen steelmaking processes were being developed in the 1970s such as IRSID continuous steelmaking, WORCRA and the Bethlehem continuous process. However, these never were commercialized, in part because most companies invested in conventional oxygen steelmaking and no new capacity was required. In recent years, other processes have been developed or proposed such as the EOF, the AISI continuous process, iron carbide continuous processes and IFCON. In this chapter, the general principles of process types are examined, selected steelmaking processes are technically evaluated and an economic analysis of selected processes is made. This has been the subject of a 1998 publication and much of this chapter comes from that publication.[1]

13.2 General Principles and Process Types

The steelmaking processes considered in this chapter can be classified into two general types: batch and continuous. The continuous processes can be classified as a continuous stirred tank reactor (CSTR) or plug flow reactors (PFR). The process types are shown in Fig. 13.1 along with their characteristics with regards to concentration and reaction rates.

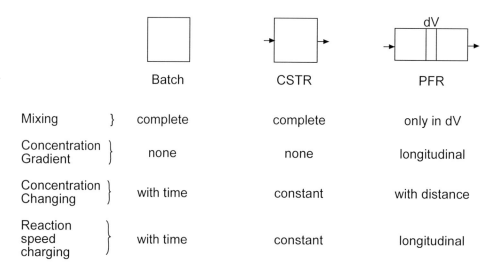

Fig. 13.1 Classification of steelmaking process types and their characteristics.

By performing a simple mass balance in a reactor, one can see that the degree of dispersion in the system defines the type of reactor where the process takes place. The contributions to backmixing of fluid flowing in the x direction can be described by the following equation[2]:

$$\frac{\partial C}{\partial t} = D \frac{\partial^2 C}{\partial x^2} \quad (13.2.1)$$

where the parameter D is called the longitudinal dispersion coefficient, which characterizes the degree of backmixing during flow. Equation 13.2.1 can be written in a dimensionless form where $z = (ut + x)/L$ and $\Theta = t/\bar{t} = tu/L$

$$\frac{\partial C}{\partial \Theta} = \left(\frac{D}{uL}\right) \frac{\partial^2 C}{\partial z^2} - \frac{\partial C}{\partial z} \quad (13.2.2)$$

where the dimensionless group (D/uL) is the inverse of Peclet number, and measures the extent of longitudinal dispersion. The Peclet number represents the ratio of material transferred by bulk flow to material transport by Eddy diffusion. L is the length of the reactor and u is the velocity in the direction of the flow. Therefore,

$$\frac{D}{uL} \to 0 \quad \text{negligible dispersion, (PFR)}$$

$$\frac{D}{uL} \to \infty \quad \text{large dispersion, (CSTR)}$$

It means that when the diffusion, plus convection inside the vessel, is small the reactor approaches an ideal plug flow reactor. On the other hand, when the diffusion and convection are large, the reactor approaches a continuous stirred tank reactor.

Szekely[3] has shown that for values of (D/uL) greater than 0.2 there would be extensive dispersion. He further concluded that furnaces of more than 100 meters in length may be required for minimal dispersion. Measurements of the (D/uL) from an actual operation of the WORCRA furnace give value of 0.1 indicating shorter furnaces may be satisfactory[4]. For plug flow Taylor[8] has shown that the length to width ratio, L/W, should be around 50 for this type of reactor to be a PFR. For highly stirred steelmaking systems it is difficult to maintain a PFR. The closest such reactor in steelmaking is a continuous casting tundish. Even for a tundish where turbulence is kept to a minimum it deviates significantly from a perfect PFR. In some PFR steelmaking processes, such as WORCRA, the slag runs counter-current to the metal. This has advantages when refining hot metal with high phosphorus contents. A further discussion of trough concurrent and counter-current steelmaking is given by G. Brooks et al[5].

The normal oxygen steelmaking processes, the EAF, and the EOF are batch processes. The AISI and IRSID processes are CSTR while the WORCRA, Bethlehem and the initial Iron Carbide processes should approach plug flow reactors; these processes are discussed later.

In steelmaking, the primary reaction is decarburization. At high carbon contents the rate is simply controlled by the oxygen input rate[6]. Below a critical carbon content, typically about 0.3%C for normal OSM, the rate is controlled by mass transfer of carbon, is first order with respect to carbon, and given by:

$$\frac{d\%C}{dt} = -k\left[\%C - \%C^e\right] \qquad (13.2.3)$$

where %C and %Ce are the carbon content and the equilibrium content with the slag respectively and k is the overall decarburization rate constant given by:

$$K = \frac{Am\rho}{W} \qquad (13.2.4)$$

where

A = reaction surface area
m = mass transfer coefficient
ρ = density of steel
W = weight of steel

It has been shown that for a CSTR the steady state or tap carbon content, $\%C_t$, is given by[7]:

$$\%C_t = \frac{F\left[\%C_o\right]}{F + k} \qquad (13.2.5)$$

where, F is the specific feed or production rate which is the production rate divided by the weight of steel in the vessel and $\%C_o$ is the initial carbon content.

In attempting to partially convert a reactor with intense stirring into one with concentration differences like a PFR, barriers are sometimes installed with a small opening for metal flow as shown schematically in Fig. 13.2. In this case, the tap carbon content is given by:

$$\%C_t = \frac{(F+\beta)\%C_1}{F+\beta+k} \qquad (13.2.6)$$

where β is the backmixing flow rate through the opening and $\%C_1$ is the carbon content of steel in Reactor I. In this case the W used in computing F is that of Reactor II.

13.3 Specific Alternative Steelmaking Processes

In this section, selected alternative steelmaking processes will be reviewed. A brief description, current status and an evaluation of the processes is given. The technical evaluation addresses issues such as productivity, yield and refining capability. First batch processes are discussed followed by CSTR and PFR. The processes considered are summarized in Table 13.1 with respect to type, typical charge make-up and current status.

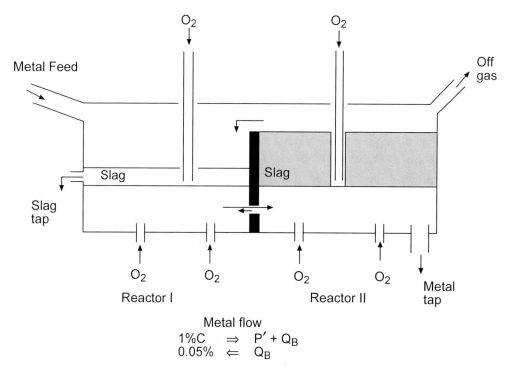

Fig. 13.2 Schematic diagram of a trough process with a barrier.

Table 13.1 Classification and Status of Steelmaking Processes.

Process	Type	Typical Charge	Current Status
BOF	batch	80% hot metal 20% scrap	65% of world production
EAF	batch	70–100% scrap 0–30% other	35% of world production
EOF (Brazil)	batch	50% hot metal 50% scrap	commercial
IRSID (France)	CSTR*	80% hot metal	pilot plant in 1970s
AISI (U.S.)	CSTR	hot metal	pilot plant in 1980s
WORCA (Australia)	PFR**	hot metal	pilot plant in 1970s
Bethlehem (U.S)	PFR	hot metal	pilot plant in 1970s
NRIM (Japan)	PFR	hot metal	pilot plant in 1970s
COSMOS (U.S.)	PFR	hot metal	pilot scale in 1980s
IFCON (South Africa)	PFR	ore, coal	pilot plant in 1990s
Carbide to steel	CSTR; PFR	iron carbide	concept

* CSTR = continuous stirred tank reactor; ** PFR = plug flow reactor.

13.3.1 Energy Optimizing Furnace (EOF)

The EOF is essentially a batch oxygen steelmaking process with high post-combustion, with coal additions and extensive scrap preheating; a schematic diagram is shown in Fig. 13.3.[9] The process theoretically can charge up to 100% scrap, but typically 40–60% scrap is used. Hot metal is charged into the vessel followed by scrap from the lower preheating chamber. In some cases, a second scrap charge is added about five minutes into the blow. The scrap is preheated in a series of preheat chambers to about 800 to 1200°C. For a 50/50 scrap to hot metal mix about 70 m³/ton of

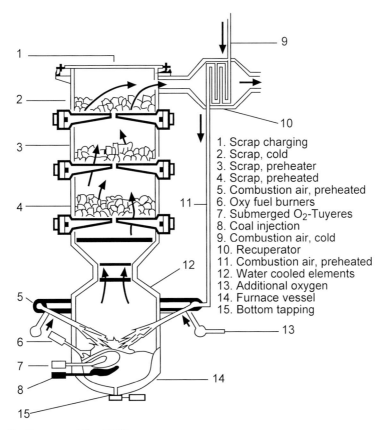

Fig. 13.3 Schematic diagram of the EOF process.

oxygen, with about one third into the metal and two thirds above the bath for post-combustion, and about 20 kg/ton of coal is used. The post-combustion gas is used to preheat the scrap and, therefore, the energy from post-combustion does not have to be completely transferred to the bath. The blowing time is about 27–32 minutes and tap-to-tap is less than one hour. Refractory consumption increases with scrap usage and is approximately 2.5 kg per ton for a 50/50 mix.

An EOF furnace has been operating in Pains Brazil for over 10 years and has produced up to 0.5 mtpy from two 30 ton furnaces. There is an 80 ton furnace at Tata Steel in India and a 60 ton furnace at AFS in Italy. The process has been reasonably well proven using a scrap charge up to 60%.

The productivity of an EOF is similar to that for a state of the art EAF and can produce a 100 ton heat in an hour. The yield with a 50/50 charge mix is 92–94% which is similar to a BOF. Nitrogen contents of 30–40 ppm can be achieved and phosphorus levels are similar to a BOF. Residuals depend on the amount and type of scrap and can be computed from a simple mass balance. Sulfur can be a significant problem as coal is used in the process. Hot metal desulfurization alone is not effective because of the high scrap rates and the use of coal. For the following conditions, per ton of steel, the final sulfur level is 0.07% for a 50/50 mix.

$$\frac{(\%S)}{[\%S]} = 4 \qquad 25 \text{ kg/ton coal}$$

70 kg slag   0.7% S in coal

0.10% S in hot metal

0.025% S in scrap

Therefore, significant steel desulfurization is required. If the hot metal is desulfurized to 0.02%, the final sulfur is 0.033% which is still high for most quality applications.

The EOF is a proven technology which allows for higher scrap melting than a BOF and reasonable flexibility with respect to the charge mix. With a 400,000 tpa COREX, or bath smelter, an 80 ton EOF with a 50/50 mix could produce over 750,000 tpa of steel.

Mention should be made of a number of converter based processes which have been developed to melt more scrap than conventional oxygen steelmaking. The first of these was developed by Klockner.[10] The KMS melted increased amounts of scrap by injecting coal into an OBM (Q-BOP) converter while the KS process could melt 100% scrap using a liquid metal heel. Nippon Steel[11] and Kawasaki Steel[12] investigated similar types of processes. The Tula or Z-BOP process which originated in Russia and used by ISCOR at its Newcastle plant and tested by Bethlehem Steel, uses lump coal in a conventional top blown BOF to melt up to 100% scrap.[13] When melting large amounts of scrap there was excessive iron oxidation. However, none of these processes continued on a commercial basis and do not appear as efficient as the EOF which maximizes post-combustion and uses scrap preheating.

13.3.2 AISI Continuous Refining

As part of the AISI-DOE direct steelmaking project a continuous steelmaking process was evaluated and a limited number of pilot scale tests were run. The process is described and evaluated in detail by Abel et al[7]. A schematic diagram is shown in Fig. 13.4. Hot metal and possibly scrap along with the required fluxes are added continuously to the top of the converter, oxygen is blown through the bottom using OBM (Q-BOP) type tuyeres and tapped continuously.

The AISI converter is essentially a CSTR and equation 13.2.5 can be used to relate the final carbon to the production rate. For example, for a 100 ton reactor processing 80% hot metal and 20% scrap producing 0.05%C could produce about 100 ton per hour. A value of 60 hr^{-1} (1.0 min.$^{-1}$) was used for k which is typical for bottom blowing processes. However, as discussed in detail elsewhere[7] since decarburization is carried out at low carbon contents a large percentage of the oxygen oxidizes iron resulting in large yield losses (>25%). For this reason, the process development was terminated.

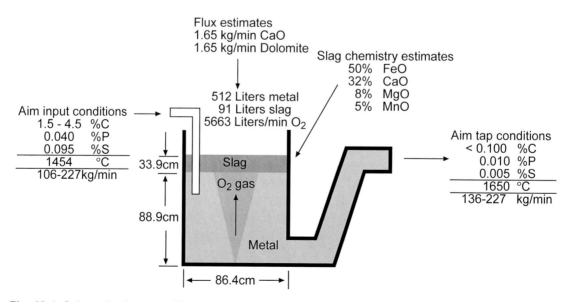

Fig. 13.4 Schematic diagram of the 4 ton AISI continuous refining process.

Alternative Oxygen Steelmaking Processes

As part of the AISI project a two zone smelter was investigated both in the pilot plant and theoretically. The basic concept is similar to that shown in Fig. 13.2. The concept is to reduce the carbon content in Reactor I to approximately 1% followed by final decarburization in Reactor II with the excess FeO in the slag flowing to Reactor I where it is reduced. A critical concern is the size of the opening in the barrier to avoid excessive back mixing. As an example, consider a 300 t/hr furnace consisting of two 100 ton CSTR separated by the barrier and the carbon content is reduced to 1% in Reactor I. To produce steel with 0.05% C, backmixing according to equation 13.2.6 must be limited to 20 t/hr ($\beta = 0.2$) Pilot plant work and water modeling by Zhang et al[14] indicates the area of the opening must be limited to about 0.1 m² or for a square opening 31.6 cm on a side. To maintain the barrier with such a small opening would be extremely difficult to do in practice.

Another problem is that there is excessive iron oxidation in Reactor II. In theory it is reduced in Reactor I supplying 75% of the oxygen required for decarburization. The reduction reaction 13.3.1 is endothermic and with the reduced oxygen gas requirement slag-metal mixing may be insufficient for rapid refining.

$$(FeO) + C = CO + Fe \tag{13.3.1}$$

Reactor I would be operating more like an open hearth type furnace. Based on the anticipated problems with the barrier, control and the nature of the reactions in Reactor I work on the development of this process was also terminated.

13.3.3 IRSID Continuous Steelmaking

Of the existing continuous steelmaking processes the IRSID process was the most extensively developed and is one of the most promising[15]. The process was tested on a 10–12 , ton per hour scale at IRSID at Maizieres-Les-Metz and up to 25 tons per hour at Hagondange. A schematic diagram of the process is shown in Fig. 13.5. The process is described elsewhere in detail.[15] Briefly, liquid hot metal is poured into the reactor at a controlled rate and an oxygen jet impinges onto the stream forming an emulsion of gas, slag and metal. The emulsion overflows into a decanter in which slag-metal separation takes place.

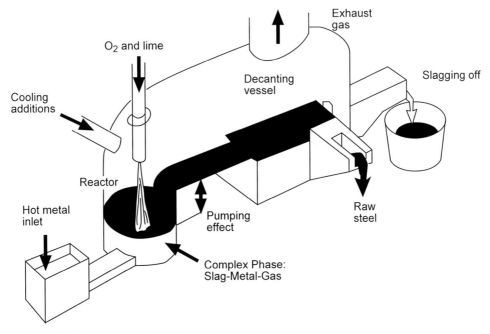

Fig. 13.5 Schematic diagram of the IRSID continuous steelmaking process.

In normal batch oxygen steelmaking, such as the BOF, the oxygen jet and reaction gases cause an extensive slag-metal emulsion with about 30–35% of the metal in the emulsion. The rate of decarburization at high carbon levels is limited simply by oxygen blowing rates. Below some critical carbon content, about 0.3%, the rate is limited by liquid phase mass transfer of carbon in the metal. The rate of mass transfer is first order with respect to carbon content. The metal drops in the emulsion are decarburized to some extent and then enter the liquid metal pool in the bottom of the furnace. The average carbon content of the metal drops is slightly less than in the bulk metal and, therefore, the rate of decarburization can be related to the bulk metal concentration. The production rate for a given furnace is limited by the rate of gas removal. That is, there is a free volume requirement for a given rate of gas generation.

The reaction mechanism for the IRSID process is not known. As the metal enters the reactor it is hit with an oxygen jet causing nearly all of the metal to be emulsified. Observations indicate there is no liquid metal pool. In this case the metal drops remain in the emulsion during their entire retention time in the reactor. The rate depends on how long the drops are in the reactor since their carbon content is changing in time. The metal which overflows into the decanter is the average carbon content of the metal in the reactor. The retention time of the metal drops in the reactor depends on the reactor size.

Obviously the IRSID process is more complex than a normal BOF and no model exists for the process. In order to extrapolate the results obtained by IRSID and evaluate the process a reasonable model is essential. The assumptions for the present model are given below:

1. The rate of decarburization of the metal drop is controlled by liquid phase mass transfer of carbon.

2. There is always sufficient oxygen present as oxygen gas or FeO in the slag.

3. The metal drops react independently and do not combine.

4. It is recognized that there is a distribution of metal drops in the emulsion that are decarburized at different rates due to differing area to volume ratios. However, it is assumed that the average rate of decarburization for the weighted average size droplets adequately describes the system. The Sauter diameter d_{32} should be used in such an analysis.

$$d_{32} = \frac{\sum_i d_L^3}{\sum_i d_i^2} \tag{13.3.2}$$

5. The carbon content of the metal leaving the reactor is the average of all the drops in the emulsion.

6. There is sufficient slag to hold all of the metal in the emulsion.

The rate of decarburization of a single droplet is controlled by mass transfer and the carbon content is given by:

$$C_D = C_o e^{-kt} \tag{13.3.3}$$

where,

C_D = carbon content of a drop after time t
C_o is the initial carbon content
k is the overall rate constant for mass transfer given by:

$$k = \frac{A_D \rho m}{W_D} \tag{13.3.4}$$

A_D = area of drop
m = mass transfer coefficient
ρ = density of the metal
W_D = weight of the drop

The reactor is not plug flow but rather a continuous stirred tank reactor (CSTR). That is, some of the particles only react for a short time and may have a high carbon content while others are very low in carbon. For a CSTR the residence time distribution f(t) is given by:

$$f(t) = \frac{1}{t_r} e^{-t/t_r} \qquad (13.3.5)$$

where f(t) is the fraction of drops in the bath for time t and t_r is the average residence time. The average or final carbon content of the metal leaving the reactor is given by:

$$C = \int_0^\infty C(t)f(t)dt \qquad (13.3.6)$$

$$C = \int_{C_0 e}^{C_0} -\left(k + \frac{1}{tr}\right)t \qquad (13.3.7)$$

$$C = \frac{C_0}{t_r k + 1} \qquad (13.3.8)$$

Equation 13.3.8 is a very simple but useful expression for evaluating the IRSID process. It is interesting to note that the rate of decarburization is reasonably fast and most of the drops are fully decarburized. The carbon in the final metal results primarily from the drops which only spend a short time in the reactor.

For typical oxygen blowing rates and slag volumes in a BOF the value of k is about 0.017 s⁻¹. In comparison to the BOF in the IRSID process all of the metal is in the emulsion. This causes the reaction area (A) to increase correspondingly. With all the metal rather than 30% in the emulsion the value of k is estimated to be a factor of about 3.3 higher or about 0.55 s⁻¹.

It is also necessary to estimate the retention time (t_r). It has been found that there is a required reactor volume to hold the slag metal emulsion in a BOF which depends on the rate of CO formation. The reactor volume at any time is filled with slag, metal and gas. In a normal BOF, 70% of the emulsion is gas. In the IRSID process, it may be even higher due to the faster rate of gas generation. If the gas occupies 70% of the volume, the retention time of the metal and slag is given by:

$$t_r = \frac{0.3 V_R}{\dot{V}_s + \dot{V}_m} \qquad (13.3.9)$$

where,

V_R = reactor volume
$\dot{V}_s$ = volume flow rate of slag
$\dot{V}_m$ = volume flow rate of metal

For example, consider a 10 m³ reactor producing 100 t/hr of steel and 75 kg of slag per ton, the retention time is about 10.75 minutes (645 s). Using equation 13.3.7, and 0.055 s⁻¹ for k and an average initial carbon of the scrap plus metal of 4%, the predicted final carbon content is about 0.1%. It should be noted that this is only a crude estimate. However, it does agree with the pilot plant results at IRSID in which in a smaller scale unit the final carbon content was about 0.08%.

The productivity of the IRSID process obviously depends on the reactor size but 200 t per hour should be easily achieved. The yield and refining for sulfur phosphorus, etc., should be similar to a BOF. Scrap can be melted but must be added continuously and sized so it melts rapidly. Metal feeding rate is critical in order to match the oxygen input. IRSID used a fairly crude but effective metal feeding system over 20 years ago. With the advent of slide gates and other metal flow control devices adequate metal flow control should be possible. Due to the high gas generation rate there could be a large amount of dust generation. However, some of the dust should settle out in the decanter. Refractory wear in the reactor may be a concern due to the violent reactions for the emulsion overflow. Refractory consumption is expected to be significantly higher than for a BOF with slag splashing.

13.3.4 Trough Process

From about 1970 to 1983 a number of continuous steelmaking processes were suggested or tested on a pilot scale in which a long trough was used in which liquid hot metal was added to one end, a series of oxygen lances or tuyeres, placed longitudinal, were used for decarburization and the metal was tapped on the other end. The slag in some cases flows counter current to the metal. The processes which were tested on a pilot scale included the Bethlehem process[16], the NRIM (Japan) process[17], the WORCRA process[18] and the COSMOS (U.S.Steel) process[19]. These processes all reached the pilot scale at production rates generally of about 5 to 12 tons per hour.

The processes were all somewhat similar and a schematic of the WORCRA process is shown in Fig. 13.6. The NRIM used a series of trough furnaces with desiliconization and partial dephosphorization carried out in the first reactor, decarburization and final dephosphorization in the second and grading and alloying in the third reactor. The slag from each furnace is removed separately. COSMOS used bottom OBM (Q-BOP) tuyeres instead of lances.

As noted earlier, to avoid backmixing the length to width ratio (L/W) should be around 50. None of the pilot facilities had this ratio and, therefore, did not act as a perfect PFR, but had significant backmixing. Even with modest values (L/W) there was excessive heat losses due to the large surface to volume ratio and relatively low production rates. Calculations indicate that even at reasonable production rates the heat losses for these processes would be significantly higher than for other processes.

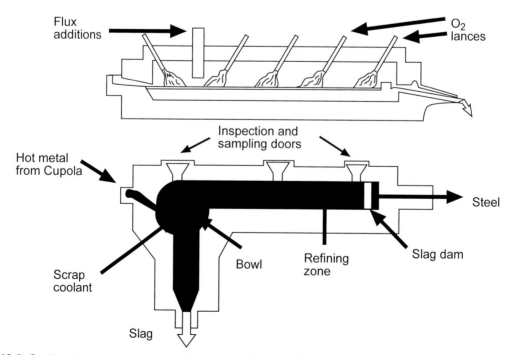

Fig.13.6 Sectional plan and elevation of L-shaped pilot plant WORCRA steelmaking furnace.

Other problems associated with these processes include metal feed control, the necessity to size scrap and other coolants, lower phosphorus removal due to inadequate slag metal mixing and the difficulty in producing low carbon steels (<0.06%).

It should be noted that these type processes had some success in nonferrous metallurgy. Noranda[20] developed a process for copper using a 21 m long reactor with 16 tuyeres. An analysis of the process indicates that the reactor worked as 6–7 CSTR reactors in series and was approaching a PFR. The QSL process[21] for lead has also been commercialized. These processes operate at lower temperatures than for steelmaking and, therefore, the high heat losses and refractory with this geometry is less critical. Also, the production rate is low compared to that required for steelmaking.

13.3.5 Other Steelmaking Alternatives

Several other alternative steelmaking processes have been proposed or reached a pilot stage. Two iron carbide continuous steelmaking processes have been proposed. In the original conceptual process, iron carbide is refined in a trough type reactor with extensive post-combustion supplying much of the energy[22]. This process has not been tested on any scale. It would suffer from all of the drawbacks of the other trough processes and the need for very high degrees of post-combustion which have not been achieved in any other process. A second process based on iron carbide, "Carbide to Steel" has been proposed by G. Geiger[23] in conjunction with U.S.Steel. They concluded that the original concept of a QSL type trough reactor would not be feasible because the CO evolution, from the unreacted FeO in the carbide and from decarburization, would be so great it would be impossible to maintain a carbon concentration gradient. They concluded it was more attractive to divide the process into two stages shown schematically in Fig. 13.7. In the first vessel, the Fe_3C would be injected and melted with carbon oxidation providing part of the energy. In the first reactor, the carbon content is reduced to about 2% and then tapped into a second reactor which is a trough reactor in which final decarburization takes place. The CO from the second reactor is transferred to the first and along with the CO from the first reactor is post-combusted to CO_2 to provide most of the energy for reduction of the remaining FeO in the carbide and melting. They

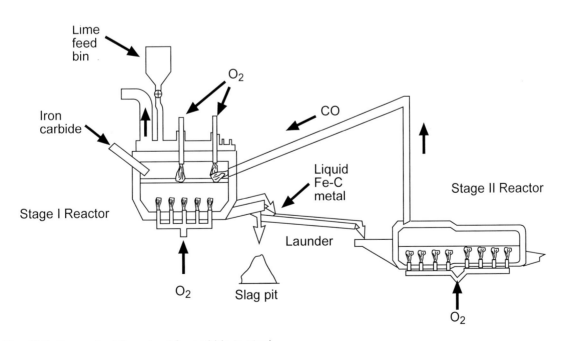

Fig. 13.7 Conceptual flow sheet for carbide to steel.

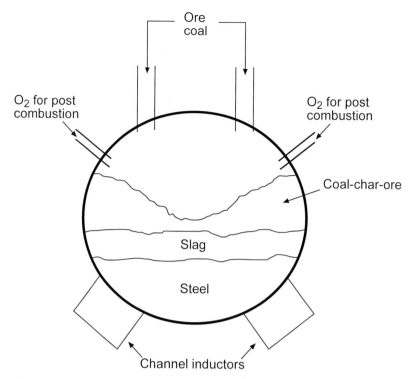

Fig 13.8 Schematic diagram of the IFCON process.

concluded that in order to avoid excessive heat losses the smallest feasible plant would be 45 tons/hour. They invited Nucor to help design such a plant. Whereas the process is theoretically attractive, it is only a concept and to date (1998) no steel has been produced. It requires higher levels of post-combustion than have even been achieved in other processes and the availability of iron carbide at a reasonable cost. If such a process was possible it could be used for other feeds such as HBI or DRI plus carbon, granulated pig iron or liquid hot metal and DRI or HBI.

Another extremely interesting, and possibly revolutionary process, is the IFCON process under development by ISCOR in South Africa. Little public information is available except for its patent.[24] A schematic diagram is shown in Fig. 13.8. The process utilizes a channel type induction furnace similar to that used in hot metal mixers. According to the patent, the process can use hot metal and partially reduced sponge iron (direct reduced iron), sponge iron only and, even more interestingly, iron ore and coal to produce steel directly. The latter version of the process would be truly revolutionary and satisfy the long time dream of steelmakers to go from ore to steel in a single reactor.

According to the ISCOR patent[24], when using ore, the ore and coal are added to the furnace continuously. Oxygen is added primarily for post-combustion of the CO from reduction and hydrogen from the coal. Post-combustion supplies a significant amount of energy for reduction. Electrical energy is supplied by the channel inductors for melting the reduced iron. ISCOR has operated a pilot plant and is considering a commercial/demonstration plant. No operating information from the pilot plant is available. However, a small pilot plant has been in operation at Pretoria and has run continuously for several campaigns of over one month each and produced steel with 0.03–0.1%C. A semi-commercial plant producing approximately 300,000 tpy is scheduled to begin operation in 1998 in South Africa. If such a process proves feasible it could drastically reduce the capital cost associated with steelmaking by eliminating the coke and sinter plants and combining iron and steelmaking in one vessel.

13.4 Economic Evaluation

For a new process to be commercialized it not only must be technically feasible, it must have a clear economic or strategic advantage over existing proven processes. Also, when comparing processes using different input materials it is the only way a meaningful comparison can be made. All economic evaluations for steelmaking processes are site specific. Furthermore, other considerations beyond cost alone, such as the availability of coke, hot metal and electricity and the type or quality of steel for the market application are important. Nevertheless, an economic evaluation is valuable as long as the assumption and unit costs used are reasonable.

The processes which appear to be the most promising are the EOF and the IRSID continuous process. All of the other processes considered have significant technical drawbacks. These two processes are compared to a BOF and an EAF.

A major concern is capital costs which are also site specific and certain assumptions must be made. The assumptions in the present case are:

1. Annual production of 800,000–1,000,000 tons.
2. The building exists.
3. Hot metal and steel ladles and cranes exist.
4. Oxygen and other utilities are at the battery boundary.

It is beyond the expertise of the authors to make accurate capital cost estimates. The capital costs for the various processes given in Table 13.2 were made by experts for the American Iron and Steel Institute (AISI)[1]. There is a relatively large range of capital cost for the EOF and IRSID processes because there is greater uncertainty since few or no units have been built. Whereas actual capital costs are very site specific the values given in Table 13.2 are for a similar site. Therefore, whereas the actual capital costs may be significantly different, the relative cost for the processes should be reasonable.

Table 13.2 Estimated Capital Cost for Selected Steelmaking Units in Dollars per Annual Ton of Production.

Process	Dollars per annual ton
BOF	80–86
EAF[1]	90–100
EOF	40–60
IRSID	40–60

[1] When using large quantities of hot metal the capital cost for the EAF will be about 10% higher due to the oxygen lances and gascleaning system.

Estimated cost of production was made for several cases. Production costs are sensitive to the unit cost of the inputs including scrap, hot metal, labor, etc. The unit costs will be a function of plant location and time and therefore can vary greatly. The analysis used was for the United States in 1998. However, as will be demonstrated, the major variables are the cost of scrap versus hot metal and the cost of capital.

For the BOF and IRSID processes, which use primarily hot metal, the charge was assumed to be 80% hot metal and 20% scrap. The EOF is for 50% scrap and 50% hot metal. The EAF estimates are for scrap plus 50% DRI/HBI or 50% hot metal. These charge balances are for the production of medium to high quality flat rolled which would be a likely product for a new plant. Materials consumption for the BOF was a typical state of the art process. For the EAF, the HBI/DRI was assumed to be 92% total iron at 92% metallization, 2% gangue, and 2% carbon. The energy and materials consumptions were calculated from a comprehensive energy and materials model for the EAF. The EOF materials consumption were from an operating furnace which were verified with an energy and materials balance. The desulfurization cost for the EOF are high because of the use of

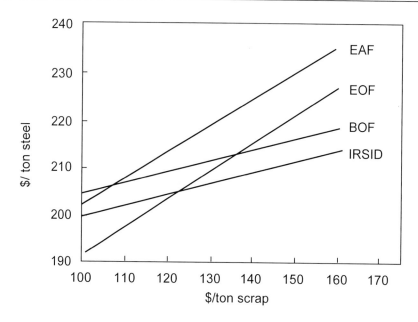

Fig.13.9 Cost of liquid steel for various processes as a function of scrap price.

coal or coke for energy to melt the scrap. The BOF refractory costs are lower due to slag splashing which extends refractory life. Other costs include dust handling and electrodes which are higher for an EAF. The BOF and IRSID processes have slightly higher labor costs due to the charging and tapping operations. The capital costs were computed assuming 8% depreciation and 8% return on investment or a total of 16% of the figures given in Table 13.2.

The most sensitive input is the cost of scrap. Therefore, a sensitivity analysis is done for the cost and amount of scrap. In the past decade the price of scrap has varied by more than 50%. The price of scrap given in Table 13.3 is the weighted average price of a typical charge mix for higher value products. A sensitivity analysis was performed in which the price of scrap varied from $100 to $160 per ton. The results are given in Fig. 13.9. As expected, the scrap based processes (EAF and EOF) costs decrease with decreasing scrap prices and at $100 per ton of scrap their costs are equal or less than those using less scrap.

The other major uncertainty is the cost of hot metal. Obviously an increase in hot metal costs will adversely affect the economics of those processes using hot metal and its effect can be easily estimated. Therefore, comparisons of processes which use differing amounts of hot metal and scrap should be made only if precise information on the cost of the charge materials are known.

The IRSID process has lower capital cost. A BOF is actually only making steel about half the time due to the time required for charging and tapping whereas the IRSID process is continuous and, therefore, has a higher productivity to volume ratio. Also, the IRSID process requires less capital because it has a permanent oxygen lance and does not require vessel rotation equipment. In the late 1970s, The United Nations[25] examined continuous steelmaking processes studied up to that time. They also concluded that the IRSID process was the most attractive alternative and potentially had significantly lower capital cost than a conventional BOF.

The EOF has lower capital cost than the EAF because it does not require a transformer, electrode mechanism, etc. It has a lower operating cost because of lower energy costs. However, it does require a source of hot metal which is currently unavailable economically from a blast furnace. It would require a COREX or new smelting ironmaking process which can produce iron economically at low production rates. The EAF could use hot metal reducing its energy cost. However, hot metal is more costly and not usually available for an EAF. In comparing an EOF and EAF with 50% hot metal, the EOF cost may be lower because of lower capital but the EOF is less flexible.

Table 13.3 Operating and Total Cost of Producing Liquid Steel.

Charge Mix	BOF 80% Hot Metal 20% Scrap		IRSID 80% Hot Metal 20% Scrap		EOF 50% Scrap 50% Hot Metal		EAF 50% Scrap 50% DRI		EAF 50% Scrap 50% Hot Metal	
Input (cost/unit)	Units	Cost	Units	Cost	Units	Cost	Units	Cost	Units	Cost
Hot Metal ($145/t)	0.85	123.30	0.85	123.30	0.55	79.70	0		0.54	64.80
Scrap ($120/t)	0.22	27.50	0.22	27.50	0.55	66.00	0.541	64.90	0.54	64.80
DRI/HBI ($130/t)							0.541	70.30		
Flux ($55/t)	0.055	3.00	0.055	3.00	0.04	2.20	0.036	2.00	0.03	1.70
Oxygen ($40/t)	0.065	2.60	0.065	2.60	0.10	4.00	0.035	1.40	0.06	2.40
Carbon ($70/t)					0.025	1.80	0.01	0.70		
Electricity ($0.04/kwh)	65	2.60	40	1.60	20	0.80	485	19.40	210	8.40
Refractories (kg/t)	1	1.00	2.5	2.50	2.5	2.50	1.5	1.50	1.5	1.50
Labor ($30/mh)	0.3	9.00	0.3	9.00	0.2	6.00	0.2	6.00	0.2	6.00
Desulfurization		7.00		7.00		12.00		7.00		7.00
Other[1]		20.00		20.00		20.00		25.00		25.00
Total Operating Cost		196.00		196.50		195.00		198.20		195.10
Cost of Capital at 8% depreciation and 8% ROI		13.30		8.00		9.00		15.20		16.70
Total Cost		209.30		204.50		204.00		213.40		211.80

[1] Other includes electrodes, gas cleaning and dust disposal.

13.5 Summary and Conclusions

Economic and strategic considerations with regards to steelmaking have changed in the past decade and will continue to change significantly. When one is considering steelmaking processes the EAF and the BOF may not necessarily be the only options. In particular the EAF will use more oxygen and virgin iron and resemble a hybrid process between the EAF and the BOF as discussed in Chapter 10.

In this chapter a technical evaluation of proposed and operating alternatives to a purely oxygen steelmaking process was conducted. The evaluation indicated that the IRSID continuous steelmaking process may be an attractive alternative to a traditional BOF and the EOF could be an attractive option when high scrap charge levels (50%) are required. The other processes considered, including the AISI continuous process and the various trough processes, have significant technical problems.

An economic analysis was performed to compare the costs of the BOF, EAF, IRSID and EOF processes. Whereas, any economic analysis is site specific and depends on the cost of the various inputs, in particular scrap, the present analysis indicates that the IRSID and EOF processes are competitive and may have an advantage over existing processes, the primary advantage being in lower capital cost. The continuous nature of the IRSID process allows for greater productivity and the fixed vessel and lance reduces its capital cost. The EOF is capable of melting scrap without the cost of the electrical energy delivery system of the EAF and results in lower operating energy costs. However, because of its high degree of development, optimization and capital investment it is doubtful these will replace existing OSM plants.

In addition to the IRSID and EOF processes other processes under development should also be considered. In particular, the IFCON process holds much promise as it purports to produce steel directly from fine raw materials, thereby significantly reducing capital cost. More development work may be required on the new processes but there is the potential of significant cost savings.

References

1. R. J. Fruehan and C. L. Nassaralla, "A Critical Review of Alternative Steelmaking Processes", *Trans ISS, Iron & Steelmaker*, Vol. 25, No.8, (1998): 59–68.
2. J. Szekely and N. J. Themelis, *Rate Phenomena in Process Metallurgy* (New York: Wiley Interscience, 1971).
3. J. Szekely, "Proc. of Heat and Mass Transfer in Process Metallurgy," I.M.M., London 1966, p. 115.
4. T. W. Jenkins, N. B. Grey, and H. K. Worner, *Metall. Trans.* 2 (1971): 1258.
5. G. A. Brooks, N. G. Ross, and H. K. Worner, 80th Steelmaking Conference proceedings, ISS-AIME, Chicago IL, 1997.
6. R. J. Fruehan: *Iron and Steelmaker* 1(1) (1976): 33.
7. C. A. Abel, R. J. Fruehan, and A. Vassilicos, *Trans. ISS, I&SM* (August 1995): 49.
8. G. Taylor, *Proces. Ray Soc.* 223A (1954): 446.
9. W. A. Tony, *Iron and Steelmaker* 15(9) (September 1988): 41.
10. K. Brotzman, 70th Steelmaking Conference Proceedings, ISS-AIME, Pittsburgh 1987, p. 3.
11. H. Ozawa et al., Proceedings of the Sixth International Iron and Steel Congress, 1990, Nagoya Japan, ISIJ, pp. 33.
12. Y. Kishimoto et al., Proceedings of the Sixth International Iron and Steel Congress, 1990, Nagoya Japan, ISIJ, pp. 41.
13. G. Golperine et al., 77th Steelmaking Conference Proceedings of ISS-AIME, Chicago IL, 1994, p. 73.
14. X. Zhang, R. J. Fruehan, and A. Vassilicos, "Modeling of the Two-Zone Smelter—Part 2: Physical Modeling and Plant Trials," Process Technology Conference Proceedings, Vol. 13, 1995, pp. 445-454.
15. A. Berthet et al., *Journal of Iron and Steel Institute* (June 1969): 790.
16. E. M. Rudzki et al., *J of Metals* 21 (1969): 57.
17. R. Nakagawa: *Trans. ISIJ* 13 (1973): 333.
18. H. K. Warner et al., *J of Metals* 21 (1969): 50.
19. R. J. Zaranek (USX Technical Center), personal communication, Monroeville PA.
20. P. Tarasoff: *Met. Trans. B* 15B (1984): 411.
21. P. Arthur, A. Siegmund, and M. Schmitt, Savard/Lee International Symposium, TMS, Montreal, 1992, TMS, Warrendale PA, p. 668.

22. G. Gieger: U.S. Patent No. 5,139,568.
23. G. Gieger: "Carbide to Steel": to be published by TMS-AIME, Warrendale, PA.
24. L. J. Fourie, Steelmaking Process, U.S. Patent 5,411,570, May 2, 1995.
25. The United Nations. *The Increasing Use of Continuous Processes in the Iron and Steel Industry.* 1979. Sales No. E79IIE-7.

Index

Adiabatic reactions 16
AISI continuous refining 748
Alternative iron, electric furnace steelmaking
    use 621
Alumina
    processed group 169
    and clay refractory group 166
Argon gas
    chemical properties 309
    production 298
    safe practices 309
    uses 295
Argon-oxygen decarburization (AOD)
    process 716, 725
    refractories for 249

Baghouses, electric furnace
    steelmaking 588
Basic oxygen furnace (see Oxygen
    steelmaking)
Bessemer process 2, 4
Bethlehem process 752
Binder types 173
BOF, BOP (see Oxygen steelmaking)

Carbide to steel process 753
Carbon dioxide gas
    chemical properties 310
    liquefaction 306
    production 305
    safe practices 310
    uses 296
Carbon refractories 170
CAS-OB process 709
Castable refractories 173
Chemical kinetics 39-45
    activation complex, activation
        energy 39-40
    equations of reaction rates 40-41
    experimental data on rates of interfacial
        reactions 42-46
    mass transfer—chemical reaction mixed controlled rates 47-49
Classification, refractories 159
Clay refractory group 166
CLU process, stainless steel 717, 727
Coal
    origin and composition 330
    chemical composition and
        classification 334
    mining 336
    preparation 339
    pulverized 342
Cold strength, refractory 182
Combustion
    calculations 322
    principles 312
Comelt furnace 649
Conarc furnace 650
Conduction, heat flow 326
Consteel process 642
Consumption, refractory 224
Containment, metal 217
Contiarc furnace 650
Continuous stirred tank reactor (CSTR) 743
Convection, heat flow 327
COSMOS process 752
Creep, refractory 181
Cryogenic gas production 298

Danarc process 634
Degassers
    CAS-OB 709
    process selection and comparison 710

refractories for 285
RH 708
tank. 705
vacuum arc. 705
Density of metals and Fe-C alloys 75-76
Density of molten slags 98-99
Density of refractories 179
Deoxidation, tapped steel in the ladle 677
Dephosphorization, hot metal. 413
Dephosphorization, hot metal. 413
Desiliconization, hot metal 413
    environmental considerations 414
    by iron oxides 414
Desulfurization
    chemical reactions 417–418
    co-injection 419
    hot metal sampling. 426
    in blast furnace runners 418
    lance designs 424
    magnesium fume capture. 418
    pneumatic injection 419
    reagent consumption 426
    tapped steel in the ladle 671, 680
Diffusion. 24-26
    chemical and self diffusivities 25
    diffusivities of ions in slags. 101
    diffusivities of solutes in iron 76-78
    Fick's diffusion laws 24-25
    gas diffusion in porous media 57-60
    types of diffusional processes 25-26
Direct reduced iron (DRI). 491, 621
Dolomite 162

EAF (see Electric furnace steelmaking)
Effluent
    limitations 385
    treatment, water 380
Electrical and thermal conductivities
    of iron alloys 76-77
    of slags. 101-102
Electric furnace steelmaking
    arc characteristics. 558
    arc regulation 555, 617
    auxiliary equipment. 537
    baghouses. 588
    bottom electrodes (DC) 541, 561
    bottom stirring 540, 615
    Comelt furnace. 649
    Conarc furnace. 650
    Consteel process 642
    Contiarc furnace. 650
    crude production share. 10, 11
    Danarc process. 634
    DC electrical considerations 560
    DC furnace operations. 618

    direct evacuation systems (DES). 579
    dust formation 590
    electrode arms 533, 617
    electrodes. 562
    environmental concerns. 590
    foamy slag practice 604
    Fuchs double shaft furnace 638
    Fuchs finger shaft furnace 640
    Fuchs single shaft furnace 635
    fume control. 577
    furnace bottom. 530
    furnace mechanical design. 243, 525
    furnace movements 532
    gas collection and cleaning 577, 586
    heat balance 628
    high voltage AC operation. 617
    IHI shaft furnace 648
    K-ES. 631
    operating cycle. 622
    operating philosophies. 550
    oxy-fuel burners. 598
    oxygen lance technology 601
    oxygen use. 597
    power supply 551
    process development 8
    reactance 557
    refractories for 243
    scrap preheating. 629
    secondary emissions control 583
    spray-cooled equipment. 530, 536
    structural support. 526
    supplementary reactance 559
    twin shell EAF. 645
    water-cooled side panels 528
Electrodes for electric furnace steelmaking
    AC vs. DC consumption 572
    continuous consumption mechanisms. . . 564
    current carrying capacity. 569
    DC special grades 575
    discontinuous consumption processes. . . 569
    manufacture 562
    properties 564
    wear mechanisms. 564
Emissivity of refractories 187
Energy optimizing furnace (EOF) 746
Energy savings, refractory use for 222
Enthalpy (heat content). 14
    standard state. 15
Enthalpy of reaction 15
Entropy 17

Ferrochromium 726
Free energy data 20-21
Fuchs shaft furnaces
    double shaft furnace. 638

Index

finger shaft furnace 640
single shaft furnace 635
Fuels
    calorific value 315
    classification 311
    economy........................... 363
    gaseous............................ 352
    liquid 344
    solid 329
Fused silica 165

Gas constant 15
Gas bubbles in oxygen steelmaking 34-38
    bubble size in steel bath.............. 35
    gas holdup and superficial velocity...... 34
    maximum rate of degassing with argon
    stirring steel bath 38-39
    rate of gas bubble reactions in steel
    bath................................ 37
Gaseous fuels
    blast-furnace gas 356
    BOF gas............................ 358
    byproduct gas....................... 356
    coke-oven gas 357
    combustion......................... 360
    manufactured gas.................... 353
    natural gas 353
    oil gas.............................. 355
    producer gas........................ 354
    water gas 354
Gas laws 13, 319
Gibbs free energy 17-18
Gunning mixes, refractory 173

Heat
    capacity 15, 316
    flow 326
    units of measurement................ 314
Hot metal............................ 489
    dephosphorization 413
    desiliconization 413
    desulfurization..................... 416
    temperature control 427
Hot strength, refractory.................. 182
Hydrogen gas
    chemical properties 309
    production 305
    safe practices 309
    steam reformer.................... 305
    uses............................... 296
Hydrogen
    pickup during tapping 662
    removal during degassing 698

IFCON process....................... 754

IHI shaft furnace...................... 648
Ionic structure of molten slags 79-80
Inclusion modification 687
Inclusion removal 672
Industrial gases
    production 297
    safety 307
    supply system options 306
    uses............................... 291
Iron–carbon alloys 64-69
    activity coefficient of carbon........... 65
    Fe-C phase diagram............... 64-69
    peritectic reaction................. 65-69
IRSID continuous steelmaking process 749

Kelly process.......................... 2
KCB-S process, stainless steel 717, 728
K-ES 631
K-OBM-S process, stainless steel 717, 726

Ladle
    alloy additions 685
    arc reheating....................... 673
    calcium treatment and inclusion
    modification....................... 687
    design............................ 265
    free open performance.............. 667
    function 262
    furnace 705
    oxygen reheating 675
    preheating......................... 665
    refractories for 262
    slidegates 278
    stirring 277, 669
Ladle refining of steel.............. 140-153
    calcium treatment.................. 150
    deoxidation equilibria 140-145
    deoxidation rates 146-147
    desulfurization..................... 147
    vacuum degassing 151-153
Laser measurements, furnace linings .. 229, 241
L-D process........................... 7
Liquidus temperature of alloy steels......... 69
Lining, refractory
    burn-in, oxygen steelmaking
    furnaces 235, 448
    construction, electric steelmaking
    furnaces 547
    construction, ladles 276
    construction, oxygen steelmaking
    furnaces 231, 446
    life, AOD furnaces............. 249, 255
    life, degassers..................... 289
    life, ladles........................ 281
    life, oxygen steelmaking furnaces.. 237, 238

life, VOD furnaces.................. 258
maintenance, electric steelmaking
    furnaces 549
maintenance, oxygen steelmaking
    furnaces 239, 241
preheating, AOD furnaces........... 254
preheating, VOD furnaces........... 261
wear, electric steelmaking furnaces..... 247
wear, oxygen steelmaking furnaces. 235, 240
zones for AOD furnaces............. 249
zones for electric steelmaking
    furnaces 244, 545
zones for oxygen steelmaking
    furnaces 230, 446
zones for ladles 268
zones for VOD furnaces 258
Liquid fuels
    burners 351
    combustion..................... 351
    grades of petroleum used as 347
    origin, composition and distribution of
    petroleum..................... 345
    properties and specifications 348

Magnesia......................... 160
Magnesia–chrome............... 163, 733
Magnesia–lime 160
Mechanisms of blast furnace reactions. 114-118
    blast furnace slag–metal data...... 115-118
    experimental work 114-115
Melting point, refractories 159
Membrane gas production 302
    membranes..................... 304
    pressure swing absorption (PSA) .. 297, 302
    vacuum pressure swing absorption
    (VPSA) 297, 302
    vacuum swing absorption (VSA) .. 297, 302
Modulus of elasticity, refractory......... 185
Modulus of rupture, refractory 182
Monolithic refractories 172
Mortars, refractory 172
MRP process, stainless steel........ 717, 727

Nitrogen gas
    chemical properties 308
    production 298
    safe practices 308
    uses........................... 294
Nitrogen
    pickup during tapping 662
    removal during degassing 701
NRIM process..................... 752

OBM
    bottom life 506
    converters....................... 476
    energy balance................... 499
    mass balance 499
    operating sequence............ 478, 506
    plant equipment 505
    powdered fluxes............... 477, 505
    process 7
    raw materials 489, 506
    refining reactions 496
    shop manning.................... 486
    slag formation 498
    tuyere coolants.................. 504
    tuyeres 504
Open hearth process 4
    crude production share........... 10, 11
Oxidation of carbon in CO_2-CO
    mixtures 28-30
Oxide activities in slags 84-88
    CaO-FeO-SiO_2 system.......... 84-85
    CaO-MnO-SiO_2 system........... 85
    MnO-Al_2O_3-SiO_2 system 87
    of MnO and FeO 88
Oxygen gas
    blowing rate, oxygen steelmaking...... 496
    chemical properties 308
    electric furnace steelmaking use 597
    production 298
    safe practices 308
    uses........................... 292
Oxygen potential diagram.......... 104-106
Oxygen steelmaking
    bottom-blown................. 453, 504
    bottom stirring 453, 510
    combination-blown 507
    crude production share........... 10, 11
    deskulling...................... 464
    dynamic process control 517
    energy balance................... 499
    furnace lining wear 228
    lance technology 465
    mass balance 499
    mixed blowing................... 507
    operating sequence............... 478
    plant layout 478
    post-combustion.................. 512
    process development 7
    raw materials 489
    refining reactions 496
    refractories for 227
    shop manning.................... 486
    slag formation 498
    static process control 515
    sub-lance equipment 471
    top-blown....................... 476
    top cone cooling................. 435

trunnion ring design. 438
vessel imbalance 445
vessel shape. 433
vessel suspensions 439

Parabolic rate of oxidation of iron 26-28
Petroleum
    grades used as fuels
    properties and specifications of liquid fuels
    origin, composition and distribution 345
Phase equilibrium diagrams for
    refractories. 194
Phosphorus
    removal 413
    reversion. 664
Plastic refractories. 172
Plug flow reactor. 743
Porosity of refractories 179
Post-combustion for electric furnace
    steelmaking 605
Post-mortem studies for refractories. 212
Preparation of refractories 172
Pressure swing absorption (PSA) 297, 302

Q-BOP (see OBM)

Radiation, heat flow 327
Reaction equilibrium constant 23
    temperature effect 24
Reactions in electric furnace
    steelmaking 132-135
    control of nitrogen. 135
    control of residuals 134
    slag chemistry 132-134
Reduction of iron oxides. 30-33
Refractories
    AOD furnaces, for 249
    binder types 173
    castables. 173
    chemical characteristics. 178
    chemical composition 178
    classification 159
    cold strength. 182
    consumption, trends and costs. 224
    creep. 181
    degassers, for 285
    density 179
    electric furnace steelmaking, for . . . 243, 545
    emissivity. 187
    general uses 215
    gunning mixes 173
    hot strength 182
    ladles, for. 262
    lining life, AOD furnaces. 249, 255
    lining life, degassers 289
    lining life, ladles 281
    lining life, oxygen steelmaking
    furnaces 237, 238
    lining life, VOD furnaces. 258
    melting point 159
    modulus of elasticity 185
    modulus of rupture (MOR) 182
    monolithics. 172
    mortars. 172
    oxygen steelmaking, for 227, 446
    phase equilibrium diagrams. 194
    physical characteristics 178
    plastic. 172
    porosity 179
    preparation. 172
    processing 176
    reactions at elevated temperatures. 194
    refractoriness 181
    selection. 206
    service conditions 178
    shapes. 172, 447
    simulated service tests. 206
    specialties. 172
    specific heat. 186
    spray mixes 173
    stress-strain behavior. 185
    thermal conductivity 190
    thermal expansion 188
    thermal shock. 194
    VOD furnaces, for. 258
    zones for AOD furnaces. 249
    zones for electric steelmaking furnaces. . 244
    zones for oxygen steelmaking furnaces. . 230
    zones for ladles 268
    zones for VOD furnaces 258
RH degasser 708

Sauter diameter. 750
Scrap 594
Selection, refractory 206
Service tests, refractory 206
Shapes, refractory. 172
Siemens, Karl Wilhelm. 4
Siliceous refractories. 164
Silica, fused. 165
Silicon carbide 165
Slag basicity 80
Slag coating, oxygen steelmaking furnace
    linings. 239, 513
Slag foaming. 102-104, 604
Slag-metal equilibrium in blast furnace
    reactions. 109-114
    manganese reaction 110
    silicon reaction. 109-110
    Si-Mn coupled reaction 111

Si-S and Mn–S coupled reaction... 112-113
Si-Ti coupled reaction 114
Slag-metal equilibrium in steelmaking. 119-122
    oxidation of carbon 120-121
    oxidation of chromium 121
    oxidation of iron 119
    oxidation of manganese............. 120
    oxidation of phosphorus 121-122
    reduction of sulfur................ 122
Slag splashing, oxygen steelmaking
    furnace linings............ 239, 241, 513
Slag structural models.................. 104
Soda ash 414
Sodium, dissolved in slag.............. 414
Solutions........................ 19, 22
Solubility of gases in liquid iron 61-64
    of H_2, N_2 and O_2 62
    of gaseous oxides................ 62-64
Solubility of gases in slags of H_2O, N_2,
    S_2 and O_2...................... 89-94
Solubility of iron oxide in liquid iron....... 69
Solubility of Pb, Ca and Mg in liquid
    iron............................ 70-72
Specialties, refractory.................. 172
Specific heat
    definition........................ 316
    of refractories.................... 186
Spray mixes, refractory................ 173
Stainless steel
    argon-oxygen decarburization (AOD)
      process....................... 716, 725
    CLU process 717, 727
    Cr reduction from slag........... 139, 715
    decarburization........ 136-137, 720, 735
    definition 715
    direct steelmaking 730
    duplex production processes 716, 721
    electric arc furnace production 715
    ferrochromium.................... 723
    gas collection systems 735
    KCB-S process................ 717, 728
    K-OBM-S process 717, 726
    lances........................... 733
    MRP process 717, 727
    nitrogen control 138
    post-combustion................... 733
    raw materials 723
    STB process.................. 717, 729
    TMBI process................. 717, 729
    triplex production processes 717, 722
    tuyeres.......................... 733
    VOD process................. 718, 730
    VODC process.................... 717
State of reactions in steelmaking 123-132
    decarburization............... 123-125
    desiliconization 127
    FeO-MnO-Mn-O relation 127-128
    hydrogen and nitrogen reaction.... 130-131
    MnO-C relation 127
    oxygen–carbon reaction............. 126
    phosphorus reaction................ 129
    sulfur reaction 129
STB process, stainless steel 717, 729
Steam reformer, hydrogen production 305
Stress-strain behavior of refractories 185
Sulfide shape control 693
Surface tension of liquid iron alloys..... 72-74
Surface tension of slags 95
    interfacial tension in iron-slag
      systems...................... 96-97

Tank degasser 705
Ternary and quaternary oxide systems... 81-84
Testing of refractories................. 206
Thermal conductivity of refractories 190
Thermal expansion of refractories 188
Thermal shock of refractories........... 194
Thermodynamic activity........ 18–19, 22–23
    activity coefficient 22–23
    Henry's law 22
    interaction coefficients........ 23, 60–61
    Raoult's law...................... 19
Thermodynamic laws 13-18
    ideal gas......................... 13
    the first law 14
    the second law.................... 16
    the third law...................... 17
Thermochemical properties of gases 49-55
    catalytic decomposition of CO 53-55
    equilibrium states in gas mixtures.... 50-52
    heat content 49
Thomas process......................... 4
TMBI process, stainless steel 717, 729
Transformers for electric furnace
    steelmaking 551
Transport properties of gases
    diffusivity, viscosity and electrical and
      thermal conductivities 55-57
Twin shell EAF....................... 645

Uses of refractories.................... 215

Vacuum arc degasser.................. 705
Vacuum degassing.................... 693
Vacuum pressure swing absorption
    (VPSA) 297, 302
Vacuum swing absorption (VSA) 297, 302
Vapor species in blast furnace
    reactions.................... 105-109
    alkali recycle 107-109

ammonia and hydrogen cyanide 109
SiO, SiS and CS. 105-107
Viscosity of liquid Fe-C alloys. 75-76
Viscosity of slags 100-101
VOD process, stainless steel. 718, 730
VODC process, stainless steel 717

Water
    boiler treatment 395
    cooling 372
    effluent limitations 385
    effluent treatment 379
    general uses for steelmaking 368
    ironmaking use and treatment 376, 386
    related problems 371
    steelmaking use and treatment 377, 390
WORCRA process 752

Z-BOP process 748
Zircon, zirconia 165
Zones, refractory
    for AOD furnaces 249
    for electric steelmaking furnaces . . . 244, 545
    for oxygen steelmaking furnaces . . . 230, 446
    for ladles 268
    for VOD furnaces 258